ENVIRONMENTAL IMPACT ASSESSMENT

McGraw-Hill Series in Water Resources and Environmental Engineering

CONSULTING EDITOR

Paul H. King, *Dean, School of Engineering, Northeastern University*

ENVIRONMENTAL IMPACT ASSESSMENT

SECOND EDITION

Larry W. Canter
University of Oklahoma

McGraw-Hill, Inc.

New York   St. Louis   San Francisco   Auckland   Bogotá   Caracas   Lisbon
London   Madrid   Mexico City   Milan   Montreal   New Delhi
San Juan   Singapore   Sydney   Tokyo   Toronto

This book was set in Times Roman by The Clarinda Company.
The editors were B. J. Clark and James W. Bradley;
the production supervisor was Denise L. Puryear.
The cover was designed by Carla Bauer.
Quebecor Printing/Fairfield was printer and binder.

ENVIRONMENTAL IMPACT ASSESSMENT

1 2 3 4 5 6 7 8 9 0 FGR FGR   9 0 9 8 7 6 5

ISBN 0-07-009767-4

Library of Congress Cataloging-in-Publication Data

Canter, Larry W.
    Environmental impact assessment / Larry W. Canter. — 2d ed.
       p.   cm. — (McGraw-Hill series in water resources and environmental engineering)
    Includes bibliographical references and index.
    ISBN 0-07-009767-4
    1. Environmental impact analysis.   I. Title.   II. Series.
TD194.6.C36   1996
333.7′14—dc20   95-935

ABOUT THE AUTHOR

LARRY W. CANTER, P.E., is the Sun Company Chair of Ground Water Hydrology, George Lynn Cross Research Professor, and Director, Environmental and Ground Water Institute, University of Oklahoma, Norman, Oklahoma. Dr. Canter received his Ph.D. in environmental health engineering from the University of Texas in 1967, M.S. in sanitary engineering from the University of Illinois in 1962, and B.E. in civil engineering from Vanderbilt University in 1961. Before joining the faculty of the University of Oklahoma in 1969, he was on the faculty of Tulane University and was a sanitary engineer in the U.S. Public Health Service. He served as Director of the School of Civil Engineering and Environmental Science at the University of Oklahoma from 1971 to 1979; and as Co-director of the National Center for Ground Water Research (a consortium of the University of Oklahma, Oklahoma State University, and Rice University) from 1979 to 1992.

Dr. Canter's research interests include environmental impact assessment (EIA) methodologies, ground water pollution source evaluation and ground water protection, soil and ground water remediation technologies, and market-based approaches for air-quality management and impact mitigation. Currently, he is conducting research on cumulative impact assessment and valuation methods for ground water resources. In 1982 he received the Outstanding Faculty Achievement in Research Award from the College of Engineering at the University of Oklahoma, and in 1983 the Regent's Award for Superior Accomplishment in Research.

Dr. Canter has written six books related to EIA; examples include *Environmental Impact Assessment* (McGraw-Hill, 1977, first edition), *Handbook of Variables for Environmental Impact Assessment* (Ann Arbor

Science, 1979), and *Environmental Impacts of Water Resources Projects* (Lewis Publishers, 1985). He is also the author or co-author of numerous book chapters, refereed papers, and research reports related to environmental impact studies. He has also written environmental assessments and environmental impact statements on projects such as power plants, gas pipelines and compressor stations, highways, wastewater treatment plants, industrial plants, and flood control dams.

Dr. Canter served on the U.S. Army Corps of Engineers Environmental Advisory Board from 1983 to 1989. Since 1979, he has taught several sessions annually of a one-week short course on EIA for the Corps. He has presented short courses, or served as advisor on EIA to several other governmental agencies in the United States and to institutions in Argentina, Brazil, Colombia, France, Germany, Greece, Hong Kong, Italy, Kuwait, Mexico, The Netherlands, Panama, the People's Republic of China, Peru, Saudi Arabia, Scotland, Sweden, Thailand, Turkey, and Venezuela. Finally, he is a member of the Consultative Expert Group on Environmental Impact Assessment of the United Nations Environment Program in Nairobi, Kenya.

To Donna, Doug and Carrie, Steve, and Greg

CONTENTS

PREFACE

The National Environmental Policy Act (NEPA) in the United States is considered to be the seminal legislation for the environmental impact assessment (EIA) process in the majority of some 100 countries that have adopted EIA legislation. EIA requirements of aid agencies and lending institutions are typically based on the principles included in the NEPA. The NEPA requires environmental impact considerations to be included in project planning along with traditional technical (engineering) and economic evaluations. The action-forcing mechanism in the NEPA is that environmental impact statements (EISs) must be prepared such that they describe the environmental consequences of major actions which significantly affect the quality of the human environment. Over 21,000 EISs have been prepared in the United States since the effective date of the NEPA (January 1, 1970), and even more will be prepared in the future. In addition, with the implementation of regulations developed in 1979 by the Council on Environmental Quality (CEQ), delineations have been made between EISs and environmental assessments (EAs). EAs are documents which are used to determine if EISs are necessary for proposed actions. It is estimated that 30,000 to 50,000 EAs are prepared on an annual basis in the United States.

This book represents an extensive revision of the 1977 edition by the same author. This author views the EIA process which culminates in either an EA or an EIS as consisting of six components: basics, impact identification, description of the affected environment, prediction and assessment of impacts, selection of proposed action, and documentation in accordance with extant guidelines. This textbook is organized according to these components. Chapters 1 and 2 encompass the basic requirements and framework of the process, including reviews of legislative requirements and information on planning impact studies. Chapter 3 highlights matrices, networks, and simple and descriptive checklists for identifying potential impacts of proposed projects or activities. Chapters 4 and 5 are related to describing the affected environment, with the latter chapter focused on the use of environmental indices. Chapters 6 through 14 address the steps for impact prediction and assessment for the physical-chemical (air, surface-water, soil and groundwater, and noise), biological (nonhabitat and habitat), cultural (historic and/or archaeological and visual resources), and socioeconomic environments, in that order. Each of these substantive-area chapters is characterized by a stepwise approach for addressing the impacts of proposed projects or activities. Chapter 15 presents various impact-assessment methodologies that can be utilized in the evaluation of alternatives and the selection of proposed actions, with the emphasis being on decision-focused checklists related to multicriteria decision making. Chapter 16 describes public participation in the EIA process, particularly as related to the selection of a proposed action. Chapter 17 discusses pertinent considerations in writing Eqs or EISs, with the basic principles of technical writing summarized. Finally, Chapter 18 presents information related to the use of monitoring in the EIA process, including baseline and post-EIS monitoring. The focus is on the use of monitoring information in impact documentation and project management.

This book is intended for use in upper-division or graduate-level courses dealing with the EIA process. It can also be used as a reference book by practitioners. The orientation is primarily for science and engineering majors; however, individuals trained in other disciplines, such as planning and geography, can also utilize this text. Information is included that is relevant for both classroom presentations and illustrations of the practice of EIA.

It is noted that this book is primarily oriented to the EIA practice in the United States, with particular emphasis to the NEPA and relevant environmental laws. It can be utilized in other countries by appropriate substitution of information related to the EIA legislation and pertinent environmental laws within the application country.

It is noted that the EIA process should be considered a part of good planning practice; it should not be viewed as an "afterthought" implemented to satisfy environmental regulatory concerns following all key decisions related to the proposed project or activity. The optimum usage of the EIA process is from the establishment of the need for a project or activity and the delineation of potential alternatives to meet that need. The primary application of the EIA process to date has been focused on proposed projects/activities. There is a current emphasis on applying the EIA process to policies, plans, and programs, with these applications being referred to as "programmatic (or strategic) environmental assessments." Representing a narrower focus, EIA process principles can also be applied in the context of the application process for permits related to water or air quality, or other waste-disposal or environmental-management activities. For example, an air-quality permit application includes an impact study related to the air-quality implications of the proposed project or activity.

This textbook has been assembled based upon continuing activities which the author has been engaged in since the initial edition in 1977. This process includes teaching university-level courses and short courses on EIA, the conduction of research related to specific methodologies or components of the process, and the actual preparation of EAs or EISs on proposed projects and activities.

This book is not meant to encompass every possible consideration in the EIA process. In fact, there are specific topics which are not addressed herein, including vibrational impacts and the potential environmental effects of electromagnetic radiation. This is a dynamic field and proper use of this textbook is as a reference for a point in time, with the understanding that it must be supplemented by additional information when technology becomes available. The following key observations are made as a result of the preparation of this book:

1. There is an enormous amount of information available for addressing different facets of the EIA process.
2. A scientific approach to impact identification, quantification, and evaluation is fundamental to the EIA process.
3. There are many tools and techniques which have been developed for usage in the EIA process, including scoping, checklists, matrices, qualitative and quantitative models, literature reviews, and decision-support systems.
4. While the EIA process can become technically complicated, it is recognized that scientifically based approaches which include simpler applications of available tools and techniques are appropriate.
5. Documentation is key to the EIA process; such documentation includes both written and verbal presentations and related environmental-monitoring data.

The author wishes to express his gratitude to a number of individuals who have participated directly or indirectly in the assemblage of information related to this book. These include former students such as Drs. Carlota Arquiaga, Sam Atkinson, Robert Knox, Mohammed Lahlou, Gary Miller, George Sammy, and Robert Westcott; and Geoff Canty, Stephen Kukoy, and Wylan Weems. These students have conducted research or participated in various EIA-related projects as part of their graduate work.

The support and assistance of current and former colleagues at the University of Oklahoma is also gratefully acknowledged; included are Drs. Loren Hill, Robert Knox, Paul Risser, James Robertson, David Sabatini, and Leale Streebin; and Professor George Reid. The author has benefitted by teaching short courses on EIA through the University of Alabama in Huntsville. Accordingly, Greg Cox, Charles Rumford, and Linda Berry are acknowledged for their continued collaboration in the planning and conduction of EIA short courses.

Numerous colleagues from the United States have also contributed indirectly to this book, including Drs. John Belshe, Jerome Delli Priscoli, Larry Leistritz, Jim Mangi, Eugene Stakhiv, Evan Vlachos, and Lee Wilson. In addition, Ray Clark of the CEQ, John Fittipaldi of the U.S. Army Construction Engineering Research Laboratory, and Carl Townsend of the U.S. Environmental Protection Agency, region VI, contributed in various ways to the development of this book.

The author has had the unique opportunity of becoming acquainted and collaborating with numerous international colleagues while participating in the EIA field. Of particular note are Professor Brian Clark and Sandra Ralston of CEMP at the University of Aberdeen and the opportunities that have been available to collaborate with them over many years in EIA training activities. Other colleagues include Drs. Gordon Beanlands, Maria Berrini, Virginio Bettini, Owen Harrop, Bindu Lohani, Barry Sadler, N. C. Thanh, and Peter Wathern; they have all contributed to this process. In addition, Ron Bisset, Robert Turnbull, and Henyk Weitzenfeld are also acknowledged.

Of major importance to the author is the positive attitude and helpfulness of Ms. Mittie Durham and Ms. Ginger Geis of the Environmental and Ground Water Institute, University of Oklahoma, in typing and retyping this manuscript. Their technical abilities and pleasant attitudes have made this book possible.

The author also expresses his gratitude to the College of Engineering, University of Oklahoma for its support during the preparation of this book. Included in this acknowledgment are Dr. Ronald Sack of the School of Civil Engineering and Environmental Science and Dean Billy Crynes of the College of Engineering. Finally, the author thanks his wife for her encouragement and continued support in the process of developing this textbook.

McGraw-Hill and the author would like to thank the following reviewers for their many helpful comments and suggestions: Samuel F. Atkinson, University of North Texas; Thomas V. Belanger, Florida Institute of Technology; Paul Chan, New Jersey Institute of Technology; Wesley Pipes, Drexel University; Frederick Pohland, Georgia Institute of Technology; Frederick G. Pohland, University of Pittsburgh; Evan Vlachos, Colorado State University; and Anthony M. Wachinski, Colorado Springs.

Larry W. Canter

ENVIRONMENTAL IMPACT ASSESSMENT

CHAPTER 1

National Environmental Policy Act and Its Implementation

The past two decades have been characterized by passage of major federal legislation dealing with the environment, including specific legislation on control of water and air pollution, solid- and hazardous-waste management, resource protection, and soil and groundwater remediation. Perhaps the most significant legislation is the National Environmental Policy Act (NEPA) of 1969 (P.L. 91-190), which became effective on January 1, 1970. This act was the first signed in the 1970s, thus signaling the importance of the environment in the decade (Kreith, 1973). It has been referred to as the "Magna Carta for the environment" in the United States (CEQ, 1993a). The thrust of this act, as well as of subsequent executive orders, Council on Environmental Quality (CEQ) guidelines and regulations, and numerous federal agency procedures and regulations, is to ensure that balanced decision making regarding the environment occurs in the total public interest. Project planning and decision making should include the integrated consideration of technical, economic, environmental, social, and other factors. The most important of these considerations can be referred to as "the three Es"

(engineering or technical, economics, and environment) in planning and decision making. Prior to the NEPA, technical and economic factors dominated the decision-making process.

TERMINOLOGY

Specialized terminology has arisen in conjunction with the process of complying with the requirements of the NEPA. Three of the most significant terms are "environmental inventory," "environmental impact assessment," and "environmental impact statement"; the latter will be described in a subsequent section.

Environmental Inventory

"Environmental inventory" is a complete description of the environment as it exists in an area where a particular proposed action is being considered. The inventory is compiled from a checklist of descriptors for the physical-chemical, biological, cultural, and socioeconomic environments. The "physical-chemical environment" includes such major areas as soils, geology, topography, surface-water and groundwater resources, water quality, air quality, and climatology. The "biological environment" refers to the flora and fauna of the area, including species of trees, grasses, fish, herpetofauna, birds, and mammals. Specific reference must be made to any threatened and/or endangered plant or animal species. General biological features such as species diversity and overall ecosystem stability should also be presented. Items in the "cultural environment" include historic and archaeological sites, and aesthetic resources such as visual quality. The "socioeconomic environment" refers to a range of considerations related to humans in the environment, including population trends and population distributions; economic indicators of human welfare; educational systems; transportation networks and other infrastructure concerns such as water supply, wastewater disposal, and solid-waste management; public services such as police and fire protection and medical facilities; and many others. The physical-chemical and biological environments can be referred to as the "natural environment," or the "biophysical environment," while the cultural and socioeconomic environments represent the "man-made environment."

The health impacts of projects, plans, programs, or policies should be considered in the decision-making process. Because of the importance of these concerns, particularly in developing countries, an environmental health impact assessment (EHIA) process has been proposed (WHO, 1987). For certain types of projects, such as nuclear power plants, it may be necessary to address psychological impacts on nearby residents ("Can Change Damage Your Mental Health," 1982).

The emphasis in environmental impact studies in the early 1970s was on the physical-chemical and biological environments; however, added attention was given to the cultural and socioeconomic environments as the decade progressed. One reason for the attention to the socioeconomic environment was the emphasis on secondary impacts mentioned in the NEPA guidelines issued by the Council on Environmental Quality in 1973. Attention to health and ecological risks is increasing and this trend is expected to accelerate.

The environmental inventory serves as the basis for evaluating the potential impacts on the environment, both beneficial and adverse, of a proposed action. It is included in an environmental impact statement (EIS) in the section referred to as "description of the affected environment" or "description of the environmental setting without the project." Development of the inventory represents an initial step in the environmental impact assessment process.

Environmental Impact Assessment

"Environmental impact assessment" (EIA) can be defined as the systematic identification and evaluation of the potential impacts (effects) of proposed projects, plans, programs, or legislative actions relative to the physical-chemical, biological, cultural, and socioeconomic components of the total environment. The primary purpose of the EIA process, also called the "NEPA process," is to encourage the consideration of the environment in planning and decision making and to ultimately arrive at actions which are more environmentally compatible. To serve as an example of incorporating environmental and other issues in decision making, the public interest factors considered by the U.S. Army Corps of Engineers in the Clean Water Act Section 404 permit program include conservation, aesthetics, wetlands, fish and wildlife values, floodplain values, navigation, recreation, water quality, food and fiber production, mineral needs, economics, general environmental concerns, historic properties, flood hazards, land use, shore erosion and accretion, water supply and conservation, energy needs, safety, and property ownership.

Barrett and Therivel (1991) have suggested that an ideal EIA system would (1) apply to all projects that are expected to have a significant environmental impact and address all impacts that are expected to be significant; (2)

compare alternatives to a proposed project (including the possibility of not developing the site), management techniques, and mitigation measures; (3) result in a clear EIS which conveys the importance of the likely impacts and their specific characteristics to nonexperts as well as experts in the field; (4) include broad public participation and stringent administrative review procedures; (5) be timed so as to provide information for decision making; (6) be enforceable; and (7) include monitoring and feedback procedures.

In the United States, the EIA process was initially perceived to be within the domain of various federal agencies because the NEPA was primarily directed toward federal actions. However, with the subsequent broadening of definitions relative to actions to encompass both permits and undertakings involving federal funding, the range of institutional and private organizations involved in the preparation of EISs has considerably expanded. In addition to federal agencies, over 30 states have the equivalent of the NEPA at the state level, or other requirements, with many of these laws being closely patterned after the EIA requirements of the NEPA. Perhaps the most stringent state law is the California Environmental Quality Act, which requires the preparation of environmental impact reports on both public and private activities. Numerous regional planning organizations, such as councils of government, and local municipalities have also implemented EIA-type requirements in conjunction with land-use planning and zoning considerations.

Practitioners associated with the EIA process include governmental agency personnel at federal, state, and local levels; engineering, planning, and environmental consulting firms; and private companies that have developed an in-house staff capability for the planning and conduction of environmental impact studies. Professionals involved in the EIA process include engineers, planners, biologists, geographers, landscape architects, and archaeologists.

FEATURES OF THE NATIONAL ENVIRONMENTAL POLICY ACT

The National Environmental Policy Act (P.L. 91-190), which is often referred to as the ''National Environmental Policy Act of 1969,'' since 1969 was the year that the law was developed, has had a profound effect on project planning and evaluation in the United States. Additionally, it could be reasonably argued that this single law has led to the adoption of environmental impact laws or policies in over 75 countries throughout the world and to the adoption of similar requirements on the part of international aid agencies and lending organizations (Sammy, 1982). NEPA is divided into two basic parts: Title I, which is a declaration of a national environmental policy, and Title II, which established the CEQ. The national goals of environmental policy as specified in Section 101 of the act are as follows (U.S. Cong., 1970, pp. 1–2):

1. Fulfill the responsibilities of each generation as a trustee of the environment for succeeding generations.
2. Assure for all Americans safe, healthful, productive, and aesthetically and culturally pleasing surroundings.
3. Attain the widest range of beneficial uses of the environment without degradation, risk to health or safety, or other undesirable and unintended consequences.
4. Preserve important historical, cultural, and natural aspects of our national heritage and maintain, where possible, an environment that supports diversity and variety of individual choice.
5. Achieve a balance between population and resource use that will permit high standards of living and a wide sharing of life's amenities.
6. Enhance the quality of renewable resources and approach the maximum attainable recycling of depletable resources.

Section 101 contains fundamental principles which encompass many issues receiving current attention (Bear, 1993); examples include pollution prevention, the importance of biological diversity, and the need for sustainable development.

Section 102 of the NEPA has three primary parts related to the environmental impact assessment process. Part A specifies that all agencies of the federal government must utilize a systematic, interdisciplinary EIA approach, which will ensure the integrated use of the natural and social sciences and environmental design arts in planning and in decision making that may have an impact on the human environment. Part B requires agencies to identify and develop methods and procedures that will ensure that presently unquantified environmental amenities and values be given appropriate consideration in decision making along with economic and technical considerations. This part has provided impetus for the development of environmental assessment methods. Part C indicates the necessity for preparing environmental statements (EISs) and identifies basic items to be included. It also indicates that agencies should include in every recommendation or report on proposals for legislation and other major federal actions significantly affecting the quality of the human environment a detailed statement that covers five major areas (U.S. Cong., 1970):

1. The environmental impact of the proposed action
2. Any adverse environmental effects that cannot be avoided should the proposal be implemented
3. Alternatives to the proposed action
4. The relationship between local short-term uses of the human environment and the maintenance and enhancement of long-term productivity
5. Any irreversible and irretrievable commitments of resources that would be involved in the proposed action should it be implemented

The requirement for preparing an EIS was not a part of the original proposed legislation that subsequently became the NEPA (Caldwell, 1973). Detailed histories of the legislative background of the NEPA have been presented by Andrews (1972) and Yannacone and Cohen (1972). Section 102 requirements were added late in the legislative review process, just prior to final action on the part of Congress. These particular requirements have been called the "action-forcing mechanism" of the NEPA and stipulate that agencies must prepare a draft statement, which is then subject to review and critique by other federal agencies as well as state and local governmental and private groups (Andrews, 1972).

One section of the NEPA that has received very little attention is Section 103, which requires that all agencies review their present statutory authority, administrative regulations, and current policies and procedures for the purpose of determining whether there are any deficiencies or inconsistencies therein that prohibit full compliance with the purposes and provisions of the NEPA. Very few written responses have been recorded with regard to action taken in conjunction with Section 103 (Caldwell, 1973; U.S. EPA, 1973a).

To aid the implementation of the EIS requirement, the NEPA also included the creation of the Council on Environmental Quality within the Executive Office of the President. This council has taken the role of providing overall coordination to the EIA process in the United States. The CEQ issued guidelines in 1971 and 1973 for federal agencies to follow in conjunction with EISs. During the decade of the 1970s, over 70 federal agencies also issued guidelines with regard to their policies and procedures in response to the NEPA requirements. Because of these multiple guidelines, there was considerable confusion as to terminology, timing requirements, and many other substantive issues. In 1978, the CEQ issued regulations which became effective in mid-1979 for responding to the requirements of the National Environmental Policy Act (CEQ, 1978). These regulations were modified slightly in 1986 (CEQ, 1987). It should be noted that these were regulations, not guidelines; therefore, they were required to be followed by all federal agencies. As a result of these regulations, federal agencies have now re-issued their previous guidelines as regulations in consonance with the CEQ regulations; a list of each agency's regulations is contained in CEQ (1992).

COUNCIL ON ENVIRONMENTAL QUALITY GUIDELINES (1971 AND 1973)

Title II of the NEPA established the CEQ, whose responsibilities in relation to the impact statement process initially included serving as a central repository for final EISs, preparing general guidelines applicable to all federal agencies in conjunction with their compliance with NEPA, reviewing draft EISs (particularly for controversial projects), and developing comparative analyses on the impact statement process. Federal agencies can request consultation and guidance from CEQ regarding NEPA compliance or the preparation of agency procedures and guidelines.

The CEQ published guidelines for the preparation of EISs on April 23, 1971, and August 1, 1973. The guidelines issued in 1971 coordinated the impact statement process, particularly with regard to review of draft EISs. Two new items were added to the five basic points specified by the NEPA for inclusion in an impact statement: a section describing the proposed action and a section discussing problems and objections raised by reviewers. The first new section precedes the basic five points, and the latter follows.

The CEQ guidelines of 1973 called for the addition of two more new sections in an impact statement, plus the expansion of a previously required section. As shown in Table 1.1, the initial section of an impact statement became a description of the proposed action, as well as a description of the existing environment (CEQ, 1973). One new section pertains to the relationship of the proposed action to existing land-use

plans, policies, and controls in the affected area, which requires a discussion of how the proposed action may conform or conflict with the objectives and specific terms of any federal, state, or local land-use matters, either approved or proposed. In addition, land-use plans developed in response to the requirements of the Clean Air Act or the Federal Water Pollution Control Act Amendments of 1972 should also be specified. The second new section calls for an indication of what other interests and considerations of federal policy are thought to offset the adverse environmental effects of the proposed action. This section is oriented to a discussion of other decision factors that the agency feels counterbalance any adverse environmental effects. Agencies that prepare cost-benefit analyses of proposed actions should summarize these analyses in this section. Where nonenvironmental costs and benefits are part of the basis for decision, it is important that the agency specify the importance of these elements in the decision. Table 1.1 lists the contents of an EIS as delineated by the CEQ (1973). Point 1 was added in 1971, whereas points 2 and 8 were added in 1973. The expansions of points 3 through 7 from the original NEPA requirements occurred primarily in 1973.

The information in Table 1.1 can be considered in relation to the five points specified in the NEPA to be addressed in an EIS. The first one is to describe "the environmental impact of the proposed action." In the early years of the preparation of impact statements, attention was focused primarily on the negative or detrimental impacts associated with a given proposed action. To be complete, the EIS should delineate both beneficial and detrimental impacts. The basic strong point of the NEPA is that it is a "full-disclosure law," implying that both the positive and negative ramifications of a given proposed action should be explored in complete detail (Best, 1972). In addition, attention must be directed toward the primary and secondary impacts associated with a proposed action. Primary

TABLE 1.1

1973 CEQ GUIDELINES FOR THE CONTENT OF ENVIRONMENTAL IMPACT STATEMENTS

From paragaph 1500.8, part (a), the following points are to be covered:

1. The EIS should include a description of the proposed action, a statement of its purposes, and a description of the environment affected, including information, summary technical data, and maps and diagrams, where relevant, adequate to permit an assessment of potential environmental impact by commenting agencies and the public. Highly technical and specialized analyses and data should be avoided in the body of the draft impact statement. Such materials should be attached as appendices or footnoted with adequate bibliographic references. The statement should also succinctly describe the environment of the area affected as it exists prior to a proposed action, including other federal activities in the area affected by the proposed action which are related to the proposed action. The interrelationships and cumulative environmental impacts of the proposed action and other related federal projects must be presented in the statement. The amount of detail provided in such descriptions should be commensurate with the extent and expected impact of the action and with the amount of information required at the particular level of decision making (planning, feasibility, design, etc.). In order to ensure accurate descriptions and environmental assessments, site visits should be made, where feasible. Agencies should also take care to identify, as appropriate, population and growth characteristics of the affected area and any population and growth assumptions used to justify the project or program or to determine secondary population and growth impacts resulting from the proposed action and its alternatives [see paragraph 3(ii)]. In discussing these population aspects, agencies should consider using the rates of growth in the region of the project contained in the projection compiled for the Water Resources Council by the Bureau of Economic Analysis of the Department of Commerce and the Economic Research Service of the Department of Agriculture (the "OBERS" projection). In any event, it is essential that the sources of data used to identify, quantify, or evaluate any and all environmental consequences be expressly noted.

2. The statement should outline the relationship of the proposed action to land-use plans, policies, and controls for the affected area. This requires a discussion of how the proposed action may conform or conflict with the objectives and specific terms of approved or proposed federal, state, and local land-use plans, policies, and controls, if any, for the area affected, including those developed in response to the *Clean Air Act* or the *Federal Water Pollution Control Act Amendments of 1972*. Where a conflict or inconsistency exists, the statement should describe the extent to which the agency has reconciled its proposed action with the plan, policy, or control, and the reasons the agency has decided to proceed notwithstanding the absence of full reconciliation.

3. The probable impact of the proposed action on the environment.
   i. This requires agencies to assess the positive and negative effects of the proposed action as it affects both the national and international environments. The attention given to different environmental factors will vary according to the nature, scale, and location of proposed actions. Among factors to consider should be the potential effect of the action on such aspects of the environment as those listed in Appendix II of the CEQ 1973 guidelines. Primary attention should be given in the statement to discussing those factors most evidently impacted by the proposed action.
   ii. Secondary or indirect, as well as primary or direct, consequences for the environment should be included in the analysis. Many major federal actions, in particular those that involve the construction or licensing of infrastructure investments (e.g., highways, airports, sewer systems, and water resource projects), stimulate or induce secondary effects in the form of associated investments and changed patterns of social and economic activities. Such secondary effects, through their impacts on existing community facilities and activities, through their role in inducing the development of new facilities and activities, or through the changes they generate in natural conditions, may often be even more substantial than the primary effects of the original action itself. For example, the effects of the proposed action on population and growth may be among the more significant secondary effects. Such population and growth impacts, if expected to be significant (using data identified as indicated in 1), should be estimated and an assessment made of the effect of any possible change in population patterns or growth upon the resource base, including land use, water, and public services, of the area in question.

TABLE 1.1

1973 CEQ GUIDELINES FOR THE CONTENT OF ENVIRONMENTAL IMPACT STATEMENTS
(Continued)

4. Alternatives to the proposed action, including, where relevant, those not within the existing authority of the responsible agency. (Section 102 (2) (D) of the act requires the responsible agency to "study, develop, and describe appropriate alternatives to recommended courses of action in any proposal which involves unresolved conflicts concerning alternative uses of available resources"). A rigorous exploration and objective evaluation of the environmental impacts of all reasonable alternative actions, particularly those that might enhance environmental quality or avoid some or all of the adverse environmental effects, is essential. Sufficient analysis of such alternatives and their environmental benefits, costs, and risks should accompany the proposed action through the agency review process in order not to foreclose prematurely options which might enhance environmental quality or have less detrimental effects. Examples of such alternatives include the alternative of taking no action or of postponing action pending further study; alternatives requiring actions of a significantly different nature which would provide similar benefits with different environmental impacts (e.g., nonstructural alternatives to flood control programs, or mass transit alternatives to highway construction); alternatives related to different designs or details of the proposed action which would present different environmental impacts (e.g., cooling ponds versus cooling towers for a power plant or alternatives that will significantly conserve energy); alternative measures to provide for compensation of fish and wildlife losses, including the acquisition of land, waters, and interests therein. In each case, the analysis should be sufficiently detailed to reveal the agency's comparative evaluation of the environmental benefits, costs, and risks of the proposed action and each reasonable alternative. Where an existing impact statement already contains such an analysis, its treatment of alternatives may be incorporated provided that such treatment is current and relevant to the precise purpose of the proposed action.

5. Any probable adverse environmental effects which cannot be avoided (such as water or air pollution, undesirable land-use patterns, damage to life systems, urban congestion, threats to health or other consequences adverse to the environmental goals set out in Section 101(b) of the act). This should be a brief section summarizing in one place those effects discussed in paragraph 3 of this section that are adverse and unavoidable under the proposed action. Included for purposes of contrast should be a clear statement of how other avoidable adverse effects discussed in paragraph 2 of this section will be mitigated.

6. The relationship between local short-term uses of man's environment and the maintenance and enhancement of long-term productivity. This section should contain a brief discussion of the extent to which the proposed action involves trade-offs between short-term environmental gains at the expense of long-term losses, or vice versa, and a discussion of the extent to which the proposed action forecloses future options. In this context short-term and long-term do not refer to any fixed time periods, but should be viewed in terms of the environmentally significant consequences of the proposed action.

7. Any irreversible and irretrievable commitments of resources that would be involved in the proposed action should it be implemented. This requires the agency to identify from its survey of unavoidable impacts in paragraph 5 of this section the extent to which the action irreversibly curtails the range of potential uses of the environment. Agencies should avoid construing the term "resources" to mean only the labor and materials devoted to an action. "Resources" also means the natural and cultural resources committed to loss or destruction by the action.

8. An indication of what other interests and considerations of federal policy are thought to offset the adverse environmental effects of the proposed action identified pursuant to paragraphs 3 and 5 of this section. The statement should also indicate the extent to which these stated countervailing benefits could be realized by following reasonable alternatives to the proposed action (as identified in paragraph 4 of this section) that would avoid some or all of the adverse environmental effects. In this connection, agencies that prepare cost-benefit analyses of proposed actions should attach such analyses, or summaries thereof, to the environmental impact statement, and should clearly indicate the extent to which environmental costs have not been reflected in such analyses.

Source: Compiled using data from CEQ, 1973.

and secondary impacts are also referred to as "direct consequences" and "indirect consequences." In general, agencies have developed methods and procedures to respond in part to direct impacts, both beneficial and adverse. However, the major impacts of a project are often from secondary or even tertiary effects, and these are much more difficult to assess because of the dearth of predictive techniques available.

The second item required by the NEPA is an identification of "any probable adverse environmental effects which cannot be avoided should the proposal be implemented." (See Table 1.1, item 5.) If a thorough approach has been utilized in describing the environmental impact of the proposed action, this section should basically be an abstract of the negative impacts, both direct and indirect, of the proposed action. This item has provided the basis for attention toward impact-mitigation measures.

The third point focuses on a discussion of "alternatives to the proposed action." This section has caused a great deal of difficulty, and many court cases have resulted from inadequate treatment of this section by the proposing agency. Implicit in the "Alternatives" section is the idea that the alternatives to the proposed action should be compared on a common basis, presumably their relative or absolute environmental impact (Hopkins, 1973). One alternative that should be discussed is the no-action, or no-project, alternative. This alternative requires the proposing agency to predict what the future environment will be without the project, and it serves as the basis against which impacts of the proposed action can be compared. Concern regarding alternatives originally arose in conjunction with the question of the retroactivity of the NEPA, and two well-known court cases that evolved from this point involve the Gillham Reservoir in southwestern Arkansas and the Cross-Florida Barge Canal (Anderson, 1973). Another point to be considered in the alternatives section is an evaluation of the alternatives through a public-participation program.

The fourth item is a description of "the relationship between local short-term uses of man's environment and the maintenance and enhancement of long-term productivity." This section is based on the principle that each generation should serve as caretakers of the environment for succeeding generations; therefore, attention must be paid to the question of whether options for future use of the environment are being eliminated or curtailed by the particular proposed action. This section is basic to current concerns regarding sustainable development. In a pragmatic sense, many impact statements have described the effects associated with the construction and the operational phases of a proposed action, considering those of the construction phase to be short-term and those of the operational phase to be long-term.

The last point is a discussion of "any irreversible and irretrievable commitments of resources which would be involved in the proposed action should it be implemented." Semantic difficulties are encountered with the terms "irreversible" and "irretrievable." Again, from a practical standpoint, most impact statements focus attention on possible changes in land usage as a result of a proposed action, loss of cultural features such as archaeological or historical sites, preclusion of development of underground mineral resources, loss of habitat for plants and animals, loss of or impact on threatened or endangered plants and/or animals or their habitats; expense of material required for project construction; energy usage required during project utilization; and even the human and monetary expenditures involved.

The 1973 CEQ guidelines also included information on federal and federal-state agencies with jurisdiction by law or de facto because of special expertise to comment on various environmental impacts (CEQ, 1973). This listing can be used in the following ways in conjunction with the EIA process: (1) to identify agencies to contact in conjunction with initial coordination and scoping activities and with subsequent public participation activities, (2) to identify agencies to contact for baseline information on the affected environment and/or for information on impact-prediction methods or approaches, and (3) to identify agencies to contact to solicit review comments on the draft EIS.

The addresses of Washington-level NEPA liaisons for various agencies are contained in CEQ (1992).

COUNCIL ON ENVIRONMENTAL QUALITY REGULATIONS

Many key concepts were included in the CEQ regulations which became effective in 1979. In May 1986, Section 1502.22 in the CEQ regulations was amended to clarify how agencies are to carry out their environmental evaluations in situations where information is incomplete or unavailable (U.S. EPA, 1989; CEQ, 1987).

A fundamental premise of the CEQ regulations is that the EIA process should be applied in the early planning considerations for a project. The Asian Development Bank defines the "project cycle" as consisting of project identification, fact-finding–preparation, preappraisal-appraisal, negotiations, implementation and supervision, completion and postevaluation. Environmental impact considerations in the project cycle are depicted in Figure 1.1. A similar project cycle with environmental considerations is used by the World Bank (1991).

A key feature of the CEQ regulations is the concept of three levels of analysis: level 1 relates to a categorical-exclusion determination, level 2 to the preparation of an environmental assessment and a finding of no significant impact, and level 3 to the preparation of an environmental impact statement (U.S. EPA, 1989). Figure 1.2 depicts the interrelationships between these three levels. Key definitions from the CEQ regulations related to Figure 1.2 include those for "federal action," "categorical exclusion," "environmental assessment," "finding of no significant impact," "significant impact," and "EIS."

"Federal actions" include the adoption of official policy (rules, regulations, legislation, and treaties) which will result in or substantially alter agency programs; adoption of formal plans; adoption of programs; and approval of specific projects, such as construction or management activities located in a defined geographic area, and actions approved by permit or other regulatory decision, as well as federal and federally assisted activities. A "categorical exclusion" refers to actions which do not individually or cumulatively have a significant effect on the human environment and which have been found to have no such effect in procedures adopted by a federal agency in implementation of the CEQ regulations. Neither an environmental assessment nor an EIS is required for categorical exclusions.

An "environmental assessment" (EA) is a concise public document that serves to briefly provide sufficient evidence and analysis for determining whether to prepare an EIS or a finding of no significant impact (FONSI), to aid an agency's compliance with the NEPA when no EIS is necessary, or to facilitate preparation of an EIS when one is necessary. A "FONSI" is a document by a federal agency which concisely presents the reasons why an action, not otherwise excluded, will not have a significant effect on the human environment and for which an EIS will not be prepared. A "mitigated FONSI" refers to a proposed action which has incorporated mitigation measures to reduce any significant negative effects to insignificant ones.

The key term in the EIA process is "significantly" or "significant impact," since a proposed action which significantly affects the human environment requires an EIS. "Significantly" as used in the NEPA requires considerations of both context and intensity. "Context" means that significance must be analyzed relative to society as a whole (human, national), the affected region, the affected interests, the locality, and the duration of effects—short- and/or long-term. "Intensity" refers to the severity

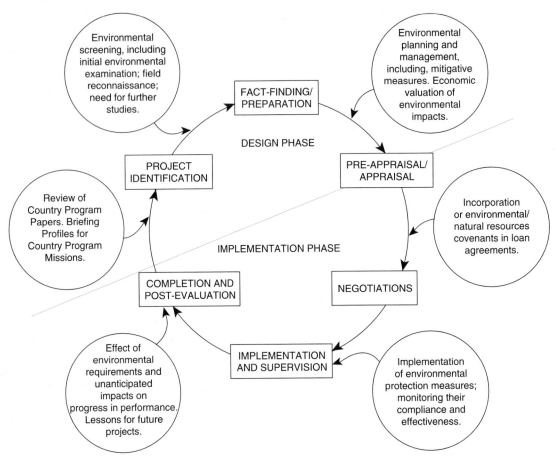

FIGURE 1.1
Project Cycle—Environmental and Natural Resources Planning and Management (Lohani, 1992).

of impact. The following should be considered in evaluating intensity:

1. The occurrence of impacts that may be both beneficial and adverse. A significant effect may exist even if the federal agency believes that on balance the effect will be beneficial.
2. The degree to which the proposed action affects public health or safety.
3. Unique characteristics of the geographic area, such as proximity to historic or cultural resources, parklands, prime farmlands, wetlands, wild and scenic rivers, or ecologically critical areas.
4. The degree to which the effects on the quality of the human environment are likely to be highly controversial.
5. The degree to which the possible effects on the human environment are highly uncertain or involve unique or unknown risks.
6. The degree to which the action may establish a precedent for future actions with significant effects or represents a decision in principle about a future consideration.

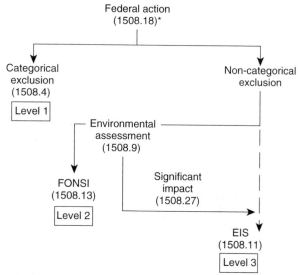

*Number denotes paragraph in CEQ regulations which contains definition (*Council on Environmental Quality*, 1987).

FIGURE 1.2
Three Levels of Analysis in the EIA Process.

7. Whether the action is related to other actions with individually insignificant but cumulatively significant impacts. Significance exists if it is reasonable to anticipate a cumulatively significant impact on the environment. Thus, significance cannot be avoided by terming an action "temporary" or by breaking it down into small component parts.

8. The degree to which the action may adversely affect districts, sites, highways, structures, or objects listed in or eligible for listing in the National Register of Historic Places or may cause loss or destruction of significant scientific, cultural, or historical resources.

9. The degree to which the action may adversely affect an endangered or threatened species or its habitat that has been determined to be critical under the Endangered Species Act of 1973.

10. Whether the action risks a violation of federal, state, or local law or requirements imposed for the protection of the environment.

An EIS is a detailed written statement which serves as an action-forcing device to ensure that the policies and goals defined in the NEPA are infused into the ongoing programs and actions of the federal government. It must provide full and fair discussion of significant environmental impacts and must inform decision makers and the public of the reasonable alternatives which would avoid or minimize adverse impacts or enhance the quality of the human environment. An EIS is more than a disclosure document; it must be used by federal officials in conjunction with other relevant material to plan actions and make decisions.

As shown in Figure 1.2, and based on the definitions above, all federal actions can be classified as either categorical exclusions or noncategorical exclusions. Each federal agency has published its list of categorical exclusions following appropriate consultation with the CEQ. For example, Table 1.2 contains categorical exclusions listed by the U.S. Army Corps of Engineers for its civil works program. A similar listing is available for the U.S. Department of the Army (1988). It should be noted that categorical exclusions do not negate other environmental considerations; for example, the U.S. Army Corps of Engineers has the following statement in its NEPA regulations (after U.S. Army Corps of Engineers, 1988, p. 3129):

> Actions listed in Table 1.2, when considered individually and cumulatively, do not have significant effects on the quality of the human environment and are categorically excluded from NEPA documentation. However, district commanders should be alert for extraordinary circumstances which may dictate the need to prepare an EA or an EIS. Even though an EA or EIS is not indicated for a federal action because of a categorical exclusion, the action is not exempt from compliance with any other federal law. For example, compliance with the Endangered Species Act, the Fish and Wildlife Coordination Act, the National Historic Preservation Act, the Clean Water Act, and so on, is always mandatory, even for actions not requiring an EA or EIS.

For federal actions which are noncategorical exclusions, it may be necessary to do an EA in order to determine whether or not an EIS will be required. Again, agencies have published lists of their typical actions which normally require an EA but not necessarily an EIS (CEQ, 1992). The listings indicate whether the regulations are in the Code of Federal Regulations (C.F.R.) or the *Federal Register*.

The key issue in Figure 1.2 is related to whether or not the federal action is a "major action significantly affecting the quality of the human environment," or MASAQHE. (This phrase is in Section 102 of the NEPA.) If a significant effect or effects (or impact or impacts) are anticipated, then an EIS is to be prepared.

If no significant effects are anticipated, or if they can be mitigated, then a FONSI is to be prepared; the FONSI was initially referred to as a "negative declaration."

Historically, defining a MASAQHE involved many quantitative and qualitative considerations. The simplest way of defining such a major action is to compare a predicted impact with an environmental quality standard for a given parameter. It is possible to do this for many substances found in air and water, for example, carbon monoxide in the atmosphere and dissolved oxygen in water. However, there are many environmental parameters for which only descriptive standards are available, such as scenic vistas and archaeological sites. Agencies may define MASAQHE by project type, indicating that certain projects require impact statements because they are major actions, and others do not because they are minor actions. In the early 1970s, the Federal Highway Administration developed guidelines of this type (U.S. Department of Transportation, 1972). Examples of major actions include a highway section entirely or generally on a new location and major upgrading of an existing highway section that requires extensive right-of-way acquisition and construction. Highway sections that may have a significant effect on the quality of the human environment include those that (1) are likely to have a significantly adverse impact on natural ecological, cultural, or scenic resources of national, state, or local significance; (2) are likely to be highly controversial regarding relocation-housing resources; (3) divide or disrupt an established community; disrupt orderly, planned development; are inconsistent with plans or goals that have been adopted by the community in which the project is located; or cause increased congestion; and (4) involve inconsistency with any national, state, or local standard relating to the environment; have a significantly detrimental impact on air or water quality or on ambient noise levels for adjoining areas; involve a possibility of contamination of a public water sup-

TABLE 1.2

CATEGORICAL EXCLUSIONS FOR THE CIVIL WORKS PROGRAM OF THE U.S. ARMY CORPS OF ENGINEERS

1. Activities at completed Corps projects which carry out the authorized project purposes. Examples include routine operation and maintenance actions, general administration, equipment purchases, custodial actions, erosion control, painting, repair, rehabilitation, replacement of existing structures and facilities such as buildings, roads, levees, groins, and utilities, and installation of new buildings, utilities, or roadways in developed areas.

2. Minor maintenance dredging using existing disposal sites.

3. Planning and technical studies which do not contain recommendations for authorization or funding for construction, but may recommend further study. This does not exclude consideration of environmental matters in the studies.

4. All operations and maintenance grants, general plans, agreements, etc., necessary to carry out land use, development, and other measures proposed in project authorization documents, project design memoranda, master plans, or reflected in the project NEPA documents.

5. Real-estate grants for use of excess or surplus real property.

6. Real-estate grants for government-owned housing.

7. Exchanges of excess real property and interests therein for property required for project purposes.

8. Real-estate grants for rights-of-way which involve only minor disturbances to earth, air, or water:
   (a) Minor access roads, streets, and boat ramps.
   (b) Minor utility distribution and collection lines, including those for irrigation.
   (c) Removal of sand, gravel, rock, and other material from existing borrow areas.
   (d) Oil and gas seismic and gravity meter surveys for exploration purposes.

9. Real-estate grants of consent to use government-owned easement areas.

10. Real-estate grants for archaeological and historical investigations compatible with the Corps' Historic Preservation Act responsibilities.

11. Renewal and minor amendments of existing real-estate grants evidencing authority to use government-owned real property.

12. Reporting excess real property to the General Services Administration for disposal.

13. Boundary line agreements and disposal of lands or release of deed restrictions to cure encroachments.

14. Disposal of excess easement interest to the underlying fee owner.

15. Disposal of existing buildings and improvements for off-site removal.

16. Sale of existing cottage-site areas.

17. Return of public domain lands to the Department of the Interior.

18. Transfer and grants of lands to other federal agencies.

Source: Compiled using data from U.S. Army Corps of Engineers, 1988.

ply system; or affect groundwater, flooding, erosion, or sedimentation.

Negative declarations, or FONSIs, can be prepared on the following types of highway improvement actions, since they are not likely to have significant impacts: (1) signing, marking, signalization, and railroad-protective devices; (2) acquisition of scenic easements; (3) modernization of an existing highway by resurfacing; less-than-lane-width widening; adding shoulders; adding auxiliary lanes for localized purposes; (4) correcting substandard curves; (5)

reconstruction of existing stream crossings where stream channels are not affected; (6) reconstruction of existing highway-highway or highway-railroad separations; (7) reconstruction of existing intersections including channelization; (8) reconstruction of existing roadbed, including minor widening, shoulders, and additional right-of-way; and (9) building rural two-lane highways on new or existing locations that are found to be generally environmentally acceptable to the public and local, state, and federal officials.

As noted earlier, the definition of "significantly" from the CEQ regulations includes the consideration of both context and intensity. "Context" is primarily related to the "when and where" of the impacts. The 10 points listed earlier for intensity can be divided into two groups, as follows: (1) those related to environmental laws, regulations, policies, and executive orders (points 2, 3, 8, 9, and 10 in the list); and (2) those related to other considerations and how they in turn may have implications for environmental laws, regulations, policies, and executive orders (points 1, 4, 5, 6, and 7 in the list). Additional considerations and approaches for significance determination are addressed later in this chapter.

Section 1502.22 of the CEQ regulations addresses necessary agency actions when there is incomplete or unavailable information on the potential environmental consequences of the proposed action. The concept of a worst-case analysis was included in the 1979 regulations which became effective on July 1, 1979 (CEQ, 1978). In April 1986, the CEQ changed the wording to a "reasonably foreseeable analysis" (CEQ, 1987). "Reasonably foreseeable" includes impacts which have catastrophic consequences, even if the probability of occurrence is low, provided that the analysis of the impacts is supported by credible scientific evidence, is not based on pure conjecture, and is within the "rule of reason." The intention is to focus attention on possible accidents and to encourage

planning for accident prevention via engineering and/or management countermeasures, and disaster management, if deemed necessary. Risk assessment can be used as a conceptual framework for this analysis.

If it is determined that a federal action has a significant impact or impacts based on the EA, or if the agency has previously listed the action as one typically requiring an EIS, then an EIS should be prepared. As an example, the listing for the civil works program of the U.S. Army Corps of Engineers includes the following actions normally requiring an EIS (U.S. Army Corps of Engineers, 1988): (1) feasibility reports for authorization and construction of major projects, (2) proposed changes in projects which increase size substantially or include additional purposes, and (3) proposed major changes in the operation and/or maintenance of completed projects. Draft EISs are circulated for review and comment, while final EISs must be filed with the U.S. EPA prior to initiating the action. This is a change from the initial years of the NEPA, when CEQ was the official recipient of EISs. The Office of Federal Activities in the EPA is the official recipient of all EISs prepared by federal agencies (U.S. EPA, 1989b).

Three types of EISs may be pertinent (draft, final, or supplemental to either draft or final); in addition to the earlier generic definition of an EIS, the following information from the CEQ guidelines or regulations is germane (after Council on Environmental Quality, 1973, 1987, p. 939):

1. *Draft EIS* The draft EIS is the document prepared by the lead agency proposing an action; it is circulated for review and comment to other federal agencies, state and local agencies, and public and private interest groups. Specific requirements with regard to timing of the review are identified in the CEQ regulations. The agency must make every effort to disclose and discuss at appropriate intervals in the draft statement all major points of view on the environmental im-

pacts of the alternatives, including the proposed action.

2. *Final EIS* The final EIS is the draft EIS modified to include a discussion of problems and objections raised by the reviewers. The final statement must be on file with the EPA for at least a 30-day period prior to initiation of construction on the project. The format for an EIS is delineated in Sections 1502.10 through 1502.18 of the CEQ regulations.

3. *Supplemental EIS* Lead agencies are to prepare supplements to either draft or final EISs if the agency makes substantial changes in the proposed action that are relevant to environmental concerns; or if there are significant new circumstances or information relevant to environmental concerns and bearing on the proposed action or its impacts. Lead agencies may also prepare supplements when the agency determines that the purposes of the act will be furthered by doing so.

The procedural aspects of filing draft, final, or supplemental EISs with the Office of Federal Activities are addressed in U.S. EPA (1989a). Several points of contact exist whereby draft or final EISs may be procured. Filed EISs are retained by the EPA for two years and then sent to the National Records Center. The EPA headquarters library in Washington, D.C., houses a microfiche collection of final EISs issued from 1970 through 1977, and all draft, final, and supplemental EISs filed from 1978 to the present time (U.S. EPA, 1989b). Regional EPA offices can also be contacted for relevant EISs. Collections of EISs are also available at the Northwestern University Transportation Library in Evanston, Illinois, Information Resources Press in Arlington, Virginia, and the Environmental Law Institute in Washington, D.C.

The approach used to prepare a draft, final, or supplemental EIS should be interdisciplinary, systematic, and reproducible. Requirements for a systematic and reproducible approach indicate

that a degree of organization and uniformity should be utilized in the assessment process. In this regard, numerous assessment methodologies have been developed since 1970, and these are discussed in more detail in Chapters 3 and 15. Requirements for an interdisciplinary approach indicate that the environment must be considered in its broadest sense; thus the input of persons trained in a number of technical fields needs to be included. The disciplines represented in a specific EIS must be oriented to the unique features of the proposed action and the environmental setting; however, at a minimum, it is necessary to have input from a physical scientist or an engineer, a biologist, and a person who can address cultural and socioeconomic impacts.

Public participation is to be achieved in the EIA process by means of (1) an early scoping process to determine the scope of issues to be addressed and to identify the significant issues related to a proposed action, (2) a public participation program during the EIA study, and (3) the review process for draft EISs. Specific activities during scoping are described in the CEQ regulations. "Scoping" refers to a process and not to an event or meeting. (See Chapter 4 for a discussion of scoping.) Some suggestions for planning and implementing the scoping process include the following (CEQ, 1981):

1. Start scoping after sufficient information is available on the proposed action.
2. Prepare an information packet.
3. Design a unique scoping process for each undertaking, project, plan, program, or policy.
4. Issue a public notice.
5. Carefully plan and conduct all public meetings.
6. Develop a plan for utilizing received comments.
7. Allocate EIS work assignments and establish a completion schedule.

To accomplish public participation (involvement), the lead agency must (CEQ, 1987) (1) make diligent efforts to involve the public in preparing and implementing the NEPA procedures; (2) provide public notice of NEPA-related hearings, public meetings, and the availability of environmental documents so as to inform those persons and agencies who may be interested or affected; (3) hold or sponsor public hearings or public meetings whenever appropriate or in accordance with statutory requirements applicable to the agency; (4) solicit appropriate information from the public; (5) explain in its procedures public sources of information or status reports on environmental impact statements and other elements of the NEPA process; and (6) make EISs, the comments received, and any underlying documents available to the public pursuant to the provisions of the Freedom of Information Act.

The "review process" refers to a process to solicit comments from various groups on the draft EIS prepared on a proposed action. Specifically, after preparing a draft EIS and before preparing a final EIS, the agency must (after Council on Environmental Quality, 1987, pp. 943–944)

1. Obtain the comments of any federal agency which has jurisdiction by law or defacto because of special expertise with respect to any environmental impact involved or which is authorized to develop and enforce environmental standards
2. Request the comments of (1) appropriate state and local agencies which are authorized to develop and enforce environmental standards, (2) Native American tribes, when the effects may be on a reservation, and (3) any agency which has requested that it receive statements on actions of the kind proposed
3. Request comments, if any, from the applicant
4. Request comments from the public, affirmatively soliciting comments from those persons or organizations who may be interested or affected

Detailed information on the review process is in the CEQ regulations. Central issues to be addressed in the review process are related to the impacts of the proposed action and the adequacy of the document. For example, Ross (1987, p. 140) listed three key questions to be addressed in the review of an EIS prepared in Canada:

1. Is the EIS suitably focused on the key questions which need to be answered to make a decision about the proposed action?
2. Is the EIS scientifically and technically sound?
3. Is the EIS clearly and coherently organized and presented so that it can be understood?

Figure 1.3 delineates time schedules and related issues associated with public participation in the three levels of analysis depicted in Figure 1.2. Planning for the integration of all environmental laws in the EIA process is depicted in Figure 1.4.

Forty frequently asked questions and answers related to the CEQ regulations are available in CEQ (1981, 1986). Their review can aid in clarifying the concepts and principles enunciated in the CEQ regulations.

"Strategic environmental assessment" (SEA) refers to the EIA process applied to policies, plans, or programs (Lee and Walsh, 1992). A programmatic EIS would be analogous to a strategic environmental assessment. Programmatic EISs (PEISs) can be used in the United States to address the environmental implications of the policies and programs of federal agencies. PEISs can also be used to address the impacts of actions that are similar in nature or broad in scope, including cases where cumulative impacts are of concern (Sigal and Webb, 1989). Site-specific or local-action EAs and/or EISs can be tiered from the PEIS.

While the concept of applying the EIA process to policies, plans, and programs was included in the NEPA, the vast majority of EISs prepared in its first 25 years have been on projects. However, the NEPA should be used in the formulation of policies and the planning of programs as well (Bear, 1993). EAs and/or EISs may be prepared for policies (including

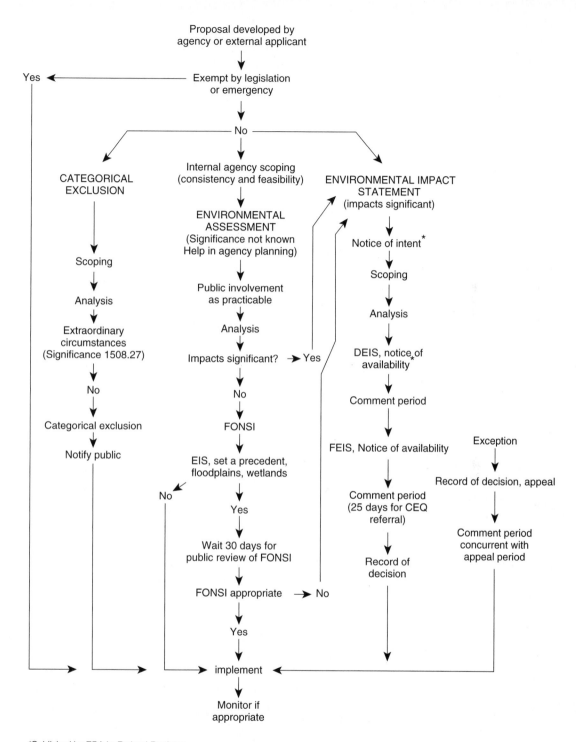

Proposal developed by
agency or external applicant

Exempt by legislation
or emergency → Yes

No

**CATEGORICAL
EXCLUSION**

Internal agency scoping
(consistency and feasibility)

**ENVIRONMENTAL IMPACT
STATEMENT**
(impacts significant)

**ENVIRONMENTAL
ASSESSMENT**
(Significance not known
Help in agency planning)

Scoping

Notice of intent*

Scoping

Analysis

Public involvement
as practicable

Analysis

Extraordinary
circumstances
(Significance 1508.27)

Analysis

DEIS, notice of
availability*

Impacts significant? → Yes

No

Comment period

No

Categorical exclusion

FONSI

FEIS, Notice of availability   Exception

Notify public

EIS, set a precedent,
floodplains, wetlands

Comment period
(25 days for CEQ
referral)

Record of decision, appeal

No

Yes

Comment period
concurrent with
appeal period

Wait 30 days for
public review of FONSI

Record of
decision

FONSI appropriate → No

Yes

implement

Monitor if
appropriate

*Published by EPA in *Federal Register*.

FIGURE 1.3
Selecting the Appropriate Level of NEPA Documentation (Shipley Associates, undated, p. 2).

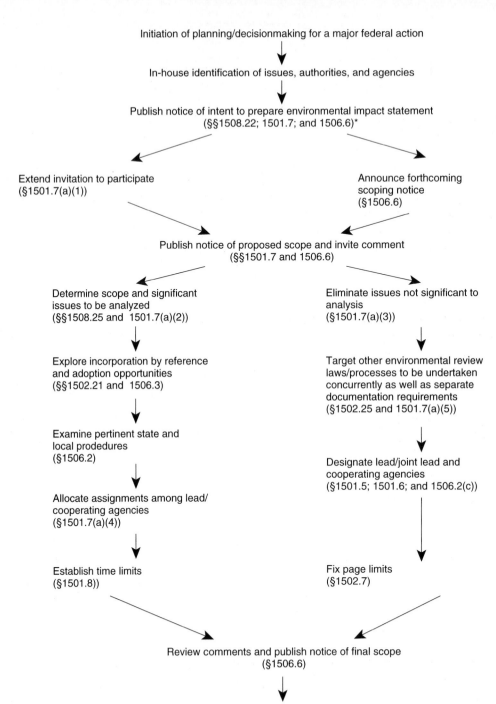

Initiation of planning/decisionmaking for a major federal action

In-house identification of issues, authorities, and agencies

Publish notice of intent to prepare environmental impact statement
(§§1508.22; 1501.7; and 1506.6)*

Extend invitation to participate
(§1501.7(a)(1))

Announce forthcoming
scoping notice
(§1506.6)

Publish notice of proposed scope and invite comment
(§§1501.7 and 1506.6)

Determine scope and significant
issues to be analyzed
(§§1508.25 and 1501.7(a)(2))

Eliminate issues not significant to
analysis
(§1501.7(a)(3))

Explore incorporation by reference
and adoption opportunities
(§§1502.21 and 1506.3)

Target other environmental review
laws/processes to be undertaken
concurrently as well as separate
documentation requirements
(§1502.25 and 1501.7(a)(5))

Examine pertinent state and
local prodedures
(§1506.2)

Allocate assignments among lead/
cooperating agencies
(§1501.7(a)(4))

Designate lead/joint lead and
cooperating agencies
(§1501.5; 1501.6; and 1506.2(c))

Establish time limits
(§1501.8))

Fix page limits
(§1502.7)

Review comments and publish notice of final scope
(§1506.6)

Proceed according to plan and schedule established in notice of final scope

* Refers to paragraphs in CEQ regulations.

FIGURE 1.4
Planning for Integration of Environmental Laws in EIA Process (Council on Environmental
Quality, 1991).

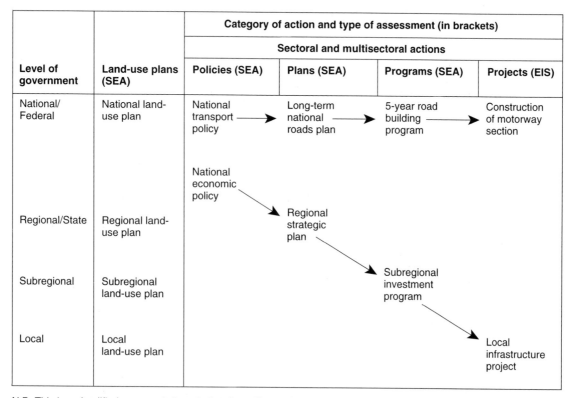

N.B. This is a simplified representation of what, in reality, could be a more complex set of relationships. In general, those actions at the highest tier level (e.g., national policies) are likely to require the broadest and least detailed form of strategic environmental assessment.

FIGURE 1.5
Actions and Assessments Within a Broad EIA Framework (Lee and Walsh, 1992).

rules, regulations, and/or new legislation), plans, or programs. Plans or programs may include operational considerations for extant single or multiple projects and repair, evaluation, maintenance, and rehabilitation of extant projects. Decommissioning of existing facilities also requires the application of the EIA process.

Figure 1.5 illustrates the relationships between policies, plans, programs, and projects, and SEAs and EISs (Lee and Walsh, 1992). The EIA process would exhibit its most far-reaching effects if it were systematically applied to pol-

icies, plans, and programs prior to being applied to specific projects.

As noted earlier, the CEQ was created by the NEPA as an office in the White House. As of August 1993 discussions were ongoing as to whether to rename the CEQ the "Office of Environmental Policy," which would remain within the White House, or to make it a part of a proposed cabinet-level office—the Department of the Environment. In either case, it is expected that the CEQ regulations will continue in effect, perhaps being renamed to match the new governmental designation.

SCREENING IN THE EIA PROCESS

"Screening" and "scoping" are terms which have been developed for usage in the EIA processes which are implemented in numerous countries. There are both subtle and significant differences in the usage of these terms from country to country. In effect, these terms are expansions of the concept of impact-significance determination described earlier. Figure 1.6 illustrates the conceptual relationship between screening and scoping. Fundamentally, if a proponent is developing a specific action, then the basic concern initially is the potential applicability of EIA requirements. Screening addresses the issue of whether or not an environmental impact study would be required for the potential action. In many countries, the resultant report on the environmental impacts is called an "environmental impact assessment" (EIA) or an "environmental assessment" (EA); in the United States, the resultant report is called an "EIS." Therefore, the fundamental issue addressed via screening is whether or not a comprehensive EIS (or EA) should be conducted. Scoping is focused primarily on determining the specific issues and impacts which may need to

be addressed in a comprehensive environmental impact study. Scoping is typically introduced after an affirmative decision has been made on the need for a comprehensive EIS, although it could also be a component of screening considerations.

It is possible to directly determine the need for a comprehensive environmental impact study, and then to address the scope of the issues or impacts which are relevant. Conversely, it may be necessary to conduct a preliminary study as a part of determining whether or not a comprehensive environmental impact study is needed. A confusing point is that this "preliminary study" is referred to by different terms in different countries. For example, in the United States the term "EA" is used, whereas in Canada and other countries the term "initial environmental evaluation" (IEE) finds application (Federal Environmental Assessment Review Office, undated).

Two fundamental approaches for determining whether or not to conduct a comprehensive environmental impact study for a proposed action are the use of policy delineations based on project type or size, or the conduction of a preliminary study.

FIGURE 1.6

Conceptual Framework for Screening and Scoping.

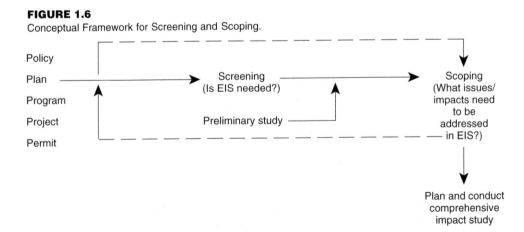

Policy Delineations

An example of a policy-based screening approach is that utilized by the World Bank in its directive on the conduction of environmental assessments (World Bank, 1989). Specifically, there are four categories of projects within the overall World Bank program, delineated by project type: those projects which would normally require an environmental impact study (Category A), those which may need some limited environmental review (Category B), those which normally do not need an environmental analysis (Category C), and environmentally beneficial projects and emergency recovery projects (Category D). Another example is the European Community directive on EIA; this directive has also categorized projects according to their level of need for a comprehensive environmental impact study (European Communities, 1985). Member states of the European Community can also develop further guidelines for projects to delineate those requiring impact studies.

In the United States, the approach has been to identify categorical exclusions from the EIA process. "Categorical exclusions" refer to those categories of action which do not in and of themselves require the preparation of even a preliminary study, much less an EIS. Examples of categorical exclusions are in Table 1.2.

Preliminary Study to Determine Impact Significance

The second fundamental approach is to conduct a preliminary study and, pending the findings of the study, either proceed to a comprehensive impact study or document that the findings were such that a comprehensive study would not be required. In the United States, the preliminary study yields an EA, with the fundamental issue being to determine whether the anticipated impacts of the action would have a significant effect on the quality of the human environment. Impacts resulting from proposed actions can be placed in one or more of the following catego-

ries: (1) beneficial or detrimental; (2) naturally reversible or irreversible; (3) repairable via management practices or irreparable; (4) short-term or long-term; (5) temporary or continuous; (6) occurring during the construction phase or the operational phase; (7) local, regional, national, or global; (8) accidental or planned (recognized beforehand); (9) direct (primary) or indirect (secondary); and (10) cumulative or single. Two key terms from the last pair of categories are "effects" (direct or indirect), and "cumulative impact"; their definitions are as follows (after CEQ, 1987, p. 955):

> *Effects, or impacts* These terms can be considered synonymous. Two broad categories of effects are "direct effects" and "indirect effects." "Direct effects" are caused by the action and occur at the same time and place. "Indirect effects" are caused by the action and are later in time or farther removed in distance, but are still reasonably foreseeable. Indirect effects may include growth-inducing impacts and other effects related to induced changes in the pattern of land use, population density or growth rate, and related effects on air, water, and other natural systems, including ecosystems. Effects may be ecological (such as the effects on natural resources and on the components, structures, and functioning of affected ecosystems), aesthetic, historic, cultural, economic, social, or health-related, whether direct, indirect, or cumulative. Effects may also include those resulting from actions which may have both beneficial and detrimental effects, even if on balance the agency believes that the overall effect will be beneficial.

> *Cumulative impact* The impact on the environment which results from the incremental impact of the action when added to other past, present, and reasonably foreseeable future actions regardless of what agency (federal or nonfederal) or person undertakes such other actions. Cumulative impacts can result from individually minor, but collectively significant actions taking place over a period of time.

Upon consideration of the above categories of impacts, several simple to more-structured options can be used to determine significance. At a minimum, the definition of "significantly" in the CEQ regulations could be applied as described in the previous section. Additional options include the application of significance criteria based on project type, structured methods, mitigation potential, and a series of sequenced questions.

To illustrate significance criteria based on project type, in the United States an option for water resources projects involves three considerations. "Principles and Guidelines" (Office of the President, 1983, pp. 9–10.) defined "significance" for environmental quality (EQ) resources based on institutional, on public, and/or on technical recognition; the relevant definitions and concepts are as follows:

1. "Significance based on institutional recognition" means that the importance of an EQ resource or attribute is acknowledged in the laws, adopted plans, and other policy statements of public agencies or private groups. Sources of institutional recognition include: public laws, executive orders, rules and regulations, treaties, and other policy statements of the federal government; plans and constitutions, laws, directives, resolutions, gubernatorial directives, and other policy statements of states with jurisdiction in the planning area; laws, plans, codes, ordinances, and other policy statements of regional and local public entities with jurisdiction in the planning area; and charters, bylaws, and formal policy statements of private groups.

2. "Significance based on public recognition" means that some segment of the general public recognizes the importance of an EQ resource or attribute. Public recognition may take the form of controversy, support, conflict, or opposition and may be expressed formally (as in official letters) or informally. Environmentally related customs and traditions should also be considered. EQ resources or attributes recognized by the public will often change over time as public awareness and perceptions change.

3. "Significance based on technical recognition" means that the importance of an EQ resource or attribute is based on scientific or technical knowledge or judgment of critical resource characteristics.

The concept of screening can lead to some proposals' being automatically excluded and some being automatically referred. Screening permits the proponent to arrive at one or more of the following conclusions relative to the potential impacts of the proposal: (1) there are insignificant adverse effects; (2) there are significant adverse effects which are mitigable; (3) there are significant adverse effects which may or may not be mitigable; (4) uncertainty exists relative to the environmental effects; (5) the effects are unknown; or (6) significant adverse effects are anticipated, and/or significant public concern exists, thus an EIS should be conducted. In accordance with the procedures in Canada where such studies are referred to as IEEs (as noted earlier), the central consideration is the significance of the anticipated impacts. Basic questions are asked to determine the significance of ecosystem components and the associated significance of anticipated impacts. Table 1.3 delineates a sequence of questions which can be used to determine the significance of the resource and anticipated impacts, with the basic presumption being that for those projects with resources or impacts which are deemed to be significant, an environmental impact study would be required. Valued ecosystem components must be determined before significant impacts can be discussed. This requires evaluation of the political, legal, public, and professional significance of environmental components. Determination of significant impact also relates to the degree of change in valued ecosystem components measured against some standard or threshold. This requires definition of the magnitude, prevalence, duration, frequency, and likelihood of potential change.

One example of a structured method for significance determination involves establishing a threshold of concern—a priority of that concern—and a probability that a potential environmental impact will, in fact, cross the threshold

TABLE 1.3

QUESTIONS RELATED TO SIGNIFICANCE DETERMINATION IN CANADA

1. Is the environmental component legally recognized as important?

- Environmental component is important if it is specifically protected by a law, policy, plan, control, or regulation; or is part of a legally defined management unit (e.g., a national park or an ecological reserve).
- Level of legal protection (i.e., federal, provincial, regional, or local) and the type of protection (i.e., law, plan, policy, control, or regulation) can affect the level of importance.
- Present legal status, the past and future predicted status.
- Environmental components legally identified as significant are commonly, also publicly, politically, and professionally, identified as important and, as such, usually rank high in relative importance.

2. Is the environmental component politically or publicly recognized as significant?

- Conditions affecting recognition of an environmental component as politically and publicly important:

  (1) Conflict over the use(s);
  (2) Resource availability and supply, and changes to that base;
  (3) Demand and changes in demand; and
  (4) Knowledge about the component and changes in that knowledge.

- Importance can be identified by any segment of the public, and the importance may be perceived rather than real.

- Assessment of the importance of an environmental component based on public input should consider:

  (1) Who and how many consider the environmental component to be important;
  (2) The past history of the use;
  (3) The public's expectations of future use;
  (4) Value of the environmental component to the public (monetary and otherwise); and
  (5) Real or perceived importance.

3. Is the environmental component professionally judged to be important?

- Professional judgment may often form the only basis for recognizing the significance of an environmental component. Careful documentation of that determination is essential.
- Key aspects evaluated by the professional in analyzing importance of an environmental component include:

  (1) Past, present, and projected future condition in the assessment area;
  (2) The condition in the context of the local area, the region, the province, the nation;
  (3) The size and extent of the environmental component;
  (4) Scarcity;
  (5) Monetary value; and
  (6) Biological, physical, and socioeconomic attributes of the environmental component.

Source: Federal Environmental Assessment Review Office, undated, pp. 18–19.

of concern (Haug et al., 1984). A "threshold of concern" is a maximum or minimum number, or other value, for an environmental impact or resource use which, if exceeded, causes that impact or use to take on new importance. Threshold priorities are defined as follows (Haug et al., 1984, p. 19):

1. *Highest priority—legal thresholds* Thresholds of impacts or resource use established by law or regulation. These may not be exceeded under any circumstances.

2. *Very high priority—functional thresholds* Thresholds established for resource use, or thresholds involving unavoidable adverse im-

pacts on the human environment, and so great that, if these thresholds are exceeded, the impacts will disrupt the functioning of an ecosystem sufficiently to destroy resources important to the nation or biosphere irreversibly and/or irretrievably.

3. *High priority—normative thresholds* Thresholds of impacts or resource use that are clearly established by social norms, usually at the local or regional level, and often tied to social or economic concerns.

4. *Moderate priority—controversial thresholds* Thresholds of impacts or resource use that are highly controversial, or which are sources of conflict between various individuals, advocacy groups, or organizations, and which do not warrant higher priority for other reasons.

5. *Low priority—preference thresholds* Thresholds of impacts or resource use that are preferences of individuals, groups, or organizations only, as distinct from society at large, and which do not warrant higher priorities for other reasons.

The probabilities of the impacts' crossing a threshold are expressed as (Haug et al., 1984, p. 24):

A *High likelihood* Greater than 1/2 chance of occurrence ($P > 0.5$)

B *Low likelihood* Less than (or equal to) 1/2, but at least a 1/20 chance of occurrence ($P = 0.05-0.5$)

C *Negligible* Less than 1/20 chance of occurrence ($P < 0.05$)

Table 1.4 summarizes the suggested integration of the three considerations. For example, if a threshold falls in categories 1A through 3A, it is significant. If it falls between 5A and 1C, it needs to be analyzed. Below 1C, thresholds are considered negligible and may be omitted from further analysis. Category 4A is a manager's option, in this example.

Thompson (1990) evaluated 24 EIA methodologies (such as matrices and various types of checklists) in terms of how they addressed impact significance determination. Wide variations were noted, with none of the methodologies providing a comprehensive framework, along with instructions, for determining the significance of anticipated impacts.

One of the things which can be done in conjunction with identification of significant negative impacts is to consider appropriate mitigation measures which could be applied to reduce these negative impacts within reasonable environmental and economic constraints. Relative to the U.S. practice, mitigation includes (CEQ, 1987) (1) avoiding the impact altogether by not taking a certain action or parts of an action; (2) minimizing the impact by limiting the degree or magnitude of the action and its implementation; (3) rectifying the impact by repairing, rehabili-

---

TABLE 1.4

PRIORITY-PROBABILITY[a] SCREEN FOR IMPACT SIGNIFICANCE DETERMINATION

Priority category	Impact likelihood category		
	A (high)	B (low)	C (negligible)
1 (highest)	Yes	No	No
2 (very high)	Yes	No	Omit
3 (high)	Yes	No	Omit
4 (moderate)	Maybe	No	Omit
5 (low)	No	No	Omit

[a]Yes = significant; no = needs to be analyzed but probably is not a significant concern; omit = no further consideration is necessary; maybe = on borderline between yes and no.
Source: Adapted from Haug et al., 1984, p. 24.

tating, or restoring the affected environment; (4) reducing or eliminating the impact over time by preservation and maintenance operations during the life of the action; and (5) compensating for the impact by replacing or providing substitute resources or environments. Usage of these measures should be attempted in sequence, beginning with avoiding the impact, or according to ease of application.

Negative impacts could be grouped in one of three categories: (1) insignificant; (2) significant but mitigable; or (3) significant but not mitigable. The mitigation concept in terms of mitigation would be that if potentially significant negative impacts are identified, and if they can be reduced to effects of lesser concern, then it would be possible to do a mitigated FONSI following an EA and without going to a comprehensive study leading to an EIS. However, it should be noted that this is feasible only if the mitigation requirements have been carefully delineated so as to ensure that the pertinent measures which have been identified are implemented.

Based upon the above discussion, a sequenced approach for impact significance determination would be appropriate. A "sequenced approach" suggests several levels of consideration in determining the potential significance of impacts from a proposed federal action (a project, plan, or program) or policy. It can be achieved by applying the series of questions as follows in the order shown (the answer to any or all of the questions can be used to determine whether an EIS should be prepared):

1. Does the proposed project, plan, program, and/or policy cause impacts that exceed the definition of "significant" as contained in pertinent laws, regulations, or executive orders?
2. Is a quantitative threshold criterion exceeded in terms of the type, size, or cost of the undertaking?
3. Is the action located in a protected habitat or land-use zone, or within an exclusionary zone relative to land usage? Is the environmental resource to be affected a significant resource?
4. Is the proposed undertaking and/or policy expected to be in compliance with pertinent environmental laws, regulations, policies, and/or executive orders?
5. What is the anticipated percentage change in pertinent environmental factors from the proposed action, and will the changes be within the normal variability of the factors? What is the sensitivity of the environment to the anticipated changes; or is the environment susceptible or resilient to changes? Will the "carrying capacity" of the resource (ability to support and maintain environmental processes) be exceeded?
6. Are there sensitive human, living, or inanimate receptors to the environmental stresses from the proposed project, plan, program, and/or policy?
7. Can the anticipated negative impacts be mitigated in a cost-effective manner?
8. What is the professional judgment of experts in the pertinent substantive areas, such as water quality, ecology, planning, landscape architecture, and archaeology?
9. Are there public concerns due to the impact risks of the proposed project, plan, program, and/or policy?
10. Are there cumulative impacts which should be considered, or impacts related to future phases of the proposed action and associated cumulative impacts?

Detailed, more-specific questions related to the above 10 can be developed.

ROLE OF THE U.S. ENVIRONMENTAL PROTECTION AGENCY

The U.S. Environmental Protection Agency was established in December 1970 as the environmental regulatory agency of the United States. It is not the chief administrative agency for EISs, although it does serve as the central repository for them. The EPA reviews environmental impact statements prepared by others, particularly with regard to water pollution, air pollution, solid-waste management, noise, radiation, and pesticides. Each statement reviewed is assigned a rating based on the nature of the proposed action and the EIS document itself (U.S. EPA, 1973b). The EPA system of rating other agency actions is as follows (after U.S. EPA, 1984, pp. 4-5–4-6):

1. Rating the environmental impact of the action:

   LO (lack of objections) The review has not identified any potential environmental impacts requiring substantive changes to the preferred alternative. The review may have disclosed opportunities for application of mitigation measures that could be accomplished with not more than minor changes to the proposed action.

   EC (environmental concerns) The review has identified environmental impacts that should be avoided in order to fully protect the environment. Corrective measures may require substantial changes to the preferred alternative or the application of mitigation measures that can reduce the environmental impact.

   EO (environmental objections) The review has identified significant environmental impacts that should be avoided in order to adequately protect the environment. Corrective measures may require substantial changes to the preferred alternative or consideration of some other project alternative (including the no-action alternative or a new alternative). The basis for environmental objections can include situations in which *(a)* an action might violate or be inconsistent with achievement or maintenance of a national environmental standard; *(b)* the federal agency violates its own substantive environmental requirements that relate to the EPA's areas of jurisdiction or expertise; *(c)* there is a violation of an EPA policy declaration; *(d)* there are no applicable standards, or applicable standards will not be violated but there is potential for significant environmental degradation that could be corrected by project modification or other feasible alternatives; or *(e)* proceeding with the proposed action would set a precedent for future actions that collectively could result in significant environmental impacts.

   EU (environmentally unsatisfactory) The review has identified adverse environmental impacts that are of sufficient magnitude that the EPA believes the proposed action must not proceed as outlined. The basis for an "environmentally unsatisfactory" determination consists of identification of environmentally objectionable impacts as defined above and fulfillment of one or more of the following conditions: *(a)* the potential violation of or inconsistency with a national environmental standard is substantive and/or will occur on a long-term basis; *(b)* there are no applicable standards, but the severity, duration, or geographical scope of the impacts associated with the proposed action warrant special attention; or *(c)* the potential environmental impacts resulting from the proposed action are of national importance because of the threat to national environmental resources or policies.

2. Adequacy of the impact statement:

   Category 1 (adequate) The draft EIS adequately sets forth the environmental impact(s) of the preferred alternative and those of the alternatives reasonably available to the action. No further analysis or data collection is necessary, but the reviewer may suggest the addition of clarifying language or information.

   Category 2 (insufficient information) The draft EIS does not contain sufficient information to completely assess environmental impacts that should be avoided in order to fully protect the environment, or the reviewer has identified new reasonably available alternatives that are within the spectrum of those analyzed in the draft EIS and which could reduce the environmental impacts of the proposal. The identified additional information, data, analyses, or discussion should be included in the final EIS.

Category 3 (inadequate) The draft EIS does not adequately assess the potentially significant environmental impacts of the proposal, or the reviewer has identified new, reasonably available alternatives which are outside of the spectrum of those analyzed in the draft EIS and which should be analyzed in order to reduce the potentially significant environmental impacts. The identified additional information, data, analyses, or discussions are of such a magnitude that they should have full public review at a draft stage. This rating indicates the EPA's belief that the draft EIS does not meet the purposes of the NEPA and/or the Clean Air Act Section 309 review, and thus should be formally revised and made available for public comment in a supplemental or reworked draft EIS.

All category 3 and EU ratings must be cleared at the EPA headquarters in order to double-check that all such ratings are consistent with policies and practices followed by EPA on a nationwide scale. The originating agency is notified of the assigned ratings at the time a comment letter is sent. The EPA also notifies the CEQ of its comments on all category 3 or EU projects so that the council can begin to follow up the project at an early stage in its development.

Although the EPA does not have official responsibility with regard to the acceptance or re-

jection of EISs, the review by this agency is critical, and the procurement of a satisfactory evaluation of the impact statement and the proposed action is imperative.

SUMMARY STATISTICAL INFORMATION ON EISs

During the 1970s, there were approximately 1,200 final EISs produced on an annual basis by the various federal agencies. During the 1980s, the number decreased and currently is in the range of 400 to 500 EISs annually. Table 1.5 lists the number of EISs produced annually by various agencies in the period 1979–1991; a detailed breakdown by agency is contained in CEQ (1993a). While it would appear that there has been a decrease in EIA emphasis in the United States (based upon the reduction in numbers of EISs), it should be noted that there has been a corresponding significant increase in the number of EAs which have been prepared. Unfortunately, no statistics are available on the number of EAs; however, one EPA official recently estimated that 30,000 to 50,000 are drafted annually (Smith, 1989). Approximately 45,000 EAs were prepared in 1992 (CEQ, 1993a). In addition, EIA concepts such as scoping and mitigation are now being included in initial project planning and decision making, thus reducing the need for subsequent preparation of EISs. Moore (1992) suggested that several factors have contributed to the decline in the number of EISs prepared annually: (1) better scoping, (2) better project planning to reduce negative impacts and avoid the need for EISs, (3) use of mitigated FONSIs, and (4) less litigation.

An issue related to the EIA process in the United States that has always been of concern is litigation. A total of 2,260 cases were filed from 1970 to 1991, with over 210 injunctions granted. In the first 14 years (1970–1983), the number of cases averaged 116 annually; from

TABLE 1.5

ENVIRONMENTAL IMPACT STATEMENTS FILED, BY FEDERAL AGENCY, 1979–1991

Agency	1979	1980	1981	1982	1983	1984	1985	1986	1987	1988	1989	1990	1991
Agriculture	172	104	102	89	59	65	117	118	75	68	89	136	145
Commerce	54	53	36	25	14	24	10	8	9	3	5	8	13
Defense	1	1	1	1	1	0	0	0	2	0	0	0	0
Air Force	8	3	7	4	6	5	7	8	9	6	11	19	20
Army	40	9	14	3	6	5	5	2	10	8	9	9	21
COE	182	150	186	127	119	116	106	91	76	69	40	46	45
Navy	11	9	10	6	4	9	8	13	9	6	4	19	9
Energy	28	45	21	24	19	14	4	13	11	9	6	11	2
EPA	84	71	96	63	67	42	16	18	19	23	25	31	16
GSA	13	11	13	8	1	0	4	0	1	3	0	4	3
HUD	170	140	140	93	42	13	15	18	6	2	7	5	7
Interior	126	131	107	127	146	115	105	98	110	117	61	68	64
Transportation	277	189	221	183	169	147	126	110	101	96	90	100	87
TVA	9	6	4	0	2	1	0	1	0	0	0	3	0
Other	98	44	76	55	22	21	26	15	17	20	23	18	24
Total	1,273	966	1,033	808	677	577	549	521	455	430	370	477	456

Notes: Years refer to calendar years. Number of EISs includes draft EISs, EIS supplements, and final EISs filed during the specified year. Some proposed projects may have several draft and final EISs filed over a period of years. COE = U.S. Army Corps of Engineers. HUD = Department of Housing and Urban Development. EPA = U.S. Environmental Protection Agency. GSA = General Services Administration. TVA = Tennessee Valley Authority.
Source: Council on Environmental Quality, 1993a, p. 169.

TABLE 1.6

TYPES OF COMPLAINTS FILED UNDER NEPA, 1991

Causes of action	Number for 1991 cases	Number for pre-1991 cases resulting in injunctions in 1991
Inadequate Environmental Impact Statement	26	
No Environmental Impact Statement, when one should have been prepared	41	
Inadequate Environmental Assessment	20	
No Environmental Assessment, when one should have been prepared	12	4
No Supplemental Environmental Impact Statement, when one should have been prepared	6	
Other	23	
Total	128	4

Source: Council on Environmental Quality, 1993a, p. 167.

1984 to 1991, the number of cases averaged 79 annually (CEQ, 1993a). The most common complaint was that no EIS was prepared when one should have been prepared; the second most common complaint was that the EIS prepared was inadequate. It should also be noted that the primary focus of the majority of the court cases has not been on technical impact issues related to proposed projects, but on procedural or administrative requirements and lack of compliance therewith; this is demonstrated upon examination of the types of 1991 complaints, as shown in Table 1.6. The plaintiffs for NEPA lawsuits in 1991 included individuals or citizen groups (41 cases), environmental groups (37 cases), property owners or residents (16 cases), local governments and business groups (5 cases

each), state governments (4 cases), and Native American tribes and other groups (2 cases each). The total of 128 lawsuits includes multiple plaintiffs on some cases (CEQ, 1993a).

STATE ENVIRONMENTAL POLICY ACTS

Beginning in 1970, several states adopted legislation equivalent to the NEPA at the state level. As of 1991, 16 states, the District of Columbia, and Puerto Rico had adopted environmental policy acts, or "little NEPAs." Eighteen states and the District of Columbia have limited-environmental-review requirements established by statute, executive order, or other administrative directives. Table 1.7 lists the states in each category.

TABLE 1.7

SUMMARY OF STATES WITH EIA REQUIREMENTS

States and districts with environmental policy acts	States and districts with limited environmental review requirements established by statute, executive order, or other administrative directives
Arkansas	Arizona
California	Arkansas
Connecticut	California
District of Columbia	Delaware
Florida	District of Columbia
Hawaii	Georgia
Indiana	Louisiana
Maryland	Massachusetts
Massachusetts	Michigan
Minnesota	New Jersey
Montana	North Carolina
New York	North Dakota
North Carolina	Oregon
Puerto Rico	Pennsylvania
South Dakota	Rhode Island
Virginia	South Dakota
Washington	Utah
Wisconsin	Washington
	Wisconsin

Source: Adapted from Council on Environmental Quality, 1992, p. 373.

EIA AT THE INTERNATIONAL LEVEL

As of the early 1980s, over 75 countries had adopted EIA legislation or regulations (Sammy, 1982). With current international lending and aid agency EIA requirements, it is estimated that over 100 countries have now participated in the EIA process.

UTILITY OF THE EIA PROCESS

One of the uncertainties in the planning and conduct of impact studies is related to appropriate costs for such studies. There has never been any systematic development of a cost algorithm which could be used for estimation purposes. Costs for impact studies occur over a wide range in terms of the cost of the study expressed as a percentage of the overall project costs. A rule of thumb is that the EIS will probably cost on the order of 1 percent or less of total project costs. However, this percentage can vary considerably, with study expenses for smaller projects being on the high side of 1 percent of project costs and, conceivably, as much as 5 to 10 percent. Conversely, for larger projects, the costs for environmental impact studies might be in the range of 0.1 to 0.5 percent of the project costs, or even less.

The usefulness of environmental impact studies has been questioned from the perspective of whether or not the preparers of the impact documentation are biased, particularly when they work for the agency proposing the action or work as consultants to agencies or private sector entities which are project proponents. There is little that can be done to completely remove this concern. Perhaps the strongest argument against the occurrence of bias is that if impact studies are conducted by professionals, such as engineers and planners, who have some training and experience, their own personal and professional codes of ethics will cause them to do the most appropriate job within the constraints of the study. Individuals should be expected to work in the most appropriate and expeditious manner and to fully disclose pertinent information which is found. However, one of the real difficulties is that, in many cases, information about the proposed action or environmental setting is unintentionally or unavoidably incomplete.

Potential bias could be introduced during any activity in an impact study; accordingly, it is necessary to carefully document the activities associated with the conduct of the study. One of the ways to avoid the introduction of bias is for the preparers of EAs or EISs to independently question information regarding the proposed action and the environmental setting. Additionally, usage of information from ''look-alike'' projects can be very helpful in the analysis.

The best approach in an environmental impact study is to do a thorough job of impact identification and analysis, to carefully document all the activities and information sources utilized in the study, and to appropriately conduct scoping (discussed later) and public information activities so as to answer questions that might exist in the minds of individuals and groups who are advocates of the particular project, as well as those who are its opponents.

It can be argued that there have been few benefits from the conduct of environmental impact studies over the past two decades. This perspective focuses upon the ''mountains'' of reports that have been prepared, many of which could be termed ''encyclopedic,'' and many of which have been forgotten as soon as a permit is received or a decision is made to go ahead with project construction and operation. On the other hand, numerous examples could be cited where projects have not been completed because of environmental concerns or where a project's design has been changed or size has been adjusted to make it more environmentally compatible. Accordingly, and in conjunction with the general spirit and intent of the NEPA, the overall goal should be use of the EIA process to facilitate incorporation of the environment as a factor in project decision making, along with engineering and economic factors, and to then develop the project and the location in such a manner as to result in an environmentally appropriate project. It is realized that this desired outcome is idealistic, and that in many cases it may not be easily accomplished. On the other hand, an open EIA process can be an aid in the development of the undertaking. Additional benefits resulting from the EIA process are that it has kept environmental issues before the public; it has altered the decision-making process of federal agencies; along with the Freedom of Information Act, it has forced public disclosure of decision making and created public participation in that decision making; and finally, it has been a model for environmental policy legislation by individual states and by a number of foreign nations (Moore, 1992).

Some potential benefits of the EIA process in developing countries are that it may encourage the inclusion of environmental considerations by developers, aid in obtaining better information about projects, help identify interests and trade-offs, help identify management and mitigation measures, bring about coordination and consultation among interested parties, increase technical expertise and experience, and facilitate better decisions (U.N. Environment Program, 1987). Constraints to the EIA process are related to fragmented authority among government agencies, the power of major development sectors, uncoordinated decision making, a lack of awareness either within the central government or at the local level, a lack of baseline data-evaluation criteria or analytical techniques; the difficulty of making judgments; the need for proven environmental-mitigation mechanisms; and the lack of technical and financial resources (U.N. Environment Program, 1987).

The EIA process can influence public policy in a variety of ways. One such policy influence might be on the stipulated mitigation or permit requirements associated with project approval. These requirements may take the form of pollution control equipment specifications, pollution control measures, operational cycles, and construction practices and timing, for examples. In addition, and again depending upon the nature and scope of the project, an impact study could lead to the delineation of long-term environmental-monitoring requirements.

EXPANDED SCOPE OF EIA

In February 1992, the United States signed the 1991 Convention on Environmental Impact Assessment in a Transboundary Context; additional signers included 28 other countries and the European Community (CEQ, 1993a; Economic Commission for Europe, 1991). The convention stipulates the obligations of parties to carry out the assessment of environmental impacts and, for certain activities likely to cause significant adverse transboundary effects, to arrange for its conduction at an early stage of planning. The convention provides procedures, in a transboundary context, for the embedding of environmental considerations in the decision-making process, thereby promoting sustainable development. It codifies the general obligation of states to notify and consult each other on all major projects under consideration, ranging from nuclear power stations to major road-construction and deforestation projects that are likely to cause significant adverse environmental impacts across boundaries. The convention also stipulates that public participation in the assessment procedure be established by each party at the project level and that foreign participants be given the same opportunities given to the local public and other nationals (Lee, Walsh, and Jones, 1991).

Application of the EIA process to a life-cycle analysis of manufactured products would incorporate considerations of the direct and indirect environmental consequences of the extraction of raw materials from the ground and of the various related processing, manufacturing, fabrication, and transportation steps, concluding with the acts of consumption, disposal, and recovery for recycling (Hunt, Sellers, and Franklin, 1992). The life-cycle viewpoint focuses on the total effects of the product system. This type of study has been referred to as a "resource and environmental profile analysis" (REPA); it has also been called a "life-cycle analysis" (LCA).

The CEQ (1989) has provided guidance to federal agencies relative to applying the EIA process to broader environmental issues, such as stratospheric ozone depletion and global warming. Climate change is a potentially relevant topic in the EIA process as a result of (1) the potential for an action to individually alter climate, (2) the cumulative impacts of the action in concert with other actions, and (3) the potential for future climate change to alter the baseline environment (and thus to affect the action or to alter the impact of the action) (Cushman et al., 1989).

Additional global issues receiving increasing international attention in EIA considerations are biological diversity and sustainable development. The CEQ (1993b) has already proposed that federal agencies address the impacts of their actions on biological diversity. Sustainable-development issues may receive greater attention in the future; an underlying principle is included as the first environmental policy listed in the NEPA—that each generation should fulfill its responsibilities as trustee of the environment for succeeding generations.

NARROWED SCOPE OF EIA

The concepts of the EIA process have become fundamental to numerous federal- and state-level environmental programs, permits, and reports. The documentation is analogous to a targeted EA or EIS. Examples of relevant permits and/or reports include (1) air quality permits and related reporting required by the Clean Air Act Amendments of 1990; (2) point-source wastewater discharge permits and related reporting required by the National Pollutant Discharge

Elimination System (NPDES) program of the Federal Water Pollution Control Act Amendments and its subsequent amendments; (3) industrial-area storm-water discharge permits and related reporting required by the NPDES program of the Clean Water Act of 1987; (4) permits for dredging and filling activities in navigable waters as required by Section 404 of the Clean Water Act of 1972; (5) remedial investigations, feasibility studies, and records of decision on uncontrolled hazardous-waste sites identified under the auspices of the Comprehensive Environmental Response, Compensation, and Liability Act (CERCLA, or the "Superfund" act) of 1981 and the Superfund Amendments and Reauthorization Act (SARA) of 1986 (Sharples and Smith, 1989); (6) replacements, permits, and reports on underground storage tanks regulated by the Resource Conservation and Recovery Act Amendments of 1984; (7) operating permits and closure plans for sanitary landfills or hazardous-waste landfills required by the Resource Conservation and Recovery Act Amendments (also called the "Hazardous and Solid Waste Act") of 1984 (Sharples and Smith, 1989); (8) reports prepared on site, or property-transfer, assessments to establish owner-buyer-lender liability for contamination; (9) reports prepared on regulatory audits; and (10) environmental-reporting requirements related to new-chemicals or -products licensing.

SUMMARY

The NEPA and the EIA process have had a profound effect on project planning and decision making in the United States. Process concepts and regulations are well-established. EIA concepts are now being applied to both broader, global issues and more-focused permit issues. However, based upon over two decades of EIA practice in the United States, CEQ annual reports, court cases, and other lessons learned, four key issues are current items of concern in relation to the NEPA requirements:

1. The need to determine the extent of mitigation-planning and mitigation-identification responsibility which an agency should undertake prior to issuance of an EIS.
2. The need for a methodology or procedure for systematically addressing cumulative impacts of proposed actions. This is a difficult issue to address; some reasons that cumulative impacts are only marginally considered in the United States include *(a)* the absence of a coordinated regional planning system (county, state, or federal), *(b)* the limited development of methods and policies, *(c)* study constraints in timing and funding, and *(d)* limited guidance from federal agencies.
3. The need for a usable methodology or procedure for conducting a reasonable foreseeability analysis (analogous to a worst-case analysis) of the consequences of a proposed action, particularly when information is incomplete or unavailable.
4. The need for follow-up environmental auditing to document experienced impacts and compare such them to preproject predicted ones. Reviews of the accuracy of predicted impacts in EISs have suggested that only a few forecasts were grossly inaccurate; however, only about 30 percent of the experienced impacts were close to their forecasts (Culhane, 1987). These findings suggest that feedback from impact-prediction auditing could be used to improve the forecasting of impacts from future projects. From an expanded perspective, it has also been suggested that agencies and federal facilities should audit their compliance with the NEPA, and a protocol for federal-facility-compliance auditing has been developed (Sigal and Cada, 1990).

Because of the need to integrate information from a number of substantive areas in the EIA process, and because of the relative newness of this field in environmental management when compared with more traditional disciplines such as biology, chemistry, and environmental engineering, there is a need for practitioners who work on the planning, conduction, and review of environmental impact studies to receive appropriate training.

Another outcome of the EIA requirements in the United States has been the planning and implementation of research programs focused on answering specific impact-related questions. Such programs have typically been associated with research conducted within specific agencies, such as the U.S. Department of Energy and the U.S. Army Corps of Engineers. An overall EIA research program to address comprehensive needs and issues across a range of agency requirements has never been developed in the United States. Creation of this type of generic research program continues to be a fundamental prerequisite to accomplishing effective environmental management in the nation.

SELECTED REFERENCES

Anderson, F. R., *NEPA in the Courts,* Resources for the Future, Washington, D.C., 1973.

Andrews, R. N. L., "Environmental Policy and Administrative Change: *The National Environmental Policy Act of 1969,* 1970–1971," Ph.D. dissertation, University of North Carolina, Chapel Hill, 1972, pp. 76–109, and app. A.

Barrett, B. F. D., and Therivel, R., *Environmental Policy and Impact Assessment in Japan,* Routledge, Chapman, and Hall, New York, 1991, p. 149.

Bear, D., "NEPA: Substance or Merely Process," *Forum for Applied Research and Public Policy,* vol. 8, no. 2, summer 1993, pp. 85–88.

Best, J. A., *"The National Environmental Policy Act* as a Full Disclosure Law," NTIS Rep. PB-227-809, U.S. Department of Commerce, National Technical Information Service, Springfield, Va., Dec. 1972.

Caldwell, L. K., *"The National Environmental Policy Act:* Status and Accomplishments," *Proceedings of the 38th North American Wildlife Natural Resources Conference,* Wildlife Management Institute, Washington, D.C., 1973.

"Can Change Damage Your Mental Health?" *Nature,* vol. 295, Jan. 21, 1982, pp. 177–179.

Council on Environmental Quality (CEQ), "Draft of Guidance to Federal Agencies Regarding Consideration of Global Climate Change in Preparation of Environmental Documents," Washington, D.C., June 21, 1989a.

———, "Environmental Quality," *Twenty-first Annual Rep.,* U.S. Government Printing Office, Washington, D.C., Apr. 1991, pp. 198–199.

———, "Environmental Quality," *Twenty-second Annual Rep.,* U.S. Government Printing Office, Washington, D.C., Mar. 1992, pp. 141–150, 352–377.

———, "Environmental Quality," *Twenty-third Annual Rep.,* U.S. Government Printing Office, Washington, D.C., Jan. 1993a, pp. 151–172.

———, Code of Federal Regulations (C.F.R.), Vol. 40, chap. V, U.S. Government Printing Office, Washington, D.C., July 1, 1987, pp. 929–971.

———, "Incorporating Biodiversity Considerations into Environmental Impact Analysis Under the *National Environmental Policy Act,"* Washington, D.C., Jan. 1993b.

———, "Memorandum: Questions and Answers About the NEPA Regulations," *Federal Register,* vol. 46, Mar. 23, 1981, p. 18026 ff., and as amended, vol. 51, Apr. 25, 1986, p. 15618 ff.

———, "Memorandum: Scoping Guidance," Washington, D.C., Apr. 30, 1981.

———, *"National Environmental Policy Act*—Regulations," *Federal Register,* vol. 43, no. 230, Nov. 29, 1978, pp. 55978–56007.

———, "Preparation of Environmental Impact Statements: Guidelines," *Federal Register,* vol. 38, no. 147, Aug. 1, 1973, pp. 20550–20562.

Culhane, P. J., "The Precision and Accuracy of U.S. Environmental Impact Statements," *Environmental Monitoring and Assessment,* vol. 8, no. 3, May 1987, pp. 217–238.

Cushman, R. M., Hunsaker, D. B., Salk, M. S., and Reed, R. M., "Global Climate Change and NEPA Analyses," conference paper no. CONF-891098-4, Oak Ridge National Laboratory, Oak Ridge, Tenn., 1989.

Economic Commission for Europe (ECE), "Convention on Environmental Impact Assessment in a Transboundary Context," E/ECE/1250, United Nations, New York, Feb. 1991.

European Communities, "Council Directive of 27 June, 1985, on the Assessment of the Effects of Certain Public and Private Projects on the Environment," *Official Journal of the European Communities,* no. L 175, Brussels, Sept. 7, 1985, pp. 40–48.

Federal Environmental Assessment Review Office, "Initial Environmental Assessment: Procedures and Practices," Toronto, undated, pp. 8–20.

Haug, P. T., Burwell, R. W., Stein, A., and Bandurski, B. L., "Determining the Significance of Environmental Issues Under the *National Environmental Policy Act,"* *Journal of Environmental Management,* vol. 18, no. 1, Jan. 1984, pp. 15–24.

Hopkins, L. D., "Environmental Impact Statements: A Handbook for Writers and Reviewers," Rep. IIEQ 73-8, Illinois Institute for Environmental Quality, Chicago, Aug. 1973.

Hunt, R. G., Sellers, J. D., and Franklin, W. E., "Resource and Environmental Profile Analysis: A Life Cycle Environmental Assessment for Products and Procedures," *Environmental Impact Assessment Review,* vol. 12, 1992, pp. 245–269.

Kreith, F., "Lack of Impact," *Environment,* vol. 15, no. 1, 1973, pp. 26–33.

Lee, N., and Walsh, F., "Strategic Environmental Assessment: An Overview," *Project Appraisal,* vol. 7, no. 3, Sept. 1992, pp. 126–136.

———, ———, and Jones, C. E., "The European Commission's EIA Activities," *EIA Newsletter 6,* winter 1991, University of Manchester, Manchester, England, p. 4.

Lohani, B. N., "Environmental Assessment and Review During the Project Cycle: The Asian Development Bank's Approach," Chap. 14 in *Environmental Impact Assessment for Developing Countries,* A. K. Biswas, and S. B. C. Agarwal, eds., Butterworth-Heinemann, Oxford, England, p. 179.

Moore, S. N., "A View of the History of *National Environmental Policy Act,"* *National Association of Environmental Professionals Newsletter,* vol. 17, no. 5, Nov. 1992, pp. 8–9.

Office of the President, "Economic and Environmental Principles and Guidelines for Water and

Related Land Resources Implementation Studies," Mar. 1983.

Ross, W. A., "Evaluating Environmental Impact Statements," *Journal of Environmental Management,* vol. 25, 1987, pp. 137–147.

Sammy, G. K., "Environmental Impact Assessment in Developing Countries," Ph.D. dissertation, University of Oklahoma, Norman, 1982.

Sharples, F. E., and Smith, E. D., "NEPA/CERCLA/RCRA Integration," CONF-891098-9, Oak Ridge National Laboratory, Oak Ridge, Tenn., 1989.

Shipley Associates, "Environmental Protection Agency *Clean Air Act* Section 309 Review Summary and EPA Policy and Procedures Manual," Bountiful, Utah, p. 2.

Sigal, L. L., and Cada, G. F., "What About Compliance with NEPA?" CONF-9004182-3, Oak Ridge National Laboratory, Oak Ridge, Tenn., 1990.

———, and Webb, J. W., "The Programmatic Environmental Impact Statement: Its Purpose and Use," *The Environmental Professional,* vol. 11, no. 1, 1989, pp. 14–24.

Smith, E. D., "Future Challenges of NEPA: A Panel Discussion," CONF-891098-10, Oak Ridge National Laboratory, Oak Ridge, Tenn., 1989.

Thompson, M. A., "Determining Impact Significance in EIA: A Review of 24 Methodologies," *Journal of Environmental Management,* vol. 30, 1990, pp. 235–250.

United Nations Environment Program, "Senior Level Expert Workshop to Evaluate Benefits and Constraints of Environmental Impact Assessment Process in SACEP Countries," Regional Office for Asia and the Pacific, Bangkok, Thailand, June 1987, pp. 6–7.

U.S. Army Corps of Engineers, "Environmental Quality: Procedures for Implementing the *National Environmental Policy Act* (NEPA)," *Federal Register,* vol. 53, no. 22, Feb. 3, 1988, pp. 3120–3137.

U.S. Congress, *National Environmental Policy Act,* P.L. 91-190, S. 1075, 91st Congress, 1970.

U.S. Department of the Army, "Environmental Effects of Army Actions," Reg. 200-2, *Federal Register,* vol. 53, no. 221, Nov. 16, 1988, pp. 46322–46361.

U.S. Department of Transportation, Federal Highway Administration, "Environmental Impact and Related Statements," app. F, Policy and Procedure Memorandum 90-1, Washington, D.C., Sept. 1972.

U.S. Environmental Protection Agency (EPA) "Application of the *National Environmental Policy Act* to EPA's Environmental Regulatory Activities," Washington, D.C., Feb. 1973a.

———, "Environmental Impact Statement Guidelines," rev. ed., Apr. 1973b, region X, Seattle, Wash. p. 120.

———, "Facts About the National Environmental Policy Act," LE-133, Washington, D.C., Sept. 1989a.

———, "Filing System Guidance for the Implementation of 1506.9 and 1506.10 of the CEQ Regulations Implementing the Procedural Provisions of the NEPA," *Federal Register,* vol. 54, no. 43, Mar. 7, 1989b, pp. 9592–9594.

———, "Policy and Procedures for the Review of Federal Actions Impacting the Environment," Manual 1640, Oct. 1984, pp. 4-5–4-7.

World Bank, "Operational Directive 4.00, Annex A3, Environmental Screening," Washington, D.C., Sept. 1989.

World Health Organization (WHO) "Health and Safety Component of Environmental Impact Assessment," Copenhagen, 1987.

Yannacone, V. J., Jr., and Cohen, B. S., *Environmental Rights and Remedies,* chap. 5, The Lawyers Co-operative Publishing Company, Rochester, N.Y., 1972.

CHAPTER 2

Planning and Management of Impact Studies

The planning and management of environmental impact assessment studies for proposed projects (or plans or programs) involves several considerations over a wide range of issues. This chapter addresses selected aspects of EIA study planning and management, including (1) a conceptual approach for planning and conducting environmental impact studies, (2) proposal development, (3) interdisciplinary-team formation, (4) team leader selection, (5) general study management, and (6) fiscal control. It is critical to recognize that the technical validity of EIA studies can be easily compromised without effective study planning and management.

CONCEPTUAL APPROACH FOR ENVIRONMENTAL IMPACT STUDIES

To provide a basis for addressing the EIA process, a 10-step or 10-activity model, shown in Figure 2.1, is suggested for the planning and conduction of the studies. This model is flexible and can be adapted to various project types by modification, as needed, to facilitate the addressing of special concerns of specific projects in unique locations. It should be noted that the

focus in this model is on projects, although it could also be applied to plans, programs, policies, or regulatory actions.

An alternative seven-phase (or -step or -activity) model for planning and conducting impact studies is shown in Figure 2.2. Table 2.1 provides a general checklist of activities associated with planning and preparing an EIS. Finally, numerous guidelines have been developed for various components of or considerations in the EIA process. For example, guidelines exist for procedural aspects, methodological considerations, generic types of projects, development aid projects, specific types of environmental settings (e.g., wetlands), assessment tasks such as screening or scoping, and various governmental organizations (EIA Centre, 1992). The review of various study planning models, checklists, and guidelines can be useful preparation for planning an impact study.

Referring to Figure 2.1, the first activity in the 10-step approach would be to determine the features of the proposed project, the need for the project, and the alternatives which either have been or could be considered for the subject project. This step is called "PDN," for "project

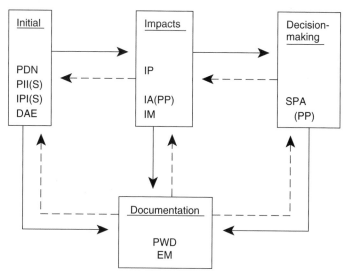

Note: Definitions of terms are in the text.

FIGURE 2.1
Conceptual Framework for Environmental Impact Studies.

description and need." Key information which would be needed relative to the proposed project includes such items as

1. A description of the type of project and how it functions or operates in a technical context.
2. The proposed location for the project and why it was chosen.
3. The time period required for project construction.
4. The potential environmental requirements or outputs (stresses) from the project during its operational phase, including land requirements, air pollution emissions, water usage and water pollutant emissions, and waste-generation and -disposal needs.
5. The identified current need for the proposed project in the particular, proposed location. This need could be related to housing, flood control, industrial development, economic development, and many other requirements. (It is important to begin to consider project need because it must later be addressed as part of the environmental document-ation).
6. Any generic alternatives which have been considered; these ought to include information such as site location, project size, project design features and pollution-control meas-ures, and project timing relative to construc-tion and operational issues; project need in relation to the proposed project size should be clearly delineated. It should be noted that the range of alternatives considered may be limited because of the individual preferences of project sponsors, primary focus on tradi-tional engineering solutions, and time pres-sures for decision making (Bacow, 1980).

It is important that the proponents of the proj-ect begin to think in terms of the above list of items and to organize information which could be utilized by individuals planning and con-ducting the impact studies, whether these indi-viduals be in-house staff, consultants, or agency personnel. To serve as an illustration, pertinent

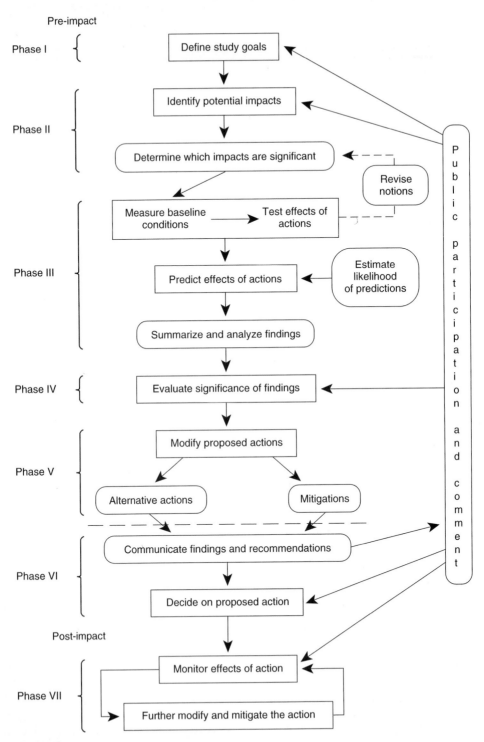

FIGURE 2.2
Phases of Impact Assessment (Westman, 1985).

TABLE 2.1

CHECKLIST FOR EIS PREPARATION

	Scheduled date	Actual date
A. Preliminary Activities		
1. Identify basic issues		
— need for action	_____	_____
— technical alternatives	_____	_____
— geographic alternatives	_____	_____
— administrative/procedural alternatives	_____	_____
2. Identify authorizations needed for action:		
— sponsor's authority & budget to proceed: legislative, presidential & judicial	_____	_____
— nonsponsor authority: budget; approval of specific parts of the action; permitting, licensing & special enabling action (as may be obtained by interagency land transfers, agreements, etc.)	_____	_____
B. Scoping		
3. Develop mailing/notification list		
— federal agencies	_____	_____
— state agencies	_____	_____
— local authorities & Indian tribes	_____	_____
— citizen & environmental groups	_____	_____
— private parties with major stake in outcome	_____	_____
4. Prepare information package		
— describe proposed action & alternatives	_____	_____
— describe potential environmental concerns	_____	_____
— describe proposed scope of DEIS	_____	_____
5. Notify interested parties & invite comments:		
— publish notice of intent in *Federal Register*	_____	_____
— mail notice and information package to selected government and private parties	_____	_____
— make information package available to public at designated locations	_____	_____
6. Obtain and consider comments:		
— collect comments (public meetings optional, if so announced in item 5)	_____	_____
— consider all comments	_____	_____
7. Develop EIS preparation strategy		
— incorporation by reference	_____	_____
— tiering of NEPA documents	_____	_____
— integration of other federal & state laws (i.e., concurrent compliance plan)	_____	_____
— participation of other federal & state agencies	_____	_____
— role of Indian tribes & local governments	_____	_____
— preliminary assessment of motivations for judicial review	_____	_____
— preliminary strategy to avoid judicial review (e.g., agreements, mitigation measures, etc.)	_____	_____
— plan to manage public communications & to respond to public concerns	_____	_____

TABLE 2.1

CHECKLIST FOR EIS PREPARATION *(Continued)*

	Scheduled date	Actual date
C. Draft EIS (DEIS) Preparation		
8. Prepare EIS implementation plan:		
— work breakdown structure (WBS)	_____	_____
— budget & schedule	_____	_____
— responsibilities for preparation	_____	_____
— page limits	_____	_____
9. Prepare prelim. DEIS (Prepare checklist per WBS)	_____	_____
10. Comply with internal agency review procedures	_____	_____
11. Finalize DEIS	_____	_____
12. Publish notice and invite comments		
— mailing list	_____	_____
— availability in public places	_____	_____
— optional scheduling of public meetings	_____	_____
13. Obtain comments		
— correspondence	_____	_____
— public meetings (optional)	_____	_____
— coordination meetings with government agencies	_____	_____
14. Respond to comments		
— make changes	_____	_____
— enlarge EIS scope (new WBS elements)	_____	_____
— negotiate & adopt mitigation measures	_____	_____
— prepare written record of response to comments	_____	_____
D. Final EIS		
15. Produce final EIS (Expand checklist per new WBS)	_____	_____
16. Comply with internal agency review procedures	_____	_____
17. Distribute final EIS & invite comments	_____	_____
18. Receive & consider comments on final EIS	_____	_____
E. Record of Decision (ROD)		
19. Prepare draft ROD	_____	_____
20. Follow internal agency review procedures	_____	_____
21. Publish ROD in *Federal Register*	_____	_____

Source: Freeman, March, and Spensley, 1992, pp. 106–107.

TABLE 2.2

PROJECT-RELATED INFORMATION FOR INDUSTRIAL AND/OR FOSSIL-FUEL-FIRED POWER PLANTS

(a) Purpose and physical characteristics of the project, including details of proposed access and transport arrangements, and of numbers and origins of employees.

(b) Land use requirements and other physical features of the project

    (i) during construction
    (ii) when operational
    (iii) after use has ceased (where appropriate).

(c) Production processes and operational features of the project

    (i) types and quantities of raw materials, energy and other resources consumed
    (ii) residues and emissions by type, quantity, composition and strength including
        discharges to water
        emissions to air
        noise
        vibration
        light
        heat
        radiation
        deposits/residues to land and soil
        others

(d) Main alternative sites and processes considered, where appropriate, and reasons for final choice

Source: Selman, 1992, p. 120.

---

project-related information, or a PDN, for industrial or fossil-fuel-fired power plants is delineated in Table 2.2.

The second activity should focus on pertinent institutional information (PII) related to the construction and operation of the proposed project. "Institutional information" refers to a multitude of environmental laws, regulations, and/or policies or executive orders related to the physical-chemical, biological, cultural, and socioeconomic environments. To completely list all pertinent legislation for any country would be difficult; however, an example list of over 50 federal statutes that may have relevance in the planning and conduction of impact studies is shown in Table 2.3. In addition to these federal laws, statutes exist at state and local levels which may have relevance for specific projects.

Several books are available which summarize key laws in the United States. One such example is a recent environmental law handbook (Arbuckle et al., 1989) which includes chapters on the Comprehensive Environmental Response, Compensation, and Liability Act (Superfund); Water Pollution Control, Air Pollution Control, Safe Drinking Water Act; Noise Control, Resource Conservation and Recovery Act; and underground storage tanks, among others.

One of the approaches which can be used to aid the PII activity is scoping, where "scoping" (S) refers to an early and open process to identify significant environmental issues and impacts relative to proposed projects (CEQ, 1978). As a part of the scoping process, which would include contacts with regulatory agencies and other interested publics, the identification of PII would be expected to occur. This information primarily serves two functions: (1) it is an aid in the interpretation of existing conditions and (2) it provides a basis for interpreting

TABLE 2.3

EXAMPLES OF FEDERAL STATUTES RELATED TO ENVIRONMENTAL PROTECTION

Title of law	U.S. code
Abandoned Shipwreck Act of 1987	43 USC 2101
Agriculture and Food Act (Farmland Protection Policy Act) of 1981	7 USC 4201 et seq.
American Folklife Preservation Act of 1976, as amended	20 USC 2101
Anadromous Fish Conservation Act of 1965, as amended	16 USC 757a et seq.
Antiquities Act of 1906, as amended	16 USC 431
Archaeological and Historic Preservation Act of 1980, as amended	16 USC 469a
Archaeological Data Conservation Act of 1974	16 USC 469
Archaeological Resources Protection Act of 1979	16 USC 470
Bald Eagle Act of 1972	16 USC 668
Clean Air Act of 1972, as amended	42 USC 7401 et seq.
Clean Water Act of 1972, as amended	33 USC 1251 et seq.
Coastal Barrier Resources Act of 1982	16 USC 3501–3510
Coastal Zone Management Act of 1972, as amended	16 USC 1451 et seq.
Comprehensive Environmental Response, Compensation and Liability Act of 1980	42 USC 9601
Conservation of Forest Lands Act of 1960	16 USC 580mn
Deepwater Port Act of 1974, as amended	33 USC 1501
Emergency Flood Control Funds Act of 1955, as amended	33 USC 701m
Endangered Species Act of 1973	16 USC 1531
Estuary Protection Act of 1968	16 USC 1221 et seq.
Federal Environmental Pesticide Control Act of 1972	7 USC 138 et seq.
Federal Water Project Recreation Act of 1965, as amended	16 USC 4601
Fish and Wildlife Coordination Act of 1958, as amended	16 USC 661
Flood Control Act of 1944, as amended, sec. 4	16 USC 460b
Food Security Act of 1985 (Swampbuster)	16 USC 3811 et seq.
Hazardous Substance Response Revenue Act of 1980, as amended	26 USC 4611
Historic Sites Act of 1935	16 USC 461
Land and Water Conservation Fund Act of 1965	16 USC 4601
Marine Mammal Protection Act of 1972, as amended	16 USC 1361
Marine Protection, Research, and Sanctuaries Act of 1972	33 USC 1401
Migratory Bird Conservation Act of 1928, as amended	16 USC 715
Migratory Bird Treaty Act of 1918, as amended	16 USC 703
National Environmental Policy Act of 1969, as amended	42 USC 4321 et seq.
National Historic Preservation Act of 1966, as amended	16 USC 470
Native American Religious Freedom Act of 1978	42 USC 1996
Noise Control Act of 1972, as amended	42 USC 4901 et seq.
Reservoir Salvage Act of 1960	16 USC 469
Resource Conservation and Recovery Act of 1976	42 USC 6901–6987
River and Harbor Act of 1888, sec. 11	33 USC 608
River and Harbor Act of 1899, secs. 9, 10, 13	33 USC 401–413
River and Harbor and Flood Control Act of 1962, sec. 207	16 USC 460d
River and Harbor and Flood Control Act of 1970, secs. 122, 209, 216	33 USC 426 et seq.
Safe Drinking Water Act of 1974, as amended	42 USC 300f
Submerged Lands Act of 1953	43 USC 1301 et seq.
Superfund Amendments and Reauthorization Act of 1986	42 USC 9601
Surface Mining Control and Reclamation Act of 1977	30 USC 1201–1328
Toxic Substances Control Act of 1976	15 USC 2601
Water Resources Development Act of 1974, as amended	88 Stat. 12
Water Resources Development Act of 1976, sec. 150	90 Stat. 2917
Water Resources Development Act of 1986	33 USC 2201 et seq.
Water Resources Planning Act of 1965, as amended	42 USC 1962a
Watershed Protection and Flood Control Act of 1954, as amended	16 USC 1001 et seq.
Wild and Scenic Rivers Act of 1968, as amended	16 USC 1271 et seq.

the anticipated impacts, or effects, of the project.

The third step or activity is the identification of potential impacts (IPI) of the subject project. This early qualitative identification of anticipated impacts can help in focusing subsequent steps; for example, it can aid in describing the affected environment and subsequent impact calculations. IPI can be an outcome of the scoping process. It should include consideration of the generic impacts related to the project type being analyzed. In this regard, there is an abundance of published information generated over the past two decades which enables planners of impact studies to more easily identify anticipated impacts. For example, Figure 2.3 displays an overall framework for identifying the direct, indirect, and ultimate impacts of a proposed action.

An appropriate task during the third activity is the conduction of computer-aided literature reviews to identify generic impacts related to the project type being analyzed. There are a number of other tools and techniques which have been developed, including the identification of potential impacts by means of the preparation of simple interaction matrices (consisting of a list of project construction and operational actions juxtaposed with a list of environmental factors). Development of the matrix involves the systematic consideration of interaction points between various project actions and environmental factors. Matrices, networks, and simple and descriptive checklists for identifying impacts are addressed in detail in Chapter 3.

The fourth activity is focused on the preparation of a description of the affected environment (DAE). This activity is placed fourth so as to enable the selective identification of pertinent baseline factors for the study in subsequent activities in the model. Early impact studies, and to some extent current impact studies, have required extensive efforts in the preparation of exhaustive descriptions of the environmental setting. A selective approach is valuable in that key environmental factors anticipated to be changed by the proposed project are identified, and appropriate, extensive descriptions of existing conditions relative to only these factors are prepared.

Sources of information for describing physical features include study-specific field investigations, topographic contour maps published by the *U.S. Geological Survey,* county soil surveys and maps prepared by the U.S. Soil Conservation Service, and aerial photographs produced at intervals by various governmental agencies, including the U.S. Soil Conservation Service, the U.S. Forest Service, and the U.S. Bureau of Land Management (Marsh, 1991). In addition, the *Earth Science Data Directory* (ESDD) of the *U.S. Geological Survey* (1989) can be used for determining the availability of specific earth science and natural resources data. GEOINDEX, a bibliographic database of published geological maps of the United States published by the *U.S. Geological Survey,* state geological surveys, geosciences societies, and others, is another useful tool; pertinent information and software can be ordered from the *U.S. Geological Survey,* Federal Center, Denver, Colorado.

Nineteen information systems related to the water environment have been described by the U.S. EPA (1990). Selected examples of the 19 systems include the Drinking Water Supply File, the Federal Reporting Data System, the Hazardous Waste Injection Well Data Base, the Industrial Facilities Discharge File, and STORET, (which includes the Biological System, the Daily Flow System, the Fish Kill File, and the Water Quality System). The Endangered Species Information System (ESIS) of the U.S. Fish and Wildlife Service (1988) is an example of a database which can provide information on nationally listed threatened or endangered plant or animal species. Examples of additional environmental data systems managed by various federal agencies are summarized in the CEQ's *Twenty-Second Annual Report* (CEQ, 1992).

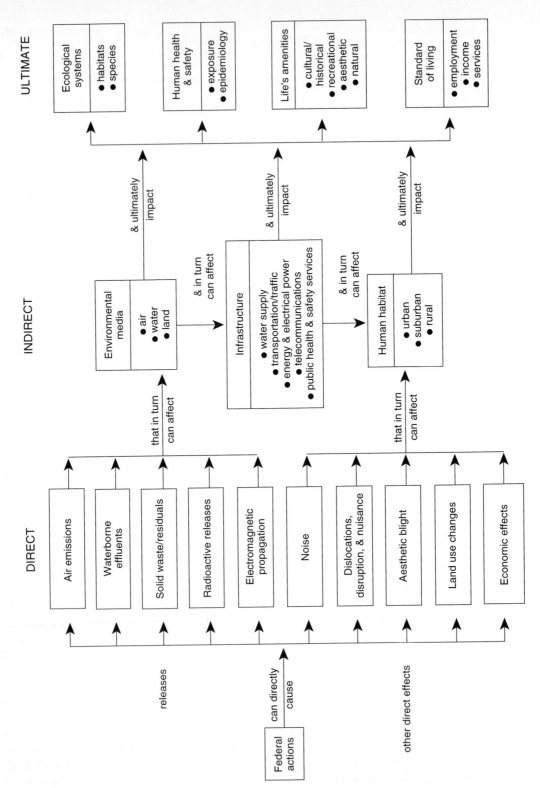

FIGURE 2.3
Applying a Systematic Approach for Impact Identification (Freeman, March, and Spensley, 1992).

44

Other information sources and environmental databases are available, including some related to emission inventories, contaminant transport and fate characteristics, the quality characteristics of environmental media, and available contaminant control technologies (U.S. EPA, 1991; Kokoszka, 1992). Chapter 4 also delineates information storage and retrieval systems, such as STORET.

Finally, the Environmental Technical Information System (ETIS), operated by the U.S. Army Construction Engineering Research Laboratory, includes the following major databases (Department of Urban and Regional Planning, 1990, pp. 1–2):

CELDS (Computer-aided Environmental Legislative Data System): contains abstracts of federal and state environmental regulations and standards;

EIFS (Economic Impact Forecast System): contains national socioeconomic data, at the county level; anticipated impacts on the local economy from a proposed activity can be assessed using one of several economic models;

SOILS (Soils Systems): contains data from the U.S. Soil Conservation Service (SCS), retrievable by soil series name or a combination of soil characteristics;

EICS (Environmental Impact Computer-aided System): an interactive system that assists the user in determining how activities such as construction or testing may affect various aspects of the environment like water, land use, and air.

Electronic bulletin boards are also part of ETIS's offerings; examples include DEEP (Discuss with Experts Environmental Problems Knowledge-based System), CRIBB (Cultural Resources Information Bulletin Board), RACE (Regulations and Compliance Expertise), and HAZE (Hazardous Expertise Knowledge-based System). These systems store and distribute information from users (Defillo, 1991).

In summary, there is considerable environmental information available in computerized information storage and retrieval systems which can aid the DAE activity. Examples of such systems in the United States include databases for air quality information, water quality information, soils information, habitat types in geographical areas, threatened and endangered species, historic and archaeological properties in geographical areas, and a variety of information related to the socioeconomic environment, including population density, income levels, and infrastructure characteristics in particular locations. Specific information sources for various substantive areas are described in Chapters 6 through 14.

The fifth and technically most difficult and challenging activity is called "impact prediction" (IP). "Impact prediction" basically refers to the quantification, where possible (or, at least, the qualitative description) of the anticipated impacts of the proposed project on various environmental factors. Depending upon the particular impact, technically demanding mathematical models might be required for impact prediction. Other approaches include the conduction of laboratory testing, such as leachate testing for dredged material and solid- or hazardous-waste materials or sludges. Still other laboratory studies might be appropriate; examples include the construction of scale models to accompany the collection of experimental data to identify anticipated impacts. Impact prediction can also be accomplished by the use of look-alike (analogous) information on actual impacts from similar types of projects in diverse or similar geographical locations. Finally, the use of environmental indexing methods or other systematic techniques for relatively addressing anticipated impacts can also be considered. It should be noted that there are many intrinsic difficulties in predicting impacts, particularly for large-scale proposals such as hydroelectric projects (Berkes, 1988). Examples of such difficulties include many unknown impacts, natural environmental variability and resiliency, and lack of adequate models. Specific prediction techniques for various substantive areas are described in Chapters 5 through 14.

It is desirable to quantify as many impacts as possible, because, in so doing, it has been frequently determined that the concerns related to anticipated changes are not as great as would be supposed, in the event of nonquantification. Also, if anticipated impacts are quantified, it would be appropriate to use specific, numerical environmental quality standards as the bases for interpretation of the anticipated changes. However, many environmental impacts are either nonquantifiable, or the monetary and personnel-resource requirements to accomplish such quantification would be beyond the scope and budget of the impact study. In many cases, it is necessary for professionals to use their best judgment in trying to anticipate impacts; useful tools include expert systems. ''Expert systems'' refer to special computer programs that are encoded with and apply the knowledge of experts to provide solutions to problems within specialized fields. Such systems have potential usage in project design and the prediction and evaluation of anticipated impacts. One example, related to environmental design and called ''ENDOW'' (for Environmental Design of Waterways), was developed by the U.S. Army Corps of Engineers (Shields, 1988). ENDOW is useful for rapidly identifying environmental alternatives for inclusion in project plans, designs, or procedures for operation and maintenance. The current version of ENDOW contains modules for streambank-protection- and flood-control channel projects, and for levee projects. When running ENDOW, a user answers queries from the program regarding the project setting and environmental objectives. ENDOW then responds with a list of project features for further study and possible inclusion in the EIA.

Lein (1988) developed a prototype expert system (IMPACT-EXPERT) consisting of 120 rules that can be used to evaluate 11 types of recognized environmental impacts. The users interact with the knowledge base through 36 screening questions that have been kept purposely broad to accommodate the review of a wide range of activities. Each question has been assigned to a specific environmental factor, consequence, attribute, or effect. The present version of IMPACT-EXPERT is limited in range because its knowledge base is comparatively small. However, it provides a useful guide for the development of more-sophisticated expert systems in this area of application (Lein, 1988).

Risk-assessment concepts could also be useful in impact prediction. The potential benefits of applying the principles of risk identification, dose-response and exposure assessment, risk interpretation, and risk management in the EIA process include (1) the encouragement of integrated thinking (such as for environmental transport pathways and associated health or ecological effects); (2) the opportunity to focus attention on risk-reduction activities such as waste minimization, pollution prevention, and mitigation measures; and (3) the inclusion of emphases on emergency response measures in the event of accidents and associated environmental perturbations (Canter, 1993).

The Senior Advisors to the Economic Commission for Europe (ECE) recently conducted a study of EIA impact prediction methods; in general, they found that quantitative and qualitative methods for the prediction of environmental impacts appear to be the two basic approaches for incorporating environmental concerns into the decision-making process. Selected technically focused recommendations resulting from this study included (after Senior Advisors to ECE Governments on Environmental and Water Problems, 1992, pp. 4–5)

1. International cooperation in the development, standardization, and application of environmental-impact prediction methods should be enhanced, particularly in relation to large-scale environmental issues.

2. In environmental predictions, the identification and quantification of the sources of uncertainty should be an important step in the application of methods. The results of environmental predictions should indicate the margin of uncertainty involved.

3. For many complex environmental problems, the prediction methods should take into account the interaction of various trophic levels and biotic-abiotic factors and be reliable and comprehensive. When possible, simple, quick, and inexpensive methods should be preferred.

4. Experience with methods related to the transfer of pollutants from one environmental compartment to another should be promoted in order to improve environmental impact prediction as a whole. Many environmental concerns, including those in a transboundary context, demonstrate the need for transcompartmental prediction methods. An intensive research effort should be undertaken to further clarify this topic.

5. Prediction methods used within EIA should be validated and verified through the use of reliable monitoring.

The next activity in the conceptual study model is entitled "impact assessment" (IA). Of necessity, impact studies represent a blend of technical information and analysis along with value judgments (Bacow, 1980). In the terminology used herein, "assessment" refers to the interpretation of the significance of anticipated changes related to the proposed project. Impact interpretation can be based upon the application of the definitions of "significant," as described in Chapter 1. A systematic EIA approach involving the careful review of the anticipated impacts relative to the definitions would provide a useful basis for impact assessment. However, it should be noted that many of the items in the definitions are general and require the use of considerable judgment in interpretation. One example of the application of such professional judgment is in the context of assessment of impacts related to the biological environment, with the biological scientist in the study team rendering decisions as to the potential significance of the loss of particular habitats, including wetland areas.

Another component of impact assessment is public input; this input could be received through a continued scoping process or through the conduction of public meetings and/or public participation (PP) programs. The general public can delineate important environmental resources and values for particular areas, and this should be considered in impact assessment. Chapter 16 addresses various facets of planning PP programs. For some types of anticipated impacts, there are specific numerical standards or criteria which can be used as bases for impact interpretation. Examples include air quality standards, environmental-noise criteria, surface- and groundwater quality standards, and wastewater discharge standards for particular facilities.

The next activity is associated with identifying and evaluating potential impact mitigation (IM) measures. Mitigation measures include (1) avoiding the impact altogether by not taking a certain action or parts of an action; (2) minimizing impacts by limiting the degree or magnitude of the action and its implementation; (3) rectifying the impact by repairing, rehabilitating, or restoring the affected environment; (4) reducing or eliminating the impact over time by preservation and maintenance operations during the life of the action; and (5) compensating for the impact by replacing or providing substitute resources or environments (CEQ, 1978). The definition of "mitigation" suggests a sequential consideration—avoidance, followed by sizing, followed by rectifying, and so on. While the IM activity is identified as the seventh step, there is no reason to wait until this point in the study to identify and evaluate impact-mitigation measures. For example, these measures could be targeted in association with the IPI (identification of potential impacts) activity.

A generic mitigation guidelines manual has been developed by the U.S. Army Corps of Engineers (1983) to summarize potential measures by various types of resources or impacts. Generic mitigation measures are included for issues related to the physical environment (topography, erosion and sedimentation, mass wasting, land subsidence, geologic hazards, soils, prime and unique agricultural lands, mineral and energy resources, paleontological resources, flooding, wild and scenic rivers, water supply, water and air quality, noise, and land use), the biological environment (terrestrial and aquatic habitats, critical habitats, wildlife refuges and sanctuaries, wildlife management areas, and threatened and endangered species), the cultural environment, and the resources and components of concern to Native Americans.

The next activity in the study model is associated with selecting the proposed action (SPA) from alternatives, which may also have been evaluated earlier. In public projects in the United States there is a great deal of emphasis on the identification and evaluation of alternatives. The CEQ regulations indicate that the analysis of alternatives represents the heart of the EIA process. Conversely, for many private developments, the range of alternatives may be limited. Even so, there are still potential alternative measures which could be evaluated, including those relating to project size and design features, even if location alternatives are not available.

There are numerous systematic procedures which can be used to compare and evaluate the environmental consequences of alternatives. Many of these can be incorporated in multiple-criteria decision-making techniques and can include PP components. These techniques represent useful tools for systematically demonstrating why the proposed action was chosen for a particular project. Chapter 15 describes several examples of decision-focused checklists which can be used in this regard.

The ninth activity is associated with preparing the written documentation (PWD) relating to the proposed project. Written documentation could involve the preparation of a preliminary report, or EA, or it could encompass the actual preparation of a complete EIS. The most important point to note about PWD is that sound principles of technical writing should be utilized. These include the development of outlines, careful documentation of data and information, the liberal usage of visual display materials, and the careful review of written materials so as to ensure effective communication to both technical and nontechnical audiences. Additional information is in Chapter 17. Preparing a report outline can be one of the first tasks in planning an impact study and developing the scope of work (Burack, 1992). A technical writer-editor is often used to facilitate written documentation of the EIA process. Information on specific functions of a writer-editor is contained in Murthy (1988).

The final activity suggested in the framework shown in Figure 2.1 is the planning and implementation of appropriate environmental-monitoring (EM) programs; this activity is particularly important for larger projects with potentially significant environmental consequences. Environmental monitoring may be necessary in establishing baseline conditions in the area of the project; however, of more relevance is longer-term monitoring in the environs of the project to carefully document impacts which are actually experienced, and the use of this information in project management. General information on planning and implementation of environmental-monitoring programs is in Chapter 18. Detailed information is in Canter and Fairchild (1986).

Referring again to Figure 2.1, it is noted that these ten steps or activities are related to each other and do not necessarily represent discrete activities which can be accomplished in a sequential fashion. It is always possible to iterate an activity to gain additional information which would be relevant to the impact study.

PROPOSAL DEVELOPMENT

Scheduling and budgeting are critical to planning an impact study; and both efforts will probably need revising one to several times during a study (Burack, 1992). Scheduling can be aided by the use of critical-path methods or project-evaluation and -review techniques. The time required to conduct an impact study and the resulting costs vary with the type, size, and complexity of the project; the characteristics of its physical, socio-cultural, and institutional settings; and the amount and quality of environmental data already available (World Bank, 1991).

One of the approaches which can be used for developing cost and time estimates for a proposal for an impact study is to think through, in a systematic fashion, the 10 activities in study conduct. These activities could be further

subdivided into cost elements, including professional person-days of effort, travel, and other related costs such as analysis and printing. Table 2.4 displays a sample worksheet for calculating the costs of an EIS organized in the fashion suggested. This approach can be useful in providing a basis for a budget and, as necessary, the results can be adjusted during the conduction of the study as different needs arise or as different issues become important.

Another concern related to the EIA process is associated with issues or situations that might arise and cause increases in the costs of such studies. Examples of these issues include (1) an extensive period of time is devoted to gathering extant information, with this time largely allocated to making multiple telephone calls and/or visits to various information sources; (2) changes in project design features which occur

TABLE 2.4

EIS BUDGETARY PLANNING WORKSHEET

Activity	Person-days			Travel	Other costs	Calendar time
	Professional[a]	Technical staff	Secretarial			
PDN						
PII						
IPI						
DAE						
IP						
IA (PP)						
IM						
SPA (PP)						
PWD						
EM						
Totals						

[a]Could be subdivided by profession.
PDN = preparation of a description of the project, the need for the project, and appropriate alternatives; PII = assemblage of pertinent institutional information; IPI = identification of potential impacts; DAE = description of affected environment; IP = impact prediction; IA = impact assessment; IM = impact mitigation; SPA = selection of proposed action; PWD = preparation of written documentation; EM = environmental monitoring.

during the conduction of the study which necessitate recalculations or reconsiderations of anticipated impacts; (3) it becomes necessary to plan and conduct a baseline EM program for critical environmental resources; (4) there is an occurrence of controversy related to the proposed project, with this controversy leading to additional meetings with regulatory and other governmental agencies and different publics, including those opposed to the project; and (5) unique risks, not identified prior to the beginning of the study, that might be related to project construction and/or operation are discovered. Because of these issues, it is desirable to include funds for contingencies. It may also be desirable to delineate options for additional costs in the event that one or more of the above issues occurs.

INTERDISCIPLINARY TEAM FORMATION

Environmental impact studies are often conducted by interdisciplinary teams. It is important that interdisciplinary and not multidisciplinary activities dominate the process. "Multidisciplinary activities" denote those in which persons versed in different disciplines work together without specific, pre-established interrelationships. The team members' findings are typically presented as individual topical reports. "Interdisciplinary activities" are characterized by interrelationships and the sharing and integration of the findings of the team members (Van Dusseldorp and Van Staveren, 1983).

An "interdisciplinary team" can be defined as a group of two or more persons trained in different fields of knowledge with different concepts, methods, and data and terms which has been organized to address a common problem with continuous communication among participants from different disciplines (Dorney and Dorney, 1989). An interdisciplinary team for a specific impact study can be considered as a temporary entity which has been assembled, and possibly specifically appointed, for meeting the identified purpose of conducting an EIS for a proposed project. The team may be assembled with formal authority, responsibility, and accountability; however, a more typical approach is the delineation of an informal authority within the team, that is, the team is basically subject to management by the team leader (Cleland and Kerzner, 1986). The roles of all team members, consultants, and advisors need to be clearly defined (Burack, 1992).

A detailed list of specialists who might be appropriate for an interdisciplinary team is in Table 2.5. A core team for an impact study could consist of the following (World Bank, 1991): (1) a project manager or team leader—often a planner, social or natural scientist, or environmental engineer—who has experience in preparing several similar studies, management skills, and sufficiently broad training and/or experience to be able to provide overall guidance and to integrate the findings of individual disciplines; (2) an ecologist or a biologist (with aquatic, marine, or terrestrial specialization, as appropriate); (3) a sociologist-anthropologist—who has experience with communities similar to that of the project; (4) a geographer or geologist–hydrologist–soils scientist; and (5) an urban or a regional planner. The core team could be supported by the specialists listed in Table 2.5.

The number of members of an interdisciplinary team can vary from as few as 2 to perhaps as many as 8 or 10 individuals, depending upon the size and complexity of the study. Typically a team comprises three to four members. In selecting the interdisciplinary team, the team leader, project proponent, or consulting company should take into consideration the following (Canter, 1991, p. 380):

1. The types of expertise needed relative to the environmental impact study (as determined in a pre-study effort and/or the scoping process);
2. The experience of the prospective team members on similar or other types of projects;

TABLE 2.5

SPECIALISTS RELATED TO THE EIA PROCESS

Natural resource	Subcomponent	Specialist
Air	Air quality Wind direction/speed Precipitation/humidity Temperature Noise	Air quality/pollution analyst Air pollution control engineer Meteorologist Noise expert
Land	Land capability Soil resources/structure Mineral resources Tectonic activity Unique features	Agronomist Soils engineer Soils scientist Civil engineer Geologist Geotechnical engineer Mineralogist Mining engineer Engineering geologist Seismologist
Water	Surface waters Groundwater regime Hydrologic balance Drainage/channel pattern Flooding Sedimentation	Hydrologist Water pollution control engineer Water quality/pollution analyst Marine biologist/engineer Chemist Civil/sanitary engineer Hydrogeologist

Flora and fauna	Environmentally sensitive areas:
	    wetlands, marshes, wildlands,
	    grasslands, etc.
	Species inventory
	Productivity
	Biogeochemical/nutrient cycling
		Ecologist
		Forester
		Wildlife biologist
		Botanist
		Zoologist
		Conservationist

Human	Social infrastructure/institutions Cultural characteristics Physiological and psychological well-being Economic resources	Social anthropologist Sociologist Archaeologist Architect Social planner Geographer Demographer Urban planner Transportation planner Economist

Source: World Bank, 1991, p. 22.

3. The orientation of the individual toward working with other individuals on a group effort;
4. The receptivity of individuals to the viewpoints of other disciplines;
5. The range of interest of the individual, with a broader range of interest being more conducive to successful work on an environmental impact study than a narrow or limited range of interest;
6. Availability within the overall work unit time schedule to work on the team;
7. Some indication of the following work traits and personal characteristics: organized, orientation to working on a time schedule, no aversion to writing, willingness to travel and make site visits, willingness to work with other individuals and serve as a team player, self-starter, creative, expertise related to the local geographical area, adequate verbal and written communication skills, credibility with other professionals in the field, and adaptability.

TEAM LEADER SELECTION AND DUTIES

A critical individual in the successful delineation and operation of an interdisciplinary team is the team leader (project manager). The team leader provides direction for the team itself in accomplishing the end purpose—the successful conduct of the impact study (Cleland and Kerzner, 1986). The team leader is expected to provide day-to-day technical direction; schedule the work and insure that deadlines are met; control costs; coordinate with various departments and disciplines; provide overall integration of the technical, scientific, and policy aspects of the project; and provide for quality control and peer review (Murthy, 1988).

The team leader should exhibit a number of specific, pertinent personal and professional qualities; examples include (after Cleland and Kerzner, 1986, p. 22)

1. Demonstrated knowledge and leadership skills in a specialized professional field
2. A positive attitude in support of the conduction of the environmental impact study
3. A rapport with individuals
4. An ability to communicate with both technical and nontechnical persons
5. Pride in his or her technical specialty area
6. Self-confidence
7. Initiative, self-starter ability
8. A reputation as a person who gets things done
9. The ability to deal successfully with the challenge of doing quality work
10. The willingness to assume responsibility for the overall study and team leadership

In summary relative to the team leader, several key characteristics should be considered in this selection process. These characteristics include, in order of priority: (1) experience in serving as team leader or project manager, (2) management and leadership skills, and (3) a substantive area of expertise.

GENERAL STUDY MANAGEMENT

A number of considerations are related to the management of an interdisciplinary team and an EIA study. The team leader should consider several management techniques and develop approaches to utilize them. For example, Cleland and Kerzner (1986) suggested six factors which are basic to the successful management of an interdisciplinary team, and which can be summarized as follows:

1. A clear, concise statement of the mission or purpose of the team
2. A summary of the goals or milestones that the team is expected to accomplish or reach in planning and conducting the environmental impact study
3. A meaningful identification of the major tasks required to accomplish the team's purposes, with each task broken down by individual assignment
4. A summary delineation of the strategy of the team relative to policies, programs, procedures, plans, budgets, and other resource-

allocation methods required in the conduction of the EIS

5. A statement of the team's organizational design, with information included on the roles, authority, and responsibility of all members of the team, including the team leader

6. A clear delineation of the human- and non-human-resource-support services available for usage by the interdisciplinary team

A fundamental technique for team operation is the conduction of periodic team meetings with planned agendas. It is a primary role of the team leader to develop schedules and to establish priorities with regard to manpower and other resources allocated to particular activities within the impact study. In addition, the team leader must allow individual team members to work in their own particular areas to carry out agreed-to assignments, and then should subject the work products, or at least the ideas resulting from the work, to team review. The pattern of meeting, individual work, and a follow-up review meeting is useful in the operation of an interdisciplinary team. While it is theoretically possible, it is unlikely that the interdisciplinary team will completely work together on every aspect of an impact study.

Periodic review meetings involving the proponent, EIA study team, and various pertinent publics are a necessity. Review of the resultant report by an environmental attorney can also be worthwhile, particularly if the project is controversial.

One of the issues related to team management is associated with the periodic necessity for special studies conducted by experts who are not members of the interdisciplinary team. An example might be the conduction of specific cultural-resources surveys by archaeologists. If special studies are required—and they are common in impact studies—then the team management process should include a meeting to discuss the requirements of the special studies, the particular terms of reference for groups or individuals to conduct such special studies, and the clear delineation of the anticipated output from the studies, with particular care given to ensuring that the output from such special studies will coincide with the needs of the overall impact study.

FISCAL CONTROL

"Fiscal control" involves conforming project staff requirements to the available budget over time. However, once a budget and timetable have been established, the project manager must not assume that they will be followed. There are a variety of graphs, charts, and computerized techniques to track expenses, to assist in comparing percent of task completion with percent of task budget expended, and to chart actual progress against scheduled progress. Use of these procedures is not unique to impact studies (Bingham, 1992).

Once reporting procedures are in place, the team leader needs to monitor the use of resources, weekly or monthly, depending upon the size of the budget and the rate of spending. Projects may require low levels of expenditures at the beginning and then build to a high peak, or vice versa. It is in the last weeks or months of a project that unanticipated expenditures seem to occur, particularly if report production is underway. Planning for contingencies and the inevitable crunch at the scheduled time for completion is wise (Bingham, 1992).

SUMMARY

The conduction of environmental impact studies in the United States has improved considerably since the passage of the NEPA. The scientific and technical validity of EISs has improved, as well, and emphasis is currently being given to mitigation measures which could preclude the need for EISs. This chapter includes detailed information on a 10-step model for planning and conducting environmental impact studies, and it also contains suggestions as to how this approach can be utilized in identifying necessary study budgets and interdisciplinary-team requirements. Team management principles and fiscal control principles are also summarized.

SELECTED REFERENCES

Arbuckle, J. G., et al., *Environmental Law Handbook,* 10th ed., Government Institutes, Rockville, Md., 1989.

Bacow, L. S., "The Technical and Judgmental Dimensions of Impact Assessment," *Environmental Impact Assessment Review,* vol. 1, no. 2, 1980, pp. 109–124.

Berkes, F., "The Intrinsic Difficulty of Predicting Impacts: Lessons from the James Bay Hydro Project," *Environmental Impact Assessment Review,* vol. 8, no. 3, Sept. 1988, pp. 201–220.

Bingham, C. S., "EIA and Project Management: A Process of Communication," paper presented at *13th International Seminar on Environmental Impact Assessment,* University of Aberdeen, Aberdeen, Scotland, June 28–July 11, 1992.

Burack, D. A., "Better Management: The Key to Successful Environmental Impact Assessment (EIA)," paper presented at *12th Annual Meeting of the International Association for Impact Assessment,* Washington, D.C., Aug. 19–22, 1992.

Canter, L. W., "Interdisciplinary Teams in Environmental Impact Assessment," *Environmental Impact Assessment Review,* vol. 11, no. 4, Dec. 1991, pp. 375–387.

———, "Pragmatic Suggestions for Incorporating Risk Assessment Principles in EIA Studies," *The Environmental Professional,* vol. 15, no. 1, 1993, pp. 125–138.

———, and Fairchild, D. M., "Post-EIS Environmental Monitoring," in *Methods and Experiences in Impact Assessment,* H. Becker and A. Porter, eds., D. Reidel Publishing Company, The Hague, The Netherlands, 1986, pp. 265–285.

Cleland, D. I., and Kerzner, H., *Engineering Team Management,* Van Nostrand Reinhold Company,

New York, 1986, pp. 1–3, 19, 22–24, 68, 268–270, and 324.

Council on Environmental Quality (CEQ), "Environmental Quality," *Twenty-second Annual Rep.,* U.S. Government Printing Office, Washington, D.C., Mar. 1992, pp. 48–55.

———, "National Environmental Policy Act—Regulations," *Federal Register,* vol. 43, no. 230, Nov. 29, 1978, pp. 55978–56007.

Defillo, R., "Electronic Bulletin Boards: An Overview," *ETIS Quarterly,* Department of Urban and Regional Planning, University of Illinois, Champaign, June 1991, pp. 1–4.

Dorney, R. S., and Dorney, L. C., *The Professional Practice of Environmental Management,* Springer-Verlag, New York, 1989, p. 93.

EIA Centre, "EIA Guidelines," Leaflet 12, Department of Planning and Landscape, University of Manchester, Manchester, England, Sept. 1992, pp. 1–8.

Freeman, L. R., March, F., and Spensley, J. W., *NEPA Compliance Manual,* Government Institutes, Rockville, Md., 1992, pp. 66–67, 104–107.

Kokoszka, L. C., "Guide to Federal Environmental Databases," *Pollution Engineering,* vol. 24, no. 3, Feb. 1992, pp. 83–92.

Lein, J. K., "An Expert System Approach to Environmental Impact Assessment," *International Journal of Environmental Studies,* vol. 33, 1988, pp. 13–27.

Marsh, W. M., *Landscape Planning: Environmental Applications,* 2d ed., John Wiley and Sons, New York, 1991, pp. 48–49.

Murthy, K. S., *National Environmental Policy Act (NEPA) Process,* CRC Press, Boca Raton, Fla., 1988, pp. 57–59.

Selman, P., *Environmental Planning: The Conservation and Development of Biophysical Resources,* Paul Chapman Publishing, London, 1992, p. 120.

Senior Advisors to ECE Governments on Environmental and Water Problems, "Methods and Techniques for Prediction of Environmental Impact," ECE/ENVWA/21, Economic Commission for Europe, United Nations, New York, Apr. 1992.

Shields, F. D., Jr., "ENDOW: An Application of an Expert System in Technology Transfer," Water Operations Technical Support Information Exchange Bulletin, vol. E-88-3, U.S. Army Corps of Engineers, Waterways Experiment Station, Vicksburg, Miss., Dec. 1988, pp. 1–5.

University of Illinois, Department of Urban and Regional Planning, "Environmental Technical Information System: A Brief Introduction," *ETIS Quarterly,* Department of Urban and Regional Planning, University of Illinois, Champaign, Sept. 1990, pp. 1–2.

U.S. Army Corps of Engineers, "Generic Environmental Mitigation Guidelines Manual," San Francisco District, San Francisco, Apr. 1983.

U.S. Department of the Interior, "Information Sheet: The Earth Science Data Directory," *Geological Survey,* Reston, Va., Mar. 1989.

U.S. Environmental Protection Agency (EPA), "EPA Information Sources," EPA 600/M-91-038, Washington, D.C., Oct. 1991.

———, "Environmental and Program Information Systems Compendium FY 1990," EPA 500/9-90-002, Aug. 1990, Washington, D.C.

U.S. Fish and Wildlife Service, "Endangered Species Information System—Project Brief," Washington, D.C., May 1988.

Van Dusseldorp, D. B., and Van Staveren, J. M., "The Interdisciplinary Procedure of Regional Planning," chap. 3 in J. M. Van Staveren and D. B. Van Dusseldorp, eds., *Framework for Regional Planning in Developing Countries,* International Institute for Land Reclamation and Improvement, Wageningen, The Netherlands, 1983, pp. 37–49.

Westman, W. E., *Ecology, Impact Assessment, and Environmental Planning,* John Wiley and Sons, New York, 1985, pp. 10–14.

World Bank, "Environmental Assessment Sourcebook—Vol. I—Policies, Procedures, and Cross-Sectoral Issues," Tech. Paper no. 139, Washington, D.C., July 1991, pp. 20–22.

CHAPTER 3

Simple Methods for Impact Identification—Matrices, Networks, and Checklists

Several activities are required in an environmental impact study, including impact identification, preparation of a description of the affected environment, impact prediction and assessment, selection of the proposed action from the alternatives evaluated to meet identified needs, and summarization and communication of information. The objectives of the various activities differ, as do the pertinent methodologies for accomplishing the activities. The term "methodology" as used herein refers to structured approaches for accomplishing one or more of the basic activities. The structured approaches encompass various substantive areas within the biophysical and socioeconomic environments, thus distinguishing them from impact prediction methods or models for specific substantive areas. Numerous methodologies have been developed to aid in achieving the various activities in the EIA process. The purpose of this chapter is to describe some simple methods for impact identification. This will be done by highlighting matrices, networks, and simple and descriptive checklists. Background information is provided on the overall purposes of the methodologies as well as a classification

scheme; this information is also relevant to decision-focused checklists and alternatives evaluation, discussed in Chapter 15.

BACKGROUND INFORMATION

EIA methodologies can be broadly categorized into interaction matrices and checklists, with networks representing variations of interaction matrices. Interaction matrices range from simple considerations of project activities and their impacts on environmental factors to stepped approaches which display interrelationships between impacted factors. Checklists range from simple listings of environmental factors to descriptive approaches which include information on measurement, prediction, and interpretation of changes for identified factors. Checklists may also involve the scaling-rating (or -ranking) of the impacts of alternatives on each of the environmental factors under consideration. Scaling or rating techniques include the use of numerical scores, letter assignments, or linear proportioning. Alternatives can be ranked from best to worst in terms of potential impacts on each factor. The most sophisticated checklists are those involving the assignment of importance weights to environmental factors and the scaling-rating of the impacts for each alternative on each factor. Resultant comparisons can be made through the development of a product matrix and overall impact index for each alternative. The index, or score, is determined by multiplying importance weights by the scale-rating value for each alternative.

Methodologies can be useful, although not specifically required, throughout the impact assessment process, with certain ones being of greater value for specific activities. Table 3.1

identifies five activities and relevant useful methodologies. For example, matrices and networks are particularly useful for impact identification, while weighting-and-scaling, -rating, or -ranking checklists find greatest application in the final evaluation of alternatives and the selection of a proposed action (Lee, 1988). It is not necessary to use a methodology in entirety in an impact study; it may be instructive to use portions of several methodologies for certain requisite activities. In that regard, methodology selection may be considered a part of an impact study. Some desirable characteristics of an EIA

TABLE 3.1

APPLICATIONS OF METHODOLOGIES IN EIA PROCESS

Process activity	Methodologies		Relative usefulness
Impact identification	Matrices	Simple	High
		Stepped	Medium
	Networks		High
	Checklists	Simple[a]	Medium
		Descriptive	Medium
Describing affected environment	Matrices	Simple	Low
		Stepped	
	Networks		
	Checklists	Simple[a]	High
		Descriptive	
Impact prediction and assessment	Matrices	Simple	Medium
		Stepped	Medium
	Networks		Medium
	Checklists	Descriptive	High
		Scaling, rating, ranking	Low
Selection of proposed action (based on evaluation of alternatives)	Matrices	Simple	Medium
		Stepped	Low
	Checklists	Scaling, rating, ranking	Medium
		Weighting-scaling, -rating, -ranking	High
Study summarization and communication	Matrices	Simple	High
		Stepped	Low
	Checklists	Simple[a]	Medium

[a]Simple checklists include questionnaire methods.

method selected for usage include the following: (1) it should be appropriate to the necessary task, such as impact identification or comparison of alternatives (not all methods are equally useful for all tasks), (2) it should be sufficiently free from assessor bias (the results should be essentially reproducible from one assessor group to another), and (3) it should be economical in terms of costs and its requirements of data, investigation time, personnel, and equipment and facilities (Lee, 1983). Additional information on methodology selection is in Chapter 15.

While numerous methodologies have been developed, and still additional methodologies are emerging, there is no "universal" methodology which can be applied to all project types in all environmental settings. It is unlikely that an all-purpose methodology will be developed, given the lack of technical information and the need for exercising subjective judgment about predicted impacts in the environmental setting wherein the project may occur. Accordingly, an appropriate perspective is to consider methodologies "tools" which can be used to aid the EIA process. In that sense, every utilized methodology should be project- and location-specific, with the basic concepts derivable from existing methodologies. These could be called "ad hoc" methods.

Methodologies do not provide complete answers to all questions related to the impacts of a potential project or set of alternatives. Methodologies are not "cookbooks" in which a successful study is achieved by meeting the detailed requirements of the methodologies. Methodologies must be selected based on appropriate evaluation and professional judgment, and they must be used with the continuous application of judgment relative to data inputs and analysis and interpretation of results.

One of the purposes of using methodologies is to insure that all pertinent environmental factors are included in the study. Most methodologies contain lists of environmental factors ranging from about 50 to 1,000 items, with the majority having between 50 and 100 items. Another purpose of using methodologies is to aid in planning baseline studies in locations where relevant environmental data is lacking. For example, if information is not available on the factors identified using appropriate methodologies, it may be determined that field studies will be necessary.

One of the most important reasons for using methodologies is that they provide a means for the synthesis of information and the evaluation of alternatives on a common basis. Within the United States, even the comparative analysis of alternatives has often been less than optimal. In many cases, alternatives have been eliminated from detailed consideration based only on economic comparisons. Usage of structured methodologies can provide the basis for evaluation of alternatives using a common framework of decision factors. Methodologies can also be useful in evaluating the cost-effectiveness of proposed impact-mitigation (IM) measures. Evaluation of a proposed project with and without mitigation will enable a clearer delineation of the effectiveness of potential IM measures.

An important element in impact studies is the communication of resultant information to other practitioners, regulatory agencies, and the general public. Some methodologies have features which are particularly useful in communicating impact information in summary form; an example is the simple interaction matrix. Finally, the NEPA requires that agencies utilize methods and procedures which ensure that unquantified environmental amenities and values be given appropriate consideration in decision making, along with more-traditional economic and technical considerations. Therefore, in order to comply with this requirement, the use of methods is strongly encouraged.

In addition to matrices and checklists, several other classifications of methodologies have been developed. For example, Warner (1973) and Warner and Bromley (1974) divided impact methodologies into five main classes: ad hoc procedures, overlay techniques, checklists, matrices, and networks. Ad hoc procedures involve assembling a team of specialists to identify impacts in their areas of expertise, with minimal guidance beyond the requirements of the NEPA. This approach was essentially utilized by all federal agencies in the period immediately following enactment of the NEPA. It is still used in the sense that as extant methodologies are adapted to specific needs, the results can be called ad hoc methods.

The term "overlay techniques" describes several well-developed approaches used in planning and landscape architecture. These techniques are based on the use of a series of overlay maps depicting environmental factors or land features (McHarg, 1971). The overlay approach is generally effective for selecting alternatives and identifying certain types of impacts; however, it cannot be used to quantify impacts or to identify secondary and tertiary interrelationships. Overlay techniques utilizing computerization for more-effective data analysis have been developed. Geographic information systems are now being used as layered overlay mapping techniques.

Comparative reviews of impact identification methods are in Canter (1979), Nichols and Hyman (1982), Bisset (1980, 1983), Lohani and Halim (1990), and ESCAP (1990). Twelve specific methodologies were systematically reviewed by Nichols and Hyman (1982); while Bisset (1980, 1983) and Lohani and Halim (1990) delineated the features of over 15 methods. Canter (1979) summarized over 100 methods and techniques for use in the EIA process.

The primary focus of this chapter will be on the use of matrix, network, and simple and descriptive checklist methods for impact identification. In using these methodologies, it is important to delineate the uncertainty associated with impact predictions (Lee, 1983). In other words, the use of scientific approaches will require the exercise of professional judgment in the interpretation of the results.

INTERACTION-MATRIX METHODOLOGIES

Interaction matrices were one of the earliest types of EIA methodologies. A "simple interaction matrix" displays project actions or activities along one axis, with appropriate environmental factors listed along the other axis of the matrix. When a given action or activity is expected to cause a change in an environmental factor, this is noted at the intersection point in the matrix and further described in terms of separate or combined magnitude and importance considerations. Many variations of the simple interaction matrix have been utilized in impact studies, including stepped matrices (Canter, 1986; ESCAP, 1990; Lohani and Halim, 1990; International Institute for Applied Systems Analysis, 1979).

Simple Matrices

The interaction-matrix method developed by Leopold et al. (1971) will be used as an example of a simple matrix. The matrix lists approximately 100 specified actions and 90 environmental items. Figure 3.1 illustrates the concept of the Leopold matrix, and Table 3.2 contains the list of the actions and environmental items. In the use of the Leopold matrix, each action and its potential for creating an impact on each environmental item must be considered. Where an impact is anticipated, the matrix is marked with a diagonal line in the appropriate interaction box.

The second step in using the Leopold matrix is to describe the interaction in terms of its magnitude and importance. The "magnitude" of an interaction is its extensity or scale and is described by the assignment of a numerical value from 1 to 10, with 10 representing a large magnitude and 1 a small magnitude. Values near 5 on the magnitude scale represent impacts of intermediate extensity. Assignment of a numerical value for the magnitude of an interaction should

Actions causing impact

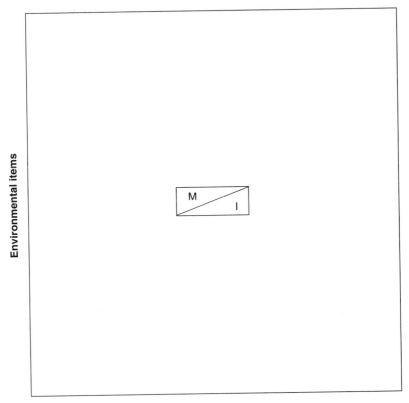

FIGURE 3.1
Leopold Interaction Matrix: M = magnitude; I = importance (Leopold et al., 1971).

be based on an objective evaluation of facts related to the anticipated impact. The ''importance'' of an interaction is related to its significance, or an assessment of the probable consequences of the anticipated impact. The scale of importance also ranges from 1 to 10, with 10 representing a very important interaction and 1 an interaction of relatively low importance. Assignment of a numerical importance value is based on the subjective judgment of the individual, small group, or interdisciplinary team working on the study.

One of the attractive features of the Leopold matrix is that it can be expanded or contracted—that is, the number of actions can be increased or decreased from the total of about 100, and the number of environmental factors can be increased or decreased from about 90. The primary advantages of using the Leopold matrix are that it is very useful as a gross screening tool for impact identification purposes, and it can provide a valuable means for impact communication by providing a visual display of the impacted items and of the major actions causing impacts.

Summation of the number of rows and columns designated as having interactions can offer insight into impact assessment. Additional refinements can be used to discuss the results of a simple interaction matrix. For example,

TABLE 3.2

ACTIONS AND ENVIRONMENTAL ITEMS IN LEOPOLD INTERACTION MATRIX

Actions		Environmental items	
Category	Description	Category	Description
A Modification of regime	a Exotic fauna introduction	A Physical and chemical characteristics	
	b Biological controls	1 Earth	a Mineral resources
	c Modification of habitat		b Construction material
	d Alteration of ground cover		c Soils
	e Alteration of ground-water hydrology		d Land form
	f Alteration of drainage		e Force fields and background radiation
	g River control and flow modification		f Unique physical features
	h Canalization	2 Water	a Surface
	i Irrigation		b Ocean
	j Weather modification		c Underground
	k Burning		d Quality
	l Surfacing or paving		e Temperature
	m Noise and vibration		f Recharge
B Land transformation and construction	a Urbanization		g Snow, ice, and permafrost
	b Industrial sites and buildings	3 Atmosphere	a Quality (gases, particulates)
	c Airports		b Climate (micro, macro)
	d Highways and bridges		c Temperature
	e Roads and trails	4 Processes	a Floods
	f Railroads		b Erosion
	g Cables and lifts		c Deposition (sedimentation, precipitation)
	h Transmission lines, pipelines, and corridors		d Solution
	i Barriers, including fencing		e Sorption (ion exchange, complexing)
	j Channel dredging and straightening		f Compaction and settling
	k Channel revetments		g Stability (slides, slumps)
	l Canals		h Stress-strain (earthquakes)
	m Dams and impoundments		i Air movements
	n Piers, seawalls, marinas, and sea terminals	B Biological conditions	
	o Offshore structures	1 Flora	a Trees
	p Recreational structures		b Shrubs
	q Blasting and drilling		c Grass
	r Cut and fill		d Crops
	s Tunnels and underground structures		e Microflora
C Resource extraction	a Blasting and drilling		f Aquatic plants
	b Surface excavation		g Endangered species
			h Barriers

TABLE 3.2

ACTIONS AND ENVIRONMENTAL ITEMS IN LEOPOLD INTERACTION MATRIX *(continued)*

Actions		Environmental items	
Category	**Description**	**Category**	**Description**
	c Subsurface excavation and retorting	2 Fauna	i Corridors
			a Birds
	d Well dredging and fluid removal		b Land animals including reptiles
	e Dredging		c Fish and shellfish
	f Clear cutting and other lumbering		d Benthic organisms
			e Insects
	g Commercial fishing and hunting		f Microfauna
			g Endangered species
D Processing	a Farming		h Barriers
	b Ranching and grazing		i Corridors
	c Feed lots	C Cultural factors	
	d Dairying	1 Land use	a Wilderness and open spaces
	e Energy generation		b Wetlands
	f Mineral processing		c Forestry
	g Metallurgical industry		d Grazing
			e Agricultural
	h Chemical industry		f Residential
	i Textile industry		g Commercial
	j Automobile and aircraft		h Industry
			l Mining and quarrying
	k Oil refining	2 Recreation	a Hunting
	l Food		b Fishing
	m Lumbering		c Boating
	n Pulp and paper		d Swimming
	o Product storage		e Camping and hiking
E Land alteration	a Erosion control and terracing		f Picnicking
			g Resorts
	b Mine sealing and waste control	3 Aesthetics and human interest	a Scenic views and vistas
	c Strip-mining rehabilitation		b Wilderness qualities
			c Open-space qualities
	d Landscaping		d Landscape design
	e Harbor dredging		e Unique physical features
	f Marsh fill and drainage		f Parks and reserves
F Resource renewal	a Reforestation		g Monuments
	b Wildlife stocking and management		h Rare and unique species or ecosystems
	c Groundwater recharge		i Historical or archaeological sites and objects
	d Fertilization application		j Presence of misfits
	e Waste recycling	4 Cultural status	a Cultural-patterns (life-style)
G Changes in traffic	a Railway		
	b Automobile		
	c Trucking		
	d Shipping		

TABLE 3.2

ACTIONS AND ENVIRONMENTAL ITEMS IN LEOPOLD INTERACTION MATRIX *(continued)*

Actions		Environmental items	
Category	**Description**	**Category**	**Description**
H Waste emplacement and treatment	e Aircraft f River and canal traffic g Pleasure boating h Trails i Cables and lifts j Communication k Pipeline a Ocean dumping b Landfill c Emplacement of tailings, spoils, and overburden d Underground storage e Junk disposal f Oil well flooding g Deep well emplacement h Cooling water discharge i Municipal waste discharge including spray irrigation j Liquid effluent discharge k Stabilization and oxidation ponds l Septic tanks, commercial and domestic m Stack and exhaust emission n Spent lubricants	5 Manufactured facilities and activities D Ecological relationships E Others	b Health and safety c Employment d Population density a Structures b Transportation network (movement, access) c Utility networks d Waste disposal e Barriers f Corridors a Salinization of water resources b Eutrophication c Disease and insect vectors d Food chains e Salinization of surficial material f Brush encroachment g Other
I Chemical treatment	a Fertilization b Chemical deicing of highways, etc. c Chemical stabilization of soil d Weed control e Insect control (pesticides)		
J Accidents	a Explosions b Spills and leaks c Operational failure		
K Others			

Source: Compiled using data from Leopold et al., 1971.

assume that a matrix incorporates the impacts of 8 actions on 20 environmental factors. Further, assume that the average action would cause 10 factors to be impacted, and the average number of impacts per factor is 6. The impacts could be grouped and discussed in terms of those actions exhibiting a greater-than-average, near-average, and fewer-than-average number of impacts. A similar approach could be used for addressing the impacted factors.

The Leopold matrix can also be utilized to identify beneficial as well as detrimental impacts through the use of appropriate designators, such as plus and minus signs. In addition, the Leopold matrix can be employed to identify impacts at various temporal phases of a project—for example, construction, operation, and post-operation phases, and to describe impacts associated with various spatial boundaries—namely, at the site and in the region.

Many uses of the Leopold matrix have involved the assignment of three levels of magnitude and importance. Major interactions would be assigned maximum numerical scores, with minor interactions being assigned minimal scores. Intermediate-level interactions would be assigned values between the major and minor scores.

In Table 3.2, there are very few items in the list of environmental factors that are oriented to the socioeconomic environment. This does not mean that these items could not be added, but rather that in 1970 and 1971, the period of time in which the matrix concept was developed, less emphasis was given to this substantive area.

Variations of the Leopold matrix have been utilized for impact analysis for many types of projects. The Federal Aviation Administration (1973) has used interaction matrices for aviation-type projects. The Oregon Highway Department (1973) has developed an interaction matrix for impact identification, and the various actions and environmental factors included in this matrix are shown in Table 3.3. Condensed versions of the Leopold matrix have been employed in EIAs for a coal mine, a generation plant, a county road and railroad project, a water supply system, and a transmission line, for examples (Chase, 1973).

Information expressed by means of ranks other than numerical values for magnitude and importance can be included in the impact scales associated with identification of an interaction. In an earth-filled–dam project, the potential impact of various actions on environmental factors has been shown in 11 categories: neutral, a range of five degrees of beneficial impact, and a range of five degrees of detrimental impact (Chase, 1973). Scales have also been used to describe the probability of occurrence of an impact, with the scale ranging from low to intermediate to high probability of impact. Impact scales can also be developed to show the extent of potential reversibility associated with a beneficial or detrimental impact.

Another approach for impact rating in a matrix involves the use of a predefined code denoting the characteristics of the impacts and whether or not certain undesirable features could be mitigated. Table 3.4 displays an example of this type of an interaction matrix for a proposed wastewater collection, treatment, and disposal project in Barbados. For this analysis, the following definitions are used for the codes (Canter, 1991):

SB = Significant beneficial impact; represents a highly desirable outcome in terms of either improving the existing quality of the environmental factor or enhancing that factor from an environmental perspective.

SA = Significant adverse impact; represents a highly undesirable outcome in terms of either degrading the existing quality of the environmental factor or disrupting that factor from an environmental perspective.

B = Beneficial impact; represents a positive outcome in terms of either improving the

TABLE 3.3

HIGHWAY INTERACTION MATRIX

Actions that may cause impact		Environmental conditions	
Category	**Action**	**Category**	**Factor**
A Elements of design and location		A Physical and chemical characteristics	
1 Modification of regime	a Modification of habitat	1 Earth	a Mineral resources, precious
	b Alteration of groundwater hydrology		b Mineral resources, common
	c Canalization		c Soils
	d Irrigation		d Land form
	e Surfacing and paving	2 Water	a Surface
2 Land transformation and construction	a Highways and bridge construction		b Ocean and estuaries
	b Road and trail construction		c Underground
	c Construction of barriers including fencing		d Snow and ice
			e Recharge-percolation
	d Channel dredging and straightening		f Quality
	e Channel revetments		g Temperature
	f Dams and impoundments	3 Atmosphere	a Quality
	g Piers and seawalls		b Climate
	h Recreational structures		c Temperature
	i Cut and fill	4 Processes	a Floods
	j Tunnels and underground-structure construction		b Erosion (air or water)
	k Erosion control		c Deposition (air or water)
	l Landscaping		d Solution
	m Harbor dredging		e Compaction and settling
	n Marsh fill and draining		f Stability (sides and slumps)
	o Scenic-wayside alteration		g Air movements
	p Junkyard and billboard removal		h Fire
			i Evaporation
3 Well drilling		B Biological characteristics	
4 Resource renewal and protection	a Reforestation	1 Flora	a Trees
	b Scenic-strip acquisition		b Shrubs
5 Changes in traffic	a Railway		c Grass
	b Automobile		d Crops
	c Trucking		e Microflora
	d River and canal traffic		f Aquatic plants
	e Pleasure boating		g Endangered species
	f Trails		h Barriers
	g Communication		i Corridors
	h Pipeline	2 Fauna	a Birds
			b Land animals

65

TABLE 3.3

HIGHWAY INTERACTION MATRIX (continued)

Actions that may cause impact		Environmental conditions	
Category	**Action**	**Category**	**Factor**
B During construction 1 Modification of regime	a Exotic flora and fauna introduction b Biological controls c Alteration of ground cover d Alteration of drainage e River control and flow modification f Burning		c Fish and shellfish d Other aquatic organisms e Insects f Microfauna g Endangered species h Barriers i Corridors
2 Land transformation and construction	a Blasting and drilling b Marsh fill and drainage c Clearing and grubbing d Dams-impoundments	C Cultural factors 1 Land use	a Wilderness b Open spaces c Wetlands d Forestry e Grazing f Agricultural g Residential h Commercial i Industrial j Lakes and rivers
3 Resource extraction	a Blasting and drilling b Surface excavation c Subsurface excavation d Well drilling and fluid removal e Dredging	2 Recreation	a Hunting b Fishing c Boating d Swimming e Camping f Hiking g Picnicking h Resorts i Winter sports j Rockhounding
4 Changes in traffic	a Railway b Automobile c Trucking d River and canal traffic e Pleasure boating f Trails g Communication h Pipeline	3 Aesthetics and human interest	a Scenic views and vistas b Wilderness qualities c Open-space qualities d Landscape design e Unique physical features f Parks and reserves
5 Waste emplacement and treatment	a Landfill b Emplacement of tailings, spoil, and overburden c Liquid and exhaust discharge d Stack and exhaust emissions e Occurrence of spent lubricants		
6 Chemical stabilization of soil			

7 Accidents
   a Explosions
   b Spills and leaks
   c Operational failure

C Operation
1 Waste emplacement and treatment
   a Liquid effluent discharge
   b Septic-tank issues
   c Stack and exhaust emissions
2 Chemical treatment
   a Fertilization
   b Chemical deicing
   c Weed control
   d Insect control
3 Accidents
   a Explosions
   b Spills, leaks
   c Operational failures

   g Monuments
   h Rare or unique species or ecosystems
   i Historical or archaeological sites and objects
   j Presence of incompatible features
4 Cultural status
   a Cultural patterns
   b Health
   c Population density
   d Institutions
   e Minority groups
   f Economic groups
5 Manufactured facilities and activities
   a Structures
   b Transportation
   c Utility networks
   d Waste disposal
   e Barriers
   f Corridors
   g Governmental activities

Source: Compiled using data from Oregon Highway Department, 1973.

TABLE 3.4

INTERACTION MATRIX FOR SOUTH COAST (BARBADOS) SEWERAGE PROJECT

Environmental factor or resource	Existing quality	Construction phase				Operation phase			
		Collection system	Treatment plant	Outfall line	Resultant quality	Collection system	Treatment plant	Outfall line	Resultant quality
Air quality	In compliance with air quality standards	A/M	A/M	a	Dusts, CO	a (odor at lift station sites)	A/M	O	Localized odor
Noise	Typical of urban residential areas	A/M	A/M	a	Increase in local noise	a (pumps)	a	a (pumps)	Small increase in noise
Groundwater	Satisfactory for area	O	O	O	Same as existing	b	b	b	Better quality due to less sheet water discharge
Graeme Hall	"Natural" biological resource	NA	a/M (no encroachment)	NA	Some disturbance, recovery expected	NA	O	NA	Same as existing
Beach erosion, coral reef, coastal water quality	Erosion of 0.1 to 0.3 m/yr. deteriorating coral reef and coastal water quality	NA	NA	a (water quality)	Turbidity increase	b	SB	NA	Improve quality
Coastal fisheries	Some decline due to deteriorating coral reef and coastal water quality	NA	NA	a	Local turbidity	b	SB	NA	Improve quality
Marine environment at outfall diffuser	Good	NA	NA	a	Some local disturbance	NA	NA	a	Small decrease in quality
Traffic	Current problem	SA/M	a	a	Increase in congestion	a	a	a	Continued problem due to tourism increase
Tourism	Important to economy	a	NA	a	Traffic congestion might cause decrease	B	B	B	Increase in economy

A = adverse impact, M = mitigation measure planned for adverse impact, a = small adverse impact, NA = environmental factor not applicable, SA = significant adverse impact, b = small beneficial impact, B = beneficial impact, SB = significant beneficial impact.
Source: Canter, 1991, pp. iii–iv.

existing quality of the environmental factor or enhancing that factor from an environmental perspective.

A = Adverse impact; represents a negative outcome in terms of either degrading the existing quality of the environmental factor or disrupting that factor from an environmental perspective.

b = Small beneficial impact; represents a minor improvement in the existing quality of the environmental factor or a minor enhancement in that factor from an environmental perspective.

a = Small adverse impact; represents a minor degradation in the existing quality of the environmental factor or a minor disruption in that factor from an environmental perspective.

O = No measurable impact is expected to occur as a result of considering the project action relative to the environmental factor.

M = Some type of mitigation measure can be used to reduce or avoid a minor adverse, adverse, or significant adverse impact.

NA = The environmental factor is not applicable or not relevant to the proposed project.

Simple interaction matrices have been used for analyzing the impacts of still other types of projects; examples include flood-control and/or hydropower, highway, transmission-line, offshore oil lease, coal mine, power plant, industrial plant, industrial park, pipeline, housing development, tourism, and coastal development projects. A generic simple interaction matrix has been developed to identify key impacts of a variety of development projects in coastal zones in Asian Development Bank (1991). ''Prompter questions'' tied to impact concerns are also included to focus the impact identification process.

Stepped Matrices

A stepped matrix, also called a ''cross-impact matrix,'' can be used to address secondary and tertiary impacts of initiating actions. A ''stepped matrix'' is one in which environmental factors

FIGURE 3.2
Concept of Stepped Matrix.

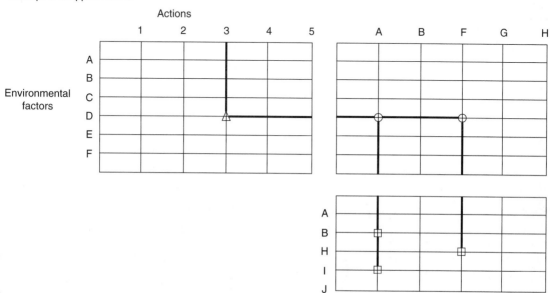

are displayed against other environmental factors. The consequences of initial changes in some factors on other factors can be displayed. Figure 3.2 displays the concept of a stepped matrix. In the figure, action 3 impacts factor D; changes in factor D then cause changes in factors A and F. Finally, changes in factor A cause changes in factors B and I, while changes in factor F cause changes in factor H. Stepped matrices facilitate the tracing of impacts and the recognition of the environment as a system; they represent an intermediate method between simple matrices and networks. Stepped matrices with multiple actions and several types and levels of impact can become visually complicated.

Johnson and Bell (1975) developed both a simple and stepped (cross-impact) interaction matrix for identifying impacts from the construction and operation of water resources reservoir projects. Project activities and the 92 included environmental attributes are listed in Table 3.5. Definitions for each were included in the method. Both letters and numbers were used in impact rating, for example, a rating could be A3, based on the following categories (Johnson and Bell, 1975, p. 3):

A = adverse, always occurs
B = adverse, usually occurs
C = adverse, sometimes occurs
N = not necessarily good or bad
X = beneficial, always occurs
Y = beneficial, usually occurs
Z = beneficial, sometimes occurs
1 = strong, permanent
2 = moderate, permanent
3 = minor, permanent
4 = strong, temporary
5 = moderate, temporary
6 = minor, temporary
Blank = no impact

A cross-impact matrix was used to display the relationships of the 92 environmental attributes to the same 92 attributes. The following codes were used to denote how changes in one attribute can affect others (Johnson and Bell, 1975, p. 3):

× = interaction between two attributes
○ = interaction between groups of attributes
Blank = no impact

As another example, the environmental impacts of petroleum development in Alaska have been identified through the use of both simple and stepped interaction matrices (Hanley, Hemming, and Morsell, 1981). Primary effects were identified based on four project phases (exploration, development, production, and termination) and seven categories of environmental perturbations (land-surface disturbances, stream or lake disturbances, noise and activity, land pollution, water pollution, air pollution, and indirect human activities). Specific activities and factors are listed in Table 3.6. An impact rating involving three levels (minor, medium, and maximum) of beneficial and detrimental impacts was used. Matrices were developed for treeless permafrost and forested nonpermafrost terrain. Descriptive rationale was presented for each assigned rating. Algebraic sums of the impact ratings were calculated for each phase or activity, thus allowing for a discussion of its relative impacts. The implications of the physical changes and disturbances were then addressed by means of a stepped matrix relating the disturbances to fish and wildlife resources; this matrix is shown in Table 3.7. An x denoted an impact and the basic rationale was described; many interaction cells were assigned no rating.

A 74-acre industrial park project in the southwestern portion of Fresno, California, also made use of a stepped matrix (U.S. Economic Development Administration, 1973). The project involved two separate and simultaneous actions. The first action was the securing of a loan to improve the processing facilities of two industries located on the project site. In addition to increased production, the loan also permitted greater control of the emission of objectionable odors. The second action was the use of a grant to the city of Fresno, to facilitate the acquisition, construction, and redevelopment of the site into

TABLE 3.5

SIMPLE AND STEPPED INTERACTION MATRIX FOR WATER RESOURCES RESERVOIR PROJECTS

Construction and operation activities (*x* axis)	Environmental attributes (*y* axis)
Clearing	Air quality
Grubbing	
Stripping	Microclimate
Excavation	A. Air movement
Stockpiling	B. Air temperature
Loading-hauling	C. Relative humidity
Placement of materials	D. Incident radiation
Grading	
Compaction	Soil conditions
Removal of materials	A. Temperature
Blasting	B. Soil moisture
Concrete placement	C. Soil structure
Surfacing	D. Soil flora
Building erection	E. Soil fauna
Building movement	
Building demolition	Ecological relationships
Pavement demolition	A. Terrestrial ecosystems
Batch and aggregate plants	1. Change in ecostructure
Temporary buildings	2. Trophic structure
Vehicle and equipment maintenance	3. Pollution of land
Restoration	4. Rare or unique ecotypes
Filing reservoir	5. Diversity of ecotypes
Flood-control operation	6. Biogeochemical cycles
	B. Aquatic ecosystems
	Fauna
	A. Terrestrial animals
	1. Mammals
	2. Birds
	3. Other vertebrates
	4. Mosquitoes
	5. Other invertebrates
	6. Rare and endangered species
	7. Species diversity, etc.
	8. Nuisance species
	B. Aquatic animals
	Flora
	A. Terrestrial plants
	1. Natural vegetation
	2. Rare and endangered species
	3. Species diversity
	4. Primary productivity
	5. Weedy species
	6. Detritus
	B. Aquatic flora
	Groundwater hydrology
	A. Depth
	B. Movement
	C. Recharge rates
	Surface-water hydrology
	A. Elevation
	B. Flow pattern
	C. Stream discharge
	D. Velocity

TABLE 3.5

SIMPLE AND STEPPED INTERACTION MATRIX FOR WATER RESOURCES RESERVOIR PROJECTS *(continued)*

Construction and operation activities (*x* axis)	Environmental attributes (*y* axis)

	Land forms and processes
	A. Compaction of soil
	B. Topography
	C. Stability of land forms
	D. Water erosion of soil
	E. Silt deposition
	F. Wave movement of soil
	G. Wind movement of soil
	
	Outdoor recreation
	A. Land-based
	B. Water-based
	
	Preservation of natural resources
	A. Fauna
	B. Flora
	C. Natural ecosystem types
	D. Open and green space
	E. Water supply
	F. Agricultural land
	
	Special-interest areas
	Aesthetics
	A. Air quality
	B. Construction scars
	C. Man-made features
	D. Scenic views
	E. Landscape diversity
	F. Vegetation
	G. Water quality
	H. Noise
	
	Surface-water quality
	A. Physical attributes
	    1. Color
	    2. Discharge
	    3. Redox potential
	    4. Turbidity
	    5. Water temperature
	B. Chemical attributes
	    1. Carbon dioxide
	    2. COD
	    3. Dissolved oxygen (DO)
	    4. Nitrate
	    5. Phosphorus
	    6. Sulfur

Source: Adapted from Johnson and Bell, 1975, p. 86.

TABLE 3.6

ENVIRONMENTAL PERTURBATIONS THAT MAY OCCUR AS A RESULT OF PETROLEUM INDUSTRY PRACTICES[a]

Environmental Perturbations[b] (y-axis)		Petroleum Development Phase and Activity (x-axis)	
		Exploration	
		Geophysical Survey	**Drilling**
Land surface disturbances	Destruction of vegetation Tree clearing Slash disposal Altered soil characteristics Thermal erosion/thermokarst Hydraulic erosion Altered surface water hydrology Fill over land surface Above ground obstructions	Overland travel Stream crossings Seismic wave production Remote-area human support Air traffic Accidents Regional population increase	Overland travel including stream crossings Access road construction Airstrip construction Drill and camp site development Gravel mining Drilling and related activities Remote area human support Air traffic Accidents Regional population increase
Stream or lake disturbances	Stream bank erosion Siltation Channel construction Altered current velocity Channel obstruction Shock wave Perched drainage structure Bottom substrate disturbance Long term channel changes Reduced water volume Altered water quality Drainage of lake basin	**Development**	
		Production Facility Construction	**Pipeline Construction**
		Overland travel Access road construction Site development (all facilities) Gravel mining Drilling Mechanical facilities Construction Gathering pipeline construction Remote area human support Air traffic Accidents Regional population increase	Access road construction Right-of-way development Stream crossings Gravel mining Pump station construction Pipe laying Restoration Remote area human support Air traffic Accidents Regional population increase

TABLE 3.6

ENVIRONMENTAL PERTURBATIONS THAT MAY OCCUR AS A RESULT OF PETROLEUM INDUSTRY PRACTICES *(continued)*

		Production	
		Field Operation	**Pipeline Operation**
Noise and activity	Loud noise (blasting, aircraft, etc.) Moderate noise Human activity	Road travel Operation of mechanical Facilities Permanent human support Air traffic Accidents Local population increase Regional population increase	Development of ancillary indust. Road travel Operation of mechanical facil. Presence of above ground pipe Gravel mining Permanent human support Air traffic Accidents Regional population increase
Land pollution	Oil and fuel spills Toxic chemical spills Drilling fluid Domestic solid waste litter Edible substance availability	**Termination**	
		Road travel Removal of equipment Gravel removal and stockpile Terrain rehabilitation Revegetation Remote area human support Air traffic Accidents	
Water pollution	Suspended sediment Oil and fuel spills Toxic chemical spills Drilling fluid Sanitary waste efflluent		
Air pollution	Dust generation Emissions from int. combustion eng. Emissions from major facilities Incinerator and burn smoke		
Indirect human activities	Hunting and fishing Intensified land use demands Incr. domestic waste processing		

[a]Assumes the area is sufficiently remote to require full human support facilities.
[b]The rating system is based on a subjective scale of 0-3 with 3 representing a maximum detrimental impact; negative numbers are used to represent beneficial impacts.
Source: Hanley, Hemming, and Morsell, 1981, p. 142.

TABLE 3.7

POTENTIAL IMPACTS TO FISH AND WILDLIFE AS A RESULT OF PETROLEUM-INDUSTRY-RELATED ENVIRONMENTAL PERTURBATIONS

Impacts to fish and wildlife resources or their habitats (columns) vs. *Environmental perturbations — Land surface disturbances* (rows).

The table is presented below with impact categories as rows and the nine land-surface-disturbance perturbations as columns:

Impact to fish and wildlife	Destruction of vegetation	Tree clearing	Slash disposal	Altered soil characteristics	Thermal erosion/thermokarst	Hydraulic erosion	Altered surface water hydrology	Fill over land surface	Above ground obstructions
Alteration of Terrestrial Habitat — Direct Habitat Loss									
Reduced primary productivity	X					X	X	X	
Wildlife range reduction	X					X	X	X	
Displacement of low mobility species	X					X	X		
Elimination of critical habitat	X					X	X	X	
Interruption of energy & nutrient flow	X					X	X		
Alteration of Terrestrial Habitat — Altered Habitat Charac.									
Reduced plant cover & productivity	X	X			X	X	X		
Altered community composition	X	X			X	X	X		
Altered wildlife use	X	X	X	X		X	X	X	
Outbreaks of pest organisms			X						
Interruption of energy & nutrient flow	X	X		X	X	X	X		
Alteration of Terrestrial Habitat — Interfer. with Movements									
Interruption of critical life history stages								X	X
Prevention of access to critical habitats								X	X
High energy expenditure—lowered survival								X	X
Injury of flying birds								X	X
Death or injury from toxic substances									
Alteration of Aquatic Habitat — Direct Habitat Loss									
Reduced productivity									
Displacement of low mobility species									
Elimination of critical fish habitat									
Interruption of energy & nutrient flow									
Alteration of Aquatic Habitat — Altered Habitat Charac.									
Reduced productivity							X		
Reduction of critical fish habitat							X		
Reduction of fish food organisms							X		
Interruption of energy & nutrient flow							X		
Alteration of Aquatic Habitat — Interfer. with Movements									
Blockage of spawning migration									
Blockage of access to feeding habitat									
Prevention of escape from adverse habitat									
High energy expenditure—lowered survival									
Interruption of energy & nutrient flow									
Death or injury to fish or eggs									
Disturbance of Wildlife (Noise, etc.)									
Short term displacement of mobile species									
Long term displacement—range reduction									
Displacement of some population segments									
Displacement from critical habitat									
Increased stress—lowered survival									
Interruption of critical life history stages									
Interruption of energy & nutrient flow									
Attraction of Wildlife									
Poor survival upon removal of food source									
Man induced mortality if nuisance animal									
Altered behavior passed on to progeny									
Imbalance between pop. size & food supply									
Breakdown in natural social structure									
Increased stress—lowered survival									
Human Exploit.									
Increased fish mortality									
Increased wildlife mortality									

75

Category	Item
Stream or lake disturbances	Stream bank erosion
	Siltation
	Channel constriction
	Altered current velocity
	Channel obstruction
	Shock wave
	Perched drainage structure
	Bottom substrate disturbance
	Long term channel changes
	Reduced water volume
	Altered water quality
	Drainage of lake basin
Noise and activity	Loud noise (blasting, aircraft, etc.)
	Moderate noise
	Human activity
Land pollution	Oil and fuel spills
	Toxic chemical spills
	Drilling fluid
	Domestic solid waste litter
	Edible substance availability
Water pollution	Suspended Sediment
	Oil and fuel spills
	Toxic chemical spills
	Drilling fluid
	Sanitary waste effluent
Air pollution	Dust generation
	Emissions from int. combustion eng.
	Emissions from major facilities
	Incinerator and burn smoke
Indirect human activities	Hunting and fishing
	Intensified land use demands
	Incr. domestic waste processing

Source: Hanley, Hemming, and Morsell, 1981, p. 154.

improved sites for use by heavier industry. The stepped matrix that was developed is shown in Figure 3.3. The steps involved in the use of the matrix are as follows:

1. Enter the matrix at the upper left-hand corner under the heading Project Elements. In this example, the element is 2 Future Improvements.
2. Read to the right. A causal factor that may result in an impact is shown at Surfacing. A dot (○) indicates that a relationship exists between Future Improvements and Surfacing.
4. Read downward from the ○ until either a ★, ⋆, □, ▯, or ∪ is encountered. If a ★ appears, a major positive impact exists. A ⋆ indicates a minor positive impact. A □ indicates a major negative impact. A ▯ indicates a minor negative impact. A ∪ indicates that an impact exists, but its magnitude or direction cannot be determined at present. Reading downward from Surfacing, a □ is encountered.

5. Read along the row, beginning at the left. A major negative impact would be a change in Subsurface Water. The 2 next to the □ indicates that the impact originates at 2 Future Improvements.
6. Continue reading to the right.
7. In the column Initial Condition is the notation "High quality," indicating that the altered element is presently of high quality.
8. In the column Mechanisms of Change is a notation "Reduced," describing the way in which the element would be altered.
9. In the column Possible Final Condition is the notation "Little," describing the altered element after the impact has taken place.
10. The Potential Corrective Measures column is reserved for those impacts against which some mitigation or effect-minimization steps have been or could be taken. These steps would be noted here.

FIGURE 3.3

Guide to Using Stepped Matrix for Industrial Park Project (U.S. Economic Development Administration, 1973).

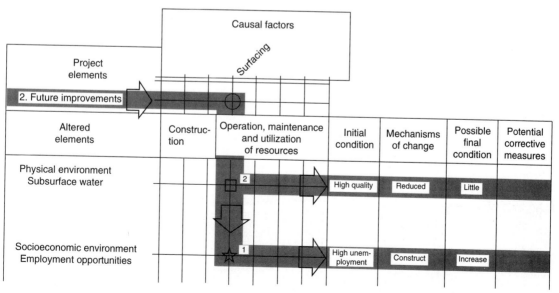

Development of a Simple Matrix

It is considered better to develop a specific matrix for the project, plan, program, or policy being analyzed, rather than using a generic matrix. The following steps can be used by an individual or an interdisciplinary team in preparing a simple interaction matrix:

1. List all anticipated project actions and group them according to temporal phase, such as construction, operation, and postoperation.
2. List all pertinent environmental factors from the environmental setting, and group them (a) according to physical-chemical, biological, cultural, and socioeconomic categories, and (b) based on spatial considerations such as site and region, or upstream, site, and downstream.
3. Discuss the preliminary matrix with study team members and/or advisors to team or study manager.
4. Decide on an impact-rating scheme (for example, numbers, letters, or colors) to be used.
5. Talk through the matrix as a team, and make ratings and notes in order to identify and summarize impacts (documentation).

Other Types of Matrices

Simple matrices can also be used for purposes other than impact identification. For example, Table 3.8 shows a matrix framework which can be used to summarize baseline environmental conditions. In this example, relative factor importance, present condition, and extent of management can be taken into consideration.

TABLE 3.8

CONCEPT OF AN ENVIRONMENTAL BASELINE MATRIX

Identification	Evaluation		
	Scale of importance	Scale of present condition	Scale of management
Environmental elements/units	1 2 3 4 5 low high	1 2 3 4 5 low high	1 2 3 4 5 low high
Biological: flora fauna ecological relationships			
Physical-chemical: atmosphere water earth			
Cultural: households communities economy communications			
Bio-cultural linkages/units: resources recreation conservation			

Source: Fischer and Davies, 1973, p. 215.

Importance weighting of impacts can also be included in interaction matrices (ESCAP, 1990; Lohani and Halim, 1990). A final example of using a matrix approach relates to the adaptive-environmental-assessment (AEA) process developed in the early 1970s (International Institute for Applied Systems Analysis, 1979). The methodology uses short, intensive workshops in which participants (resource specialists, managers, and policy makers) are assisted by a workshop staff in the construction of an interactive, computerized simulation model of the resource system being studied. The modeling exercise is used to promote communication and understanding among participants, to identify data gaps and research priorities, and to examine possible results of various management alternatives.

An AEA workshop generally consists of 5 days of meetings. It begins with an introduction to the AEA process and to the use of computer models in systems analysis and resource allocation. Basic modeling terms are defined, and preliminary goals for the week are discussed. The next step is to have workshop participants define clearly which components of the resource system are to be included in the simulation model. Important variables, potential management actions, and performance measures are identified, and decisions are made concerning the time interval to be covered by the model and the spatial extent and resolution to be included. Components so identified are then grouped into five or six major categories for further consideration by workshop participants. The next process, construction of a "looking-outward" matrix, is used to initiate interdisciplinary communication. The purpose of this activity is to identify those pieces of information needed to connect the five or six major components of the system. Each group of specialists is asked to articulate what they need to know about all of the other components in order to predict how their subsystem will behave under various environmental conditions. In other words, they are asked to look outward at the kinds of inputs that affect their subsystem. As a result of this exercise, each subgroup receives a list of inputs that will be provided by other subgroups and a list of outputs that they must in turn provide. The complexity of each submodel (or matrix) is thus limited by these two lists of variables. The resultant looking-outward matrix is then used in the development of one or more pertinent simulation models.

The AEA process can be based on the activities of one workshop. However, additional workshops may be desirable. The first workshop provides a good beginning to environmental analysis. Although the first version of the matrix and model may be incomplete, it serves to clarify information gaps and to provide a framework for the integration of existing and proposed studies. Ideally, additional workshops follow as new information becomes available. The periods between workshops are used for research, data collection, and model refinement. Each subsequent workshop produces a more credible model and/or matrix that is more useful in evaluating management alternatives. At all times, the model and/or matrix is used as a focal point for discussion, as a mechanism for integration of research results, and as a tool for testing the probable consequences of various management alternatives.

Summary Observations on Matrices

Based upon the above examples of matrices and other experiences in using such matrices, the following nonprioritized observations can be made:

1. It is critical to carefully define the spatial boundaries associated with environmental factors, as well as each environmental factor; the temporal phases and specific actions associated with the proposed project; and the impact rating or summarization scales used in the matrix.

2. A matrix should be considered a tool for purposes of analysis, with the key need being to clearly state the rationale utilized for the impact ratings assigned to a given temporal phase and project action, and a given spatial boundary and environmental factor.

3. The development of one or more preliminary matrices can be a useful technique in discussing a proposed action and its potential environmental impacts. This can be helpful in the early stages of a study to assist each team member in understanding the implications of the project and developing detailed plans for more-extensive studies on particular factors and impacts.

4. The interpretation of impact ratings should be very carefully considered, particularly when realizing that there may be large differences in spatial boundaries, as well as temporal phases, for a proposed project.

5. Interaction matrices can be useful for delineating the impacts of the first and second or multiple phases of a two-phase or multiphase project; the cumulative impacts of a project when considered relative to other past, present, and reasonably foreseeable future actions in the area; and the potential positive effects of mitigation measures. Creative codes can be used in the matrix to delineate this information; examples are shown in Figure 3.4.

6. If interaction matrices are used to display comparisons between different alternatives, it is necessary to use the same basic matrix in terms of spatial boundaries and environmental factors, and temporal phases and project actions for each alternative being analyzed. Completion of such matrices can provide a basis for trade-off analysis.

7. Impact quantification and comparisons to relevant standards can provide a valuable basis for the assignment of impact ratings to different project actions and environmental factors.

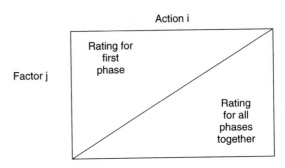

(a) Two-phase to multiphase project

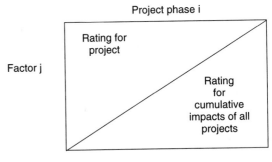

(b) Cumulative impacts

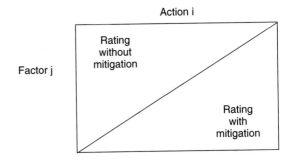

(c) Mitigation measures and effectiveness

FIGURE 3.4
Examples of Codes for Displaying Impacts in Simple Leopold Interaction Matrices.

8. Color codes can be used to display and communicate information on anticipated impacts. For example, beneficial impacts could be shown by using green or shades of green; whereas, detrimental or adverse effects could be depicted with red or shades of red. Impact matrices can be used without the incorporation of number, letter, or color ratings. For example, circles of varying size could be used to denote ranges of impacts.

9. One of the concerns relative to interaction matrices is that project actions and/or environmental factors are artificially separated, when they should be considered together. It is possible to use footnotes in a matrix to identify groups of actions, factors, and/or impacts which should be considered together. This would allow the delineation of primary and secondary effects of projects.

10. The development of a preliminary interaction matrix does not mean that it would have to be included in a subsequent EA or EIS. The preliminary matrix could be used as an internal working tool in study planning and development.

11. It is possible to utilize importance weighting for environmental factors and project actions in a simple interaction matrix. If this approach is chosen, it is necessary to carefully delineate the rationale upon which differential importance weights have been assigned. Composite indices could be developed for various alternatives by summing the products of the importance weights and the impact ratings.

12. Usage of an interaction matrix forces the consideration of actions and impacts related to a proposed project within the context of other related actions and impacts. In other words, the matrix will aid in preventing overriding attention being given to one particular action or environmental factor.

NETWORK METHODOLOGIES

"Networks" are those methodologies which integrate impact causes and consequences through identifying interrelationships between causal actions and the impacted environmental factors, including those representing secondary and tertiary effects. Several illustrations of networks, also known as "sequence diagrams," will be shown. A linear network display for an impoundment project is in Figure 3.5. Figure 3.6 shows a network diagram for a dredging project. In both networks, the initiating action is shown on the left, with other causal actions and impacted factors shown in the phases of the network. Figure 3.7 displays a portion of a type of network diagram (sometimes called an "impact tree") of the aerial application of herbicides. Figure 3.8 displays a variation of network presentations, with the diagram related to coastal development projects.

Network analyses are particularly useful for identifying anticipated impacts associated with potential projects. Networks can also aid in organizing the discussion of anticipated project impacts. Network displays are useful in communicating information about an environmental impact study to interested publics. The primary limitation of the network approach is the minimal information provided on the technical aspects of impact prediction and the means for comparatively evaluating the impacts of alternatives. In addition, networks can become very visually complicated.

Directed graphs, or "digraphs," represent a variation of networks. Figure 3.9 depicts a directed graph showing the primary impacts of a residential housing project. The numerical relationships can be developed via questionnaire surveys of selected relevant professionals (Hepner, 1981a). Digraphs are useful in depicting relationships between biophysical and socioeconomic systems; a relevant concern is their visual complexity and the validity of established numerical relationships.

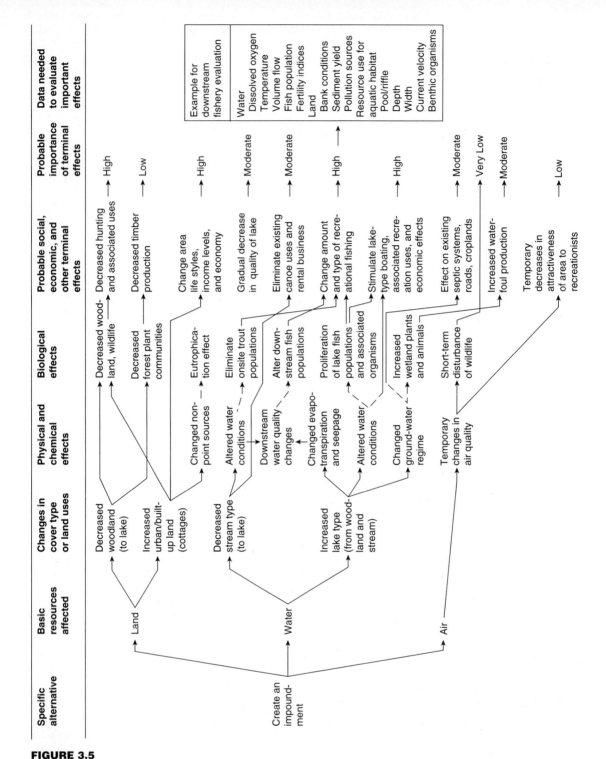

FIGURE 3.5
An Example of a Network Diagram for Analyzing Probable Environmental Impacts (U.S. Soil Conservation Service, 1977).

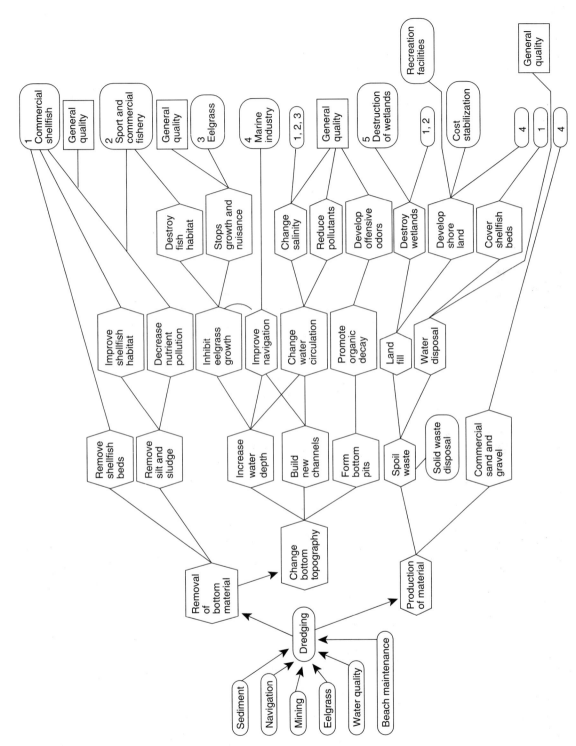

FIGURE 3.6
Network Diagram for Dredging Project (Sorensen, 1971).

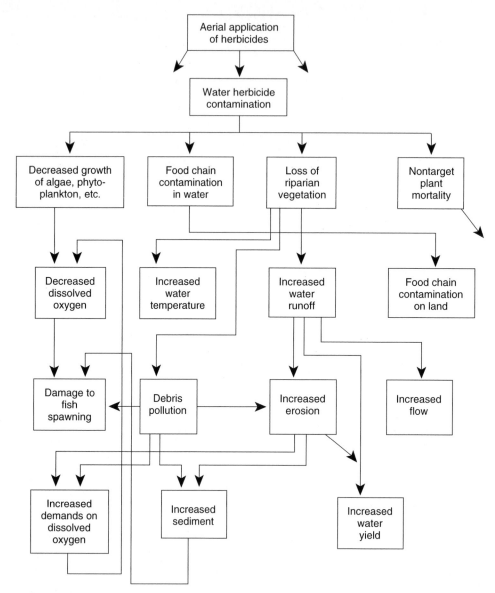

FIGURE 3.7
A Section of an Impact Tree (Bisset, 1983).

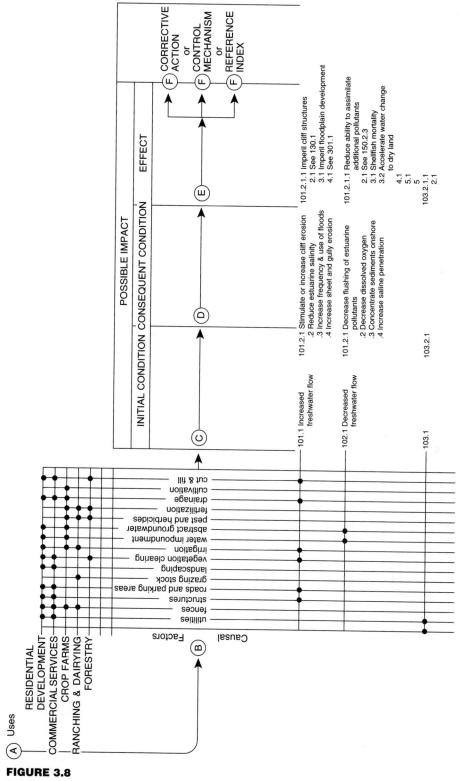

FIGURE 3.8
The Sorensen Network (Bisset, 1983).

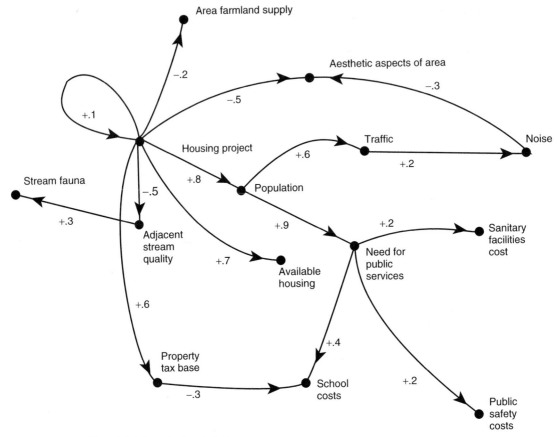

The digraph is interpreted such that:

(+) indicates an augmenting effect; i.e., an increase in vertex factor x leads to an increase in vertex factor y, and a decrease in x leads to a decrease in y.

(−) indicates an inhibiting effect; i.e., an increase in x leads to a decrease in y, and a decrease in x leads to an increase in y.

FIGURE 3.9
Example of a Directed Graph, or Digraph, of the Primary Impacts of a Residential Housing Project (Hepner, 1981b).

CHECKLIST METHODOLOGIES

Checklist methodologies range from listings of environmental factors to highly structured approaches involving importance weightings for factors and the application of scaling techniques for the impacts of each alternative on each fac-

tor. "Simple checklists" represent lists of environmental factors which should be addressed; however, no information is provided on specific data needs, methods for measurement, or impact prediction and assessment. "Descriptive checklists" refer to methodologies that include lists of environmental factors along with information on measurement and impact prediction and assessment.

Simple Checklists

Simple checklists were extensively used in the initial years following the passage of the NEPA, and they represent a valid approach for providing systemization to an EIS. For example, Table 3.9 contains a listing of the environmental factors in a simple checklist utilized for gas pipeline projects in the United States. Another example of a checklist is that developed by the Cooperative Research Service of the U.S. Department of Agriculture (USDA) for use with projects that might impact agricultural lands (USDA, 1990). This extensive checklist, shown in Table 3.10, can be used in both planning and summarizing an environmental impact study.

Another example of a simple checklist is that developed by the Asian Development Bank for usage on major dam, reservoir, and hydropower projects (Asian Development Bank, 1987). The checklist also includes mitigation information; additional, supportive information is available on the details of this checklist in the original guidelines publication (Asian Development Bank, 1987). A comprehensive questionnaire checklist developed by ESCAP (1990) for small reservoir projects is in Table 3.11.

Many other checklists could be cited, with some focused on particular categories of impacts, such as health impacts (U.S. Agency for International Development, 1980; World Bank, 1982; WHO Regional Office for Europe, 1983). Table 3.12 includes a questionnaire checklist of the potential health impacts of water resource developments and irrigation projects.

Descriptive Checklists

Descriptive checklists are widely used in environmental impact studies. For example, Carstea et al. (1976) developed a descriptive-checklist approach for projects in coastal areas. The methodology addressed the following issues, actions, and projects: riprap placement; bulkheads;

groins and jetties; piers, dolphins, mooring piles, and ramp construction; dredging (new and maintenance); outfalls, submerged lines, and pipes; and aerial crossings. For each of the items, environmental impact information was provided on potential changes in erosion, sedimentation, and deposition; flood heights and drift; water quality; ecology; air quality; noise; safety and navigation; recreation; aesthetics; and socioeconomics.

Several descriptive checklists have been developed for water resources projects. For example, Canter and Hill (1979) suggested a list of about 65 environmental factors related to the environmental quality account used for project evaluation in the United States; the list is in Chapter 4. For each factor, information is included on its definition and measurement, prediction of impacts, and functional curves for data interpretation (where one was available or easily developed).

Descriptive checklists are also used for transportation projects and land development projects. The transportation methodology addresses social, economic, and physical impacts of highway construction and operation (U.S. Department of Transportation, 1975). "Social impacts" include effects related to community cohesion, accessibility of facilities and services, and displacement of people. "Economic impacts" relate to those on employment, income and business activity, residential activity, property taxes, regional and community plans and growth, and resources. "Physical impacts" address changes on aesthetics and historic values, terrestrial and aquatic ecosystems, air quality, noise, and vibration. For each of the identified environmental factors, workable state-of-the-art methods and techniques for impact identification, data collection, analysis, and evaluation are included. A portion of a descriptive checklist containing several factors for housing and other land development projects is shown in Table 3.13. The Bases for Estimates column presents a simplified, brief

TABLE 3.9

SAMPLE OF SIMPLE CHECKLIST METHODOLOGY FOR GAS PIPELINE PROJECTS

Category	Comments
Land features and uses	Identify present uses and describe the characteristics of the land area.
	Land uses Describe the extent of present uses, as in agriculture, business, industry, recreation, residence, wildlife, and other categories, including the potential for development; locate major nearby transportation corridors, including roads, highways, ship channels, and air traffic patterns; locate transmission facilities and their placement (underground, surface, or overhead); identify water resources.
	Topography, physiography, and geology Provide a detailed description of the topographical, physiographical, and geological features within the area of the proposed action. Include *U.S. Geological Survey* Topographic Maps, aerial photographs (if available), and other such graphic material.
	Soils Describe the physical characteristics and chemical composition of the soils, including the relationship of these factors to land slope.
	Geological hazards Indicate the potential for the occurrence of geological hazards in the area, such as earthquakes, "slumping," landslides, subsidence, permafrost, and erosion.
Species and ecosystems	Identify those species and ecosystems that will be affected by the proposed action.
	Species List in general categories, by common and scientific names, the plant and wildlife species found in the area of the proposed action, and indicate those having commercial and recreational importance.
	Communities and associations Describe the dominant plant and wildlife communities and associations located within the area of the proposed action. Provide an estimate of the population densities of major species. If data are not available for the immediate area of the proposed action, data from comparable areas may be used.
	Unique and other biotic resources Describe unique ecosystems or rare or endangered species and other biotic resources that may have special importance in the area of the proposed action.
Socioeconomic considerations	If the proposed action could have a significant socioeconomic effect on the local area, discuss the socioeconomic future of the area without the implementation of the proposed action; describe the economic development in the vicinity of the proposed action, particularly the lcoal tax base and per capita income; and identify trends in economic development and/or land use of the area, from both an historical and a prospective viewpoint. Describe the population densities of both the immediate and generalized area. Include distances from the site of the proposed action to nearby residences, cities, and urban areas, and list the populations of these areas. Indicate the number and types of residences, businesses, and industries that will be directly affected and those that would require relocation if the proposed action occurs.
Air and water environment	Describe the prevailing climate and the quality and quantity of the air and water resources of the area.
	Climate Describe the cl'matic conditions that have prevailed in the vicinity of the proposed action: extremes and means of monthly temperatures, precipitation, and wind speed and direction. In addition, indicate the frequency of temperature inversions, fog, smog, and destructive storms such as hurricanes and tornadoes.
	Hydrology and hydrography Describe surface waters (fresh, brackish, or saline) in the vicinity of the proposed action, and discuss drainage basins, physical and chemical characteristics, water use, water supplies, and circulation. Describe the groundwater situation, water uses and sources, aquifer systems, and flow characteristics.
	Air, noise, and water quality Provide data on the existing quality of the air and water [indicate the distance(s) from the proposed action site to monitoring stations] and the mean and maximum noise levels at the site boundaries.
Unique features	Identify unique or unusual features of the area, including historical, archaeological, and scenic sites and values.

Source: Abstracted from Federal Power Commission, 1973.

TABLE 3.10

USDA CHECKLIST FOR ADDRESSING AND/OR SUMMARIZING ENVIRONMENTAL IMPACTS

Topical Issue	Yes	Maybe	No	Comments

Land form. Will the project result in:

- Unstable slopes or embankments?
- Extensive disruption to or displacement of the soil?
- Impact to land classified as prime or unique farmland?
- Changes in ground contours, shorelines, stream channels, or river banks?
- Destruction, covering or modification of unique physical features?
- Increased wind or water erosion of soils?
- Foreclosure on future uses of site on a long-term basis?

Air/climatology. Will the project result in:

- Air pollutant emissions which will exceed Federal or State standards or cause deterioration of ambient air quality (e.g., radon gas)?
- Objectionable odors?
- Alteration of air movements, humidity, or temperature?
- Emissions of hazardous air pollutants regulated under the Clean Air Act?

Water. Will the project result in:

- Discharge to a public water system?
- Changes in currents or water movements in marine or fresh water?
- Changes in absorption rates, drainage patterns, or the rate and amount of surface water runoff?
- Alterations to the course or flow of flood waters?
- Impoundment, control, or modifications of any body of water equal to or greater than 10 acres in surface area?
- Discharges into surface waters or alteration of surface water quality including, but not limited to, temperature or turbidity?
- Alteration of the direction or rate of flow of groundwaters?
- Alterations in groundwater quality?
- Contamination of public water supplies?
- Violation of State Stream Quality Standards, if applicable?
- Location in a riverine or coastal floodplain?
- Exposure of people or property to water-related hazards such as flooding?
- Location in a State's coastal zone and subject to consistency with the State Coastal Zone Management Plan?
- Impact on or construction in a wetland or inland floodplain?

Solid Waste. Will the project:

- Generate significant solid waste or litter?

Noise. Will the project:

- Increase existing noise levels?
- Expose people to excessive noise?

Plant life. Will the project:

- Change the diversity or productivity of species or number of any species of plants (including trees, shrubs, grass, crops, microflora, and aquatic plants)?
- Reduce the numbers or affect the habitat of any State or federally designated unique, rare, or endangered species of plants? (Check State and Federal lists of endangered species.)
- Introduce new species of plant into the area or create a barrier to the normal replenishment of existing species?
- Reduce acreage or create damage to any agricultural crop?

TABLE 3.10

USDA CHECKLIST FOR ADDRESSING AND/OR SUMMARIZING ENVIRONMENTAL IMPACTS *(continued)*

Topical Issue	Yes	Maybe	No	Comments

Animal life. Will the project:

- Reduce the habitat or numbers of any State or federally designated unique, rare, or endangered species of animals? (Check State and Federal lists and Migratory Bird Treaty Act.)
- Introduce new species of animals into an area or create a barrier to the migration or movement of animals or fish?
- Cause attraction, entrapment, or impingement of animal life?
- Harm existing fish and wildlife habitats?
- Cause emigration resulting in human-wildlife interaction problems?

Land use. Will the project:

- Substantially alter the present or planned use of an area?
- Impact a component of the National Park system, the National Wildlife Refuge system, the National Wild and Scenic River system, the National Wilderness system, or National Forest land?

Natural resources. Will the project:

- Increase the rate of use of any natural resources?
- Substantially deplete any nonreusable natural resources?
- Be located in an area designated as or being considered for wilderness, wild and scenic river, national park, or ecological preserve?

Energy. Will the project:

- Use substantial amounts of fuel or energy?
- Substantially increase the demand on existing sources of energy?

Transportation and traffic circulation. Will the project result in:

- Movement of additional vehicles?
- Effects on existing parking facilities or demands for new parking?
- Substantial impact on existing transportation system(s)?
- Alterations to present patterns of circulation or movement of people and/or goods?
- Increased traffic hazards to motor vehicles, bicyclists, or pedestrians?
- Construction of new roads?

Public service. Will the project have an effect on, or result in, a need for new or altered governmental services in any of the following areas:

- Fire protection?
- Schools?
- Other governmental services?

Utilities. Will the project result in a need for new systems or alterations to the following utilities:

- Power and natural gas?
- Communications systems?
- Water?
- Sewer or septic tanks?
- Storm sewers?

TABLE 3.10

USDA CHECKLIST FOR ADDRESSING AND/OR SUMMARIZING ENVIRONMENTAL IMPACTS *(continued)*

Topical Issue	Yes	Maybe	No	Comments

Population.   Will the project:

- Alter the location or distribution of human population in the area?

Accident risk.   Does the project:

- Involve the risk of explosion or release of potentially hazardous substances including, but not limited to, oil, pesticides, chemicals, radiation, or other toxic substances, in the event of an accident or "upset" condition?

Human health.   Will the project:

- Create any health hazard or potential health hazard?
- Expose people to potential health hazards?

Economic.   Will the project:

- Have any adverse effect on local or regional economic conditions, e.g., tourism, local income levels, land values, or employment?

Community reaction.   Is the project:

- Potentially controversial?
- In conflict with locally adopted environmental plans and goals?

Aesthetics.   Will the project:

- Change any scenic vista or view open to the public?
- Create an aestheticaly offensive site open to the public view (e.g., out of place with character or design of surrounding area)?
- Significantly change the visual scale or character of the vicinity?

Archaeological, cultural, and historical.   Will the project:

- Alter archaeological, cultural, or historical sites, structures, objects, or buildings, either in or eligible for inclusion in the National Register (e.g., be subject to the Historic Preservation Act of 1974)?

Hazardous waste.   Will the project:

- Involve the generation, transport, storage or disposal of any regulated hazardous waste (e.g., asbestos, if demolition or building alterations are involved)?

Source: U.S. Department of Agriculture (USDA), 1990, Attachment B, pp. 1–7.

TABLE 3.11

SAMPLE MODIFIED CHECKLIST FOR SMALL RESERVOIR PROJECTS IN OREGON

Instructions

Answer the following questions by placing an "×" in the appropriate YES/NO space; consider activity, construction, operational, as well as indirect impacts.

Use the "explanation" section to clarify points or add information.

A. NATURAL BIOLOGICAL ENVIRONMENT

1. Might the proposed activity affect any natural feature or water resource adjacent to or near the activity areas? _____ NO __×__ YES

If YES, specify natural feature affected:

	Direct	Indirect	Synergistic	Short Term	Long Term	Reversible	Irreversible	Severe	Moderate	Insignificant
(1) Surface water hydrology	(×)	( )	( )	( )	(×)	( )	(×)	( )	( )	(×)
(2) Surface water quality	(×)	( )	( )	( )	(×)	( )	(×)	(×)	( )	( )
(3) Soil/erosion	(×)	( )	( )	( )	(×)	( )	(×)	(×)	( )	( )
(4) Geology	(×)	( )	( )	( )	(×)	( )	(×)	(×)	( )	( )
(5) Climate	(×)	( )	( )	( )	(×)	( )	(×)	(×)	( )	( )

2. Might the activity affect wildlife or fisheries? _____ NO __×__ YES

If YES, specify wildlife or fisheries affected:

	Direct	Indirect	Synergistic	Short Term	Long Term	Reversible	Irreversible	Severe	Moderate	Insignificant
(1) Wildlife habitat	(×)	( )	( )	( )	(×)	( )	(×)	(×)	( )	( )
(2) Ecology of fisheries	(×)	( )	( )	( )	(×)	( )	(×)	(×)	(×)	( )

3. Might the activity affect natural vegetation? _____ NO __×__ YES

If YES, specify vegetation and acreage(s) affected.

B. ENVIRONMENTAL HAZARDS

1. Might the activity involve the use, storage, release of, or disposal of any potentially hazardous substances? __×__ NO _____ YES

If YES, specify substance and potential effect.

2. Might the activity cause an increase or probability of increase of environmental hazards? __×__ NO _____ YES

If YES, specify type.

3. Might the activity be susceptible to environmental hazard due to its locations? __×__ NO _____ YES

If YES, specify type.

C. RESOURCE CONSERVATION AND USE

1. Might the activity affect or eliminate land suitable for agricultural or timber production? _____ NO __×__ YES

TABLE 3.11

SAMPLE MODIFIED CHECKLIST FOR SMALL RESERVOIR PROJECTS IN OREGON *(continued)*

If YES, specify acres and soil class which will be affected:

(1) Inundation area   (×)   ( )   ( )   ( )   (×)   ( )   (×)   ( )   (×)   ( )
(If the project might result in inundation)

2. Might the activity affect commercial fisheries or aquacultural resources or production?   _____ NO   __×__ YES

If YES, specify type affected.

3. Might the activity affect the potential use or extraction of an indispensable or scarce mineral or energy resource?   __×__ NO   _____ YES

If YES, specify resource affected and approximate amount.

D. WATER QUALITY AND QUANTITY

1. Might the activity affect the quality of water resources within, adjacent to, or near the activity area?   _____ NO   __×__ YES

If YES, specify water resource affected and amount in gallons/day.

	Direct	Indirect	Synergistic	Short Term	Long Term	Reversible	Irreversible	Severe	Moderate	Insignificant
(1) Quality and quantity	(×)	( )	( )	( )	(×)	( )	(×)	(×)	( )	( )

2. Might the activity result in a deleterious effect on the quality of any water resource areas or watersheds?   _____ NO   __×__ YES

If YES, specify water resource that might be affected.

	Direct	Indirect	Synergistic	Short Term	Long Term	Reversible	Irreversible	Severe	Moderate	Insignificant
(1) Water supply downstream irrigation (if applicable)	(×)	( )	( )	( )	(×)	(×)	( )	( )	( )	(×)

If YES, specify possible substances causing effects.

	Direct	Indirect	Synergistic	Short Term	Long Term	Reversible	Irreversible	Severe	Moderate	Insignificant
(1) Human and animal toxics	(×)	( )	( )	( )	(×)	(×)	( )	( )	( )	(×)
	Impact type			Duration		Reversibility		Severity		

E. AIR QUALITY/ATMOSPHERIC ENVIRONMENT

1. Might the activity affect the air quality in the projection area, immediately adjacent areas, or the regional airshed?   _____ NO   __×__ YES

If YES, specify possible substance affecting air quality.

	Direct	Indirect	Synergistic	Short Term	Long Term	Reversible	Irreversible	Severe	Moderate	Insignificant
(1) Atmospheric environment	( )	(×)	( )	( )	(×)	( )	(×)	( )	( )	(×)

F. NOISE/SONIC ENVIRONMENT

1. Might the activity result in the generation of noise?   __×__ NO   _____ YES

If YES, specify noise source.

TABLE 3.11

SAMPLE MODIFIED CHECKLIST FOR SMALL RESERVOIR PROJECTS IN OREGON *(continue)*

G. COMMUNITY FACILITIES/SERVICES

1. Might the proposed activity result in changes in community facilities, services or institutions? _____ NO __×__ YES

2. Are any mitigation or enhancement measures foreseen to compensate for the above stated impacts? _____ NO __×__ YES

   Explain below.

3. Will the activity create new opportunities for recreational experiences? _____ NO __×__ YES

   If YES, specify.

(1) Reservoir-related recreation   ( ) (×) ( ) ( ) (×) ( ) (×) (×) ( ) ( )

J. HISTORIC RESOURCES

1. Might any site or structure of historic significance be affected? __×__ NO _____ YES

2. Might any known archaeologic or palaeontological site be affected by the activity? __×__ NO _____ YES

K. VISUAL RESOURCES

1. Might the activity cause a change in the visual character in or near the activity area through alteration of natural or cultural features? _____ NO __×__ YES

   If YES, specify natural and cultural features that may be changed.

(1) Natural   (×) ( ) ( ) ( ) (×) ( ) (×) (×) ( ) ( )

2. Might the activity affect views or access to views of natural or cultural landscape features? _____ NO __×__ YES

   If YES, specify viewshed affected.

(1) Natural   (×) ( ) ( ) ( ) (×) ( ) ( ) ( ) ( ) (×)

3. Might the activity introduce new materials, colours, and forms to the immediate landscape? _____ NO __×__ YES

   If YES, specify.

(1) Reservoirs   (×) ( ) ( ) ( ) (×) ( ) (×) ( ) ( ) (×)

	Impact type	Duration	Reversibility	Severity

L. ECONOMICS AND ENVIRONMENT

1. Might the proposed activity cause elimination or relocation of existing commercial and industrial enterprises? __×__ NO _____ YES

2. Might the activity cause generation of or reduction in employment? _____ NO __×__ YES

3. Might the proposed activity affect property values and local tax revenues? _____ NO __×__ YES

   If YES, specify potential effects.

TABLE 3.11

SAMPLE MODIFIED CHECKLIST FOR SMALL RESERVOIR PROJECTS IN OREGON *(continued)*

	Direct	Indirect	Synergistic	Short Term	Long Term	Reversible	Irreversible	Severe	Moderate	Insignificant
(1) Irrigation district	(×)	( )	( )	( )	(×)	(×)	( )	(×)	( )	( )

4. Might the proposed activity affect local expenditures for infrastructural services (sewer, water, etc.)?   __×__ NO   _____ YES

5. Might the proposed activity affect the local and regional economics?   _____ NO   __×__ YES

If YES, to what extent, how and at what scale(s)?

	Direct	Indirect	Synergistic	Short Term	Long Term	Reversible	Irreversible	Severe	Moderate	Insignificant
(1) Increase of the local revenues	( )	(×)	( )	( )	(×)	(×)	( )	(×)	( )	( )

6. Might the activity cause an increase or decrease in seasonality of employment?   _____ NO   __×__ YES

If YES, indicate which and state occupation and groups affected.

	Direct	Indirect	Synergistic	Short Term	Long Term	Reversible	Irreversible	Severe	Moderate	Insignificant
(1) Farmers—irrigation	(×)	( )	( )	( )	(×)	( )	(×)	(×)	( )	( )

M. PLANNING CO-ORDINATION AND GROWTH

1. Will the activity require a variance from or result in a potential violation of any statute, ordinance, by-law, regulation, or prevent or minimize damage to the environment?   __×__ NO   _____ YES

If YES, specify variance and or statutes.

2. Might the activity stimulate additional local and regional land use development?   _____ NO   __×__ YES

If YES, specify extent and scale(s).

	Direct	Indirect	Synergistic	Short Term	Long Term	Reversible	Irreversible	Severe	Moderate	Insignificant
(1) Land use—irrigation	(×)	( )	( )	( )	(×)	( )	(×)	( )	(×)	( )

3. Are there any other developments planned which are or will be impacted by the proposed activity including those beyond the control of the submitting agency?   __×__ NO   _____ YES

If YES, specify other development(s) affected.

Source: Economic and Social Commission for Asia and the Pacific (ESCAP), 1990, pp. 22–26.

TABLE 3.12

QUESTIONNAIRE CHECKLIST OF POTENTIAL HEALTH IMPACT OF WATER RESOURCE DEVELOPMENTS AND IRRIGATION PROJECTS

A. DIRECT IMPACTS ON PEOPLE IN THE PROJECT AREA:

- Will new diseases or new strains of the disease be introduced by immigration of construction workers or new settlers? Will these affect new settlers or residents or both?
- Will relocated communities be exposed to diseases to which they have little or no immunity?
- Will new settlers be exposed to locally endemic diseases to which they have little or no immunity?
- Will food, waste or water cycles aggravate sanitation and disease problems?
- Will housing and sanitary facilities become overburdened, misused or not used at all, leading to conditions conducive to increases in water washed diseases and spread of communicable diseases by the faecal-oral route?
- Will soil and water be contaminated by excreta, facilitating spread of communicable disease?
- Will introduction of migrant workers cause increases in venereal disease among workers and subsequently residents?
- Will new settlers and relocated communities be exposed to physical, social and cultural changes leading to psychological strains and traumas? These may include changes in lifestyles and employment.
- Will changes in food supplies lead to possibilities of malnutrition, nutritional deficiencies or toxic effects? These effects may occur because of:

  - introduction of Western-style convenience foods;
  - changes in staple foods—possibly using unfamiliar toxic plants as substitutes for usual foods;
  - contamination of soil or agricultural water supplies with toxic substances;
  - reduced productivity of soils caused by hydrological changes (waterlogging, etc.), mineralisation or pollution of ground and surface waters;
  - reduced productivity of fisheries caused by hydrological changes or water pollution;
  - change in availability of trace metals in soils caused by hydrological changes (lowering or raising of water table etc.).

- Will effluents and emissions, or substances released intentionally into the environment (e.g. pesticides) pollute air or water or soil presenting a threat to human health?
- Will irrigation of fields increase opportunities for human contact with water borne, water based and water related disease?
- Will traffic in the area, and therefore road accidents, increase as a result of the development?
- Will new industries and similar activities attracted to the area by growth, result in pollution of air, soil or water or noise, with subsequent impacts on human health?

B. INDIRECT IMPACTS THROUGH EFFECTS ON DISEASE VECTORS:

- Will new vectors be introduced into the area from upstream as a result of hydrological changes?
- Will new vectors be introduced into the area on vehicles, animals, transplanted plants, soil, etc.?
- Will existing vectors be infected or reinfected by contact with infected humans coming into the area?
- Will the prevalence and distribution of existing infected vectors be changed by changes in the availability of suitable habitats for breeding and survival? These changes may result from hydrological changes (water velocities, temperature, depth, standing water, etc.), morphological changes (bank slopes, cover, etc.), climate changes (rainfall, humidity) and biological changes (vegetation, predators, etc.). They may affect presently infected or uninfected areas.

C. DIRECT IMPACTS ON WORKERS:

- Will migrant workers be exposed to locally endemic diseases to which they have little or no immunity?
- Will migrant workers be exposed to psychological strains and traumas from changes in living and working conditions?
- Will workers be exposed to physical threats to their safety (injuries, deaths) or chemical and physical hazards to health (toxic substances, noise, vibration, radiation, high pressures, etc.)?
- Will workmen be particularly exposed to contact with water and thus with water associated disease during their work?
- Will workmen be exposed to dangerous animals during their work (snakes, scorpions, etc.)?
- Will adequate supplies of food be provided to prevent malnutrition and minimise spread of disease (e.g. by use of itinerant food vendors)?

D. IMPACT ON HEALTH SERVICES:

- Will health and other social services be overburdened with consequent effects on health of residents and workers?

Source: World Health Organization (WHO) Regional Office for Europe, 1983, p. 13.

TABLE 3.13

PORTION OF A DESCRIPTIVE CHECKLIST FOR LAND DEVELOPMENT PROJECTS

Factor	Bases for Estimates
I. Local economy	
Public fiscal balance Net change in government fiscal flow (revenue less expenditures)	Public revenues: expected household income, by residential housing type; added property values. Public expenditures: analysis of new-service demand, current costs, available capacities, by service
Employment Change in numbers and percent employed, unemployed, and underemployed, by skill level	Direct from new business, or estimated from floor space, local residential patterns, expected immigration, current unemployment profiles
Wealth Change in land values	Supply and demand of similarly zoned land, environmental changes near property
II. Natural environment	
Air quality Health Change in air pollution concentrations by frequency of occurrence, and number of people at risk	Current ambient concentrations, current and expected emissions, dispersion models, population maps
Nuisance Change in occurrence of visual (smoke, haze) or olfactory (odor) air quality nuisances, and number of people affected	Baseline citizen survey, expected industrial processes, traffic volumes
Water quality Changes in permissible or tolerable water uses, and number of people affected for each relevant body of water	Current and expected effluents, current ambient concentrations, water quality model
Noise Change in noise levels and in frequency of occurrence, and number of people bothered	Changes in nearby traffic or other noise sources and in noise barriers; noise-propagation model or nomographs relating noise levels to traffic, barriers, etc.; baseline citizen survey or current satisfaction with noise levels

Source: Abstracted from Schaenman, 1976.

---

listing of key data and models (if any) needed for the factor.

An environmental impact computer system (EICS) has been developed by the U.S. Army Construction Engineering Research Laboratory (Lee et al., 1974). This system uses computer techniques to identify potential environmental impacts from 9 functional areas of Army activities on 11 broad environmental categories (Jain et al., 1973). The nine functional areas are construction; operation, maintenance, and repair; training; mission change; real estate; procurement; Army industrial activities; research, development, testing, and evaluation; and administration and support. Each of these functional areas has a number of additional basic activities. Examples of basic activities in the construction functional area include clearing trees, removing broken concrete, backfilling foundations, curing bituminous pavement, cleaning used concrete forms, installing insulation, and landscaping sites. A total of approximately 2,000 basic activities have been identified in the nine functional areas. In the EICS system, the environment is divided into 11 topical areas: ecology, health science, air quality, surface water, groundwater, sociology, economics, earth science, land use, noise, and transportation. Within each of these categories additional parameters are defined. Approximately 1,000 specific environmental factors are defined for the 11 environmental categories. On this basis, it is possible to have a checklist that addresses the impacts of approximately 2,000 basic Army activities on 1,000 environmental factors.

The computer system is used to identify potential impacts associated with various types of activities. In a sense, this method is similar to a computerized interaction matrix. It is considered here a descriptive checklist because each of the environmental factors is described in detail, with information given on actual measurement and data interpretation. The system codes each interaction as belonging to one of four categories: the first category indicates that the potential impact must be assessed every time the activity is carried out; the second, that the impact is usually present but may be absent depending upon individual circumstances; the third, that the impact arises in a small but predictable number of cases and its presence should be considered in the individual circumstance; and, finally, if there is no indication of potential impact, then the particular activity is considered to have no impact upon this environmental factor. Variations of this descriptive checklist have been devel-

oped in U.S. Army Construction Engineering Research Laboratory (1989).

Summary Observations on Simple and Descriptive Checklists

Simple and descriptive checklists of environmental factors and/or impacts to consider can be helpful in planning and conducting an EIS, particularly if one or more checklists for the specific project type can be utilized. The following nonprioritized comments are pertinent summaries of simple and descriptive checklists:

1. Published agency checklists and/or project-specific checklists represent the collective professional knowledge and judgment of their developers; hence, they have professional credibility and usability.
2. Checklists provide a structured approach for identifying key impacts and/or pertinent environmental factors for consideration in impact studies. More-extensive lists of factors or impacts do not necessarily represent better lists, since relevant factors or impacts will need to be selected. Checklists can be easily modified (items can be added or deleted) to make them more pertinent to particular project types in given locations.
3. Checklists can be used to stimulate or facilitate interdisciplinary-team discussions during the planning, conduction, and/or summarization of EISs.
4. In using a checklist it is important to carefully define the utilized spatial boundaries and environmental factors. Any special impact codes or terminology used within the checklist should also be defined.
5. Documentation of the rationale basic to identifying key factors and/or impacts should be accomplished. In this regard, factor-impact quantification and comparison to pertinent standards can be helpful.
6. Factors and/or impacts from a simple or descriptive checklist can be grouped together to demonstrate secondary and tertiary im-

pacts and/or environmental system interrelationships.

7. Importance weights could be assigned to key environmental factors or impacts; the rationale and methodology for such importance-weight assignments should be clearly delineated.

8. Key impacts which should be mitigated can be identified through the systematic usage of a simple or descriptive checklist.

SUMMARY

Numerous EIA methodologies have been developed within the last two decades. These methodologies can be useful in identifying anticipated impacts, determining appropriate environmental factors for inclusion in a description of the affected environment, providing information on prediction and assessment of specific impacts, allowing for systematic evaluation of alternatives and the selection of a proposed action, and summarizing and communicating impact study results. The most-used methodologies can be categorized as interaction matrices, networks, or checklists. Interaction matrices are of greatest value in impact identification and the display of comparative information on alternatives. Network methodologies provide useful information for impact identification, as well as valuable approaches for communicating information on interrelationships between environmental factors and anticipated project impacts.

Checklist approaches range from simple listings of environmental factors to complex methods involving assignment of relative importance weights to environmental factors and the scaling of environmental impact factors for each of a series of alternatives. Simple and descriptive-checklist approaches, including questionnaire checklists, are useful for identifying environmental factors and providing information on impact prediction and assessment.

SELECTED REFERENCES

Asian Development Bank, "Environmental Evaluation of Coastal Zone Projects: Methods and Approaches," Manila, The Philippines, June 1991, pp. 13–41.

———, "Environmental Guidelines for Selected Industrial and Power Development Projects," Manila, The Philippines, 1987.

Bisset, R., "A Critical Survey of Methods for Environmental Impact Assessment," *An Annotated Reader in Environmental Planning and Management,* T. O'Riordan and R. K. Turner, eds., Pergamon Press, Oxford, England, 1983, pp. 168–186.

———, "Methods for Environmental Impact Analysis: Recent Trends and Future Prospects," *Journal of Environmental Management,* vol. 11, 1980, pp. 27–43.

Canter, L. W., *Environmental Health Impact Assessment,* Pan American Health Organization, Metepec, Mexico, 1986, pp. 232–247.

———, "Environmental Impact Assessment of South Coast Sewerage Project," report submitted to Inter-American Development Bank, Washington, D.C., July 1991.

———, *Water Resources Assessment—Methodology and Technology Sourcebook,* Ann Arbor Science Publishers, Ann Arbor, Mich., 1979.

——— and Hill, L. G., *Handbook of Variables for Environmental Impact Assessment,* Ann Arbor Science Publishers, Ann Arbor, Mich., 1979.

Carstea, D., et al., "Guidelines for the Environmental Impact Assessment of Small Structures and Related Activities in Coastal Bodies of Water," HTR-6916, rev. 1, Aug. 1976, The MITRE Corporation, McLean, VA.

Chase, G. B., "Matrix Analyses in Environmental Impact Assessment," paper presented at the *Engineering Foundation Conference on Preparing Environmental Impact Statements,* Henniker, N.H., July 29–Aug. 3, 1973.

Economic and Social Commission for Asia and the Pacific (ESCAP), "Environmental Impact Assessment—Guidelines for Water Resources Development," ST/ESCAP/786, United Nations, New York, 1990, pp. 19–48.

Federal Aviation Administration (FAA), "Procedures for Environmental Impact Statement Preparation,"

FAA Order 1050.1A, Washington, D.C., June 19, 1973.

Federal Power Commission, "Implementation of the *National Environmental Policy Act of 1969—Order 485*," Washington, D.C., June 1973.

Fischer, D. W., and Davies, G. S., "An Approach to Assessing Environmental Impact," *Journal of Environmental Management,* vol. 1, no. 3, 1973, pp. 207–227.

Hanley, P. T., Hemming, J. E., and Morsell, J. W., "Natural Resource Protection and Petroleum Development in Alaska," FWS/OBS-80/22, U.S. Fish and Wildlife Service, Washington, D.C., Aug. 1981, pp. 140–162.

Hepner, G. F., "A Directed Graph Approach to Locational Analysis of Fringe Residential Development," *Geographical Analysis,* vol. 13, no. 3, July 1981a, pp. 276–283.

———, personal communication to author, Department of Geography, Western Michigan University, Kalamazoo, Aug. 1981b.

International Institute for Applied Systems Analysis, "Expect the Unexpected—an Adaptive Approach to Environmental Management," Executive Rep. 1, Laxenburg, Austria, 1979.

Jain, R. K., et al., "Environmental Impact Assessment Study for Army Military Programs," Tech. Rep. D-13, U.S. Army Construction Engineering Research Laboratory, Champaign, Ill., Dec. 1973.

Johnson, F. L., and Bell, D. T., "Guidelines for the Identification of Potential Environmental Impacts in the Construction and Operation of a Reservoir," Forestry Research Rep. 75-6, Department of Forestry, University of Illinois, Champaign, June 1975.

Lee, E. Y. S., et al., "Environmental Impact Computer System," Tech. Rep. E-37, U.S. Army Construction Engineering Research Laboratory, Champaign, Ill., Sept. 1974.

Lee, N., "Environmental Impact Assessment: A Review," *Applied Geography,* vol. 3, 1983, pp. 5–27.

———, "An Overview of Methods of Environmental Impact Assessment," Environmental Impact Assessment Workshop, Seville, Spain, Nov. 1988.

Leopold, L. B., et al., "A Procedure for Evaluating Environmental Impact," *Circular 645, U.S. Geological Survey,* Washington, D.C., 1971.

Lohani, B. N., and Halim, N., "Environmental Impact Identification and Prediction: Methodologies and Resource Requirements," background papers for course "Environmental Impact Assessment of Hydropower and Irrigation Projects," International Center for Water Resources Management and Training (CEFIGRE), Bangkok, Thailand, Aug. 13–31, 1990, pp. 152–182.

McHarg, I., *Design with Nature,* Doubleday and Company, Garden City, N.Y., 1971.

Nichols, R., and Hyman, E., "Evaluation of Environmental Assessment Methods," *Journal of the Water Resources Management and Planning Division,* American Society of Civil Engineers, vol. 108, no. WR1, Mar. 1982, pp. 87–105.

Personal communication to author, Oregon Highway Department, July 1973.

Schaenman, P. S., "Using an Impact Measurement System to Evaluate Land Development," URI 15500, The Urban Institute, Washington, D.C., Sept. 1976.

Sorensen, J., "A Framework for Identification and Control of Resource Degradation and Conflict in the Multiple Use of the Coastal Zone," Master's thesis, Department of Landscape Architecture, University of California, Berkeley, 1971.

U.S. Agency for International Development, "Environmental Design Considerations for Rural Development Projects," Washington, D.C., 1980.

U.S. Army Construction Engineering Research Laboratory, "Environmental Review Guide for USAREUR," 5 vols., Champaign, Ill., 1989.

U.S. Department of Agriculture (USDA), "Checklist for Summarizing the Environmental Impacts of Proposed Projects," Cooperative State Research Service, Stillwater, Okla., 1990.

U.S. Department of Transportation, Federal Highway Administration, "Environmental Assessment Notebook Series," Washington, D.C., 1975.

U.S. Economic Development Administration, "Final Environmental Statement, Fruit/Church Industrial Park, Fresno, California," Washington, D.C., Feb. 1973.

U.S. Soil Conservation Service, "Guide for Environmental Assessment," Washington, D.C., Mar. 1977.

Warner, M. L., "Environmental Impact Analysis: An Examination of Three Methodologies," Ph.D. dissertation, University of Wisconsin, Madison, 1973. p. 28.

———— and Bromley, D. W., "Environmental Impact Analyses: A Review of Three Methodologies," Tech. Rep., Wisconsin Water Resources Center, University of Wisconsin, Madison, 1974.

World Bank, "The Environment, Public Health, and Human Ecology: Considerations for Economic Development," Washington, D.C., 1982.

World Health Organization (WHO), Regional Office for Europe, "Environmental Health Impact Assessment of Irrigated Agricultural Development Projects, Guidelines and Recommendations: Final Report," Copenhagen, 1983.

CHAPTER 4

Description of Environmental Setting (Affected Environment)

The description of the environmental setting (also referred to as "baseline," "existing," "background," or "affected environment") is an integral part of an environmental impact study. The Council on Environmental Quality regulations contain the following comments relative to this issue (CEQ, 1978, p. 55996):

> The environmental impact statement shall succinctly describe the environment of the area(s) to be affected or created by the alternatives under consideration. The descriptions shall be no longer than is necessary to understand the effects of the alternatives. Data and analyses in a statement shall be commensurate with the importance of the impact, with less important material summarized, consolidated, or simply referenced. Agencies shall avoid useless bulk in statements and shall concentrate effort and attention on important issues. Verbose descriptions of the affected environment are themselves no measure of the adequacy of an environmental impact statement.

The principles enunciated in the CEQ regulations have applicability to both environmental assessments (EAs) and environmental impact statements (EISs). There are two major purposes of describing the environmental setting of the proposed project area in an impact study, namely, (1) to assess existing environmental quality, as well as the environmental impacts of the alternatives being studied, including the no-action or no-project alternative, and (2) to identify environmentally significant factors or geographical areas that could preclude the development of a given alternative or alternatives (this could be referred to as "identifying any 'fatal flaws' in the proposed project setting"). Examples of environmentally significant factors or areas include the presence of stream segments with poor water quality, geographical areas with "marginal" air quality, habitat for threatened or endangered plant or animal species, and significant historical or archaeological sites.

Additional purposes of describing the setting include to provide sufficient information so that decision makers and reviewers unfamiliar with the general location can develop an understanding of the project need, as well as the environmental characteristics of the study area, and to serve as a basis for establishing project need, whether the project involves construction of a highway, reservoir, or sewage-treatment plant;

expansion and/or modification of an airport facility; or development of an industrial park. Even though there may be a section in the EA or EIS that deals specifically with project need, the basis can be briefly delineated in the description of the environmental setting. A final example of one purpose of describing the setting is to allow development of information for a given geographical area, which can then be used for multiple EAs or EISs; this use would be applicable for projects involving military installations and large land holdings of various agencies. An example is the baseline characterization of the Hanford site of the U.S. Department of Energy in the state of Washington (Cushing, 1990).

This chapter describes a conceptual framework for preparing a description of the environmental setting. Techniques for identifying an initial list of environmental factors are described, along with several approaches for screening this list, or reducing it to a selected list. Examples related to different types of projects are included. Finally, some special issues and concerns are highlighted.

CONCEPTUAL FRAMEWORK

Figure 4.1 depicts a conceptual framework which can be used for preparing a description of the environmental setting. The methodology involves (1) the identification of from one to

several lists of environmental factors, (2) the application of a screening process leading to a selected list of environmental factors, (3) the procurement of relevant data for the selected factors and/or the conduction of pertinent baseline studies, and (4) the preparation of the description of the setting. The primary emphasis in this chapter is on the first two steps of the procedure; the latter steps are primarily addressed in the more-substantive area chapters (Chapters 5 through 14). Selection is the central issue, since one of the key issues in describing the environmental setting is to ensure that all environmental factors that need to be considered are included, while excluding those items that require extensive identification and interpretation effort but that have little relevance to the environmental impact of the proposed action or any of its alternatives. There are no specific generic requirements with regard to the number and type of environmental factors that must be included in a description of the environmental setting.

INITIAL LIST OF FACTORS

Four approaches which can be used to identify an initial list of environmental factors of potential relevance to a proposed project are (1) use of pertinent agency guidelines or regulations, (2) use of professional knowledge regarding the anticipated impacts of similar projects, (3) re-

FIGURE 4.1

Conceptual Framework for Preparing a Description of the Environmental Setting.

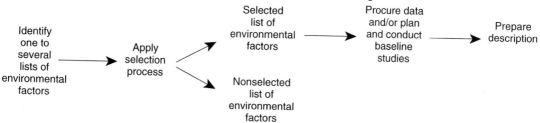

view of other recent EAs or EISs on similar projects or projects in the same geographical area as the proposed project, and (4) use of lists of factors in EIA methodologies. Examples of these four approaches will be discussed.

Agency Guidelines or Regulations

Two generic examples of the use of pertinent agency guidelines will be presented. In the first, a list of the environmental factors associated with describing the environmental setting for a nuclear power plant will be given. One of the advantages of this list is that it is not exhaustive, yet it is comprehensive. The factors are (Atomic Energy Commission, 1973)

1. *Site location and topography*
2. *Regional demography, land, and water use*   Determine population distribution for 10- and 50-mi radius; identify present and projected land-use and zoning restrictions for 5-mi radius; indicate water use and surface- and groundwater sources within 50-mi radius; note locations of any stack discharges in area.
3. *Regional historical, scenic, cultural, and natural landmarks*   Check national and state registries of historical places; perform an archaeological survey.
4. *Geology (topography, stratigraphy, soils, and rock types)*   The geology is relevant to earthquake potential and cooling ponds.
5. *Hydrology*   Describe physical, chemical, biological, and hydrological characteristics (and their seasonal variations) of the surface waters and groundwater of the site and its immediate environs; note existing pollution sources; note applicable low flow; cite any applicable water quality standards.
6. *Meteorology*   Describe diurnal and monthly averages and extremes of temperature, dew point, and humidity, wind speed and direction, atmospheric stability, mixing heights, precipitation, and storms such as hurricanes and tornadoes; also include air quality data,

and information regarding sources of air pollution and applicable air quality standards.
7. *Ecology*   Identify important flora and fauna of the region and their habitats, distribution, and relationships to other species; note rare and endangered species; show distribution maps; discuss ecological succession; and define any preexisting environmental stresses.
8. *Background radiological characteristics*

It should be noted that while the above list was developed for a nuclear power plant, its applicability is not limited to that type of project. It would also be generally useful for industrial plants and coal-fired power plants.

A second example of a generic list is the one developed for gas pipeline projects by the Federal Power Commission (1973), now called the Federal Energy Regulatory Commission. The listing is as shown in Table 3.9.

Although, as stated earlier, the list in Table 3.9 was developed for gas pipeline projects, it could also be used for other "corridor-type" projects such as highways and transmission lines.

Site assessments related to establishing seller-buyer-lender liability have been a recent trend in the United States. Guidelines developed for describing the site environmental setting as it applies to such cases can also be useful in the EIA process. For example, the following topics and environmental factors are to be addressed in environmental baseline surveys conducted at U.S. Army installations (U.S. Department of the Army, 1990b, p. 75):

1. *History.*   Description of past and present activities on/in the property.
2. *Location.*   A map should show the subject property in its geographical context.
3. *Physiography/surface hydrology.*   Topography, flood plain and wetland locations, low and minimum receiving stream flow, water supply capabilities, and flood potential of existing and proposed use.
4. *Soils.*   Type, depth, erosion, and contaminant migration potentials.

5. *Geology.* Summary of the geology of the region and subject property, emphasizing the potential for migration of contaminants.
6. *Hydrogeology.* Depth to uppermost aquifer, ground water quality, rate and direction of flow, water supply capabilities, potential for contaminant migration, and potential for contaminating deeper aquifers.
7. *Meteorology.* Precipitation and evaporation rates, prevailing wind speed and direction, temperatures.
8. *Contaminated structures, buildings, or fixtures.* An identification of the structure and the kind of contamination potential, for example, asbestos, radon, PCB transformers, pesticides/rodenticides/herbicides, chemical agents, explosives.
9. *Land use patterns.* Residential, commercial, industrial, agricultural, etc., compatibility of proposed use with existing neighboring usage.
10. *Noise.* An assessment of the ambient noise level contours.
11. *Existing ecological baseline.* Subject areas include but are not limited to all existing and former sites involved in generating, transporting, storing, treating, or disposing of hazardous materials/substances/wastes, wastewaters, solid wastes, fuels, explosives, ordnance, and other potential hazards, such as excessive noise, asbestos, or radon gas.
12. *Population.* This element summarizes the existing and potential human populations on the property and in the region.
13. *Environmental compliance.* This element summarizes the status of compliance with existing environmental requirements, any closure requirements if pollution control facilities would have to be abandoned due to the transaction, and any anticipated future regulatory requirements.

An important thing to do in conjunction with agency guidelines or regulations is to attempt to procure pertinent requirements from agencies that do projects of the type which is being analyzed. For example, if the project type to be analyzed is a hydroelectric power project, then agencies which build and operate or regulate such projects should have guidelines or regulations which list pertinent environmental factors. In the United States, such federal agencies include but are not limited to the Army Corps of Engineers, Bureau of Reclamation, Tennessee Valley Authority, Soil Conservation Service, and Federal Energy Regulatory Commission.

Professional Knowledge

One of the better ways of identifying environmental factors is to utilize professional knowledge related to the anticipated impacts of specific types of projects. The example given below is related to the general effects of impoundments on water quality. Considering water quality only, it is known that impoundment of water will lead to the following beneficial effects:

Turbidity reduction

Hardness reduction

Oxidation of organic material

Coliform reduction

Flow equalization

The detrimental effects of impoundment include the following:

Lower reaeration

Buildup of inorganics

Algae blooms

Stratified flow

Thermal stratification

Perhaps the most significant impact on water quality is due to thermal stratification, which leads to the following additional changes in water quality: (1) decreased dissolved oxygen in hypolimnion, (2) anaerobic conditions in hypolimnion, and (3) dissolution of iron and manganese from bottom deposits. Thermal stratification also results in changes in mixing patterns. A "stratified, or density, current" is the movement, without loss of identity by turbulent mixing at the bounding surfaces, of a stream of fluid under, through, or over a body of fluid with which it is miscible; the ''density difference'' is a function of the differences of temperature, solids content, and/or salt content of the two fluids. In the case of thermal stratification, the density difference is a function of differences in temperature. Water exhibits maximum density at 4°C. Thermal stratification can result in ''overflow'' (warmer water flowing over the surface of colder water), ''interflow'' (cool water flowing between upper layers of warmer water and lower layers of colder water), or ''underflow'' (cooler water flowing underneath warmer surface water). These conditions of thermal stratification prevent complete mixing from occurring in normal time frames.

An additional concern related to water impoundment is the reduction in waste assimilative capacity of the body of water being impounded. ''Waste assimilative capacity'' is the ability of a body of water to receive organic wastes and purify itself through natural reaeration. In general, water impoundment decreases the reaeration ability of a body of water, thus reducing the waste loading that the body of water can receive without having the dissolved oxygen concentration fall below a prescribed water quality standard.

Utilization of professional knowledge related to the impacts of many types of projects can be aided by resorting to organized lists of environmental factors of concern for specific types of projects. An example of a simple checklist of environmental factors for impoundment projects is in Table 4.1. Other methods which can aid in identifying impacts are described in Chapter 3; examples of such systematic methods include interaction matrices, networks, and descriptive checklists. Weighting-scaling checklists developed for specific types of projects are described in Chapter 15; these checklists include environmental factors which should be addressed in describing the environmental setting.

Another example of the use of professional knowledge to identify anticipated impacts is re-lated to construction activities. Most proposed actions involve a construction phase, with the activities being similar for many types of projects. Table 4.2 contains an example list of potential environmental impacts resulting from construction activities—in this case, associated with nuclear power plant site preparation and plant- and transmission-facilities construction. Although this list was developed for the nuclear industry, many of the construction activities are relevant to numerous other types of projects. Consideration of the areas listed under Potential Environmental Impacts in Table 4.2 would be another way of developing a list of environmental factors for consideration. This detailed approach could be used for major construction projects.

To serve as a final illustration, another issue of concern in many projects is that of potential impacts on health. One particular example relates to the health impacts of water projects such as dams for flood control and/or hydroelectric power development, and irrigation systems in agricultural areas. The following questions are relevant to identifying pertinent environmental factors and obtaining baseline information on the potential health impacts of projects:

1. What are the major health problems in the surrounding population? What is the level of infection and degree of resistance?
2. What are the environmental transmission pathways for existing diseases?
3. What habitats are important for breeding and feeding of disease organisms and vectors?
4. What are the life cycles of important disease organisms and vectors?
5. What are the numbers, locations, and characteristics of existing populations?
6. What information is available on local environmental and social conditions?
7. What time period and geographical coverage are necessary for a survey of existing conditions to encompass important temporal and spatial variations?

TABLE 4.1

CHECKLIST OF BIOPHYSICAL- AND CULTURAL-ENVIRONMENT FACTORS FOR IMPOUNDMENT PROJECTS

Category	Subcategory	Factor
Terrestrial	Population	Crops
		Natural vegetation
		Herbivorous mammals
		Carnivorous mammals
		Upland game birds
		Predatory birds
	Habitat, land use	Bottomland forest[a]
		Upland forest[b]
		Open (nonforest) lands[c]
		Drawdown zone
		Land use
	Land quality, soil erosion	Soil erosion
		Soil chemistry
		Mineral extraction
	Critical community relationships	Species diversity
Aquatic	Population	Natural vegetation
		Wetland vegetation
		Zooplankton
		Phytoplankton
		Sport fish
		Commercial fisheries
		Intertidal organisms
		Benthos, epibenthos
		Waterfowl
	Habitat	Stream[d]
		Freshwater lake[e]
		River swamp[f]
		Nonriver swamp[g]
	Water quality	pH
		Turbidity
		Suspended solids
		Water temperature
		Dissolved oxygen
		Biochemical oxygen demand
		Dissolved solids
		Inorganic nitrogen
		Inorganic phosphate
		Salinity
		Iron and manganese
		Toxic substances
		Pesticides
		Fecal coliforms
		Stream assimilative capacity

TABLE 4.1

CHECKLIST OF BIOPHYSICAL- AND CULTURAL-ENVIRONMENT FACTORS FOR IMPOUNDMENT PROJECTS *(continued)*

Category	Subcategory	Factor
	Water quantity	Stream-flow variation Basin hydrologic loss
	Critical community relationships	Species diversity
Air	Quality	Carbon monoxide Hydrocarbons Oxides of nitrogen Particulates
	Climatology	Diffusion
Human interface	Noise	Noise
	Aesthetics	Width and alignment Variety within vegetation type Animals, domestic Native fauna Appearance of water Odor and floating materials Odor and visual quality Sound
	Historical	Historical internal and external packages
	Archaeological	Archaeological internal and external packages

[a]"Bottomland forest" represents a composite consideration of the following 11 parameters: species associations, percent mastbearing trees, percent coverage by understory, diversity of understory, percent coverage of ground cover, diversity of ground cover, number of trees greater than 16 in (or 18 in) dbh/acre, percent of trees greater than 16 in (or 18 in) dbh, frequency of inundation, quantity edge, and quality edge.

[b]"Upland forest" represents a composite consideration of the following 10 parameters: species associations, percent mastbearing trees, percent coverage by understory, diversity of understory, percent coverage of ground cover, diversity of ground cover, number of trees greater than or equal to 16 in dbh/acre, percent of trees greater than or equal to 16 in dbh, quantity of edge, and mean distance to edge.

[c]"Open (nonforest) lands" represent a composite consideration of the following four parameters: land use, diversity of land use, quantity of edge, and mean distance to edge.

[d]"Stream" represents a composite consideration of the following eight parameters: sinuosity, dominant centrarchids, mean low-water width, turbidity, total dissolved solids, chemical type, diversity of fishes, and diversity of benthos.

[e]"Freshwater lake" represents a composite consideration of the following 10 parameters: mean depth, turbidity, total dissolved solids, chemical type, shore development, spring flooding above the vegetation line, standing crop of fishes, standing crop of sport fish, diversity of fishes, and diversity of benthos.

[f]"River swamp" represents a composite consideration of the following six parameters: species associations, percent forest cover, percent flooded annually, ground cover diversity, percent coverage by ground cover, and days subject to river overflow.

[g]"Nonriver swamp" represents a composite consideration of the following five parameters: species associations, percent forest cover, percent flooded annually, ground cover diversity, and percent coverage by ground cover.

Source: Adapted from Canter and Hill, 1979.

Table 4.2

POTENTIAL ENVIRONMENTAL IMPACTS RESULTING FROM CONSTRUCTION PRACTICES

Construction phase	Construction practice	Potential environmental impacts
Preconstruction	Site inventory	Short-term and nominal
	Vehicular traffic	Dust, sediment, and tree injury
	Test pits	Tree root injury, sediment
	Environmental monitoring	Negligible if properly done
	Temporary controls	Short-term and nominal
	Storm water	Vegetation, water quality
	Erosion and sediment	Vegetation, water quality
	Vegetative	Fertilizers in excess
	Dust	Negligible if properly done
Site work	Clearing and demolition	Short-term
	Clearing	Decrease in the area of protective tree, shrub, and ground covers, stripping of topsoil; increased soil erosion, sedimentation, and storm-water runoff; increased stream water temperatures; modification of stream banks and channels, water quality
	Demolition	Increased dust, noise, solid wastes
	Temporary facilities	
	Shops and storage sheds	Increased surface areas impervious to water infiltration, increased water runoff, petroleum products
	Access roads and parking lots	Increased surface areas impervious to water infiltration, increased water runoff, generation of dust on unpaved areas
	Utility trenches and backfills	Increased visual impacts, soil erosion, and sedimentation for short periods
	Sanitary facilities	Increased visual impacts, soild wastes
	Fences	Barriers to animal migration
	Lay-down areas	Visual impacts, increased runoff
	Concrete batch plant	Increased visual impacts; disposal of wastewater, increased dust and noise
	Temporary and permanent pest control (termites, weeds, insects)	Nondegradable or slowly degradable pesticides accumulated by plants and animals, then passed up the food chain to humans; degradable pesticides having short biological half-lives preferred for use
	Earthwork	Long-term
	Excavation	Stripping, soil stockpiling, and site grading;
	Grading	increased erosion, sedimentation, and
	Trenching	runoff, soil compaction; increased soil levels
	Soil treatment	of potentially hazardous materials; side effects on living plants and animals, and the incorporation of decomposition products into food chains; water quality
	Site drainage	Long-term
	Foundation drainage	Decrease in the volume of underground water
	Dewatering	for short and long time periods, increased
	Well points	stream flow volumes and velocities,
	Stream channel relocation	downstream damages, water quality

Table 4.2

POTENTIAL ENVIRONMENTAL IMPACTS RESULTING FROM CONSTRUCTION PRACTICES
(continued)

Construction phase	Construction practice	Potential environmental impacts

	Landscaping
	   Temporary seeding
	   Permanent seeding and sodding
		Decreased soil erosion and overland flow of storm water, stabilization of exposed cut and fill slopes, increased water infiltration and underground storage of water, visual impacts

Permanent facilities	Transmission lines and heavy traffic areas
	   Parking lots
	   Switchyard
	   Railroad spur line
	Buildings
	   Warehouses
		Long-term
		Storm-water runoff, petroleum products
		Visual impacts, sediment, runoff
		Storm-water runoff
		Long-term
		Impervious surfaces, storm-water runoff, solid wastes, spillages

	Sanitary waste treatment
	   Cooling towers
	Related facilities
	   Reactor intake and discharge channel
		Odors, discharges, bacteria, viruses
		Visual impacts
		Long-term
		Shoreline changes, bottom topography changes, fish migration, benthic fauna changes

	Water supply and treatment
	   Storm-water drainage
	   Wastewater treatment
	   Dams and impoundments
	   Breakwaters, jetties, etc.
	   Fuel-handling equipment
	   Oil-storage tanks, controls, and piping
	   Conveying systems (cranes, hoists, chutes)
		Waste discharges, water quality
		Sediment, water quality
		Sediment, water quality, trace elements
		Dredging, shoreline erosion
		Circulation patterns in the waterway
		Spillages, fire, and visual impacts
		Visual impacts
		Visual impacts

	Waste-handling equipment (incinerators, wood chippers, trash compactors)	Noise, visual impacts

	Security fencing
	   Access road
	   Fencing
		Long-term
		Increased runoff
		Barriers to animal movements

Project closeout	Removal of temporary offices and shops
	   Demolition
	   Relocation
		Short-term
		Noise, solid waste, dust
		Storm-water runoff, traffic blockages, soil compaction

	Site restoration
	   Finish grading
	   Topsoiling
	   Fertilizing
	   Sediment controls
	Preliminary start-up
	   Cleaning
	   Flushing
		Short-term
		Sediment, dust, soil compaction
		Erosion, sediment
		Nutrient runoff, water quality
		Vegetation
		Short-term
		Water quality, oils, phosphates, and other nutrients

Source: Adapted from Hittman Associates, 1974, pp. B13–B-18.

Review of EISs

Over 20,000 draft and final EISs have been prepared in the United States since the 1970 effective date of the National Environmental Policy Act (NEPA). In addition, many tens of thousands of EAs and other related issues reports have been prepared. For example, in a recent conference on the NEPA, the deputy director of the Office of Federal Activities of the U.S. EPA observed that some 30 to 50 programmatic and 250 to 450 project EISs are issued each year, whereas many more decisions are covered in EAs and NEPA categorical exclusions. The estimate of the annual number of EAs was 30,000 to 50,000 (Smith, 1989). For certain project types, there may be several hundred impact statements or EAs that could be examined for the environmental factors included in the description of the environmental setting, as well as the environmental impacts discussed in the impact statement proper. For example, the documented impacts for 55 EISs dealing with U.S. Army Corps of Engineers dams and reservoirs were reported on by Ortolano and Hill (1972). The presumption is that existing conditions should be described for those environmental factors expected to be changed.

Another example of documented impacts for dam and reservoir projects is the list of 23 typical impacts in Table 4.3. Also, Goodland (1989) noted that the area of influence, or boundaries, of dam and reservoir projects in-

---

TABLE 4.3

TYPICAL IMPACTS OF DAMS AND RESERVOIRS

1. Change in quality of impounded water (seasonal)
2. Water loss due to evaporation (seasonal)
3. Downstream effects in terms of decreased (and more-uniform) flow into estuaries, thus causing changes in saltwater intrusion patterns and changes in estuarine fisheries
4. Changes in local groundwater levels and quality
5. In-reservoir landslides and/or increased seismic activity in the area due to water pressure
6. Changes in microclimate of area—more wind, humidity, and/or precipitation
7. Inundation of mineral resources
8. Changes in number and types of fish—from coldwater to warmwater fishery
9. Preclusion of movement of migratory fish (salmon on Columbia River are an example)
10. Fish destruction in turbines and pumps (use protective screens)
11. Possible creation of "new reservoir fishery" as positive impact
12. Increase areas for breeding of mosquitoes and related insects—and their public health implications (e.g., malaria and schistosomiasis)
13. Promote growth of aquatic weeds such as water hyacinths
14. Changes to habitat in inundated area and wildlife associated with habitat
15. Changes to waterfowl habitat from shallow, flowing habitat to deeper lakes; possible impact on migratory birds
16. Impacts on rare, threatened, endangered, unique floral and faunal species
17. Decrease in waste assimilative capacity of river segment
18. Inundation of historical, cultural, archaeological, or religious resources
19. People relocation-resettlement (and possible change in style of life)
20. Influx of construction workers and associated social, infrastructure, and health impacts
21. Increased tourism around reservoir
22. Downstream effects on traditional floodplain cultivation; reduced flood delivery of nutrients to downstream fields
23. Developments in catchment area resulting from roads and from other associated increases in sediment and nutrients into reservoir

Source: Compiled using data from Carpenter and Maragos (1989), Goodland (1989), Dixon, Talbot, and Le Moigne (1989), and Liao, Bhargava, and Das (1988).

cludes (1) the catchments contributing to the reservoir or project area and the area below the dam down to the estuary, coastal zone, and offshore region; (2) all ancillary aspects of the project, such as power-transmission corridors, pipelines, canals, tunnels, relocation and access roads, borrow and disposal areas, and construction camps, as well as unplanned developments arising from the project (e.g., logging or shifting agriculture along access roads); (3) off-site areas required for resettlement or compensatory tracts; (4) the airshed—for example, locations where airborne pollution (smoke, dust) may enter or leave the area of influence; and (5) migratory routes of humans, wildlife, or fish, particularly where related to public health, economics, or environmental conservation.

Two examples related to military projects will be cited. First, regarding the U.S. Army Base Realignment and Closure (BRAC) program, the environmental factors addressed in the setting description for the ''Fort Dix'' EIS included the following (U.S. Department of the Army, 1990a):

A. Physical environment
  1. Geology, minerals, soils
  2. Climate
  3. Water resources (surface water and groundwater)
  4. Water quality
  5. Air quality and emissions
  6. Noise
  7. Hazardous-waste management
B. Biological environment
  1. Terrestrial ecosystems
    a. Natural vegetation
    b. Natural fauna
  2. Aquatic ecosystems
  3. Threatened and endangered species
C. Human resources
  1. Demography
  2. Development and economy
  3. Land use
  4. Infrastructure
    a. Housing
    b. Schools
    c. Health care
    d. Transportation
    e. Recreation
    f. Utilities
  5. Archaeological and historic resources

The second military example is a questionnaire checklist developed for use at Fort Jackson, South Carolina; this checklist, too, could be used as a basis for describing the environmental setting (Knight, 1991). The checklist, or slight modifications thereof, has been used for numerous Army installations. It is similar to the one shown in Table 3.10, which was used in impact identification for agricultural projects.

A final example of the use of other EISs is related to wastewater treatment plants. Canter (1978) conducted a detailed review of 28 draft or final EISs prepared by the U.S. Environmental Protection Agency or their contractors on wastewater facility plans. At the time of the review, the sample group represented approximately 15 percent of all EISs prepared on wastewater-treatment-plant projects. The geographic distribution included 17 states. The sample EISs encompassed a wide variety of projects ranging from those designed to meet the wastewater-management needs for populations as low as 1,600 persons (in Jacksonville, Oregon) to 3.6 million persons (in Los Angeles County, California). Projects consisted of sixteen involving collection or conveyance, treatment, and disposal; four, treatment and disposal only; six, collection or conveyance only; and two, sludge disposal only.

To prepare a description of the environmental setting, quantitative information should be assembled on as many of the identified pertinent factors as possible. Existing environmental standards, such as water or air quality standards, should be included, to provide for assessing existing environmental quality. The description of the environmental setting should focus primarily on the broad study area for the project; however, specific information should be included for proposed interceptor routes or treatment plant sites. In the case of wastewater treatment projects, the boundaries or study areas can be defined by the sewer service area and by the area of influence of the wastewater discharge.

While the majority of the 28 EISs in the sample group had adequate descriptions of their environmental setting, all did not evaluate the same environmental factors. This is appropriate, since emphasis should be given to those factors that are critical or important in a given locale. The environment generally was described in two broad categories: the natural environment, consisting of physical-chemical and biological factors, and the man-made environment, consisting of cultural resources and socioeconomic concerns. Examples of the factors in each category and sensitive environmental areas included are

1. *Natural environment*—Air quality and odor, water quantity and quality, groundwater quantity and quality, saltwater intrusion, water usage and water rights, meteorology and climatology, catastrophic meteorological events, benthal deposits, noise, soils, geology, seismicity, topography, aquatic and terrestrial biology, threatened or endangered plant or animal species, visual characteristics, and aesthetic features of recreational areas, open spaces, and natural areas.
2. *Man-made environment*—Historical, archaeological, and paleontological sites; land use; population (permanent versus temporary or recreational); skiing and other recreational uses of area; infrastructure; resource consumption in terms of energy and minerals; and regulatory and planning frameworks for the area, including water laws and regulations and planning groups and agencies.
3. *Sensitive environmental areas*—Prime agricultural lands, floodplains, groundwater recharge areas, wetlands, shellfishing areas,

watershed-protection areas, steeply sloping land, and recreational open spaces.

One way of improving information communication relative to the environmental setting involves the liberal usage of figures, tables, maps, and photographs of the area. Several EISs in the wastewater-project study group incorporated photographs to describe the environmental setting, as well as new approaches for presenting biological information. These presentation approaches contrast with the more-typically undertaken assemblage of complete species lists of flora and fauna within the study area. One such approach is to identify habitat types within the study area and to include a map showing their geographical distribution. Another approach is to use a map of plant communities on which are delineated the geographical distribution. Summary tables identifying community type, value of the community type, vegetation and environmental characteristics, geographic distribution, and representative species are also useful (Canter, 1978).

Environmental Impact-Assessment Methodologies

Another approach that could be used for preparing an initial list of environmental factors for consideration is to select factors utilized in environmental impact-assessment (EIA) methods. Only one method will be cited at this point, although several hundred have been developed since the passage of the NEPA. Additional methods are described in Chapters 3 and 15. A list of about 90 environmental items incorporated in the Leopold interaction matrix was prepared for use by the *U.S. Geological Survey* (Leopold et al., 1971) and is shown in the right-hand column of Table 3.2.

SELECTION PROCESS

Selecting environmental factors from an initial list could involve site visits, interdisciplinary-

TABLE 4.4

SOME REASONS FOR CONDUCTING SITE VISITS

- Develop sense of existing conditions and improve writing
- Check existing information (ground truthing)
- Check surrounding area in terms of other developments and secondary impacts
- Aid in identifying unknown factors and data (new information; some cursory data gathering; videotapes or photographs of area)
- Verify proposal
- Review or plan work by environmental contractors
- Coordinate with other agencies
- Provide credibility
- Determine applicability of specific requirements (e.g., Sec. 404)
- Determine current status of project—see if anyone (applicant) has "jumped the gun"
- Meet with proponent
- Facilitate joint discussion through team visit
- Have interagency discussions on site
- Provide independent review of site away from proponent
- Check changes in project and environment due to time lapse
- Obtain field experience which adds to cumulative knowledge over time and makes a person a better professional
- Help to plan environmental-monitoring program
- Talk to locals
- Verification of possibly erroneous information provided by various interest groups

team discussions, scoping, the application of criteria questions, and/or professional judgment.

Site Visits

It is important for each member of the study team to visit the proposed area for the project. A site visit can provide familiarization with the area and enable more-effective review of extant environmental data. Some reasons for site visits are listed in Table 4.4.

Exact delineations of what to look for on a site visit will be dependent on the type of proposed project and anticipated impacts, and on the particular features of the proposed location. Site assessments as a precursor to the sale or purchase of property have become commonplace in recent years. Such assessments typically include site visits, and several site-visit checklists have been developed; one such list is in Table 4.5. The list could also be used as is or modified for site visits for the EIA process.

Interdisciplinary-Team Discussions

The study team should engage in group discussions of relevant environmental factors appearing on one to several initial lists. These discussions can lead to a greater familiarity with project impacts and, possibly, to the identification of pertinent environmental factors not included on any initial list. These discussions can take place during and following site visits; they are most productive once all team members have an understanding of the features of the proposed project.

Scoping

Scoping can be used in the selection of pertinent environmental factors for inclusion in the description of the environmental setting. "Scoping" refers to an early and open process for determining the scope of issues to be addressed and for identifying the significant issues related to a proposed action (CEQ, 1978). The environmental setting needs to be described in terms of

TABLE 4.5

CHECKLIST OF ITEMS TO ADDRESS IN A SITE ASSESSMENT

Physical Features	**Process and Wastewater**
Parking areas	Sizes and construction
Roads	Function
Power lines	Contents
Public buildings	Inside or outside
Dwellings	Above or below grade
Structural improvements	Leaks
Easements and rights-of-way	Piping
	Chemical transfer points
	Pumps
	Spill evidence
Facilities	**Effluents Discharge**
Number, location, size	Manholes
Prior use	Catch basins
Type of construction	Process sewer
Insulation material	Fill pipes
Ages	Leach fields
Flooring material	
Floor drains	
Spill evidence	
Odors	
Internal contamination	
Adjacent Land Use	**Pits/Ponds/Lagoons**
Surface water	Sizes and location
Roads and utilities	Materials of construction
Residential property	Purpose
Industrial sites	Contents
Vacant land	Lined or unlined
Types of vegetation	Above or below grade
	General condition
	Freeboard
	Leaks
	Fill and drain pipes

(continued)

any factors anticipated to be impacted by the proposed action.

Criteria Questions

It is suggested that only the environmental factors that are pertinent to the specific study be described; environmental factors that require extensive data procurement and interpretation but have little relevance to the study should be excluded. The following three criteria questions are suggested for evaluating the factors to be considered; if any of these criteria apply to a given factor, then that factor should be described in the environmental setting:

1. Will the environmental factor be affected, either beneficially or adversely, by any of the alternatives (including the no-action alternative) under study?
2. Will the environmental factor exert an influence on construction scheduling or a subsequent operational phase of any of the alternatives?
3. Is the factor of particular public interest or controversy within the local community?

Professional Judgment

Professional judgment should be relied upon in the selection of environmental factors for inclu-

TABLE 4.5

CHECKLIST OF ITEMS TO ADDRESS IN A SITE ASSESSMENT *(continued)*

Utilities Electricity Natural gas Oil Telephone Sewer to publicly-owned treatment works (POTW) Water Storm drains **Water Features** Wells Springs Seeps Swamps and wetlands Ponds Streams Evidence of flooding Direction of runoff Surface erosion **Geological Features** Topography and slope Soil characteristics Rock outcrops Sink holes Excavation Mining activity Diversion ditches Soil stockpiles	**Disposal Areas** Sizes, location, ages Contents General condition Debris, sludge, rubble Stressed vegetation Surface contours Monitoring and recovery wells **Chemical/Fuel Drum Storage** General condition Security and access Spill control Emergency response plan Spill evidence **Waste Evidence** Drums, barrels, containers Waste materials Construction and demolition Debris Discolored soil Leachate seeps Discolored surface water

Source: Bowman, 1989, pp. 19–20.

sion in a description of the environmental setting for a project. The judgment can be a part of interdisciplinary-team discussions, or it can be solicited during conversations with recognized experts in particular, relevant disciplines for a project.

DOCUMENTATION OF SELECTION PROCESS

For potentially controversial projects, it may be necessary to carefully document the rationale used for inclusion-exclusion of environmental factors in the description of the environmental setting. Additionally, careful documentation is a good practice to be used for all projects. Table 4.6 displays one approach which can be used to document the selection-exclusion process. The Table 4.6 information could be included in an appendix in the resultant EA or EIS.

DATA SOURCES

Relevant data for describing the physical-chemical, biological, cultural, and socioeconomic environments can be obtained from reports and unpublished data from numerous regulatory agencies at the local, state, regional, and federal government levels. A list of agencies which might have information on various environmental factors would be helpful. Such a list is available for the environmental factors included in the Leopold interaction matrix (Canter, 1977). Additionally, a list of federal agencies and fed-

TABLE 4.6

DOCUMENTATION OF FACTOR-SELECTION PROCESS

| | Basis for selection | | | | | Basis for exclusion | |
Factor	S^a	Q	SV	IDT	PJ	Nonoccurrence in area	No basis for selection
1	×		×	×			
2						×	
3							×
4	×						
.							
.							
.							
n							

^{a}S = scoping; Q = criteria questions; SV = site visits; IDT = interdisciplinary-team discussion; PJ = professional judgment

eral-state agencies with jurisdiction by law or qualified by special expertise to comment on various environmental matters was in the 1973 CEQ guidelines. This list could be expanded to include the state and local government corollary agencies for each federal entity, thus extending the list of contacts for environmental setting information. Remote-sensing information can also be useful for describing large-scale features of the environment; remote sensing includes aerial photography, infrared photography, and satellite data. Environmental data in given geographical areas might be obtained from various monitoring programs being conducted by local universities and/or private industry. While this information may not have general public distribution, use of appropriate contacts may lead to its procurement.

Extensive environmental data in the United States is found in computerized information storage and retrieval systems. Examples include air quality data (U.S. EPA Storage and Retrieval of Aerometric Data System—SAROAD), meteorological data (U.S. National Oceanic and Atmospheric Administration Climatic Center), water quality data (U.S. EPA Storage and Retrieval of Water Quality Data System—STORET), water quality and quantity data (*U.S. Geological Survey* National Water Data Storage

and Retrieval System—WATSTORE), terrestrial and aquatic biological data (U.S. Forest Service Wildlife and Fish Assessment System—WAFA, and Brookhaven National Laboratory Endangered Species of the United States of America System—ESUSA). The Oak Ridge National Laboratory's GEOCOLOGY System includes information on terrain, land use, wildlife, and vegetation. Socioeconomic data can be procured from the U.S. Bureau of the Census, the U.S. Army Corps of Engineers Socioeconomic and Environmental Data Information System (SEEDIS), and the U.S. Army Construction Engineering Research Laboratory (CERL) Economic Impact Forecast System (EIFS). In addition to providing baseline data, the EIFS can be used for impact predictions. Environmental laws and regulations are basic to the conduction of an environmental impact study, and the CERL Computer-aided Environmental Legislative Data System (CELDS) provides summary information on applicable laws and regulations.

An emerging information source being developed by the U.S. EPA is the Ecological Monitoring and Assessment Program (EMAP). EMAP will have integrated ecological-monitoring networks to estimate current extent, and changes in indicators of the condition of the nation's ecological resources on a regional basis with a known confidence level; to monitor indicators of pollutant exposure and habitat condition and seek associations between human-induced stresses and ecological condition; and to provide periodic statistical summaries and estimates and interpretive reports on status and trends of indicators (Hunsaker and Carpenter, 1990). EMAP will also provide a framework for cooperative planning and implementation with other agencies and organizations that have active monitoring programs in the ecological and natural resource areas.

Six ecological resource categories have been defined within EMAP: near-coastal waters, inland surface waters, wetlands, forests, arid lands, and agroecosystems. Within each of these categories, several ecological resource classes have been identified (e.g., large estuaries, small lakes, emergent estuarine wetlands, sagebrush-dominated shrubland, and orchard cropland). In addition to measures of extent (numbers, length, area), EMAP will make routine measurements of environmental indicators on resource-sampling units selected from resource classes. Several types of indicators will be used, as shown in Figure 4.2. "Response indicators" are characteristics measured to provide evidence of the biological condition of a resource at the organism, population, community, or ecosystem level of organization. "Exposure indicators" and "habitat indicators" are diagnostic indicators measured in conjunction with response indicators. "Exposure indicators" are characteristics measured to provide evidence of the occurrence or magnitude of a response indicator's contact with a physical, chemical, or biological stress. "Habitat indicators" are physical attributes which characterize conditions necessary to support an organism, population, or community in the absence of pollutants. "Stressor indicators" are characteristics to quantify a natural process, environmental-hazard indicators, or a management activity that effects changes in exposure and habitat (Hunsaker and Carpenter, 1990). Information generated by EMAP will be useful for describing existing conditions and trends, and for addressing broader issues such as biological diversity.

Several key federal agencies have multiple environmental data storage and retrieval systems; such systems can include media quality data, pollution sources, and pollutant toxicity and environmental mobility. For example, the U.S. EPA has information systems which contain nationwide statistics concerning the quality of treated drinking water, industrial and municipal point-source discharges from facilities possessing National Pollutant Discharge Elimination System (NPDES) permits, surface- and groundwater quality, ambient air quality, air pollutant emissions inventories, hazardous-waste sites, toxics releases, health risks of chemicals, and many other issues (U.S. EPA 1990, 1991). Other agencies with extensive environmental databases include the *U.S. Geological Survey,* U.S. Soil Conservation Service, U.S. Fish and Wildlife Service, and U.S. Bureau of the Census. A partial guide to federal environmental databases is available in Kokoszka (1992).

Environmental data systems are also available through international organizations such as the United Nations Environment Program and the World Health Organization (WHO). Many countries already have or are in the process of establishing environmental databases.

Even though there are numerous environmental databases, they are not of equal reliability. Quality control–quality assurance programs must be associated with data collection, data input and associated calculations, and data output. Comparative evaluations of environmental data and databases may be needed (Ramamoorthy and Baddaloo, 1990).

SPECIAL ISSUES AND CONCERNS

Several special issues and concerns can be raised in conjunction with the process of describing the environmental setting. These in-

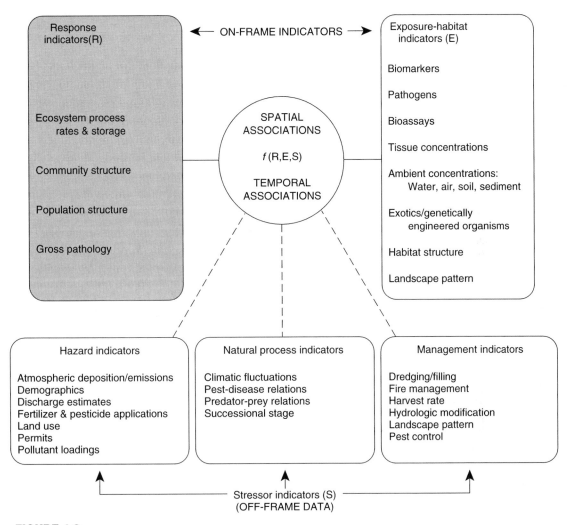

FIGURE 4.2
EMAP Conceptual Indicator Strategy (Hunsaker and Carpenter, 1990).
Note: Indicators are objective, quantifiable surrogates for assessment endpoints and stressors
(environmental hazards, management actions, and natural phenomena). The circle indicates that
analysis is by statistical association, rather than by explicit (causal) mathematical relationships.

clude determining the project boundaries to be addressed in the setting description, review of existing data for correctness, dealing with temporal variability in data, and cost-effective alternative information-gathering approaches to use in the complete absence of data. The usage of environmental indices and indicators for de-

scribing the environmental setting is addressed in Chapter 5.

Project boundaries should be established based on consideration of the expected locations of the impacts. Boundaries can differ as a function of whether physical-chemical, biological, cultural, or socioeconomic consequences are being considered. Project boundaries could also be defined based on agency policy; as a special case, it may be appropriate to use project exclusionary zones as the boundaries. Obviously, the boundaries established will influence the level of effort associated with describing the environmental setting.

All procured data, whether from information storage and retrieval systems, reports, or files should be reviewed for general correctness. Professionals in various substantive areas of expertise can review published data and information and quickly determine if obvious errors exist.

Normal daily, seasonal, and/or annual fluctuations in environmental data should be recognized and addressed in EAs or EISs. Of particular importance may be extreme conditions, since pertinent standards may exist, or since the impacts may be greatest at these times. For example, water quality standards are typically most critical at low-flow conditions in a river segment.

For many situations there may be a complete absence of site-specific data for identified factors. Two approaches could be used to address this absence of information: (1) plan and implement a target-monitoring program focused on key indicators and/or (2) consider the usage of pertinent available information from nearby areas that are similar in characteristics to the project study area.

SUMMARY

The preparation of an appropriate description of the environmental setting is one of the key steps in the EIA process. Proper delineation of a list of environmental factors to be addressed, as well as documentation of information and data utilized, is a necessary component of this phase. A summary of detailed technical information and data should be provided, with the referenced information appropriately contained in tables, maps, footnotes, and appendices. Several approaches can be used for developing a list of

pertinent environmental factors for a given project. No single approach is universal, and utilization of a combination of approaches leads to the best results. The information and data sources for relevant environmental factors are manifold; however, these can generally be narrowed to a minimal number of key sources.

SELECTED REFERENCES

Atomic Energy Commission, "Preparation of Environmental Reports for Nuclear Power Plants," Regulatory Guide 4.2, Washington, D.C., Mar. 1973.

Bowman, V. A., *Preacquisition Assessment of Commercial and Industrial Property,* Pudvan Publishing Company, Northbrook, Ill., 1989, pp. 19–20.

Canter, L. W., *Environmental Impact Assessment,* McGraw-Hill Book Company, New York, 1977, pp. 287–293.

———, *Environmental Impact Statements on Municipal Wastewater Programs,* Information Resources Press, Washington, D.C., 1978.

———, and Hill, L. G., *Handbook of Variables for Environmental Impact Assessment,* Ann Arbor Science Publishers, Ann Arbor, Mich., 1979.

Carpenter, R. A., and Maragos, J. E., eds., "How to Assess Environmental Impacts on Tropical Islands and Coastal Areas," *Training Manual for South Pacific Regional Environment Programme* (SPREP), Environment and Policy Institute, East-West Center, Honolulu, Oct. 1989.

Council on Environmental Quality (CEQ), *"National Environmental Policy Act—Regulations," Federal Register,* vol. 43, no. 230, Nov. 29, 1978, pp. 55978–56007.

Cushing, C. E., "Hanford Site *National Environmental Policy Act* (NEPA) Characterization," PNL-6415-Rev. 3, Battelle Pacific Northwest Laboratories, Richland, Wash., Sept. 1990.

Dixon, J. A., Talbot, L. M., and Le Moigne, G. J., "Dams and the Environment—Considerations in World Bank Projects," Tech. Paper No. 110, World Bank, Washington, D.C., 1989, pp. 12–20.

Federal Power Commission, "Implementation of the *National Environmental Policy Act of 1969,* Order 485, Order Amending Part 2 of the General Rules to Provide Guidelines for the Preparation of Applicants' Environmental Reports Pursuant to Order 415-C," Washington, D.C., June 1973.

Goodland, R., "The World Bank's New Policy on the Environmental Aspects of Dam and Reservoir Projects," World Bank reprint ser. no. 458, World Bank, Washington, D.C., 1989.

Hittman Associates, "General Environmental Guidelines for Evaluating and Reporting the Effects of Nuclear Power Plant Site Preparation, Plant and Transmission Facilities Constructions," prepared for Atomic Industrial Forum, Washington, D.C., Feb. 1974, pp. B-13–B-18.

Hunsaker, C. T., and Carpenter, D. E., "Environmental Monitoring and Assessment Program—Ecological Indicators," EPA 600/3-90/060, U. S. Environmental Protection Agency, Research Triangle Park, N.C., Sept. 1990, pp. xii–xvii.

Knight, J. B., personal communication, Ft. Jackson, S.C., Aug. 13, 1991.

Kokoszka, L. C., "Guide to Federal Environmental Databases," *Pollution Engineering,* vol. 24, no. 3, Feb. 1992, pp. 83–92.

Leopold, L. B., et al., "A Procedure for Evaluating Environmental Impact," *Circular 645,* U.S. Geological Survey, Washington, D.C., 1971.

Liao, W., Bhargava, D. S., and Das, J., "Some Effects of Dams on Wildlife," *Environmental Conservation,* vol. 15, no. 1, 1988, pp. 68–70.

Ortolano, L., and Hill, W. W., "An Analysis of Environmental Statements for Corps of Engineers' Water Projects," Rep. 72-3, U.S. Army Corps of Engineers, Institute for Water Resources, Alexandria, Va., 1972.

Ramamoorthy, S., and Baddaloo, E., *Evaluation of Environmental Data for Regulatory and Impact Assessment, Studies in Environmental Science No. 41,* Elsevier Science Publishers, Amsterdam, 1990.

Smith, E. D., "Future Challenges of NEPA: A Panel Discussion," CONF-891098-10, Oak Ridge National Laboratory, Oak Ridge, Tenn., 1989.

U.S. Department of the Army, "Draft Environmental Impact Statement—Fort Dix Realignment, Including forts Bliss, Dix, Jackson, Knox, Lee, and Leonardwood," Washington, D.C., Jan. 1990a.

———, "Environmental Protection and Enhancement," AR 200-1, Washington, D.C., Apr. 1990b, pp. 73–78.

U.S. Environmental Protection Agency (EPA), Office of Water, "Environmental and Program Information Systems Compendium," EPA 500/9-90-002, Office of Water, Washington, D.C., Aug. 1990.

———, "EPA Information Sources," EPA 600/M-91-038, Washington, D.C., Oct. 1991.

CHAPTER 5

Environmental Indices and Indicators for Describing the Affected Environment

An "environmental index" in its broadest concept is a numerical or descriptive categorization of a large quantity of environmental data or information, with the primary purpose being to simplify such data and information so as to make it useful to decision makers and various publics. Selected indicators can also be used in impact studies. The primary focus of this chapter is on several types of indices which have been or could be used in environmental impact studies. The chapter is organized into sections addressing background information, examples of environmental-media indices (air quality, water quality, noise, ecological sensitivity and diversity, archaeological resources, visual quality, and quality of life), and the necessary steps in the development of an index. It is interesting to note that examples of indices can be cited for all of the typical biophysical and socioeconomic components of an impact study. Additional information on index methods is contained in Chapters 6 through 14 dealing with various environmental components.

In terms of EISs, environmental indices can be useful in accomplishing one or more of the following objectives:

1. To summarize existing environmental data
2. To communicate information on the quality of the affected (baseline) environment
3. To evaluate the vulnerability or susceptibility of an environmental category to pollution
4. To focus attention on key environmental factors
5. To serve as a basis for the expression of impact by forecasting the difference between the pertinent index with the project and the same index without the project

BACKGROUND INFORMATION

It should be noted that an environmental index is not the same as an environmental indicator. "Indicators" refer to single measurements of factors or biological species, with the assumption being that these measurements are indicative of the biophysical or socioeconomic system. Ecological indicators have been used for many decades (Hunsaker and Carpenter, 1990). For example, in the western United States, plants have been much used as indicators of water and soil conditions, especially as these conditions affect grazing and agricultural potentials

(Odum, 1959). The use of vertebrate animals, as well as plants, as indicators of temperature zones has also been practiced. Odum (1959) suggested that some of the important considerations which should be borne in mind in dealing with ecological indicators are as follows:

1. In general, "steno-" species make much better indicators than "eury-" species. *Steno* means "narrow" and *eury* means "wide." Steno- species are often not the most abundant ones in the community.
2. Large species usually make better indicators than small species, because a larger and more-stable biomass or standing crop can be supported with a given energy flow. The turnover rate of small organisms may be so great that the particular species present at any one moment may not be very instructive as an ecological indicator.
3. Before relying on a single species or groups of species as indicators, there should be abundant field evidence, and, if possible, experimental evidence that the factor in question is limiting. Also, the species ability to compensate or adapt should be known.
4. Numerical relationships between species, populations, and whole communities often provide more-reliable indicators than single species, since a better integration of conditions is reflected by the whole than by the part.

Relative to pollution effects, an "indicator organism" is a species selected for its sensitivity or tolerance (more frequently sensitivity) to various kinds of pollution or its effects—for example, metal pollution or oxygen depletion (Chapman, 1992). Relative to water quality, the different groups of organisms which have been used as indicators include bacteria, protozoa, algae, macroinvertebrates, macrophytes, and fish. For example, mussels in San Diego Bay have been used as bioindicators of the environmental effects of tributyltin (TBT) antifouling coatings applied to ships, marine structures, and vessels (Salazar and Salazar, 1990).

As described in Chapter 4, the EPA's Ecological Mapping and Assessment Program (EMAP) relies on the use of indicators for a variety of purposes (Hunsaker and Carpenter, 1990). Environmental indicators have also been suggested as useful tools for monitoring the state of the environment in relation to sustainable development and associated environmental threats (Organization for Economic Cooperation and Development, 1991). Indicators are being considered which would enable the measurement of environmental performance with respect to the level of (and changes in the level of) environmental quality; the integration of environmental concerns in sectoral policies; and the integration of environmental concerns in economic policies more generally through environmental accounting, particularly at the macro level. Table 5.1 contains the preliminary set of 25 indicators; included are 18 environmental indicators per se, followed by 7 key indicators reflecting economic and social changes of environmental significance.

While some environmental indices are fairly complicated from a mathematical perspective, it should be remembered that simple comparisons of data can be useful. For example, the following ratios yield relative indices that can be useful in an EIS:

$$\frac{\text{Existing quality}}{\text{Environmental quality standard}}$$

$$\frac{\text{Emission quantity or quality}}{\text{Emission standard}}$$

$$\frac{\text{Existing quality}}{\text{Temporal average}}$$

$$\frac{\text{Existing quality}}{\text{Spatial (geographical) average}}$$

Before proceeding to the examples of indices, it should be noted that conceptual concerns related to environmental indices have been identified, with the primary concern being the dis-

---

TABLE 5.1

PRELIMINARY SET OF NATIONAL ENVIRONMENTAL INDICATORS

Biophysical environment indicators

1. CO_2 emissions
2. Greenhouse gas emissions
3. SOx emissions
4. NOx emissions
5. Use of water resources
6. River quality
7. Wastewater treatment
8. Land use changes
9. Protected areas
10. Use of nitrogenous fertilizers
11. Use of forest resources
12. Trade in tropical wood
13. Threatened species
14. Fish catches
15. Waste generation
16. Municipal waste
17. Industrial accidents
18. Public opinion

Social and economic environment indicators

19. Growth of economic activity
20. Energy intensity
21. Energy supply
22. Industrial production
23. Transport trends
24. Private fuel consumption
25. Population

Source: Organization for Economic Cooperation and Development, 1991, p. 9.

---

tortion that can occur in the simplification process implied by aggregating environmental variables into one single value (Alberti and Parker, 1991). However, with the careful selection of indices and their systematic usage and a comparative interpretation of results, it is considered that the risk of distortion can be minimized.

ENVIRONMENTAL-MEDIA INDEX— AIR QUALITY

Indices of air pollution or air quality have been used for about 25 years. For example, Thom and Ott (1975) summarized a number of indices which represented various combinations of air quality factors. Because of the wide diversity in the indices, a common pollutant standards index (PSI) was developed for use in the United States (Ott, 1978). Ten criteria were delineated for the PSI and used in its promulgation; these criteria were that the PSI should (1) be easily understood by the public, (2) include major pollutants and be capable of including future pollutants, (3) relate to ambient air quality standards, (4) relate to air pollution episode criteria, (5) be calculated in a simple manner using reasonable assumptions, (6) be based on a reasonable scientific premise, (7) be consistent with perceived air-pollution levels, (8) be spatially meaningful, (9) exhibit day-to-day variation, and (10) enable forecasting a day in advance (Ott, 1978). Based upon these criteria, Table 5.2 was developed to represent information for the direct determination of the PSI. Five pollutants (total suspended particulates, sulfur dioxide, carbon monoxide, oxidants, and nitrogen dioxide) are considered individually in the PSI; combination effects such as those from sulfur dioxide and particulates are not addressed. Additional pollutants may be added in the future.

The PSI is established by defining an index value of 100 as the equivalent of the short-term (24 hours or less), national, primary ambient air quality standards. These short-term primary standards represent the concentration below which adverse health effects have not been observed, thus the PSI is based on health effects. The procedure is to calculate a simple ratio subindex value for each of the five pollutants considered, and then to report the PSI as the maximum subindex for the five pollutants. The subindex is calculated as follows:

$$\text{Subindex}_i = \frac{\text{concentration of pollutant}}{\text{short-term primary standard}}(100)$$

The reported daily PSI is the maximum subindex for the five pollutants considered, with the pollutant involved being identified. On days when two or more pollutants have subindices greater than 100, each pollutant having a subindex value of greater than 100 is reported, along with the maximum subindex value for all pollutants. Historical information on the PSI can be used in describing ambient air quality in an air impact study.

ENVIRONMENTAL-MEDIA INDEX— WATER QUALITY

There are numerous water quality indices which have been developed over the last 25 years. One example, called simply the "water quality index" (WQI), developed in 1970 by the U.S. National Sanitation Foundation (NSF), will be described. The WQI was based on the Delphi

approach, using a panel of 142 persons from throughout the United States with expertise in various aspects of water quality management (Table 5.3). A series of three questionnaires was mailed to the members of this panel. In questionnaire no. 1, the respondents were asked to consider 35 water-pollutant variables for possible inclusion in a water quality index (Table 5.4). Respondents were permitted to add any variables to the list which they felt should be included in the WQI. They were asked to designate each variable as follows: "do not include," "undecided," or "include." Respondents also were asked to rate each "include" variable according to its significance to overall water quality. This rating was done on a scale of 1 (highest relative significance) to 5 (lowest relative significance).

When respondents returned questionnaire no. 1, the results were tabulated and returned to the respondents for their further consideration, along with questionnaire no. 2. In questionnaire no. 2, each member was asked to review their original ratings and to modify the response if desired. Each member was instructed to note his or her replies for each variable and to compare them with those of the entire group. Following the receipt of results from questionnaire no. 2, the nine individual variables of greatest importance were identified as dissolved oxygen (DO), fecal coliforms, pH, 5-day biochemical oxygen demand (BOD_5), nitrates (NO_3), phosphates (PO_4), temperature deviation, turbidity (in JTU), and total solids (TS). The resultant importance weights based on the ratings for each variable are listed in Table 5.5. The weights have a public health focus based on using the water for human consumption.

In questionnaire no. 3, the respondents were asked to develop a rating curve for each of the included variables (Ott, 1978). This was accomplished by providing blank graphs to each respondent. Levels of Water Quality from 0 to 100 were indicated on the ordinate of each graph, while various levels (or strengths) of the

TABLE 5.2

COMPARISON OF PSI VALUES WITH POLLUTANT CONCENTRATIONS, DESCRIPTOR WORDS, GENERAL HEALTH EFFECTS, AND CAUTIONARY STATEMENTS

Index value	Air quality level	Pollutant levels					Health effect descriptor	General health effects	Cautionary statements
		TSP (24-hr) ($\mu g/m^3$)	SO$_2$ (24-hr) ($\mu g/m^3$)	CO (8-hr) (mg/m^3)	O$_3$ (1-hr) ($\mu g/m^3$)	NO$_2$ (1-hr) ($\mu g/m^3$)			
500	Significant harm	1000	2620	57.5	1200	3750	Hazardous	Premature death of ill and elderly. Healthy people will experience adverse symptoms that affect their normal activity.	All persons should remain indoors, keeping windows and doors closed. All persons should minimize physical exertion and avoid traffic.
400	Emergency	875	2100	46.0	1000	3000	Hazardous	Premature onset of certain diseases in addition to significant aggravation of symptoms and decreased exercise tolerance in healthy persons.	Elderly and persons with existing diseases should stay indoors and avoid physical exertion. General population should avoid outdoor activity.
300	Warning	625	1600	34.0	800	2260	Very unhealthful	Significant aggravation of symptoms and decreased exercise tolerance in persons with heart or lung disease, with widespread symptoms in the healthy population.	Elderly and persons with existing heart or lung disease should stay indoors and reduce physical activity.
200	Alert	375	800	17.0	400[a]	1130	Unhealthful	Mild aggravation of symptoms in susceptible persons, with irritation symptoms in the healthy population.	Persons with existing heart or respiratory ailments should reduce physical exertion and outdoor activity.
100	NAAQS	260	365	10.0	160	b	Moderate		
50	50% of NAAQS	75[c]	80[c]	5.0	80	b	Good		
0		0	0	0	0	b			

[a] 400 $\mu g/m^3$ was used instead of the O$_3$ alert level of 200 $\mu g/m^3$ (see text).
[b] No index values reported at concentration levels below those specified by "alert level" criteria.
[c] Annual primary NAAQS.
Source: Ott, 1978, pp. 149–150.

TABLE 5.3

PROFESSIONS OF NSF WQI PANEL PARTICIPANTS

Regulatory officials (federal, interstate, state, territorial and regional)	101
Managers of local public utilities	5
Consulting engineers	6
Academicians	26
Others (industrial waste control engineers and representatives of professional organizations)	4
Total	142

Source: Ott, 1978, p. 203.

TABLE 5.4

35 CANDIDATE VARIABLES CONSIDERED FOR THE NSF WQI IN QUESTIONNAIRE NO. 1

Dissolved oxygen	Oil and grease
Fecal coliforms	Turbidity
pH	Chlorides
Biochemical oxygen demand (5-day)	Alkalinity
Coliform organisms	Iron
Herbicides	Color
Temperature	Manganese
Pesticides	Fluorides
Phosphates	Copper
Nitrates	Sulfates
Dissolved solids	Calcium
Radioactivity	Hardness
Phenols	Sodium and potassium
Chemical oxygen demand	Acidity
Carbon chloroform extract	Bicarbonates
Ammonia	Magnesium
Total solids	Aluminum
	Silica

Source: Ott, 1978, p. 203.

particular variable were arranged along the abscissa. Each respondent was asked to draw a curve on each graph which, in their judgment, represented the variation of water quality produced by the various quantities of each pollutant variable. The resultant relationships are called "functional relationships" or "functional curves."

The investigators subsequently averaged the curves from the respondents to produce a set of "average curves," one for each pollutant variable. The resulting curves are shown in Figures 5.1 through 5.9. In each figure, the solid line represents the arithmetic mean of all respondents' curves, while the dotted lines bounding the shaded area represent the 80 per-

TABLE 5.5

EXAMPLE CALCULATIONS FOR WATER QUALITY INDEX

Variable	Measurement	I_i	W_i	I_iW_i	$I_i^{W_i}$
DO	60%	60	0.17	10.2	2.01
Fecal coliforms	10^3	20	0.15	3.0	1.57
pH	7	90	0.12	10.8	1.72
BOD$_5$	10	30	0.10	3.0	1.41
NO$_3$	10	50	0.10	5.0	1.48
PO$_4$	5	10	0.10	1.0	1.26
Temperature deviation	5	40	0.10	4.0	1.45
Turbidity	40 JTU	44	0.08	3.5	1.35
Total solids (TS)	300	60	0.08	4.8	1.39
				WQI$_a$ = 45.3	WQI$_m$ = 38.8

Subindex values are from Figures 5.1 through 5.9.
Importance weights assigned to variables = W_i.

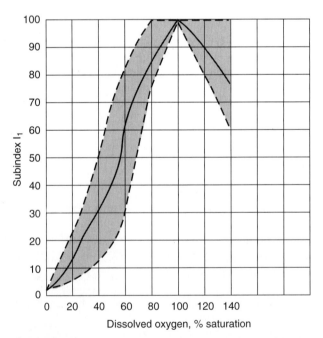

FIGURE 5.1
Subindex Function for DO in the NSF WQI (For DO
> 140%, I_1 = 50) (Ott, 1978).

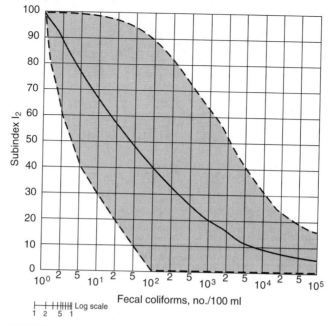

FIGURE 5.2
Subindex Function for Fecal Coliforms (average number
of organisms per 100 ml) in the NSF WQI (For fecal
coliforms > 10^5/100 ml, $I_2 = 2$) (Ott, 1978).

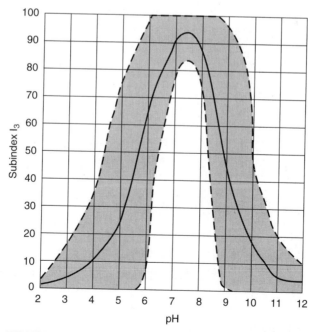

FIGURE 5.3
Subindex Function for pH in the NSF WQI (Ott, 1978).

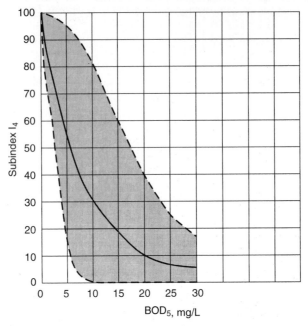

FIGURE 5.4
Subindex Function for BOD$_5$ in the NSF WQI (For BOD$_5$
> 30 mg/l, I$_4$ = 2) (Ott, 1978).

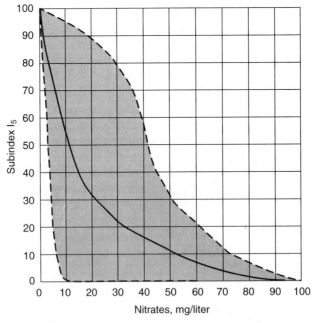

FIGURE 5.5
Subindex Function for Nitrates in the NSF WQI (For
nitrates > 100 mg/l, I$_5$ = 1) (Ott, 1978).

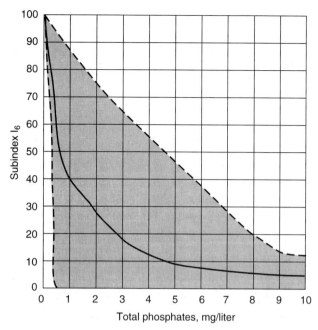

FIGURE 5.6
Subindex Function for Total Phosphates in the NSF WQI
(For total phosphates > 10 mg/l, I_6 = 2) (Ott, 1978).

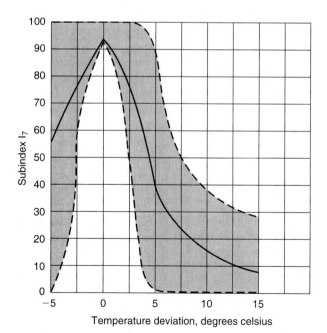

FIGURE 5.7
Subindex Function for Temperature Deviation from
Equilibrium (ΔT) in the NSF WQI (For ΔT > 15°C,
I_7 = 5) (Ott, 1978).

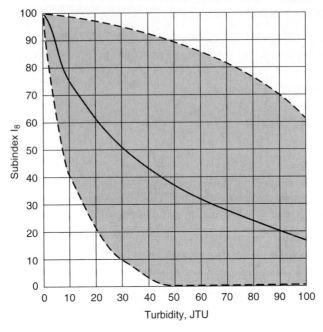

FIGURE 5.8
Subindex Function for Turbidity (Jackson Turbidity Units) in the
NSF WQI (For turbidity > 100 JTU, $I_8 = 5$) (Ott, 1978).

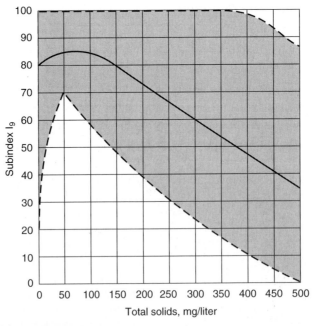

FIGURE 5.9
Subindex Function for Total Solids in the NSF WQI (For
total solids > 500 mg/l, $I_9 = 20$) (Ott, 1978).

TABLE 5.6

DESCRIPTOR WORDS AND COLORS SUGGESTED FOR REPORTING THE EXAMPLE WQI

Descriptor Words	Numerical Range	Color
Very Bad	0–25	Red
Bad	26–50	Orange
Medium	51–70	Yellow
Good	71–90	Green
Excellent	91–100	Blue

Source: Ott, 1978, p. 212.

cent confidence limits. Approximately 80 percent of the respondents' curves lie within the shaded zone. A narrow band of shading, such as the one for DO (Figure 5.1), denotes greater agreement among respondents than does a wide band, such as the one for turbidity (Figure 5.8).

To calculate the aggregate WQI, either a weighted linear sum of the subindices (WQI_a) or a weighted product-aggregation function (WQI_m) can be used. These are expressed mathematically as follows (Ott, 1978):

$$\text{NSF WQI}_a = \sum_{i=1}^{n} w_i I_i$$

$$\text{NSF WQI}_m = \prod_{i=1}^{n} I_i^{w_i}$$

An example calculation using both the WQI_a and the WQI_m is shown in Table 5.5. The interpretation of the resultant index could be based on the descriptors suggested in Table 5.6.

In summary, the steps in the application of the WQI in an impact study are as follows: (1) assemble average and extreme data for each parameter (published or monitoring); (2) use Figures 5.1 through 5.9 to determine I_i for average and extreme conditions; and (3) calculate WQI_a and/or WQI_m for average and extreme conditions, and interpret the results as appropriate.

Some general comments on the WQI are as follows: (1) it has been used in at least 17 states

in the United States (Ott, 1978); (2) conceptually similar methods are used in other countries for calculating water quality indices; and (3) the focus is on "conventional" pollutant indicators, not on toxics.

Environmental indices which are indirectly related to groundwater quality have also been developed, including indices expressing the vulnerability of aquifer systems to pollution and to the transport of pesticides through the subsurface to water-bearing zones. Simple examples of these indices are presented in Chapter 8; more-detailed information is included in Canter, Knox, and Fairchild (1987) and in Knox, Sabatini, and Canter (1993).

ENVIRONMENTAL-MEDIA INDEX— NOISE

Von Gierke et al. (1977) developed guidelines for addressing noise in environmental impact statements. In addition to generic audible-noise environments, the guidelines cover, separately, environments of high-energy impulse noise and special noises such as ultrasound and infrasound, and the environmental impact of structure-borne vibration. Whenever feasible and practical, a single-number noise-impact characterization should be used, and should be based on the concept of level-weighted population: that is, the summation over the total population of the product of each residential person times a weighting factor that varies with the yearly day-night average sound level (L_{dn}) outside the residence of that person. This single-number approach is analogous to use of a noise index. A discussion of additional means of expressing noise in relation to land usage and human-population density is in Chapter 9.

ENVIRONMENTAL-MEDIA INDEX—ECOLOGICAL SENSITIVITY AND DIVERSITY

Cooper and Zedler (1980) described an index method (basically a classification system) for evaluating the relative sensitivity to perturbations of the ecosystems in a region. The ecological sensitivity of each area or ecosystem in a region to perturbations is assessed in terms of (Cooper and Zedler, 1980) (1) significance of the ecosystem both regionally and globally, (2) rarity or abundance of the ecosystem relative to others in the region or elsewhere, and (3) the resilience of the ecosystem. Following the application and evaluation of these components, a map of ecologically sensitive areas, with accompanying narrative descriptions, is carefully prepared. The descriptive text should point out the specific features that make each area ecologically important and environmentally sensitive and should indicate the kinds of ecological disruptions that might be expected from proposed projects (Cooper and Zedler, 1980).

"Ecosystem significance" represents a subjective valuation of the biological importance of species and of the ecosystem. Table 5.7 lists the characteristics considered in the significance determination.

Rarity or abundance is usually the easiest element to measure in the sensitivity model. The area occupied by each major ecosystem, or the number of plants and animals of interest, is usually known within some acceptable limit of error or can be estimated from satellite imagery, small-scale aerial photographs, and field surveys. Other things being equal, ecosystems covering larger areas offer greater scope and flexibility in project location and design. Conversely, an ecosystem otherwise relatively insignificant may acquire greater importance simply because of its small size and the accompanying greater probability of elimination. Ecosystem sensitivity is thus related to area inversely, but not linearly (Cooper and Zedler, 1980).

"Resilience" is a measure of an ecosystem's ability to absorb environmental stress without changing to a recognizably different ecological state. It implies the ability of a system to reorganize itself under stress and to establish alternative energy-flow pathways that enable it to remain viable after perturbation, although per-

---

TABLE 5.7

CHARACTERISTICS CONSIDERED IN DETERMINING THE SIGNIFICANCE OF SPECIES AND ECOSYSTEMS

1. Role of the local ecosystem in regional ecosystem function, or importance of the species in ecosystem function.
2. Uniqueness and isolation.
3. Actual and potential aesthetic value.
4. Actual and potential scientific value.
5. Actual and potential economic value.
6. Relative size or rarity.
7. Prospects for continued persistence.

Source: Cooper and Zedler, 1980, p. 289.

haps with a modified species structure (Cooper and Zedler, 1980). As shown in Table 5.8, the degree of ecosystem or species resilience measured as a response to a given environmental stress is a composite of many partially independent reactions. Perhaps the most important indicator of a species' or ecosystem's resilience is its birthrate, or reestablishment rate. The ecological significance of a given level of mortality, from whatever cause, has to be considered in the light of the species' ability to repopulate depleted areas (Cooper and Zedler, 1980).

Four levels of ecological sensitivity were defined in the index approach of Cooper and Zedler (1980). These four levels are convenient divisions of a continuous gradient. "Minimally sensitive areas" were those already extensively disturbed by man, where additional human interference was not thought likely to induce much measurable ecological change. "Maximum sensitivity" was limited to areas where ecologically significant plants or animals are likely to be highly responsive even to slight intrusion by man, and the consequences of this impact could probably not be reduced by any practical measure to a level generally regarded as acceptable. The other two sensitivity levels ("major" and "moderate") lie between the extremes. The ecological-sensitivity levels were tested in a regional study in southern California (Cooper and Zedler, 1980).

Biological indices for water quality and/or pollution assessments are typically based on aquatic-community data and related information. An early example was the saprobic system assessment and associated saprobic indices for describing the self-purification of river systems receiving domestic wastewater discharges, described in Chapman (1992). Four types of zones associated with gradual self-purification were defined: (1) the polysaprobic (extremely severe pollution), (2) the α-mesosaprobic (severe-pollution), (3) the β-mesosaprobic (moderate-pollution), and (4) the oligosaprobic (no-pollution or very slight pollution). These zones are characterized by indicator species, certain chemical conditions, and the general nature of the bottom of the water body and of the water itself (Chapman, 1992).

Alternative approaches to the saprobic index for water pollution assessments include methods based on the presence or absence of certain indicator groups and/or indicator species at selected sampling points. Examples include the Trent biotic index and the Chandler biotic index, which are based on invertebrates (Chapman, 1992).

The aquatic-community-structure approach examines the numerical abundance of each species in a community. Methods focused on pollution are typically based on indices of community structure, either a diversity index or a similarity index. A "diversity index" expresses

---

TABLE 5.8

ECOLOGICAL RESPONSES TO ENVIRONMENTAL STRESS CONSIDERED IN EVALUATING RESILIENCE OF ECOSYSTEMS OR SPECIES POPULATIONS

1. Mortality.
2. Changes in birth rates.
3. Displacement (emigration or immigration).
4. Change in coverage, growth, or vitality of individuals.
5. Changes in behavior.
6. Disruption of ecosystem interrelationships (e.g., predator-prey interactions).

Source: Cooper and Zedler, 1980, p. 290.

TABLE 5.9

EXAMPLES OF AQUATIC DIVERSITY INDICES

Index	Calculation
Simpson Index (D)	$D = \dfrac{\sum\limits_{i=1}^{s} n_i (n_i - 1)}{n(n - 1)}$
Species deficit according to Kothé	$\dfrac{A_1 - A_x}{A_1} \times 100$
Margalef Index (D)	$D = \dfrac{S - 1}{\ln N}$
Shannon and Weaver: Shannon Index (H')	$H' = -\sum\limits_{i=1}^{s} \dfrac{n_i}{n} \ln \dfrac{n_i}{n}$
Evenness (E)	$E = \dfrac{H'}{H'_{max}}$

S The number of species in either a sample or a population
A_1 The number of species in a control sample
A_x The number of species in the sample of interest
N The number of individuals in a population or community
n The number of individuals in a sample from a population
n_i The number of individuals of species (i) in a sample from a population
Source: Chapman, 1992, p. 197.

the data on species abundance in a community as a single number. A "similarity index" is obtained by comparing two samples, one of which is often a control. Table 5.9 lists some currently used diversity indices and their models. As a general principle, greater diversity indicates a more stable aquatic community.

Habitat-based methods for biological impact assessment are basically index methods; two examples are described in Chapter 11—namely, the habitat evaluation system (HES) of the U.S. Army Corps of Engineers and the habitat evaluation procedures (HEP) of the U.S. Fish and Wildlife Service.

ENVIRONMENTAL-MEDIA INDEX— ARCHAEOLOGICAL RESOURCES

An index approach to evaluate the potential for occurrence of and impact on significant archaeological resources along eight alternative water-way-navigation routes was developed for a project in northeast–north central Oklahoma and southeast–south central Kansas (Canter, Risser, and Hill, 1974). The project involved the consideration of extending waterway navigation from Tulsa, Oklahoma, to Wichita, Kansas, and eight routes were evaluated. Detailed archaeological investigation of the eight routes was not possible; thus, 19 parameters (or factors) were used to evaluate the potential for archaeological resources. Each route was evaluated in terms of the 19 parameters. Additional information on this approach is in Chapter 12, along with descriptions of predictive models that are analogous to index approaches.

ENVIRONMENTAL-MEDIA INDEX— VISUAL QUALITY

Index methods have also been developed for evaluating both existing visual quality and the potential impact of proposed projects on visual resources (Bagley, Kroll, and Clark, 1973; Leopold, 1969; Smardon, Palmer, and Felleman, 1986). Because of the uniqueness of the terminology, no examples are included here; however, detailed information on several visual-quality indices is in Chapter 13.

ENVIRONMENTAL-MEDIA INDEX— QUALITY OF LIFE

"Quality of life" (QOL) is a term which has been developed to indicate the overall characteristics of the socioeconomic environment in a given area. In many instances, structured approaches (including indices) have been developed to describe QOL, and these approaches are included in the EIA process. Canter, Atkinson, and Leistritz (1985) provided a comprehensive review of nine such approaches (or methodologies), with the nine divided into three groups:

1. *Structured checklists*—Those approaches wherein the QOL or social well-being considerations are organized into categories and associated factors.
2. *Structured checklists with importance weights*—Those approaches wherein the categories and factors have been previously assigned relative importance weights. The purpose of the importance weighting is to allow calculation of a numerical QOL index.
3. *Structured checklists with interpretation information*—Those approaches wherein information is provided on whether the information on a given factor should be interpreted as positive or negative in terms of improving QOL. Further, it should be noted that increases in the numerical information for some factors denotes a QOL improvement (+), while for other factors increases may be negative (−), or denoting a lowered QOL.

It is beyond the scope of this chapter to present a detailed review of the nine methodologies; however, two related to water resources projects

will be briefly highlighted. For example, Guseman and Dietrich (1978) developed a structured checklist for addressing the social well-being account for water resources projects. The checklist includes 6 categories (real-income distribution; life, health, and safety; educational, cultural, and recreational opportunities and other community services; emergency preparedness; community cohesion; and other population characteristics), 20 subcategories (subdivisions of the 6 categories), and 68 specific factors. In this approach each category, subcategory, and factor is addressed for without- and with-project conditions.

As a second example, Fitzsimmons, Stuart, and Wolff (1975) described a structured checklist with importance weights for water resources projects. This approach was presented as part of a comprehensive methodology for addressing the social well-being account for such projects. Table 5.10 shows the QOL index as a weighted composite for 29 QOL dimensions. The 29 dimensions, or factors, address psychological well-being and situational concerns (economic, social, leisure, and political). Usage of this approach involves exercising considerable judgment in assigning numerical scores to the 29 dimensions. Information on the acceptability of the relative-importance weights for different types of projects in different geographical areas is not available.

Based upon a review of the nine approaches, and considering the availability of information, Canter, Atkinson, and Leistritz (1985) proposed a generic structured checklist for QOL, with this checklist based on the application of the following recommendations:

1. The approach (list of QOL factors) should be comprehensive in the use of "life domains"—QOL consists of many dimensions.
2. The approach should incorporate both perceptual and objective QOL factors (indicators)—these two basic types of indicators essentially account for different phenomena.

TABLE 5.10

QUALITY OF LIFE (QOL) INDEX

Weights	Quality of life dimensions
.10	**I. Psychological Well Being**
	1. Love, companionship
	2. Self respect
	3. Peace of mind
	4. Stimulation, challenge
	5. Other: popularity, accomplishment, individuality, sexual satisfaction, involvement, comfort, novelty, dominance, privacy, etc.
.90	**II. Situational Descriptors**
.40	Economic
.50	6. Standard of living (income per capita, discretionary income)
.10	7. Unemployment
.10	8. Financial dependency (welfare, dependency ratio)
.20	9. Housing (persons per room, home ownership, % substandard)
.05	10. Supply and distribution bottlenecks (food, fuel, commodity, etc., shortages)
.05	11. Transportation convenience (including commuting time)
.25	Social
.17	12. Family relations
.13	13. Friendships
.20	14. Job satisfaction
.05	15. Crime and violence (crime index and exposure to civil strife and political violence)
.35	16. Health, safety and nourishment (calorie and protein consumption, infant mortality, disability rate, life expectancy)
.20	17. Education (reading achievement, enrollment ratios, graduation rates)
.20	Leisure
.30	18. Media entertainment (radio, TV, movies)
.15	19. Entertainment: other (spectator sports, "night life," live programs)
.15	20. Cultural opportunities (performing arts, art objects, reading, museums, historical sights)
.20	21. Recreational facilities and areas (sports, strolling, play, picnics, etc.)
.10	22. Areas of natural beauty (park lands, landscapes, access to)
.10	23. Exposure to pollution (air, water, radiation, noise)
.15	Political
.15	24. Political participation (voting, campaigning, activism)
.15	25. News coverage
.30	26. Freedoms and civil rights
.15	27. Government responsiveness
.05	28. Public services (shortages, blackouts, interruptions, unavailability)
.20	29. Equality (income, opportunity, justice)

Source: Fitzsimmons, Stuart, and Wolff, 1975, p. 39.

TABLE 5.11

GENERIC STRUCTURED CHECKLIST FOR QOL

Category	Domain	Indicators (factors)
Basic Life Needs	Income	Household income distribution Real income Income per capita Cost of living index
	Housing	Percent of owner occupied housing units Median value, owner occupied, SFDU's Net housing starts Vacancy rate Satisfaction with housing
Well-being Needs	Employment	Unemployment rate Percent of labor force employed Satisfaction with employment and job opportunities
	Health	Infant mortality rate Communicable disease index Number of physicians per 1000 people Hospital beds per 1000 people Death rate per 1000 people Satisfaction with health care
	Safety	Crime seriousness index (per 1000 people) Number of police per 1000 population Percentage of crimes cleared by arrest Percentage of recovered stolen property Perceived safety
Opportunity Needs	Education	Public school expenditures per capita Public school tax base School enrollment Continuing education opportunity Satisfaction with education opportunities

3. Specific factors should be chosen based on local conditions—no one set of indicators should be applied across all conditions.

Table 5.11 contains the generic structured checklist proposed by Canter, Atkinson, and Leistritz (1985). The framework in Table 5.11 is adaptable to a variety of conditions. This framework is based on three components: (1) categories of quality-of-life needs, (2) domains, and (3) specific indicators (factors). Thus, any methodology to predict and assess QOL impacts should address four categories of needs—basic, well-being, opportunity, and amenity. Within each of these categories, several domains must be included. For example, basic life needs should include income and housing components; well-being needs should include employment, health, and safety components. An approach which fails to include this diversity of domains is likely to distort QOL. Within each domain, a variety of indicators can be used. Selection of these indicators should be based on local conditions, the time and resources available for data collection, and the data already available. Typically, information on most of the

TABLE 5.11

GENERIC STRUCTURED CHECKLIST FOR QOL *(continued)*

Category	Domain	Indicators (factors)
	Transportation Mobility	Ratio of miles of surfaced streets to miles of unsurfaced streets Number of traffic accidents per 1000 people Motor vehicle registrations per 1000 population Percent of workers who use public transportation to work Satisfaction with transportation
	Information	Number of books in public library per 1000 people Local Sunday newspaper circulation per 1000 population Local radio stations per 1000 people
	Equality	Ratio of white to non-white unemployment rates Ratio of male to female unemployment rates Percent of families with incomes below poverty level Perceived inequality among residents
	Participation	Percent of eligible voters Vote turnout in local elections Satisfaction with opportunity to participate
Amenity Needs	Recreation	Acres of parks and recreational areas per 1000 population Miles of trails per 1000 population Satisfaction with recreational opportunities
	Environmental Quality	Air pollution index Mean annual inversion frequency Water pollution index Noise pollution index Satisfaction with environmental quality
	Cultural Opportunities	Cultural events per 1000 people (dance, drama, music events) Fairs and festivals, annual rate Sports events Satisfaction with cultural opportunities

Source: Canter, Atkinson, and Leistritz, 1985, pp. 254–256.

objective indicators is routinely available from census records, institutional databases, chambers of commerce, and the like. However, this is not true for perceptual indicators, which may require data collection activities using tools such as new surveys or questionnaires.

Simpler indices for analyzing the social and/or socioeconomic implications of development projects have been advocated (Asian Development Bank, 1991). One example is the human development index (HDI) for developing countries. The HDI combines the three factors considered to best represent the human condition—that is, life expectancy, literacy, and income (Asian Development Bank, 1991). Many other indices of this type are used by international development banks and aid agencies.

DEVELOPMENT OF INDICES

Several generic steps are associated with the development of numerical indices or classifications of environmental quality, media vulnerability, or pollution potential of human activities. These include factor identification, assignment of importance weights, establishment of scaling functions or other methods for factor evaluation, determination and implementation of the appropriate aggregation approach, and application and field verification. Several techniques discussed in Chapter 15 can aid in the accomplishment of each of these steps, including unranked-pairwise comparisons, the nominal-group process, standards of practice, and the Delphi approach.

''Factor identification'' basically consists of delineating key factors that can be used as indicators of environmental quality, susceptibility to pollution, or the pollution potential of the source type. To serve as an example, the characterization of groundwater pollution sources—such as sanitary or chemical landfill disposal sites and liquid-waste pits, ponds, and lagoons—will be discussed. Existing methodolo-

gies generally entail consideration of the characteristics of the waste materials, or components thereof, which may be transported to the aquifer, and the characteristics of the local surface and subsurface environment relative to the inducement of attenuation of pollutant movement from the source to the aquifer. Some methodologies give consideration to the atmospheric transport of waste materials or components, and the associated human-population exposure. For example, Table 5.12 summarizes the waste-rating factors included in three empirical assessment methodologies—two for landfills and one for surface impoundments. ''Waste-rating factors'' are biological, chemical, or physical parameters which address toxicity, persistence, and mobility. To continue the example, Table 5.13 lists the site-rating factors for the three methodologies which take into account soil, groundwater, and air parameters. If a methodology is being developed for a new source type, the pollution characteristics of the source must be con-

---

TABLE 5.12

SUMMARY OF WASTE-RATING FACTORS IN EMPIRICAL ASSESSMENT METHODOLOGIES[a]

Waste rating factor	Method
1. Human toxicity	HPH, PNM
2. Ground water toxicity	HPH, PNM
3. Disease transmission potential	HPH, PNM
4. Biological persistence	HPH, PNM
5. Chemical persistence	PNM
6. Sorption properties	PNM
7. Viscosity	PNM
8. Acidity/basicity	PNM
9. Waste application rate	PNM
10. Waste mobility	HPH
11. Waste hazard potential (source/type)	LeG

[a]HPH = Hagerty, Pavoni, and Heer (1973); LeG = LeGrand (1964), and LeGrand and Brown (1977); PNM = Phillips, Nathwani and Mooij (1977).
Source: Canter, Knox, and Fairchild, 1987, p. 318.

TABLE 5.13

SUMMARY OF SITE-RATING FACTORS IN EMPIRICAL ASSESSMENT METHODOLOGIES[a]

Site rating factors	Method
A. Soil Parameters	
1. Soil permeability	HPH, LeG, PNM
2. Filtering capacity (sorption)	HPH, PNM, LeG
3. Adsorptive capacity	HPH
4. Depth to ground water table (unsaturated soil thickness)	HPH, PNM, LeG
5. Infiltration	PNM
6. Infiltration capacity (field capacity)	HPH
B. Ground Water Parameters	
7. Ground water gradient	LeG, PNM
8. Ground water velocity	HPH
9. Distance from source to point of use	LeG, PNM
10. Organic content	HPH
11. Buffering capacity	HPH
12. Ground water quality	LeG
13. Travel distance	HPH
C. Air Parameters	
14. Prevailing wind direction	HPH
15. Population factor	HPH

[a]HPH = Hagerty, Pavoni, and Heer (1973); LeG = LeGrand (1964), and LeGrand and Brown (1977); PNM = Phillips, Nathwani, and Mooij (1977).
Source: Canter, Knox, and Fairchild, 1987, p. 318.

sidered, along with the factors which could be chosen as indicators of the groundwater pollution potential.

Factor identification should be based on the collective professional judgment of knowledgeable individuals relative to the environmental-media or pollution-source category. Organized procedures such as the Delphi approach can be used to aid in the solicitation of this judgment and the aggregation of the results (Linstone and Turoff, 1975).

The second step in the development of an index is the assignment of relative-importance weights to the environmental-media and/or source-transport factors, or at least the ranking of these factors in order of importance. Some techniques which could be used to achieve this step include the Delphi approach, unranked-

pairwise comparisons, multiattribute utility measures, rank ordering, rating against a predefined scale, and the nominal-group process. Information on these techniques is included in Chapter 15; one technique will be briefly described here.

The ''nominal-group process'' technique has been used in many environmental studies. In the case of importance-weight assignments, four steps can be identified (Canter, Knox, and Fairchild, 1987): (1) nominal (silent and independent) generation of importance-weight ideas in writing by a panel of professionals, (2) round-robin listing of ideas generated by participants on a flip chart during a serial discussion, (3) discussion of each recorded idea by the group for the purposes of clarification and evaluation, and (4) independent voting on priority ideas (or importance weights), with the group decision determined by mathematical rank ordering.

Several approaches have been used to scale or evaluate the data associated with factors in index methodologies. Examples of techniques for scaling or evaluation for this purpose include the use of (1) linear scaling or categorization based on the range of data, (2) letter or number assignments designating data categories, (3) functional curves, or (4) the unranked-pairwise-comparison technique. Again, detailed information on these and other techniques is contained in Chapter 15. The development of scaling or evaluation approaches should be based on the collective professional judgment of individuals knowledgeable in areas related to the environmental-media or pollution-source category. The techniques used can be based on published approaches used by others, or on the application of structured techniques such as the nominal-group process or the Delphi approach.

Aggregation of the information on the weighted and scaled (or evaluated) factors into a final numerical index (or classification) is the important next-to-last step in the development of the index. The aggregation may include simple additions, multiplication, and/or the use of power functions. Details on the features of various aggregation approaches are in Ott (1978). As a minimum, the collective professional judgment of individuals knowledgeable in one or more substantive areas should be used.

A final step in the development of an index-classification should include field verification of its applicability. This may involve data collection and statistical testing ranging from simple to complex. At a minimum, the usability of the index should be explored in terms of data needs and availability.

SUMMARY

Environmental indicators and/or environmental indices can be useful tools in preparing a description of the environmental setting for a proposed project. These tools can aid in gathering and summarizing extant data, in communicating information on existing environmental quality, and in providing a structured basis for impact prediction and assessment. Examples of indicators and indices for various categories of the biophysical and socioeconomic environments are described in this chapter in terms of their usage and limitations.

SELECTED REFERENCES

Alberti, M., and Parker, J. D., "Indices of Environmental Quality—The Search for Credible Measures," *Environmental Impact Assessment Review,* vol. 11, no. 2, June 1991, pp. 95–101.

Asian Development Bank, "Guidelines for Social Analysis of Development Projects," Manila, The Philippines, June 1991, pp. 31–35.

Bagley, M. D., Kroll, C. A. and Clark, C., "Aesthetics in Environmental Planning," EPA 600/5-73-009, U.S. Environmental Protection Agency, Washington, D.C., Nov. 1973.

Canter, L. W., Atkinson, S. F., and Leistritz, F. L., *Impact of Growth,* Lewis Publishers, Chelsea, Mich., 1985, pp. 235–258.

———, Knox, R. C., and Fairchild, D. M., *Ground Water Quality Protection,* Lewis Publishers, Chelsea, Mich., 1987, pp. 317–321.

———, Risser, P. G., and Hill, L. G., "Effects Assessment of Alternate Navigation Routes from Tulsa, Oklahoma to Vicinity of Wichita, Kansas," University of Oklahoma, Norman, June 1974, pp. 295–341.

Chapman, D., ed., *Water Quality Assessments: A Guide to the Use of Biota, Sediments and Water in Environmental Monitoring,* Chapman and Hall, London, 1992, pp. 183–198.

Cooper, C. F., and Zedler, P. H., "Ecological Assessment for Regional Development," *Journal of Environmental Management,* vol. 10, 1980, pp. 285–296.

Fitzsimmons, S. J., Stuart, L. I., and Wolff, P. C., "A Guide to the Preparation of the Social Well-Being Account," U.S. Bureau of Reclamation, Washington, D.C., July 1975.

Guseman, P. K. and Dietrich, K. T., "Profile and Measurement of Social Well-Being Indicators for

Use in the Evaluation of Water and Related Land Management Planning," Misc. Paper Y-78-2, U.S. Army Engineer Waterways Experiment Station, Vicksburg, Miss., 1978.

Hagerty, D. J., Pavoni, J. L., and Heer, J. E., Jr., *Solid Waste Management,* Van Nostrand Reinhold, New York, 1973, pp. 242–262.

Hunsaker, C. T., and Carpenter, D. E., "Environmental Monitoring and Assessment Program—Ecological Indicators," EPA 600/3-90-060, U.S. Environmental Protection Agency, Research Triangle Park, N.C., Sept. 1990.

Knox, R. C., Sabatini, D. A., and Canter, L. W., *Subsurface Transport and Fate Processes,* Lewis Publishers, Ann Arbor, Mich., 1993, pp. 243–282.

LeGrand, H. E., "System of Reevaluation of Contamination Potential of Some Waste Disposal Sites," *Journal American Water Works Association,* vol. 56, Aug. 1964, pp. 959–974.

———, and Brown, H. S., "Evaluation of Ground Water Contamination Potential from Waste Disposal Sources," Office of Water and Hazardous Materials, U.S. Environmental Protection Agency, Washington, D.C., 1977.

Leopold, L. B., "Quantitative Comparison of Some Aesthetic Factors Among Rivers," *Geological Survey Circular 620,* U.S. Geological Survey, Washington, D.C., 1969.

Linstone, H. A. and Turoff, M., *The Delphi Method—Techniques and Applications,* Addison-Wesley Publishing Company, Reading, Mass., 1975.

Odum, E. P., *Fundamentals of Ecology,* W. B. Saunders Company, Philadelphia, 1959, pp. 142–143.

Organization for Economic Cooperation and Development, "Environmental Indicators," Organization for Economic Cooperation and Development, Paris, 1991, pp. 8–10.

Ott, W. R., *Environmental Indices: Theory and Practice,* Ann Arbor Science Publishers, Ann Arbor, Mich., 1978, pp. 2, 5, 135–169, 202–213.

Phillips, C. R., Nathwani, J. D., and Mooij, H., "Development of a Soil-Waste Interaction Matrix for Assessing Land Disposal of Industrial Wastes," *Water Research,* vol. 11, Nov. 1977, pp. 859–868.

Salazar, M. H., and Salazar, S. M., "Mussels as Bioindicators: A Case Study of Tributyltin Effects in San Diego Bay," *Proceedings of the 17th Annual Aquatic Toxicity Workshop,* Scripps Institution of Oceanography, La Jolla, Calif., Nov. 1990, pp. 47–75.

Smardon, R. C., Palmer, J. F., and Felleman, J. P., *Foundations for Visual Project Analysis,* John Wiley and Sons, New York, 1986.

Thom, G. C., and Ott, W. R., "Air Pollution Indices: A Compendium and Assessment of Indices Used in the United States and Canada," Council on Environmental Quality and U.S. Environmental Protection Agency, Washington, D.C., Dec. 1975, p. 24.

Von Gierke, H. E., et al., "Guidelines for Preparing Environmental Impact Statements on Noise," rep. of Working Group 69, National Research Council, National Academy of Sciences, Washington, D.C., June 1977.

CHAPTER 6

Prediction and Assessment of Impacts on the Air Environment

This chapter addresses basic concepts of and a methodological approach for conducting a scientifically based analysis of the potential air quality impacts of proposed projects and activities. Projects exhibiting air quality impacts include the construction and operation of fossil-fuel-fired power plants, petroleum refineries, petrochemical operations, iron and steel mills, hazardous-waste incinerators, major highways or freeways, and airports. Military training activities, controlled burning of forest areas, and land disposal of hazardous wastes can also cause air quality impact concerns.

The primary focus of this chapter is on a six-step methodological approach for air quality impact quantification. Several levels of analysis could be incorporated within the steps, with these levels primarily relating to the extent of potential air quality effects and the necessary attention which must be given to them in an EA or EIS. Information on air quality concerns, relevant ambient air quality standards, and the key requirements of the 1990 Clean Air Act is also included in this chapter.

BASIC INFORMATION ON AIR QUALITY ISSUES

To systematically address the air quality impacts of potential projects or activities, it is necessary to be familiar with basic information regarding air pollution. Accordingly, this section includes a definition of air pollution and information on the types and effects of specific air pollutants, along with summary information on air pollution sources.

Air Pollution

"Air pollution" can be defined as the presence in the outdoor atmosphere of one or more contaminants (pollutants) in such quantities and of such duration as may be (or may tend to be) injurious to human, plant, or animal life, or to property (materials), or which may unreasonably interfere with the comfortable enjoyment of life or property, or the conduct of business. It should be stressed that the attention in this definition is on the outdoor, or ambient, air, as opposed to the indoor, or work environment, air. This chapter is focused upon air quality analysis with regard to the ambient atmosphere.

Air pollution can be caused by the presence of one or more contaminants. Examples of traditional contaminants include sulfur dioxide, oxides of nitrogen, carbon monoxide, hydrocarbons, ozone, oxidants, hydrogen sulfide, particulate matter, smoke, and haze. This list of air pollutants can be divided into two categories: gases and particulates. Gases, such as sulfur dioxide and oxides of nitrogen, exhibit diffusion properties and are normally formless fluids which can be changed to the liquid or solid state only by a combined effect of increased pressure and decreased temperature. Particulates represent any dispersed matter, solid or liquid, in which the individual aggregates are larger than single small molecules (about 0.0002 micrometers, μm, in diameter) but smaller than about 500 μm. (One micrometer is 10^{-4} cm.) Of recent attention is particulate matter which is equal to or less than 10 μm in size, with this size range of concern relative to potential human health effects. Additionally, particles in the atmosphere can have a lifetime ranging from only a few minutes to several months; larger particles settle more rapidly than smaller particles.

Of recent importance is the focus on "air toxics," or hazardous air pollutants. Air toxics are a class of compounds which may be present in the atmosphere and exhibit potentially toxic effects not only to humans but also to the overall ecosystem. In the 1990 Clean Air Act, the air toxics category includes 189 specific chemicals which may be of relevance in an air quality impact study (Quarles and Lewis, 1990). This group of compounds represents typical compounds of concern in the industrial air environment, with these compounds and associated relevant quality standards being adjusted for outdoor atmospheric conditions.

The above definition also mentions the quantity or concentration of the contaminant in the atmosphere, and its associated duration or period of occurrence. This is an important concept in that pollutants that are present at extremely low concentrations and for short time periods may be insignificant in terms of the planning and conduction of an impact study.

Additional air pollutants or atmospheric effects which have become of concern include photochemical smog, acid rain, and global warming. "Photochemical smog" refers to the formation of oxidizing constituents, such as ozone, in the atmosphere as a result of the photo-induced reaction of hydrocarbons (or volatile organic chemicals) and oxides of nitrogen. This phenomenon was first recognized in Los Angeles, California, in the period following World War II, and it has become a major air-pollution concern throughout the United States. "Acid rain" refers to atmospheric reactions which can lead to precipitation which exhibits a pH value less than the normal pH of rainfall (which is approximately 5.7 when carbon dioxide equilibrium is considered). Within recent years, in central Europe and in several Scandinavian countries, along with Canada and the northeastern United States, attention has been directed to potential environmental consequences of acid precipitation. Causative agents in acid-rain formation are typically associated with sulfur dioxide emissions and, possibly, nitrogen oxide emissions, along with gaseous hydrogen chloride. From a worldwide perspective, sulfur dioxide emissions are the dominant precursor of acid-rain formation.

Another issue of importance from a global perspective is the influence of air pollution on atmospheric heat balances and associated absorption or reflection of incoming solar radiation. As a result of increasing levels of carbon dioxide and other carbon-containing compounds in the atmosphere, there is a growing concern that the earth's surface has already started exhibiting increased temperature levels, and this, in turn, can have major implications in terms of shifting climatic conditions throughout the world.

Sources of Air Pollutants

Air pollutant sources can be categorized from several perspectives, including the type of source, their frequency of occurrence and spatial distribution, and the types of emissions. Characterization by source type can be delineated as arising from natural sources or from man-made sources. ''Natural sources'' include plant pollens, windblown dust, volcanic eruptions, and lightning-generated forest fires. ''Man-made sources'' can include transportation vehicles, industrial processes, power plants, construction activities, and military training activities.

Source characterization according to number and spatial distribution can include such categories as single or point sources (stationary), area or multiple sources (stationary or mobile), and line sources. ''Point sources'' are characteristic of pollutant emissions from industrial-process stacks, as well as fuel-combustion facility stacks. ''Area sources'' include vehicular traffic, fugitive-dust emissions from resource-material stockpiles or construction, or military training activities over large geographical areas. Figure 6.1 includes one delineation of source categories which can be used for analyzing air pollutant sources in a given geographical area.

Effects of Air Pollutants

Air pollution effects may also be divided into several categories, with such effects encompassing those that are health-related as well as those associated with damage to property or materials or which cause decreases in atmospheric aesthetic features. Examples of effects on human health include eye irritation, headaches, and aggravation of respiratory difficulties. Plants and crops have been subjected to the undesirable consequences of air pollution, including abnormal growth patterns, leaf discoloration or spotting, and death. Animals such as cattle have been subjected to undesirable consequences of atmospheric fluorides. Property and materials damages include property devaluation because of odors, deterioration of materials such as concrete statuary, and discoloration of painted surfaces on cars, buildings, and bridge structures.

The aesthetic effects include reductions in visibility, discoloration of air, photochemical–smog–related traffic disruptions at airports, and the general nuisance aspects of odors and dust.

KEY FEDERAL LEGISLATION AND REGULATIONS

In November 1990, the Clean Air Act was amended and passed by the U.S. Congress with the concurrence of the President. Examples of detailed standards, regulations, and information developed from the Clean Air Act of 1990, or its precursors, and included in the Code of Federal Regulations (C.F.R.) are the national primary and secondary air quality standards (Part 50), state implementation plans (Part 51), ambient air monitoring (Parts 53 and 58), national emission standards for hazardous air pollutants (Part 61), regulation of fuels and fuel additives (Part 80), control of air pollution from motor vehicles and motor vehicle engines (Part 85), and control of air pollution from aircraft and aircraft engines (Part 87) (U.S. EPA, 1991c). Part 60 contains standards of performance for new stationary sources, including emissions standards for a large number of source types, fossil-fuel-fired steam generators, incinerators, nitric acid plants, petroleum refineries, primary and secondary lead smelters, sewage-treatment plants, phosphate fertilizer plants, Kraft pulp mills, bulk gasoline terminals, and many others (U.S. EPA, 1991c).

Table 6.1 includes the list of national ambient air quality standards that are in effect in the United States as of 1992. The primary standards are focused on preventing any adverse impacts on human health. Additional standards related to ambient air quality in the United States include those being developed for hazardous air pollutants.

Emission standards, also called ''new source performance standards,'' have also been developed and promulgated for more than 60 cate-

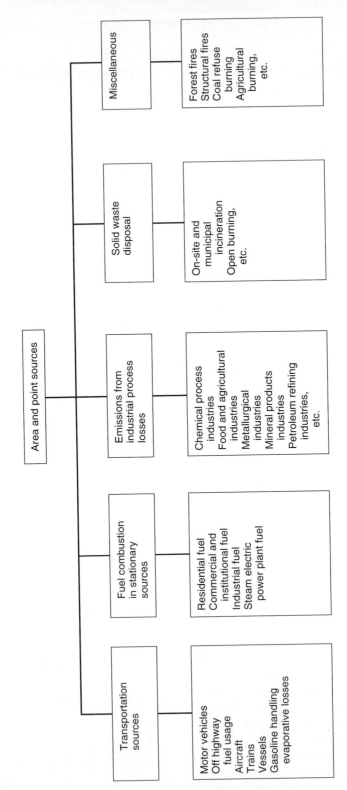

FIGURE 6.1
Source Categories for Emission Inventories (U.S. Environmental Protection Agency, 1972).

TABLE 6.1

NATIONAL AMBIENT AIR QUALITY STANDARDS (NAAQS)

Pollutant	Primary (health related)		Secondary (welfare related)	
	Type of average	Standard level concentration[a]	Type of average	Standard level concentration[a]
CO	8-hour[b]	9 ppm (10 mg/m³)	No Secondary Standard	
	1-hour[b]	35 ppm (40 mg/m³)	No Secondary Standard	
NO_2	Annual Arithmetic Mean	0.053 ppm (100 µg/m³)	Same as Primary Standard	
O_3	Maximum Daily 1-hour Average[c]	0.12 ppm (235 µg/m³)	Same as Primary Standard	
Pb	Maximum Quarterly Average	1.5 µg/m³	Same as Primary Standard	
PM-10	Annual Arithmetic Mean[d]	50 µg/m³	Same as Primary Standard	
	24-hour[d]	150 µg/m³	Same as Primary Standard	
SO_2	Annual Arithmetic Mean	80 µg/m³ (0.03 ppm)	3-hour[b]	1300 µg/m³ (0.50 ppm)
	24-hour[b]	365 µg/m³		

[a]Parenthetical value is an approximately equivalent concentration.
[b]Not to be exceeded more than once per year.
[c]The standard is attained when the expected number of days per calendar year with maximum hourly average concentrations above 0.12 ppm is equal to or less than 1, as determined according to Appendix H of the Ozone NAAQS in the Code of Federal Regulations.
[d]Particulate standards use PM-10 (particles less than 10µ in diameter) as the indicator pollutant. The annual standard is attained when the expected annual arithmetic mean concentration is less than or equal to 50 µg/m³; the 24-hour standard is attained when the expected number of days per calendar year above 150 µg/m³ is equal to or less than 1, as determined according to Appendix K of the PM NAAQS in the Code of Federal Regulations.
Source: U.S. Environmental Protection Agency, 1992, p. 3.

gories of industrial sources, with this information being relevant in EISs related to these source categories. The majority of these new source performance standards were developed in the time period from approximately 1977 through 1990. The Clean Air Act of 1977 was the basic air-pollution legislation in force in the United States over a 13-year period, until it was superseded by the 1990 amendments. Tables 6.2 and 6.3 illustrate emissions standards for fossil-fuel-fired steam generators and petroleum refineries, respectively.

An additional relevant issue is directed toward regulation of hazardous air pollutants, also

referred to as "air toxics." This program, created by Section 112 of the act, represents a major new emphasis in air pollution regulation; it has relevance to many projects and activities in that a number of specific hazardous air pollutants are addressed by the new law. Table 6.4 includes a list of 189 substances regulated by the U.S. Environmental Protection Agency as air toxics under Section 112. Regulation includes the establishment of emission standards, ambient air quality standards, and control technologies. The Clean Air Act of 1990 also includes a number of additional source categories for which the U.S. Environmental Protection

TABLE 6.2

EMISSION STANDARDS* FOR FOSSIL FUEL-FIRED STEAM GENERATORS

Pollutant	Standards

Particulate matter	1. No owner or operator shall cause to be discharged into the atmosphere from any affected facility any gases which
	   a. Contain particulate matter in excess of 43 nanograms per joule, (ng/J) heat input (0.10 lb per million Btu) derived from fossil fuel or fossil fuel and wood residue
	   b. Exhibit greater than 20 percent opacity except for one 6-minute period per hour or not more than 27 percent opacity

Sulfur dioxide	1. No owner or operator shall cause to be discharged into the atmosphere from any affected facility any gases which contain sulfur dioxide in excess of
	   a. 340 ng/J heat input (0.80 lb per million Btu) derived from liquid fossil fuel or liquid fossil fuel and wood residue
	   b. 520 ng/J heat input (1.2 lb per million Btu) derived from solid fossil fuel or solid fossil fuel and wood residue
	2. When different fossil fuels are burned simultaneously in any combination, the applicable standard (in nanograms per joule) shall be determined by proration using the following formula: $$PS_{so_2} = [y(340) + z(520)]/(y + z)$$ where PS_{so_2} = prorated standard for sulfur dioxide when burning different fuels simultaneously, heat input derived from all fossil fuels fired or from all fossil fuels and wood residue fired, ng/J
	
	y = total heat input derived from liquid fossil fuel, %
	
	z = total heat input derived from solid fossil fuel, %

Nitrogen oxides	1. No owner or operator shall cause to be discharged into the atmosphere from any affected facility any gases which contain nitrogen oxides, expressed as NO_2 in excess of
	   a. 86 ng/J heat input (0.20 lb per million Btu) derived from gaseous fossil fuel
	   b. 129 ng/J heat input (0.30 lb per million Btu) derived from liquid fossil fuel, liquid fossil fuel and wood residue, or gaseous fossil fuel and wood residue
	   c. 300 ng/J heat input (0.70 lb per million Btu) derived from solid fossil fuel or solid fossil fuel and wood residue (except lignite or a solid fossil fuel containing 25%, by weight, or more of coal refuse)
	   d. 260 ng/J heat input (0.60 lb per million Btu) derived from lignite or lignite and wood residue (except as provided below) [item 1(e)]
	   e. 340 ng/J heat input (0.80 lb per million Btu) derived from lignite which is mined in North Dakota, South Dakota, or Montana and which is burned in a cyclone-fired unit
	2. Except as provided in items 3 and 4, below, when different fossil fuels are burned simultaneously in any combination, the applicable standard (in nanograms per joule) is determined by proration using the following formula: $$PS_{NOx} = \frac{w(260) + x(86) + y(130) + z(300)}{w + x + y + z}$$ where PS_{NOx} = prorated standard for nitrogen oxides when burning different fuels simultaneously, in ng/J heat input derived from all fossil fuels fired or from all fossil fuels and wood residue fired
	
	w = total heat input derived from lignite, %
	
	x = total heat input derived from gaseous fossil fuel, %
	
	y = total heat input derived from liquid fossil fuel, %
	
	z = total heat input derived from solid fossil fuel (except lignite), %
	
	3. When a fossil fuel containing at least 25%, by weight, of coal refuse is burned in combination with gaseous, liquid, or other solid fossil fuel or wood residue, the standard for nitrogen oxides does not apply.
	4. Cyclone-fired units which burn fuels containing at least 25% of lignite that is mined in North Dakota, South Dakota, or Montana remain subject to 1 (e) above, regardless of the types of fuel combusted in combination with that lignite.

*Standards are abstracted from Sections 60.42, 60.43, and 60.44 of the Code of Federal Regulations (C.F.R.) Vol. 40, Subpart C.
Source: Compiled using data from U.S. Environmental Protection Agency, 1991c.

TABLE 6.3

EMISSION STANDARDS* FOR PETROLEUM REFINERIES

Pollutant	Standards

Particulate matter	1. No owner or operator shall discharge or cause the discharge into the atmosphere from any fluid catalytic-cracking-unit catalyst regenerator of
	   a. Particulate matter in excess of 1.0 kg/1,000 kg (1.0 lb/1,000 lb) of coke burn-off in the catalyst regenerator
	   b. Gases exhibiting greater than 30% opacity, except for one 6-min average opacity reading in any 1-hr period
	2. Where the gases discharged by the fluid catalytic-cracking-unit catalyst regenerator passed through an incinerator or waste-heat boiler in which auxiliary or supplemental liquid or solid fossil fuel is burned, particulate matter in excess of that permitted by item 1 (a) above may be emitted to the atmosphere, except that the incremental rate of particulate-matter emissions shall not exceed 43.0 g/MJ (0.10 lb/million Btu) of heat input attributable to such liquid or solid fossil fuel

Carbon monoxide	No owner or operator shall discharge or cause the discharge into the atmosphere from any fluid catalytic-cracking-unit catalyst regenerator any gases that contain carbon monoxide (CO) in excess of 500 ppm by volume (dry basis).

Sulfur oxides	1. No owner or operator shall
	   a. Burn in any fuel-gas combustion device any fuel gas that contains hydrogen sulfide (H_2S) in excess of 230 mg/dscm [dry standard cubic meter] (0.10 gr/dscf) [dry standard cubic foot]. The combustion in a flare of process-upset gases or fuel gas that is released to the flare as a result of relief-valve leakage or other emergency malfunctions is exempt
	   b. Discharge or cause the discharge of any gases into the atmosphere from any Claus sulfur-recovery plant containing in excess of
	      i. For an oxidation-control system or a reduction-control system followed by incineration, 250 ppm by volume (dry basis) of sulfur dioxide (SO_2) at 0% excess air
	      ii. For a reduction-control system not followed by incineration, 300 ppm by volume of reduced sulfur compounds and 10 ppm by volume of hydrogen sulfide (H_2S), each calculated as ppm SO_2, by volume (dry basis) at 0% excess air
	2. Each owner or operator shall comply with one of the following conditions for each affected fluid catalytic-cracking-unit catalyst regenerator:
	   a. With an add-on control device, reduce sulfur dioxide emissions to the atmosphere by 90 percent or maintain sulfur dioxide emissions to the atmosphere less than or equal to 50 ppm by volume (vppm), whichever is less stringent.
	   b. Without the use of an add-on control device, maintain sulfur oxides emissions calculated as sulfur dioxide to the atmosphere less than or equal to 9.8 kg/1,000 kg coke burn-off.
	   c. Process in the fluid catalytic-cracking-unit fresh feed that has a total sulfur content no greater than 0.30% by weight.

*Standards are abstracted from Sections 60.102, 60.103, and 60.104 of the Code of Federal Regulations, Part 40, Subpart C.
Source: Compiled using data from U.S. Environmental Protection Agency, 1991c.

TABLE 6.4

LIST OF HAZARDOUS AIR POLLUTANTS TO BE REGULATED UNDER SECTION 112 OF THE CLEAN AIR ACT

Chemical name		
Acetaldehyde	2,4-D, salts and esters	Hexachlorobenzene
Acetamide	DDE	Hexachlorobutadiene
Acetonitrile	Diazomethane	Hexachlorocyclopentadiene
Acetophenone	Dibenzofurans	Hexachloroethane
2-Acetylaminofluorene	1,2-Dibromo-3-chloropropane	Hexamethylene-1,6-diisocyanate
Acrolein	Dibutylphthalate	Hexamethylphosphoramide
Acrylamide	1,4-Dichlorobenzene(*p*)	Hexane
Acrylic acid	3,3-Dichlorobenzidene	Hydrazine
Acrylonitrile	Dichloroethyl ether	Hydrochloric acid
Allyl chloride	[Bis(2-chloroethyl)ether]	Hydrogen fluoride
4-Aminodiphenyl	1,3-Dichloropropene	(hydrofluoric acid)
Aniline	Dichlorvos	Hydroquinone
o-Anisidine	Diethanolamine	Isophorone
Asbestos	*N,N*-Diethylaniline	Lindane (all isomers)
Benzene	(*N,N*-dimethylaniline)	Maleic anhydride
(including benzene from gasoline)	Diethyl sulfate	Methanol
Benzidine	3,3-Dimethoxybenzidine	Methoxychlor
Benzotrichloride	Dimethyl aminoazobenzene	Methyl bromide (bromomethane)
Benzyl chloride	3,3'-Dimethyl benzidine	Methyl chloride (chloromethane)
Biphenyl	Dimethyl carbamoyl chloride	Methyl chloroform
Bis(2-ethylhexyl)phthalate (DEHP)	Dimethyl formamide	(1,1,1-trichloroethane)
Bis(chloromethyl)ether	1,1-Dimethyl hydrazine	Methyl ethyl ketone (2-butanone)
Bromoform	Dimethyl phthalate	Methyl hydrazine
1,3-Butadiene	Dimethyl sulfate	Methyl iodide (iodomethane)
Calcium cyanamide	4,6-Dinitro-*o*-cresol, and salts	Methyl isobutyl ketone (hexone)
Caprolactam	2,4-Dinitrophenol	Methyl isocyanate
Captan	2,4-Dinitrotoluene	Methyl methacrylate
Carbaryl	1,4-Dioxane (1,4-diethyleneoxide)	Methyl tert butyl ether
Carbon disulfide	1,2-Diphenylhydrazine	4,4-Methylene bis (2-chloroaniline)
Carbon tetrachloride	Epichlorohydrin	Methylene chloride
Carbonyl sulfide	(1-chloro-2,3-epoxypropane)	(dichloromethane)
Catechol	1,2-Epoxybutane	Methylene diphenyl diiocyanate
Chloramben	Ethyl acrylate	(MDI)
Chlordane	Ethyl benzene	4,4'-Methylenedianiline
Chlorine	Ethyl carbamate (urethane)	Naphthalene
Chloroacetic acid	Ethyl chloride (chloroethane)	Nitrobenzene
2-Chloroacetophenone	Ethylene dibromide	4-Nitrobiphenyl
Chlorobenzene	(dibromoethane)	4-Nitrophenol
Chlorobenzilate	Ethylene dichloride	2-Nitropropane
Chloroform	(1,2-dichloroethane)	*N*-Nitroso-*N*-methylurea
Chloromethyl methyl ether	Ethylene glycol	*N*-Nitrosodimethylamine
Chloroprene	Ethylene imine (aziridine)	*N*-Nitrosomorpholine
Cresols or cresylic acid	Ethylene oxide	Parathion
(isomers and mixture)	Ethylene thiourea	Pentachloronitrobenzene
o-Cresol	Ethylidene chloride	(quintobenzene)
m-Cresol	(1,1-dichloroethane)	Pentachlorophenol
p-Cresol	Formaldehyde	Phenol
Cumene	Heptachlor	*p*-Phenylenediamine

TABLE 6.4

LIST OF HAZARDOUS AIR POLLUTANTS TO BE REGULATED UNDER SECTION 112 OF THE CLEAN AIR ACT (continued)

Chemical name		
Phosgene	Tetrachloroethylene	Xylenes (isomers and mixture)
Phosphine	(perchloroethylene)	o-Xylenes
Phosphorus	Titanium tetrachloride	m-Xylenes
Phthalic anhydride	Toluene	p-Xylenes
Polychlorinated biphenyls	2,4-Toluene diamine	Antimony compounds
(aroclors)	2,4-Toluene diisocyanate	Arsenic compounds (inorganic,
1,3-Propane sultone	o-Toluidine	including arsine)
beta-Propiolactone	Toxaphene (chlorinated	Beryllium compounds
Propionaldehyde	camphene)	Cadmium compounds
Propoxur (baygon)	1,2,4-Trichlorobenzene	Chromium compounds
Propylene dichloride (1,2-	1,1,2-Trichloroethane	Cobalt compounds
dichloropropane)	Trichloroethylene	Coke oven emissions
Propylene oxide	2,4,5-Trichlorophenol	Cyanide compounds[a]
1,2-Propylenimine (2-methyl	2,4,6-Trichlorophenol	Glycol ethers[b]
aziridine)	Triethylamine	Lead compounds
Quinoline	Trifluralin	Manganese compounds
Quinone	2,2,4-Trimethylpentane	Mercury compounds
Styrene	Vinyl acetate	Fine mineral fibers[c]
Styrene oxide	Vinyl bromide	Nickel compounds
2,3,7,8-Tetrachlorodibenzo-p-	Vinyl chloride	Polycylic organic matter[d]
dioxin	Vinylidene chloride	Radionuclides (including radon)[e]
1,1,2,2-Tetrachloroethane	(1,1-dichloroethylene)	Selenium compounds

[a]X'CN where X = H' or any other group where a formal dissociation may occur.
[b]Includes mono- and di-ethers or ethylene glycol, diethylene glycol, and triethylene glycol R-(OCH2CH2) n-OR'
where n = 1, 2, or 3; R' = alkyl or aryl groups; R' = R, H, or groups which, when removed, yield glycol ethers with the structure
R-(OCH2CH)n-OH. Polymers are excluded from the glycol category.
[c]Includes mineral fiber emissions from facilities manufacturing or processing glass, rock, or slag fibers (or other mineral-derived fibers) of
average diameter 1 μm or less.
[d]Includes organic compounds with more than one benzene ring, and which have a boiling point greater than or equal to 100°C.
[e]A type of atom which spontaneously undergoes radioactive decay.
Source: Adapted from Quarles and Lewis, 1990, pp. 86–88.

Agency may establish maximum-achievable control technology (MACT) standards under Section 112.

Additional provisions of the Clean Air Act of 1990 which may have relevance for impact studies include the nonattainment provisions (Title I), reductions of vehicular emissions through the usage of alternative fuels (Title II), control of acid rain, sulfur dioxide, and/or nitrogen dioxide through the use of appropriate technologies and allowance trading (Title IV), permitting requirements (Title V), and reductions in chlorofluorocarbon emissions to protect the ozone layer (Title VI).

The concluding point with regard to legislation related to air quality is that this subject area is dynamic; regulations are forthcoming to supplement those which are already in force. State programs must be as stringent as the federal requirements, and they can even be more stringent depending upon the pollutants and circumstances. Therefore, for the appropriate conduction of air quality impact studies, it is necessary to keep up-to-date in this subject area.

Step 1:   Identification of air quality impacts of proposed project

                    ↓

Step 2:   Description of existing air environment conditions

                    ↓

Step 3:   Procurement of relevant air quality standards and/or guidelines

                    ↓

Step 4:   Impact prediction

                    ↓

Step 5:   Assessment of impact significance

                    ↓

Step 6:   Identification and incorporation of mitigation measures

FIGURE 6.2
Conceptual Approach for Study Focused on Air Environment Impacts.

CONCEPTUAL APPROACH FOR ADDRESSING AIR ENVIRONMENT IMPACTS

To provide a basis for addressing air environment impacts, a six-step or six-activity model is suggested for the planning and conduct of impact studies. This model is flexible and can be adapted to various project types by modification, as needed, to enable the addressing of specific concerns of specific projects in unique locations. The identified steps are typical of air impact studies. It should be noted that the focus in this model will be on projects and their air quality impacts; however, the model could also be applied to plans, programs, policies, and regulatory (or permit) actions.

The six generic steps associated with air environment impacts are (1) identification of air pollutant emissions and impact concerns related to the construction and operation of the development project; (2) description of the environmental setting in terms of existing ambient air quality, emission inventory, and meteorological data in the project study area; (3) procurement of relevant laws, regulations, or criteria related to ambient air quality and/or pollutant emission standards; (4) conduction of impact prediction activities, including the use of mass balances, simple dilution calculations, comprehensive mathematical models, and/or qualitative predictions based on case studies and professional judgment; (5) use of pertinent information from step 3, along with professional judgment and public input, to assess the significance of anticipated beneficial and detrimental impacts; and (6) identification, development, and incorporation of appropriate mitigation measures for the adverse impacts. Figure 6.2 delineates the relationship between the six steps or activities in the suggested conceptual approach.

The six-step model can be used to plan a study focused on air quality impacts, to develop the scope of work for such a study, and/or to review air quality impact information in EAs or EISs.

STEP 1: IDENTIFICATION OF THE TYPES AND QUANTITIES OF AIR POLLUTANTS AND OF THEIR IMPACTS

An appropriate initial step when analyzing any proposed project-activity is to consider what types of air pollutants might be emitted during the construction and/or operational phases of the proposed project-activity, and the quantities in which such air pollutants are expected to occur. Use of emission-factor information organized according to project type or activity is a suggested approach. An ''emission factor'' is the average rate at which a pollutant is released into the atmosphere as a result of some activity, such as combustion or industrial production, divided by the level of that activity (U.S. EPA, 1973). "Emission factors" relate the types and quantities of pollutants emitted to indicators such as production capacity, quantity of fuel burned, or vehicle-miles traveled by an automobile.

A considerable body of information exists on emission factors for a variety of projects and associated activities. Emission-factor information has been developed using techniques (such as source testing) involving many measurements related to multiple-process variables or single measurements not clearly defined in relationship to process operating conditions, the preparation of process material balances, and engineering appraisals and professional judgment (U.S. EPA, 1973). To indicate the accuracy of the factors assessed for a specific process, each factor-process variable is rated as A, B, C, D, or E. For a process with an A rating, the emission factor is considered excellent, based on field measurements of a large number of sources. A process rated B is considered above-average, based on a limited number of field measurements. A rating of C is considered average; D, below-average; and E, poor.

The key information source for emission factors is Publication AP-42 of the U.S. Environmental Protection Agency; this publication was originally distributed in 1973. Several subsequent editions and supplements to the original AP-42 compendium have been published. Volume 1 of the most recent edition addresses stationary point and area sources such as fuel combustion, combustion of solid waste, evaporation of fuels and solvents, industrial processes, and miscellaneous sources (U.S. EPA, 1985). Supplements to Volume 1 were issued in 1986, 1988, 1990, 1991, and 1992 (it is important to utilize the most-recent information in delineating pertinent emission factors for a project). Volume 2 addresses mobile sources, with the

1991 edition revising previous emission factors for highway mobile sources (U.S. EPA, 1991a). Emission-factor information is provided for eight vehicle types and the highway vehicle fleet as a whole for several conditions, such as calendar year, average speed, temperature, fuel volatility, and operating modes. The primary point of contact for acquisition of the AP-42 publications is the U.S. Environmental Protection Agency, located in Research Triangle Park, North Carolina. Other sources of emission-factor information are publications of the U.S. Army, several of which have been developed by and/or for the U.S. Army Construction Engineering Research Laboratory; one specific illustration is described in Schanche et al. (1976).

Information on several types of emission factors will be presented as illustrations. In order to know the types of emission factors to look for, an initial task should be the systematic identification and listing of pertinent construction and/or operational activities that would generate air pollutants. Following the development of this list, specific emission-factor information for each pertinent source type could be procured. Examples of potential air pollution sources associated with U.S. Army installations are delineated in Table 6.5.

An air pollution source common to many projects and activities are unpaved roads. Dust plumes behind vehicles moving along unpaved roads represent a typical occurrence, since as the vehicle travels over an unpaved road, the force of the wheels on the road surface causes pulverization of the surface material. Particles are lifted and dropped from the rolling wheels, and the road surface is exposed to strong air currents in turbulent shear with the surface. The turbulent wake (behind the vehicle) continues to act on the road surface after the vehicle has passed (U.S. EPA, 1975). As an approximation, fugitive dust (dust generated from unpaved roads is termed "fugitive dust" because it is not discharged into the atmosphere in a confined-flow stream) from unpaved roads can be considered

TABLE 6.5

EXAMPLES OF RELEVANT AIR POLLUTANT SOURCES ASSOCIATED WITH U.S. ARMY INSTALLATIONS

1. External Combustion Sources
   1.1 Bituminous Coal Combustion
   1.2 Anthracite Coal Combustion
   1.3 Fuel Oil Combustion
   1.4 Natural Gas Combustion
   1.5 Liquified Petroleum Gas Combustion
   1.6 Wood Waste Combustion in Boilers
   1.7 Lignite Combustion

2. Solid Waste Disposal
   2.1 Refuse Incineration
   2.4 Open Burning
   2.5 Sewage Sludge Incineration

3. Internal Combustion Engine Sources
   3.1 Highway Vehicles
   3.2 Off-Highway, Mobile Sources
   3.3 Off-Highway, Stationary Sources

4. Evaporation Loss Sources
   4.1 Dry Cleaning
   4.2 Surface Coating
   4.3 Petroleum Storage
   4.4 Gasoline Marketing

8. Mineral Products Industry
   8.1 Asphaltic Concrete Plants
   8.9 Coal Cleaning
   8.10 Concrete Batching
   8.19 Sand and Gravel Processing
   8.20 Stone Quarrying and Processing

11. Miscellaneous Sources
    11.1 Forest Wildfires
    11.2 Fugitive Dust Sources

Appendix A. Miscellaneous Data
Appendix B. Projected Emission Factors for Highway Vehicles

Note: Numbers correspond to sections in U.S. Environmental Protection Agency Publ. AP-42 (1973) and supplements.
Source: Schanche et al., 1976, p. 9.

to average about 75 lb per vehicle-mile of travel (Hesketh and Cross, 1981).

The specific quantity of dust emissions from a given segment of unpaved road varies linearly with the volume of traffic. In addition, emissions depend upon the correction parameters (average vehicle speed, vehicle mix, surface texture, and surface moisture) that characterize the condition of a particular road and the associated vehicular traffic (U.S. EPA, 1975). In the typical speed range on unpaved roads—that is, 30 to 50 mi/hr (50 to 80 km/hr), the results of field measurements indicate that emission quantity is directly proportional to vehicle speed. Limited field measurements further indicate that vehicles produce dust from an unpaved road in proportion to the number of wheels. For roads with a significant volume of vehicles with six or more wheels, the traffic volume should be adjusted to equal the volume of an equal number of four-wheeled vehicles.

In addition, dust emissions from unpaved roads have been found to vary in direct proportion to the fraction of "silt" (that is, particles smaller than 75 μm in diameter, as defined by the American Association of State Highway Officials) in the road surface material. The silt fraction is determined by measuring the proportion of loose, dry surface dust that passes a 200-mesh screen. The silt content of gravel roads averages about 12 percent, and the silt content of a dirt road may be estimated based on the silt content of the parent soil in the area. Unpaved roads have a hard, nonporous surface that dries quickly after a rainfall. The temporary reduction in emissions because of rainfall may be accounted for by neglecting emissions on "wet days"—that is, days with more than 0.01 in (0.254 mm) of rainfall. The quantity of fugitive-dust emissions from an unpaved road, per vehicle-mile of travel, may be estimated (within ±20 percent) using the following empirical expression (U.S. EPA, 1975):

$$E = (0.81s)\left(\frac{S}{30}\right)\left(\frac{365 - w}{365}\right)$$

where E = emission factor, lb per
    vehicle-mile
  s = silt content of road
    surface material, %
  S = average vehicle speed, mi/hr
  w = mean annual number of
    days with 0.01 in
    (0.254 mm) or more of rainfall

The equation is valid for vehicle speeds in the range of 30 to 50 mi/hr (50 to 80 km/hr).

Emission factors for refuse incinerators, along with several other types of incinerators,

are shown in Table 6.6. The emission factors in Table 6.6 are rated A, with this being considered excellent, since the data are from field measurements of a large number of sources. In using emission-factor information it is desirable to state the rating category (if it is known).

Emission-factor information is also available for six Superfund-site remediation technologies (thermal treatment, air stripping, soil vapor extraction, solidification and stabilization, physical and chemical treatment, and biotreatment and land treatment). The pollutants addressed include volatile organic compounds, metals, particulate matter, sulfur dioxide, nitrogen oxides, carbon monoxide, hydrochloric acid, and hydrofluoric acid (Thompson, Inglis, and Eklund, 1991). An emissions model for surface impoundments is also available in Watkins (1989).

A topic of growing importance is emission-factor information for toxic (hazardous) air pollutants (U.S. EPA, 1988c). Reference texts for this topic and other example pollution sources are emerging. Of relevance is the reference by Pope, Cruse, and Most (1988), which provides emission factors for a number of toxic air pollutants and sources. The emission factors in the report were compiled from a review of the literature for more than 200 air toxics compounds. Also included are brief descriptions of emission-factor derivations, notes on control measures associated with factors, and references (Pope, Cruse, and Most, 1988). A software system containing the air-toxics emission factors in the compilation was developed to enable easy access and updating of the data (Radian Corporation, 1988; Pope et al., 1990).

Smokes and obscurants are often used in military training activities. Shinn et al. (1987) have suggested 13 major components that should be included in an environmental impact report associated with field tests of smokes and obscurants. These are (1) an introduction, (2) a statement of the proposed action, (3) a description of the environmental setting, (4) a discussion of the physical, chemical, and biological properties

TABLE 6.6

EMISSION FACTORS FOR REFUSE INCINERATORS WITHOUT CONTROLS[a]—EMISSION-FACTOR RATING: A

Incinerator type	Particulates		Sulfur oxides[b]		Carbon monoxide		Hydrocarbons[c]		Nitrogen oxides[d]	
	lb/ton	kg/MT	lb/ton	kg/MT	lb/ton	kg/MT	lb/ton	kg/MT	lb/ton	kg/MT
Municipal										
Multiple chamber, uncontrolled	30(8 to 70)	15	1.5	0.75	35(0 to 233)	17.5	1.5	0.75	2	1
With settling chamber and water-spray system	14(3 to 35)	7	1.5	0.75	35(0 to 233)	17.5	1.5	0.75	2	1
Industrial-commercial										
Multiple chamber	7(4 to 8)	3.5	1.5	0.75	10(1 to 25)	5	3(0.3 to 20)	1.5	3	1.5
Single chamber	15(4 to 31)	7.5	1.5	0.75	20(4 to 200)	10	15(0.5 to 50)	7.5	2	1
Controlled air	1.4(0.7 to 2)	0.7	1.5	0.75	Neg.	Neg.	Neg.	Neg.	10	5
Flue-fed	30(7 to 70)	15	0.5	0.25	20	10	15(2 to 40)	7.5	3	1.5
Flue-fed (modified)	6(1 to 10)	3	0.5	0.25	10	5	3(0.3 to 20)	1.5	10	5
Domestic single chamber										
Without primary burner	35	17.5	0.5	0.25	300	150	100	50	1	0.5
With primary burner	7	3.5	0.5	0.25	Neg.	Neg.	2	1	2	1
Pathological	8(2 to 10)	4	Neg.	Neg.	Neg.	Neg.	Neg.	Neg.	3	1.5

[a]Average factors given based on EPA procedures for incinerator stack testing. Use high side of particulate, HC, and CO emission ranges when operation is intermittent and combustion conditions are poor.
[b]Expressed as SO_2.
[c]Expressed as methane.
[d]Expressed as NO_2.
Source: Adapted from U.S. Environmental Protection Agency, 1973, p. 2.1–3.

TABLE 6.7

AIR POLLUTION EMISSION FACTORS FOR PASSENGER VEHICLES

Transport mode	Carbon dioxide (pounds/passenger-mile)	Organic compounds	Carbon monoxide	Nitrogen oxides	Sulfur dioxide
		(grams/passenger-mile)			
Truck (gasoline):					
Single occupancy	1.55	3.20	27.46	2.05	0.23
Average occupancy	0.81	1.68	14.45	1.08	0.12
Car:					
Single occupancy	1.12	2.57	20.36	1.61	0.14
Average occupancy	0.68	1.51	11.98	0.95	0.08
Vehicle rideshare:					
3-person carpool	0.37	0.86	6.79	0.54	0.05
4-person carpool	0.28	0.64	5.09	0.40	0.03
9-person vanpool	0.17	0.36	3.05	0.23	0.03
Bus (diesel):					
Transit	0.39	0.25	1.21	1.82	n/a
Rail:					
Amtrak/intercity					
Diesel	0.43	1.12	0.6	0.9	0.51
Electric	0.26	neg	0.05	1.1	2.07
Commuter (diesel)	0.53	1.04	1.44	4.10	0.63
Transit (electric)	0.37	neg	0.06	1.48	2.89
Aircraft	0.57	0.5	0.52	1.08	0.08
Bicycle	0	0	0	0	0
Walk	0	0	0	0	0

Source: World Resources Institute, 1992, p. 70.

of the smokes and obscurants being tested, (5) a discussion of impact criteria, (6) an identification of environmental effects, (7) a consideration of environmental consequences, (8) a discussion of the cumulative, long-term effects of repeated tests, (9) a discussion of short-term effects versus effects on long-term productivity, (10) a statement of recommended alternatives, (11) a consideration of mitigation measures, (12) recommendations for the next step in the NEPA process, and (13) references. It should be noted that inclusion of these 13 elements can be used in studies addressing a broader range of impacts than those on air quality. An example EA for smokes and obscurants has

been prepared by Shinn, Sharmer, and Novo (1987).

Emission factors for passenger vehicles are shown in Table 6.7. Because of the importance of highway and motor vehicles, a computer program that calculates emission factors for hydrocarbons (HCs), carbon monoxide (CO), and oxides of nitrogen (NOx) from gasoline-fueled and diesel-fueled highway motor vehicles has been developed (U.S. EPA, 1989a). The program, MOBILE4.1, calculates emission factors for eight individual vehicle types in two classes of regions (high and low altitude) in the United States. MOBILE4.1 emission estimates depend on various conditions such as ambient temperature, speed, and mileage accrual rates (U.S. EPA, 1991d). Toxic air pollutants such as benzene, formaldehyde, 1,3-butadiene, acetaldehyde, diesel particulate matter, gasoline particulate matter, and gasoline vapors from motor vehicles have been identified, and information is available, or being developed, on pertinent emission factors (U.S. EPA, 1993).

In summary relative to step 1, information should be aggregated by source type and construction and/or operational phase of the proposed project or activity, which could be used to delineate both the types and quantities of air pollutants of concern. For conventional air pollutants (such as particulates, carbon monoxide, hydrocarbons, oxides of nitrogen, and carbon monoxide) emission-factor information should be readily available. If the air quality issue of concern is related to toxic air pollutants, it should be recognized that the availability of emission-factor information for this type of pollutant is somewhat limited. If emission standards have been developed for the pollutants of interest, they could be used in lieu of emission factors to represent worst-case conditions.

STEP 2: DESCRIPTION OF EXISTING AIR QUALITY CONDITIONS

Existing air quality conditions can be described in terms of ambient air quality data, emission inventories, and meteorological information which relates to atmospheric dispersion.

Compilation of Air Quality Information

As the next step, information should be assembled on the existing air quality, for the pollutants of concern delineated in step 1. In the United States, the sources of information on air-quality-monitoring data include the relevant county, regional, and/or state air pollution control agencies, and private industries in the area that might be maintaining air-quality-monitor-

ing programs for their particular interests. The pertinent state, regional, or county agencies may also operate monitoring stations that are a part of a national system coordinated by the U.S. Environmental Protection Agency. As of 1990, almost 4,100 monitoring sites reported air quality data to the U.S. EPA; approximately 30 percent reported PM_{10} data, 20 percent each reported on ozone and sulfur dioxide, and 10 percent each reported on carbon monoxide, nitrogen dioxide, and lead (U.S. EPA, 1992).

The Aerometric Information Retrieval System (AIRS) is a computer-based repository of air quality–related information in the United States. Included in AIRS is the National Emissions Data System (NEDS), containing data related to emission inventories for criteria pollutants [PM_{10}, carbon monoxide, sulfur dioxide, nitrogen dioxide, lead, reactive volatile organic compounds (VOCs), and ozone]; and the National Air Toxics Information Clearinghouse (NATICH), which includes regulatory-program descriptions, acceptable ambient concentrations, ambient-air-monitoring information, and emissions inventory data. Finally, SAROAD (storage and retrieval of aerometric data) is a system for editing, storing, summarizing, and reporting ambient air quality data throughout the United States (Kokoszka, 1992). Historical air-quality-monitoring data from governmental programs throughout the United States is also available from SAROAD.

Appropriate interpretation of air quality data should include consideration of historical trends, as well as information about the monitoring station. If possible, it is desirable to examine the complete history of air quality for the sampling stations in the particular locale. To utilize this information appropriately, one must carefully describe the characteristics of each sampling site, including any unique factors about the site, such as surrounding land usage, height of the sampling device above the ground surface, and the type and calibration history of the sampling equipment. Graphical presentation of air quality information may be useful, particularly if there appear to be trends, either upward or downward, in the air quality levels of any of the air pollutants.

One of the tasks in summarizing the existing air quality may be to express the collected raw data in accordance with the averaging times in pertinent ambient-air-quality standards. For example, it may be necessary to calculate annual average concentrations, along with the pertinent statistical distributions. Determination of data distributions may also be necessary for shorter averaging-time data (such as 8 hr or 24 hr). The pollutant standards index (PSI) described in Chapter 5 could be used for presenting air quality information.

A common problem in addressing baseline air quality in an area of a proposed project or activity is the absence of data for the specific site. One solution is to explore the availability of data from nearby areas that exhibit similar characteristics in terms of land usage (and possible air pollution sources) and climatological features. If appropriately qualified, data from such look-alike areas can be used.

The fundamental usage of existing ambient-air-quality data is in determining whether the air quality exceeds, attains, or does not comply with relevant standards. In addition to standards for traditional pollutants (such as sulfur dioxide, particulates, and carbon monoxide), it may be necessary to determine if ambient air quality criteria or standards have been established for pertinent toxic air pollutants (see the list of 189 chemicals in Table 6.4). Greater significance should be attached to, and greater attention should be given to, those pollutants which do not meet or barely meet the allowable ambient air concentrations. Additional air quality management measures may be required if one or more pollutants are in a nonattainment area. Examples of such measures include source-reduction programs and emissions trading; the non-

attainment provisions of the 1990 Clean Air Act address this subject.

Procurement or Development of Emission Inventory

In analyzing the potential air quality impacts of a proposed project or activity, it is necessary to consider the study area (potential area or region of influence) associated with the air pollution emissions. The delineation of a study area can be made using the boundaries of the land associated with the project-activity, or the delineation can include a larger area by considering the atmospheric dispersion patterns within the vicinity of the proposed project or activity. In the continental United States, a relevant approach for defining the study area is to consider the county or parish in which the proposed project or activity is located.

The primary item of information which should be procured or developed is an emission inventory for the air pollutants within the study area. An "emission inventory" is the compiled information on the quantities of air pollution from all the sources in a defined geographical area entering the atmosphere in a given time period (typically a 1-yr period is used). A properly developed emission inventory provides information concerning all source emissions and defines the location, magnitude, frequency, duration, and relative contribution of these emissions. It can be used as a baseline marker against which previous and anticipated future air pollutant emissions and their increases in the geographical study area as a result of the activity can be judged. Additionally, it can be used in a comparative context with regard to data from other nearby geographical areas.

An emission inventory, therefore, provides a summary of the air pollutant emissions under current conditions in the vicinity of the proposed project-activity. It should be noted that emission inventories have limitations, since they do not give consideration to atmospheric reactions, nor do they account for unequal effects of air pollutants on a mass basis.

In the United States, emission inventories have been developed for all 50 states, for approximately 250 air quality control regions within and between states, and for several thousand counties or parishes within the air quality control regions. The primary information source for these inventories is the National Emissions Data System, or NEDS (U.S. EPA, 1988a). NEDS issues annual reports which summarize annual cumulative estimates of source emissions of five criteria pollutants: particulates, sulfur oxides, nitrogen oxides, VOCs, and carbon monoxide. Summary data are presented for the nation as a whole, for individual states, for air quality control regions, and for individual portions thereof (U.S. EPA, 1988a). Nationwide emissions-trend information for the period from 1940 to 1989 is available for six major pollutants: particulates (PM/TSP and PM_{10}), sulfur oxides, nitrogen oxides, reactive VOCs, carbon monoxide, and lead (U.S. EPA, 1991b).

Table 6.8 presents an example of the summarization of an emission inventory. This summarization approach is basic to emission inventories developed in the United States; source categories are considered in terms of fuel combustion, process losses, solid-waste disposal, transportation, and miscellaneous area sources. Examination of the information in such a summary can indicate the major sources of particular pollutants in the geographical area. This information is useful for planning air quality management programs and air-quality-monitoring networks. To date, the emphasis has been on conventional air pollutants. However, with the passage of the 1990 Clean Air Act, there is a growing focus upon the development of emission inventories for toxic air pollutants. The U.S. Environmental Protection Agency maintains a toxics-release inventory wherein information on atmospheric emissions of air toxics is aggregated. If the impact study project involves atmospheric emissions of various toxics, then this inventory can serve as a useful basis for comparison.

If an emission inventory does not exist or is not specific enough for the study area, it may be necessary to develop one. The steps associated with compiling a comprehensive emission inventory are as follows (U.S. EPA, 1972):

1. Classification of all pollutants and sources of emissions in the geographical area being addressed
2. Identification and aggregation of information on emission factors for each of the identified pollutants and sources
3. Determination of the daily quantity of materials handled, processed, or burned; or other unit-production information, depending upon the individual identified sources
4. Computation of the rate at which each pollutant is emitted to the atmosphere, with this rate typically expressed on an annual basis
5. Summation of the specific pollutant emissions from each of the identified source categories

Even if an emission inventory is available for the geographical study area, it may be somewhat outdated; thus it may be desirable to update the inventory as emission-factor and -production information change, as sources are eliminated, or as new sources become viable. Again, if the inventory is being maintained by a governmental agency, this could probably be accomplished by making only periodic checks with the agency. On the other hand, if a military installation is maintaining the inventory, it would probably be necessary to conduct periodic reviews and recalculations because of either changes in source types, or changes in unit-production data for pertinent source categories. Fagin (1988) has described an approach for preparing emission inventories for military installations (in particular, for U.S. Air Force bases). Information is included on the use of

TABLE 6.8

SUMMARY CATEGORIES AND SELECTED SUBCATEGORIES IN AN EMISSION INVENTORY FORM

Source category	Tons of pollutant/year				
	Particulates	SO$_2$	CO	HC	NO$_x$
I. Fuel combustion					
A. Residential fuel, area source					
2. Distillate oil					
3. Natural gas					
6. Total					
B. Commercial—institutional and Industrial					
1.b. Bituminous coal, point source					
3.a. Distillate oil, area source					
b. Distillate oil, point source					
4.a. Residual oil, area source					
b. Residual oil, point source					
5.a. Natural gas, area source					
b. Natural gas, point source					
8.a. Other (specify), area source					
b. Other (specify), point source					
9. Total					
C. Steam—electric power plant					
2. Bituminous coal					
3. Distillate oil					
4. Residual oil					
5. Natural gas					
7. Total					
D. Total fuel combustion					
II. Process losses					
A. Area sources					
B. Point sources					
III. Solid-waste disposal					
A. Incineration					
2. Municipal, etc., point source					
B. Open burning					
1.a. Onsite, area source					
D. Total solid-waste disposal					
IV. Transportation, area source					
A. 1. Motor vehicles—gasoline					
2. Motor vehicles—diesel					
B. Off-highway fuel usage					
C. Aircraft					
D. Railroad					
E. Vessels					
F. Gasoline-handling evaporation losses					
G. Other (specify) petroleum storage loss					
H. Total transportation					
V. Miscellaneous, area sources					
B. Other (specify)					
C. Total miscellaneous					
VI. Grand total					
A. Area source					
B. Point source					
C. Total					

Note: The numbers and letters refer to predesignated source categories.

emission factors for manual calculation of emissions, as well as data collection procedures necessary for completion of the emissions inventory. The report provides guidelines and example calculations, blank data sheets, emission factors, and a completed example.

Summary of Key Meteorological Data

The key meteorological data for the study area should be summarized, with the emphasis being on those parameters which are indicative of limited-dispersion conditions for pollutants emitted to the atmosphere. The necessary data can be grouped in three categories: (1) data which is indicative of the general air-pollution-dispersion characteristics of the study area, (2) data which can be used to qualitatively describe the atmospheric dispersion of air pollutants from a project-activity, and (3) data which is necessary for the use of mathematical models for determining actual pollutant dispersion.

The ability to describe general atmospheric dispersion conditions demonstrates a fundamental understanding of atmospheric transport; more importantly, limiting times, months, or seasons can be identified during this process, and this information can be used in construction-phase planning and operational-phase decision making. Data indicative of the general characteristics of the area with regard to air-pollution dispersion include mixing height, inversion height, and mean annual wind speeds. "Mixing height" refers to the vertical distance available above the earth's surface at a given location and at a given time (or during a given period) for the mixing of pollutants. Mixing heights vary daily, seasonally, and topographically. Figures 6.3 and 6.4 indicate the mean annual morning and mean annual afternoon mixing heights, respectively, for the continental United States. Larger numerical values for mixing heights are desirable, since they indicate larger dilution potential.

"Inversions" occur when temperature increases with height above the earth's surface (Hosler, 1961). Inversions typically are present during the night or early morning hours because of the heating and cooling pattern at the earth's surface. In general, inversions are more frequent during the fall than during any other season. One of the characteristics of inversions is that they are often accompanied by wind speeds less than 7 mi/hr; thus, they often represent time periods when there is limited horizontal and vertical dispersion. Figure 6.5 shows seasonal maps of the percentage of total hours of the occurrence of inversions or isothermal conditions below 500 ft during the winter and the summer. Larger numerical values for inversion heights are also desirable, since they indicate larger dilution potential.

Finally, "mean annual wind speeds" can be used as general indicators of dispersion conditions, with larger numerical values being more desirable as they signify the more rapid dispersion of air pollutants from the study area. Figures 6.6 and 6.7 represent isopleths of the mean annual wind speed averaged through the morning and the afternoon mixing layers, respectively, in the United States. Some areas are characterized by speeds as low as 3 m/sec in the morning, whereas others by speeds as high as 9 m/sec in the afternoon.

The suggested approach regarding the general dispersion characteristics is to summarize data on one or more of the above three indicators, giving particular emphasis to the limiting conditions (daily, monthly, and/or seasonally). The implications of these limiting conditions in relation to the types and quantities of air pollutants to be emitted from the proposed project-activity should be considered. One other thing that should be considered is the historical record of air pollution episodes in the study area. An "episode" refers to the occurrence of high concentrations of air pollutants with potentially damaging effects on humans, plants, and animals; the occurrence is associated with limiting meteorological conditions, generally temperature inversions, that reduce the effective volume

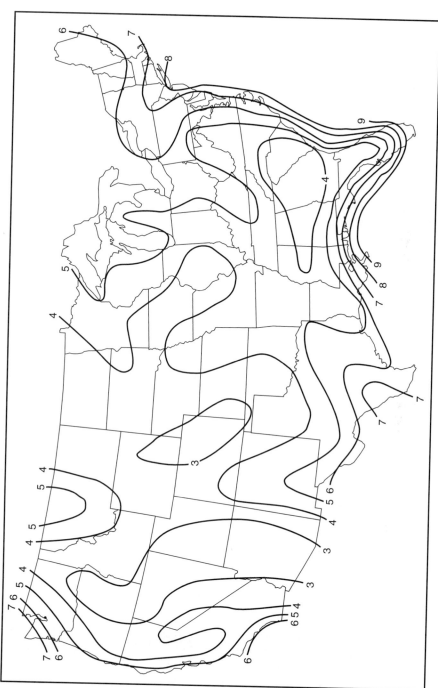

FIGURE 6.3
Isopleths (m $\times$ 10^2) of Mean Annual Morning Mixing Heights (Holzworth, 1972).

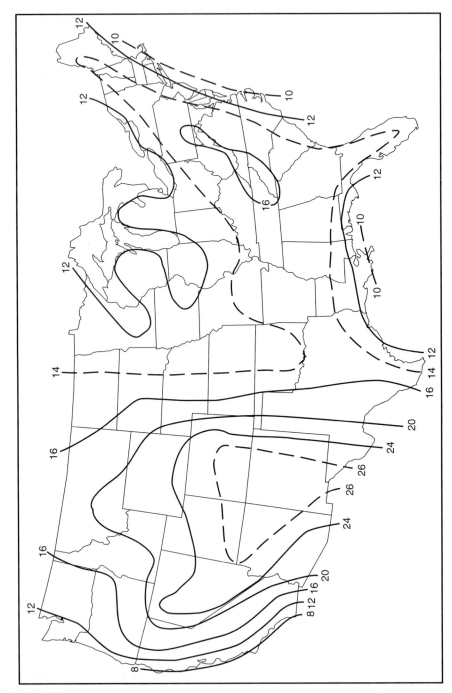

FIGURE 6.4
Isopleths (m × 10^2) of Mean Annual Afternoon Mixing Heights (Holzworth, 1972).

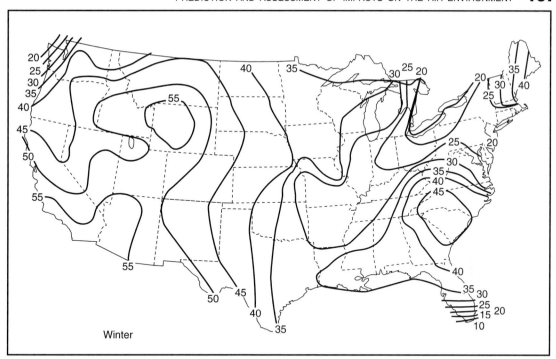

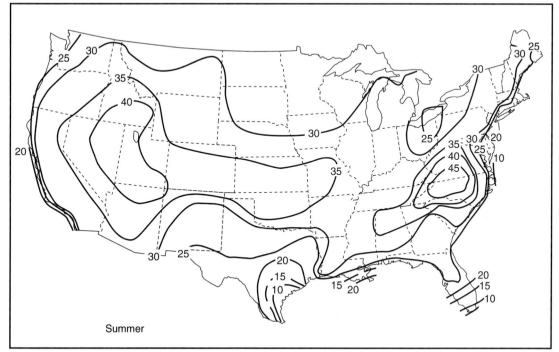

FIGURE 6.5
Isopleths of Percentage Frequency (percent of total hours) of the Occurrence of Inversions or
Isothermal Conditions Based Below 500 ft during the Winter and Summer (Hosler, 1961, p. 322).

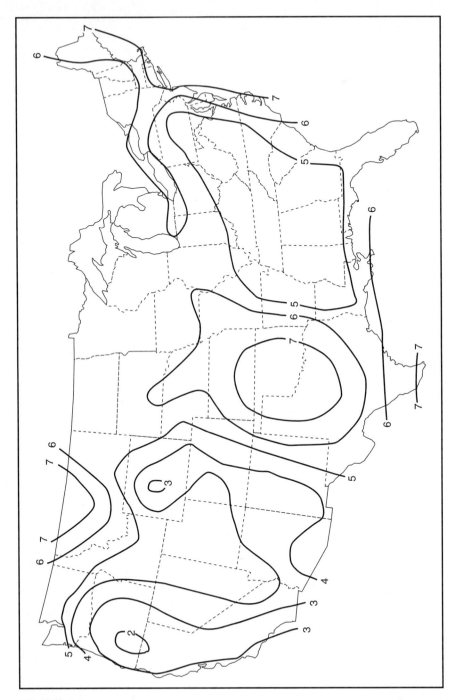

FIGURE 6.6
Isopleths (m/sec) of Mean Annual Wind Speed Averaged Through the Morning Mixing Layer
(Holzworth, 1972).

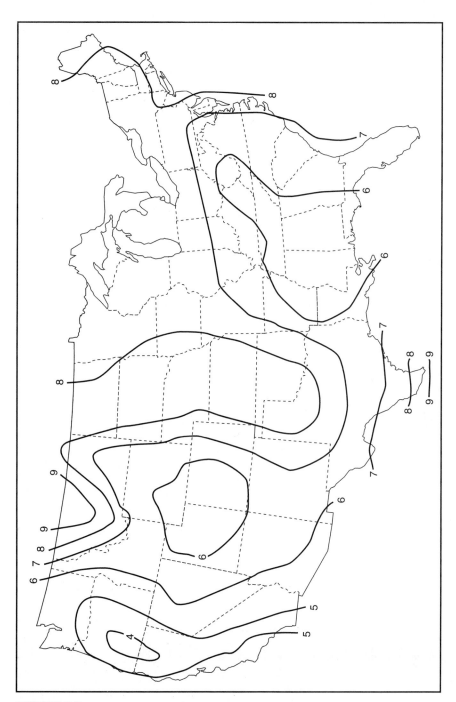

FIGURE 6.7
Isopleths (m/sec) of Mean Annual Wind Speed Averaged Through the Afternoon Mixing Layer
(Holzworth, 1972).

of air in which pollutants are emitted and subsequently diluted. As inversion conditions persist and emission sources continue to be discharged to the ambient air, concentrations of specific air pollutants will increase. Any previous air pollution episodes in the study area should be documented.

Any unique meteorological phenomena that occur in the area should be noted, particularly as related to the occurrence of tornadoes or characteristics such as fog formation and persistence. Some agencies require a discussion of the probability of a tornado occurring in an area, and this probability can be calculated by using the approach suggested by Thom (1963).

Wind-rose information can be used to qualitatively describe the atmospheric dispersion of air pollutants from a project-activity. A "wind rose" is a diagram designed to show the distribution of wind direction at a given location over a considerable period of time (Hesketh, 1972). It is a pictorial graph showing the prevailing wind direction and speed, with the wind direction being the direction from which the wind is blowing. In Figure 6.8, two wind roses

FIGURE 6.8
January and July Wind Roses, Cincinnati. The Monthly Distributions of Wind Direction and Wind Speed Are Summarized on Polar Diagrams. The Positions of the Spokes Show the Direction from which the Wind Was Blowing; the Length of the Segments Indicates the Percentage of the Speeds in Various Groups (Smith, 1968).

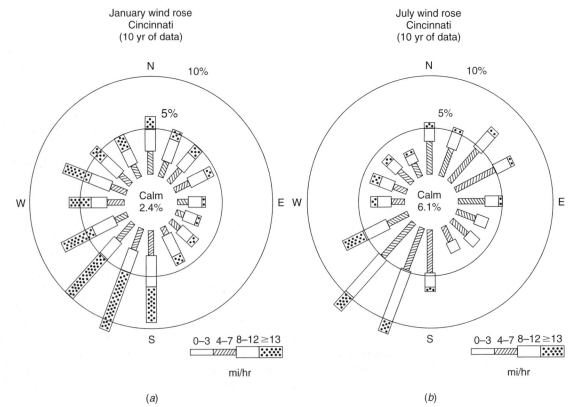

(a)   (b)

representing the average of 10 yr of data are shown. Wind roses should be presented for the study area or the nearest weather station. Based upon pertinent wind-rose data, qualitative statements can be made about the directions of air pollutant movement from a project or activity.

Specific meteorological data necessary for mathematical modeling of air pollutant dispersion may include hourly, daily, and/or monthly records of precipitation, temperature, wind speed and direction, solar radiation, atmospheric stability, humidity, and other items. The two basic factors that influence the movement of pollutants from their points of origin to some other location are "horizontal wind speed and direction" and the "vertical temperature structure" of the atmosphere. These two parameters influence the vertical and horizontal motion of pollutants released to the atmosphere. The influence of these two parameters can be combined and the joint parameter is then called "atmospheric stability" (some representative values for each of five categories are shown in Table 6.9.)

Class A indicates the greatest amount of spreading under the most unstable atmospheric conditions, whereas class F indicates the least amount of spreading under the most stable atmospheric conditions. Most mathematical models involve the use of a stability classification. Usage of this specific meteorological data will probably not be necessary unless a mathematical modeling analysis is utilized.

Sources of meteorological data for a study area can include local airports (commercial and/ or at military installations), local or state meteorological-climatological offices, and federal agencies such as the National Oceanic and Atmospheric Administration or the Federal Aviation Administration (FAA). State or regional air quality agencies may also have pertinent meteorological data. In the absence of pertinent existing meteorological information, it is possible to generate the data; however, the cost and time considerations and the inability to collect the data over multiple years may be constraining factors. Specific data collection would only be warranted on an as-needed basis.

TABLE 6.9

ATMOSPHERIC STABILITY CATEGORIES FOR USE IN DISPERSION MODELING

Surface wind speed at 10 m height (m/sec)	Insolation stability classes[a]				
	Day			Night	
	Strong[b]	Moderate[c]	Slight[d]	Thinly overcast or > ½ cloud[e]	Clear to < ½ cloud
>2 (4.5 mi/hr)	A[f]	A-B	B	—	—
2–3 (4.5–6.7)	A-B	B	C	E	F
3–5 (6.7–11)	B	B-C	C	D	E
5–6 (11–13.5)	C	C-D	D	D	D
>6 (>13.5 mi/hr)	C	D	D	D	D

[a]Insolation, amount of sunshine.
[b]Sun >60° above horizontal; sunny summer afternoon; very convective.
[c]Summer day with few broken clouds.
[d]Sunny fall afternoon; summer day with broken low clouds; or summer day with sun from 15 to 35° with clear sky.
[e]Winter day.
[f]Class A indicates greatest amount of spreading and most unstable atmospheric conditions, and Class F indicates least spreading and most stable atmospheric conditions.
Source: Hesketh, 1972, p. 61.

Baseline Monitoring

"Ambient-air-quality monitoring" refers to appropriate sampling and analysis to establish the ambient concentrations of specific pollutants. Targeted monitoring might be desirable in order to verify the experienced changes in air quality concentrations for those pollutants determined in earlier steps to be of concern. Any planned monitoring should be coordinated with existing monitoring programs conducted by local, regional, state, or federal agencies. Detailed information on planning an air-quality-monitoring program is beyond the scope of this chapter. General information on monitoring-program planning is in Chapter 18.

STEP 3: PROCUREMENT OF RELEVANT AIR QUALITY STANDARDS AND REGULATIONS

The primary sources of information on air quality standards, criteria, and policies will be the relevant local, state, and federal agencies which have a mandate for overseeing the air resources of the study area. Documentation of this information will allow the determination of the significance of air quality impacts incurred during projects or activities and will aid in deciding between alternative actions or in assessing the need for mitigating measures for a given alternative. Pertinent institutional information and sources of data related to the air environment have been described earlier. Also, specific air-quality-management policies or requirements may be in existence for particular areas, and the particular requirements of such policies may need to be ascertained. It is beyond the scope of this chapter to completely address this issue; however, some examples of such policies are prevention of significant deterioration (U.S. EPA, 1989b), emissions trading and banking (U.S. EPA, 1986; Tietenberg, 1985), and alternative-fuel considerations stipulated in the 1990 Clean Air Act (Quarles and Lewis, 1990; U.S. EPA, 1988b).

STEP 4: IMPACT PREDICTION

Air quality impact prediction can be based on one to several approaches, including mass balances, the use of simple to detailed mathematical models, and other considerations.

Mass-Balance Approaches

Air pollutant emissions from the construction and/or operational phase of a project-activity can be considered in relation to the existing emission inventory for the study area. This approach will necessitate the development of an inventory representing a mass balance of the total air pollutant emissions from all sources for a proposed project or activity entering the atmosphere during the construction and/or operational phase. The basic steps associated with compiling an emission inventory for a proposed project or activity consist of the following (U.S. EPA, 1972):

1. Classification of all pollutants and sources of emissions in the geographical study area that would be derived from the proposed project or activity. Consideration should be given to emissions during both the construction and the operational phases.

2. Identification and aggregation of information on the emission factors for each of the identified sources for each of the pertinent pollutants.

3. Determination of appropriate unit-production information, which, when multiplied by the emission-factor information, results in an overall mass-balance value. The specific unit-production information varies depending upon the source and its type; it may range from vehicle-miles traveled to tons of coal consumed to the geographical extent of a construction site.

4. Computation of the rate at which each pollutant is emitted into the atmosphere, with this rate typically being extended to an annual basis. The annual basis is chosen so as to enable the systematic comparison of the

emissions from the proposed project or activity with the existing emission inventory for the area.

5. Summation of the specific pollutant emissions from each of the identified source categories associated with the proposed project or activity.

The next aspect of this approach can be described as involving a "mesoscale impact calculation"; it requires the consideration of the increase in the existing emission inventory for one or more pollutants as the result of the construction and/or operational phase of the proposed project or activity. The basic mathematical relationship is as follows:

$$\begin{array}{l}\text{Percentage} \\ \text{increase} \\ \text{in inventory}\end{array} = \frac{\begin{array}{c}\text{project-activity emission} \\ \text{inventory information (100)}\end{array}}{\begin{array}{c}\text{existing emission} \\ \text{inventory information}\end{array}}$$

Percentage increases can be calculated for each pertinent pollutant and each project-activity phase. A total percentage increase could be calculated by summing the values of all the pollutants in the inventory.

Since the basic output from this approach is a figure representing a percentage increase in the current inventory for one or more air pollutants, the issue then becomes focused upon how to interpret this percentage change information. No air quality criteria or standards provide a delineation of an appropriate interpretation. Instead, interpretations can be based on professional judgment and consideration of the following: (1) the existing air quality for the pollutant(s) of interest, (2) the quantity of emissions and the magnitude of the percentage change, (3) the time period of the expected percentage change, (4) the potential for visibility reduction, and (5) any local, sensitive receptors to damage from the pollutant(s).

In addition to these factors, it is necessary to examine the anticipated emissions from the proposed action in light of applicable emission

standards. It is presumed that the proposed action will be in compliance with pertinent emission standards; however, the extent of compliance involved in the proposed action should be discussed.

Box-Model Approaches

A simple atmospheric dispersion model, called a "box model," can be used to calculate ground-level concentrations of specific air pollutants of concern emitted from the project-activity. A box model is based on the assumption that pollutants emitted to the atmosphere are uniformly mixed in a volume, or "box," of air (Canter, 1985). The most critical aspect of the usage of the box model is to establish, with rationale, the downwind, cross-wind, and vertical dimensions of the box.

In addition, the time period over which pollutant emissions will be considered must be established; a typical time period is 1 hr. The time and physical dimension considerations are based on the assumption of steady state conditions; that is, it is supposed that the emissions, wind speed, and characteristics of air available for dilution will not vary over time (Ortolano, 1985). A box model is also based on the assumption that discharges mix completely and instantaneously with the air available for dilution and the released material is chemically stable and remains in the air.

Box models have been frequently used to analyze the air quality impacts of airports (line sources of emissions from aircraft-landing and -takeoff cycles) (Nelson and LaBelle, 1975). The dimensions of a box model for an airport runway, as an example, are established as follows: (1) the height is determined to be 1,100 m based on emissions during the landing-takeoff cycle of an aircraft, (2) the length is a function of the type of aircraft and its flight angles through the 1,100-m height, and (3) the width is 1,600 m (representing calm wind conditions).

The box model can be used for single-point, multiple-point, area, or line, or "hybrid-type" sources of air pollutants. It can also be used in valley settings. The basic box model is depicted in Figure 6.9 and mathematically expressed as follows (Ortolano, 1985; Canter, 1985):

$$C = \frac{Qt}{xyz}$$

where C = average concentration of gas or particulate < 20 μm in size, throughout box, including at ground level, μg/m^3

Q = release rate of gas, or particulates < 20 μm in size, from source type(s), μg/sec (related to emission-factor information described in step 1)

t = time period over which assumption of uniform mixing in box holds valid, sec (typical period, 1 hr)

x = downwind dimension of box, m (chosen based on average wind speed and physical aspects of terrain)

Y = crosswind dimension of box, m (chosen based on average wind speed, source configuration, and physical aspects of terrain)

z = vertical dimension of box, m (chosen based on limiting inversion heights in area and physical aspects of terrain)

The box-model approach could be focused on the key pollutants of concern identified using mass-balance approaches. One method of establishing the dimensions of the box and the time period for the emissions is to use the data on limiting meteorological conditions as developed in step 2. In this approach, the box-model calculations are focused on worst-case conditions.

The box-model results can be interpreted on a pollutant-by-pollutant basis, in relation to existing ambient air quality and the relevant standard. It is important to compare the sum of the existing pollutant concentration and the concentration from the proposed project or activity, as

FIGURE 6.9
Air Available for Dilution in a Simple Box Model (Ortolano, 1985).

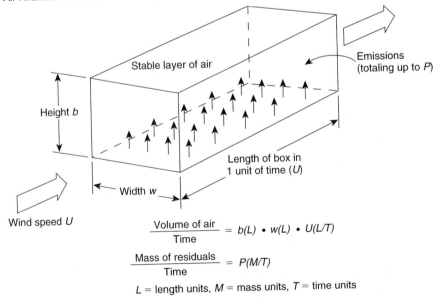

$\dfrac{\text{Volume of air}}{\text{Time}} = b(L) \bullet w(L) \bullet U(L/T)$

$\dfrac{\text{Mass of residuals}}{\text{Time}} = P(M/T)$

L = length units, M = mass units, T = time units

calculated from the box model, to the value given in the applicable standard. Greater significance should be attached to, and greater attention should be given to, those pollutants which do not meet or are near the allowable ambient air concentrations.

Air-Quality Dispersion-Modeling Approaches

From a user perspective, air-quality dispersion models can be classified according to source type [elevated point (stacks), ground-level point, ground-level area, or line], pollutant type (gases or particulates), averaging times (short-term, 24-hr, monthly, or annual), and atmospheric reactions (deposition, photochemical-smog formation, or acid-rain formation).

It is beyond the scope of this chapter to derive the various mathematical models for gaseous and particulate dispersion. Instead, this section will present several mathematical models for use in air quality impact calculations and identify reference sources that will support a more detailed study for an individual project or activity. Detailed information on the theory and practice of air-quality dispersion modeling is found in Turner (1970, 1979), U.S. EPA, (1978), Ortolano (1985), Szepesi (1989), Lyons and Scott (1990), and Zannetti (1990). The last reference is a text which encompasses dispersion theory, computational methods, and available computer software.

Each of the mathematical models for prediction of microscale impact involves the use of a stability classification (see Table 6.9). The models to be described herein can be considered in two groups: (1) manual-calculation (or calculator) models and (2) computer models. Three basic manual-calculation models will be presented as illustrative of a large number of potentially usable models for calculating the air quality impacts of projects or activities. All three models are useful for calculating short-term (on the order of hours), average concentrations of air pollutants at specific locations.

Several categories of projects-activities have stack emissions (elevated point sources); examples include chemical plants, and heat- and steam-generation facilities. The following model, known as the "Pasquill model, as modified by Gifford," is frequently used to analyze the air quality impacts of single, elevated point sources (Turner, 1970):

$$C_{x,y,o} = \frac{Q}{\Pi\sigma_y\sigma_z\overline{u}} \exp\left[-\left(\frac{H^2}{2\sigma_z^2} + \frac{y^2}{2\sigma_y^2}\right)\right]$$

where $C_{x,y,o}$ = ground-level concentration of gas, or particulate < 20 μm in size, at distance x in m downwind from source, and distance y in m crosswind (90° from wind direction) from source, in μg/m^3

Q = release rate of gas, or particulate < 20 μm in size, from elevated point source, μg/sec

σ_y = horizontal dispersion coefficient which represents amount of plume spreading in crosswind direction at distance x downwind from source, and under a given atmospheric-stability condition, m (determine stability class from Table 6.9 and read σ_y from Figure 6.10)

σ_z = vertical dispersion coefficient which represents amount of plume spreading in vertical direction at distance x downwind from source, and under a given atmospheric-stability condition, m [determine stability class from Table 6.9 and read σ_z from Figure 6.11 (Turner, 1970)]

$\overline{u}$ = mean wind speed, m/sec

H = effective stack height (actual physical height plus any rise of plume as it leaves the stack), m; plume rise is result of momentum effect caused by vertical velocity of gas leaving stack, and buoy-

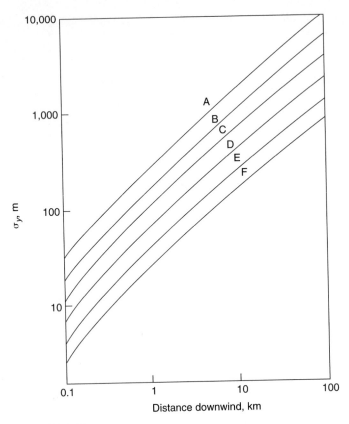

FIGURE 6.10
Horizontal Dispersion Coefficient as a Function of Downwind
Distance from the Source (Turner, 1970).

ancy effect, which is related to warm stack gases tending to rise in a cooler surrounding atmosphere; plume rise can be calculated from the Holland or Briggs equations (not presented herein)

Several categories of projects or activities have stack or point air-pollutant emissions which would be located at (or near enough to) ground level to be considered ground-level point sources. Examples include industrial areas with small incinerators. The following model can be used for ground-level point sources:

$$C_{x,y,o} = \frac{Q}{\Pi \sigma_y \sigma_z \bar{u}} \exp\left[-\left(\frac{y^2}{2\sigma_y^2}\right)\right]$$

where all the terms are as previously defined for the elevated point-source model.

A frequently occurring type of air pollution source associated with many projects and activities is the area source, with the second most frequently occurring source being the line source. Examples of area sources include air pollutants from agricultural operations, open burning, wind erosion, and pesticide applications. Line-source examples include unpaved roads and vehicular traffic. The following model

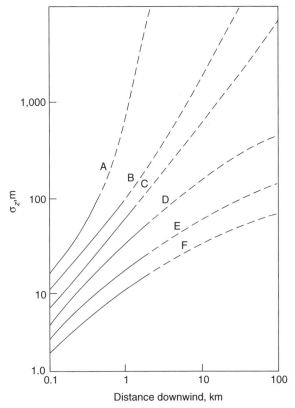

FIGURE 6.11
Vertical Dispersion Coefficient as a Function of
Downwind Distance from the Source (Turner, 1970).

can be used for ground-level area or line sources:

$$C_{x,o,o} = \frac{Q}{\Pi(\sigma_y^2 + \sigma_{yo}^2)^{1/2}\sigma_z\bar{u}}$$

where $C_{x,o,o}$ = ground-level concentration of gas, or particulate less than 20 μm in size, directly downwind and at distance x in m downwind from source, $\mu g/m^3$

σ_{yo} = one-fourth of emission width of area or line source along axis which coincides with wind direction, m

and Q, σ_y, σ_z, u are as previously defined for the elevated point-source model.

Several examples of computer-based air quality simulation models are summarized in Table 6.10. Many of these models are available for mainframe-computer applications, or in personal computer software. The models are a part of the U.S. Environmental Protection Agency's UNAMAP (User's Network for Applied Modeling of Air Pollution) system. All the listed models in Table 6.10 are based upon gaussian plume concepts. Brief information on several additional models available from the U.S. Environmental Protection Agency is in Table 6.11.

TABLE 6.10

EXAMPLES OF COMPUTER-BASED AIR QUALITY DISPERSION MODELS

Model	Description
APRAC	APRAC is Stanford Research Institute's urban carbon monoxide model. Computes hourly averages for any urban location. Requires an extensive traffic inventory for the city of interest.
CDM	The climatological dispersion model (CDM) determines long-term (seasonal or annual), quasi-stable pollutant concentrations at any ground-level receptor using average emission rates from point and area sources and a joint frequency distribution of wind direction, wind speed, and stability for the same period.
CRSTER	This algorithm estimates ground-level concentrations resulting from up to 19 colocated, elevated stack emissions for an entire year and prints out the highest and second-highest 1-hr, 3-hr, and 24-hr concentrations, as well as the annual mean concentrations at a set of 180 receptors (5 distances by 36 azimuths). The algorithm is based on a modified form of the steady state gaussian-plume equation, which uses empirical dispersion coefficients and includes adjustments for plume rise and limited mixing. Terrain adjustments are made as long as the surrounding terrain is physically lower than the lowest stack-height input. Pollutant concentrations for each averaging time are computed for discrete, nonoverlapping time periods (no running averages are computed), using measured hourly values of wind speed and direction, and estimated hourly values of atmospheric stability and mixing height.
HIWAY	HIWAY computes the hourly concentrations of nonreactive pollutants downwind of roadways. It is applicable for uniform wind conditions and level terrain. Although best suited for at-grade highways, it can also be applied to depressed highways (cut sections).
PAL	PAL stands for "point, area, line" source algorithm. This short-term, gaussian steady state algorithm estimates concentrations of stable pollutants from point, area, and line sources. Computations from area sources include effects from the edge of the source. Line-source computations can include effects from a variable emission rate along the source. The algorithm is not intended for application to entire urban areas but for smaller-scale analysis of such sources as shopping centers, airports, and single plants. Hourly concentrations are estimated, and average concentrations from 1 hr to 24 hrs can be obtained.
PTMAX[a]	PTMAX performs an analysis of the maximum short-term concentrations from a single point source as a function of stability and wind speed. The final plume height is used for each computation.
PTDIS[a]	Estimates short-term concentrations directly downwind of a point source at distances specified by the user. The effect of limiting vertical dispersion by a mixing height can be included, and gradual plume rise to the point of final rise is also considered. An option allows the calculation of isopleth half-widths for specific concentrations at each downwind distance.
PTMTP[a]	Estimates the concentration from a number of point sources for a number of arbitrarily located receptor points at or above ground-level. Plume rise is determined for each source. Downwind and crosswind distances are determined for each source-receptor pair. Concentrations at a receptor from various sources are assumed to be additive. Hourly meteorological data is used; both hourly concentrations and averages over any averaging time from 1 to 24 hr can be obtained.
Valley	This algorithm is a steady state, univariate gaussian-plume dispersion algorithm designed for estimating either 24-hr or annual concentrations resulting from emissions from up to 50 (total) point and area sources. Calculations of ground-level pollutant concentrations are made for each frequency designed in an array defined by six stabilities, 16 wind directions, and six wind speeds for 112 program-designed receptor sites on a radial grid of variable scale. Empirical dispersion coefficients are used, and the calculation includes adjustments for plume rise and limited mixing. Plume height is adjusted according to terrain elevations and stability classes.

[a]Model uses Briggs plume-rise equation and Pasquill-Gifford dispersion methods, as given in Turner (1970).
Source: Adapted from Turner, 1979, pp. 518–519.

TABLE 6.11

EXAMPLES OF AIR QUALITY MODELS AVAILABLE FROM U.S. ENVIRONMENTAL PROTECTION AGENCY

Model	Brief description
CALINE 3 (California line-source model)	Line-source dispersion model that can be used to predict carbon monoxide concentrations near highways and arterial streets given traffic emissions, site geometry, and meteorology
CDM 2 (climatological dispersion model)	Updated version of CDM listed in Table 6.10
HIWAY-ROADWAY	Two models which compute the hourly concentrations of nonreactive pollutants downwind of roadways and predict pollutant concentrations within 200 m of a highway, respectively
INPUFF	Gaussian integrated puff model which is capable of addressing the accidental release of a substance over several minutes or of modeling the more typical continuous plume from a stack
ISCLT (industrial source complex long-term)	Steady state gaussian plume model which can be used to access pollutant concentrations from an industrial source complex
ISCST (industrial source complex short-term	Steady state gaussian plume model which can be used to access pollutant concentrations from an industrial source complex
LONGZ, SHORTZ	Designed to calculate the long-term and short-term pollutant concentration produced at a large number of receptors by emissions from multiple stack, building, and area sources
MESOPUFF	Lagrangian model suitable for calculating the transport, diffusion, and removal of air pollutants from multiple point and area sources at transport distances beyond 10 to 50 km
MPTER	Multiple-point-source gaussian model with optional terrain adjustments
MPTDS	Modification of MPTER that explicitly accounts for gravitational settling and/or deposition loss of a pollutant
PEM (pollutant episodic model)	Urban-scale air pollution capable of predicting short-term average surface concentrations and deposition fluxes of two gaseous or particulate pollutants
PBM (photochemical box model)	Simple, stationary single-cell model with a variable-height lid designed to provide volume-integrated hourly averages of ozone and other photochemical-smog pollutants for an urban area for a single day of simulation

Detailed information on the models listed in Tables 6.10 and 6.11 can be obtained by contacting the National Technical Information Service (NTIS), U.S. Department of Commerce, Springfield, Virginia.

A PC-compatible SCREEN model has been developed by the U.S. Environmental Protection Agency to facilitate an initial evaluation of the air quality impacts of stationary sources (Brode, 1988). The SCREEN model can be used in a "screening mode" to calculate ground-level concentrations under limited-dispersion conditions. If the results are of concern when compared to standards, more-detailed modeling can be used. Table 6.12 summarizes some recommended models for the detailed analysis.

There are numerous specific models of potential relevance; examples related to fugitive-

TABLE 6.12

SELECTED U.S. EPA–RECOMMENDED PLUME MODELS AND APPROPRIATE APPLICATIONS

Model	Averaging Period	Source Type[a]	Terrain[b]	Land Use[c]
SCREEN	hourly, daily	point, area	simple, complex	rural, urban
ISCST2	hourly to annual	point, area, volume	simple	rural, urban
ISCLT2	monthly, seasonal, annual	point, area, volume	simple	rural, urban
MPTER	hourly to annual	point	simple	rural
COMPLEX I	hourly to annual	point	complex	rural
SHORTZ	hourly to annual	point, area	complex	urban
LONGZ	seasonal to annual	point, area	complex	urban

[a]Source types include "point" sources such as stacks and vents, "area" sources such as evaporating liquid spills, and "volume" sources such as raw material storage piles.
[b]Complex terrain is defined by the EPA as "terrain exceeding the height of the stack being modeled." Locations where the terrain is at or lower than the source heights are modeled as "simple terrain."
[c]Land use is classified by the EPA as either "rural" or "urban." Rural conditions exist if less than 50 percent of the total area within a 3-kilometer radius of the source contains heavy and light-moderate industry, commercial establishments and compact residential areas, with less than 35 percent vegetation. Otherwise, an area impacted by air pollution emissions should be considered urban. By far, most areas in the United States would be classified as rural.
Source: Sadar, 1993, p. 20.

dust emissions, military air bases, heavy gases (gases with densities greater than air), evaporation of toxic spills, and hazardous-waste sites will be mentioned.

The "fugitive-dust model" (FDM) is a computerized air-quality model specifically designed for computing concentration and deposition impacts from fugitive-dust sources (Carey, 1990). The sources may be point, line, or area. Emissions for each source are apportioned by the user into a series of particle size classes. A gravitational-setting velocity and a deposition velocity are calculated by FDM for each class. Concentration and deposition can be computed at all user-selectable receptor locations. The model can be implemented on an IBM PC–compatible using the DOS operating system.

Segal (1988a and b) has developed a microcomputer-based pollution model for civilian airports and U.S. Air Force bases. The user-friendly model is called the "emissions and dispersion modeling system" (EDMS); it can be used to produce (1) an emissions inventory of all sources at an airport or air base and (2) an

estimate of the concentrations produced by these sources at four airport locations. An inexperienced user should be able to process the example problem included in the software in a few hours.

Fuel and other chemical usage in rockets and jet aircraft may be hazardous when emitted to the ambient atmosphere. These fuels and chemicals are also stored in tanks and transported within and between military installations. Several dispersion models have been developed for potential accidental releases of toxic and/or combustible gases that could occur during transportation or routine maintenance operations. Ermak and Merry (1989) have developed a methodology for evaluating the effectiveness of mathematical models in predicting the atmospheric dispersion of heavier-than-air vapor releases. The methodology is based upon ratio comparisons of the model-predicted value to the observed value in field-scale experiments involving continuous releases of denser-than-air gases. Plume characteristics used in the ratio comparisons include maximum concentration, centerline concentration, plume half-width, and plume height, all as functions of downwind distance from the source.

One specific model which addresses the atmospheric dispersion of denser-than-air vapors is called "SLAB" (Ermak, 1988). The types of releases treated by the model include an evaporating pool (ground-level area source), an elevated horizontal jet, and an instantaneous volume source. Subsequent dispersion of the released vapor is modeled as either a steady state plume, a transient puff, or a combination of the two, depending upon the duration of the release. In the case of a finite-duration evaporating-pool or horizontal-jet release, cloud dispersion is initially described using the plume model while the source is active. Once the source is shut off, dispersion is calculated using the puff model. In cases of an instantaneous release, the puff dispersion model is used for the entire calculation.

Vossler (1989) has developed a model for estimating evaporation which can occur from a toxic-chemical spill. The model is based on calculating the temperature of the spilled chemical pool based on the net energy input into the pool from all possible sources. The pool temperature is then used to calculate the evaporation rate under steady state conditions.

The final example of a site-specific model type addresses area-source emissions at hazardous-waste sites. Touma (1989) has described the area-source emission characteristics associated with Superfund sites in the United States and provides a review of existing, available techniques for modeling area sources. Five short-term and three long-term area-source models were applied to a number of example applications, using one field-database, in order to compare the magnitude of concentration predictions and test the conformance of the resulting concentration estimates with mathematical and physical principles.

Utilization of dispersion models allows the development of isopleths of equal concentration around the source of emission for various types of air pollutants. These isopleths should be calculated bearing in mind the frequency of occurrence based on the meteorological data assembled earlier. The calculated ground-level concentrations at various positions in the project area should then be compared to the applicable ambient air standards. If the calculated ground-level concentrations are less than the standards, then fractional proportions should be indicated to present the degree of safety. If the calculated ground-level concentrations are greater than the ambient air quality standards, then control measures or abatement strategies must be developed. If there are other sources of air pollution in the area of the proposed action that would significantly contribute to the anticipated ground-level concentrations from the proposed action, these should also be included in the calculations so as to examine the possible additive or even potentiating effect of various air pollutant sources. As always, greater significance should be attached to, and greater attention should be given to, those pollutants which do not meet or are near the maximum allowable ambient air concentrations. Stringent mitigation measures may be required in such cases.

In summary, models can be useful in estimating the impacts of the project or activity on air quality. Also, various modifications of the proposed project-activity can be evaluated to assess the effectiveness of mitigating efforts to minimize the impacts of the project-activity. Selection of an appropriate model to meet a given need typically involves consideration of both the technical capabilities of models and model-related managerial issues such as economic considerations, and necessary training and experience of model users.

Other Considerations

Air quality impact analyses of emissions from nuclear power plants, fossil-fueled power plants, and industrial plants should address the issue of accidental releases. A risk or safety analysis of the involved processes could form the basis for identifying the probabilities of releases of atmospheric pollutants. Atmospheric-dispersion modeling should be done for selected release scenarios involving different types and quantities of air pollutants. Particular attention should be given to releases of air toxics and the resultant comparison of maximum ground-level concentrations to appropriate ambient air quality guidelines or standards. In conjunction with these analyses, it is important to identify prevention, control, and mitigation measures for accidental releases of air toxics. Details on process- and plant-design considerations, operating procedures and practices, emergency planning, and real-time countermeasures is in Davis et al. (1989). This type of information could be incorporated in mitigation measures in both project design and operational planning.

Finally, human health effects can occur as a result of atmospheric emissions of pollutants. A quantitative health-risk-assessment approach for setting ambient air quality standards has been described by Merkhofer (1985); this approach could be used in analyzing potential health impacts of projects, plans, programs, or policies. The key feature of the approach is the combining of an emissions-and-atmospheric-dispersion model, an exposure model, health dose-response models, and a significance-weighting model.

STEP 5: ASSESSMENT OF IMPACT SIGNIFICANCE

The next activity entails impact-significance assessment. In the terminology used herein, "significance assessment" refers to the interpretation of the significance of anticipated changes related to the proposed project. One basis for impact assessment is public input; this input could be received through a continued scoping process or through the conduction of public meetings and/or public participation programs. The general public can delineate important en-

vironmental resources and values for particular areas, and this input should be considered in impact assessment. Professional judgment can also be used in relation to the percentage changes from baseline conditions in terms of air-pollutant emission levels and/or exposed human population, or the PSI; such changes could be considered during both the construction and operational phases of a project.

For some types of projects or air-pollutant prediction methods, there are numerical standards or criteria which can be used as a basis of interpretation. This information should have been accumulated in step 3.

A final consideration could include the identification of specific effects of the types of air pollutants from a proposed project or activity. This may include the identification of sensitive receptors in the study area (for example, elderly persons, specific crops, or terrestrial vegetation) (Barker and Tingey, 1991).

As an example of information on effects, a study described by Cataldo et al. (1990) for determining the terrestrial effects of the use of smokes and obscurants will be discussed. The primary objective was to characterize the fate and response of the soil and biotic components of the terrestrial environment to aerosols, deposited brass, and brass in combination with fog oil. Important physical, chemical, and biotic aspects were investigated using five different types of terrestrial vegetation representative of that found on U.S. Army training sites and in surrounding environments. No significant foliar-contact toxicity was observed for brass. The weathering and chemistry of brass aerosols deposited and amended to soils was assessed, along with the impacts of acid precipitation and moisture regimes on weathering rates. Soil-weathering processes and brass solubilization on seed germination produced no detectable effects of brass. However, moderate toxicity effects were noted after 160 days of soil incubation. The toxicity effects were proportional to soil-loading levels.

Fox et al. (1989) have developed a screening procedure to evaluate air pollution effects in wilderness areas, which will serve as an example of receptor-focused considerations. The procedure involves an initial estimate of susceptibility to critical loadings for sulfur, nitrogen, and ozone. It also provides criteria for requesting necessary additional information where potential adverse impacts are identified.

STEP 6: IDENTIFICATION AND INCORPORATION OF MITIGATION MEASURES

"Mitigation measures" refer to project-activity design or operational features that can be used to minimize the magnitude of the air quality impacts. The key approach is to revise the design as needed in order to reduce the air pollutants expected to be emitted from the project-activity. The revised project or activity can then be reassessed to determine if the mitigation measures have eliminated or sufficiently minimized the deleterious air quality impacts. Examples of mitigation measures for reducing air pollutant emissions include

1. Limitations on the practice of open burning of agricultural crop residues; such limitations might include a permitting program, delineation of specified times for burning to occur, and the establishment of distance requirements between residences and open-burning areas (Canter, 1985).

2. Wind erosion from open land can be controlled by use of three basic techniques (watering, use of chemical stabilizers, and windbreaks) in addition to a vegetative cover. Watering, the most common method, provides only temporary dust control of about 50 percent. The use of chemicals to treat exposed surfaces provides longer-term dust suppression of up to 70 percent but may be costly, have adverse impacts on plant and animal life, or contaminate the treated material.

Windbreaks and source enclosures are often impractical because of the size of fugitive-dust sources; the emissions may be reduced by only about 30 percent (Canter, 1985). Planting of rapidly growing vegetation in construction areas to reduce dust generation can also be useful (U.S. Army Construction Engineering Research Laboratory, 1989).

3. Common techniques for controlling air pollutant emissions from unpaved roads are paving, surface treating with penetration chemicals, working of soil-stabilization chemicals into the roadbed, watering, and traffic-control regulations (Canter, 1985). Paving as a control technique is often not practical because of its high cost. Surface chemical treatments and watering can be accomplished with moderate to low costs, but frequent retreatments are required for such techniques to be effective. Traffic controls, such as speed limits and traffic-volume restrictions, provide moderate emission reductions, but such regulations may be difficult to enforce. Watering, because of the frequency of treatments required, is generally not feasible for public roads and is effectively used only when watering equipment is readily available and roads are confined to a single site, such as a construction location (U.S. Army Construction Engineering Research Laboratory, 1989).

4. Fugitive-dust-control measures can be also used for open-waste piles and staging areas, dry surface impoundments, landfills, land treatment systems, and waste-stabilization measures (Cowherd et al., 1990). The development of control plans for open dust sources of PM_{10}, storage piles, construction-demolition activities, open-area wind erosion, and agricultural tilling has been described by Cowherd, Muleski, and Kinsey (1988). An overview of the source category describing emission characteristics and mechanisms is included, followed by available emission factors for analyzing the op-

erative nature of control measures. Demonstrated control techniques are discussed for estimating efficiency and determining costs of implementation.

5. Several options are available for controlling or minimizing airborne pesticide residues resulting from or following pesticide usage. The spraying of pesticides can create liquid particulate pollutants which could consist of hazardous chemicals, hydrocarbons, and other materials. This type of fugitive-particulate emission would not travel very far, except after the material had evaporated. To reduce this type of emission, low-pressure spray nozzles (less than or equal to 20 psig) can be used to minimize the generation of fine particles. Also, pesticide spraying could be planned to coincide with periods of low wind velocities.

6. Several recent studies have been conducted on the use of methanol as an alternative automotive fuel, and the 1990 Clean Air Act includes specific emphases in this regard. Two example studies will be mentioned. In the first, the U.S. EPA (1988b) has conducted an economic and environmental effects analysis of several alternative fuels, including methanol, ethanol, compressed natural gas, liquified petroleum gas, electricity, and reformulated gasoline. In the second, Santini, Saricks, and Sekar (1988) described the environmental-quality changes which would occur from the replacement of diesel-oil-fueled buses by methanol-fueled buses. Information from both studies may be pertinent to an analysis of the use of alternative fuels in a proposed project-activity.

7. Information on exhaust emission control technologies for gasoline- and diesel-fueled cars, trucks, and buses, including the use of alternative fuels, has been compiled by the OECD (Organization for Economic Cooperation and Development, 1988). A comparative summary of regulations on exhaust emissions and lead in gasoline for OECD countries having such regulation is available; the pertinent countries include the member countries of the European Community, Switzerland, Sweden, the United States, Japan, Australia, and Canada. The report has chapters or subchapters on exhaust emission control technologies for gasoline- or diesel-fueled cars, trucks, buses, and two-wheeled vehicles; and fuel modifications related to unleaded fuel, fuel extenders, and alternative fuels. Car-pooling regulations can also be implemented as a part of a mitigation program.

8. Air pollution control equipment can be used for point sources of emissions. Examples include cyclones, scrubbers, fabric filters, and/or electrostatic precipitators for control of particulate emissions. Carbon adsorption or combustion (incineration) can be used for gaseous organics; and absorption systems such as flue-gas desulfurization can be used for SO_2 control. Detailed information on control technologies and their applications for numerous industries is in Buonicore and Davis (1992). Finally, Sink (1991) summarized information on various control technologies for hazardous air-pollutant emissions from point and fugitive sources. The technologies addressed included thermal incineration, catalytic incineration, flares, carbon adsorption, absorption, condensers, fabric filters, electrostatic precipitators, and Venturi scrubbers.

SUMMARY

This chapter has presented a six-step methodology for addressing the impacts of proposed projects (or plans, programs, or policies) on the air environment. These steps provide a general framework which can be used (1) as a guide to study planning and conduction, (2) as an indication of areas for which more-detailed information will be necessary, (3) to discuss a study with a contractor (or sponsor) and develop appropriate terms of reference, and (4) to review impact study work done by others.

It will be necessary to document the six-step methodology in the resulting air-environment-impact report, which could be a section or chapter in either an EA or EIS. The report should address potential air quality impacts identified for the project or activity (step 1), the existing characteristics of the air environment and the emission inventory and meteorological characteristics of the study area (step 2), the applicable air quality standards and regulations (step 3), the quantification of the anticipated air impacts due to the project or activity (step 4), the rationale for the assessment of the predicted impacts (step 5), and the considered and included mitigation measures (step 6).

Numerous relevant topics are only briefly mentioned here; they may need to be expanded depending on the study. Examples of these topics include (1) specific details of the Clean Air Act of 1990 and resultant programs or policies; (2) receptors of impacts at different air quality levels; (3) the planning and implementing of source and/or ambient air monitoring; (4) emission and ambient air quality standards for air toxics; (5) point- and area-source control measures; (6) formation and effects of acid rain; (7) broader issues, such as global warming and destruction of the ozone layer; and (8) market-based air quality management strategies such as allowance or emissions trading. Another issue which may need to be addressed in an impact study is the proposed undertaking's impact on acid rain and/or global warming.

SELECTED REFERENCES

Barker, J. R., and Tingey, D. T., eds., *Air Pollution Effects on Biodiversity,* Van Nostrand Reinhold, New York, 1991.

Brode, R. W., "Screening Procedures for Estimating the Air Quality Impact of Stationary Sources," EPA 450/4-88-010, U.S. Environmental Protection Agency, Research Triangle Park, N.C., Aug. 1988.

Brophy, R., "Software Programs Model Air Emissions," *Environmental Protection,* Nov./Dec., 1991, pp. 13–21.

Buonicore, A. J., and Davis, W. T., *Air Pollution Engineering Manual,* Van Nostrand Reinhold, New York, 1992.

Canter, L. W., *Environmental Impacts of Agricultural Production Activities,* Lewis Publishers, Chelsea, Mich., 1985, pp. 169–209.

Carey, P., "Fugitive Dust Model (FDM) (for Microcomputers)," EPA SW/DK-90-941, U.S. Environmental Protection Agency, Seattle, May 1990.

Cataldo, D. A., et al., "Evaluation Characterization of Mechanisms Controlling Fate and Effects of Army Smokes (Transport, Transformations, Fate and Terrestrial Ecological Effects of Brass Obscurants)," AD-A227 134/4/WEP, Battelle Pacific Northwest Laboratory, Richland, Wash., Aug. 1990.

Cowherd, C., Englehart, P., Muleski, G. E., Kinsey, J. S., and Rosbury, K. D., *Control of Fugitive and Hazardous Dusts,* Noyes Data Corporation, Park Ridge, N.J., 1990.

———, Muleski, G. E., and Kinsey, J. S., "Control of Open Fugitive Dust Sources," EPA 450/3-88-008, U.S. Environmental Protection Agency, Research Triangle Park, N.C., Sept. 1988.

Davis, S. D., DeWolf, G. B., Ferland, K. A., Harper, D. L., Keeney, R. C., and Quass, J. D., *Accidental Releases of Air Toxics: Prevention, Control, and Mitigation,* Noyes Data Corporation, Park Ridge, N.J., 1989.

Ermak, D. L., "SLAB: A Denser-than-Air Atmospheric Dispersion Model," UCRL-99882, Lawrence Livermore National Laboratory, Livermore, Calif., Oct. 1988.

——— and Merry, M., "Methodology for Evaluating Heavy Gas Dispersion Models," AFESC/ESL-TR-88-37, Lawrence Livermore National Laboratory, Livermore, Calif., June 1989.

Fagin, G. T., "Manual Calculation Methods for Air Pollution Inventories," USAFOEHL-88-070 EQ 0111 EEB, U.S. Air Force Occupational and Environmental Health Laboratory, Brooks AFB, Tex., May 1988.

Fox, D. G., et al., "Screening Procedure to Evaluate Air Pollution Effects on Class I Wilderness Areas," FSGTR-RM-168, Rocky Mountain Forest and Range Experiment Station, Ft. Collins, Colo., Mar. 1989.

Hesketh, H. E., *Understanding and Controlling Air Pollution,* Ann Arbor Science Publishers, Ann Arbor, Mich., 1972, p. 50.

———— and Cross, Jr., F. L., "The Magnitude of the Agricultural Air Pollution Problem," Paper no. 81-3070, *Summer Meeting of American Society of Agricultural Engineers,* June 1981.

Holzworth, G. C., "Mixing Heights, Wind Speeds, and Potential for Urban Air Pollution throughout the Contiguous United States," Publ. AP-101, U.S. Environmental Protection Agency, Research Triangle Park, N.C., Jan. 1972, pp. 3–6, 22, 26, 31, 36, 41.

Hosler, C. R., "Low-Level Inversion Frequency in the Contiguous United States," *Monthly Weather Review,* vol. 89, Sept. 1961, pp. 319–339.

Kokoszka, L. C., "Guide to Federal Environmental Databases," *Pollution Engineering,* vol. 24, no. 3, Feb. 1992, pp. 83–92.

Lyons, T. J., and Scott, W. D., *Principles of Air Pollution Meteorology,* Belhaven Press, London, 1990.

Merkhofer, M. W., "An Approach for Assessing Health Risks Associated with Alternative Ambient Air Quality Standards," *Environmental Impact Assessment, Technology Assessment, and Risk Analysis,* NATO ASI Series, vol. G4, Covello, V. T., et al., eds., Springer-Verlag, Berlin, 1985, pp. 691–722.

Nelson, K. E., and LaBelle, S. J., "Handbook for the Review of Airport Environmental Impact Statements," ANL/ES-46, Argonne National Laboratory, Argonne, Ill., July 1975.

Organization for Economic Cooperation and Development (OECD), "Transport and the Environment," Organization for Economic Cooperation and Development, Paris, 1988.

Ortolano, L., "Estimating Air Quality Impacts," *Environmental Impact Assessment Review,* vol. 5, no. 1, Mar. 1985, pp. 9–35.

Pope, A. A., Brooks, G. R., Carfagna, P. F., and Lynch, S. K., "Toxic Air Pollutant Emission Factors—A Compilation of Selected Air Toxic Compounds and Sources," 2d ed., EPA 450/2-90-011, U.S. Environmental Protection Agency, Research Triangle Park, N.C., Oct. 1990.

————, Cruse, P. A., and Most, C. C., "Toxic Air Pollutant Emission Factors: A Compilation for Selected Air Toxic Compounds and Sources," EPA 450/2-88/006A, U.S. Environmental Protection Agency, Research Triangle Park, N.C., Oct. 1988.

Quarles, J., and Lewis, W. H., *The New Clean Air Act,* Morgan, Lewis, and Bockius, Washington, D.C., 1990.

Radian Corporation, "Toxic Air Pollutant Emission Factors: Information Storage and Retrieval System—User's Manual," EPA 450/2-88-006B, U.S. Environmental Protection Agency, Research Triangle Park, N.C., Oct. 1988.

Sadar, A. J., "Dispersion Modeling," Environmental Protection, vol. 4, no. 3, 1993, pp. 16–20.

Santini, D. J., Saricks, C. L., and Sekar, R., "Environmental Quality Changes Arising from the Replacement of Diesel Oil-Fueled Buses by Methanol-Fueled Buses," Argonne National Laboratory, Argonne, Ill., 1988.

Schanche, G. W., et al., "Pollution Estimation Factors," CERL TR-N-12, U.S. Army Construction Engineering Research Laboratory, Champaign, Ill., Nov. 1976, pp. 8–11.

Segal, H. M., "Microcomputer Pollution Model for Civilian Airports and Air Force Bases—Model Application and Background," FAA-EE-88-5, Federal Aviation Administration, Washington, D.C., Aug. 1988a.

————, "Microcomputer Pollution Model for Civilian Airports and Air Force Bases—User's Guide—Issue 2," FAA-EE-88-6, Federal Aviation Administration, Washington, D.C., Aug. 1988b.

Shinn, J. H., et al., "Smokes and Obscurants: A Guidebook of Environmental Assessment, vol. 1: Method of Assessment and Appended Data," UCRL-21004-VOL-1, Lawrence Livermore National Laboratory, Livermore, Calif., Sept. 1987.

————, Sharmer, L., and Novo, M., "Smokes and Obscurants: A Guidebook of Environmental Assessment, vol. 2: A Sample Environmental Assessment," UCRL-21004-VOL-2, Lawrence Livermore National Laboratory, Livermore, Calif., Sept. 1987.

Sink, M. K., "Handbook: Control Technologies for Hazardous Air Pollutants," EPA 625/6-91-014, U.S. Environmental Protection Agency, Cincinnati, June 1991.

Smith, M., ed., "Recommended Guide for the Prediction of the Dispersion of Airborne Effluents," American Society of Mechanical Engineers, New York, May 1968, p. 2.

Szepesi, D. J., *Compendium of Regulatory Air Quality Simulation Models,* Akademiai Kiado, Budapest, 1989.

Thom, H. C. S., "Tornado Probabilities," *Monthly Weather Review,* vol. 91, no. 10-12, 1963, pp. 730–736.

Thompson, P., Inglis, A., and Eklund, B., "Air/Superfund National Technical Guidance Study Series—Emission Factors for Superfund Remediation Technologies," EPA 450/1-91-001, U.S. Environmental Protection Agency, Research Triangle Park, N.C., Mar. 1991.

Tietenberg, T. H., *Emissions Trading—An Exercise in Reforming Pollution Policy,* Resources for the Future, Washington, D.C., 1985.

Touma, J. S., "Review and Evaluation of Area Source Dispersion Algorithms for Emission Sources at Superfund Sites," EPA 450/4-89-020, U.S. Environmental Protection Agency, Research Triangle Park, N.C., Nov. 1989.

Turner, D. B., "Atmospheric Dispersion Modeling—A Critical Review," *Journal of the Air Pollution Control Association,* vol. 29, no. 5, 1979, pp. 502–519.

———, "Workbook of Atmospheric Dispersion Estimates," Publ. AP-26, U.S. Environmental Protection Agency, Research Triangle Park, N.C., 1970.

U.S. Army Construction Engineering Research Laboratory, "Environmental Review Guide for USAREUR," 5 vols., Champaign, Ill., 1989.

U.S. Environmental Protection Agency (EPA), "Air Quality Atlas," EPA 400/K-92-002, Office of Air Quality Planning and Standards, U.S. Environmental Protection Agency, Research Triangle Park, N.C., May 1992.

———, "Analysis of the Economic and Environmental Effects of Methanol as an Automotive Fuel," U.S. Environmental Protection Agency, Ann Arbor, Mich., Sept. 1988b.

———, "Compilation of Air Pollutant Emission Factors," 2d ed., Publ. AP-42, U.S. Environmental Protection Agency, Research Triangle Park, N.C., Apr. 1973, p. 1.

———, "Compilation of Air Pollutant Emission Factors, vol. 1: Stationary Point and Area Sources," AP-42-SUPPL-B, U.S. Environmental Protection Agency, Research Triangle Park, N.C., Sept. 1985.

———, "Compilation of Air Pollutant Emission Factors, vol. 2: Mobile Sources," AP-42-SUPPL-A,

Office of Mobile Sources, U.S. Environmental Protection Agency, Ann Arbor, Mich., Jan. 1991a.

———, "Emissions Trading Policy Statement—General Principles for Creation, Banking and Use of Emission Reduction Credits," *Federal Register,* vol. 51, no. 233, Dec. 4, 1986, pp. 43814–43860.

———, "Guide for Compiling A Comprehensive Emission Inventory," Publ. no. APTD-1135, U.S. Environmental Protection Agency, Research Triangle Park, N.C., June 1972.

———, "Guidelines on Air Quality Models," EPA 450/2-78-027, U.S. Environmental Protection Agency, Research Triangle Park, N.C., Apr. 1978, pp. 16–27, A-12–A-24.

———, "MOBILE 4.1: Highway Vehicle Mobile Source Emission Factor Model (IBM PC Compatible) (for Microcomputers)," EPA SW/DK-91-099, Office of Mobile Sources, U.S. Environmental Protection Agency, Ann Arbor, Mich., Aug. 1991d.

———, "Motor Vehicle-Related Air Toxics Study," EPA 420/R-93-005, Office of Mobile Sources, U.S. Environmental Protection Agency, Ann Arbor, Mich., Apr. 1993.

———, "National Air Pollutant Emission Estimates—1940–1989," EPA 450/4-91-004, Office of Air Quality Planning and Standards, U.S. Environmental Protection Agency, Research Triangle Park, N.C., Mar. 1991b.

———, "National Emissions Report, 1985—National Emissions Data System (NEDS) of the Aerometric and Emissions Reporting System (AEROS)," EPA 450/4-88-018, U.S. Environmental Protection Agency, Research Triangle Park, N.C., Sept. 1988a.

———, "Parts 50, 51, 53, 58, 60, 61, 80, 85, and 87 of Air Programs," *Code of Federal Regulations,* vol. 40, subchap. C, Washington, D.C., July 1, 1991c.

———, "Prevention of Significant Deterioration for Particulate Matter," *Federal Register,* vol. 54, no. 192, Oct. 5, 1989b, pp. 41218–41232.

———, "Supplement No. 5 for Compilation of Air Pollutant Emission Factors," AP-42-SUPPL-5, U.S. Environmental Protection Agency, Research Triangle Park, N.C., Dec. 1975.

———, "Toxic Air Pollutant Emission Factors—Information Storage and Retrieval System—User's Manual," EPA 450/2-88-006B, U.S. Environmental Protection Agency, Research Triangle Park, N.C., Oct. 1988c.

———, "User's Guide to MOBILE 4 (Mobile Source Emission Factor Model)," EPA AA/TEB-89/01, U.S. Environmental Protection Agency, Office of Mobile Sources, Ann Arbor, Mich., Feb. 1989a.

Vossler, T. L., "Comparison of Steady State Evaporation Models for Toxic Chemical Spills: Development of a New Evaporation Model," GL-TR-89-0319, Geophysics Laboratory (AFSC), Hanscom AFB, Mass., Nov. 1989.

Watkins, S., "Background Document for the Surface Impoundment Modeling System (SIMS)," EPA SW/DK–EPA 450/4-90-009B–89-013B, U.S. Environmental Protection Agency, Research Triangle Park, N.C., Sept. 1989.

World Resources Institute, *The 1992 Environmental Almanac,* Houghton Mifflin Company, Boston, 1992, p. 70.

Zannetti, P., *Air Pollution Modeling—Theories, Computational Methods, and Available Software,* Van Nostrand Reinhold, New York, 1990.

CHAPTER 7

Prediction and Assessment of Impacts on the Surface-Water Environment

Many types of projects, plans, programs, or policies have impact implications for the surface-water environment (rivers, lakes, estuaries, or oceans). Effects can be represented by quantity and/or quality changes; these changes can, in turn, have aquatic faunal or floral species and aquatic ecosystem implications. Examples of projects which create impact concerns for the surface-water environment include (1) industrial plants or power plants withdrawing surface water for use as cooling water (this may be of particular concern during low-flow conditions); (2) power plants discharging heated wastewater from their cooling cycles; (3) industries discharging process wastewaters from either routine operations or as a result of accidents and spills; (4) municipal wastewater treatment plants discharging primary, secondary, or tertiary treated wastewaters; (5) dredging projects in rivers, harbors, estuaries, and/or coastal areas (increased turbidity and releases of sediment contaminants may occur); (6) projects involving "fill" or creation of "fast lands" along rivers, lakes, estuaries, and coastal areas; (7) surface-mining projects with resultant changes in surface-water hydrology and nonpoint pollution;

(8) construction of dams for purposes of water supply, flood control, or hydropower production; (9) river channelization projects for flow improvements; (10) deforestation and agricultural development resulting in nonpoint-source pollution associated with nutrients and pesticides, and irrigation projects leading to return flows laden with nutrients and pesticides; (11) commercial hazardous-waste disposal sites, and/or sanitary landfills, with resultant runoff water and nonpoint-source pollution; and (12) tourism projects adjacent to estuaries or coastal areas, with concerns related to bacterial pollution.

An additional issue of potential concern is the transboundary nature of surface-water systems (Economic Commission for Europe, 1991). For example, rivers can flow from region to region within a country, or from one country to another. Impacts caused by projects in one location may be experienced in distant locations. Therefore, surface-water impact studies may need to address the implications of projects in a transboundary or transnational context. This is particularly important in that the major effects of projects are often experienced downstream.

189

This chapter is focused on a methodological approach for addressing surface-water quantity and/or quality impacts from proposed undertakings (projects, plans, programs, or policies). The initial section summarizes some basic information on quantity-quality considerations. The next section discusses key federal legislation pertaining to surface-water quality. The main portion, comprising the next seven sections, highlights a six-step methodology. The final section consists of a summary.

BASIC INFORMATION ON SURFACE-WATER QUANTITY AND QUALITY

Surface-Water Hydrology

When considering surface-water quantity or quality it is important to understand the processes that create bodies of surface water (rivers, streams, lakes, and the like). Surface water is replenished by rainfall (or snowfall) that ends up in runoff and by groundwater that discharges into it, as shown in Figure 7.1. Rainfall can infiltrate the subsurface, be intercepted by foliage (initial abstraction), or result in runoff. The rainfall may subsequently "evapotranspirate" (evaporate naturally or through vegetative growth), enter the groundwater, and/or result in surface-water flow. The runoff flows downgradient (usually from a higher to a lower elevation) into creeks, streams, lakes, and rivers, and eventually into the oceans (unless it evaporates, infiltrates the subsurface, or is withdrawn along the way). The rainfall that infiltrates the subsurface and becomes groundwater may discharge into a surface water at some other location; in this case, the surface water is referred to as a "receiving stream" and the discharging groundwater is referred to as "base flow." Base flow

FIGURE 7.1
Hydrologic Cycle (*Source:* Adapted from Linsley and Franzini, 1979, p. 10).

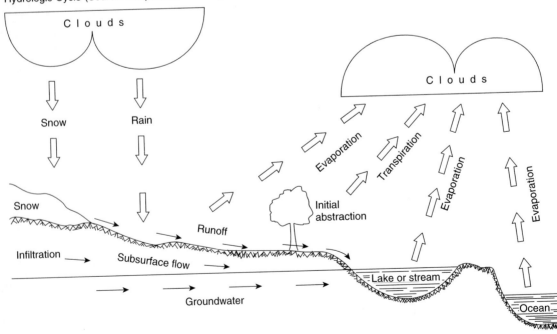

accounts for the flow in streams, rivers, and so on, between rainfall-runoff events. Once the rainfall has reached the oceans (or anytime between), evaporation can return the runoff to the atmosphere for a future rainfall event (likely at some other location) and the cycle starts over again; thus, the system is referred to as the "hydrologic cycle." Figure 7.1 demonstrates the cycle of both surface-water and groundwater hydrology, ("hydrology" refers to the distribution of water throughout the hydrologic cycle). Because of the dynamic nature of both the quantity- and quality-influencing processes, natural variations occur in the flow and quality characteristics, respectively.

Surface-Water Quality Parameters

Surface water comprises rainfall, runoff, base flow, and so on. Each of these inputs to the surface-water system can contribute natural compounds of relevance to water quality. For example, rainfall in highly industrialized regions may consist of acidic precipitation which is introduced to the surface water; runoff may bring with it natural organics, sediments, and so on; and base flow may have elevated levels of hard-

ness from the flow of the water through the subsurface. Human activities may increase the concentration of existing compounds in a surface water or may cause additional compounds to enter the surface water. For examples, discharge of wastewater (treated or otherwise) greatly adds to the organic loading of the surface water and clearing of land (for construction, farming, etc.) can result in increased erosion and sediment load in the surface water. Thus, it is important to recognize the natural (background) quality of surface waters and the existing impacts of human activities on this water quality.

"Surface-water pollution" can be defined in a number of ways; however, most definitions address excessive concentrations of particular substances for sufficient periods of time to cause identifiable effects. "Water quality" can be defined in terms of the physical, chemical, and biological characterization of the water. Physical parameters include color, odor, temperature, solids (residues), turbidity, oil content, and grease content. Each physical parameter can be broken down into subcategories. For example, the characterization of solids can be further subdivided

FIGURE 7.2
Relationships of Various Solids Tests Used for Water Quality Characterization.

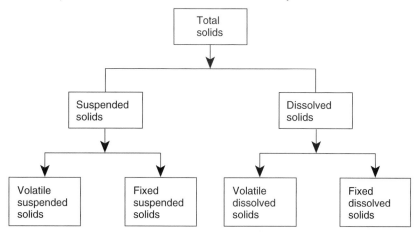

into suspended and dissolved solids (size and settleability) as well as organic (volatile) and inorganic (fixed) fractions. Figure 7.2 depicts a hierarchy of different solids tests. Chemical parameters associated with the organic content of water include biochemical oxygen demand (BOD), chemical oxygen demand (COD), total organic carbon (TOC), and total oxygen demand (TOD). It should be noted that BOD is a measure of the organics present in the water; it is determined by measuring the oxygen necessary to biostabilize the organics (the oxygen equivalent of the biodegradable organics present). A BOD response curve is shown in Figure 7.3. Inorganic chemical parameters include salinity; hardness; pH; acidity; alkalinity; and the presence of substances including iron, manganese, chlorides, sulfates, sulfides, heavy metals (mer-

cury, lead, chromium, copper, and zinc), nitrogen (organic, ammonia, nitrite, and nitrate), and phosphorus. Biological properties include bacteriological parameters such as coliforms, fecal coliforms, specific pathogens, and viruses. Table 7.1 provides an overview of various types and sources of water quality characteristics.

In evaluating surface-water pollution impacts associated with the construction and operation of a potential project, two main sources of water pollutants should be considered: nonpoint and point. Nonpoint sources are also referred to as "area" or "diffuse" sources. "Nonpoint pollutants" refer to those substances which can be introduced into receiving waters as a result of urban-area, industrial-area, or rural runoff—for example, sediment, pesticides, or nitrates enter-

FIGURE 7.3
BOD Response Curve.

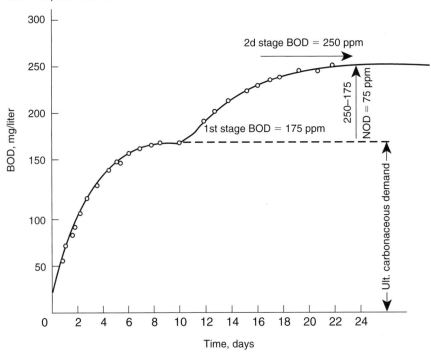

NOD = Nitrogenous oxygen demand

TABLE 7.1

PHYSICAL, CHEMICAL, AND BIOLOGICAL WATER-QUALITY CHARACTERISTICS AND THEIR SOURCES

Characteristic	Sources
Physical properties:	
Color	Domestic and industrial wastes, natural decay of organic materials
Odor	Decomposing wastewater, industrial wastes
Solids	Domestic water supply, domestic and industrial wastes, soil erosion, inflow/ infiltration
Temperature	Domestic and industrial wastes
Chemical constituents:	
Organic:	
Carbohydrates	Domestic, commercial, and industrial wastes
Fats, oils, and grease	Domestic, commercial, and industrial wastes
Pesticides	Agricultural wastes
Phenols	Industrial wastes
Proteins	Domestic, commercial, and industrial wastes
Priority pollutants	Domestic, commercial, and industrial wastes
Surfactants	Domestic, commercial, and industrial wastes
Volatile organic compounds	Domestic, commercial, and industrial wastes
Other	Natural decay of organic materials
Inorganic:	
Alkalinity	Domestic wastes, domestic water supply, groundwater infiltration
Chlorides	Domestic wastes, domestic water supply, groundwater infiltration
Heavy metals	Industrial wastes
Nitrogen	Domestic and agricultural wastes
pH	Domestic, commercial, and industrial wastes
Phosphorus	Domestic, commercial, and industrial wastes; natural runoff
Priority pollutants	Domestic, commercial, and industrial wastes
Sulfur	Domestic water supply; domestic, commercial, and industrial wastes
Gases:	
Hydrogen sulfide	Decomposition of domestic wastes
Methane	Decomposition of domestic wastes
Oxygen	Domestic water supply, surface-water infiltration
Biological constituents:	
Animals	Open watercourses and treatment plants
Plants	Open watercourses and treatment plants
Protists:	
Eubacteria	Domestic wastes, surface-water infiltration, treatment plants
Archaebacteria	Domestic wastes, surface-water infiltration, treatment plants
Viruses	Domestic wastes

Source: Metcalf and Eddy, 1991, p. 57.

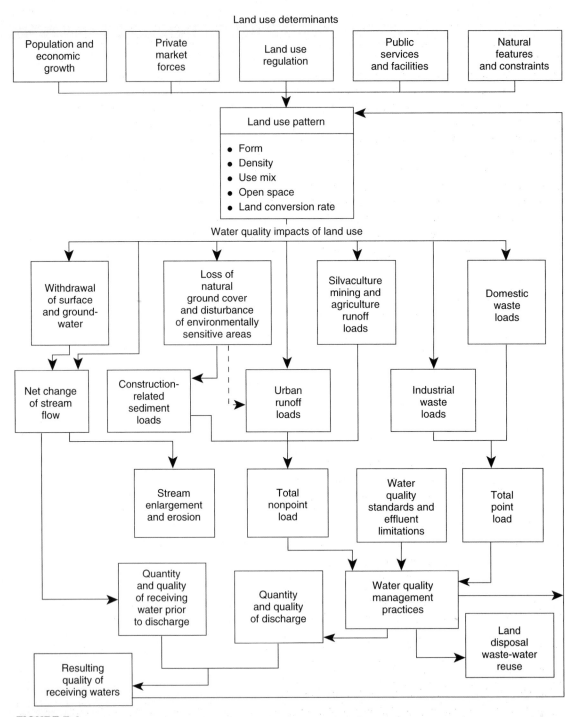

FIGURE 7.4
Schematic Diagram of the Land-Use—Water-Quality Relationship (Shubinski and Tierney, 1973).

TABLE 7.2

IMPORTANT SURFACE-WATER CONTAMINANTS AND THEIR IMPACTS

Contaminants	Reason for importance
Suspended solids	Suspended solids can lead to the development of sludge deposits and anaerobic conditions when untreated wastewater is discharged in the aquatic environment.
Biodegradable organics	Composed principally of proteins, carbohydrates, and fats, biodegradable organics are measured most commonly in terms of BOD (biochemical oxygen demand) and COD (chemical oxygen demand). If discharged untreated to the environment, their biological stabilization can lead to the depletion of natural oxygen resources and to the development of septic conditions.
Pathogens	Communicable diseases can be transmitted by the pathogenic organisms in wastewater.
Nutrients	Both nitrogen and phosphorus, along with carbon, are essential nutrients for growth. When discharged to the aquatic environment, these nutrients can lead to the growth of undesirable aquatic life. When discharged in excessive amounts on land, they can also lead to the pollution of groundwater.
Priority pollutants	Organic and inorganic compounds selected on the basis of their known or suspected carcinogenicity, mutagenicity, teratogenicity, or high acute toxicity. Many of these compounds are found in wastewater.
Refractory organics	These organics tend to resist conventional methods of wastewater treatment. Typical examples include surfactants, phenols, and agricultural pesticides.
Heavy metals	Heavy metals are usually added to wastewater from commercial and industrial activities and may have to be removed if the wastewater is to be reused.
Dissolved inorganics	Inorganic constituents such as calcium, sodium, and sulfate are added to the original domestic water supply as a result of water use and may have to be removed if the wastewater is to be reused.

Source: Metcalf and Eddy, 1991, p. 58.

ing a surface water because of runoff from agricultural farms. Point sources are related to specific discharges from municipalities or industrial complexes—for example, organics or metals entering a surface water as a result of wastewater discharge from a manufacturing plant. In a given body of surface water, nonpoint-source pollution can be a significant contributor to the total pollutant loading, particularly with regard to nutrients and pesticides. Figure 7.4 illustrates the relationship between land usage and these pollution sources relative to the resulting quality of receiving waters.

Some general characteristics of nonpoint-source pollution are as follows (Novotny and Chesters, 1981): (1) nonpoint-source discharges enter surface waters in a diffuse manner and at intermittent intervals that are related mostly to the occurrence of meteorological events; (2) pollution arises over an extensive area of land and is in transit overland before it reaches surface waters; (3) nonpoint-source discharges generally cannot be monitored at the point of origin, and the exact source is difficult or impossible to trace; (4) elimination or control of these pollutants must be directed at specific sites; and (5) in general, the most effective and economical controls are land-management techniques and conservation practices in rural zones and architectural or hydrological control in urban zones.

The effects of pollution sources on receiving-water quality are manifold and dependent upon the type and concentration of pollutants (Nemerow and Dasgupta, 1991). Soluble organics, as represented by high BOD wastes, cause depletion of oxygen in the surface water. This can result in fish kills, undesirable aquatic life, and undesirable odors. Even trace quantities of certain organics may cause undesirable taste and odors, and certain organics may be biomagnified in the aquatic food chain. Suspended solids decrease water clarity and hinder photosynthetic processes; if solids settle and form sludge deposits, changes in benthic ecosystems result. Color, turbidity, oils, and floating materials are of concern because of their aesthetic undesirability and possible influence on water clarity and photosynthetic processes. Excessive nitrogen and phosphorus can lead to algal overgrowth,

with concomitant water treatment problems resulting from algae decay and interferences with treatment processes. Chlorides cause a salty taste to be imparted to water, and, in sufficient concentration, limitations on water usage can occur. Acids, alkalies, and toxic substances have the potential for causing fish kills and creating other imbalances in stream ecosystems. Thermal discharges can also cause imbalances, as well as reductions in stream waste-assimilative capacity. Stratified flows from thermal discharges minimize normal mixing patterns in receiving streams and reservoirs. Table 7.2 provides an overview of important surface-water contaminants and their impacts. Table 7.3 summarizes the impacts of certain pollutants in relation to potential impairment of water usage.

TABLE 7.3

LIMITS OF WATER USES DUE TO WATER QUALITY DEGRADATION

Pollutant	Use						
	Drinking water	Aquatic wildlife, fisheries	Recreation	Irrigation	Industrial uses	Power and cooling	Transport
Pathogens	xx	0	xx	x	xx[a]	na	na
Suspended solids	xx	xx	xx	x	x	x[b]	xx[c]
Organic matter	xx	x	xx	+	xx[d]	x[e]	na
Algae	x[e,f]	x[g]	xx	+	xx[d]	x[e]	x[h]
Nitrate	xx	x	na	+	xx[a]	na	na
Salts[i]	xx	xx	na	xx	xx[j]	na	na
Trace elements	xx	xx	x	x	x	na	na
Organic micropollutants	xx	xx	x	x	?	na	na
Acidification	x	xx	x	?	x	x	na

xx Marked impairment causing major treatment or excluding the desired use
x   Minor impairment
0   No impairment
na Not applicable
+   Degraded water quality may be beneficial for this specific use
?   Effects not yet fully realized
[a] Food industries
[b] Abrasion
[c] Sediment setting in channels
[d] Electronic industries
[e] Filter clogging
[f] Odor, taste
[g] In fish ponds higher algal biomass can be accepted
[h] Development of water hyacinth (*Eichhomia crassipes*)
[i] Also includes boron, fluoride, etc.
[j] Ca, Fe, Mn in textile industries, etc.

Source: Chapman, 1992, p. 9.

KEY FEDERAL LEGISLATION

Surface-water-quality legislation can be considered from pollution-control and water-usage perspectives. Selected examples of major surface-water-quality legislation in the United States with a focus on pollution control are in Table 7.4. The Federal Water Pollution Control Act Amendments of 1972 (P.L. 92-500) established basic water quality goals and policies for the United States (U.S. Congress, 1972). The primary objective of the Clean Water Act of 1987 (also known as the "Water Quality Act of 1987") is to restore and maintain the chemical, physical, and biological integrity of the nation's waters; this was also an objective of the precursor laws (Dennison, 1992).

The federal laws have promulgated state requirements and programs. In terms of water usage, it may be necessary to consider different quality standards for different uses. For example, the Safe Drinking Water Act of 1986, which included amendments to the original law passed

TABLE 7.4

SELECTED EXAMPLES OF FEDERAL SURFACE-WATER-QUALITY LEGISLATION

Law	Date	Selected features
Rivers and Harbors Act	1899	Prohibited discharge of refuse into waterways that would interfere with navigation without a permit from the U.S. Army Corps of Engineers
Water Pollution Control Act	1948	Provided limited federal financial assistance to local governments for construction of municipal wastewater treatment facilities
Federal Water Pollution Control Act	1956	Increased federal financial assistance for municipal wastewater treatment facilities
Water Quality Act	1965	Required states to develop state water quality standards for interstate waters, and created the Federal Water Pollution Control Administration to establish broad guidelines and approve state standards. Increased federal financial assistance for municipal wastewater treatment facilities
Federal Water Pollution Control Act Amendments	1972	Greatly increased federal financial assistance for municipal wastewater treatment faciltiies. Instituted uniform technology-based effluent limitations for industrial dischargers and a national permit system for all point-source dischargers. Designated the U.S. Army Corps of Engineers as the permit-granting authority over discharge of dredged or fill material into U.S. waters
Clean Water Act	1977	Encouraged states to accept delegation of the national permit system and assume management of the construction grants program. Added control of priority toxic pollutants to the federal program
Municipal Wastewater Treatment Construction Grant Amendments	1981	Reduced federal financial assistance for municipal wastewater treatment facilities
Clean Water Act or "Water Quality Act"	1987	Phased out federal grants for construction of municipal wastewater treatment facilities; provided capitalization grants to state revolving funds. Required the EPA to develop regulations for storm-water runoff control. Required states to prepare nonpoint-source management programs

Source: After "News Update," 1992, p. 8.

in 1974, focused on setting minimum national standards for drinking water. Examples of maximum contaminant levels for categories including inorganic chemicals, organic chemicals, turbidity, and radioactivity are in Table 7.5. "Primary standards" are protective of public health, while "secondary standards" protect against nonhealth effects (Table 7.5).

Three key components of the Clean Water Act of 1987 of relevance to impact studies include water quality standards and planning, discharge permits, and effluent limitations.

Water Quality Standards and Planning

Typically, state water quality standards represent statewide goals for individual water bodies and provide a legal basis for decision making. The standards designate the use or uses to be made of the water and set criteria necessary to protect the uses (U.S. EPA, 1991c). Key relevant terms in water quality management include (after U.S. EPA, 1991b, pp. 260–261):

1. *Load allocation (LA)* The portion of a receiving water's loading capacity that is attributed either to one of its existing or future nonpoint sources of pollution or to natural (background) sources.
2. *Wasteload allocation (WLA)* The portion of a receiving water's loading capacity that is allocated to one of its existing or future point sources of pollution. WLAs constitute a type of water-quality-based effluent limitation.
3. *Total maximum daily load (TMDL)* The sum of the individual WLAs for point sources and LAs for nonpoint sources and background sources. If a receiving water has only one point-source discharger, the TMDL is the sum of that point-source WLA plus the LAs for any nonpoint sources of pollution and for background sources, tributaries, or adjacent segments. TMDLs can be expressed in terms of mass per time, toxicity, or other appropriate measures. If best management practices (BMPs) or other nonpoint-source

pollution controls make more-stringent load allocations practicable, then wasteload allocations can be made less stringent. Thus, the TMDL process provides for nonpoint-source-control trade-offs.

4. *Water-quality-limited segment* Any segment of which the water quality does not meet applicable standards, and/or is not expected to meet applicable standards, even after the application of technology-based effluent limitations.
5. *Water quality management (WQM) plan* A state- or areawide waste-treatment management plan developed and updated in accordance with the provisions of Sections 205(j), 208, and 303 of the Clean Water Act of 1977 and of 1987 and subsequent regulations. The biennial report required by Section 305(b) is related to the WQM plan.
6. *Best management practice (BMP)* Methods, measures, or practices (or combination of practices) determined by a state or designated areawide planning agency to be the most effective practicable means (including technological, economic, and institutional considerations) of preventing or reducing the amount of pollution generated by nonpoint sources to a level compatible with water quality goals—that is, the best means of meeting the particular nonpoint-source-control needs (Novotny and Chesters, 1981). BMPs include, but are not limited to, structural and nonstructural controls, and operation and maintenance procedures, as well as schedules of activities, prohibition of practices, and other management practices to prevent or reduce runoff pollution. BMPs can be applied before, during, and after pollution-producing activities to reduce or eliminate the introduction of pollutants into receiving waters.

Water-quality-limited segments cannot be expected to meet established water quality standards even if all point sources achieve effluent

TABLE 7.5

MAXIMUM CONTAMINANT LEVELS IN COMMUNITY WATER SYSTEMS

Contaminant category	Contaminant	Maximum contaminant level
Primary standards		
Inorganic chemicals	Arsenic	0.05 mg/L
	Barium	1
	Cadmium	0.010
	Chromium	0.05
	Fluoride	4.0
	Lead	0.05
	Mercury	0.002
	Nitrate (as N)	10
	Selenium	0.01
	Silver	0.05
Organic chemicals	Chlorinated hydrocarbons	
	Endrin	0.0002 mg/L
	Lindane	0.004
	Methoxychlor	0.1
	Toxaphene	0.005
	Chlorophenoxys	
	2,4-D (2,4-dichlorophenoxyacetic acid)	0.1
	2,4,5-TP Silvex (2,4,5-trichlorophenoxypropionic acid)	0.01
	Total trihalomethanes (the sum of the concentrations of bromodichloromethane, dibromochloromethane, tribromomethane (bromoform), and trichloromethane (chloroform)	0.10
Turbidity	Turbidity	1.0 JTU (turbidity units)
Radioactivity	Combined radium 226 and radium 228	5 pCi/L
	Gross alpha-particle activity (including radium 226 but excluding radon and uranium)	15 pCi/L
Bacteriological	Total coliform	1/100 mL
Secondary standards		
Miscellaneous	Aluminum	0.05 to 0.2 mg/L
	Chloride	250 mg/L
	Color	15 CU (color units)
	Copper	1.0 mg/L
	Corrosivity	Noncorrosive
	Fluoride	2.0 mg/L
	Foaming agents	0.5 mg/L
	Iron	0.3 mg/L
	Manganese	0.05 mg/L
	Odor	3 Ton[a]
	pH	6.5 to 8.5
	Silver	0.1 mg/L
	Sulfate	250 mg/L
	Total dissolved solids (TDS)	500 mg/L
	Zinc	5 mg/L

[a]Threshold odor number

Source: Compiled using data from U.S. Environmental Protection Agency, 1991e, and 1991f.

limitations such as secondary treatment for publicly owned treatment works and best practicable treatment for industrial discharges. "Effluent-limited segments" are those where water quality standards can be achieved once all point sources meet effluent limitations. For water-quality-limited stream segments, the individual states are to establish total maximum daily loads (TMDLs), with the concurrence of the U.S. Environmental Protection Agency. The TMDL for a stream segment indicates the greatest amount of a pollutant that can be received daily without violating the state's water quality standards (The Clean Water Act Requires A Water-Quality-Based Approach . . . , 1989). Depending upon the situation, controls for storm-water discharges or advanced levels of wastewater treatment may be necessary. If a proposed project is in a water-quality-limited segment, it would be desirable to ascertain the pertinent TMDL (if one has been developed) and use it in assessing the daily load from the project.

Section 319 of the Clean Water Act of 1987 requires states to develop nonpoint-source pollution control programs; such programs are to include an assessment of problems and the development of a management program.

State-level water quality management plans have been developed through the requirements of Section 208 of the Federal Water Pollution Control Act Amendments of 1972 and its successors, and through river basin planning efforts. Information on existing point and nonpoint loadings for organics, metals, nutrients, and other pollutants, organized by stream segment, should be available from the pertinent state agency or the regional office of the U.S. Environmental Protection Agency.

Water quality standards vary from state to state, river basin to river basin, and among various segments within river basins. As an example, Oklahoma water quality standards will be used. Most state standards include a statement regarding "antidegradation." In Oklahoma, the statement is as follows (Oklahoma Water Resources Board, 1982, p. 2):

> It is recognized that certain waters of the State possess an existing water quality which exceeds those levels necessary to support propagation of fish, shellfish, wildlife, and recreation in and on the water. These high quality waters shall be maintained and protected unless the State decides, after full satisfaction of the intergovernmental coordination, and public participation provisions of the State's continuing planning process, to allow lower water quality as a result of necessary and justifiable economic or social development. Furthermore, where limited degradation is justified, the State shall require that any new point source of pollution, or increased load from an existing point source, protect all existing and attainable beneficial uses through the highest statutory and regulatory requirements, and feasible management or regulatory programs pursuant to Section 208 of Public Law 92-500 as amended by PL 95-217 for nonpoint sources.

State standards also include consideration of present and potential beneficial uses of water. Beneficial-use designations in Oklahoma are as follows:

Public and private water supplies
Emergency public and private water supplies
Fish and wildlife propagation
Agriculture
Hydroelectric power generation
Industrial and municipal process and cooling water
Primary body-contact recreation
Secondary body-contact recreation
Navigation
Aesthetics
Smallmouth bass fisheries (excluding lake waters)
Trout fisheries (put and take)

Table 7.6 summarizes the stream quality requirements for public and private drinking water supplies in Oklahoma.

TABLE 7.6

STREAM QUALITY REQUIREMENTS FOR PUBLIC AND PRIVATE DRINKING WATER SUPPLIES IN OKLAHOMA

Characteristics	Standards
Physical	
Color	Color-producing substances from other than natural sources must be limited to concentrations equivalent to 75 color units (CUs).
Odor	Taste- and odor-producing substances from other than natural origin must be limited to concentrations that will not interfere with the production of a potable water supply by modern treatment methods.
Temperature	At no time is heat to be added to any stream in excess of the amount that will raise the temperature of the receiving water more than 5°F. In streams, temperature determinations are to be made by averaging representative temperature measurements of the cross-sectional area of streams at the end of the mixing zone. The normal daily and seasonal variations that were present before the addition of heat from other than natural sources is to be maintained. The maximum temperature due to man-made causes must not exceed 68°F in trout streams, 84°F in smallmouth bass streams, and 90°F in all other streams and lakes.
Inorganic elements, mg/L	
Arsenic	0.05
Barium	1.0
Cadmium	0.01
Chromium	0.05
Copper	1.0
Fluoride, at 90°F	1.6
Lead	0.05
Mercury	0.002
Nitrates	0
pH	pH values must be between 6.5 and 8.5; pH values less than 6.5 or greater than 8.5 must not be due to water discharge(s).
Selenium	0.01
Silver	0.05
Zinc	5.0
Organic chemicals, mg/L	
Cyanide	0.2
Detergents, total	0.2
Methylene blue active substances	0.5
Oil and grease	All waters must be maintained free of oil and grease to prevent a visible film of oil or globules of oil or grease on or in the water. Oil and grease must not be present in quantities that adhere to stream banks and coat bottoms of watercourses or that cause deleterious effects to the biota. For public and private water supplies, the water must be maintained free from oil, grease, taste, and odors that emanate from petroleum products.
Phthalate esters	0.003
Microbiological	
Coliform organisms	The bacteria of the fecal coliform group must not exceed a monthly geometric mean of 200/100 mL, as determined by multiple tube fermentation or membrane filter procedures based on a minimum of not less than five (5) samples taken over not more than a thirty- (30-) day period. Further, in no more than 10% of the total samples during any thirty- (30-) day period is the bacteria of the fecal coliform group to exceed 400/100 mL.

Source: Compiled using data from Oklahoma Water Resources Board, 1979.

Discharge Permits

Operators of new point sources of wastewater discharge have to apply for National Pollutant Discharge Elimination System (NPDES) permits under Section 402 of the Clean Water Act of 1987 and its precursors (dating back to 1972). Applications for permit renewals at five-year intervals will be necessary over the operational life of the project. The permits may be issued by state agencies with primacy, or by the regional office of the U.S. Environmental Protection Agency. Permits typically address pertinent effluent limitations (discharge standards) for conventional and toxic pollutants, monitoring and reporting requirements, and schedules of compliance (Miller, Taylor, and Monk, 1991). Effluent limitations may be technology-based or water-quality-based, with the former involving consideration of best practicable control technology currently available (BPT), best available technology economically achievable (BAT), best conventional-pollutant control technology (BCT), or new source performance standards (NSPSs). BPT emphasizes "end-of-pipe" controls and is reflective of the average of the best for the industry category; it deals primarily with conventional pollutants such as BOD, oil and grease, solids, pH, and some metals. BAT may include pollution prevention through process control and end-of-pipe technology; it deals primarily with toxics such as organics and heavy metals. BCT is to be used with BAT. NSPSs are to be based on the best available demonstrated control technology (BADCT); they are typically similar to implementations of BAT with BCT. Water-quality-based effluent limitations may require greater levels of treatment as dictated by a waste-load-allocation scheme based on the TMDL for the relevant stream segment. Detailed information on NPDES permits is contained in U.S. Environmental Protection Agency (1991a).

Section 402(p) of the Clean Water Act of 1987 requires NPDES permits for storm-water (runoff) discharges associated with industrial

---

TABLE 7.7

SECONDARY TREATMENT STANDARDS FOR POTWs

Parameter	Standards
BOD_5	1. The 30-day average must not exceed 30 mg/L.
	2. The 7-day average must not exceed 45 mg/L.
	3. The 30-day average % removal is not to be less than 85%
	4. At the option of the NPDES permitting authority, in lieu of the parameter BOD_5 and the levels of the effluent quality specified above, the parameter $CBOD_5$ may be substituted with the following levels of the $CBOD_5$ effluent quality, provided
	i. The 30-day average does not exceed 25 mg/L.
	ii. The 7-day average does not exceed 40 mg/L.
	iii. The 30-day average % removal is not less than 85%.
Suspended solids	1. The 30-day average must not exceed 30 mg/L.
	2. The 7-day average must not exceed 45 mg/L.
	3. The 30-day average % removal must not be less than 85%.
pH	The effluent values for pH must be maintained within the limits of 6.0 to 9.0, unless the publicly owned treatment works demonstrate both of the following:
	1. Inorganic chemicals are not added to the waste stream as part of the treatment process.
	2. Contributions from industrial sources do not cause the pH of the effluent to be less than 6.0 or greater than 9.0.

Notes: BOD_5 = The 5-day measure of the biochemical oxygen demand (BOD) pollutant parameter.
    $CBOD_5$ = The 5-day measure of the carbonaceous biochemical oxygen demand (CBOD) pollutant parameter.
Source: Compiled using data from U.S. Environmental Protection Agency, 1991d.

activity; discharges from large municipal separate storm-water systems (those serving a population of 250,000 or more); and discharges from medium-size municipal separate storm-water systems (those serving a population of 100,000 or more, but less than 250,000) (Government Institutes, 1991). Nonpoint-source pollution in storm water from industrial-type land uses and other areas at military installations also

needs to be addressed. Information on application requirements and procedures, including required technical supporting data, is contained in Government Institutes (1991). (Technical supporting data includes drainage maps and outfall lines, estimates of discharge flow rates and volumes, and estimates of pollutant loadings based on nonpoint-source-unit waste-generation factors.)

TABLE 7.8

BPT EFFLUENT LIMITATIONS GUIDELINES FOR CRACKING-SUBCATEGORY POINT SOURCES FROM PETROLEUM REFINING

Pollutant or pollutant property	BPT effluent limitations	
	Maximum for any 1 day	Maximum average of daily values for 30 consecutive days
	Metric units, kg/1,000 m^3 of feedstock[a]	
BOD$_5$	28.2	15.6
TSS	19.5	12.6
COD[b]	210	109
Oil and grease	8.4	4.5
Phenolic compounds	0.21	0.10
Ammonia as N	18.8	8.5
Sulfide	0.18	0.082
Total chromium	0.43	0.25
Hexavalent chromium	0.035	0.016
pH	(c)	(c)
	English units, (lb/1,000 bbl of feedstock)[a]	
BOD$_5$	9.9	5.5
TSS	6.9	4.4
COD[b]	74.0	38.4
Oil and grease	3.0	1.6
Phenolic compounds	0.074	0.036
Ammonia as N	6.6	3.0
Sulfide	0.065	0.029
Total chromium	0.15	0.088
Hexavalent chromium	0.012	0.0056
pH	(c)	(c)

[a]"Feedstock" denotes the crude oil and natural gas liquids fed to the topping units.
[b]In any case in which the applicant can demonstrate that the chloride ion concentration in the effluent exceeds 1,000 mg/L (1,000 ppm), the regional administrator may substitute TOC as a parameter in lieu of COD. Effluent limitations for TOC shall be based on effluent data from the plant correlating TOC to BOD$_5$. If, in the judgment of the regional administrator, adequate correlation data are not available, the effluent limitations for TOC shall be established at a ratio of 2.2 to 1 to the applicable effluent limitations on BOD$_5$.
[c]Within the range of 6.0 to 9.0.
Source: Compiled using data from U.S. Environmental Protection Agency, 1991g.

NPDES permits for storm water from industrial areas require the development of a pollution prevention plan to reduce pollution at the source. The plan is to include five major phases: (1) planning and organization, (2) assessment, (3) best management practice (BMP) selection and plan design, (4) implementation, and (5) evaluation and site inspection (U.S. EPA, 1992c). Detailed information on developing pollution prevention plans and BMPs for construction activities and industrial activities is also available in U.S. Environmental Protection Agency (1992a, 1992b).

Effluent Limitations

Effluent limitations are typically established by the NPDES permitting process. Two examples will be cited, one for publicly owned treatment works (POTWs) and one for a petroleum industry application. Secondary treatment standards for POTWs are delineated in Table 7.7.

Table 7.8 summarizes the effluent limitations based on BPT for point sources associated with the Cracking subcategory of petroleum refining. Adjustment factors for the effluent limitations, which account for facility size and specific processes, are in the regulations (U.S. EPA, 1991g). Effluent-limitations information is also available for BAT, BCT, and NSPS for the Cracking subcategory, and for BPT, BAT, BCT, and NSPS for the Topping, Petrochemical, Lube, and Integrated subcategories in U.S. Environmental Protection Agency (1991g).

CONCEPTUAL APPROACH FOR ADDRESSING SURFACE-WATER-ENVIRONMENT IMPACTS

To provide a basis for addressing surface-water-environment impacts, a six-step or six-activity model is suggested for the planning and conduction of impact studies. This model is flexible and can be adapted to various project types by modification, as needed, to enable the addressing of specific concerns of specific projects in unique locations. It should be noted that the focus in this model will be on projects, and their surface-water impacts; however, the model could also be applied to plans, programs, and regulatory actions. Although this chapter is mainly related to river systems, the principles could also be applied to the examination of impacts on lakes, estuaries, coastal zones, and the ocean.

The six generic steps are (1) identification of the types and quantities of water pollutants to be introduced, water quantities to be withdrawn, or other impact-causing factors related to the development project; (2) description of the environmental setting in terms of river, lake, or estuarine flow patterns; water quality characteristics; existing or historical pollution problems; pertinent meteorological factors (examples are precipitation, evaporation, and temperature); relationships to area groundwater resources; existing point and nonpoint sources of pollution; and pollution loadings and existing water withdrawals; (3) procurement of relevant laws, regulations, or criteria related to water quality and/or water usage, and any relevant compacts (agreements) between states, countries, or other entities related to relevant transnational waters; (4) conduction of impact prediction activities, including the use of mass balances in terms of water quantity and/or pollutant-loading changes, mathematical models for relevant pollutant types (conservative, nonconservative, bacterial, nutrient, and thermal), aquatic ecosystem models to account for floral and faunal changes and nutrient-pollutant cycling or qualitative predictions based on case studies and professional judgment; (5) use of pertinent information from step 3, along with professional judgment and public input, to assess the significance of anticipated beneficial and detrimental impacts; and (6) identification, development, and incorporation of appropriate mitigation measures for the adverse impacts. Figure 7.5 delineates the relationship between the six steps or activities in the suggested conceptual approach.

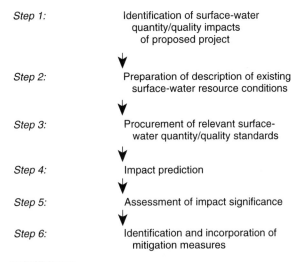

Step 1:   Identification of surface-water
           quantity/quality impacts
           of proposed project

Step 2:   Preparation of description of existing
           surface-water resource conditions

Step 3:   Procurement of relevant surface-
           water quantity/quality standards

Step 4:   Impact prediction

Step 5:   Assessment of impact significance

Step 6:   Identification and incorporation of
           mitigation measures

FIGURE 7.5
Conceptual Approach for Study Focused on Surface-
Water Environment Impacts.

STEP 1: IDENTIFICATION OF SURFACE-WATER QUANTITY OR QUALITY IMPACTS

The first activity is to determine the features of the proposed project, the need for the project, and the potential alternatives which either have been or could be considered. Key information relative to the proposed project includes such items as (1) the type of project and how it functions or operates in a technical context, particularly with regard to water usage and waste-water generation, or to the creation of changes in water quantity or quality; (2) the proposed location of the project; (3) the time period required for project construction; (4) the potential environmental outputs from the project during its operational phase, including information relative to water usage and water-pollutant emissions, and waste-generation and -disposal needs; (5) the identified need for the proposed project in the particular location (this need could be related to flood control, industrial development, economic development, and many other requirements; it is important to begin to consider proj-

ect need because it will be addressed as part of the subsequent related environmental documentation); and (6) any alternatives which have been considered, with generic alternatives for factors including site location, project size, project design features and pollution-control measures, and project timing relative to construction and operational phases.

The focus of this step is on identifying potential impacts of the subject project. This early qualitative identification of anticipated impacts can help in refining subsequent steps; for example, it can aid in describing the affected environment and in calculating potential impacts. The identification of potential impacts can be an outcome of the scoping process, and step 1 should also include consideration of the generic impacts related to the project type. There is an abundance of published information generated over the past two decades which enables planners of impact studies to more easily identify anticipated impacts.

Several methods have been developed as aids for identifying generic impacts related to the project type being analyzed; one approach

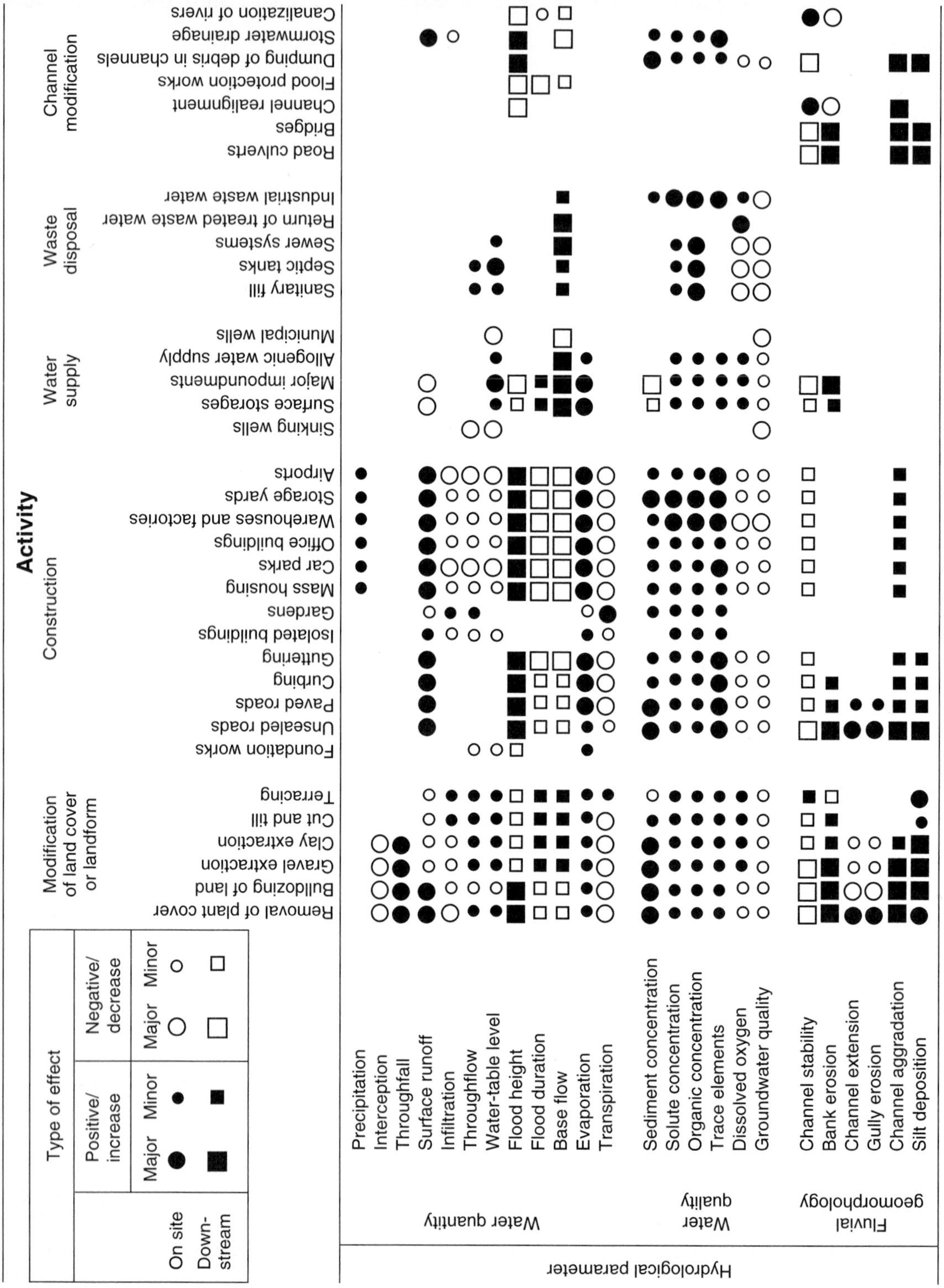

FIGURE 7.6
Impacts of Land Uses and Changes on Selected Hydrological Parameters (Source: Douglas, 1983, p. 58).

makes use of simple interaction matrices. For example, Figure 7.6 displays an impact matrix which depicts various land-use changes and their consequences on selected hydrological parameters. Careful examination of Figure 7.6 indicates that both positive and negative effects might occur either at the site of the land-use change (development) or downstream of the development. Additional methods for impact identification, such as networks and simple and descriptive checklists, are discussed in Chapter 3.

Identifying potential impacts may require the delineation of the quantities of surface-water usage, the types and quantities of potential surface-water pollutants to be utilized or generated during the project, and/or the activities that will alter the amount and quality of runoff that results from a precipitation episode. A first consideration might be to develop a list of the materials to be utilized during the project and of those materials which will require disposal. The materials utilized during the project may contaminate surface water during storage as the result of precipitation-runoff events. Materials which are disposed of during the project may contaminate surface water if not properly managed. Materials that may result in surface-water contamination include fuels and oils, preservatives, bituminous products, insecticides, fertilizers, various other chemicals, and solid and liquid wastes. (See Tables 7.2 and 7.3 for a summary of important contaminants and impacts.)

It is often beneficial to conduct a review of literature on projects similar to the one under consideration. Since there are over 20 years of experience in the conduct of impact studies in the United States (as well as in numerous other countries), it is likely that published literature and/or project reports exist that will provide some insight into the surface-water impacts resulting from many types of proposed projects. Computer-based literature searches of the National Technical Information Service (NTIS),

Pollution Abstracts, Compendex (engineering index), Biosis (biological abstracts), and other information systems can be helpful. Following a literature review, a listing of the typical impacts of proposed projects can be assembled. For example, a listing of the typical impacts of dam and reservoir projects identified through a computer-based literature search was shown earlier in Table 4.3. Considerable information on the water quality impacts of dam and reservoir, channelization, levee, and dredging projects is available through the Waterways Experiment Station of the U.S. Army Corps of Engineers. Finally, the physical failure of large dams can cause large losses of life and productive land in addition to transient shifts in water quality. For examples, the filling of the Vaiont Dam reservoir in Italy led to a landslip in 1963 that resulted in 2,000 deaths; 13 dam failures in the United States in the period 1874–1977 caused the loss of over 3,500 lives (Roberts, Liss, and Saunders, 1990).

Information on the characteristics of wastewater discharges from municipalities and various industries can also be assembled. For example, the typical composition of untreated domestic wastewater in the United States is shown in Table 7.9. Several books are available which summarize the characteristics of wastewaters from various industries; examples include Nemerow (1978), Corbitt (1990), and Nemerow and Dasgupta (1991). Information on the types and quantities of water pollutants from major private sector categories such as the apparel, food, materials, chemical, and energy industries is available in Nemerow and Dasgupta (1991). To serve as an example of industrial wastewaters, brief information will be presented on the pollutant discharges from petroleum refineries. The total amount of water used in a petroleum refinery has been estimated to be 770 gal/bbl of crude oil (Nemerow, 1978). Approximately 80 to 90 percent of the water is used for cooling purposes only and is not contaminated except by leaks in the lines. Process

TABLE 7.9

TYPICAL COMPOSITION OF UNTREATED DOMESTIC WASTEWATER

Contaminants	Unit	Concentration		
		Weak	Medium	Strong
Solids, total (TS)	mg/L	350	720	1,200
Dissolved solids, total (TDS)	mg/L	250	500	850
Fixed	mg/L	145	300	525
Volatile	mg/L	105	200	325
Suspended solids (SS)	mg/L	100	220	350
Fixed	mg/L	20	55	75
Volatile	mg/L	80	165	275
Settleable solids	mL/L	5	10	20
Biochemical oxygen demand 5-day, 20°C (BOD$_5$, 20°C)	mg/L	110	220	400
Total organic carbon (TOC)	mg/L	80	160	290
Chemical oxygen demand (COD)	mg/L	250	500	1,000
Nitrogen, total, as N	mg/L	20	40	85
Organic	mg/L	8	15	35
Free ammonia	mg/L	12	25	50
Nitrites	mg/L	0	0	0
Nitrates	mg/L	0	0	0
Phosphorus, total, as P	mg/L	4	8	15
Organic	mg/L	1	3	5
Inorganic	mg/L	3	5	10
Chlorides[a]	mg/L	30	50	100
Sulfate[a]	mg/L	20	30	50
Alkalinity as CaCO$_3$	mg/L	50	100	200
Grease	mg/L	50	100	150
Total coliform	no./100 mL	10^6–10^7	10^7–10^8	10^7–10^9
Volatile organic compounds (VOCs)	μg/L	<100	100–400	>400

[a]Values should be increased by amount present in domestic water supply.
Note: 1.8(°C) + 32 = °F.
Source: Adapted from Metcalf and Eddy, 1991, p. 64.

wastewaters, constituting 10 to 20 percent of the total, may include free and emulsified oil from leaks, spills, tank drawoff, and other sources; waste caustic, caustic sludges, and alkaline waters; acid sludges and acid waters; emulsions incident to chemical treatment; condensate waters from distillate separators and tank drawoff; tank-bottom sludges; coke from equipment tubes, towers, and other locations; acid gases; waste catalyst and filtering clays; and special chemicals from by-product chemical manufacturing.

Both conventional and toxic pollutants may be found in the wastewaters from petroleum refineries. As noted earlier, "conventional pollutants" refer to those which have received

TABLE 7.10

SUBCATEGORIZATION OF THE PETROLEUM-REFINING INDUSTRY TO SIGNIFICANT DIFFERENCES IN WASTEWATER CHARACTERISTICS

Subcategory	Characteristics
Topping	Topping and catalytic reforming, whether or not the facility includes any other process.
	This subcategory is not applicable to facilities which include thermal processes (coking, visbreaking, and so on) or catalytic cracking
Cracking	Topping and cracking, whether or not the facility includes any other processes, except as specified in any of the subcategories listed below
Petrochemical	Topping, cracking, and petrochemical operations,[a] whether or not the facility includes any other process, except lube-oil manufacturing operations
Lube	Topping, cracking, and lube-oil manufacturing processes, whether or not the facility includes any other process, except petrochemical operations[a] (see above)
Integrated	Topping, cracking, lube-oil manufacturing processes, and petrochemical operations, whether or not the facility includes any other processes

[a]The term "petrochemical operations" means the production of second-generation petrochemicals (alcohols, ketones, cumene, styrene, and so on) or first-generation petrochemical and isomerization products (BTX, olefins, cyclohexane, and so on) when 15 percent or more of refinery production is as first-generation petrochemicals and isomerization products.
Source: Compiled from data in U.S. Environmental Protection Agency, 1980.

historical attention, while "toxic pollutants" relate to parameters being given increasing attention because of their potential environmental toxicity (U.S. EPA, 1980). In delineating information on wastewater characteristics, five refinery subcategories, based on throughputs and process capacities, can be used; these are shown in Table 7.10. Table 7.11 presents ranges and median loadings in raw wastewater of conventional pollutants for the petroleum-refining industry subcategories. "Raw wastewater" has been defined as the effluent from the oil separator, which is an integral part of refinery process operations for product and raw material recovery prior to wastewater treatment.

Quality characteristics of industrial wastes vary considerably depending upon the type of industry. A useful parameter in describing industrial wastes is population equivalent:

$$PE = \frac{(A)(B)(8.34)}{0.17}$$

where PE = population equivalent based on organic constituents in industrial waste

A = industrial waste flow, mgd
B = industrial waste BOD, mg/L
8.34 = lb/gal
0.17 = lb BOD per person-day

A similar type of population-equivalent calculation could be made for suspended solids, nutrients, and other pertinent constituents. To express all waste loadings on a similar basis, population-equivalent calculations can be made for various pollutants from both point and nonpoint sources.

Nonpoint sources of water pollution have been recognized as potential major contributors to the total waste load within the aquatic environment, and it is vitally important to consider nonpoint sources of water pollution along with point sources. Information on unit-waste-generation factors for nonpoint sources is increasing; Table 7.12 presents information on the rates

TABLE 7.11

RAW WASTEWATER[a] LOADINGS BY SUBCATEGORY IN PETROLEUM-REFINING INDUSTRY (IN NET KILOGRAMS PER 1,000 M^3 OF FEEDSTOCK THROUGHPUT)

Characteristics	Topping subcategory		Cracking subcategory		Petrochemical subcategory	
	Range[b]	Median	Range[b]	Median	Range[b]	Median
Flow[c]	8.00–558	66.6	3.29–2,750	93.0	26.6–443	109
BOD_5	1.29–217	3.43	14.3–466	72.9	40.9–715	172
COD	3.43–486	37.2	27.7–2,520	217	200–1,090	463
TOC	1.09–65.8	8.01	5.43–320	41.5	48.6–458	149
TSS	0.74–286	11.7	0.94–360	18.2	6.29–372	48.6
Sulfides	0.002–1.52	0.054	0.01–39.5[d]	0.94[d]	0.009–91.5	0.86
Oil and grease	1.03–88.7	8.29	2.86–365	31.2	12.0–235	52.9
Phenols	0.001–1.06	0.034	0.19–80.1	4.00	2.55–23.7	7.72
Ammonia	0.077–19.5	1.20	2.35–174	28.3	5.43–206	34.3
Chromium	0.0002–0.29	0.007	0.0008–4.15	0.25	0.014–3.86	0.234

Characteristics	Lube subcategory		Integrated subcategory	
	Range[b]	Median	Range[b]	Median
Flow[c]	68.6–772	117	40.0–1,370	235
BOD_5	62.9–758	217	63.5–615	197
COD	166–2.290	543	72.9–1,490	329
TOC	31.5–306	109	28.6–678	139
TSS	17.2–312	71.5	15.2–226	59.1
Ammonia	6.5–96.2	24.1		
Phenols	4.58–52.9	8.29	0.61–22.6	3.78
Sulfides	0.00001–20.0	0.014	0.52–7.87[d]	2.00[d]
Oil and grease	23.7–601	120	20.9–269	74.9
Chromium	0.002–1.23	0.046	0.12–1.92	0.49

[a]After refinery oil separator.
[b]Probability of occurrence less than or equal to .10 or .90, respectively.
[c]1,000 m^3/1,000 m^3 of feedstock throughput.
[d]Sulfur.
Source: U.S. Environmental Protection Agency, 1980, p. II.14–17.

of erosion from various land uses. More-detailed or site-specific information could be developed by using the "universal soil loss equation"; however, the values in Table 7.12 are useful for generating an approximation of the sediment load. For urban developments, Table 7.13 summarizes the relationship between six land uses and four pollutants based on the percentage of land covered by impervious material, or "density." Similar information regarding storm-water pollution loading based on the units per acre of residential development is in Table 7.14.

"Loading functions" refer to simple mathematical expressions that have been developed to evaluate either the production and/or the transport of a given pollutant in a specific area under a given usage. Table 7.15 summarizes (for various land uses) the availability of loading functions for different pollutants. General information on loading functions is in McElroy et al. (1976).

The primary point of Tables 7.9 through 7.14 is to provide information to enable calculation of the total quantity of water pollution anticipated from a given activity. This information is basic to prediction of "mesoscale" and "microscale" water quality impacts. Information is also available on unit-waste-generation factors for nutrients, metals, and organics

TABLE 7.12

REPRESENTATIVE RATES OF EROSION FROM VARIOUS LAND USES

Land use	Erosion rate		
	Metric tons/km²-yr	Tons/mi²-yr	Relative to forest = 1
Forest	8.5	24	1
Grassland	85	240	10
Abandoned surface mines	850	2,400	100
Cropland	1,700	4,800	200
Harvested forest	4,250	12,000	500
Active surface mines	17,000	48,000	2,000
Construction	17,000	48,000	2,000

Note: Rainfall is approximately 30 in/yr.
Source: U.S. Environmental Protection Agency, 1973, p. 6.

TABLE 7.13

STORMWATER POLLUTION FOR SELECTED URBAN LAND USES

Land use	Density[a]	Nitrogen[b]	Phosphorus[b]	Lead[b]	Zinc[b]
Residential, large lot (1 acre)	12%	3.0	0.3	0.06	0.20
Residential, small lot (0.25 acre)	25%	8.8	1.1	0.40	0.32
Townhouse apartment	40%	12.1	1.5	0.88	0.50
High-rise apartment	60%	10.3	1.2	1.42	0.71
Shopping center	90%	13.2	1.2	2.58	2.06
Central Business District	95%	24.6	2.7	5.42	2.71

[a]Based on percentage of the land covered by impervious (hard surface) material.
[b]Pounds per acre of land per year.
Source: Marsh, 1991, p. 161.

TABLE 7.14

ANNUAL STORM-WATER POLLUTION LOADING FOR RESIDENTIAL DEVELOPMENT

Residential density	Phosphorus[a]	Nitrogen[a]	Lead[a]	Zinc[a]	Sediment[b]
0.5 unit/ac (1.25 person)	0.8	6.2	0.14	0.17	0.09
1.0 unit/ac (2.5 persons)	0.8	6.7	0.17	0.20	0.11
2.0 units/ac (5 persons)	0.9	7.7	0.25	0.25	0.14
10.0 units/ac (25 persons)	1.5	12.1	0.88	0.50	0.27

[a]Pounds per acre per year.
[b]Tons per acre per year.
Source: Marsh, 1991, p. 162.

TABLE 7.15

LAND-USE–POLLUTANTS MATRIX AND AVAILABLE LOADING FUNCTIONS

Land use	Major pollutant	Loading functions; base data
Agriculture	Sd, N, Ph, P, BOD, M	***
Irrigation return flow	TDS	**
Silviculture	Sd, N, Ph, BOD, M	*
Feedlots	Sd, N, Ph, BOD	**
Urban runoff	Sd, N, Ph, P, BOD, TDS, M, coliform	***
Highways	Sd, N, Ph, BOD, TDS, M	*
Construction	Sd, M	*
Terrestrial disposal	N, Ph, TDS, M, others	*
Background	Sd, N, Ph, BOD, TDS, M, radiation	**
Mining	Sd, M, radiation, acidity	*

Key: Sd = sediment, N = nitrogen, Ph = phosphorus, P = pesticides, BOD = biochemical oxygen demand, TDS = total dissolved solids, M = heavy metals

***Wide range of data is available.
**Less data is available
*A little data is available
Source: Canter et al., 1990, p. 32.

from agricultural and urban nonpoint sources in Novotny and Chesters (1981).

In addition to information on pollutant types and quantities, it may also be necessary to assemble information on the transport and fate of specific pollutant materials. For example, information may be needed on the fate of petroleum products, other organics, nutrients, metals, and the like, in the water environment. It is important to know whether the pollutant will partition between the water and sediment phases, or become associated with aquatic flora and fauna. Metals can occur in surface-water systems as both dissolved and particulate constituents. Bio-

geochemical partitioning of metals can yield a diversity of forms, including hydrated or "free" ions, colloids, precipitates, adsorbed phases, and coordination complexes with dissolved organic and inorganic ligands. An excellent review related to aluminum, arsenic, cadmium, chromium, copper, iron, lead, manganese, mercury, nickel, and zinc is in Elder (1988). Controlling factors for biogeochemical partitioning include pH, oxidation-reduction potential, hydrologic features, sediment grain size, clay minerals, organic matter, and biological processes. Some illustrations of biogeochemical cycles are in Chapter 10.

Chromium can be used to illustrate changes within aqueous systems, since it is a transition metal that exhibits various oxidation states and behavior patterns (Canter and Gloyna, 1968). Trivalent chromium is generally present as a cation, $Cr(OH)^{2+}$, and is very reactive chemically, tending to sorb on suspended materials and subsequently settle from the liquid phase. Hexavalent chromium is anionic (CrO_4^{2-}) and chemically unreactive, thus tending to remain in solution. Changes can occur in the chromium oxidation state as a result of changes in stream water quality. For example, hexavalent chromium can be chemically reduced to trivalent chromium under anaerobic conditions, whereas trivalent chromium can be oxidized to hexavalent chromium under aerobic conditions. This information could be used to qualitatively predict the impacts of chromium discharges into river systems. Of particular concern is the potential reconcentration of pollutant materials (e.g., heavy metals, pesticides) into aquatic organisms and their subsequent harvesting and consumption by humans. Again, conduction of literature searches and review of reports from similar projects may be helpful in assessing the transport and fate characteristics of the pollutants of concern.

Contaminant partitioning will be illustrated in the context of surface-water reservoir projects, since such undertakings are often evaluated in terms of the potential buildup of contaminants in sediments. Examples include the association of pesticides, polychlorinated biphenyls (PCBs), other organics, and iron and manganese with sediments, and their related interchanges with the water column. A recent survey has determined that many U.S. Army Corps of Engineers reservoir projects are experiencing problems resulting from contaminants in project waters and sediments (Gunnison, 1990). Eighteen percent of 442 Corps of Engineers reservoir projects were reported to have problems with organic contaminants. An identical percentage of these projects reported difficulties with pH

and acidity. Thirty-three percent had problems resulting from contamination by metals other than iron and manganese—primarily cadmium, copper, mercury, and lead; problems with manganese and iron occurred in 24 and 25 percent, respectively. Actual percentages of projects experiencing problems with each of these groups of contaminants may be much higher, since between 29 and 54 percent of the projects were not evaluated for problems caused by each of the various groups of contaminants. Many fundamental processes are associated with these partitioning problems; examples include adsorption, sedimentation, desorption, precipitation, and microbial transformations.

It may also be necessary to consider joint toxicity effects when several chemicals are present in the surface-water environment, and toxicity testing using juvenile fathead minnows or other appropriate organisms may be desirable (Broderius, 1990).

If the proposed project is to involve usage of a surface-water resource (such as a river or lake) as a water supply, information should be gathered on the quantities of water to be used and the distribution of this water demand over time (and location, if appropriate). If the project involves irrigation, a water balance should be developed to determine the percentage of the applied water that will recharge groundwater or appear in runoff to the surface-water resource. For projects entailing construction activities that require surface-water usage, the impact on the resource will be a function of the period of time and the season of the year during which these activities occur.

STEP 2: DESCRIPTION OF EXISTING SURFACE-WATER RESOURCE CONDITIONS

Step 2 involves describing existing (background) conditions of the surface-water resource(s) potentially impacted by the project. Pertinent activities include assembling information on water quantity and quality, identifying unique pollution problems, highlighting key climatological information, conducting baseline monitoring, and summarizing information on point- and nonpoint-pollution sources and on water users and uses.

Compilation of Water Quantity-Quality Information

Information should be assembled on both the quantity (flow variations) and quality of the surface water in the river reach of concern, and, potentially, in relevant downstream reaches. The quality emphasis should be on those water pollutants expected to be emitted during the construction and operational phases of the project. If possible, consideration should be given to historical trends in surface-water quantity and quality characteristics in the study area. Information for a given geographical area may already have been collected, and may be available in one or more computerized databases (an example is STORET). The procurement of water quality and quantity information from data storage systems must be coupled with an examination of the data for reliability. A professional interface is necessary in order to effectively utilize the information obtained. For example, hydrologic atlases prepared by state water-geological agencies contain information on stream, river, and lake locations; precipitation data and hydrographs; flow-duration curves for selected streams; quality of surface water; and water usage and supplies for larger cities. Additional sources of information include relevant city, county, and state water resources agencies and private industries that have monitoring programs. A summary of existing information on quantity and quality for the pertinent surface waters should be developed. When data is incomplete or not normally distributed, nonparametric testing of water-quality time-series data has been suggested as a means of identifying and analyzing trends (Hipel, 1988).

One approach for summarizing baseline quality-quantity information is through the use of environmental indices. Chapter 5 includes a discussion of a water quality index, or WQI, and its development (Ott, 1978). Multivariate statistical techniques have been used to aggregate water quality parameters into an index which permits examination of the effects of low-level increments in mining, grazing, and logging (Mahmood and Messer, 1982). Four suitability-for-use WQIs have been developed for use in New Zealand; the indices are for general use, recreational use, water supply, and fish spawning (salmonid) (Smith, 1987).

Water quality standards can be used to interpret assembled data. Since water quality standards vary according to the specific beneficial uses assigned for particular streams or stream segments, it is necessary to evaluate existing water quality relative to various standards. This step is important for projects that may have impacts over large distances in a single stream and for other projects, such as pipelines, that may cross numerous streams in several states.

Current and historical stream-flow information can be obtained from the *U.S. Geological Survey,* as well as from local and state agencies dealing with water resources. For example, statistical analysis of stream flows for the Mountain Fork River near Eagletown, Oklahoma, are shown in Figure 7.7; this indicates that 10 percent of the time the flow in the Mountain Fork River is $\geq$ 3,500 cfs, 50 percent of the time it is $\geq$ 350 cfs, and 90 percent of the time it is $\geq$ 10 cfs (Oklahoma Water Resources Board, 1969). An issue which may need to be addressed is the relationship between surface- and groundwater resources, particularly if groundwater provides the base flow for the stream segment.

One of the key concerns with regard to stream flow is the flow-frequency information, which is utilized for calculation of compliance with water quality standards. In some instances, the 7-day, 2-yr low flow must be utilized; in

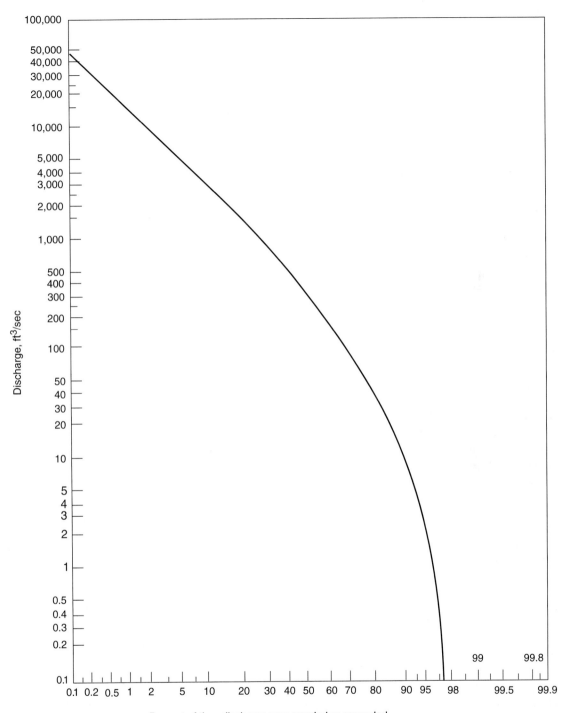

FIGURE 7.7
Duration Curve of Daily Discharge, Mountain Fork River, Eagletown, Oklahoma. *Location:* latitude 34°02′30″, longitude 94°37′15″, on downstream side of pier of bridge on U.S. Highway 70, 2 miles west of Eagletown and at mile 8.9. *Drainage area:* 787 mi². *Average discharge:* 37 yr (water years 1925, 1930–1965) 1,291 cfs (934,600 acre-ft/yr). *Daily discharge:* max., 62,100 cfs: median, 350 cfs: min., no flow. (Oklahoma Water Resources Board, 1969).

other cases the 7-day, 10-yr low-flow data is required. The phrase "7-day, 2-yr low flow" indicates that this is the minimum flow that occurs over a 7-day period at a frequency of once every 2 yr. Flow-frequency information used for water quality management is available from local and state water resources agencies.

Identification of Unique Pollution Problems

The identification of any unique pollution problems that have occurred in the project study area is a prerequisite if one is to adequately describe the environmental setting, to indicate a familiarity with the area and establish credibility, and to focus on environmentally sensitive parameters for "fatal flaws." Examples of unique pollution problems that should be identified include fish kills, excessive algal growth, and thermal discharges causing stratified flow. Many sources can be used to obtain information on unique pollution problems. Local and state water resources agencies constitute one source, conservation groups, another. Local newspapers can also provide historical documentation of pollution concerns.

Highlighting of Key Climatological Information

Meteorological data are required in order to predict and assess air quality impacts associated with proposed actions. In addition, certain climatological factors—such as precipitation, evaporation, and air temperature—are important for predicting and assessing water quality impacts. The primary sources of information include local and state water resources agencies, as well as the National Oceanic and Atmospheric Administration. Precipitation information will be useful for addressing hydraulic balances (flows), delineating acid rainfall, evaluating nonpoint sources of pollution, and scheduling construction. Evaporation information, including the loss of water from impoundments, is also relevant to hydraulic balances. Air temper-

ature can be related to water temperature, and thus, in turn, to water quality modeling; in addition, this information may be necessary for construction scheduling.

Baseline Monitoring

It may be necessary to plan and conduct specific baseline flow and quality studies to collect original background data. For example, various water quality parameters which ought to be included in riverwater-quality-monitoring programs are listed in Table 7.16. Biological surveys should be fully integrated with toxicity and chemical-specific assessment methods when evaluating attainment-nonattainment of water quality standards. Such surveys can detect impacts caused by (1) pollutants that are difficult to identify chemically or characterize toxicologically, (2) complex or unanticipated exposures from spills, and (3) habitat degradation due to channelization, sedimentation, or historical contamination. Aquatic life uses, biological integrity, and biological criteria also need to be considered (U.S. EPA, 1991h).

Detailed information on the planning and conduction of such studies is contained in *U.S. Geological Survey* (1982), Canter and Fairchild (1986), and Canter (1985). Some general information on the planning of monitoring activities is in Chapter 18. Detailed information related to selecting variables and biota and sediment sampling is in Chapman (1992). The type of background data to be assembled for baseline monitoring will depend on the potential impacts of the project, identified in step 1, and the sensitivity and usage of the surface water. Also, surface-water characteristics which are especially sensitive and which may be influenced by the proposed project should be evaluated.

Summary of Pollution Sources and Water Uses

It is appropriate, in any evaluation of the potential surface-water impacts of a proposed project, to consider what other potential and actual

TABLE 7.16

SELECTION OF PARAMETERS FOR RIVER WATER-QUALITY-MONITORING SURVEYS

		Chemical parameters			Biological parameters	
Type of survey	Physical parameters	Inorganics	Organics	Nutrients	Microbiological	Hydrobiological
Proposed for inclusion in all surveys	Color pH Specific conductance Suspended solids Total solids		Chemical oxygen demand (COD) Total organic carbon (TOC)		Coliforms, total and fecal	
Recommended for collection of baseline data	Odor	Acidity Alkalinity Calcium (Ca) Chlorides (Cl) Dissolved oxygen Hardness Iron (Fe) Magnesium (Mg) Manganese (Mn) Potassium (K) Selenium (Se) Silver (Ag) Sodium (Na)	Biochemical oxygen demand (BOD) immediate 5-day ultimate	Nitrate nitrogen, NO_3	Total plate count	
Recommended additional parameters where municipal and/or industrial pollution are expected	Floating solids	Arsenic (As) Barium (Ba) Beryllium (Be) Boron (B) Cadmium (Cd) Chromium (Cr) Copper (Cu) Dissolved carbon dioxide (CO_2) Fluorides (F) Hydrogen sulfide (H_2S) Lead (Pb) Mercury (Hg) Nickel (Ni) Vanadium (V) Zinc (Zn)	Cyanide (CN) Dissolved organic carbon Methylene blue active substances (MBAS) Oil and grease Pesticides Phenolics	Ammonia nitrogen (NH_3) Nitrite nitrogen (NO_2) Organic nitrogen Soluble phosphorus Total phosphorus	Fecal streptococci Salmonella	Benthos Plankton counts
Optional parameters for surveys of special purpose	Bed load Light penetration Particle size Sediment concentrations Settleable solids	Aluminum (Al) Sulfates	Carbon alcohol extract (CAE) Carbon chloroform extract (CCE) Chlorine demand	Organic phosphorus Orthophosphates Polyphosphates Reactive silica	Shigella Viruses Coxsackie A and B Polio Adenoviruses Echoviruses	Chlorophylls Fish Periphyton Taxonomic composition

Source: Compiled using data from IDH-WHO Working Group on the Quality of Water, 1978.

TABLE 7.17

SUMMARY OF SPECIFIC QUALITY CHARACTERISTICS OF SURFACE WATERS USED AS SOURCES FOR INDUSTRIAL WATER SUPPLIES[a]

Characteristics	Boiler makeup water: Industrial, 0–1,500 psig	Boiler makeup water: Utility, 700–5,000 psig	Cooling water Fresh: Once through	Cooling water Fresh: Makeup recycle	Cooling water Brackish[b]: Once through	Cooling water Brackish[b]: Makeup recycle	Process water: Textile industry, SIC-22	Process water: Lumber industry, SIC-24	Process water: Pulp and paper industry, SIC-26	Process water: Chemical industry, SIC-28	Process water: Petroleum industry, SIC-29	Process water: Primary metals industry, SIC-33	Process water: Food and kindred products, SIC-20	Process water: Leather industry, SIC-31
Silica (SiO_3)	150	150	50	150	25	25	—	—	50	—	50	—		
Aluminum (Al)	3	3	3	3	—	—	—	—	—	—	—	—		
Iron (Fe)	80	80	14	80	1.0	1.0	0.3	—	2.6	5	15	—		
Manganese (Mn)	10	10	2.5	10	0.02	0.02	1.0	—	—	2	—	—		
Copper (Cu)	—	—	—	—	—	—	0.5	—	—	—	—	—		
Calcium (Ca)	—	—	500	500	1,200	1,200	—	—	—	200	220	—		
Magnesium (Mg)	—	—	—	—	—	—	—	—	—	100	85	—		
Sodium and potassium (Na + K)	—	—	—	—	—	—	—	—	—	—	230	—		
Ammonia (NH_3)	—	—	—	—	—	—	—	—	—	—	—	—		
Bicarbonate (HCO_3)	600	600	600	600	180	180	—	—	—	600	480	—		
Sulfate (SO_4)	1,400	1,400	680	680	2,700	2,700	—	—	—	850	570	—		
Chloride (Cl)	19,000	19,000	600	500	22,000	22,000	—	—	200[c]	500	1,600	500		
Fluoride (F)	—	—	—	—	—	—	—	—	—	—	1.2	—		
Nitrate (NO_2)	—	—	30	30	—	—	—	—	—	—	8	—		
Phosphate (PO_4)	—	50	4	5	5	5	—	—	—	—	—	—		
Dissolved solids	35,000	35,000	1,000	1,000	35,000	35,000	150	—	1,080	2,500	3,500	1,500		
Suspended solids	15,000	15,000	5,000	15,000	250	250	1,000	[d]	—	10,000	5,000	3,000		
Hardness ($CaCO_3$)	5,000	5,000	850	850	7,000	7,000	120	—	475	1,000	900	1,000		
Alkalinity ($CaCO_3$)	500	500	500	500	150	150	—	—	—	500	500	200		
Acidity ($CaCO_3$)	1,000	1,000	0	200	0	0	—	—	—	—	—	75		
pH, units	—	—	5.0–8.9	3.5–9.1	5.0–8.4	5.0–8.4	6.0–8.0	5–9	4.6–9.4	5.5–9.0	6.0–9.0	3–9		
Color, units	1,200	1,200	1,200	1,200	—	25	—	—	360	500	25	—		
Organics														
Methylene blue active substances	2[e]	10	1.3	1.3	1.3	1.3	—	—	—	—	—	—		
Carbon tetrachloride extract	100	100	[f]	100	[f]	100	—	—	—	—	—	30		
Chemical oxygen demand (O_2)	100	500	—	100	—	200	—	—	—	—	—	—		
Hydrogen sulfide (H_2S)	—	—	—	—	4	4	—	—	—	—	—	—		
Temperature, °F	120	120	100	120	100	120	—	—	95[g]	—	—	100		

For the above two categories (Food and kindred products, SIC-20; Leather industry, SIC-31) the quality of raw surface supply should be that prescribed by the National Technical Advisory Subcommittee on Water Quality Requirements for Public Water Supplies.

[a] Unless otherwise indicated, units are mg/L and values are maximums. No one water will have all the maximum values shown.
[b] Water containing in excess of 1,000 mg/L dissolved solids.
[c] May be ≤1,000 for mechanical pulping operations.
[d] No large particles ≤3 mm diameter.
[e] 1 mg/L for pressures up to 700 psig.
[f] No floating oil.
[g] Applies to bleached chemical pulp and paper only.

Source: Federal Water Pollution Control Administration, 1968a, p. 189.

sources of surface-water pollution already exist in the study area and also to consider current and potential usage of the surface-water resource(s) for purposes of water supply. The objective should be to assemble sufficient information to enable the determination of the level of other sources of pollution and of the types and degrees of water uses in the study area. This does not always require the conduction of a detailed surface-water-monitoring program. Pertinent information could be assembled on items such as the types of land usage (and size of each land use), storage areas (stockpiles, tailing piles), and the like, within the study area. For example, if nitrates or pesticides are of concern as pollutants from the project, the land area within the study region associated with the agricultural usage of these pollutants should be established, and data (including the types, quantities, and timing of fertilizer and/or pesticide applications) should be compiled.

A detailed review of the current numbers of surface-water users and the quantities associated with such uses should also be assembled. The types of information that may be accumulated include general estimates of the number of users of the surface water (private, public, industrial), types of water uses (drinking water, recreation, cooling water, etc.), the location and rates of existing surface-water withdrawals, and the location, quantity, and quality of existing discharges into the surface water, and so on. Water quantity concerns are of major importance in water-deficient areas. The types of water uses are also important, since quality requirements vary for different uses. Table 7.17 summarizes specific quality characteristics of surface waters that have been used as sources for industrial water supplies. Additionally, consideration should be given to potential increases in usage over time without the proposed project. If this information is not available, it may be necessary to conduct a survey to gather the most-pertinent information.

Sources of information include appropriate water supply and/or surface-water pollution studies conducted by relevant governmental agencies. An example would be the updated "208 study" or basinwide water quality management study prepared by the pertinent governmental entity. If these studies do not address both point and nonpoint sources in the stream segment of interest, it may be necessary to assemble such information. Of particular importance would be load allocations to nonpoint sources and waste-load allocations to point sources. Depending on the types of pollutants identified in step 1, it may be necessary to obtain information regarding organics, nutrients, metals, thermal discharges, and/or specific toxicants. Toxics-release inventories are being conducted by the U.S. Environmental Protection Agency under requirements of the Pollution Prevention Act of 1990 and Section 313 of the Emergency Planning and Community Right-to-Know Act of 1986 ("News Update," 1992).

In addition, special studies are often conducted in conjunction with the development of water supply sources; these studies may include inventory information of the type desired in this step. Local, state, regional, or federal agencies in the respective locations with a mandate to oversee surface-water resources will likely know about the existence of such study reports. In the absence of such information it is possible to generate the data; however, the requirements in terms of cost and time and the inability to collect the data over multiple years may be constraining factors.

STEP 3: PROCUREMENT OF RELEVANT SURFACE-WATER QUANTITY-QUALITY STANDARDS

To determine the severity of the impacts that may result from a project, it is necessary to make use of institutional measures for determining impact significance. Surface-water quantity and quality standards, regulations, or policies are examples of these measures. Thus, determination of the specific requirements for a given surface water will require contacting governing agencies in one or several regions, states, and/or countries with jurisdiction for that surface water. (Relevant institutional information from U.S. government agencies has been described earlier.) The intended use of the surface water will also affect the standards to be applied for that surface water, with use as a drinking water supply typically resulting in the most stringent standards. Effluent limitations regulating the permissible quality of discharged wastewater from domestic and industrial sources may also be pertinent, along with regulations concerning nonpoint discharges from industrial areas. In some cases, there may be limitations on the amount and timing of water usage from a given body of water. Water quality management policies may also be pertinent; examples of such policies include antidegradation goals, cleanup or remediation goals, and/or goals for preservation of aquatic ecosystems and scenic beauty.

It should be noted that most water quality standards and water-use restrictions are related to low-flow periods in the river system. For example, dissolved oxygen (DO) minima may be applicable during the 7-day, 2-yr low-flow conditions or the 7-day, 10-yr low-flow conditions. These conditions need to be considered in interpreting existing quality-quantity, and in impact calculations and their resultant interpretation. In addition, it may be necessary to consider low-flow requirements for the maintenance of the aquatic ecosystem. One such procedure for ascertaining requisite flows is the "instream flow incremental methodology"; it will be described later in this chapter, along with other aquatic-ecosystem-modeling approaches.

One of the approaches which can be used for identifying and procuring institutional information is scoping (CEQ, 1978). Contacts with regulatory agencies and other interested publics can aid in the identification of pertinent institutional information. This information can primarily serve two functions: (1) as an aid in the interpretation of existing quality and/or quantity conditions relative to the environment and (2) as a basis for interpreting the anticipated quality and/or quantity impacts of the resultant project. In the complete absence of water quality-quantity and aquatic-ecosystem standards, it might be appropriate to utilize generic criteria or standards developed for nearby geographical areas. National and/or international water-management and environmental agencies could also be contacted regarding pertinent institutional information.

STEP 4: IMPACT PREDICTION

"Impact prediction" basically refers to the quantification (or, at least, the qualitative description), where possible, of the anticipated impacts of the proposed project on various surface-water environment factors. Depending upon the particular impact, technically demanding mathematical models might be required for prediction. Other approaches include the conduction of laboratory testing, such as leachate testing for dredged material (Brannon, 1978), and for solid- or hazardous-waste materials (U.S. EPA, 1986) or sludges (Deeley and Canter, 1986). Still other laboratory studies might be appropriate; examples include chronic-toxicity testing (Canter, Robertson, and Hargrave, 1990). Still other techniques include the use of look-alike or analogous information on actual impacts from similar types of projects in other, similar geographical locations. Finally, environmental indexing methods such as the WQI or other types of systematic techniques for relatively addressing anticipated impacts can also be considered.

It is desirable to quantify as many impacts as possible, because, in so doing, it has been frequently determined that the concerns related to anticipated changes are not as great as they would appear to be in the event of nonquantification. Also, if anticipated impacts are quantified, it would be appropriate to use specific numerical standards as the basis for interpretation (assessment) of the anticipated changes.

However, many impacts cannot be quantified, or the requirements in terms of monetary and personnel resources to accomplish such quantification would be beyond the scope and budget of the impact study. In many cases, it is necessary for professionals to use their best judgment in qualitatively describing impacts.

Impact prediction will involve considering whether the pollutants are conservative, nonconservative, bacterial, or thermal. Conservative pollutants are not biologically degraded in a stream, nor will they be lost from the water phase as a result of precipitation, sedimentation, or volatilization. The basic approach for prediction of downstream concentrations of conservative pollutants is to consider the dilution capacity of the stream and use a mass-balance approach with appropriate assumptions. "Nonconservative pollutants" refer to organic materials that can be biologically decomposed by bacteria in aqueous systems. Nutrients are also nonconservative, since they can be involved in biochemical cycling and plant uptake. Prediction of impacts resulting from nonconservative and bacterial pollutants and thermal discharges typically require mathematical-modeling approaches.

Mass-Balance Approaches

One mathematical-modeling prediction approach uses mass-balance calculations to determine average downstream concentrations resulting from point- or nonpoint-source discharges or to determine percentage changes in stream flow or pollutant loadings. The results can be compared to pertinent effluent limitations, quality-quantity standards, or baseline flow and quality characteristics.

If several discharge streams are mixed prior to final discharge, or if the quantity of water and mass of constituent are calculated separately, it will be necessary to conduct a mass-balance analysis to determine the quality of the final discharge stream. This is accomplished by determining the mass contribution of a constituent

from each stream, summing these masses for all streams, and dividing this total mass by the total flow from all streams. In equation form, this can be expressed as shown below:

$$C_{\text{avg}} = \frac{\Sigma \, C_i Q_i}{\Sigma \, Q_i} = \frac{\Sigma \, M_i}{\Sigma \, Q_i}$$

where C_{avg} = average concentration of constituent for combined discharge stream

C_i = concentration of constituent in ith discharge stream

Q_i = flow for ith discharge stream

M_i = mass of constituent in ith discharge stream

A mass-balance calculation can also be used to estimate erosional impacts on water quality which may occur during either the construction or operational phases of a project. The quantity of erosion per time period is divided by the stream flow in the immediate stream segment for the same time period to estimate the suspended solids (or turbidity). The results can be compared to the normal variations of suspended solids in the stream. This calculation assumes uniform erosion, no transport losses as the eroded materials move to the stream, and complete mixing within the stream. Adjustments in the calculations can be made to account for sedimentation in overland flow and the receiving stream. Similarly, calculations and interpretation for the construction phase can be based on the time period of construction, water users in the immediate area, and the use of construction-phase specifications for erosion minimization. Mass-balance calculations could also be used for estimating the suspended-solids impacts of pipeline river crossings when trenching is performed.

Mass-balance calculations can be used for stream segments, "cells" in lakes and estuaries, and the ocean, as well. Mass-balance calculations involving a defined "cell" can be referred to as "box models." Such models, which couple hydrodynamic and water quality considerations, may be useful for studies involving river segments, lakes, estuaries, coastal areas, and oceans (Bird and Hall, 1988). For example, Canale and Auer (1987) described a mass-balance box model for the water quality conditions in 19 cells in Green Bay, Wisconsin. The model can be used for the constituents, or variables, listed in Table 7.18. Mass-balance considerations include exchanges among adjacent model cells (horizontal and vertical mass transport) and all sources and sinks of material. The phosphorus concentration controls algal activity and the production of organic carbon. The breakdown of organic carbon in the water column and in sediment influences the dissolved oxygen mass balance. The general form of the mass-balance equation for oxygen in each model cell is (Canale and Auer, 1987)

$$V_i \frac{dC_i}{dt} = \sum_j \left[\frac{E_i A_i}{L_i} (C_j - C_i) + Q_j C_j \right] + W_i - Q_i C_i + K_i (C_s - C_i) + P_i - S_i - R_i$$

where A_i = area of interface between cell i and adjacent cell j

C_i = oxygen concentration in cell i

C_j = oxygen concentration in adjacent cell j

C_s = saturation oxygen concentration

E_i = coefficient for dispersion across boundary of cell i

K_i = atmospheric oxygen exchange coefficient for cell i

L_i = distance between the centers of cells i and j

P_i = photosynthetic production of oxygen in cell i

Q_i = flow leaving cell i

Q_j = flow entering from cell j

R_i = water column respiration in cell i

S_i = sediment respiration in cell i

t = time

V_i = volume of cell i

W_i = tributary oxygen loading to cell i

Mass-balance equations have also been developed for chloride, total phosphorus, and total organic carbon, with the relevant source and sink characteristics for each included, as appropriate (Canale and Auer, 1987).

An example of the use of a mass-balance equation developed for chloride will be given here (Canter, 1973). Figure 7.8 represents a schematic drawing of a power plant located adjacent to a stream, with the power plant using a 205-acre cooling pond. Data about the stream at withdrawal point x are shown in Table 7.19.

TABLE 7.18

SOURCE AND SINK CHARACTERISTICS FOR MASS-BALANCE MODEL

Variable	Source	Sinks
Chloride	Tributary loads	None (conservative)
Total phosphorus	Tributary loads Atmospheric loads Sediment release	Settling
Total organic carbon	Tributary loads Primary production	Settling Water column respiration
Dissolved oxygen	Atmospheric exchange Photosynthesis	Atmospheric exchange Water column respiration Sediment respiration

Source: Canale and Auer, 1987, p. 93.

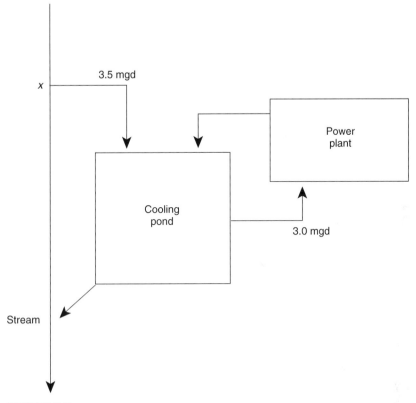

FIGURE 7.8
Schematic Drawing of Power Plant and Water Usage (Canter, 1973).

TABLE 7.19		
SUMMARY OF TOTAL DISSOLVED SOLIDS IN STREAM AT WITHDRAWAL POINT		
Stream flow, mgd	**Frequency, days per year**	**Total dissolved solids, mg/L**
10.0	37	720
17.3	219	520
80.0	109	305

Source: Adapted from Canter, 1973, pp. 85 and 88.

These data are expressed as a frequency distribution relative to flow and concentration of total dissolved solids. The total dissolved solids are composed primarily of chlorides. Water withdrawal from the stream into the cooling pond is 3.5 mgd, and cooling-water usage in the power plant is 3.0 mgd. Water losses in the cooling cycle occur from consumptive losses in the plant cooling cycle and from pond evaporation. Consumptive losses average 1.6 mgd, and evaporation losses vary from 0 mgd in the winter to 1.4 mgd during the period from May to October.

TABLE 7.20

SUMMARY OF CALCULATED TOTAL DISSOLVED SOLIDS DOWNSTREAM FROM COOLING-POND DISCHARGE

Condition[a]	Flow, mgd	Total dissolved solids, mg/L
Worst	7.0	1,030 (720)[b]
Average	14.7	610 (520)
Best	78.4	310 (305)

[a]Worst = Stream flow of 6.5 mgd (3.5 mgd withdrawn from 10 mgd), pond effluent flow of 0.5 mgd, occurs 37 days/yr; average = stream flow of 13.8 mgd (3.5 mgd withdrawn from 17.3 mgd), pond effluent flow of 0.9 mgd, occurs 219 days/yr; best = stream flow of 76.5 mgd (3.5 mgd withdrawn from 80 mgd), pond effluent flow of 1.9 mgd, occurs 109 days per yr.
[b]Numbers in parentheses represent stream water quality just upstream from confluence with cooling-pond discharge.
Source: Adapted from data in Canter, 1973, p. 94.

On this basis, total water losses in the cooling pond vary from 1.6 to 3.0 mgd. If the concentration factor in the pond is defined as " 'water in' divided by 'water out'," while maintaining the pond water level constant on a daily basis, the highest concentration factor is 7.0 (3.5 mgd/ 0.5 mgd) and the lowest is 1.8 (3.5 mgd/1.9 mgd). Using these concentration factors, calculations can be made for the total dissolved solids in the receiving stream; the results are summarized in Table 7.20. The calculated values in the receiving stream can then be made compared to those for the applicable water quality standard.

The impact of a project on surface-water quantity can be determined by quantifying the expected water usage and expressing this as a percentage of the average-, high-, and low-flow conditions in the river or lake. Determination of percentage changes under different flow conditions can also be made for quality parameters.

Determination of total pollutant loadings on the receiving-stream segment as a function of duration and timing may provide additional insights into the surface-water impacts. For instance, if the discharges from the project are seasonal and the surface-water quality is also seasonal (naturally or as a result of other seasonal variations in discharges), the cumulative

effect of these variations is important. For example, if the project discharges will occur during periods when the surface-water quality is naturally elevated, then the resulting water quality will be better than that estimated for a typical period. Also, if discharges from other sources are also seasonal and occur in different seasons, then the net effect will be less severe than if the discharges occurred simultaneously. Similar considerations would also apply to water quantity impact concerns.

Mesoscale impact calculations based on mass-balance considerations can be useful. "Mesoscale" refers to a large scale; such calculations could include the stream segment where the project is located, that segment plus the next downstream segment, those two segments plus the next downstream segment, and so on, possibly to include the entire river basin, for example. The mesoscale approach consists of multiplying unit waste-generation factors (from step 1) by their appropriate production quantities during construction and operation and then comparing these calculated daily pollutant loads with existing pollutant loads in the study area (step 2). Both point and nonpoint loads need to be considered. One means of assessing the impact is to calculate percentage changes in pollutant loads, by comparing these loads for the alternatives to existing pollutant loads in the study area. Depending on the type of project, pollutant loads should be considered for organic (nonconservative), inorganic (conservative), solid, nutrient, and bacterial pollutants, and for thermal discharges. Interpretation of mesoscale impact calculations can be based on comparisons of existing and projected load allocations and waste-load allocations, considerations of the time periods of the anticipated changes, and evaluations of percentage changes in relation to existing flow conditions and the quality of relevant parameters.

It is also possible that the load generated by the project—even though a small fraction of the total waste load discharged into the surface water—will happen to be just enough to cause the water quality standards to be exceeded. Given this scenario, it would be helpful to quantify the total mass loading to the surface water and determine what fraction of this total contribution could be attributed to the project of concern. If the waste-loading contribution of the project is minor (say, less than 1 to 3 percent), if one or more major contributors are identified (accounting for, say, greater than 40 to 50 percent of the waste load), and the major contributor(s) can readily reduce their waste loading to the surface water, then it can be argued that the project should be allowed to discharge at some (lower) level and the other contributor(s) encouraged to reduce their contributions. Similar considerations would also apply to water-quantity impact concerns.

Data obtained from the use of mass-balance approaches may be sufficient to conclude that surface-water quality-quantity impacts are not significant for a proposed project, particularly if mitigation measures are to be included (see step 6); if so, detailed mathematical modeling for water quality or aquatic ecology may not be necessary. In addition, it should be noted that certain types of contaminants, such as BOD_5 and thermal discharges, do not lend themselves to mass-balance calculations. These types of contaminants are not conserved in water but can change as a result of processes other than dilution.

Mathematical-Modeling Approaches

A more advanced analysis will likely require the use of mathematical models to estimate surface-water impacts. Such calculations can be termed "microscale impact prediction" in that the emphasis is on the small scale in the immediate environs of the project.

In all surface-water models the physical system must be described using mathematical expressions (typically, partial differential equations). In the process of going from physical reality to differential equations, simplifying assumptions are always made. Every assumption is a potential source of error. The set of equations can be solved exactly (analytically) for simple conditions; however, as the complexity of the surface-water system increases, analytical solutions become less practical (an example would be the set of equations describing estuaries with tidal flows). In these cases, it is possible to solve the equations using numerical approximations to the governing differential equations. While this approach allows solution of the governing equations, it also introduces the potential for numerical errors. However, by simplifying the dimensionality of the system in the solution technique, the risk of error introduction is minimized. The greater the dimensionality of the model (three-dimensional versus two- or one-dimensional), the more complicated is the data input required and the techniques necessary to solve the equations. Thus, quite commonly the physical system is reduced to one or two dimensions to simplify the data requirements and solution techniques.

The temporal, or transient, nature of the problem is another element that may be simplified. All processes are time-dependent; however, many systems will approach a time-independent solution (an equilibrium condition) rather quickly and/or this equilibrium condition may be the case of interest (e.g., a worst-case scenario). Determination of equilibrium conditions is commonly much simpler than determination of the time-dependent results leading up to them. However, it may be necessary to analyze the temporal results to determine if the limiting case is realized under these conditions. This is especially pertinent if not all of the inputs are constant (e.g., if wastewater discharges are discontinuous). Model inputs may vary spatially (at various locations) and temporally (at a given location with time transient). Any of the above conditions will greatly complicate the data input needs and the complexity of the solution technique.

In summary, models can be classified according to their underlying basis (either descriptive of fundamental processes or statistical), flow regime (steady state or dynamic), dimensional considerations (one to three dimensions), type of water body (river, lake, estuarine, coastal, or ocean), and number of parameters (single to multiple parameters—up to about 15).

Modeling approaches for predicting impacts of organic materials in rivers must include consideration of the changes in dissolved oxygen (DO) resulting from bacterial demand for oxygen in the decomposition process, and the supply of oxygen by means of natural reaeration. The saturation DO concentration is a function of temperature, pressure, and salt content. DO in a stream is deficient when the actual concentration is less than the saturation concentration for existing conditions of temperature, pressure, and salt content. There are certain demand and supply forces relative to dissolved oxygen in a stream. Demand for oxygen is exerted by bacteria in the decomposition of organic materials, both in the liquid phase and in bottom deposits. Oxygen is supplied both from natural reaeration and as the net effect of photosynthesis. A clas-

FIGURE 7.9
Characteristic Oxygen Sag Curve Obtained Using the Streeter-Phelps Equation (Metcalf and Eddy, 1991, p. 84).

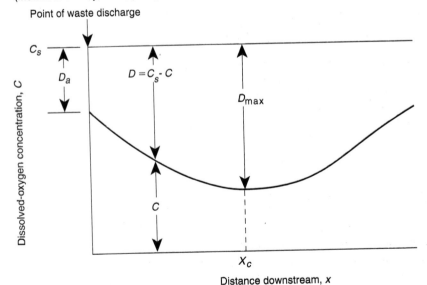

sical DO model addressing liquid-phase demand and natural reaeration was developed by Streeter and Phelps in 1925; it is as follows:

$$D_t = \frac{K_1 L_a}{K_2 - K_1} (10^{-K_1 t} - 10^{-K_2 t}) + D_a 10^{-K_2 t}$$

where D_t = DO deficit at any flow time t or distance x downstream, days

= saturation DO concentration (C_s) − actual DO concentration (C)

K_1 = coefficient of deoxygenation, day^{-1}

K_2 = coefficient of reaeration, day^{-1}

L_a = ultimate BOD in stream following mixing, mg/L

D_a = DO deficit upstream of or at point of waste discharge, mg/L

The above model can be used to calculate the "oxygen-sag" curve, as shown in Figure 7.9. It should be noted that K_1, K_2, and L_a are influenced by temperature. Specific mathematical relationships describing the temperature influence are as follows:

$$K_{1(T)} = K_{1(20)} (1.047)^{T-20}$$
$$K_{2(T)} = K_{2(20)} (1.016)^{T-20}$$
$$L_{a(T)} = L_{a(20)} (0.02T + 0.6)$$

where T = water temperature, °C, and $K_{1(20)}$, $K_{2(20)}$, $L_{a(20)}$ = values at 20°C.

Other factors that may be important in predicting the DO impact are related to critical conditions in terms of the location and value of the minimum point on the oxygen-sag curve, and as the maximum permissible BOD loading that can be introduced without exceeding the dissolved oxygen standard. Equations for critical time and deficit are as follows:

$$t_c = \frac{1}{K_2 - K_1} \log \left(\frac{K_1 L_a - K_2 D_a + K_1 D_a}{K_1 L_a} \frac{K_2}{K_1} \right)$$
$$D_c = \frac{K_1}{K_2} L_a 10^{-K_1 t_c}$$

where t_c = critical time (time of flow) to point of occurrence of minimum DO concentration,

days, and D_c = critical (maximum) deficit (mg/L) at time of flow t_c.

The equation that can be used for determining the maximum permissible BOD loading is as follows:

$$\log L_a =$$
$$\log D_{all} + \left[1 + \frac{K_1}{K_2 - K_1} \left(1 - \frac{D_a}{D_{all}} \right)^{0.418} \right] \log \frac{K_2}{K_1}$$

where D_{all} = allowable deficit, in mg/L = saturation DO concentration − DO standard.

Instream water temperature can be predicted, based on hydrometeorological data, through the use of the "instream flow and aquatic systems group" (IFG) model (Theurer, Voos, and Prewitt, 1982). The IFG model can be used to predict average daily water temperature and diurnal fluctuations within a stream network.

Numerous other DO models have been developed based on the classical Streeter-Phelps approach, including models which incorporate oxygen demand by sediments and oxygen supply by photosynthesis. In addition, there are lake and reservoir models which address chemical constituents as well as temperature fluctuations, stream and estuary models related to dissolved oxygen, and models relating to specific types or sources of water pollution such as thermal, nonpoint, and irrigation return flows. Comprehensive reviews of surface-water quality models are in Biswas (1981); Grimsrud, Finnemore, and Owen (1976); and U.S. Army Corps of Engineers (1987). A state-of-the-art book on water quality modeling is James (1993). It includes chapters on modeling water quality in rivers, estuaries, lakes, and reservoirs; in addition, modeling water quality in sewage outfalls in the marine environment is also addressed. Models for predicting the water quality impacts of oil or chemical spills are also available.

One surface-water quality model in the United States which is being used by govern-

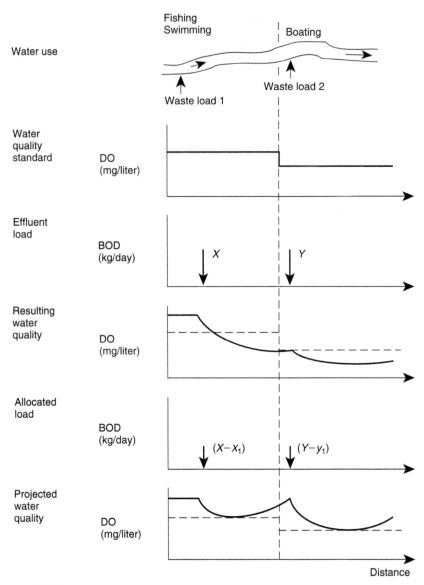

FIGURE 7.10
A Simple Waste-Load Allocation Problem for Dissolved Oxygen (Ray, 1990, p. 66).

mental regulatory agencies and consulting engineers is QUAL-IIE (Loucks, 1981). QUAL-IIE can be used to develop or evaluate waste-load allocations to rivers (Ray, 1990). The major steps in the WLA process are the designation of desirable water use and the corresponding water quality standard, a cause-effect analysis of projected waste inputs and the water quality response, and a projection analysis for achieving water quality standards under various levels of waste-load input. Figure 7.10 illustrates the principal steps in WLA modeling.

The basic equation solved by QUAL-IIE is the one-dimensional advection-dispersion–mass-transport equation, which is numerically integrated over space and time for each water quality constituent. For any constituent C, this equation can be written as (Ray, 1990)

$$\frac{\partial C}{\partial t} = \frac{\partial\left(A_x D_L \frac{\partial c}{\partial x}\right)}{A_x \partial x} - \frac{\partial(A_x \bar{u} C)}{A_x \partial x} + \frac{dC}{dt} + \frac{s}{V}$$

where x = distance (L; L = length)
$\quad\quad\; t$ = time (T, T = time)
$\quad\quad\; C$ = concentration (ML^{-3}; M = mass)
$\quad\quad\; A_x$ = cross-sectional area (L^2)
$\quad\quad\; D_L$ = dispersion coefficient ($L^2 T^{-1}$)
$\quad\quad\; u$ = mean velocity (LT^{-1})
$\quad\quad\; s$ = external source or sinks (MT^{-1})
$\quad\quad\; V$ = (M/C) = $A_x dx (L^3)$

The right-hand side of the equation represents the dispersion, advection, constituent changes, and external source-sink terms, respectively. The term dC/dt represents constituent changes such as growth and decay. The term $\partial C/\partial t$ represents the local concentration gradient. The term s/V expresses the effect of advection, dispersion, constituent interactions, sources and sinks, and dilution. The constituents modeled by the above equation in QUAL-IIE include (Ray, 1990) (1) dissolved oxygen, (2) carbonaceous biochemical oxygen demand, (3) temperature, (4) algae as chlorophyll-a, (5) nitrogen species (organic, ammonia, nitrate, and nitrite), (6) phosphorus species (organic and dissolved), (7) coliforms, (8) an arbitrary (variable) nonconservative constituent, and (9) three conservative constituents.

To give some perspective on the detailed nature of QUAL-IIE, Table 7.21 contains a summary listing of the features, data requirements, and output. The QUAL-IIE model is being continuously updated and adapted for usage by a

---

TABLE 7.21

SUMMARY FEATURES OF QUAL-IIE MODEL

Description	One-dimensional longitudinal stream water quality model
Major features	Simulates up to 15 water quality constituents
	Addresses steady state or time-varying water quality
	Addresses steady or slowly varying flow
	Accounts for flow augmentation
	Addresses branching lateral inflows and withdrawals
	Can specify or roughly simulate hydraulics resulting from instream structures
Data requirements	River-reach identification and riverine data
	Flow-augmentation data (optional)
	Hydraulic or channel data
	Reaction-rate coefficients
	Initial conditions
	Inflow data
	Meteorological data
Output	Input-data echo
	Tabular output of hydraulic data, meteorologic data, fluxes, productivity data, and concentrations
	Longitudinal graphical profiles for constituents

Source: Compiled using data from U.S. Army Corps of Engineers, 1987.

variety of software for personal computers. There are numerous examples of the usage of QUAL-IIE for many applications, and two illustrations will be cited. Ray (1990) described its use for evaluating the impact of a food-processing plant's wastewater on the water quality of a small river which subsequently flowed into a larger river. Several conditions and scenarios were studied by Ray (1990). Summers, Kazyak, and Weisberg (1991) described the use of a version of QUAL-IIE to model the receiving-water quality impacts of the discharges from several wastewater-treatment facilities and a large kraft (paper) mill.

Finally, Ray (1990) noted that in order to apply QUAL-IIE (or modified versions thereof) to the prediction of water quality responses to existing and altered waste-load situations, the models must be calibrated and verified. Model verification should be conducted at various flow and waste-load situations to determine the validity of the model for the purpose of application. Periodic verification and calibration ensures that predicted parameters are representative of the system simulated. The kinetic coefficients should be properly selected and determined to represent the reaction mechanisms occurring in the water system.

In addition to the use of the QUAL-IIE model, impacts of bacterial pollution can also be addressed through the application of a specific model relating the aquatic environment characteristics and the bacterial death rate. This approach involves the prediction of "bacterial self-purification," which is defined as the decrease of bacteria of all types, and especially those of fecal origin, as a function of flow distance or flow time in a river (Phelps, 1944). The mathematical relationship that describes bacterial self-purification is as follows:

$$B_t = B_0 10^{-Kt}$$

where B_t = bacterial residual after any time t, days

B_0 = initial number of bacteria in stream

K = bacterial death rate, day^{-1}

Modeling the WLA process for toxic chemicals requires a different approach than that used for conventional pollutants. Present WLAs for toxicants generally rely on simple dilution calculations for low stream-flow conditions. However, new water quality criteria for toxicants and new modeling strategies to implement the criteria have been developed. The new criteria specify an acute threshold concentration and a chronic-no-effect concentration for each toxicant, as well as tolerable durations and frequencies of exposure at or above the two concentrations. Methods for using steady state and dynamic models to derive WLAs based on the new criteria were reviewed by Ambrose et al. (1988). Physical, chemical, and biological processes affecting toxicants in the mixing-zone and far-field regimes of streams have also been summarized by Ambrose et al. (1988).

Several models have been developed to account for the water pollution impacts of nonpoint pollution. For example, the CREAMS model can be used for chemicals, runoff, and erosion from agricultural management systems; the GLEAMS model is used for addressing the groundwater loading effects of agricultural management systems (Knizel and Nicks, 1980; Williams and Nicks, 1982; Leonard et al., 1988). The "storm water management model" (SWMM) of the U.S. Environmental Protection Agency simulates urban runoff-water quality and quantity in storm and combined sewer systems. The simulations address surface and subsurface runoff, transport through the drainage network, storage, and treatment (Huber and Dickinson, 1988). Summary information on these three models is in Table 7.22. A detailed review of models to calculate water pollution impacts for both nonpoint and point sources is in Stefan, Ambrose, and Dortch (1990). Water quality modeling for impacts from agricultural nonpoint sources is addressed in DeCoursey (1990).

Thermal-pollution modeling approaches are also available, and the extant models range from empirical relationships to sophisticated and sci-

TABLE 7.22

SUMMARY OF THREE NONPOINT-SOURCE MODELS

Development	Process	Scale	Components	Other characteristics
CREAMS: chemicals, runoff, and erosion from agricultural management systems				
Developed by USDA in 1979. It is the first USDA model that accounts for sediment, nutrient, and pesticides (Knizel and Nicks, 1980).	Continuous formulation	Field-size areas	Hydrology Based on the U.S. Soil Conservation Service Curve Numbers (SCSCN) method. It has also an infiltration-based subroutine as a secondary option. Erosion and sediment transport Based on overland flow transport capacity, interrill and rill detachment, and impoundment deposition. Quality subroutines Simulate pesticides and nutrient losses, and account for mineralization, nitrification immobilization, leaching, adsorption-desorption, volatilization, and degradation.	A modified version of this model is called "GLEAMS."
GLEAMS: ground water loading effects of agricultural management systems				
Developed as an extension of CREAMS to evaluate the impact of management practices on potential pesticide leaching below the root zone, as well as surface runoff and sediment losses.	Storm events	Field-size area	Hydrology Uses daily climatic data to calculate the water balance in the root zone. Precipitation is partitioned between surface runoff and infiltration into the soil surface. Uses the curve number method of estimating runoff as modified by Williams and Nicks (1982). A seasonally frozen-soil representation was added to better estimate snowmelt runoff. A storage-routing technique is used to simulate redistribution of infiltrated water within the root zone, and percolation out of the bottom of the root zone is estimated. Soil evaporation and plant transpiration are estimated with a modified Penman equation. Erosion and sediment transport Uses a modified "universal soil loss equation" (USLE) for storm-by-storm estimates of rill and interrill erosion in overland flow areas. Channel and pond elements were added to calculate erosion or deposition in the field delivery system to estimate sediment yield at edge of field. Eroded soil is routed with runoff by particle size, which enables calculation of storm-by-storm sediment-enrichment ratios for use in estimating adsorbed pesticide transport.	Has been shown to be effective in assessing potential pesticide leaching below the root zone (Leonard et al., 1988).

TABLE 7.22

SUMMARY OF THREE NONPOINT-SOURCE MODELS (continued)

Development	Process	Scale	Components	Other characteristics
			Quality subroutines	
			Retains concepts of CREAMS pesticide component for surface losses in runoff and with sediment. The same adsorption characteristics were coupled with the water-storage-routing technique to route pesticides within and through root zone. Upward movement of pesticides in soil by evaporation and plant uptake by transpiration were included in modification.	
		SWMM: stormwater management model		
Developed for the U.S. Environmental Protection Agency to simulate overland water quantity and quality produced by storms in urban watersheds.	Continuous single events	Urban watershed	Hydrology Uses a distributed-parameters submodel (RUNOFF) which simulates runoff based on the concept of surface storage balance and the use of small, homogenous subcatchments (up to 200). A transport routine (TRANSPORT) uses storm-water runoff data generated by RUNOFF and distributed among the various routes and accounts for infiltration (INFIL) and the effect of natural and/or man-made storages and dampening of storm-runoff peaks. Sediment/pollution load For impervious areas, is computed from the daily or hourly increase in particle accumulation based on a linear formulation. For pervious areas, sediment load is determined based on the USLE (a modified form). Pollutants other than sediment are computed using the concept of potency factors.	The application of this model is limited to drainage areas ranging from 5 to 2000 ha (U.S. EPA, 1984).

entifically derived fundamental relationships (Krenkel and Parker, 1969; James, 1993). A useful reference with example calculations is in Federal Water Pollution Control Administration (1968b). The QUAL-IIE model can also be used for modeling temperature changes in river systems.

The models used to simulate only stream conditions are least complex because of the one-dimensional nature of flow. Models for simulating stratified lakes and reservoirs fall next in line in terms of complexity, followed by estuarine models. Quantitative models for describing lakes or reservoirs typically involve the consideration of multiple influencing factors and processes (Henderson-Sellers, 1991). Quantitative models also typically consider reservoir dynamics in lieu of steady state conditions; models for transient conditions can express continuous time or discrete time conditions. As an example, the oxygen relationships in a reservoir can be modeled by considering decomposition, reaeration, benthic oxygen demand, and photosynthesis. Quantitative models can be one-dimensional (x direction), two-dimensional (x and y directions), or three-dimensional (x, y, and z directions); the multidimensional models are more complicated from a mathematical perspective. In addition, they are more data-intensive. Table 7.23 summarizes the key characteristics of several quantitative models which can be used for reservoir water-quality studies.

Released tailwater from hypolimnetic zones in a reservoir can exhibit DO depletion and increased concentrations of dissolved nutrients (ammonium and inorganic phosphorus), sulfide, reduced metals (iron and manganese), and specific organic substances (simple organic acids and methane). A PC-based model for addressing tailwater quality, called the "TWQM" (tailwater quality model), has been developed based on modifications to the one-dimensional QUAL-IIE model (Tillman, Dortch, and Bunch, 1992). Water quality constituents addressed in TWQM include DO, carbonaceous BOD, temperature, algae as chlorophyll-a, total organic nitrogen, ammonium nitrogen, nitrite plus nitrate nitrogen, total organic phosphorus, dissolved inorganic (orthophosphate) phosphorus, dissolved (reduced) iron, dissolved (reduced) manganese, total dissolved sulfide (HS^- and H_2S), iron sulfide, one arbitrary nonconservative constituent, and two conservative constituents. Figure 7.11 illustrates the system relationships between the model constituents. The dissolved oxygen concentration from water releases by gated or ungated spillways, sharp-crested overflow weirs, and hollow cone valves have been described, along with a spreadsheet calculational procedure to simulate blending the flows from various release patterns at a reservoir (Wilhelms, 1993).

Estuary models are more complex because the prototype flow is usually in at least two dimensions, and the boundary conditions, such as tides, vary rapidly compared with those in lakes. The costs of model application tend to be proportional to their complexity (Grimsrud, Finnemore, and Owen, 1976). Selected information on this category of models is in James (1993).

Aquatic-Ecosystem-Modeling Approaches

Several methods have been developed to quantify and assess biological impacts on aquatic resources (Brookes, 1988). One example is the "instream flow incremental methodology" (IFIM); the steps are in Table 7.24. The IFIM was originally developed in the United States as a means of assessing how much water could be extracted from a river at various times of the year without adversely affecting the fishery resource. The approach is based on the concept that a particular species can be correlated with a set of particular habitat requirements, such as specific water quality, velocity, depth, substrate, temperature, and cover conditions. If these requirements are known, then an assessment of the habitat suitability for a particular species can be made by measuring the quality of the available habitat in terms of these factors. The advantage of this approach is that measuring the

TABLE 7.23

SUMMARY COMPARISONS OF SEVERAL WATER QUALITY MODELS

Model name	Description	Major features	Data requirements	Output
CE-THERM-1	1-D vertical reservoir model for temperature	Temperature, total dissolved solids (TDS), suspended solids (SS) coupled to density. Specify outflow ports or ports based on temperature objective. Reregulation pool, pumped-storage, and/or peaking hydropower options.	Inflow rates and constituent values Outflow rates, operations Structural configuration and hydraulic constraints of outlets Initial constituent profiles Morphometric data Meteorological data Process and rate coefficients Release flow and temperature targets if using outflow-port decision routine	Vertical profiles and outflow values for constituents over time (printed and/or plotted) Statistics of predicted and observed values Flux information Operations schedules for multilevel outlet configurations
CE-QUAL-R1	1-D vertical reservoir model for water quality	All CE-THERM-R1 features. Allow simulation of most major physical, chemical, and biological processes and associated water quality constituents. Simulates anaerobic processes. Monte Carlo simulations.	Same as CE-THERM-R1 plus additional water quality data and coefficients	Same forms as CE-THERM-R1
CE-QUAL-W2	2-D longitudinal, vertical hydrodynamic and water quality model for reservoir, estuarine, and other 2-D waterbodies	Solves 2-D hydrodynamics. Head of flow boundary conditions. Allows multiple branches. Simulates temperature, salinity, and up to 19 other water quality variables.	Basically same as CE-QUAL (THERM)-R1 Tidal boundary conditions for estuarine applications Morphometric data, including widths for each cell	Velocities and water quality constituents at all points on 2-D grid (printed) 2-D vector plots and 2-D constituent concentration contour or shading plots Time-series data and plots Statistical output Restart files for subsequent hot restart simulations

Model	Description	Input	Output
SELECT	1-D vertical steady-state model of selective withdrawal from a reservoir	Computation of withdrawal zone distribution from a density-stratified reservoir. Release temperature, density, conservative constituents computed. Multiple outflow types (spillway, water quality gate, flood-control outlet, etc.) handled internally. User specifies ports operating or selects ports internally based on quality objective (e.g., temperature). Reaeration of hydropower and flood-control releases.	Reservoir profiles for temperature (density) and conservative constituents Outflow rate, operation Structural configuration and hydraulic constraints or outlet(s) Quality targets if deciding port operation	Vertical profile of withdrawal zone Release qualities Appropriate port operations to meet quality targets
STEADY	1-D longitudinal steady-state stream temperature and DO model	1-D steady state. Steady flow. Allows branches, loops, and lateral inflows and withdrawals. Flow can be piecewise nonuniform.	Flows, depths, and velocities Average equilibrium temperature and heat-exchange coefficient Inflow temperature and DO Rate coefficients	Printed output for predicted temperature and DO at each node
CE-QUAL-RIV1	1-D, dynamic flow, time-varying stream hydraulic (RIV1H), and water quality (RIV1Q) models	Simulates dynamic (highly unsteady) flows. Simulates up to 10 time-varying water quality constituents. Allows branching systems. Allows multiple control structures. Stream, structural, and wind reaeration options. Direct energy balance or equilibrium temperature approach for temperatures.	Physical data, cross-section geometry, elevations, and locations of nodes; lateral inflows and tributaries; control structures Initial conditions Boundary conditions for flow and water quality Rate coefficients and other parameters Meteorological data or equilibrium temperatures and exchange coefficients	Hydraulic information and water quality constituent values printed for all nodes at specified print intervals Time-series plots of selected variables at selected nodes

TABLE 7.23

SUMMARY COMPARISONS OF SEVERAL WATER QUALITY MODELS (continued)

Model name	Description	Major features	Data requirements	Output
HEC-5Q	Reservoir system simulation/ optimization model for multiple water-resource purposes including water quality, water supply, hydropower, and flood control	Balanced reservoir system regulation determination. Optimum gate regulation for multiple water quality constituents.	Inflow quantity and quality Initial water quality conditions System configuration and physical description Reservoir regulation manual operation criteria System diversions Water quantity and quality targets at system control points	Reservoir and river water quality profiles Reservoir and river discharge rates, elevations, and travel time

Source: Compiled using data from U.S. Army Corps of Engineers, 1987.

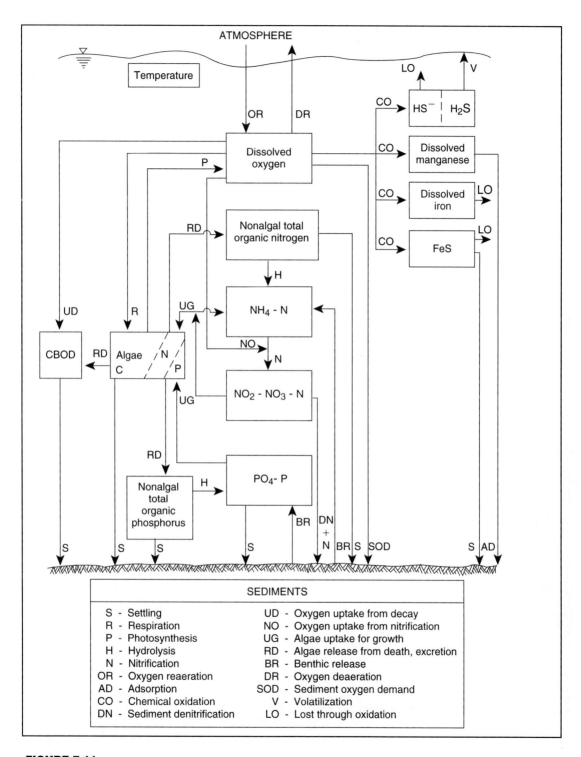

FIGURE 7.11
TWQM Compartmental Diagram (Tillman, Dortch, and Bunch, 1992, p. 4.).

TABLE 7.24

COMPONENTS OF THE INSTREAM FLOW INCREMENTAL METHODOLOGY

1. Characteristics of the stream are assessed (flow, sediment, water chemistry, temperature).
    Need to determine whether the system is stable or responding to change.
2. Stream is divided into reaches which have constant morphology, chemistry, and temperature.
    Shallow areas such as riffles which may present difficulties for fish migration should be noted.
3. Representative sampling reaches are taken and sampling is made through a range of discharge conditions.
    Transects are then taken so that each variation in morphology (pool, riffle, undercut bank, debris jam) is represented. The elevation of the water surface, depth, velocity, substrate, and cover is measured for each transect.
4. The measurements are taken to be representative of the habitat conditions within the 'cell' bounded laterally by the spot sampling points across the transect and longitudinally by the point midway between adjacent transects.
5. The measured habitat conditions within each cell are then compared with data on the habitat requirements for each species or life stage. A selection of species is taken, which can include those which are sensitive to change or those which are economically valuable.
6. Models, based on probability of use (POU) curves, are available to assess the habitat requirements of many species. The suitability of each habitat attribute is assessed between 0 (unsuitable) and 1 (optimum). Combinations of attributes are obtained by multiplying POU values, thereby giving an overall POU for each cell.
7. The product of the POU value and surface area of the cell gives a weighted useable area (WUA). The sum of the WUA value for all the cells provides a measure of the available habitat for each species/life stage under prevailing discharge conditions.
8. Calculations can be made over a range of discharges to produce a relationship between discharge and the available habitat for each species/life stage. This forms the output from the IFIM simulation programme.

Source: Brookes, 1988, p. 70.

existing habitat conditions by assigning values is easier than determining the actual fish population. It is also easier to predict the changes in habitat conditions that will be produced by a proposed water-extraction project (Brookes, 1988). Numerous aquatic-productivity models are available for use in impact studies.

A second example is the "habitat evaluation procedure" (HEP). The HEP was developed by the U.S. Fish and Wildlife Service, in conjunction with other agencies and research organizations, so that terrestrial and aquatic wildlife resources could be considered as part of an environmental impact assessment (U.S. Fish and Wildlife Service, 1980). A third example of a methodology for biological impacts on aquatic systems is the "habitat evaluation system" (HES) developed by the U.S. Army Corps of Engineers (1980). Detailed information on the usage of these latter two methodologies is in Chapter 11.

Other Considerations

Several other considerations may be relevant to the prediction of surface-water quantity-quality impacts. Examples include (1) frequency distribution of decreased quality and quantity; (2) effects of sedimentation on the stream-bottom ecosystem; (3) fate of nutrients by incorporation into biomass; (4) reconcentration of metals, pesticides, or radionuclides into the food web; (5) chemical precipitation or oxidation-reduction of inorganic chemicals; and (6) anticipated distance downstream of decreased water quality and the implications for water users and related raw-water-quality requirements.

STEP 5: ASSESSMENT OF IMPACT SIGNIFICANCE

The next activity is impact assessment. In the terminology used herein, "assessment" refers to the *interpretation* of the significance of anticipated changes related to the proposed project. Impact interpretation can be based upon the systematic application of a definition of "significance"; one example is the definition as included in CEQ (1978) and cited in Chapter 1. For some types of anticipated impacts, there are specific numerical standards or criteria which can be used as a basis for impact interpretation. Examples include the application of surface- and coastal-water quality standards and waste-water-discharge standards (effluent limitations) from particular facilities. Professional judgment can also be used in evaluating the percentage changes from baseline conditions in terms of pollutant loading and selected parameters or a water quality index; such changes could be anticipated during both the construction and operational phases of a project.

Another example is the application of professional judgment in the context of assessing impacts related to the biological environment; for example, the biological scientist in the study team would render judgments as to the applicability of various laws and the potential significance of the loss of particular habitats, including wetland areas. Additional information concerning prediction and assessment of biological-environment impacts is in Chapter 10. Another basis for impact assessment is public input; this input could be received through a continued scoping process or through the conduction of public meetings and/or public participation programs. The general public can often delineate important environmental resources and values for particular areas, and these should be considered in impact assessment.

STEP 6: IDENTIFICATION AND INCORPORATION OF MITIGATION MEASURES

The next activity is that of identifying and evaluating potential impact-mitigation measures. Mitigation measures may need to be added to the project proposal to make it acceptable.

These mitigation measures might consist of decreasing the magnitude of the surface-water impacts or including features that will compensate for the surface-water impacts. The specific mitigation measures will be dependent upon the particular project type and location; however, examples of some things which could be considered as mitigation or control measures, depending on the type of project, are listed below:

1. Decrease surface-water usage and waste-water generation through the promotion of water conservation and wastewater treatment and reuse. Pretreat wastewaters prior to discharging into receptor.

2. Minimize erosion during the construction and operational phases of the project; this could be facilitated by the use of on-site sediment-retention basins and by planting rapidly growing vegetation.

3. In projects involving the use of agricultural chemicals, consider measures that could be used to better plan the timing of chemical applications, the rate of application, and the extent of such applications in an effort to minimize erosion and chemical transport to surface-water systems. "Integrated pest management" (IPM) could also be used to decrease the pesticide loading from agricultural areas. "IPM" is an approach which combines biological, chemical, cultural, physical, and/or mechanical means, as appropriate, to deal with unwanted insects, weeds, and other pests (Franck and Brownstone, 1992). The net result of IPM should be a decrease in pesticide usage and thus a decrease in the nonpoint-source-pollution contribution to the surface-water environment.

4. Manage nonpoint-source pollution through the application of best management practices (BMPs) as determined by a state or a designated areawide planning agency to be the most effective practicable means of achieving pollutant levels compatible with

TABLE 7.25

SUMMARY OF WATER-QUALITY-ENHANCEMENT-TECHNIQUE APPLICATIONS AT U.S. ARMY CORPS OF ENGINEERS PROJECTS

Water quality concern	Technique applied	Total number of projects	Degree of success		
			Successful	Marginal	Unknown
Tailwater technique applications					
Temperature	Selective withdrawal	9	2	3	4
	Weir	3	3	—	—
	Operational modification	2	1	1	—
	Under study	7	—	—	7
Low DO	Operational modification	6	6	—	—
	Under study	2	—	—	2
Trace metals	Operational modification	3	3	—	—
	Localized mixing	1	1	—	—
Acid mine drainage	Operational modification	2	2	—	—
N supersaturation	Structural modification	3	—	3	—
Miscellaneous	Operational modification	2	1	—	1
In-lake enhancement technique applications					
Turbidity	Drawdown and planting	12	1	—	11
	Dredging	1	—	1	—
	Operational modification	2	2	—	—
	Inflow diversion	1	1	—	—
	Under study	4	—	—	4
Eutrophication	Olszenski tube	1	—	—	1
	Destratification	2	1	1	—
	Hypolimnetic aeration	1	—	—	1
	Copper sulfate	1	—	1	—
	Pool fluctuation	1	—	1	—
	Structural modification	1	—	—	1
	Wetland creation	1	—	—	1
	Best management practice	1	—	—	1
	Under study	2	—	—	2

Macrophytes	Herbicides	6	4	2
	Grass carp	1	—	1
Acid mine drainage	Operational modification	4	—	1
	Selective withdrawal	1	—	—
	Liming	2	—	—
Total dissolved solids (TDS)	Operational modification	1	—	1
Point source (oil wells)	Plug wells	1	—	—
Mercury (Hg)	Under study	1	—	1
Lock and dam enhancement technique applications				
Low DO	Operational modification	7	—	7

Source: Price, 1990, pp. 8–9.

water quality goals (Novotny and Chesters, 1981). This determination should be made after a process of problem assessment, examination of alternative practices, and appropriate public participation.

5. Develop a nonpoint-pollution-control program for coastal waters; information is available on management measures for agricultural sources, forestry, urban areas, marinas, recreational boating, hydromodification projects (channelization and channel modification, dams, and streambank and shoreline erosion), wetlands riparian areas, and vegetated treatment systems (U.S. EPA, 1990, 1993).

6. Use constructed wetlands to control nonpoint-source pollution involving nutrients, pesticides, and sediments (Olson and Marshall, 1991). As an example, a constructed system might include, in hydraulic order, a sediment basin, grassy filter, wetland, and deep pond (Wengrzynek, 1991).

7. Consider alternative wastewater-treatment schemes to achieve treatment goals in a cost-effective manner. For point sources, the treatment schemes could include primary, secondary, and/or tertiary processes involving physical, biological, and/or chemical principles of pollutant removal. For thermal effluents the use of cooling ponds or towers might be appropriate.*

---

*Note: The wastewater from "central vehicle wash facilities" (CVWFs) at military installations may be used as a treatment example. These facilities are used to allow the exterior of tactical vehicles to be cleaned in an efficient and environmentally safe manner. Research has been conducted to evaluate alternatives or modifications to existing design guidance for CVWF secondary treatment (Gerdes et al., 1991). The findings revealed that intermittent sand filtration, lagoons, and constructed wetlands are acceptable for secondary treatment. All three alternatives function with little attention from the CVWF operator. Lagoons and constructed wetlands require little maintenance. Sand on the surface of intermittent sand filters needs to be removed periodically. All three types of secondary treatment provide water that should be of discharge quality in most states.

8. Use discharge credit trading within watersheds to enable the trading of permitted pollution credits between parties responsible for both point- and nonpoint-source discharges.

9. Consider project operational modes that would minimize detrimental impacts. One example is related to operating dam-reservoir projects. In a recent survey of water resources projects operated by the U.S. Army Corps of Engineers, a number of enhancement techniques were identified for minimizing both in-reservoir and downstream water quality problems. Table 7.25 summarizes the technique applications and their success. Techniques for improving the dissolved oxygen in turbine-hydropower reservoir releases include in-reservoir, in-structure, and tailwater alternatives. In-reservoir options include diffused air or oxygen injection in the project forebay and the withdrawal of higher-quality surface water; in-structure options include air aspiration or air-oxygen injection, and operational changes to enhance reaeration; and tailwater options include diffused air or oxygen injection in the project tailwater and overflow weirs to promote reaeration (Wilhelms, 1992). A DO-monitoring system should be incorporated into the reservoir water quality improvement program. Finally, Price (1989) has reported on evaluations of commercially available destratification devices involving water pumping and hydraulic mixing.

10. Use techniques such as sediment removal and macrophyte (weed) harvesting for restoring lakes and reservoirs from water quality deterioration and eutrophication. These techniques have been described in terms of their scientific basis, method of application, effectiveness, beneficial and detrimental impacts, and costs in Cooke et al. (1986).

SUMMARY

This chapter has presented a six-step methodology for addressing the impacts of proposed projects, plans, programs, or policies on the surface-water environment. It is important to note that these steps provide a general framework which can be used: (1) to guide the study planning and conduction, (2) to indicate sources for which more-detailed information will be necessary, (3) to discuss a study with a contractor (or sponsor) and develop appropriate terms of reference, and (4) to review impact study work done by others. NPDES permit applications for point sources and for storm-water discharges from industrial and/or municipal areas with separate storm-water systems require the preparation of information analogous to a surface-water impact-focused impact study. Therefore, the six steps enumerated herein can also serve as the basis for development of permit applications for either point or nonpoint sources of pollution.

It will be necessary to document the six-step methodology via the preparation of a report, which could later serve as a section or chapter in either an EA or EIS. The report should address the potential surface-water impacts (quantity and/or quality) identified for this activity (step 1), the existing characteristics of the surface-water resource(s) in question (step 2), the applicable surface-water standards (step 3), the quantification or qualitative description of the surface-water impacts due to the project-activity (step 4), the assessment of the predicted impacts (step 5), and the considered and included mitigation measures, if any (step 6).

Legislation and regulations related to the surface-water environment continue to become more extensive and stringent. Key federal laws such as the Clean Water Act and Safe Drinking Water Act are reauthorized on an approximate five-year-cycle basis, frequently with additional source-control requirements and pollutants of emphasis. Additional water quality standards are being developed for toxic pollutants, and baseline data for such pollutants may be minimal. Therefore, these changes must be considered in the application of the six-step methodology.

SELECTED REFERENCES

Ambrose, R. D., et al., "Waste Allocation Simulation Models: A State-of-the-Art Review," *Journal Water Pollution Control Federation,* vol. 60, no. 9, Sept. 1988, pp. 1646–1655.

Bird, S. L., and Hall, R., "Environmental Impact Research Program: Coupling Hydrodynamics to a Multiple-Box Water Quality Model," WES/TR/EL-88-7, U.S. Army Engineer Waterways Experiment Station, Vicksburg, Miss., Mar. 1988.

Biswas, A. K., ed., *Models for Water Quality Management,* McGraw-Hill Book Company, New York, 1981.

Brannon, J. M., "Evaluation of Dredged Material Pollution Potential," Rep. no. WES-TR-DS-78-6, U.S. Army Engineer Waterways Experiment Station, Vicksburg, Miss., Aug. 1978.

Broderius, S. J., "Modeling the Joint Toxicity of Xenobiotics to Aquatic Organisms: Basic Concepts and Approaches," EPA 600/D-90-063, Environmental Research Laboratory, U.S. Environmental Protection Agency, Duluth, Minn., 1990.

Brookes, A., *Channelized Rivers—Perspectives for Environmental Management,* John Wiley and Sons, Chichester, England, 1988, pp. 67–71.

Canale, R. P., and Auer, M. T., "Personal Computers and Environmental Engineering: Part II—Applications," *Environmental Science and Technology,* vol. 21, no. 10, 1987, pp. 936–942.

Canter, L. W., "Environmental Impact of Comanche Power Station in Lawton, Oklahoma," report prepared for Public Service Company of Oklahoma, Tulsa, Feb. 1973.

———, *River Water Quality Monitoring,* Lewis Publishers, Chelsea, Mich., 1985.

———, et al., "Methods for Assessment and Prioritization of the Water Pollution Potential of Nonpoint Discharges from Navy Facilities," University of Oklahoma, Norman, Feb. 1990.

——— and Fairchild, D. M., "Post-EIS Environmental Monitoring," in *Methods and Experiences* in *Impact Assessment,* H. Becker and A. Porter, eds., D. Reidel Publishing Company, The Hague, 1986, pp. 265–285.

———— and Gloyna, E. F., "Transport of Chromium-51 in an Organically Polluted Environment," *Proc. 23d Purdue Industrial Waste Conference,* Purdue University, Lafayette, Ind., 1968, pp. 374–387.

————, Robertson, J. M., and Hargrave, S. L., "Use of Biomonitoring for Mitigating Biological Impacts During Construction and Postconstruction Phases of Selected Projects," presented at the *Eleventh International Seminar on Environmental Impact Assessment and Management,* University of Aberdeen, Scotland, July 13–14, 1990.

Chapman, D., ed., *Water Quality Assessments,* Chapman and Hall, London, 1992, p. 9.

"The Clean Water Act Requires a Water Quality-Based Approach When Technology-Based Pollution Controls Are Not Sufficient," *Water Newsletter,* vol. 31, no. 7, Apr. 15, 1989, p. 3.

Cooke, G. D., Welch, E. B., Peterson, S. A., and Newroth, P. R., *Lake and Reservoir Restoration,* Butterworth Publishers, Stoneham, Mass., 1986.

Corbitt, R. A., *Standard Handbook of Environmental Engineering,* McGraw-Hill Publishing Company, New York, 1990.

Council on Environmental Quality (CEQ), "National Environmental Policy Act—Regulations," *Federal Register,* vol. 43, no. 230, Nov. 29, 1978, pp. 55978–56007.

DeCoursey, D. G., ed., *Proceedings of the International Symposium on Water Quality Modeling of Agricultural Non-point Sources—Parts 1 and 2,* ARS-81, U.S. Department of Agriculture, Washington, D.C., June 1990.

Deeley, G. M., and Canter, L. W., "Distribution of Heavy Metals in Waste Drilling Fluids under Conditions of Changing pH," *Journal of Environmental Quality,* vol. 15, Apr.–June, 1986, pp. 108–112.

Dennison, M. S., "Updated Water Act Faces Capitol Debate," *Environmental Protection,* vol. 3, no. 6, July-Aug. 1992, pp. 13–17.

Douglas, I., *The Urban Environment,* Edward Arnold Publishers, Baltimore, 1983, pp. 58–61.

Economic Commission for Europe (ECE), "Convention on Environmental Impact Assessment in a Transboundary Context," E/ECE/1250, United Nations, New York, Feb. 1991.

Elder, J. F., "Metal Biogeochemistry in Surface-Water Systems—A Review of Principles and Concepts," *U.S. Geological Survey Circular 1013,* U.S. Geological Survey, Denver, 1988.

Federal Water Pollution Control Administration, "Industrial Waste Guide on Thermal Pollution," Washington, D.C., Sept. 1968b.

————, "Water Quality Criteria," Washington, D.C., Apr. 1, 1968a, p. 189.

Franck, I., and Brownstone, D., *The Green Encyclopedia,* Prentice-Hall General Reference, New York, 1992, pp. 167–168.

Gerdes, G. L., et al., "Evaluation of Alternatives for Secondary Treatment at Central Wash Facilities," CERL-TR-N-91/29, U.S. Army Construction Engineering Research Laboratory, Champaign, Ill., Aug. 1991.

Government Institutes, *Storm Water—Guidance Manual for the Preparation of NPDES Permit Applications for Storm Water Discharges Associated with Industrial Activity,* Rockville, Md., June 1991.

Grimsrud, G. P., Finnemore, E. J., and Owen, H. J., "Evaluation of Water Quality Models—A Management Guide for Planners," EPA 600/5-76-004, U.S. Environmental Protection Agency, Washington, D.C., July 1976, pp. 21–35.

Gunnison, D., "Interactions of Contaminants with Sediment and Water in Reservoirs," Information Exchange Bulletin, vol. E-90-1, U.S. Army Engineer Waterways Experiment Station, Vicksburg, Miss., June 1990.

Henderson-Sellers, B., *Water Quality Modeling—Vol. IV—Decision Support Techniques for Lakes and Reservoirs,* CRC Press, Boca Raton, Fla., 1991.

Hipel, K. W., "Nonparametric Approaches to Environmental Impact Assessment," *Water Resources Bulletin,* vol. 24, no. 3, June 1988, pp. 487–492.

Huber, W. C., and Dickinson, R. E., "Storm Water Management Model, Version 4, Part A: User's Manual," EPA 600/3-88-001A, University of Florida, Gainesville, June 1988.

IHD-WHO (World Health Organization) Working Group on the Quality of Water, "Water Quality Surveys," *Studies and Reports in Hydrology,* no. 23, World Health Organization, Geneva, 1978.

James, A., ed., *An Introduction to Water Quality Modeling,* John Wiley and Sons, West Sussex, England, 1993.

Knizel, G. W., and Nicks, A. D., "CREAMS: A Field Scale Model for Chemicals, Runoff, and Erosion from Agricultural Management Systems," Conservation Research Rep. no. 26, U.S. Department of Agriculture, Washington, D.C., 1980.

Krenkel, P. A., and Parker, F. L., "Engineering Aspects of Thermal Pollution," *Proceedings of National Symposium on Thermal Pollution,* Vanderbilt University Press, Nashville, 1969.

Leonard, R. A., et al., "Modeling Pesticides Metabolite Transport with GLEAMS," *Proceedings on Planning Now for Irrigation and Drainage,* IR DIV/ASCE, Lincoln, Neb., 1988.

Linsley, R. K., and Franzini, J. B., *Water-Resources Engineering,* 3d ed., McGraw-Hill Book Company, New York, 1979.

Loucks, D. P., "Water Quality Models for River Systems," Chap. 1 in *Models for Water Quality Management,* A. K. Biswas, ed., McGraw-Hill Book Company, New York, 1981, pp. 17–25.

Mahmood, R., and Messer, J. J., "Multivariate Water Quality Index for Land Use in Management of a Wildland Watershed," Water Resources Planning Series UWRL/p-82/08, Utah Center for Water Resources Research, Utah State University, Logan, Nov. 1982.

Marsh, W. M., *Landscape Planning—Environmental Applications,* 2d ed., John Wiley and Sons, New York, 1991, pp. 161–162.

McElroy, A. D., et al., "Loading Functions for Assessment of Water Pollution from Nonpoint Sources," EPA 600/2-76-151, U.S. Environmental Protection Agency, Washington, D.C., 1976.

Metcalf and Eddy, *Wastewater Engineering-Treatment, Disposal and Reuse,* 3d ed., McGraw-Hill Book Company, New York, 1991.

Miller, L. A., Taylor, R. S., and Monk, L. A., *NPDES Permit Handbook,* Government Institutes, Rockville, Md., June 1991.

Nemerow, N. L., *Industrial Water Pollution: Origin, Characteristics and Treatment,* Addison-Wesley Publishing Company, Reading, Mass., 1978, pp. 529–549.

———, and Dasgupta, A., *Industrial and Hazardous Waste Treatment,* Van Nostrand Reinhold Company, New York, 1991.

"News Update," *Environmental Protection,* vol. 3, no. 6, July-Aug. 1992, p. 8.

Novotny, V., and Chesters, G., *Handbook of Nonpoint Pollution,* Van Nostrand Reinhold Company, New York, 1981.

Oklahoma Water Resources Board, "Appraisal of the Water and Related Land Resources of Oklahoma—Regions Five and Six," Publ. 27, Oklahoma Water Resources Board, Oklahoma City, 1969, pp. 76–77.

———, "Oklahoma's Water Quality Standards—1982," no. 111, Oklahoma City, 1982, pp. 1–2.

———, "Oklahoma's Water Quality Standards—1979," no. 101, Oklahoma City, 1979.

Olson, R. K., and Marshall, K., *Workshop Proceedings: The Role of Created and Natural Wetlands in Controlling Nonpoint Source Pollution,* EPA 600/9-91-042, ManTech Environmental Technology, Corvallis, Ore., Nov. 1991.

Ott, W. R., *Environmental Indices—Theory and Practice,* Ann Arbor Science Publishers, Ann Arbor, Mich., 1978, pp. 202–213.

Phelps, E. B., *Stream Sanitation,* John Wiley and Sons, New York, 1944, pp. 201–221.

Price, R. E., "Evaluating Commercially Available Destratification Devices," Information Exchange Bulletin, vol. E-89-2, U.S. Army Engineer Waterways Experiment Station, Vicksburg, Miss., Dec. 1989.

———, "Water Quality Enhancement Techniques Used within the Corps of Engineers," Miscellaneous Paper W-90-1, U.S. Army Engineer Waterways Experiment Station, Vicksburg, Miss., Oct. 1990.

Ray, C., "Use of QUAL-I and QUAL-II Models in Evaluating Waste Loads to Streams and Rivers," Chap. 7 in A. K. Biswas, T. N. Khoshoo, and A. Khosla, eds., *Environmental Modeling for Developing Countries,* Tycooly Publishing, London, 1990, pp. 65–79.

Roberts, L. E. J., Liss, P. S., and Saunders, P. A. H., *Power Generation and the Environment,* Oxford University Press, Oxford, 1990, pp. 180–181.

Shubinski, R. P., and Tierney, G. F., "Effects of Urbanization on Water Quality," *Proceedings of ASCE Urban Transportation Div. Specialty Conference on Environmental Impact,* American Society of Civil Engineers, New York, 1973, p. 180.

Smith, D. G., "Water Quality Indexes for Use in New Zealand's Rivers and Streams," Publication no. 12, Water Quality Centre, Ministry of Works and Development, Hamilton, New Zealand, Nov. 1987.

Stefan, H. G., Ambrose, R. B., and Dortch, M. S., "Surface Water Quality Models: Modeler's Perspective," *Proceedings of the International Symposium on Water Quality Modeling of Agricultural Non-point Sources, Part 1,* ARS-81, U.S. Department of Agriculture, Washington, D.C., June 1990, pp. 329–379.

Summers, J. K., Kazyak, P. F., and Weisberg, S. B., "A Water Quality Model for a River Receiving Paper Mill Effluents and Conventional Sewage," *Ecological Modeling,* vol. 58, 1991, pp. 25–54.

Theurer, F. D., Voos, K. A., and Prewitt, C. G., "Application of IFG's Instream Water Temperature Model in the Upper Colorado River," *Proceedings of International Symposium on Hydrometeorology,* American Water Resources Association, Denver, June 1982, pp. 287–292.

Tillman, D. H., Dortch, M. S., and Bunch, B. W., "Tailwater Quality Model (TWQM) Development," vol. E-92-4, U.S. Army Engineer Waterways Experiment Station, Vicksburg, Miss., Oct. 1992.

U.S. Army Corps of Engineers, "A Habitat Evaluation System for Water Resources Planning," U.S. Army Corps of Engineers, Lower Mississippi Valley Division, Vicksburg, Miss., Aug. 1980.

———, "Water Quality Models Used by the Corps of Engineers," Information Exchange Bulletin, vol. E-87-1, U.S. Army Engineer Waterways Experiment Station, Vicksburg, Miss., Mar. 1987.

U.S. Congress, *"Federal Water Pollution Control Act Amendments of 1972,"* P.L. 92-500, 92nd Congress, Oct. 18, 1972.

U.S. Environmental Protection Agency (EPA), "Coastal Nonpoint Pollution Control Program," U.S. Environmental Protection Agency, Office of Wetlands, Oceans, and Watersheds, Washington, D.C., Jan. 1993.

———, "Guidance Specifying Management Measures for Sources of Nonpoint Pollution in Coastal Waters," U.S. Environmental Protection Agency, Office of Wetlands, Oceans, and Watersheds, Washington, D.C., 1990.

———, "Hazardous Waste Management System," *Federal Register,* vol. 51, no. 114, June 13, 1986, pp. 21648–21693.

———, "Methods for Identifying and Evaluating the Nature and Extent of Non-Point Sources of Pollutants," EPA 430/9-73-014, U.S. Environmental Protection Agency, Washington, D.C., Oct. 1973, p. 6.

———, "Part 125—Criteria and Standards for the National Pollutant Discharge Elimination System," *Code of Federal Regulations,* Title 40, Ch. 1, July 1, 1991, pp. 211–215.

———, "Part 130—Water Quality Planning and Management," *Code of Federal Regulations,* Title 40, Ch. 1, July 1, 1991b, pp. 259–272.

———, "Part 131—Water Quality Standards," *Code of Federal Regulations,* Title 40, Ch. 1, July 1, 1991c, pp. 272–277.

———, "Part 133—Secondary Treatment Regulation," *Code of Federal Regulations,* Title 40, Ch. 1, July 1, 1991d, pp. 285–289.

———, "Part 141—National Primary Drinking Water Standards—Subpart B—Maximum Contaminant Levels," *Code of Federal Regulations,* Title 40, Ch. 1, July 1, 1991e, pp. 585–589.

———, "Part 143—National Secondary Drinking Water Regulations," *Code of Federal Regulations,* Title 40, Ch. 1, July 1, 1991f, pp. 758–762.

———, "Part 419—Petroleum Refining Point Source Category," *Code of Federal Regulations,* Title 40, Ch. 1, July 1, 1991g, pp. 419–457.

———, "Policy on the Use of Biological Assessments and Criteria in the Water Quality Program," U.S. Environmental Protection Agency, Office of Science and Technology, Washington, D.C., May 1991h.

———, "Storm Water Management for Construction Activities—Developing Pollution Prevention Plans and Best Management Practices," EPA 832/R-92-005, U.S. Environmental Protection Agency, Office of Water, Washington, D.C., Sept. 1992a.

———, "Storm Water Management for Industrial Activities—Developing Pollution Prevention Plans and Best Management Practices," EPA 832/R-92-006, U.S. Environmental Protection Agency, Washington, D.C., Sept. 1992b.

———, "Storm Water Management for Industrial Activities—Developing Pollution Prevention Plans and Best Management Practices—Summary Guidance," EPA 833-R-92-002, U.S. Environmental Protection Agency, Office of Water, Washington, D.C., Oct. 1992c.

————, "Storm Water Management Model User's Manual, Version 3," EPA 600/2-84/109a, b, Washington, D.C., 1984.

————, "Treatability Manual, Vol. II, Industrial Descriptions," EPA 600/8-80-042b, Section II.14 (Petroleum Refining), U.S. Environmental Protection Agency, Washington, D.C., July 1980.

U.S. Fish and Wildlife Service, "Habitat Evaluation Procedures (HEP)," ESM 102, U.S. Fish and Wildlife Service, Washington, D.C., Mar. 1980.

U.S. Geological Survey, "National Handbook of Recommended Methods for Water Data Acquisition," U.S. Geological Survey, Reston, Va., March 1982.

Wengrzynek, R. L., "Constructed Wetlands to Control Nonpoint Source Pollution," PAT-APPL-7-764 924/WEP, U.S. Department of Agriculture, Agricultural Research Service, Washington, D.C., Sept. 1991.

Wilhelms, S. C., "Improvement of Low-head Hydropower Releases," vol. E-92-3, U.S. Army Engineer Waterways Experiment Station, Vicksburg, Miss., September 1992.

————, "Improvement of Reservoir Releases—Alternative Release Strategies," vol. E-93-1, U.S. Army Engineer Waterways Experiment Station, Vicksburg, Miss., June 1993.

Williams, J. R., and Nicks, A. D., "CREAMS Hydrology Model—Option I," in *Applied Modeling in Catchment Hydrology, Proceedings of the International Symposium on Rainfall-Runoff Modeling,* Water Resources Publication Company, Littleton, Colo., 1982.

CHAPTER 8

Prediction and Assessment of Impacts on the Soil and Groundwater Environments

Land development, resource extraction, and waste-disposal projects can cause certain undesirable impacts on soil and/or groundwater resources. These impacts may be in the form of either quantity or quality changes. In addition, activities that may result from development projects, such as urban growth near a new water-supply reservoir, can cause soil and/or groundwater effects as a result of urban-waste-disposal leachates moving through the subsurface system. Other projects which may have impacts on groundwater quantity-quality include those in which groundwater is used for purposes of water supply. Therefore, in considering the potential impacts of projects of various types on soil and groundwater resources, attention should be given to both quality and quantity concerns.

The purpose of this chapter is to present a conceptual framework for addressing potential soil or groundwater quantity-quality impacts associated with proposed projects or activities. The first sections contain background information on soil and groundwater, and relevant regulations, standards, and policies. The bulk of the chapter focuses on a six-step methodology for addressing soil or groundwater impacts.

BACKGROUND INFORMATION ON THE SOIL ENVIRONMENT

Soil systems have developed over many millions of years, and they can be influenced by numerous factors. Some of these factors involve natural environmental influences, and others are related to man-induced influences. Table 8.1 delineates the influence of man on five classic factors associated with soil formation. Human activities can influence each fundamental factor either beneficially or detrimentally. The primary point is that the soil characteristics in a given geographical area at a given point in time are a function of both natural influences and human activities. Accordingly, this should be considered both in describing the affected environment and interpreting the potential significance of the changes resulting from a project-activity.

The geological features in an area are also a function of natural forces and processes which have existed and occurred over millions of years. While these factors are influential in terms of a specific location, it is unlikely that many human-caused significant changes will have occurred in the geological environment; an

TABLE 8.1

POSSIBLE HUMAN-INDUCED EFFECTS ON FIVE CLASSIC FACTORS OF SOIL FORMATION

Factors of soil formation	Type of effect	Nature of effect
Climate	Beneficial	Adding water by irrigation; rainmaking by seeding clouds; removing water by drainage; diverting winds, etc.
	Detrimental	Subjecting soil to excessive insolation, to extended frost action, to wind, etc.
Organisms	Beneficial	Introducing and controlling populations of plants and animals; adding organic matter including 'nightsoil'; loosening soil by ploughing to admit more oxygen; fallowing; removing pathogenic organisms as by controlled burning
	Detrimental	Removing plants and animals; reducing organic matter content of soil through burning, ploughing, over-grazing, harvesting, etc; adding or fostering pathogenic organisms; adding radioactive substances
Topography	Beneficial	Checking erosion through surface roughening, land forming and structure-building; raising land level by accumulation of material; land levelling
	Detrimental	Causing subsidence by drainage of wetlands and by mining; accelerating erosion; excavating
Parent material	Beneficial	Adding mineral fertilizers; accumulating shells and bones; accumulating ash, locally; removing excess amounts of substances such as salts
	Detrimental	Removing, through harvest, more plant and animal nutrients than are replaced; adding materials in amounts toxic to plants or animals; altering soil constituents in a way to depress plant growth.
Time	Beneficial	Rejuvenating the soil through adding of fresh parent material or through exposure of local parent material by soil erosion; reclaiming land from under water
	Detrimental	Degrading the soil by accelerated removal of nutrients from soil and vegetation cover; burying soil under solid fill or water

Source: Goudie, 1984, p. 246.

exception is human-induced seismic activity. Accordingly, addressing the geological environment in an impact study typically involves the utilization of fundamental resources-oriented information relative to the features in the project study area. Excellent background information on the soil and geological environments can be found in several textbooks. For example, Strahler and Strahler (1987) contains both descriptive and quantitative information related to geographic features of the environment, and Keller (1982) and Rahn (1986) do similarly from a geological perspective. Examples of projects of many types, as well as pertinent environmental consequences, are included in all three textbooks.

The soil and geological environments are typically associated with the physical-chemical environment. However, they also exhibit fundamental relationships to other environmental components. For example, the habitat types and associated vegetation found in an area will be a function of the soil characteristics. Additionally, cultural resources may be related to soil characteristics or, possibly, to unique geological features in an area. Another example of a relationship between the soil environment and other environmental components is associated with nonpoint-source water pollution. The quantity and chemical characteristics of the nonpoint-source pollution are related to the fundamental soil characteristics of an area, along with its history of erosional patterns and applications of fertilizers and pesticides. This topic is addressed in Chapter 7.

In many cases, soil and geological information is fundamental to the engineering design, construction, and, sometimes, the operation of proposed projects. For example, soil characteristics must be considered in foundation design for buildings. Soil and geological information is taken into account in the design of major hydraulic works such as multipurpose dams. In addition, a major area of concern in the design of nuclear power plants is associated with the geological characteristics in the area, including earthquake potential and fault zones. Accordingly, considerable information may have already been developed in conjunction with project planning and design which could be used, as appropriate, in an impact study. Finally, soil and geological information is frequently necessary in site selection or delineation of specific design features for certain types of projects. For example, the site determination process for hazardous-waste landfills should incorporate consideration of local soil and geological features (Kiang and Metry, 1982). Additionally, the design of liners and covers for hazardous-waste landfills should be based on the permeability characteristics of the site soil (U.S. EPA, 1986b).

Numerous types of projects could have detrimental impacts on the soil or geological environment, or both, and the resultant environmental conditions could affect the functioning of such projects, once implemented. Examples of types of projects and associated impacts include

1. Land subsidence which can occur as a result of overpumping of groundwater resources (Schumann and Genualdi, 1986) or oil or gas resources in a given geographical area or which can occur as a result of surface or subsurface mining activities associated with mineral extraction (Toy and Hadley, 1987c, 1987d).

2. The impacts associated with the identification and usage of construction material for major projects, with such material coming from identified borrow areas. (There may be resultant changes in local surface-water hydraulics and erosional patterns as a result of the removal of construction material.)

3. Construction practices in general can create some concerns related to the potential for increased soil erosion in the construction area. This increase in soil erosion could lead to specific mitigation requirements, such as the creation of sediment-retention basins or the planting of rapidly growing vegetation (Toy and Hadley, 1987e; Westcott, 1989).

4. Landslides, caused by inappropriate slope stability, which can occur as a result of overdevelopment on particular soil types within areas having certain topographic features. An example is the landslides which have occurred in recent years in southern California as a result of home-building in steep topographic areas on certain types of soils.

5. The potential concerns associated with constructing and operating nuclear power plants, chemical production plants, waste-disposal facilities, and/or large storage tank facilities in areas characterized by seismic instability and excessive earthquake potential. (This can influence siting decisions and decisions associated with construction and operation activities.)

6. Strip-mining operations for coal extraction, or other mineral resource extraction, wherein the land surface is removed and relocated, with the normal practice being to restore the original landscape, possibly in some type of alternative topographic arrangement (U.S. EPA, 1973; U.S. Department of Energy, 1988).

7. The construction of jetties along coastal areas in order to control beach erosion and littoral drift.

TABLE 8.2

EXAMPLES OF HUMAN-INDUCED EFFECTS ON SOIL CHARACTERISTICS

Soil factor	Beneficial change	Neutral change	Adverse change
Soil chemistry	Mineral fertilizers (increased fertility) Adding trace elements Desalinize (irrigation) Increase oxidation (aeration)	Altering exchangeable ion balance Altering pH (lime) Alter via vegetation changes	Chemical imbalance Toxic herbicides and herbicides Salinize Over-removal of nutrients
Soil physics	Induce crumb structure (lime and grass) Maintain texture (organic manure or conditioner) Deep plowing, alter soil moisture (irrigation or drainage)	Alter structure (plowing, harrowing) Alter soil microclimate (mulches, shelter belts, heating, albedo change)	Compaction/plow pan (poor structure) Adverse structure due to chemical changes (salts) Remove perennial vegetation
Soil organisms	Organic manure Increase pH Drain/moisten Aerate	Alter vegetation and soil microclimate	Remove vegetation and plow (less worms and microorganisms) Pathogens (e.g., slurry) Toxic chemicals
Time (rate of change)	Rejuvenate (deep plowing, adding new soil, reclaiming land)		Accelerated erosion Overuse of nutrients Urbanizing land

Source: Drew, 1983, p. 32.

8. Projects associated with military training activities, wherein the training itself leads to excessive compaction of the soil and subsequent detrimental effects on soil erosion or drainage patterns.

9. Projects which may create acid rain in localized areas, with the acid rain, in turn, having an impact on soil chemistry and, potentially, on subsurface groundwater resources; examples of this have been noted in southern Sweden and on Long Island, New York (Andersson and Stokes, 1988; Proios, 1985).

10. Projects wherein the site characteristics in terms of soil and geological features are incorporated as components in the selection process. Examples of such site-selection-oriented projects include waste-disposal projects, sludge-disposal projects, and upland locations for dredged-material disposal.

11. Projects that involve developments along coastal areas wherein coastal erosion problems may either be increased by the project, or may influence the proposed project itself. Examples of such projects include coastal marinas and associated secondary developments, industrial development projects with associated port and boat-mooring facilities, and projects which involve the development of ports and harbors.

12. The construction and operation of surface-water reservoir projects, with the purposes of the projects ranging from the single purpose of providing flood control to multiple purposes, including hydroelectric power development, provision of water supply, and so on. There are two key environmental concerns relative to soils and geological issues: the first is related to sedimentation within the reservoir and the provision of appropriate sediment-storage capacity in terms of the project lifetime; the second is related to the potential effects of such surface-water reservoir projects on the subsurface environment, including changes in soil, groundwater, and geological features that lie underneath the water pool of the reservoir.

13. Projects associated with permits for grazing leases or other leases related to agricultural uses, where the subsequent grazing or agricultural developments could lead to changes in soil characteristics such as erosion patterns and soil chemistry (Toy and Hadley, 1987a). Examples of such changes are in Table 8.2.

14. The potential effects of buried pipelines on soils such as the concerns related to the Alaska pipeline and its associated effects on the soil freeze-thaw cycle.

15. The potential effects of soil characteristics on buried pipelines, with examples including the potential loss of the physical integrity of the pipeline as a result of acid or corroding soils (Sorrell et al., 1982).

16. Projects involving the purchase or selling of land which may have been previously contaminated with hazardous-waste materials. Examples include the excess lands program within the U.S. Department of Defense, as well as numerous private-property transactions. [A type of environmental impact study which is typically associated with property transactions is referred to as a "site assessment" or an "environmental baseline survey" (Canter, 1990).]

17. Land disturbance associated with the use of recreational vehicles, such as snowmobiles or motorcycles, with the primary soil and geological concern being the inducement of additional soil erosion (Toy and Hadley, 1987b).

BACKGROUND INFORMATION ON GROUNDWATER QUANTITY AND QUALITY

Groundwater quantity concerns are typically related to groundwater usage in connection with

TABLE 8.3

POSSIBLE SOURCES OF GROUNDWATER CONTAMINATION

Category I   Sources designed to discharge substances

Subsurface percolation (e.g., septic tanks and cesspools)

Injection wells
   Hazardous waste
   Nonhazardous waste (e.g., brine disposal and drainage)
   Nonwaste (e.g., enhanced recovery, artificial recharge, solution mining, and in-situ mining)

Land application
   Wastewater (e.g., spray irrigation)
   Wastewater by-products (e.g., sludge)
   Hazardous waste
   Nonhazardous waste

Category II—Sources designed to store, treat, and/or dispose of substances; discharge through unplanned release

Landfills
   Industrial hazardous waste
   Industrial nonhazardous waste
   Municipal sanitary

Open dumps, including illegal dumping (waste)
Residential (or local) disposal (waste)
Surface impoundments
   Hazardous waste
   Nonhazardous waste

Waste tailings
Waste piles
   Hazardous waste
   Nonhazardous waste

Materials stockpiles (nonwaste)
Graveyards
Animal burial
Aboveground storage tanks
   Hazardous waste
   Nonhazardous waste
   Nonwaste

Underground storage tanks
   Hazardous waste
   Nonhazardous waste
   Nonwaste

Containers
   Hazardous waste
   Nonhazardous waste
   Nonwaste

its availability. The study of developable groundwater resources in relation to water supply needs involves the consideration of hydrogeological factors and quality characteristics. If groundwater is withdrawn at a rate greater than its natural replenishment rate, then the depth to groundwater increases, and the resource is "mined." In addition, excessive usage of groundwater in coastal areas can cause saltwater intrusion (this problem can also occur in inland

---

TABLE 8.3

POSSIBLE SOURCES OF GROUNDWATER CONTAMINATION *(continued)*

Open burning and detonation sites
Radioactive disposal sites

Category III—Sources designed to retain substances during transport or transmission

---

Pipelines
  Hazardous waste
  Nonhazardous waste
  Nonwaste

Materials transport and transfer operations
  Hazardous waste
  Nonhazardous waste
  Nonwaste

Category IV—Sources discharging substances as consequence of other planned activities

---

Irrigation practices (e.g., return flow)
Pesticide applications
Fertilizer applications
Animal feeding operations
Deicing salt applications
Urban runoff
Percolation of atmospheric pollutants
Mining and mine drainage

  Surface mine-related
  Underground mine-related

Category V—Sources providing conduit or inducing discharge through altered flow patterns

---

Production wells
  Oil (and gas) wells
  Geothermal and heat recovery wells
  Water supply wells

Other wells (nonwaste)
  Monitoring wells
  Exploration wells

Construction excavation

Category VI—Naturally occurring sources whose discharge is created and/or exacerbated by human activity

---

Ground water–surface water interactions
Natural leaching
Salt-water intrusion/brackish water upconing (or intrusion of other poor-quality natural water)

Source: Office of Technology Assessment, 1984, p. 45.

---

states where freshwater-bearing zones are underlain by saline aquifers). Finally, because of the hydraulic influences, the relationship between shallow, alluvial aquifers and the flow of surface streams and rivers may need to be explored.

Groundwater quality concerns other than inorganic quality issues, primarily focused on chlorides, have been increasing within the last decade. This increased attention is a result of greater emphasis upon hazardous-waste sites in the United States and Europe, and the realization that many global environmental problems may also have implications for groundwater resources, including the potential influence of acid

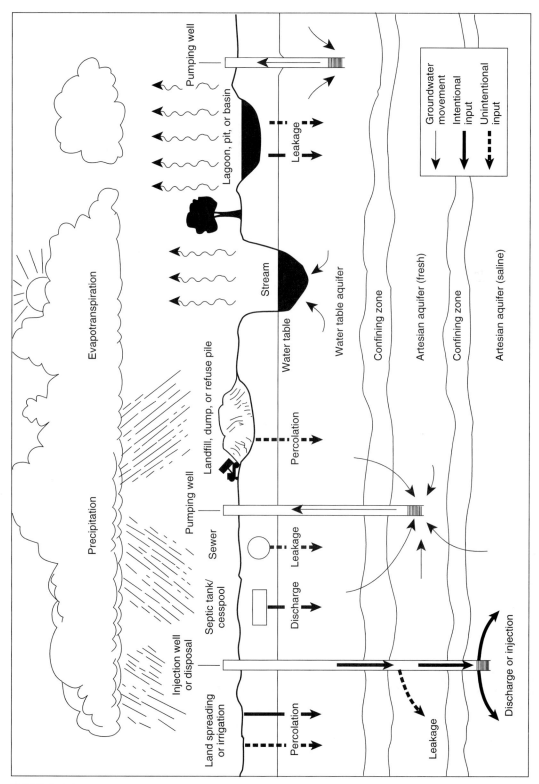

FIGURE 8.1
Examples of Sources of Ground Water Contamination (Franck and Brownstone, 1992).

TABLE 8.4

CHARACTERISTICS OF PRINCIPAL ACTIVITIES POTENTIALLY CAUSING GROUNDWATER POLLUTION

Activity	Principal characteristics of pollution					Stage of development[a]			Impact of water use		
	Distribution	Category	Main types of pollutant	Relative hydraulic surcharge	Soil zone bypassed	A	B	C	Drinking	Agricultural	Industrial
Urbanization											
Unsewered sanitation	ur	P-D	pno	x	★	xxxx	xx	x	xxxx		x
Land discharge of sewage	ur	P-D	nsop	x		x	x	x	xx	x	x
Stream discharge of sewage	ur	P-L	nop	xx	★	x	x		xx	x	x
Sewage oxidation lagoons	u	P	opn	xx	★	x	xx	x	xx		x
Sewer leakage	u	P-L	opn	x	★			xx	x		x
Landfill, solid waste disposal	ur	P	osnh		★	x	xx	xxx	x		x
Highway drainage soakaways	ur	P-L	so	xx	★	x	xx	xx	xx	x	x
Wellhead contamination	ur	P	pn		★	xxx	x		xxx		
Industrial development											
Process water/effluent lagoons	u	P	ohs	xx	★	x	xx	xx	xx		x
Tank and pipeline leakage	u	P	oh		★	x	xx	xxx	xx		xx
Accidental spillages	ur	P	oh	xx		x	xx	xxx	xxx		xx
Land discharge of effluent	u	P-D	ohs	x		x	xx	xx	x	x	x
Stream discharge of effluent	u	P-L	ohs	xx	★	x	x	x	x	x	x
Landfill disposal residues and waste	ur	P	ohs		★	x	xxx	xxx	xx		x
Well disposal of effluent	u	P	ohs	xx	★		x	xx	xx		x
Aerial fallout	ur	D	a					xx	x	x	x
Agricultural development											
Cultivation with:											
Agrochemicals	r	D	no			x	xx	xxx	xxx	x	x
Irrigation	r	D	sno	x		xx	xx	x	xxx	xxxx	x
Sludge and slurry	r	D	nos			x	x	x	xx	x	x
Wastewater irrigation	r	D	nosp	x			xx	x	xx	xx	

Livestock rearing/crop processing:

Source	Category	Distribution	Types of pollutant	* / na	A	B	C
Unlined effluent lagoons	P	r	pno	*	x	xx	x
Land discharge of effluent	P-D	r	nsop	*	x	xx	x
Stream discharge of effluent	P-L	r	onp	*	x	xx	x

Mining development

Source	Category	Distribution	Types of pollutant	* / na	A	B	C
Mine drainage discharge	P-L	ru	sha	*	xx	xx	x
Process water/sludge lagoons	P	ru	hsa	*	xx	xx	x
Solid mine tailings	P	ru	hsa	*	xx	xx	x
Oilfield brine disposal	P	r	s	*	x	xx	x
Hydraulic disturbance	D	ru	s	na	x	xx	x

Groundwater resource management

Source	Category	Distribution	Types of pollutant	* / na	A	B	C
Saline intrusion	D-L	ur	s	na	xx	xxx	xx
Recovering water levels	D	u	so	na	x	x	x

Distribution
u Urban
r Rural

Category
P Point
D Diffuse
L Line

Types of pollutant
P Fecal pathogens
N Nutrients
O Organic micropollutants
H Heavy metals
S Salinity
A Acidification

x to xxxx Increasing importance or impact
na Not applicable

[a] Stages of development: A Highly industrialized; B Newly industrializing; C Low development

Source: Chapman, 1992, p. 403–404.

rain upon groundwater quality. Table 8.3 includes a listing of over 30 potential sources of groundwater contamination. These sources are divided among six categories, with the categories ranging from sources designed to discharge substances to sources which represent naturally occurring phenomena whose discharge may be created and/or exacerbated by human activities. One of the points to note from Table 8.3 is that there are numerous potential sources of groundwater pollution which could exist in a given geographical area. Some of the sources are naturally occurring; however, the vast majority are man-made or human-caused. Figure 8.1 also depicts numerous potential sources of groundwater contamination. Finally, Table 8.4 summarizes the principal anthropogenic activities which can cause groundwater pollution.

KEY FEDERAL LEGISLATION

Soil-Environment Emphasis

While not as targeted as environmental media laws, there are several laws at the federal level that address various features of the soil environment. Specific examples of such federal laws will be mentioned; these are only representative of additional laws and regulations which may exist at the state or local level. One example is the Coastal Zone Management Act (CZMA) of 1972. The primary purpose of the CZMA was to encourage the development of voluntary programs for land and water use in coastal areas. The CZMA focused on targeting and prohibiting certain types of land uses within the coastal zone so as to protect the zone from unnecessary losses and disruptions related to man-made projects. As a result of the requirements of the CZMA of 1972, coastal states have also developed laws and regulations, often in conjunction with state coastal zone commissions, that focus upon coordinating and limiting land uses in certain ecologically fragile areas. In 1976, the CZMA was amended, and three significant points were included. The first was that state management programs were required to address shoreline erosion, shorefront access, and the presence of energy facilities. The second was that unavoidable adverse environmental impacts associated with new or expanded onshore energy facilities in coastal areas were to be mitigated with measures financed by federal grants. The third raised the federal funding to 90 percent for coastal management and authorized 50 percent federal funding for land acquisition for estuarine sanctuaries (Freedman, 1987).

The Coastal Zone Management Improvement Act was passed by the U.S. Congress in 1980 to update and strengthen various aspects of the CZMA (1972, 1976). As directed by the act, an inventory of geographical areas of primary concern was developed; it resulted in the creation of eight critical areas, as follows (Freedman, 1987): (1) areas of unique, scarce, fragile, or vulnerable natural habitat, physical features, historical significance, cultural value, and scenic importance; (2) areas of high natural productivity or essential habitat for living resources, including fish, wildlife, and the various tropic levels in the food web critical to their well-being; (3) areas of substantial recreational value or opportunity; (4) areas where developments and facilities are dependent upon the utilization of, or access to, coastal water; (5) areas of unique geologic or topographic significance to industrial or commercial development; (6) areas of urban concentration where shoreline utilization and other water uses are highly competitive; (7) areas that, if developed, would pose significant hazards, because of storms, slides, floods, erosions, seismic conditions, or other circumstances; and (8) adjacent areas needed to protect, maintain, or replenish coastal lands or resources, including coastal floodplains, aquifer recharge areas, sand dunes, coral and other reefs, beaches, offshore sand deposits, and mangrove stands.

Another major law focused upon the soil and geological environment is the Surface Mining

Control and Reclamation Act (SMCRA) of 1977. The law requires operators of surface coal mines affecting greater than 2 acres of ground to obtain permits; the permit application must be based on and include a reclamation plan and general environmental-protection provisions, through which the operator must prevent environmental damage and restore the affected lands (Freedman, 1987). The regulatory program mandated by Section 501(b) of the SMCRA has resulted in the identification of impacts and possible mitigation measures which are commonly associated with mining activities. There are four general categories of biological impacts which mitigation measures must address: (1) loss of wildlife and wildlife habitat, (2) disturbance of aquatic organisms and aquatic habitat, (3) erosion and sedimentation, and (4) destruction of vegetation (U.S. Department of the Interior, 1979). Mitigation of erosion impacts will indirectly mitigate all other impact types and consequently, increase the quality of the environment. Biological mitigation measures which were suggested for impacts to vegetation included restoration of lands to premining conditions and implementation of revegetation programs which store topsoil for reapplication. Replacing or restoring the vegetation in the mining areas is the most critical of all mitigation activities if environmental impacts to the biota are to be minimized. Vegetation has many essential roles in the ecosystem. Primarily, it is the first level in the terrestrial food chain upon which all organisms depend. Secondly, vegetation provides habitat and cover for organisms, as well as providing stability of soils with its root systems. Additional information is in Chapter 10.

Another example of a law related to the soil environment is the Clean Water Act (CWA) of 1987, and, more specifically, Section 319 of the CWA, which is primarily focused on nonpoint sources of pollution. Section 319 requires each state to develop a pertinent management and control program for nonpoint sources of pollu-

tion. The primary objective of such programs is to minimize nonpoint-source discharges into both surface waters and groundwaters (U.S. EPA, 1987c). (This law is discussed in more detail in Chapter 7.) As an example of a state-developed nonpoint-source control program, the relationship between the major pollution categories and the BMPs for the State of Delaware is shown in Table 8.5.

Several waste-disposal-related laws have been developed that focus upon delineating appropriate environmental controls for such waste disposal. Examples of these laws include the Resource Conservation and Recovery Act (RCRA) of 1976, the Hazardous and Solid Waste Amendments (HSWA) to the RCRA, the Comprehensive Environmental Response, Control, and Liability Act (CERCLA), and the Superfund Amendments and Reauthorization Act (SARA). The latter two laws are focused upon Superfund requirements, while the first two are related to sanitary- and hazardous-waste landfilling operations. The RCRA and HSWA regulations stipulate environmental performance standards for hazardous-waste landfills and other hazardous-waste-disposal facilities (U.S. EPA, 1986b). The standards give emphasis to the prevention of adverse effects on groundwater quality, surface-water quality, air quality, and the subsurface environment and associated system components.

Because of the attention directed toward the cleanup (or remediation) of contaminated soil, identification of the necessary constituent concentrations requisite to establishing that cleanup has been achieved is a key issue. No uniform soil concentration levels have been established; many states are developing standards for generic and/or site-specific levels. A number of multimedia environmental goals (MEGs) have been proposed for use in the United States (Fitchko, 1989). Discharge MEGs (DMEGs) represent approximate concentration maxima in source emissions to receiving water, atmosphere, or soil (through solid waste) which should be tol-

TABLE 8.5

EXAMPLES OF BMPs USED IN DELAWARE IN RELATION TO POLLUTION CATEGORIES

Major nonpoint source pollution categories	Resource management system	Conservation practice(s)	Samples of best management practice(s)
Agriculture	Cropland	Fertilizer Management[a]	soil test for nutrient needs calibrate equipment fertilize for yield goals	time application split application
		Pesticide Management[a]	use pest-specific chemicals apply chemical at proper time apply chemical at label rates integrated pest management	watch weather conditions dispose of container properly calibrate spray equipment manage spills and wastewater
		Conservation Cropping Sequence[a]	crop rotation	cover and green manure crop
		Conservation Tillage System[a]	reduced tillage no tillage	minimum tillage crop residue use
		Irrigation Water Management[a]		
		Structure for Water Control[a]		
		Terrace[a]		
		Grassed Waterway[a]		
		Water and Sediment Control Basin[a]		
	Farmstead/ Headquarters	Waste Management System[a]	waste storage structure waste utilization waste storage pond waste treatment lagoon	roof runoff management filterstrips effluent spray irrigation
	Pastureland	Pasture/Hayland Planting	fertilizer management soil test for nutrient needs	pesticide management
		Pasture/Hayland Management	intensive grazing pesticide management	manure management livestock exclusion
Silviculture	Forest Land	Woodland Improved Harvesting	preharvest assessment and planning proper harvesting practices buffer strip	regeneration cutting residual tree protection
		Forest Land Erosion Control System	proper harvesting practices outsloping haul roads	environmental exclusion, i.e., harvesting during dry season only
		Tree Planting	woods/haul road maintenance site preparation	skid trail and haul road closure species selection

Category	Land	Management System		
		Wooded Corridor Management	timber stand improvement interplanting	
Construction (Highway/Road/Bridges)	Transportation Services Land		proper storage facility of deicing salts vegetative erosion and sediment control measures alternate snow removal techniques water quality inlets creation of traffic-free zones public education heavy-use area protection	density restrictions in residential developments reduction of number of vehicle trips in sensitive area incorporation of open spaces regional and multiuse management measures land-use planning
Construction	Urban Land	Runoff Management System[a]	infiltration structures vegetated swales grassed waterways filterstrips artificial wetlands porous pavement	streambank/shoreline protection slope stabilization stilt fence stabilized construction entrance sediment traps critical area planting
Resource Extraction/ Exploration/ Development	Mined Land	Land Reconstruction	land reclamation sediment basins guides to include water quality requirements	diversion dike critical area planting well-head protection
Urban Runoff			artificial wetlands flood prevention measures catch basin cleaning pet litter control	critical area planting urban forestry fertilizer management pesticide management
Land Disposal	Community Services Land		public information and education state-of-the-art landfill operations resource reclamation plant energy generating facilities erosion and sediment control	composting operations density of on-site systems rural wastewater policy reduction of land application rates
Hydrologic/Habitat Modification			erosion and sediment control	public education and information
Other (Atmospheric Deposition)			restriction of certain pollutant containing building materials reduction of smoking in public facilities emission control on heating devices public information and education regional pollution reduction programs	gasoline vapor capture devices on automobiles restriction on certain paints, adhesives, and solvents restriction on water and wastewater treatment and disposal building code uniformity as to ventilation and circulation

[a]Indicator practices.
Source: Delaware Department of Natural Resources and Environmental Control, 1989, pp. 8–9.

erable for short-term exposures. Values based on acute human health effects and short-term (reversible) effects on natural biological communities are specified. The use of DMEGs in evaluating emissions is ultraconservative in that no dilution of contaminants in the environment is assumed prior to exposure. "Ambient-level MEGs" (AMEGs) are approximate levels of contaminants in water, air, or soil below which unacceptable negative effects in human populations or in natural biological communities should not occur, even with continuous exposure.

MEGs for terrestrial environments (expressed in milligrams per kilogram) are based on a simple leachate model for solid waste (in this case, DMEGs) and for contaminated soil (here, AMEGs are used), and are equal to the liquid-emission MEGs (expressed in micrograms per liter) for the chemical of concern, multiplied by a factor of 0.2. This model assumes that all contaminants in 1 kg of soil or solid waste would be leached by 2 L of water. The major human exposure route to contaminants from soil or solid waste is assumed to be consumption of contaminated drinking water. Similarly, the major exposure route for aquatic life is through the leaching of contaminated soil or solid waste by surface waters. It is further assumed that the concentrated leachate entering a body of water (groundwater or surface water) will be diluted almost instantaneously by an arbitrary factor of 100. While this model is very simplistic and, in most situations, conservative (e.g., it considers almost no retention or attenuation of contaminants before reaching surface water), it does provide a set of rough guideline limits for a broad range of contaminants without the need to consider hydrogeological and other environmental variables in detail. Table 8.6 provides selected DMEG (solid waste) and AMEG (soil) values (Fitchko, 1989).

Other relevant institutional requirements at the federal level include executive orders associated with prime and unique agricultural lands,

as well as executive orders and policy delineations associated with wetlands and wetland losses. These requirements are often focused on limiting certain types of projects, or the extent of certain projects, in given geographical areas. Accordingly, it is important that the relevance of such executive orders and/or policies be considered in addressing the potential impacts of a given project in a given geographical location.

A special category of relevant laws are those that relate to site assessments (Canter, 1990). The central purpose of a site assessment is to establish the environmental liabilities of the proper owner, potential buyer, and lending institution. Owner-operator liability for contaminated property has become one of the major issues negotiated among lenders, insurers, real estate developers, lawyers, financial brokers, purchasers, and other asset managers involved in real-property transactions. The regulatory basis for liability related to property transfer can be considered from both federal and state perspectives (Funderburk, 1990; Duvel, 1986; Severns, 1988). The key federal laws include the Resource Conservation and Recovery Act of 1976, and the Superfund laws of the 1980s.

By the mid-1980s, laws addressing site-assessment investigations had already been passed or were under consideration in several states. The Environmental Cleanup and Responsibility Act (ECRA) passed in New Jersey in 1984 required that industrial properties used to manufacture, store, or dispose of toxic waste be inspected and cleaned up before sale. This was the first law making environmental site assessments mandatory in certain real estate transactions (Cahill and Kane, 1989). The majority of the states have environmental legislation affecting property transfers (Funderburk, 1990). Because of these still-developing state and federal regulations and growing private sector liability concerns, environmental site assessments are fast becoming a requirement in many types of real estate transactions.

TABLE 8.6

SELECTED MULTIMEDIA ENVIRONMENTAL GOALS FOR SOLID WASTE AND SOIL, BY CONTAMINANT CLASS

Contaminants, by class	Concentration, mg/kg	
	Solid waste (DMEG)	Soil (AMEG)
Aliphatic and cyclic hydrocarbons		
Cyclohexane	200	100
Cyclopentane	20,000	10,000
Dicyclopentadiene	20	10
Heptanes	20,000	10,000
Heptenes	20,000	10,000
Hexanes	20,000	10,000
Hexenes	20,000	10,000
Octanes	2,000	1,000
Pentanes	200	100
Organics with carbon-oxygen bonds		
Ethanol	20,000	10,000
Methanol	20,000	10,000
Ethylene glycol	2,000	1,000
Acetone	20,000	10,000
Acetophenone	3,100	100
Acrolein	20	10
Butyraldehyde	20	10
Formaldehyde	200	100
Methyl ethyl ketone	20,000	10,000
Acetic acid	200	100
Organics with carbon-nitrogen bonds		
Acetonitrile	20,000	10,000
Aniline	200	100
Benzidine	20	10
PCBs and organohalogen pesticides		
Polychlorinated biphenyls	0.001	0.0002
Lindane	0.004	0.0008
Other halogenated organics		
Low molecular weight, volatile		
Carbon tetrachloride	2,000	100
Chloroform	200	100
Methylene chloride	2,000	1,000
Tetrachloroethylene	200	100
Trichloroethylene	2,000	1,000
Vinyl chloride	20,000	10,000
High molecular weight, halogenated aromatic		
Chlorobenzene	20	10

TABLE 8.6

SELECTED MULTIMEDIA ENVIRONMENTAL GOALS FOR SOLID WASTE AND SOIL, BY CONTAMINANT CLASS *(continued)*

	Concentration, mg/kg	
Contaminants, by class	Solid waste (DMEG)	Soil (AMEG)
High molecular weight, halogenated phenolic		
Chlorinated cresols	100	0.06
Pentachlorophenol	5	2.5
Polynuclear aromatic hydrocarbons		
Naphthalene	20	10
Other substituted aromatics		
Benzene	200	100
Toluene	200	50
Xylenes	200	100
Phenol	100	20
Organometallics		
Mercury (alkyl)	0.6	0.002
Nickel carbonyl	2.0	0.4
Tetraethyl lead	20	10
pH-volatile inorganics		
Ammonia	10	7
Cyanide	3.6	—
Hydrogen sulfide	2.0	—
Reactive inorganics		
Bromine	200	—
Chlorine	2.0	—
Chromates	0.3	—
Magnesium oxide	20,000	10,000
Heavy metals		
Antimony	620	8
Barium	180	100
Beryllium	5.4	2.2
Boron	220	—
Cadmium	0.30	0.08
Cerium	2,200	—
Cesium	2,200	—
Chromium	44	10
Cobalt	130	10
Copper	2.4	2.0
Gold	20	—
Iron	1,000	—
Lanthanum	3,200	—

TABLE 8.6

SELECTED MULTIMEDIA ENVIRONMENTAL GOALS FOR SOLID WASTE AND SOIL, BY CONTAMINANT CLASS *(continued)*

Contaminants, by class	Concentration, mg/kg	
	Solid waste (DMEG)	Soil (AMEG)
Lead	0.76	—
Lithium	75	15
Manganese	104	4.0
Molybdenum	940	2.0
Nickel	56	0.4
Rubidium	280	—
Silver	0.12	—
Strontium	2	—
Thallium	40	—
Thorium	2,200	—
Tin	120	—
Titanium	170	—
Tungsten	2,200	—
Uranium	56	20
Vanadium	20	15
Zinc	20	4.0
Major nutrients		
Phosphorus	200	—
Macroelements		
Aluminum	1.4	—
Bromide	128,000	—
Calcium	20,000	—
Fluoride	46	—
Magnesium	10,000	8,660
Potassium	8,600	4,320
Silica	26,000	—
Sulfate	8,600	—
Acid gases		
Hydrogen chloride	22,000	—
Nitric acid	90	—
Sulfuric acid	90	—

Source: Adapted from data in Fitchko, 1989, pp. 2.5a–2.5d.

Groundwater Emphasis

In the United States, there are at least 16 identifiable statutes at the federal level that have some focus on groundwater resources management; however, there is no comprehensive law that encompasses the entirety of management issues related to this resource. In fact, other environmental laws that require more-stringent pollution controls on air or industrial-wastewater emissions have caused the generation of sludges which, in turn, when they are land applied, can lead to groundwater contamination.

TABLE 8.7

SUMMARY OF FEDERAL STATUTES RELATED TO THE PROTECTION OF GROUNDWATER QUALITY

	Investigations/detection				Correction		Prevention				
Statutes	Inventories of sources[a]	Ambient groundwater monitoring	Groundwater monitoring related to sources[a]	Water supply monitoring	Federally funded remedial actions	Regulatory requirements for sources[a]	Regulate chemical production	Standards for new/existing sources[a]	Aquifer protection	Standards	Other
Atomic Energy Act	x		x		x	x		x		x	x
Clean Water Act		x	x		x	x		x		x	x
Coastal Zone Management Act									x		
Comprehensive Environmental Response, Compensation, and Liability Act	x		x		x						
Federal Insecticide, Fungicide, and Rodenticide Act			x				x	x			
Federal Land Policy and Management Act (and associated mining laws)			x					x			
Hazardous Liquid Pipeline Safety Act	x							x			
Hazardous Materials Transportation Act	x							x			
National Environmental Policy Act											x

Reclamation Act

Resource Conservation and Recovery Act

Safe Drinking Water Act

Surface Mining Control and Reclamation Act

Toxic Substances Control Act

Uranium Mill Tailings Radiation Control Act

Water Research and Development Act

[a] Programs and activities under this heading relate directly to specific sources of groundwater contamination. Table 8.3 summarizes the sources addressed by the statues.

[b] This category includes items such as research and development and grants to the states to develop groundwater related programs.

Source: Office of Technology Assessment, 1984, p. 65.

Table 8.7 lists 16 federal statutes in the United States related to the protection of groundwater quality. Examination of this table reveals a wide-ranging perspective in terms of the federal laws, and differing approaches depending upon what groundwater issue is addressed by the particular statute. Table 8.8 delineates the 16 federal statutes in accordance with the sources of groundwater pollution identified in Table 8.3. Table 8.8, through the use of a letter code, delineates the specific ways in which the statutes relate to the individual sources of pollution.

Leaking underground storage tanks (USTs) containing petroleum or other liquid products have been identified as a significant groundwater pollution source in the United States. The Hazardous and Solid Waste Amendments of 1984 (amendments to the Solid Waste Disposal Act of 1965 and the Resource Conservation and Recovery Act of 1976) included requirements related to the development of regulations to protect public health and the environment from leaking USTs (U.S. EPA, 1988). The U.S. Environmental Protection Agency issued the regulations in 1988; as a result, numerous tanks have been removed and corrective actions taken at contaminated sites (Gangadharan et al., 1988). Depending upon the extent of contamination and specific permitting requirements, EAs on corrective actions may be necessary.

Remedial action projects involving Superfund sites or other contaminated sites are subject to the preparation of EAs or EISs, or their functional equivalents. The U.S. Environmental Protection Agency is developing standards for land disposal restrictions (LDRs) for contaminated soil and debris removed from uncontrolled hazardous-waste-disposal sites and Superfund sites (Davis and Chou, 1992). The standard may be based on the delineation of best demonstrated available technologies (BDATs) such as destruction, extraction, and immobilization. EAs or EISs may be required in the decision-making process.

In addition to federal laws, within the United States there may be numerous state and local laws and regulations related to groundwater protection, groundwater quality standards, or groundwater resources management. For many countries throughout the world, applicable laws and regulations may be based on or adapted from those developed by the World Health Organization or various United Nations organizations such as the United Nations Environment Program and the United Nations Development Program.

The delineation of pertinent groundwater quality-quantity standards for given geographical areas may be difficult because of the multiplicity of laws and regulations at different governmental levels. It would be appropriate to initially determine if there were any pertinent groundwater quality standards or criteria that would be applicable for the study area. Additionally, if there are laws related to the quantity of groundwater which could be used in a given geographical area, they should also be identified and summarized.

Within the United States, there are additional groundwater management policies and programs being developed (on both federal and state levels), with one example being the usage of a groundwater- (aquifer-) classification system. Groundwater-classification systems are primarily based upon providing greater protection to those groundwater resources that are of greater importance because of their usage for public water supply, or because of their provision of unique ecological benefits in given geographical areas. Accordingly, if there is a groundwater-classification system which has been developed for the study area, this information should be identified. In the United States, the U.S. Environmental Protection Agency has suggested a three-level classification scheme as follows:

Class I—A "special" or sole-source aquifer, or one providing ecological benefits; the

TABLE 8.8

RELATIONSHIP BETWEEN SOURCES OF CONTAMINATION AND FEDERAL STATUTES

Sources	Federal statutes[a]															
	AEA	CWA	CZMA	CERCLA	FIFRA	FLPMA	HLPSA	HMTA	NEPA[b]	RA	RCRA	SDWA	SMCRA	TSCA	UMTRCA	WRDA[c]
Category I																
Subsurface percolation	E[a]											A				
Injection wells (waste)				F								A				
Injection wells (nonwaste)												A				
Land application	D			F							A					
Category II																
Landfills				F							A, B			A		
Open dumps (including illegal dumping)				F							B					
Residential (or local) disposal																
Surface impoundments				F	C	A					A		A			
Waste tailings						A					A		A		A	
Waste piles				F	C	A					A		A			
Materials stockpiles					C											
Graveyards																
Animal burial																
Aboveground storage tanks	A			F							A			A		
Underground storage tanks	A			F							A			A		
Containers				F	C						A					
Open-burning/detonation sites				F	C						A					
Radioactive disposal sites	A			F											A, F	
Category III																
Pipelines				F			A									
Materials transport/transfer operations				F				A								

TABLE 8.8

RELATIONSHIP BETWEEN SOURCES OF CONTAMINATION AND FEDERAL STATUTES (continued)

Sources	AEA	CWA	CZMA	CERCLA	FIFRA	FLPMA	HLPSA	HMTA	NEPA	RA	RCRA	SDWA	SMCRA	TSCA	UMTRCA	WRDA[c]
Category IV																
Irrigation practices	C, E															
Pesticide applications	C, E				A											
Fertilizer applications	C, E															
Animal feeding operations	C, E															
Deicing salts applications																
Urban runoff	C, E															
Percolation of atmospheric pollutants																
Mining and mine drainage	C, E					A							A, F			
Category V																
Production wells												A				
Other wells (nonwaste)					A	A										
Construction excavation	C, E															
Category VI																
Groundwater-surface water interactions	C, E															
Natural leaching										F						
Salt-water intrusion/ brackish-water upconing	C, E	E														

Note: The federal statutes are abbreviated here and coincide, in order, with the list in Table 8.7.

[a]Key: A = Requires compliance with specified federal requirements (some programs in this group may be implemented by states if they meet certain federal criteria).
B = Authorizes funding of optional state programs that address specific sources.
C = Establishes best management practices (BMPs) or recommended procedures for certain sources.
D = Establishes federal criteria that must be met in order to receive funds for specific projects related to a source of contamination.
E = Establishes a gift program to states (funds may be used at the state or local level to address contaminants or sources).
F = Funds federal cleanup of contaminated groundwater and associated sources.

[b]The NEPA does not apply to any particular source. The environmental impacts of projects involving the use of federal funds may be subject to federal agency review.
[c]The WRDA does not apply to any particular source. The set amount represents funds to states.

Source: Office of Technology Assessment, 1984, p. 76.

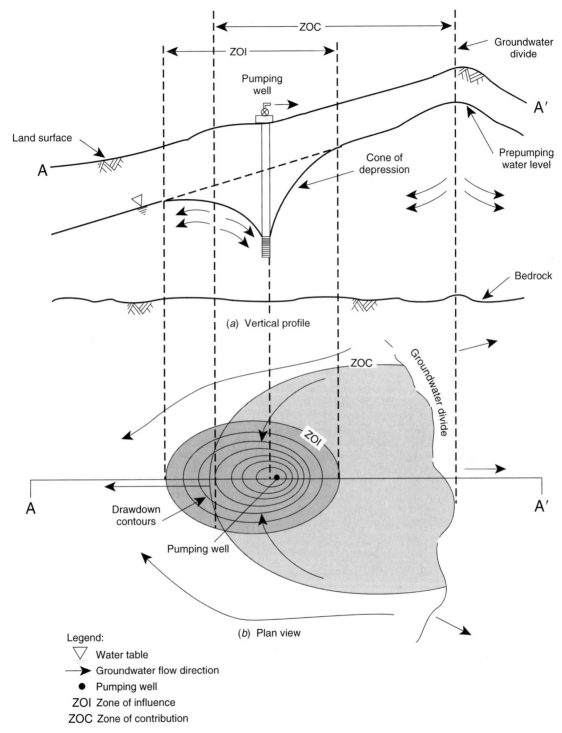

FIGURE 8.2
Terminology for Wellhead Protection Area Delineation (Hypothetical Pumping Well in Porous Media) (U.S. Environmental Protection Agency, 1987a, p. 2–20).

greatest protection should be provided for class I aquifers; the U.S. Environmental Protection Agency has designated 56 sole-source aquifers nationwide that serve more than 22 million people (U.S. General Accounting Office, 1992).

Class II—The majority of aquifers are in class II; moderate protection should be provided.

Class III—Limited-usage aquifers because of their poor natural or man-caused quality; the least protection should be provided.

Many states have developed classification systems patterned after the above three-level approach; other states have incorporated additional categories in their classification systems (Canter, Knox, and Fairchild, 1987).

Another example of a current program emphasis in the United States, as well as in many European countries, is the delineation and usage of groundwater protection zones. In the United States, this program is called the "wellhead protection program," and it was established by the Safe Drinking Water Act Amendments of 1986. A "wellhead protection area" (WHPA) means the surface and subsurface area surrounding a water well or wellfield, supplying a public water system, through which contaminants are reasonably likely to move toward and reach such water well or wellfield. In designating a WHPA, consideration needs to be given to natural groundwater flow, local hydrogeological characteristics, the influence of pumping and associated rates of water movement, and the transport and transformation characteristics of key contaminants. Figure 8.2 illustrates the terminology related to a WHPA.

The fundamental concept of a groundwater protection zone (or WHPA) is that certain limitations will be imposed upon man-made activities within a delineated zone of protection, with these limitations being for purposes of maintaining groundwater quality. Several techniques are available for delineating the boundaries of a groundwater protection zone, such as the one shown in Figure 8.2, including the use of arbitrary fixed radii, calculated fixed radii, simplified variable shapes, analytical methods, hydrogeologic mapping, and numerical flow-transport models; however, the specific discussion of these techniques is beyond the scope of this text. The key point to note is that if there are groundwater protection zones within the general vicinity of the proposed project, these should be so identified and the necessary requirements for meeting the protection zone stipulations should be summarized.

In addition to programs related to groundwater protection, there may also exist performance standards pertinent to particular activities or projects. For example, there are increasing federal requirements in terms of the provision of liners and leachate-collection and -treatment systems for sanitary- and/or hazardous-waste landfills within the United States. The basic issue is to determine whether such performance or design standards have actually been developed for the potential source category being addressed in the impact study.

CONCEPTUAL APPROACH FOR ADDRESSING SOIL- AND GROUNDWATER-ENVIRONMENT IMPACTS

To provide a basis for addressing soil- and/or groundwater-environment impacts, a six-step or six-activity model is suggested for the planning and conduct of impact studies. This model is flexible and can be adapted to various project types by modification, as needed, to enable the addressing of specific concerns of specific projects in unique locations. It should be noted that the focus in this model will be on projects and their soil and/or groundwater impacts; however, the model could also be applied to plans, programs, and regulatory actions. Although this chapter is mainly related to soil and groundwater systems, the principles could also be ap-

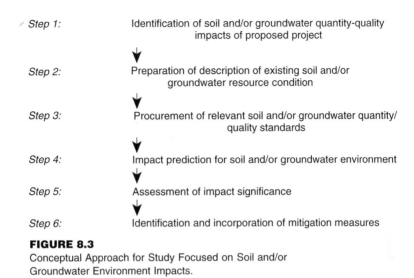

FIGURE 8.3
Conceptual Approach for Study Focused on Soil and/or
Groundwater Environment Impacts.

plied to the examination of impacts on geological systems.

The six generic steps associated with soil and/or groundwater environment impacts are (1) identification of the types and quantities of soil and/or groundwater pollutants to be introduced or groundwater quantities to be withdrawn as a result of the project, or other impact-causing factors related to the development project; (2) description of the environmental setting in terms of soil types, characteristics, and quality; groundwater flows and quality; hydrogeological characteristics; relationships to existing surface-water resources; groundwater classifications and/or WHPAs; existing point and nonpoint sources of pollution; and pollution loadings and existing groundwater withdrawals; (3) procurement of relevant laws, regulations, or criteria related to soil erosion control, soil quality, groundwater quality and/or groundwater usage, and any relevant compacts (agreements) between states, countries, or other entities related to the groundwater system; (4) conduction of impact prediction activities, including the use of qualitative approaches based on analogs and professional judgment, simple mass-balance cal-

culations, empirical index methods, and/or quantitative modeling; (5) use of pertinent information from step 3, along with professional judgment and public input, to assess the significance of anticipated beneficial and detrimental impacts; and (6) identification, development, and incorporation of appropriate mitigation measures for the adverse impacts. Figure 8.3 delineates the relationship between the six steps or activities in the conceptual approach.

STEP 1: IDENTIFICATION OF SOIL AND/OR GROUNDWATER QUANTITY-QUALITY IMPACTS

Soil Quantity-Quality Impacts

An appropriate initial activity when analyzing a proposed project or activity is to consider what types of soil and/or geological disturbances might be associated with the construction and/or operational phases of the proposed project, and what quantities of potential soil contaminants are expected to occur. There are no generic methodologies which can be uniformly applied for all projects in order to identify their

TABLE 8.9

INFORMATION ON FOUR SOILS-RELATED DATABASES

AGRICOLA (AGRICultural OnLine Access) is an extensive bibliographic database consisting of records for literature citations of journal articles, monographs, theses, patents, translations, microforms, audiovisuals, software, and technical reports. Available since 1970, AGRICOLA serves as a document locator and bibliographic access and control system for the National Agricultural Library (NAL) collection, but since 1984 the database has also included some records produced by cooperating institutions for documents not held by NAL.

CAB Abstracts is a comprehensive file of agricultural information containing all records in the 22 journals published by the Commonwealth Agricultural Bureaux (CAB) in the United Kingdom. CAB has long been recognized as a leading scientific information service in agriculture and certain fields of applied biology.

Enviroline draws material from approximately 5,000 primary and secondary sources of information, including the following: more than 3,500 scientific, technical, trade, professional, and general periodicals; papers and proceedings from environment- and resource-related conferences; documents and research reports from private and governmental agencies; congressional hearing transcripts; environmental project reports; and newspaper articles. Enviroline includes information abstracts approximately 100 words long for records from 1975 forward; it currently has about 150,000 records.

Pollution Abstracts is a leading resource for references to environmentally related technical literature on pollution, its sources, and its control. Abstracts are included for most records from the beginning of 1978 forward. References in Pollution Abstracts are drawn from approximately 2,500 primary sources from around the world, including books, conference papers and proceedings, periodicals, research papers, and technical reports. The database currently has about 150,000 records.

Source: Compiled using data from Dialog Information Services, 1989.

potential impacts on the soil and geological environments. However, one approach is to consider relevant case studies and their typical impacts (17 examples of projects or activities were listed earlier). This is a qualitative approach in that professional knowledge and case study information could be used to delineate anticipated soil- and geological-environment impacts.

One of the things that can be done to help identify soil quantity-quality impacts for the project types listed above, or for other projects or activities that could have potential effects, is to conduct surveys of published literature. The scope of computerized bibliographic searching is expanding, with the primary requisite to a successful search being the identification of descriptor words for the project-activity type; the searching can be done in various databases. Four examples of databases in the commercially available Dialog system

which might be searched are summarized in Table 8.9.

"Impact trees" or "networks" can also be used to delineate potential impacts on the soil and geological environments. For example, Figure 8.4 shows different pathways and effects of acid precipitation on different components of an ecosystem.

Regarding the identification of potential soil pollutants, a list of the materials to be utilized during the project and those materials which will require disposal could be developed. Examples of materials that may result in soil contamination include fuels and oils, bituminous products, insecticides, fertilizers, chemicals, and solid and liquid wastes. As an initial step, a simple checklist of the types and quantities of chemicals associated with each activity could be prepared and utilized. Transport and effects information on key chemicals could also be included. It may also be appropriate to consider

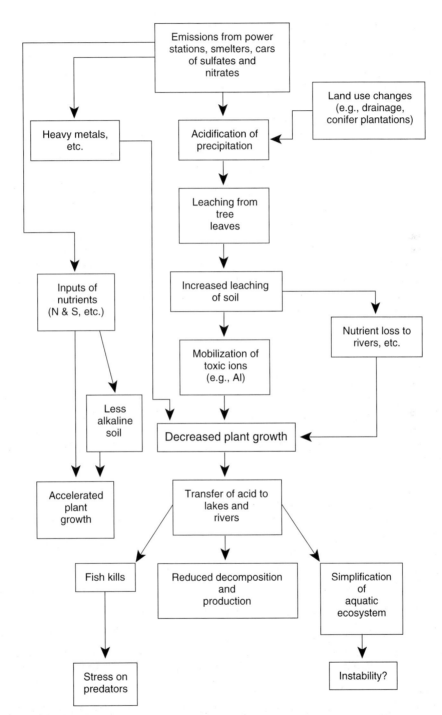

FIGURE 8.4
Pathways and Effects of Acid Precipitation on Different Components of Ecosystem (Goudie, 1986).

the quality of leachates from waste materials disposed on land; this topic is addressed in the next section.

Groundwater-Quantity and -Quality Impacts

The consideration of groundwater-quantity and -quality impacts consists of identifying the types and quantities of groundwater pollutants and/or groundwater quantity changes anticipated to be associated with the construction and operational phases of a proposed project. This activity should also be performed for any alternatives to the project or proposed plans or programs.

The specific things to be done to identify the types and quantities of groundwater pollutants associated with a proposed project will vary depending upon the project type. As an example, for a project that involves the generation of leachate, groundwater pollutants might come from dredged material to be disposed of upland or solid or hazardous waste to be disposed of on land. The basic issue of concern is associated with precipitation and other runoff waters that might move through the disposal site and thus appear as leachate in the subsurface (soil and groundwater) environment. Additionally, leachate quality can be a concern for projects involving the use of construction materials containing municipal-solid-waste (MSW) incinerator residues—for example, those using MSW incinerator fly ash in asphalt concrete or MSW incinerator bottom ash in concrete paving blocks (Van der Sloot and De Groot, 1990).

One approach for addressing leachate quality concerns is to conduct leachate testing. "Leachate testing" typically involves subjecting the material to be leached to appropriate environmental conditions relative to precipitation volumes and mixing conditions, and subsequently collecting the leachate which moves through the material and analyzing the leachate water. Several tests of this type have been developed in the United States, including the Elutriate Test for Dredged Material (Brannon, 1978), the Ex-

traction Procedure (EP) Toxicity Test for solid or hazardous waste (U.S. EPA, 1980), and the more recent Toxicity Characteristic Leaching Procedure (TCLP) Test for solid or hazardous waste (U.S. EPA, 1986b). Comparisons of the TCLP and EP tests have been made in terms of the extraction of metals and volatile organic contaminants; the results were variable (Bricka, Holmes, and Cullinane, 1991). In addition, another type of leaching test called "Sequential Extraction Tests" has been developed for muds from oil- and gas-well drilling and for municipal and industrial wastewater treatment sludges (Deeley and Canter, 1986).

Another approach to identify the types and quantities of groundwater pollutants pertinent to the project type would be to conduct appropriate literature reviews for such data compiled for projects of similar type. It might also be appropriate to actually visit the locations of similar types of projects to determine from these look-alike (analogous) conditions what types and quantities of groundwater pollutants might be relevant.

In terms of the quantity of leachate that might be produced, estimates can be developed using water-balance methods, with one example being the method developed by the U.S. EPA (Fenn, Hanley, and DeGeare, 1975). Although this water-balance method was developed for sanitary landfills, it can be also used for hazardous-waste landfills, dredged-material disposal sites, and agricultural operations. A special, computerized water-balance method for hazardous-waste landfills is called the "hydrologic evaluation of landfill performance" (HELP) model (Schroeder and Peyton, 1992). The basic advantage of using water-balance methods is that they account for precipitation, runoff, evapotranspiration, infiltration, cover-soil moisture-storage capacity, and the moisture-storage capacity of the waste material, whether it be dredged material or solid or hazardous waste. Careful accounting of these components will yield information on the quantities of leachate produced as a function of different seasons and precipitation levels. This information on quantity, when coupled with information resulting from leachate testing, represents the basis for the delineation of source factors in terms of groundwater pollution.

If the proposed project involves usage of groundwater for purposes of water supply, then information should be procured on the quantities of water to be used over time as a result of the development and implementation of the project. [One implication of groundwater extraction which will not be explored herein is land subsidence, which can be of concern depending on local hydrogeological factors (Schumann and Genualdi, 1986).] If the project involves irrigation systems, then information could be accumulated on the quantities of water to be used for purposes of irrigation, and, with appropriate adjustments, the water-balance method mentioned earlier could be used to estimate the potential quantity impact on the local groundwater system (Fenn, Hanley, and De-Geare, 1975).

Finally, for projects that may involve septic tank systems or land application of wastewater, information would be needed on the wastewater quality and application rates. When coupled with use of the water-balance method mentioned earlier, this information could be used to develop data on the potential quantitative impacts of such a project on groundwater resources in the area (Fenn, Hanley, and DeGeare, 1975). More-sophisticated quantitative models are described in the section outlining step 4.

Finally, regarding step 1 activities (for either soil or groundwater impacts), the proponent of the project, project design considerations being developed by the proponent, the conduction of pertinent testing of materials to anticipate leachate quality characteristics, and/or the conduction of pertinent literature reviews on projects of similar type are the best sources of and methods for obtaining relevant information.

STEP 2: DESCRIPTION OF EXISTING SOIL AND/OR GROUNDWATER RESOURCES

Soil Characteristics

Many impact studies have been characterized by thorough presentations of information on existing geological information in the project area, and to a lesser degree, on pertinent soil information. Geological information is often included because it is readily available, even though the relevance of this information to the specific proposed project is always not thoroughly explored. The key principle in describing the affected soil and geological environment is to consider potential impacts, and then to address the relevant soil and geological features that might be changed by the proposed project or that might exert some influence on project design, construction, or operation.

In terms of baseline conditions, a major source of information for soils in the United States are county-level soil surveys which have been developed over the last several decades. Some soil surveys are fairly old; however, because of the relatively slow change of soil characteristics over time, the older information can often be used with confidence. Information on the planning, conduction, and interpretation of soil survey results is available in Olson (1981, 1984) and can be useful for orientation purposes in impact studies. Another soils information source is the Soil Conservation Service (SCS) of the U.S. Department of Agriculture (USDA). The SCS has extensive information on soils and soil characteristics, and frequently it is the primary agency in the development of county-level soil surveys.

The *U.S. Geological Survey* (USGS) is a major source of information for geological, cartographic, and geographic data for different environmental settings (Dodd, Fuller, and Clarke, 1989). In addition, the USGS has extensive information on groundwater and on the relationship between geological features and ground-

TABLE 8.10

SOME SOURCES OF EARTH-SCIENCE-RELATED INFORMATION

Federal agencies
U.S. Department of the Interior
  Geological Survey
  Bureau of Mines
  Bureau of Reclamation
  Bureau of Land Management
U.S. Department of Agriculture
  Extension Service
  Soil Conservation Service
  Forest Service
U.S. Department of Commerce
  National Oceanic and Atmospheric Administration
U.S. Department of the Army
  Army Corps of Engineers
Energy Research and Development Administration
Environmental Protection Agency
National Aeronautics and Space Administration
Tennessee Valley Authority

State divisions or departments
Agriculture
Conservation
Forestry
Geological Surveys
Oil and Gas
Soil Conservation
Fish and Game
Water Resources
Water Quality
Mineral Resources
Colleges and Universities

County or city departments or special districts
Planning
Water
Flood Control
Agriculture
Parks and Recreation
Engineering
Building and Safety
Public Works

Private producers
Consulting Firms
Private Colleges and Universities
Professional Societies
Industries with In-house Capabilities

Source: William Spangle and Associates, 1976, p. 5.

water resources. Additionally, each state typically has its own geological survey which can serve as a valuable source of both soil and geological data for a given geographical area. Examples of information which might be procured from a state geological survey include listings of specific geological features organized by county or geological basin, hydrologic atlases delineating surface and subsurface environmental features, historical geology reports, studies related to waste-disposal siting and environmental features enhancing siting in given geographical areas, and other specific studies related to proposed projects in a given geographical area.

It may be desirable to organize the soil information according to topographical features of the area. Again, the pertinent state geological survey or the regional or area office of the USGS may be expected to have basic topographic maps. Other relevant federal agencies which may have appropriate land-use, land- or mineral-resource-, and/or soils information include the Army Corps of Engineers, the Bureau of Reclamation, and the Bureau of Land Management. In addition, many states have corollary agencies that address mining, land usage and management, and soils concerns, and these may also be sources of information. An example of a composite listing of sources of earth-science-related information is in Table 8.10.

A particularly important source of information for addressing surface soils and associated land uses may be historical aerial photographs. Aerial photography in many areas of the United States exists for periods dating back in excess of 50 years. These photographs can be valuable in examining historical trends and changes in land use and associated soil features and erosional patterns. Aerial photographic information can be procured from the SCS or other agencies within the USDA. From a more recent perspective, remote-sensing information in terms of infrared or satellite imagery can also be utilized in impact studies. This information can enable

a retrospective review of changing situations in a given geographical area; it can also serve as a basis for projecting the potential impacts of proposed projects. A particular computer-based tool which is growing in importance in impact studies is the geographic information system (GIS). GIS will be discussed in a subsequent section.

Several information-storage and -retrieval systems have been developed which incorporate soil and/or geological data. One example is the Soils data system operated by the U.S. Army Construction Engineering Research Laboratory in Champaign, Illinois (Majerus, 1988, 1989). For example, the Soils-6 database focuses on the geographic location of soils at the county, state, and regional MLRA (major land resource area) level. The Soils-5 database focuses on soil characteristics (which are compiled using USDA SCS Soil Interpretation Records) such as erosion, soil texture, pH, and available water capacity (Majerus, 1988). Table 8.11 presents a brief example of the use of the Soils system to obtain data upon which to base a landfill-siting decision.

Groundwater Quantity and Quality

In describing groundwater quantity and quality, specific indicator parameters can be utilized. For example, the following represent some of the information which could be compiled and issues which could be addressed:

1. Descriptions should be assembled on groundwater systems in the study area, indicating whether they are confined or unconfined, with the obvious pollution relevance being that unconfined groundwater systems tend to be more susceptible to groundwater contamination. Of particular importance would be the description of karst aquifer systems, since these areas can exhibit unique and rapid groundwater flow patterns (Schuknecht and Mikels, 1986).

2. Many areas are characterized by the presence of multiple groundwater systems;

TABLE 8.11

USAGE OF THE SOILS SYSTEM IN LANDFILL SITING

Example COMUUF Output
(For soils in New Jersey with soil permeability 0 to 0.06 inches/hour

Acres (of soil map units)	County
5,240	Hunterdon, NJ
7,170	Somerset, NJ
27,880	Bergen, NJ
13,920	Middlesex, NJ
2,350	Morris, NJ
15,570	Passaic, NJ
18,860	Union, NJ
2,850	Sussex, NJ
Grand total 93,840	

Source: Majerus, 1989, p. 2.

Note: Permeability is one important soil characteristic for a landfill site. Permeability is the quality of a soil which enables water to move through the soil pore space. For landfills, nonpermeable soils are best. The SOILS system can search for soils with very slow permeability and identify counties in which these soils occur.

The data listed above provides an example output from a Multiple Parameter Series Search (MPSS) for soils in th State of New Jersey with a permeability of 0 to 0.06 inches/hour followed by a search using the County Map Unit Use File (COMUUF) which lists counties and acreages for these nonpermeable soils.

Given the counties listed above as possible locations, field testing with deep-well borings could then be conducted for the soils identified for potential landfill sites. SOILS has many other applications to the site selection process for different projects.

accordingly, it would be appropriate to describe those geographical areas characterized by multiple aquifer systems.

3. If information exists on the quantitative aspects of the groundwater resource in terms of potentially useable supplies which could be extracted, it should be summarized.

4. Information should be summarized on the uses of groundwater within the study area, with a more detailed study of this subject to be conducted later.

5. A description of the relationships between local groundwater systems and surface streams, lakes, estuaries, or coastal areas may be important, since mutual quantitative or qualitative influences can occur.

6. Groundwater-pollution vulnerability is associated with whether or not the project area is in a recharge zone for a given groundwater system; thus, this should be determined because there is greater pollution potential if in recharge zone. (It should be noted that for confined aquifer systems the recharge area may be located long distances from the actual segment of the groundwater system being used for purposes of water supply.)

7. Depth to groundwater is a fundamental parameter which could be identified, with the pertinent issue being that the greater the depth of groundwater, the greater the degree of natural protection.

8. Unsaturated-zone permeability should be described. Here, the "unsaturated zone" refers to that segment of the subsurface environment between the land surface and the water table of an unconfined aquifer system. The unsaturated-zone permeability can in-

fluence the attenuation of contaminants as they move away from a source of pollution and toward the groundwater system.

9. Aquifer transmissivity should be described. This parameter represents information on the water-carrying capacity of the ground-water system.

10. Any existing data on groundwater quality should be summarized. If no such data exists, it may be necessary to appropriately plan and conduct a groundwater-monitoring program. In some unique cases, the quality data may need to be described in terms of aquatic ecosystems. For example, several threatened or endangered aquatic species have been found in springs associated with the Edwards aquifer in central Texas (Pennisi, 1993).

Describing groundwater quality may be difficult because most historical data on groundwater quality has tended to focus on inorganic constituents such as chlorides, while essentially minimal information is available on bacteriological or organic quality of groundwater systems. Furthermore, because of the slow rate of movement of most groundwater systems and the possibility of incomplete mixing, there can be considerable variations in groundwater quality, both spatially and vertically, in a given study area.

If possible, consideration should be given to historical trends in groundwater quality and quantity characteristics in the study area. In many countries, information has already been developed on the quantitative features of groundwater resources in specific geographical areas. If this information does exist, it would be pertinent to summarize it as a part of this step. It is possible that only minimal information would be available, with the need thus being to conduct pertinent baseline studies to gather additional information. The planning and conduction of baseline studies relative to groundwater-resource characterization is beyond the scope of this chapter; however, reference documents ex-

ist on this subject [see ASCE (1987); Canter, Knox, and Fairchild (1987); and U.S. EPA (1987b)].

Depending upon the type of project being examined, certain indicators of groundwater quality might be used. For example, Passmore (1989) described three indicators used for public drinking water supplies, areawide sources of contamination, and areawide sources of pesticide contamination. The indicators used were the nitrate maximum-concentration-level (MCL) violations at public water supplies, nitrate concentrations in all types of wells, and soluble-pesticide use. The three groundwater quality indicators were used to identify counties in Pennsylvania which have documented or potential groundwater quality problems caused by nonpoint sources of nitrate or by heavy pesticide use. The three indicators were consistent in identifying several counties in southeast Pennsylvania where the groundwater quality has been degraded or where there is a high potential for contamination.

One of the things which can aid in organizing and summarizing information relative to the groundwater system in a given study area is the use of an aquifer-vulnerability-mapping technique. This subject is addressed in Knox, Sabatini, and Canter (1993). The key concept in aquifer-vulnerability mapping is the aggregation of information on selected parameters so that a relative evaluation of the contamination potential of groundwater systems can be made. This approach will be described in "Step 4: impact predictions."

Sources of the information which are needed in order to accomplish step 2 can include specific governmental agencies which have been assigned responsibility for describing groundwater resources and have thus already gathered pertinent data. For example, the U.S. Environmental Protection Agency has published an aggregated report on 20 years of groundwater monitoring for pesticides (Cohen, 1993). Also, hydrologic atlases prepared by state water or geological agencies contain information on the geological setting; principal aquifers and well locations, depths, and water yields; groundwater quality in terms of dissolved solids and inorganic constituents such as bicarbonate, calcium, chloride, magnesium, potassium, sodium, and sulfate; and groundwater usage for municipal supplies. Additional sources of this type of information include relevant city, county, and state water resources agencies. Also, the USGS is currently conducting studies on regional aquifer systems. This regional aquifer systems analysis (RASA) program focuses on two types of regional systems: (1) aquifers that extend over large areas, such as those underlying the Great Plains, High Plains, Gulf Coastal Plain, and Atlantic Coastal Plain, and (2) a group of virtually independent aquifers that have so many characteristics in common that studies of a few of these aquifers can establish common principles and identify hydrogeologic factors that control the occurrence, movement, and quality of groundwater throughout these types of aquifers (*U.S. Geological Survey*, 1988).

Unique Soil or Groundwater Problems

Many geographical areas exhibit special or unique problems that should be addressed in the description of baseline conditions for the soil or groundwater resources in the study area. Examples of these problems include saline seeps, groundwater ''mining,'' saltwater intrusion, condemned groundwater supplies relative to existing bacteriological or other quality constituents, poor natural quality, and the presence of hazardous-waste sites. Dryland-farming practices involving irrigation often lead to salt accumulation in surface soils and shallow, unconfined aquifers. A ''saline seep'' refers to an intermittent or continuous saline-water discharge at or near the soil surface and downslope from recharge areas under dryland conditions; such seeps reduce or eliminate crop growth in the affected area as a result of increased soluble-salt concentrations in the root zone (Brown et

al., 1982). Saline seeps frequently occur in the northern Great Plains area of the United States.

One of the purposes of this activity is to encourage the identification of critical issues relative to the project being located in a proposed area, and to use this information to either determine that a new location would be more suitable to the project, or to develop appropriate design and/or mitigation measures to reduce any undesirable additional impacts on those groundwater characteristics that may already be of concern.

Sources of information related to these special or unique issues include local regulatory agency personnel, local newspapers from the study area, professionals working for the government or private sector who have conducted such studies or who have information on the soil- or groundwater-resource characteristics in the study area, and well drillers who have experience in the development of water supplies in the study area.

Pollution Sources and Groundwater Users

It is appropriate to consider what other potential and actual sources of soil and/or groundwater pollution may exist in the study area, and also to consider current and potential future usage of the groundwater resource for purposes of water supply. Table 8.3 lists over 30 potential soil and/or groundwater contamination sources. Some of the sources can be considered point sources in that they represent defined points of introduction of contaminants to the subsurface environment. Examples of such ''point'' sources include landfills, underground storage tanks, and the like. Conversely, many of the sources in Table 8.3 can be characterized as nonpoint or area sources—for example, large geographical areas served by septic tank systems, and agricultural lands.

The key focus should be on the assemblage of sufficient information to enable the determination of the level of existing sources of pollution in the study area. This does not necessarily require the conduction of a detailed soil- and/or groundwater-monitoring program. Pertinent information could be assembled on items such as the number, size, and materials contained within landfills in the study area or within wastewater impoundments or underground storage tanks within the study area. Pertinent information for nonpoint sources should include the number and distribution of septic tank systems in the study area, the acreage associated with agricultural usage, and the types, quantities, and timing of fertilizer or pesticide applications.

Regarding current groundwater usage, while this topic was mentioned earlier, a more detailed review of the current number of groundwater users and the quantities associated with such uses might be appropriate. Again, specific, detailed information may not be necessary; pertinent information perhaps could include general estimates of the number of individual users of groundwater and of the number of industrial and municipal areas using the groundwater resource. Additionally, consideration should be given to potential increases in usage over time without the proposed project.

Sources of information for pollution sources and groundwater users include pertinent water supply and/or groundwater pollution studies that may have been conducted by governmental agencies. An example would be an updated 208 study, mentioned in Chapter 7. In addition, special studies are often conducted in conjunction with the development of sources of water supply; it is possible that these studies may include relevant inventory information.

STEP 3: PROCUREMENT OF RELEVANT SOIL AND/OR GROUNDWATER QUANTITY-QUALITY STANDARDS

Land-use restrictions, soil quality standards, soil reclamation requirements, and groundwater quantity-quality standards, regulations, or policies are examples of institutional measures which can be used to determine impact significance and required mitigation measures. Thus, to determine the specific requirements for a given project area will require contacting appropriate governmental agencies with jurisdiction. Relevant institutional information available from U.S. sources has been described earlier.

The primary sources of information needed for step 3 will be pertinent governmental agencies—federal, state, and/or local agencies. In addition, international environmental agencies may have information pertinent to this step. The primary reason for step 3 is that the information obtained provides a basis for interpreting existing quality-quantity information in the study area; it can also be used as a basis for assessing the anticipated impacts resulting from the proposed project.

STEP 4: IMPACT PREDICTION

The prediction of the impacts of a project-activity on the soil and/or groundwater environment(s), or, conversely, the potential influence of the environment(s) on a proposed project, can be approached from three perspectives: (1) qualitative, (2) simple quantitative, and (3) specific quantitative. In general, efforts should be made to quantify the anticipated impacts; however, in many cases this will be impossible and reliance must be given to qualitative techniques. Qualitative impact prediction is typically associated with the use of look-alikes, or analogous projects for which knowledge and information is available, and/or the utilization of relevant case studies.

Qualitative Approaches—Soil Impacts

One example of qualitative impact prediction using look-alikes would be the prediction of acid-rain impacts on soils as a result of a proposed project based on data obtained from certain extensive studies which have been conducted over several years in southern Sweden (Andersson and Stokes, 1988). While it is recognized that soil types are a determinant in the effects, certain principles can be delineated from these studies, which can be applied to impact prediction in other geographical areas. Another example is related to the effects of military training on soil compaction; here, the information developed from studies conducted by the U.S. Army Construction Engineering Research Laboratory at military facilities could be used to delineate soil compaction effects of nonmilitary training activities elsewhere. While such studies are qualitatively oriented, they can provide a useful basis for impact prediction when interpreted using professional judgment.

Another example of a qualitative approach for soil-impact prediction and mitigation planning is related to pipeline construction. There are four potential impacts of pipeline construction on drainage and soils: (1) contamination of topsoil with excavated subsoil, (2) soil compaction, (3) soil erosion, and (4) disruption of drainage lines or natural drainage patterns (Sorrell et al., 1982). In most soils, the top several inches are relatively high in organic matter, nutrients, and soil biota. This "topsoil" provides a more-fertile growing medium than the relatively inorganic and nutrient-poor subsoil. Pipeline construction can result in the mixing of subsoil with topsoil in several ways: through the initial grading of the right-of-way, through the excavation and backfilling of the pipeline trench, and through the spreading of excess subsoil over the right-of-way during cleanup. In general, the mixing of subsoil with topsoil will have an adverse impact on soil fertility and soil structure. The severity of the impact will depend on the nature of the subsoil.

To minimize contamination of topsoil with excavated subsoil, several measures can be taken; the applicability and design of these measures will vary with specific soil conditions. Examples of such steps include (Sorrell et al., 1982) (1) making proper applications of lime and fertilizer to restore much of the lost productivity almost immediately and (2) removing excess subsoil after the trench has been back-filled for reuse as fill rather than spreading it over the right-of-way.

The use of heavy equipment on the right-of-way will compact the soil, resulting in reduced aeration, infiltration capacity, and permeability and, thereby, reduced plant growth and increased surface runoff and erosion. The removal of vegetation and the churning and breaking up of soil particles by construction equipment will expose soils in the right-of-way to accelerated erosion by water. The amount of erosion will depend on several factors, including slope, soil erodibility, time needed for revegetation, and amount and intensity of rainfall. Pipeline trenches may be foci of water erosion for several reasons: the looseness of the backfill, the tendency of the trench to act as a drain, and the concentration of runoff by the crown left above the trench to compensate for natural settlement of the backfill (or conversely, by the depression left where settlement exceeds the excess materials provided) (Sorrell et al., 1982).

The effect of accelerated erosion is twofold: a decline in the fertility of the soil as the more fertile topsoil is carried away and sedimentation in watercourses receiving the eroded material. There are a variety of measures commonly used for controlling erosion on pipeline rights-of-way. By far the most important is the rapid reestablishment of a vegetative cover. Revegetation can be accomplished naturally (by artificial seeding) or by plantings and, in most cases, can be substantially accelerated by the application of fertilizers.

Qualitative Approaches—Groundwater Impacts

A qualitative approach for groundwater-impact prediction involves considering the fundamental subsurface-environment processes and using this information to infer where groundwater contamination might occur and the potential extent of such an occurrence. The fundamental processes in the subsurface environment can be

---

TABLE 8.12

NATURAL PROCESSES THAT AFFECT SUBSURFACE CONTAMINANT TRANSPORT

Physical processes
Advection (porous media velocity)
Hydrodynamic dispersion
Molecular diffusion
Density stratification
Immiscible phase flow
Fractured media flow

Chemical processes
Oxidation-reduction reactions
Radionuclide decay
Ion-exchange
Complexation
Cosolvation
Immiscible phase partitioning
Sorption

Biological processes
Microbial population dynamics
Substrate utilization
Biotransformation
Adaption
Cometabolism

Source: U.S. Environmental Protection Agency, 1987a, p. 152.

---

examined relative to their hydrodynamic (physical), abiotic (chemical), and biotic (biological) aspects. Table 8.12 lists various natural subsurface processes that affect subsurface contaminant transport. Table 8.13 summarizes processes which may affect constituents of groundwater and, thereby, the subsurface transport and transformation of pollutants to or within water-bearing zones. With regard to hydrodynamic processes, if fundamental information is known on the groundwater flow direction and velocity, then the direction of groundwater contamination could be predicted by inference and, by making simple assumptions relative to the velocity and the movement of contaminants with the water phase, dilution-type calculations could be used for estimating groundwater concentrations of particular constituents.

Abiotic processes include adsorption, ion exchange, and precipitation. Information on the characteristics of various contaminants in terms of their adsorbability, their tendency to participate in ion-exchange processes, and the possibilities for precipitation occurring in the subsurface environment could be assembled from published literature. Once the types of contaminants that might be of concern have been identified, relevant information could be assembled on their potential attenuation in the subsurface environment by considering pertinent abiotic processes.

Biotic processes include biological degradation and the potential uptake of various constituents in plant material. It has been determined in recent years that a considerable amount of natural biological degradation of organic

TABLE 8.13

PROCESSES WHICH MAY AFFECT CONSTITUENTS OF GROUNDWATER

Constituent	Physical		Geochemical					Biochemical		
	Dispersion	Filtration	Complexation	Ionic strength	Acid-base	Oxidation-reduction	Precipitation-solution	Adsorption-desorption	Decay, respiration	Cell synthesis
Cl^-, Br^-	xx					xx				
NO_3^-	xx					xx		x	xx	xx
SO_4^{2-}	xx		x	x	x	xx			x	
HCO_3	xx		x	x	xx		xx		xx	xx
PO_4^{3-}	xx		xx	xx	xx		xx	xx	xx	xx
Na^+	xx			x				xx		
K^+	xx			x				xx		
NH_4^+	xx		xx	x	xx	xx		xx	xx	xx
Ca^{2+}	xx		x	xx			x	xx		
Mg^{2+}	xx		x	xx			x	xx		
Fe^{2+}	xx		xx	xx	xx	xx	xx	xx		
Mn^{2+}	xx		xx	xx	xx	xx	xx	xx		
Fe^{3+} and Mn^{4+} oxyhydroxides	xx	xx			xx	xx	xx			
Trace elements	xx		xx	xx		xx	xx	xx	xx	xx
Organic solutes	xx		xx	x	xx	xx	x	x	xx	xx
Microorganisms	xx	xx				xx			xx	xx

xx Major control
x Minor control
Source: Chapman, 1992, p. 402.

materials will take place in the unsaturated and saturated zones of a groundwater system. Accordingly, information compiled on the biodegradation potential of various types of contaminants could be used to again make qualitative inferences relative to groundwater contamination.

Simple Quantitative Approaches—Soil Impacts

Another approach for addressing impacts on the soil environment is to use simple quantitative techniques, with a range of such techniques having been developed. One example of a simple quantitative technique is the use of "overlay mapping," which was developed to delineate various land-use compatibilities in given geographical areas (McHarg, 1971). Overlay mapping basically consists of utilizing a base map of the project study area and overlaying on this map different soil or geological features or particular impact concerns of the proposed project; impact prediction involves the identification of where overlaps of particular concerns occur. Overlay mapping can be achieved through the development of handdrawn maps or the use of computer-generated maps.

Geographic information systems (GISs) represent a technological advance in terms of overlay mapping techniques. At the heart of the GIS is a database, which may contain multiple "layers" of data for the same area. Examples of possible layers are topographic data, land-use–land-cover information, hydrologic data, and erodibility indices. All layers are referenced to a common ground-datum point and orientation, allowing them to be, in essence, overlaid.

Data can be input to a GIS by either analytical or digital means. An example of the former would be the use of map digitizing and of the latter, the use of satellite imagery tapes. One of the great benefits of using a GIS is its ability to collate data from diverse sources into a consistent form, so that paper maps, aerial photographs, satellite multiband images, and so on, may be input by the most convenient method. Regardless of original scale and format, the data, once in the GIS, are consistent. They may be output in different forms for checking, and they are available for a variety of analyses.

One example of a GIS system is the Geographic Resources Analysis Support System (GRASS) developed by the U.S. Army Construction Engineering Research Laboratory. GRASS is composed of three subsystems: (1) Grid—for analyzing, overlaying, and modeling grid-cell-type maps, as well as for displaying both grid-cell and line maps; (2) Imagery—for displaying, georeferencing, comparing, and classifying satellite and aerial-photograph imagery; and (3) MAP-DEV—for digitizing and integrating landscape data generated from hardcopy maps, digital elevation files, or other sources into a form suitable for analysis in Grid. GRASS data files can be developed for large or small geographic regions at any scale desired, within the limits of the original source documents and the storage capacity of the hardware (U.S. Army Construction Engineering Research Laboratory, 1989).

GIS is beginning to be used in impact studies, since it can be a valuable tool for assessing cumulative impacts. GIS can also be used to quantify rates of regional resource loss by comparing data layers representing different years. In addition, GIS can be used to develop empirical relationships between resource loss and environmental degradation. For example, a cumulative-impact-evaluation method involving aerial photointerpretation, multivariate statistical analysis, and GIS techniques was developed and used to relate past and present soil types and wetland abundance with stream water quality in the Minneapolis–St. Paul metropolitan area (Johnson et al., 1988).

Another simple quantitative technique which can be used to address soil-impact concerns involves mass-balance calculations to quantify nonpoint-source pollution; this subject is addressed in Chapter 7. Such calculations can be used to quantify soil erosion from given geographical areas by considering different erosion rates related to specific types of land uses. Additionally, nutrient pollution can also be determined by the use of mass-balance calculations involving factors relating the quantity of nutrient to a unit area and unit time.

Examples of simple quantitative models include the "universal (or unified) soil loss equation" and models for land subsidence. The universal soil loss equation is as follows:

$$A = (R)\,(K)\,(LS)\,(C)\,(P)$$

where A = computed soil loss for a given storm, tons/ha
R = rainfall-energy factor
K = soil-erodibility factor
LS = slope-length factor
C = cropping-management (vegetative-cover) factor
P = erosion-control-practice factor

Detailed information on each factor in the equation is given in Novotny and Chesters (1981). The equation can be used to determine soil loss per unit area due to erosion by rain. Models for land subsidence are being developed for situations where such subsidence has been experienced as a result of excessive removal of groundwater resources and/or oil-extraction areas. Such models account for selected physical and engineering properties of the soil and subsurface as they enable the prediction of topographic changes caused by extraction programs.

The final suggested simple quantitative approach to be discussed here is conduction of laboratory studies, and possibly scale-model studies, to examine impact concerns. One illustration of this technique is the use of soil lysimeters for examining subsurface movement of nutrient materials; another involves the use of laboratory microcosms (columns) to study the transport and fate of specific pollutants in the subsurface environment. For example, one study involved the use of several laboratory microcosms to examine the influence of acid rain on the leachability of soil constituents, as well as the leachability of components associated with solid waste which might be placed in a sanitary landfill (Liu, 1983).

Simple Quantitative Approaches— Groundwater Impacts

Mesoscale impacts on groundwater can be calculated using mass balances. The word "mesoscale" simply denotes a large scale and could refer to the total groundwater system, or to a subset of the system within the geographical study area.

The basic approach for determining mesoscale impacts uses a materials-based, or mass-balance, technique, which could be applied during the construction and/or operational phases of the proposed project. For example, in the case of projects that involve the usage of groundwater for purposes of water supply, a simple calculation could be made to establish the percent change (increase) over current groundwater usage within the study area. Another example technique for such a project would be to determine the percentage change in the pollution-source-type inventory within the study area.

In order to determine mesoscale impacts, it is necessary to have delineated the types and quantities of water pollutants and/or the water quantity requirements of the proposed project (step 1), and to have determined existing conditions in terms of groundwater pollution sources and/or groundwater usage (step 2).

Construction-Phase Impacts on Groundwater

During the construction phase of major projects, large areas of land may be cleared, and surface runoff may be impounded in order to reduce erosion problems. In addition, for major water resources projects such as hydropower dams, or for surface-mining projects, it may be necessary to dewater the groundwater in the vicinity of the project so that the construction or resource extraction can proceed (Powers, 1981). Typically, construction-phase impacts are short-term and associated with the time period related to particular construction activities.

One potential construction-phase impact is the loss of groundwater as a resource to local-area water users when the construction-phase dewatering practices are in progress. Anticipation of this type of impact should be readily achievable—for example, by using simple well-drawdown calculations to enable the determination of the extent of the impact (Freeze and Cherry, 1979). Another issue often related to construction-phase impacts when dewatering practices are used is the need for disposal of the pumped groundwater. On occasion, this can represent a significant environmental impact in terms of the water quality of surficial receiving bodies.

Sources of information for construction-phase impacts include the project proponent, the construction industry, and additional information gathered by conducting searches of litera-ture pertinent to construction-phase drawdown practices.

Site-Selection Approaches

For many projects or activities, a key decision may be the selection of an appropriate site from several potential sites. Several methodologies that are based on considering both baseline conditions and anticipated impacts on soil and groundwater have been developed for site comparisons. Such methods are often called "empirical-index methods." For example, the siting of hazardous-waste landfills should incorporate consideration of the soil and hydrogeological features in various areas (Kiang and Metry, 1982). Site-related information may be useful in terms of project design; for example, information pertinent to the design of liners and covers for hazardous-waste landfills may be derived from the permeability characteristics of the site soil (U.S. EPA, 1986a). Additional information on site-selection methodologies is available in Bolton and Curtis (1990) and Canter, Knox, and Fairchild (1987).

TABLE 8.14

APPLICATIONS OF NINE EMPIRICAL ASSESSMENT METHODOLOGIES

Methodology	Examples of application
Surface Impoundment Assessment	Evaluation of extant liquid impounds (pits, ponds, and lagoons).
Landfill Site Rating	Evaluation of extant or new sanitary landfill sites.
Waste-Soil-Site Interaction Matrix.	Evaluation of new industrial solid or liquid waste disposal sites.
Site Rating System	Chemical landfill site selection or evaluation.
Hazard Ranking System	Ranking of hazardous-waste sites for remedial actions.
Site Rating Methodology	Ranking of hazardous-waste sites for remedial actions.
Brine Disposal Methodology	Evaluation of extant or planned practices for brine disposal from oil and gas field activities.
Pesticide Index	Ranking of pesticides based on their groundwater pollution potential.
DRASTIC	Evaluation of the potential for groundwater pollution at a specific site given its hydrogeological setting.

Source: Canter, Knox, and Fairchild, 1987, p. 278.

TABLE 8.15

ASSIGNED IMPORTANCE WEIGHTS FOR FACTORS IN TWO DRASTIC MODELS

	Importance weight	
Factor	Generic model	Pesticide model
Depth to water (D)	5	5
Net recharge (R)	4	4
Aquifer media (A)	3	3
Soil media (S)	2	5
Topography (T)	1	3
Impact of the vadose-zone media (I)	5	4
Hydraulic conductivity of the aquifer (C)	3	2

Source: After Aller et al., 1987, p. 3.

Index Methods for Source- and/or Environmental-Vulnerability Analysis

"Microscale impacts" refer to those on a small scale; in the context of this discussion, they would be the potential impacts of the proposed project on the local soil and/or groundwater within the project boundary or study area. One approach for microscale-impact prediction involves the usage of empirical index methods. Table 8.14 lists nine such methods. One example of such a method is the surface-impoundment-assessment (SIA) technique developed by the U.S. EPA and used in the United States to evaluate the groundwater-contamination potential of liquid holding ponds. In this context, the term "surface impoundments" refers to liquid-waste holding ponds associated with municipal and/or industrial wastewater treatment facilities (U.S. EPA, 1978).

The SIA method has been used in a national survey of pits, ponds, and lagoons by the U.S. Environmental Protection Agency (1978). The results provide a perspective on the nature and potential magnitude of the groundwater-pollution potential of this source category. In addition, this methodology has been used in a study of several central Oklahoma areas containing septic tank systems (Canter, 1985).

An aquifer-vulnerability-mapping technique (or numerical rating scheme), called "DRASTIC," has been developed for evaluating the potential for groundwater pollution in given areas based on their hydrogeological setting (Aller et al., 1987). This rating scheme (listed in Table 8.15) is based on seven factors chosen by over 35 groundwater scientists from throughout the United States. Information on these factors is presumed to exist for all locations in the United States. In addition, the scientists also established relative-importance weights and a point-rating scale for each factor. The acronym "DRASTIC" is derived from the seven factors in the rating scheme:

D = depth to groundwater
R = recharge rate (net)
A = aquifer media
S = soil media
T = topography (slope)
I = impact of the vadose zone
C = conductivity (hydraulic) of aquifer

Determination of the DRASTIC index for a given area involves multiplying each factor weight by its point rating and summing the total. The higher sum values represent greater potential for groundwater pollution, or greater aquifer vulnerability. For a given area being evaluated, each factor is rated on a scale of 1 to 10, indicating the relative pollution potential of the given factor for that area. Once all factors have been assigned a rating, each rating is multiplied by the assigned weight, and the resultant numbers are summed as follows:

$$D_r D_w + R_r R_w + A_r A_w + S_r S_w + T_r T_w$$
$$+ I_r I_w + C_r C_w = \text{pollution potential}$$

where r = rating for area being evaluated and w = importance weight for factor.

Weight values of from 1 to 5 express the relative importance of the factors with respect to each other. Ratings are obtained from tables or graphs for each factor, while the importance weights can be found in generic DRASTIC tables listing either weights for factors having general applicability or for factors related to the potential pollution from pesticide applications from the modified "Pesticide DRASTIC"; factor weights are shown in Table 8.15.

Depth to groundwater, or the water table, is an important factor primarily because it determines the depth of material through which a contaminant must travel before reaching the aquifer; this factor can also be used to determine the contact time with the surrounding materials. Table 8.16 contains the ranges and ratings for

TABLE 8.16

EVALUATION OF THE DEPTH-TO-GROUNDWATER FACTOR IN DRASTIC

Depth to groundwater	
Range (feet)	Rating
0–5	10
5–15	9
15–30	7
30–50	5
50–75	3
75–100	2
100+	1

Source: Aller et al., 1987, p. 21.

TABLE 8.17

EVALUATION OF THE NET-RECHARGE FACTOR IN DRASTIC

Net annual recharge	
Range (inches)	Rating
0–2	1
2–4	3
4–7	6
7–10	8
10+	9

Source: Aller et al., 1987, p. 21.

the depth to groundwater. The ranges were determined based on what the study group of groundwater professionals considered to be depths where the potential for pollution significantly changed.

"Net recharge" refers to the total quantity of water which infiltrates from the ground surface to reach the aquifer. Net recharge includes the average annual amount of infiltration and does not take into consideration the distribution, intensity, or duration of recharge events. The ranges and corresponding ratings for net recharge are given in Table 8.17. The attenuation capacity of the aquifer media is evaluated on the basis of grain sizes, fractures, and solution openings. Ranges and ratings for the aquifer-media factor in DRASTIC are illustrated in Figure 8.5. The "soil media" is considered to be the upper, weathered zone of the earth, which averages a depth of 6 ft or less from the ground surface. The soil media is evaluated on the basis of the type of clay present, the shrink-swell potential of that clay, and the grain size of the soil. The ranges and ratings for the soil-media factor are in Table 8.18.

As used in the DRASTIC methodology, "topography" refers to the slope and slope variability of the land surface. Table 8.19 contains the slope ranges which were chosen as significant relative to groundwater-pollution

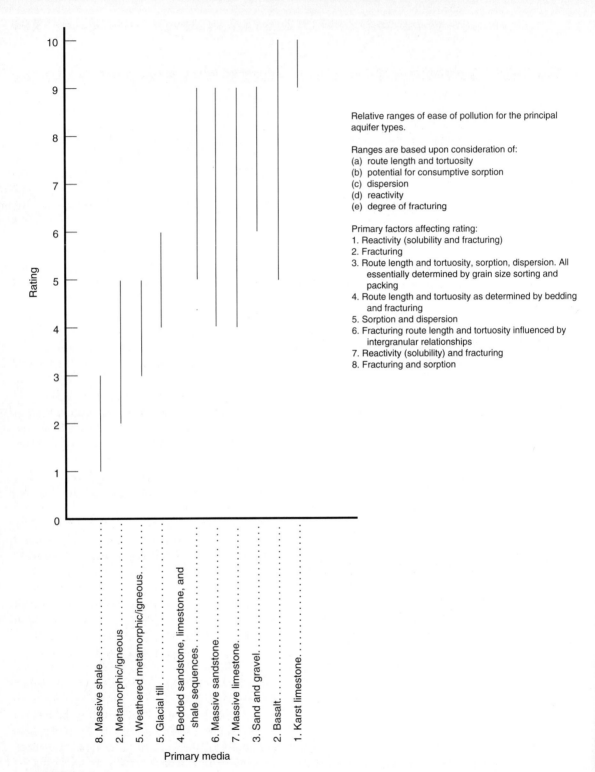

FIGURE 8.5
Ranges and Ratings for the Aquifer-Media Factor in DRASTIC (Aller et al., 1987, p. 28).

TABLE 8.18

EVALUATION OF THE SOIL-MEDIA FACTOR IN DRASTIC

Soil media	
Range	Rating
Thin or Absent	10
Gravel	10
Sand	9
Shrinking and/or Aggregated Clay	7
Sandy Loam	6
Loam	5
Silty Loam	4
Clay Loam	3
Nonshrinking and Nonaggregated Clay	1

Source: Aller et al., 1987, p. 22.

TABLE 8.19

EVALUATION OF THE TOPOGRAPHY FACTOR IN DRASTIC

Topography	
Range (percent slope)	Rating
0–2	10
2–6	9
6–12	5
12–18	3
18+	1

Source: Aller et al., 1987, p. 23.

TABLE 8.20

EVALUATION OF THE HYDRAULIC-CONDUCTIVITY FACTOR IN DRASTIC

Hydraulic conductivity	
Range (gpd/ft^2)	Rating
1–100	1
100–300	2
300–700	4
700–1000	6
1000–2000	8
2000+	10

Source: Aller et al., 1987, p. 25.

potential. The "vadose zone" is defined as that zone above the water table which is unsaturated or discontinuously saturated. The vadose zone is evaluated on the basis of grain sizes, fracturing, solution openings, and sorption potential. Figure 8.6 displays the ranges and ratings for the impact of the vadose-zone-media factor in DRASTIC. Finally, values for hydraulic conductivity are calculated from aquifer pumping tests. Information on hydraulic conductivity is typically available from published hydrogeological reports for given geographical areas. The ranges and ratings of the hydraulic-conductivity factor are given in Table 8.20.

The DRASTIC methodology was developed to enable the groundwater in the United States to be systematically evaluated. This system has been prepared to assist planners, managers, and administrators in the task of evaluating the relative vulnerability of areas to groundwater contamination from various sources of pollution (Aller et al., 1987). The DRASTIC methodology has also been used in Sweden (Swanson, 1990).

Groundwater vulnerability, precipitation distribution, population density, potential well yield, and aquifer sensitivity for each of the 48 conterminous states is also addressed in a largely graphic format in Pettyjohn, Savoca, and Self (1991). A classification scheme based on an assessment of the vulnerability of surficial and relatively shallow aquifers to contamination was developed. Aquifers having a high degree of vulnerability in areas of high population density are considered to be the most sensitive. About 46 percent of the land area of the conterminous United States consists of vulnerable aquifers; moderately vulnerable aquifers cover about 14 percent and the least vulnerable aquifers make up about 20 percent. The undefined systems, class U, account for an additional 20 percent.

It should be noted that considerable debate exists regarding the use of aquifer-vulnerability

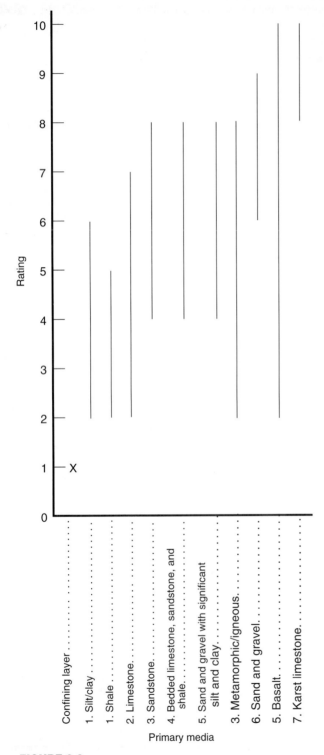

Relative impact of the Vadose Zone Media type
Range based upon:

(a) path length and tortuosity
(b) potential for dispersion and consequent dilution
(c) reactivity (solubility)
(d) consumptive sorption
(e) fracturing

Primary factors affecting rating:
1. Consumptive sorption and fracturing
2. Fracturing and reactivity
3. Fracturing path length as influenced by intergrating relationships
4. Fracturing path length and tortuosity as influenced bedding planes, sorption, and reactivity
5. Path length and tortuosity as impacted by bedding grain size, sorting and packing, sorption
6. Path length and tortuosity as influenced by grain size, sorting, and packing
7. Reactivity and fracturing

FIGURE 8.6
Ranges and Ratings for the Impact of the Vadose-Zone-Media Factor in DRASTIC (Aller et al., 1987, p. 31).

indices for groundwater protection, particularly with regard to vulnerability to pesticide contamination, and the use of this information to establish differential protection strategies regarding geographical location (U.S. General Accounting Office, 1991a).

Pesticide contamination of soil or groundwater has recently become an issue of concern, particularly in areas used for agricultural purposes. A large number of physical, chemical, and biological processes have been found to influence pesticide behavior, transport, and fate in the subsurface environment. These pesticide characteristics and relationships which may have an influence include aqueous solubility, melting point, vapor pressure, Henry's constant, octanol-water partition coefficient, sorption coefficient, and degradation half-life (Rao, Hornsby, and Jessup, 1985). Detailed evaluation of the soil- and groundwater-pollution potential of a pesticide at a site would require the following site-specific information: climatological data, including daily records of rainfall, evapotranspiration, temperature, and net radiation; irrigation, crop, and pesticide-management practices; soil profile characteristics including depth to groundwater, total porosity, volumetric soil-water content at field capacity and permanent wilting point, soil bulk density, soil organic carbon content, and groundwater net-recharge rate; and crop parameters such as rooting depth and rooting density.

Rao, Hornsby, and Jessup (1985) have suggested a simple scheme for ranking the relative potential of different pesticides to intrude into groundwater. This ranking scheme, which was listed in Table 8.14 as "pesticide index," does not require the detailed pesticide and site-characteristics information which would be necessary in a complete mathematical model of the system. The scheme addresses pesticide transport through the crop root zone and the intermediate vadose zone. The following equations and definitions are used in the ranking scheme (Rao, Hornsby, and Jessup, 1985):

$$AF = \frac{M_2}{M_0} = \exp(-B)$$

where AF = attenuation factor between 0 and 1
= index for pesticide mass emission from vadose zone
M_2 = amount of pesticide entering groundwater
M_0 = amount of pesticide applied at soil surface
B = term related to travel time and degradation time

or

$$B = \frac{0.693 t_r}{t^{1/2}}$$

where $t^{1/2}$ = degradation half-life of pesticide
t_r = time required for pesticide to travel through root zone and intermediate vadose zone

or

$$t_r = \frac{(L)\,(RF)\,(FC)}{q}$$

where L = distance from soil surface to groundwater
FC = volumetric soil-water content at field capacity
q = net-recharge rate
RF = retardation factor

or

$$RF = 1 + \frac{(BD)\,(OC)\,(K_{oc})}{FC} + \frac{(AC)\,(Kh)}{FC}$$

where BD = soil bulk density
OC = soil organic carbon content
K_{oc} = sorption coefficient of pesticide on soil
AC = air-filled porosity for soil
Kh = Henry's constant for pesticide

Rao, Hornsby, and Jessup (1985) have suggested that the attenuation factor (AF) index can be used by regulatory agencies in the preliminary evaluation of pesticides to be monitored in geographical areas with soil and/or groundwater susceptible to pesticide pollution. Usage of this index is based on the following simplifying assumptions: (1) vadose zone properties are independent of depth; (2) an average groundwater-recharge rate can be computed given local rainfall, irrigation, and evapotranspiration data; (3) a K_{oc} value can be estimated for each pesticide, based on the assumption that hydrophobic interactions are dominant; and (4) an average $t^{1/2}$ value can be estimated for each pesticide.

Transport-Modeling Approaches

At the most sophisticated level, microscale impact prediction involves the use of subsurface flow and solute-transport models. Detailed information on groundwater modeling is contained in Knox, Sabatini, and Canter (1993). The focus herein will be on examples of modeling software.

Nofziger and Hornsby (1985) have developed a model for describing chemical movement in soils. The model estimates the location of the leading edge of a nonpolar organic chemical as it moves downward in the soil. The model also determines the relative amount of the applied organic chemical remaining in the soil as a function of time. The model is available in IBM PC–compatible software. This software is based on fundamental water and solute-transport principles. It also incorporates the work of others who have shown that the partition coefficient for a particular organic chemical in a soil divided by the organic carbon content of that soil is nearly constant for a wide range of soils. Information is also included on the relationships between sorption, degradation, and loss of pesticides in soil-water systems. The model is intended to illustrate chemical transport principles; it is interactive with graphical and tabular output. The software was written to allow the user to make repeated simulations of different soil-chemical systems with a minimum of effort.

To illustrate the complexity of groundwater flow and solute-transport modeling, brief information will be presented on one simplified model. The hydrodynamic processes of advec-tion and dispersion are involved in the subsurface transport of contaminants. "Advection" refers to the transport of a solute at a velocity equivalent to that of the groundwater movement (Roberts, Reinhard, and Valocchi, 1982). This transport occurs as a result of the bulk motion of the flowing groundwater. Advection is also referred to as "convection." "Dispersion" refers to the spreading of a solute concentration front as a result of spatial variation in aquifer permeability, fluid mixing, and molecular diffusion. This spreading results in the dilution of the solute concentration. Microscale dispersion can result from hydraulic drag in pore channels, differences in pore size, and pore-channel tortuosity, branching, and interfingering.

The following transport equation can be used to describe one-dimensional, horizontal, single-phase flow in a saturated, unconsolidated, homogeneous medium (Roberts, Reinhard, and Valocchi, 1982):

$$-u\frac{\delta C}{\delta x} + D\frac{\delta^2 C}{\delta x^2} - \frac{\rho_b}{\Sigma}\frac{\delta s}{\delta t} + \frac{\delta C}{\delta t_{rn}} = \frac{\delta C}{\delta t} \quad (8.1)$$

where u = average fluid velocity, m/sec
C = solute concentration in aqueous phase, g/m^3
x = distance in flow direction, m
D = dispersion coefficient, m^2/sec
ρ_b = bulk density of soil, g/m^3
Σ = soil-void fraction, unitless
s = mass of solute adsorbed per unit dry mass of soil, g/g
t = time, sec
rn = chemical reactions and/or biological degradation

The first term in Equation (8.1) relates to advection or convection, and the second term addresses dispersion. Adsorption is incorporated in the third term, and other abiotic reactions and biological degradation are accounted for in the fourth term.

Selection of an appropriate groundwater model to meet a given need typically involves consideration of the technical capabilities of models, as well as managerial issues. Technical comparisons of a number of models are contained in Van der Heijde et al. (1985). Managerial issues include economic considerations, necessary training and experience of model users, and information communication.

Groundwater-modeling information can be divided into two key categories: (1) information on types and features of models and (2) modeling software. Two organizations that provide modeling information are the International Ground Water Modeling Center (IGWMC) at the Colorado School of Mines in Golden, Colorado, and the National Water Well Association (NWWA) in Dublin, Ohio. In addition, the U.S. EPA's Center for Subsurface Modeling Support at its research laboratory in Ada, Oklahoma, provides groundwater- and vadose-zone-modeling software and technical support to public agencies and private companies.

STEP 5: ASSESSMENT OF IMPACT SIGNIFICANCE

Several approaches can serve as a basis for interpreting the anticipated project-induced changes to the soil and groundwater environments. One approach is to consider the percentage and direction of change from existing conditions for a particular soil or groundwater environmental factor. While this can be helpful, it does presume that quantitative information is available for the baseline conditions for such factors, and that anticipated changes in the factors as a result of a project can be quantified.

Another approach for impact assessment is to apply the provisions of pertinent federal, state, or local laws and regulations related to the soil and groundwater environments to the expected with-project conditions. In many cases, these institutional requirements are qualitative; however, they can be used as a "yardstick" in evaluating the project and any features the project might incorporate to minimize environmental

damage. One example of such requirements is the land reclamation stipulations of the Surface Mining Control and Reclamation Act of 1977. In terms of assessing calculated concentrations of contaminants in soil or groundwater systems, any existing quality standards, as well as information that has been assembled in conjunction with step 3 above, could be used as a basis for significance determination and impact interpretation.

A third approach for interpreting anticipated changes relies upon professional judgment and knowledge. The anticipated changes could be interpreted in relation to existing information on natural changes; next, the expected impacts could be placed in an historical context. A professional-judgment-based interpretation of anticipated changes may consist of applying rules of thumb. As an example, concerning soil erosion, the current and anticipated soil erosion patterns from a project area could be compared to regional averages or historical trends. It is generally agreed that a certain amount of soil loss is inevitable. Ideally, the loss should not exceed the rate of soil formation from parent rock and decomposed vegetation, but there is no agreement in the literature on the rate of soil formation (Carpenter and Maragos, 1989). A commonly cited, generalized upper limit of permissible, or tolerable, soil loss is about 11 ton/ha/yr, but the "permissibility" of such a loss depends on many local factors, such as the fertility and drainage characteristics of the subsoil. Many soils are vulnerable to a decline in productivity at a rate of loss from erosion considerably lower than 11 ton/ha/yr. It has been estimated that, in the United States, soil on agricultural land erodes at a rate of about 30 ton/ha/yr, which is approximately 8 times the rate at which topsoil is formed (Goudie, 1984). Downstream damage from sedimentation must also be considered in setting tolerance limits for the rate of soil erosion.

One final approach to impact assessment is to consider the anticipated changes in relation to look-alikes or the application of information from relevant case studies. In addition, this method can be a useful tool for focusing upon key issues and interpreting their potential significance.

STEP 6: IDENTIFICATION AND INCORPORATION OF MITIGATION MEASURES

The final step in this methodology is to consider and adopt pertinent mitigation measures for negative impacts, as appropriate. Relevant measures would need to be developed for specific projects; therefore, the information to be gathered here should focus primarily on delineations of pertinent mitigation measures for soil and/or groundwater impacts. For example, some of the following measures might be included:

1. Use of techniques to decrease soil erosion during either the construction or operational phase of a project. Examples of such techniques include minimization of the exposed time during the construction phase by planting rapidly growing vegetation and the use of sediment-catchment basins. Additionally, as various types of grasses and vegetation have relatively greater or less potential for minimizing soil erosion, the selection of pertinent vegetation for usage should take this characteristic into account. As an example, the specific land-restoration requirements of the Surface Mining Control and Reclamation Act of 1977 rely on such considerations.

2. Use of BMPs focused on minimizing nonpoint-source pollution. For example, Table 8.5 listed some BMPs which can be utilized to minimize nonpoint-source pollution to both surface and groundwater systems.

3. Rotation of land-use practices in the project area can be used to permit natural recovery without the continuing stress related to any one land-use practice. Examples include the rotation of military training areas, agricultural crops in given geographical areas, and grazing patterns in areas permitted by pertinent governmental agencies.

4. The project can be designed to exhibit greater earthquake resistance if this is a potential concern for the project area. Examples include structural designs for withstanding shocks associated with the occurrence of earthquakes.

5. For projects involving usage of the groundwater resource, groundwater usage could be decreased.

6. If the potential impact of concern is land subsidence, management techniques to minimize groundwater usage in the area where subsidence is expected to occur could be implemented. These could encompass water conservation measures so as to reduce groundwater extraction requirements, for example.

7. Development of comparative information to enable more-systematic site selections, which will, in turn, make maximum usage of the natural attenuation capacity of given environmental settings to prevent groundwater contamination; an example for solid-waste disposal sites is in Bolton and Curtis (1990).

8. For projects which may be of concern because of leachate generation, measures could be taken to immobilize the constituents and prevent their occurrence in leachate through the use of waste-solidification techniques.

9. Liners could be used to provide a physical barrier to limit the movement of contaminant materials from waste-disposal sites into and through the subsurface environment. Liners can be composed of either natural or man-made materials. Leachate collection systems could also be used, along with surface-water runon-runoff control systems. Design standards for liners and leak-detection systems for hazardous-waste

land-disposal units have been promulgated by the U.S. Environmental Protection Agency (1992) and address units such as surface impoundments, landfills, and waste piles.

10. If the project involves the use of agricultural chemicals, consideration should be given to techniques that could be used to better plan the timing of chemical applications, the rate of application, and the extent of such applications; in addition, nitrification inhibitors could be used to minimize nitrate contamination of groundwater.

SUMMARY

This chapter addresses a variety of considerations associated with predicting and assessing a proposed project's (or activity's) potential impacts on the soil and groundwater environments. The six-step methodological framework represents a logical approach which begins with the identification of potential impacts and progresses through a sequence of steps associated with environmental-setting information, institutional information, and impact prediction. The approaches for impact prediction are described in terms of qualitative, simple quantitative, and quantitative techniques. Impact interpretation is assumed to be primarily based upon professional judgment and the suitable application of relevant institutional information. Finally, impact-mitigation measures should be identified and then incorporated in project construction and operational characteristics so as to minimize undesirable effects on the soil and groundwater environments.

A key issue that may influence EIS findings and subsequent interpretation is that the subsurface environment is typically characterized by very nonuniform conditions. Accordingly, consideration of such nonuniformity must be included in project evaluation and in interpretation of anticipated impacts. Another issue of concern is the difficulty of taking a consistent approach for impact interpretation given the rapidly changing regulatory programs and policies. Management of soil and groundwater quality represents a new field, and considerable changes are occurring relative to the development and implementation of pertinent legislation, standards, and regulations.

It will be necessary to document the six-step methodology in the resulting soil-environment- or groundwater-environment-impact report, which could be a section or chapter in either an EA or EIS. The report should address the potential impacts identified for the project or activity (step 1); the key features of the environmental setting in the study area (step 2); the pertinent laws, regulations, standards, and/or policies (step 3); the predicted impacts of the proposed project-activity, whether based on the use of qualitative or of quantitative approaches (step 4); the assessment of the predicted impacts (step 5) and its rationale; and the mitigation measures considered and incorporated (step 6).

It is of significance to note that a major new emphasis in the U.S. Environmental Protection Agency is on groundwater-pollution prevention (U.S. EPA, 1990). Accordingly, the EPA will give increased attention to prevention of groundwater contamination and strive to achieve a greater balance between prevention and remediation activities. The importance of preproject analysis of potential groundwater impacts is expected to increase as a result of the agency's increasing efforts in this area, which are expected to parallel or even exceed its remediation efforts in the coming years (U.S. General Accounting Office, 1991a and b).

SELECTED REFERENCES

Aller, L., et al., "DRASTIC: A Standardized System for Evaluating Ground Water Pollution Potential Using Hydrogeologic Settings," EPA 600/2-87-035, U.S. Environmental Protection Agency, Ada, Okla., May 1987.

American Society of Civil Engineers (ASCE), "Ground Water Management," ASCE Manual no. 40, 3d ed., 1987, ASCE, New York, pp. 138–152.

Andersson, I., and Stokes, J., "Investigation of Potential Groundwater Acidification in Sweden Using Principal Component Analysis and Regression Analysis," SNV-3417, National Swedish Environment Protection Board, Solna, Sweden, Feb. 1988.

Bolton, K. F., and Curtis, F. A., "An Environmental Assessment Procedure for Siting Solid Waste Disposal Sites," *Environmental Impact Assessment Review,* vol. 10, 1990, pp. 285–296.

Brannon, J. M., "Evaluation of Dredged Material Pollution Potential," Rep. no. WES-TR-DS-78-6, U.S. Army Engineer Waterways Experiment Station, Vicksburg, Miss., Aug. 1978.

Bricka, R. M., Holmes, T. T., and Cullinane, M. J., "Comparative Evaluation of Two Extraction Procedures: The TCLP and the EP," EPA 600/2-91-049, U.S. Army Engineer Waterways Experiment Station, Vicksburg, Miss., Sept. 1991.

Brown, P. L., et al., "Saline-Seep Diagnosis, Control, and Reclamation," Conservation Research Rep. no. 30, U.S. Department of Agriculture, Washington, D.C., 1982.

Cahill, L. B., and Kane, R. W., *Environmental Audits,* 6th ed., Government Institutes, Rockville, Md., 1989, pp. III-1–III-27.

Canter, L. W., "Methods for Assessment of Ground Water Pollution Potential," Chap. 13 in *Ground Water Quality,* C. H. Ward, W. Giger, and P. L. McCarty, eds., John Wiley and Sons, New York, 1985, pp. 270–306.

———, "Planning and Conducting Site Assessments," Environmental and Ground Water Institute, University of Oklahoma, Norman, June 1990.

———, Knox, R. C., and Fairchild, D. M., *Ground Water Quality Protection,* Lewis Publishers, Chelsea, Mich., 1987, pp. 277–323, 499–555.

Carpenter, R. A., and Maragos, J. E., eds., "How to Assess Environmental Impacts on Tropical Islands and Coastal Areas," *Training Manual for South Pacific Regional Environment Programme (SPREP),* Environment and Policy Institute, East-West Center, Honolulu, Oct. 1989, pp. 258–266.

Chapman, D., ed., *Water Quality Assessments—A Guide to the Use of Biota, Sediments and Water in Environmental Monitoring,* Chapman and Hall, London, 1992, pp. 402–404.

Cohen, S., "EPA Releases National Ground Water Database," *Ground Water Monitoring and Remediation,* vol. 13, no. 3, summer 1993, pp. 99–102.

Davis, M. L., and Chou, G. P., "EPA's Approach to Development of LDR Standards for Contaminated Soil and Debris," *Journal of the Air and Waste Management Association,* vol. 42, no. 2, Feb. 1992, pp. 145–151.

Deeley, G. M., and Canter, L. W., "Distribution of Heavy Metals in Waste Drilling Fluids Under Conditions of Changing pH," *Journal of Environmental Quality,* vol. 15, Apr.–June 1986, pp. 108–112.

Delaware Department of Natural Resources and Environmental Control, "State of Delaware Nonpoint Source Pollution Management Program," Dover, Dela., July 1989.

Dialog Information Services, "Guide to DIALOG Searching," Lockheed Information Systems, Palo Alto, Calif., 1989.

Dodd, K., Fuller, H. K., and Clarke, P. F., "Guide to Obtaining USGS Information," *U.S. Geological Survey Circular 900, U.S. Geological Survey,* Washington, D.C., 1989.

Drew, D., *Man-Environment Processes,* George Allen and Unwin Publishers, London, 1983, p. 32.

Duvel, W. A., "Trends and Future Concerns Emerging from the Massachusetts and New Jersey Site Assessment Experience," paper presented at the *American Institute of Chemical Engineers National Meeting,* Boston, Aug. 1986.

Fenn, D. G., Hanley, K. J., and DeGeare, T. V., "Use of the Water Balance Method for Predicting Leachate Generation from Solid Waste Disposal Sites," EPA-SW-168, U.S. Environmental Protection Agency, Cincinnati, Ohio, Oct. 1975.

Fitchko, J., *Criteria for Contaminated Soil/Sediment Cleanup,* Pudvan Publishing Company, Northbrook, Ill., 1989, pp. 2.1–2.12.

Franck, I., and Brownstone, D., *The Green Encyclopedia,* Prentice-Hall General Reference, New York, 1992, p. 327.

Freedman, W., *Federal Statutes on Environmental Protection,* Quorum Books, New York, 1987, pp. 1–171.

Freeze, R. A., and Cherry, J. A., *Groundwater,* Prentice-Hall, Englewood Cliffs, N.J., 1979.

Funderburk, J. R., "Site Assessments Call for Variety of Approaches," *HAZMAT World,* vol. 3, no. 3, Mar. 1990, pp. 40–48.

Gangadharan, A. C., et al., *Leak Prevention and Corrective Action Technology for Underground Storage Tanks,* Noyes Data Corporation, Park Ridge, N.J., 1988.

Goudie, A., *The Human Impact on the Natural Environment,* Basil Blackwell, Oxford, England, 1986, p. 280.

————, *The Nature of the Environment,* Basil Blackwell, Oxford, England, 1984, p. 246.

Johnson, C. A., et al., "Geographic Information Systems for Cumulative Impact Assessment," *Photogrammetric Engineering and Remote Sensing,* vol. 54, no. 11, Nov. 1988, pp. 1609–1615.

Keller, E. A., *Environmental Geology,* 3d ed., Charles E. Merrill Publishing Company, Columbus, Ohio, 1982.

Kiang, Y. H., and Metry, A. A., *Hazardous Waste Processing Technology,* Ann Arbor Science Publishers, Ann Arbor, Mich., 1982.

Knox, R. C., Sabatini, D. A., and Canter, L. W., *Subsurface Transport and Fate Processes,* Lewis Publishers, Boca Raton, Fla., 1993, pp. 243–352.

Liu, Y. M., "The Effects of Acid Rain and Cover Soil on Leachate Quality from Simulated Solid Waste Sites," MSCE thesis, University of Oklahoma, Norman, 1983.

Majerus, K., "Identifying Sites Using SOILS Searches," *ETIS Quarterly,* University of Illinois at Urbana-Champaign, ETIS Program Office, Urbana, Apr. 1989, p. 2.

————, "SOILS Offers Data in Detail," *ETIS Quarterly,* University of Illinois at Urbana-Champaign, Urbana, ETIS Program Office, Dec. 1988, p. 2.

McHarg, I. L., *Design with Nature,* Doubleday/Natural History Press, Doubleday and Company, Garden City, N.Y., 1971.

Nofziger, D. L., and Hornsby, A. G., "Chemical Movement in Soil: IBM PC User's Guide," Circular 654, Institute of Food and Agricultural Sciences, University of Florida, Gainesville, Jan. 1985.

Novotny, V., and Chesters, G., *Handbook of Nonpoint Pollution,* Van Nostrand Reinhold Company, New York, 1981.

Office of Technology Assessment, "Protecting the Nation's Groundwater from Contamination," OTA-0-233, U.S. Congress, Washington, D.C., Oct. 1984.

Olson, G. W., *Field Guide to Soils and the Environment—Applications of Soil Surveys,* Chapman and Hall, New York, 1984.

————, *Soils and the Environment—A Guide to Soil Surveys and Their Applications,* Chapman and Hall, New York, 1981.

Passmore, M. E., "Development and Use of Three Ground Water Quality Indicators," EPA 101/F-90-011, Environmental Studies Institute, Drexel University, Philadelphia, 1989.

Pennisi, E., "Saving Hades' Creatures," *Science News,* vol. 143, Mar. 13, 1993, pp. 172–174.

Pettyjohn, W. A., Savoca, M., and Self, D., "Regional Assessment of Aquifer Vulnerability and Sensitivity in the United States," EPA 600/S2-91-043, U.S. Environmental Protection Agency, Robert S. Kerr Environmental Research Laboratory, Ada, Okla., Oct. 1991.

Powers, J. P., *Construction Dewatering: A Guide to Theory and Practice,* John Wiley and Sons, New York, 1981.

Proios, G., "Potential Impacts of Acidic Precipitation on a Sole-Source Aquifer," *Proceedings of the Second Annual Eastern Regional Ground Water Conference,* National Water Well Association, Dublin, Ohio, 1985, pp. 451–463.

Rahn, P. H., *Engineering Geology—An Environmental Approach,* Elsevier Science Publishing Company, New York, 1986.

Rao, P. S., Hornsby, A. G., and Jessup, R. E., "Indices for Ranking the Potential for Pesticide Contamination of Groundwater," *Proceedings of Soil and Crop Science Society of Florida,* vol. 44, 1985, pp. 1–8.

Roberts, P. V., Reinhard, M., and Valocchi, A. J., "Movement of Organic Contaminants in Groundwater: Implications for Water Supply," *Journal of the American Water Works Association,* Aug. 1982, pp. 408–413.

Schroeder, P. R., and Peyton, R. L., "HELP Modeling Workshop for Landfill Design Evaluation," University of Wisconsin, Milwaukee, Aug. 1992.

Schuknecht, M. R., and Mikels, J. K., "Hydrogeologic Impact Assessment of Proposed Urbanization Atop a Karst Aquifer," *Proceedings of Conference on Environmental Problems in Karst Terrains and Their Solutions,* National Water Well Association, Dublin, Ohio, 1986, pp. 435–451.

Schumann, H. H., and Genualdi, R. B., "Land Subsidence, Earth Fissures, and Water Level Change in Southern Arizona," Map 23, Arizona Bureau of Geology and Mineral Technology, Tucson, 1986.

Severns, J. J., "Controlling Environmental Liability in Property Transactions Through the Use of Environmental Assessments in the State of California," *Journal of Hazardous Materials,* vol. 18, no. 3, June 1988, pp. 245–254.

Sorrell, F. Y., et al., "Oil and Gas Pipelines in Coastal North Carolina: Impacts and Routing Considerations," CEIP Rep. no. 33, North Carolina State University, Raleigh, Dec. 1982.

Strahler, A. N., and Strahler, A. H., *Modern Physical Geography,* 3d ed., John Wiley and Sons, New York, 1987.

Swanson, G. J., "Sweden Finds Answers in United States," *Water Well Journal,* Apr. 1990, pp. 60–61.

Toy, T. J., and Hadley, R. F., Chap. 6, "Lands Disturbed by Grazing," in *Geomorphology and Reclamation of Disturbed Lands,* Academic Press, Orlando, Fla., 1987a, pp. 152–162.

———, Chap. 7, "Lands Disturbed by Recreational Use," in *Geomorphology and Reclamation of Disturbed Lands,* Academic Press, Orlando, Fla., 1987b, pp. 163–176.

———, Chap. 8, "Lands Disturbed by Surface Mining of Coal," in *Geomorphology and Reclamation of Disturbed Lands,* Academic Press, Orlando, Fla., 1987c, pp. 177–255.

———, Chap. 9, "Lands Disturbed by Surface Mining of Uranium," in *Geomorphology and Reclamation of Disturbed Lands,* Academic Press, Orlando, Fla., 1987d, pp. 256–343.

———, Chap. 10, "Lands Disturbed by Construction Activities," in *Geomorphology and Reclamation of Disturbed Lands,* Academic Press, Orlando, Fla., 1987e, pp. 344–430.

U.S. Army Construction Engineering Research Laboratory, "Geographic Resources Analysis Support System (GRASS)," EN 48, U.S. Army Construction Engineering Research Laboratory, Champaign, Ill., Jan. 1989.

U.S. Department of Energy, "Energy Technologies and the Environment—Environmental Information Handbook," DOE/EH-0077, Washington, D.C., Oct. 1988.

U.S. Department of the Interior, "Final Environmental Impact Statement: Permanent Regulatory Program Implementing Sec. 501(b) of the Surface Mining Control and Reclamation Act of 1977," Washington, D.C., Jan. 1979, pp. BIII-71–BIII-81.

U.S. Environmental Protection Agency, "EPA Ground Water Task Force Report," Draft Final, U.S. Environmental Protection Agency, Washington, D.C., Sept. 1990, pp. 1–8.

———, "Extraction Procedure Toxicity Characteristic," *Code of Federal Regulations,* vol. 40, no. 261, May 19, 1980.

———, "Guidelines for Delineation of Wellhead Protection Areas," U.S. Environmental Protection Agency, Office of Ground Water Protection, Washington, D.C., June 1987a.

———, "Handbook: Ground Water," EPA/625/6-87-016, U.S. Environmental Protection Agency, Center for Environmental Research Information, Cincinnati, Ohio, Mar. 1987b.

———, "Hazardous Waste Management System," *Federal Register,* vol. 51, no. 114, June 13, 1986a, pp. 21648–21693.

———, "Liners and Leak Detection Systems for Hazardous Waste Land Disposal Units," *Federal Register,* vol. 57, no. 19, Jan. 29, 1992, pp. 3462–3497.

———, "A Manual for Evaluating Contamination Potential of Surface Impoundments," EPA 500/9-78-003, U.S. Environmental Protection Agency, Office of Drinking Water, Washington, D.C., June 1978.

———, "Musts for USTs," EPA 530/UST-88/008, U.S. Environmental Protection Agency, Washington, D.C., Sept. 1988.

———, "Nonpoint Source Guidance," U.S. Environmental Protection Agency, Office of Water Regulations and Standards, Washington, D.C., Dec. 1987c.

———, "Processes, Procedures, and Methods to Control Pollution from Mining Activities," EPA 430/9-73-011, U.S. Environmental Protection Agency, Washington, D.C., Oct. 1973.

———, "Subchapter I—Solid Wastes" and Subchapter J—Superfund Programs," *Code of Federal Regulations,* vol. 4, parts 264–300, July 1986b, U.S. Government Printing Office, Washington, D.C., pp. 425–814.

U.S. General Accounting Office, "Drinking Water—Projects That May Damage Sole Source Aquifers Are Not Always Identified," GAO/RCED-93-4, Washington, D.C., Oct. 1992.

———, "Groundwater Protection—Measurement of Relative Vulnerability to Pesticide Contamination," GAO/PEMD-92-8, Washington, D.C., Oct. 1991a.

————, "Water Pollution—More Emphasis Needed on Prevention in EPA's Efforts to Protect Ground Water," GAO/RCED-92-47, Washington, D.C., Dec. 1991b.

U.S. Geological Survey, "Regional Aquifer System Analysis (RASA) Program," Open File Rep. 88-118, U.S. Geological Survey, Reston, Va., 1988.

Van der Heijde, P., et al., "Groundwater Management: The Use of Numerical Models," Water Resources Monograph 5, 2d ed., American Geophysical Union, Washington, D.C., 1985, pp. 13–51.

Van der Sloot, H. A., and De Groot, G. J., "Characterization of Municipal Solid Waste Incinerator Residues for Utilization: Leaching Properties," Netherlands Energy Research Foundation, Petten, The Netherlands, Apr. 1990.

Westcott, R. M., "Policies and Practices for Mitigating Biological Impacts," MES thesis, University of Oklahoma, Norman, 1989.

William Spangle and Associates, "Earth Science Information in Land-Use Planning—Guidelines for Earth Scientists and Planners," *U.S. Geological Survey Circular* 721, U.S. Geological Survey, Washington, D.C., 1976.

CHAPTER 9

Prediction and Assessment of Impacts on the Noise Environment

This chapter addresses basic noise-environment concepts and a six-step methodological approach for conducting a scientifically based analysis of the potential noise effects of proposed projects and activities. Noise impacts can be of concern during the construction and the operational phases of projects. Noise should also be considered in relation to present and future land-use zoning and policies.

Construction noise can be a significant source of community noise. Of concern are impacts on people near the construction site performing activities which are totally unrelated to construction activities (e.g., area residents, office workers, schoolchildren, and hospital residents and staff). Factors which are important in determining noise levels that will potentially impact such populations include distance from the noise source; natural or man-made barriers between the source and the impacted population; weather conditions which could potentially absorb, reflect, or focus sound (such as wind speed and direction and temperature inversions); and the scale and intensity of the particular construction phase (e.g., excavation, erection, or finishing) (U.S.

Army Construction Engineering Research Laboratory, 1989).

Examples of operational-phase impacts include noise emissions from compressor stations on gas pipelines, pumping stations for water distribution or wastewater treatment systems, highways and freeways, industrial plants, fossil-fueled power plants, and military training activities. For example, noise emissions from military training can directly impact surrounding, sensitive land uses. Two types of noise emissions are of concern: (1) impulse noise— that is, noise of short duration and high density such as explosions, sonic booms, and artillery fire, and (2) continuous noise—that is, longer-duration and lower-intensity noise such as that from construction or traffic. The military creates noise from artillery and rifle fire, low-flying jets and helicopters, tank and other heavy-vehicle traffic, whether such vehicles are traveling alone or in convoys. In addition, the military often must train at night and at other times that are normally considered quiet hours.

The primary focus of this chapter is a six-step methodological approach for noise-impact quantification. For each step, several levels of analysis are possible, depending on the extent of potential noise effects for the particular project or activity. Information is also included on noise terminology and relevant noise laws, standards, guidelines, and/or requirements in the United States.

BASIC INFORMATION ON NOISE

"Noise" can be defined as unwanted sound or sound in the wrong place at the wrong time. It can also be defined as any sound that is undesirable because it interferes with speech and hearing, is intense enough to damage hearing, or is otherwise annoying (U.S. EPA, 1972). The definition of "noise" as *unwanted* sound implies that it has an adverse effect on human beings and their environment, including land, structures, and domestic animals. Noise can also disturb natural wildlife and ecological systems.

"Sound" is mechanical energy from a vibrating surface and is transmitted by a cycling series of compressions and rarefactions of the molecules of the materials through which it passes (Chanlett, 1973). Sound can be transmitted through gases, liquids, and solids. A vibrating source which produces sound has a "total power output," and the sound results in a sound pressure wave that alternately rises to a maximum level (compression) and drops to a minimum level (rarefaction). (Noise level is related to total power output.) The number of compressions and rarefactions of the air molecules in a unit of time is described as its "frequency." Frequency is expressed in hertz (Hz), which is the same as the number of cycles per second. Humans can detect sounds with frequencies ranging from about 16 to 20,000 Hz (U.S. EPA, 1973).

Sound power (total power output or sound pressure) does not provide practical units for sound or noise measurement for two basic reasons (U.S. EPA, 1973). First, a tremendous range of sound power (or sound pressure) can be produced. Expressed in microbars (μbar,

one-millionth of 1 atm pressure), the range is from 0.0002 to 10,000 μbar for peak noises within 100 ft of large jet- and rocket-propulsion devices. Second, the human ear does not respond linearly to increases in sound pressure. The human response is essentially logarithmic. Therefore noise measurements are expressed by the term "sound-pressure level" (SPL), which is the logarithmic ratio of the sound pressure to a reference pressure and is expressed as a dimensionless unit of power, the decibel (dB). The reference level is 0.0002 μbar, the threshold of human hearing. The equation for sound-pressure level is as follows:

$$SPL = 20 \log_{10}\left(\frac{P}{P_0}\right)$$

where SPL = sound-pressure level, dB
P = sound pressure, μbar
P_0 = reference pressure, 0.0002 μbar

Table 9.1 contains a summary of various sound pressures and corresponding A-weighted decibel levels, with examples of recognized noise sources cited. Figure 9.1 lists some common and easily recognized sounds along with a subjective evaluation scale.

In most noise considerations, the "A-weighted sound-level" scale is used. This scale is appropriate because the human ear does not respond uniformly to sounds of all frequencies, being less efficient in detecting sounds at low and high frequencies than at medium, or speech, frequencies (Chanlett, 1973). To obtain a single number representing a sound level containing a wide range of frequencies and yet representative of the human response, it is necessary to weight the low and high frequencies with respect to average, or "A," frequencies. Thus, the resultant SPL is "A-weighted," and the units are A-weighted decibels (dBA). The A-weighted sound level is also called the "noise level." Sound-level meters have an A-weighting network, thus yielding A-weighted dB, or dBA, readings.

TABLE 9.1

SPL, SOUND PRESSURE, AND RECOGNIZED SOURCES OF NOISE IN OUR DAILY EXPERIENCES

Sound pressure, μbar	SPL, dBA	Example
0.0002	0	Threshold of hearing
0.00063	10	
0.002	20	Studio for sound pictures
0.0063	30	Studio for speech broadcasting
0.02	40	Very quiet room
0.063	50	Residence
0.2	60	Conventional speech
0.63	70	Street traffic at 100 ft
1.0	74	Passing automobile at 20 ft
2.0	80	Light trucks at 20 ft
6.3	90	Subway at 20 ft
20	100	Looms in textile mill
63	110	Loud motorcycle at 20 ft
200	120	Peak level from rock and roll band
2,000	140	Jet plane on the ground at 20 ft

Source: Chanlett, 1973, p. 525.

Some key definitions related to noise-impact prediction and assessment are as follows (Federal Interagency Committee on Urban Noise, 1980, pp. A-1–A-2):

1. Day-Night Average Sound Level (DNL; scientific notation, L_{dn})

   Day-Night average sound level, abbreviated as DNL and symbolized as L_{dn}, is the 24 hour average sound level, in decibels, for the period from midnight to midnight, obtained after addition of 10 decibels to sound levels in the night from midnight to 7 a.m. and from 10 p.m. to midnight. The nighttime penalty is based on the fact that many studies have shown that people are much more disturbed by noise at night than at any other time (Chanlett, 1973). DNL is a measurable quantity and can be measured directly at a specific location, using portable monitoring equipment.

2. Equivalent sound level (L_{eq})

   L_{eq} is the energy equivalent sound level, in decibels, for any time period under consideration. It is the equivalent steady noise level that, in a stated period of time, would contain the same noise energy as the time-varying noise during the same time period. In connection with its highway noise standards featuring design noise levels, the Federal Highway Administration (FHWA) uses an L_{eq} for the high "design hour" as an alternative to the L_{10} descriptor. (The design hour is normally the 30th highest traffic volume occurring during the year.) Noise levels are predicted for the design year, which is normally 20 years from construction of the highway, and the noisiest hour of the day (usually the design hour). Under typical conditions the L_{eq} (design hour) approximately equals DNL.

3. L_{10}

   While this descriptor applies to any noise source, FHWA is the only federal agency using it (as an alternative to L_{eq}). L_{10} is defined as the sound level that is exceeded 10 percent of the time for the period under consideration, which, in the case of FHWA, is the design hour. DNL under typical conditions approximately equals $L_{10} - 3$ decibels.

4. Community Noise Equivalent Level (CNEL)

   The CNEL, developed for the State of Califor-

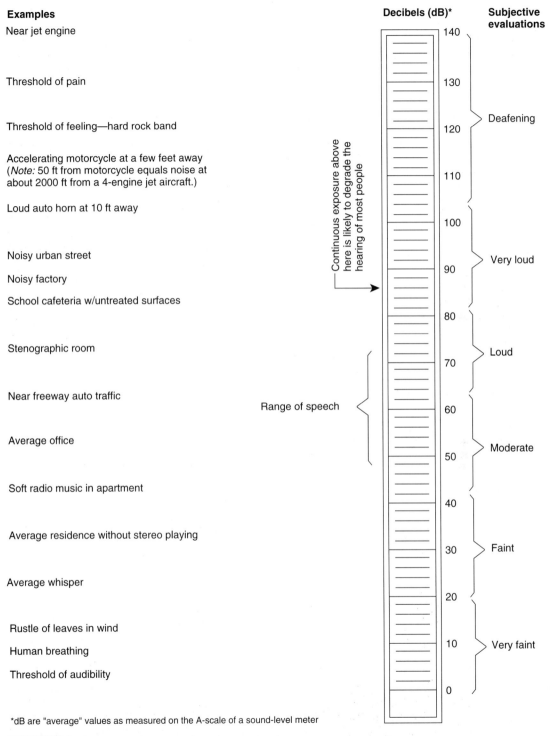

Examples	Decibels (dB)*	Subjective evaluations

Near jet engine — 140

Threshold of pain — 130

Threshold of feeling—hard rock band — 120

Accelerating motorcycle at a few feet away
(*Note:* 50 ft from motorcycle equals noise at
about 2000 ft from a 4-engine jet aircraft.) — 110

Loud auto horn at 10 ft away — 100

Noisy urban street — 90

Noisy factory

School cafeteria w/untreated surfaces — 80

Stenographic room — 70

Near freeway auto traffic — 60

Average office — 50

Soft radio music in apartment — 40

Average residence without stereo playing — 30

Average whisper — 20

Rustle of leaves in wind — 10

Human breathing

Threshold of audibility — 0

Continuous exposure above here is likely to degrade the hearing of most people

Range of speech

Deafening

Very loud

Loud

Moderate

Faint

Very faint

*dB are "average" values as measured on the A-scale of a sound-level meter

FIGURE 9.1
Examples of Common Sounds in Decibels (U.S. Department of Housing and Urban Development, 1985, p. 1).

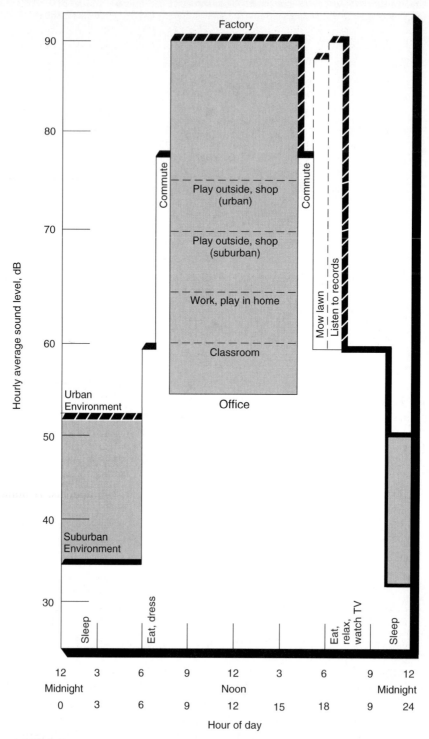

FIGURE 9.2
Generalized Individual Noise Exposure Patterns (U.S. Environmental Protection Agency, 1978, p. 16).

nia, is almost identical to the DNL, except that it introduces an intermediate weighting for the early evening hours between 7:00 p.m. and 10:00 p.m. in addition to the weighting for the nighttime hours (10:00 p.m. to 7:00 a.m.). CNEL, like DNL, is a measurable quantity and can be measured directly. DNL is approximately equal to CNEL in almost all situations.

5. Noise Exposure Forecast (NEF)

   The NEF was developed in 1967 as a refinement of the composite noise rating (CNR) for airports. It takes into account the factors considered by the CNR plus the additional exposure factors of the duration of aircraft flyovers and of discrete (pure) tones such as turbine "whine." The NEF cannot be directly measured and requires a computer for noise contour development. DNL approximately equals NEF + 35.

Detailed information for converting one expression of noise to another is in Magrab (1975). As an example of such a conversion, L_{dn} can be calculated as follows:

$$L_{dn} = 10 \log [0.625(10^{(L_d/10)})$$
$$+ 0.375(10^{(L_n + 10)/10})] \quad \text{dB}$$

where $L_d = L_{eq}$ for the daytime (0700–2200 hr) and $L_n = L_{eq}$ for the nighttime (2200–0700 hr). A typical daily pattern of noise exposure received by an individual is in Figure 9.2.

KEY FEDERAL LEGISLATION AND GUIDELINES

Examples of federal legislation related to the noise environment include the Noise Control Act of 1972, the Quiet Communities Act of 1978, and various highway and aviation laws (U.S. Department of Housing and Urban Development, 1985). The Noise Control Act of 1972 directed the EPA to promote an environment for all Americans that is free from noise that jeopardizes their health and welfare. It also required the EPA to set a noise-level criterion designed to protect health and welfare with an adequate margin of safety but without regard to cost or feasibility. The Quiet Communities Act of 1978 amended the Noise Control Act of 1972 to encourage noise-control programs at the state and community levels. The Federal Aid Highway Act of 1970 and subsequent amendments established the requirement that noise control be a part of the planning and design of all federally aided highways. Finally, the Aviation Safety and Noise Abatement Act of 1979 required the Federal Aviation Administration (FAA) to develop a single system for measuring noise at airports and under certain conditions, to prepare and publish noise maps; emphasis is given to airport noise reduction.

Information on the physiological effects of noise has been developed primarily from case studies on industrial exposures and on the psychological effects of noise conducted as a result of legal actions concerning airport noise exposure. Some typical noise effects are hearing changes and losses, interference with speech communication, and annoyance. Additional effects of noise include disruption of sleep and rest, reduction in work performance, property devaluation resulting from sonic booms, and interference with normal patterns of behavior of domestic and wild animals. A number of reference materials with detailed presentations on these manifold effects of noise are available—for example, U.S. EPA (1972, 1973, 1974), Haber and Nakaki (1989), and Morey (1990).

Table 9.2 summarizes the effects of noise on humans in residential areas. Several factors other than the magnitude of exposure have been found to influence community reaction to noise; these factors include (U.S. EPA, 1978) (1) duration of intruding noises and frequency of occurrence, (2) time of year (windows open or closed), (3) time of day of noise exposure, (4) outdoor noise level in community when intruding noises are not present, (5) history of prior exposure to the noise source, (6) attitude toward the noise source, and (7) presence of pure tones or impulses.

TABLE 9.2

EFFECTS OF NOISE ON PEOPLE (Residential Land Uses Only)

Day-night average sound level in decibels	Effects[a] Hearing loss Qualitative description	Speech interference[b] Indoor % sentence intelligibility	Speech interference[b] Outdoor Distance in meters for 95% sentence intelligibility[c]	Annoyance[b] % of population highly annoyed[c]	Average community reaction[d]	General community attitude towards area
75 and above	May begin to occur	98	0.5	37	Very Severe	Noise is likely to be the most important of all adverse aspects of the community environment.
70	Will not likely occur	99	0.9	25	Severe	Noise is one of the most important adverse aspects of the community environment.
65	Will not occur	100	1.5	15	Significant	Noise is one of the important adverse aspects of the community environment.
60	Will not occur	100	2.0	9	Moderate to Slight	Noise may be considered an adverse aspect of the community environment.
55 and below	Will not occur	100	3.5	4		Noise considered no more important than various other environmental factors.

Note: Research implicates noise as a factor producing stress-related health effects such as heart disease, high-blood pressure and stroke, ulcers and other digestive disorders. The relationships between noise and these effects, however, have not as yet been quantified.
[a]"Speech Interference" data are drawn from other U.S. Environmental Protection Agency studies.
[b]Depends on attitudes and other factors.
[c]The percentages of people reporting annoyance to lesser extents are higher in each case. An unknown small percentage of people will report being "highly annoyed" even in the quietest surroundings. One reason is the difficulty all people have in integrating annoyance over a very long time.
[d]Attitudes or other non-acoustic factors can modify this. Noise at low levels can still be an important problem, particularly when it intrudes into a quiet environment.
Source: Federal Interagency Committee on Urban Noise, 1980, p. D-2.

General Noise Criteria

Table 9.3 summarizes noise criteria developed by the EPA for the protection of public health and welfare with an adequate margin of safety. The phrase "public health and welfare" is defined as complete physical, mental, and social well-being, and not merely the absence of disease and infirmity. Table 9.3 is useful for noise-impact assessment in the absence of specific noise standards for a given area. The two key terms in Table 9.3 are L_{dn} (or DNL) and L_{eq}.

Noise Levels and Land Use

Several key federal agencies have agreed to joint efforts to incorporate noise considerations in development planning; the agencies include the Environmental Protection Agency, Department of Transportation (FAA and FHWA), Department of Housing and Urban Development, Department of Defense, and Veterans Administration (Federal Interagency Committee on Urban Noise, 1980). This cooperation has resulted in the development of noise-impact-related data such as noise-zone classifications and land-use-compatibility guidelines. Table 9.4 classifies noise levels into a common set of noise zones, and Table 9.5 contains land-use-compatibility guidelines.

In Table 9.4, noise zones are identified in order of increasing noise levels by the letters *A* through *D*. The day-night average sound level (DNL) descriptor L_{dn} can be used for all noise sources. The equivalent sound level term L_{eq} is included because some highway noise data is expressed in terms of an equivalent sound level for the highway design hour. The L_{eq} descriptor itself is not unique to highways and can be applied to any noise source. The NEF descriptor is used for aircraft noise only and is being superseded by DNL. The community noise equivalent level (CNEL) descriptor (for the state of California) uses values similar to DNL. Older descriptors unique to airport noise environments, such as the composite noise rating (CNR), may also be encountered (Federal Interagency Committee on Urban Noise, 1980).

In 1973, the U.S. Department of Defense initiated the "Air Installation Compatible Use Zone" (AICUZ) program. The information in Table 9.5 was originally available as suggested land-use guidelines developed for the AICUZ program (U.S. Department of Transportation, 1992).

In 1981 the FAA published regulations (C.F.R., Part 150) establishing a single system of noise measurement appropriate to human response to aviation noise and identifying compatible land uses near airports. Land-use-compatibility designations similar to those in Table 9.5 were also established (FAA, 1988). These guidelines represent recommendations to local authorities for determining acceptability and permissibility of land uses. The noise levels are derived from case histories involving aircraft noise problems at civilian and military airports and the resultant community response. Note that residential land use is deemed acceptable for noise exposures up to 65 L_{dn} (Table 9.5). The FAA has also developed guidelines (Order 5050.4A) for the environmental analysis of airports. Federal requirements now dictate that increases in noise levels in noise-sensitive land uses of over 1.5 L_{dn} are to be considered significant.

Noise Emissions Standards

Standards for noise emissions from various sources have been established by the EPA, in accordance with the requirements of the Noise Control Act of 1972; for example, noise emissions standards now exist for train locomotives and rail cars, trucks or buses engaged in interstate commerce, construction equipment (including air compressors), medium and heavy trucks, and motorcycles (street, competition, and off-road) (U.S. EPA, 1991).

TABLE 9.3

YEARLY AVERAGE[a] EQUIVALENT SOUND LEVELS IDENTIFIED AS REQUISITE TO PROTECT THE PUBLIC HEALTH AND WELFARE WITH AN ADEQUATE MARGIN OF SAFETY

Land use	Measure	Indoor Activity interference	Indoor Hearing loss consideration[b]	Indoor To protect against both effects[c]	Outdoor Activity interference	Outdoor Hearing loss consideration[b]	Outdoor To protect against both effects[c]
Residential with outside space and farm residences	L_{dn}	45		45	55		55
	$L_{eq(24)}$		70			70	
Residential with no outside space	L_{dn}	45		45			
	$L_{eq(24)}$		70			70	
Commercial	$L_{eq(24)}$	d	70	70[e]	d	70	70[e]
Inside transportation	$L_{eq(24)}$	d	70	d			
Industrial	$L_{eq(24)}$[f]	d	70	70[e]	d	70	70[e]
Hospitals	L_{dn}	45		45	55		55
	$L_{eq(24)}$		70			70	
Educational	$L_{eq(24)}$	45		45	55		55
	$L_{eq(24)}$[f]		70			70	
Recreational areas	$L_{eq(24)}$	d		70[e]	d	70	70[e]
Farm land and general unpopulated land	$L_{eq(24)}$				d	70	70[e]

[a] Refers to energy rather than arithmetic averages.
[b] The exposure period that results in hearing loss at the identified level is 40 yr.
[c] Based on lowest level.
[d] Since different types of activities appear to be associated with different levels, identification of a maximum level for activity interference may be difficult except in those circumstances where speech communication is a critical activity.
[e] Based only on hearing loss.
[f] An $L_{eq(8)}$ of 75 dB may be identified in these situations so long as the exposure over the remaining 16 hr/day is low enough to result in a negligible contribution to the 24-hr average, i.e., no greater than an L_{eq} of 60 dB.
Source: U.S. Environmental Protection Agency, 1974, p. 29.

TABLE 9.4

NOISE-ZONE CLASSIFICATION

| Noise zone | Noise exposure class | Noise descriptor | | | HUD noise standards |
		DNL[a] day-night average sound level	L_{eq} (hour)[c] equivalent sound level	NEF[d] noise exposure forecast	
A	Minimal Exposure	Not Exceeding 55	Not Exceeding 55	Not Exceeding 20	"Acceptable"
B	Moderate Exposure	Above 55[b] But Not Exceeding 65	Above 55 But Not Exceeding 65	Above 25 But Not Exceeding 30	
C-1	Significant Exposure	Above 65 Not Exceeding 70	Above 65 Not Exceeding 70	Above 30 But Not Exceeding 35	"Normally Unacceptable"[e]
C-2		Above 70 But Not Exceeding 75	Above 70 But Not Exceeding 75	Above 35 But Not Exceeding 40	
D-1	Severe Exposure	Above 75 But Not Exceeding 80	Above 40 But Not Exceeding 80	Not Exceeding 45	"Unacceptable"
D-2		Above 80 But Not Exceeding 85	Above 80 But Not Exceeding 85	Above 45 But Not Exceeding 50	
D-3		Above 85	Above 85	Above 50	

[a]CNEL—Community Noise Equivalent Level (California only) uses the same values.
[b]HUD, DOT and EPA recognize $L_{dn} = 55$ dB as a goal for outdoors in residential areas in protecting the public health and welfare with an adequate margin of safety.
However, it is not a *regulatory* goal. It is a level defined by a negotiated scientific consensus without concern for economic and technological feasibility or the needs and desires of any particular community.
[c]The Federal Highway Administration (FHWA) noise policy uses this descriptor as an alternative to L_{10} (noise level exceeded ten percent of the time) in connection with its policy for highway noise mitigation. The L_{eq} (design hour) is equivalent to DNL for planning purposes under the following conditions: (1) heavy trucks equal ten percent of total traffic flow in vehicles per 24 hours; (2) traffic between 10 p.m. and 7 a.m. does not exceed fifteen percent of the average daily traffic flow in vehicles per 24 hours. Under these conditions DNL equals $L_{10} - 3$ decibels.
[d]For use in airport environs only; is now being superceded by DNL.
[e]The HUD Noise Regulation allows a certain amount of flexibility for non-acoustic benefits in zone C-1. Attenuation requirements can be waived for projects meeting special requirements.
Source: Federal Interagency Committee on Urban Noise, 1980, p. 5.

TABLE 9.5

SUGGESTED LAND-USE-COMPATIBILITY GUIDELINES

					Noise zones/DNL levels in L_{dn}				
			A	B	C-1	C-2	D-1	D-2	D-3
SLUCM no.	Land use / Name		0–55	55–65	65–70	70–75	75–80	80–85	85+
10	**Residential**								
11	Household units								
11.11	Single units—detached		Y	Y*	25[a]	30[a]	N	N	N
11.12	Single units—semidetached		Y	Y*	25[a]	30[a]	N	N	N
11.13	Single units—attached row		Y	Y*	25[a]	30[a]	N	N	N
11.21	Two units—side-by-side		Y	Y*	25[a]	30[a]	N	N	N
11.22	Two units—one above the other		Y	Y*	25[a]	30[a]	N	N	N
11.31	Apartments—walk up		Y	Y*	25[a]	30[a]	N	N	N
11.32	Apartments—elevator		Y	Y*	25[a]	30[a]	N	N	N
12	Group quarters		Y	Y*	25[a]	30[a]	N	N	N
13	Residential hotels		Y	Y*	25[a]	30[a]	N	N	N
14	Mobile home parks or courts		Y	Y*	N	N	N	N	N
15	Transient lodgings		Y	Y*	25[a]	30[a]	35[a]	N	N
16	Other residential		Y	Y*	25[a]	30[a]	N	N	N
20	**Manufacturing**								
21	Food and kindred products—manufacturing		Y	Y	Y	Y[b]	Y[c]	Y[d]	N
22	Textile mill products—manufacturing		Y	Y	Y	Y[b]	Y[c]	Y[d]	N
23	Apparel and other finished products made from fabrics, leather, and similar materials—manufacturing		Y	Y	Y	Y[b]	Y[c]	Y	N
24	Lumber and wood products (except furniture)—manufacturing		Y	Y	Y	Y[b]	Y[c]	Y[d]	N
25	Furniture and fixtures—manufacturing		Y	Y	Y	Y[b]	Y[c]	Y[d]	N
26	Paper and allied products—manufacturing		Y	Y	Y	Y[b]	Y[c]	Y[d]	N
27	Printing, publishing, and allied industries		Y	Y	Y	Y[b]	Y[c]	Y[d]	N
28	Chemicals and allied products—manufacturing		Y	Y	Y	Y[b]	Y[c]	Y[d]	N
29	Petroleum refining and related industries		Y	Y	Y	Y[b]	Y[c]	Y[d]	N
31	Rubber and misc. plastic products—manufacturing		Y	Y	Y	Y[b]	Y[c]	Y[d]	N
32	Stone, clay, and glass products—manufacturing		Y	Y	Y	Y[b]	Y[c]	Y[d]	N
33	Primary metal industries		Y	Y	Y	Y[b]	Y[c]	Y[d]	N
34	Fabricated metal products—manufacturing		Y	Y	Y	Y[b]	Y[c]	Y[d]	N
35	Professional, scientific, and controlling instruments; photographic and optical goods; watches and clocks—manufacturing		Y	Y	Y	25	30	N	N
39	Miscellaneous manufacturing		Y	Y	Y	Y[b]	Y[c]	Y[d]	N
40	**Transportation, communication, and utilities**								
41	Railroad, rapid rail transit and street railway transportation		Y	Y	Y	Y[b]	Y[c]	Y[d]	Y
42	Motor vehicle transportation		Y	Y	Y	Y[b]	Y[c]	Y[d]	Y
43	Aircraft transportation		Y	Y	Y	Y[b]	Y[c]	Y[d]	Y

Code	Land use category						
44	Marine craft transportation	Y	Y	Yb	Yc	Yd	Y
45	Highway and street right-of-way	Y	Y	Yb	Yc	Yd	Y
46	Automobile parking	Y	Y	Yb	Yc	Yd	Y
47	Communication	Y	Y	25e	30e	N	N
48	Utilities	Y	Y	Yb	Yc	Yd	Y
49	Other transportation, communication and utilities	Y	Y	25e	30e	N	N
50	**Trade**						
51	Wholesale trade	Y	Y	Yb	Yc	Yd	N
52	Retail trade—building materials, hardware and farm equipment	Y	Y	Yb	Yc	Yd	N
53	Retail trade—general merchandise	Y	Y	25	30	N	N
54	Retail trade—food	Y	Y	25	30	N	N
55	Retail trade—automotive, marine craft, aircraft and accessories	Y	Y	25	30	N	N
56	Retail trade—apparel and accessories	Y	Y	25	30	N	N
57	Retail trade—furniture, home furnishings and equipment	Y	Y	25	30	N	N
58	Retail trade—eating and drinking establishments	Y	Y	25	30	N	N
59	Other retail trade	Y	Y	25	30	N	N
60	**Services**						
61	Finance, insurance and real estate services	Y	Y	25	30	N	N
62	Personal services	Y	Y	25	30	N	N
62.4	Cemeteries	Y	Y	Yb	Yc	Yd,k	Yf,k
63	Business services	Y	Y	25	30	N	N
64	Repair services	Y	Y	Yb	Yc	Yd	N
65	Professional services	Y	Y	25	30	N	N
65.1	Hospitals, nursing homes	Y*	25*	30*	N	N	N
65.1	Other medical facilities	Y	Y	25	30	N	N
66	Contract construction services	Y	Y	25	30	N	N
67	Governmental services	Y*	Y	25*	30*	N	N
68	Educational services	Y*	25*	30*	N	N	N
69	Miscellaneous services	Y	Y	25	30	N	N
70	**Cultural, entertainment and recreational**						
71	Cultural activities (including churches)	Y*	25*	30*	N	N	N
71.2	Nature exhibits	Y*	Y*	N	N	N	N
72	Public assembly	Y	Y	N	N	N	N
72.1	Auditoriums, concert halls	Y	Y	25	30	N	N
72.11	Outdoor music shells, amphitheaters	Y*	N	N	N	N	N
72.2	Outdoor sports arenas, spectator sports	Y	N	Yg	Yg	N	N
73	Amusements	Y	Y	Y	Y	N	N
74	Recreational activities (incl. golf courses, riding stables, water recreation)	Y*	25*	30*	N	N	N
75	Resorts and group camps	Y*	Y*	Y*	N	N	N
76	Parks	Y*	Y*	Y*	N	N	N
79	Other cultural, entertainment and recreation	Y*	Y*	Y*	N	N	N

TABLE 9.5

SUGGESTED LAND-USE-COMPATIBILITY GUIDELINES (continued)

SLUCM no.	Name	Noise zones/DNL levels in L_{dn}						
		A 0–55	B 55–65	C-1 65–70	C-2 70–75	D-1 75–80	D-2 80–85	D-3 85+
80	**Resource production and extraction**							
81	Agriculture (except livestock)	Y	Y	Y^h	Y^i	Y^i	$Y^{j,k}$	$Y^{j,k}$
81.5 to 81.7	Livestock farming and animal breeding	Y	Y	Y^h	Y^i	N	N	N
82	Agricultural related activities	Y	Y	Y^h	Y^i	Y^i	$Y^{j,k}$	$Y^{j,k}$
83	Forestry activities and related services	Y	Y	Y^h	Y^i	Y^i	$Y^{j,k}$	$Y^{j,k}$
84	Fishing activities and related services	Y	Y	Y	Y	Y	Y	Y
85	Mining activities and related services	Y	Y	Y	Y	Y	Y	Y
89	Other resource production and extraction	Y	Y	Y	Y	Y	Y	Y

*The designation of these uses as "compatible" in this zone reflects individual Federal agencies' consideration of cost and feasibility factors as well as program objectives. Localities, when evaluating the application of these guidelines to specific situations, may have different concerns or goals to consider.

aAthough local conditions may require residential use, it is discouraged in C-1 and strongly discouraged in C-2. The absence of viable alternative development options should be determined and an evaluation indicating that a demonstrated community need for residential use would not be met if development were prohibited in these zones should be conducted prior to approvals.

Where the community determines that residential uses must be allowed, measures to achieve outdoor to indoor Noise Level Reduction (NLR) of at least 25 dB (Zone C-1) and 30 dB (Zone C-2) should be incorporated into building codes and be considered in individual approvals. Normal construction can be expected to provide an NLR of 20 dB, thus the reduction requirements are often stated as 5, 10 or 15 dB over standard construction and normally assume mechanical ventilation and closed windows year round. Additional consideration should be given to modifying NLR levels based on peak noise levels.

NLR criteria will not eliminate outdoor noise problems. However, building location and site planning, design and use of berms and barriers can help mitigate outdoor noise exposure particularly from ground level sources. Measures that reduce noise at a site should be used wherever practical in preference to measures which only protect interior spaces.

bMeasures to achieve NLR of 25 must be incorporated into the design and construction of portions of these buildings where the public is received, office areas, noise sensitive areas or where the normal noise level is low.

cMeasures to achieve NLR of 30 must be incorporated into the design and construction of portions of these buildings where the public is received, office areas, noise sensitive areas or where the normal noise level is low.

dMeasures to achieve NLR of 35 must be incorporated into the design and construction of portions of these buildings where the public is received, office areas, noise sensitive areas or where the normal noise level is low.

eIf noise sensitive use indicated NLR; if not use is compatible.

fNo buildings.

gLand use compatible provided special sound reinforcement systems are installed.

hResidential buildings require an NLR of 25.

iResidential buildings require an NLR of 30.

jResidential buildings not permitted.

kLand use not recommended, but if community decides use is necessary, hearing protection devices should be worn by personnel.

Key:

SLUCM	Standard Land Use Coding Manual
Y (Yes)	Land Use and related structures compatible without restrictions.
N (No)	Land Use and related structures are not compatible and should be prohibited.
NLR (Noise Level Reduction)	Noise Level Reduction (outdoor to indoor) to be achieved through incorporation of noise attenuation into the design and construction of the structure.
Y* (Yes with restrictions)	Land Use and related structures generally compatible; see notes b through d.
25, 30, or 35	Land Use and related structures generally compatible; measures to achieve NLR of 25, 30 or 35 must be incorporated into design and construction of structure.
25*, 30* or 35*	Land Use generally compatible with NLR; however, measures to achieve an overall noise level do not necessarily solve noise difficulties and additional evaluation is warranted.

Source: Federal Interagency Committee on Urban Noise, 1980, pp. 6–11.

Occupational Noise-Exposure Limits

The Occupational Safety and Health Administration (OSHA) noise exposure limits for the work environment are listed in Table 9.6.

Installation Compatible Use Zone (ICUZ) Program for Military Installations— A Special Example

Noise incompatibility (in relation to land use) at a military installation is addressed by means of a land-use planning effort known as the "Installation Compatible Use Zone" (ICUZ) program. The ICUZ zones are delineated in Table 9.7. An ICUZ study involves the identification and control of noise impacts through the preparation of noise-zone maps of the installation's existing and future noise environment; analysis of land-use-compatibility problems and solutions, including (1) identification of existing in-

compatible land uses within zones II and III; (2) identification of possibly incompatible land uses within zones II and III; and (3) identification of desirable land uses within zones II and III; review of installation master plans to ensure that existing and future facility siting is consistent with the noise environment, and the identification of noise sources that create impact, along with the investigation of possible mitigations, and the programming of resources for noise-impact reduction (U.S. Department of the Army, 1990; Fittipaldi et al., 1988).

Housing, schools, medical facilities, and other noise-sensitive land uses should be sited according to the following principles: (1) ICUZ zone I, acceptable; (2) ICUZ zone II, normally unacceptable; or (3) ICUZ zone III, unacceptable.

Large-weapon firing (for example, artillery, armor, or demolition) at military installations

TABLE 9.6

OSHA NOISE EXPOSURE LIMITS FOR THE WORK ENVIRONMENT
(Noise exposures in dBA)

Noise	Permissible exposure (hours and minutes)
85	16 hrs
87	12 hrs 6 min
90	8 hrs
93	5 hrs 18 min
96	3 hrs 30 min
99	2 hrs 18 min
102	1 hr 30 min
105	1 hr
108	40 min
111	26 min
114	17 min
115	15 min
118	10 min
121	6.6 min
124	4 min
127	3 min
130	1 min

Note: Exposures above or below the 90 dB limit have been "time weighted" to give what OSHA believes are equivalent risks to a 90 dB eight-hour exposure.
Source: Marsh, 1991, p. 322.

TABLE 9.7			
ICUZ NOISE ZONES			
ICUZ zone	Percentage population highly annoyed	A-weighted day-night sound level ADNL (dBA)	C-weighted day-night sound level CDNL (dBC)
I	<15	<65	<62
II	15–39	65–75	62–70
III	>39	>75	>70

Source: U.S. Department of the Army, 1990, p. 44.

should be assessed using "C-frequency weighting," for impulse noise. All other noise not meeting the criteria for high-energy impulsive sounds should be assessed using A-frequency weighting. Assessment of helicopter noise should include a distance factor and another helicopter-specific factor to account for the special character of the sound. Helicopters are an example of a wider class of noise sources which can be described either as impulsive or as having high-level, low-frequency energy strong enough to rattle windows or other building elements, or as both. Evidence shows that the degree of increased annoyance generated by this class of noise is quite variable. Rattles and other nonauditory effects can be induced by some helicopters when the helicopters are loud enough and near enough (U.S. Department of the Army, 1990).

The primary means of noise assessment at a military installation should be by mathematical modeling and computer simulation. Simulations will normally be summarized using the annual average DNL. Separate, overall A-weighted noise-zone maps expressed in A-weighted DNL (ADNL) and C-weighted maps expressed in CDNL should normally be prepared for each major noise source. All maps should be labeled using the zone designations I, II, and III. In locations where the DNL is determined by a few infrequent, very high-level noise sources (for example, blasts with C-weighted sound exposure levels in excess of 110 to 115 dB), the contour map should be supplemented with descriptions of these single events and potential community reaction (U.S. Department of the Army, 1990).

CONCEPTUAL APPROACH FOR ADDRESSING NOISE-ENVIRONMENT IMPACTS

To provide a basis for addressing noise-environment impacts, a six-step or six-activity model is suggested for the planning and conduction of impact studies. This model is flexible and can be adapted to various project types by modification, as needed, to enable the addressing of specific concerns of specific projects in unique locations. However, the identified steps are typical of noise impact studies (Von Gierke et al., 1977). It should be noted that the focus in this model will be on projects and their noise impacts; however, the model could also be applied to plans, programs, and regulatory actions.

The six generic steps associated with noise-environment impacts are (1) identification of levels of noise emissions and impact concerns related to the construction and operation of the development project; (2) description of the environmental setting in terms of existing noise levels and noise sources, along with land-use information and unique receptors in the project area; (3) procurement of relevant laws, regula-

Step 1:　　　　　　　Identification of noise impacts of proposed project

Step 2:　　　　　　　Preparation of description of existing noise environment conditions

Step 3:　　　　　　　Procurement of relevant noise standards and/or guidelines

Step 4:　　　　　　　Impact prediction

Step 5:　　　　　　　Assessment of impact significance

Step 6:　　　　　　　Identification and incorporation of mitigation measures

FIGURE 9.3
Conceptual Approach for Study Focused on Noise-Environment
Impacts.

tions, or criteria related to noise levels, land-use compatibility, and noise emission standards; (4) conduction of impact prediction activities, including the use of simple noise-attenuation models, simple noise-source-specific models, comprehensive mathematical models, and/or qualitative-prediction techniques based on the examination of case studies and the exercise of professional judgment; (5) use of pertinent information from step 3, along with professional judgment and public input, to assess the significance of anticipated beneficial and detrimental impacts; and (6) identification, development, and incorporation of appropriate mitigation measures for the adverse impacts. Figure 9.3 delineates the relationship between the six steps or activities in the suggested conceptual approach.

STEP 1: IDENTIFICATION OF NOISE IMPACTS

The first step in the methodology is to determine the potential impacts of the proposed project (or activity) on the noise environment. This requires the identification of the noise levels associated with the project. A considerable body of information exists on noise levels associated with a variety of projects and associated activities (Planning and Management Consultants, 1990),

and several general examples will be mentioned here.

Construction activities generally generate noise levels in excess of those typically found in the project environs. Construction sites can be categorized into four major types: domestic housing, including residences for from one to several families; nonresidential buildings, including offices, public buildings, hotels, hospitals, and schools; industrial, including buildings, religious and recreational centers, stores, and repair facilities; and public works, including roads, streets, water mains, and sewers (U.S. EPA, 1972). Noise from construction of major civil works, such as dams, generally affects relatively few people other than those employed at or near construction sites; thus such sites are not included in these categories.

Noise at a construction site varies relative to the particular operation in progress. Operations can be divided into five consecutive phases: ground clearing, including demolition and removal of structures, trees, and rocks; excavation; placing foundations, including the reconditioning of old roadbeds and the compacting of trench floors; erection, including framing; the placing of walls, floors, windows, and pipe installation; and finishing, including filling, paving, and cleanup. Table 9.8 shows typical en-

TABLE 9.8

TYPICAL RANGES OF ENERGY-EQUIVALENT NOISE LEVELS (IN dBA) AT CONSTRUCTION SITES

Phase	Domestic housing		Office building, hotel, hospital, school, public works		Industrial parking garage, religious amusement and recreations, store, service station		Public works roads and highways, sewers, and trenches	
	I[a]	II[b]	I	II	I	II	I	II
Ground clearing	83	83	84	84	84	83	84	84
Excavation	88	75	89	79	89	71	88	78
Foundations	81	81	78	78	77	77	88	88
Erection	81	65	87	75	84	72	79	78
Finishing	88	72	89	75	89	74	84	84

[a]I, all pertinent equipment present at site.
[b]II, minimum required equipment present at site.
Source: U.S. Environmental Protection Agency, 1972, p. 2-104.

ergy-equivalent noise levels at construction sites. Table 9.9 contains information on noise levels observed 50 ft from various types of construction equipment. These levels range from 72 to 96 dBA for earth-moving equipment, from 75 to 88 dBA for materials-handling equipment, and from 68 to 87 dBA for stationary equipment; impact equipment may generate noise levels up to 115 dBA (U.S. EPA, 1972).

Noise from project operations includes sound emissions from highway vehicles, aircraft, recreation vehicles, internal-combustion engines, and industrial machinery. Noise produced by highway vehicles can be attributed to three major generating systems: rolling stock such as tires and gearing, propulsion systems related to engine accessories and other accessories, and aerodynamic and body systems. Noise levels produced by highway vehicles are a function of vehicle speed, as illustrated in Figure 9.4).

Operational-noise data for OH-58D Army helicopters and other military aircraft is available in Benson, White, and Murphy (1992). Finally, information on noise levels from power

lines and their attenuation with distance, along with related examples, is available in Berglund and Berglund (1986).

Specific noise-impact guides have been developed for certain key sources; one example is the guide developed for electric power plants (Edison Electric Institute, 1984). This guide addresses prediction, evaluation, measurement, and control of power plant noise emissions. It is based on actual noise output data from equipment for any form of power plant—coal, oil, gas, or nuclear; open or enclosed—and the expected longtime average sound levels over a range of community positions for a wide range of weather and geographic conditions. A "modified composite noise rating" procedure is recommended for estimating the expected response of a community to power plant noise. The procedure provides a means for estimating the expected response of a typical community to power plant noise emissions that contain long-term, steady broadband sounds (e.g., those from cooling towers, condensers, and boilers), intermittent high-level sounds (e.g., those from safety vents, outdoor public-address systems,

TABLE 9.9

CONSTRUCTION-EQUIPMENT NOISE RANGES

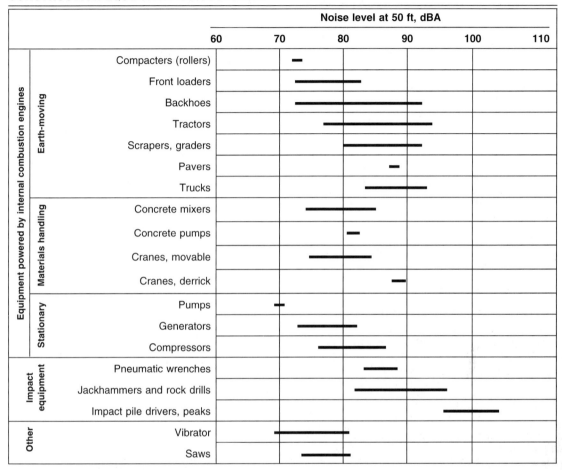

			Noise level at 50 ft, dBA					
			60	70	80	90	100	110
Equipment powered by internal combustion engines	Earth-moving	Compacters (rollers)		▬				
		Front loaders		▬▬▬▬▬				
		Backhoes		▬▬▬▬▬▬▬▬				
		Tractors			▬▬▬▬▬▬			
		Scrapers, graders			▬▬▬▬▬▬			
		Pavers				▬		
		Trucks			▬▬▬▬			
	Materials handling	Concrete mixers		▬▬▬▬▬				
		Concrete pumps			▬			
		Cranes, movable			▬▬▬			
		Cranes, derrick				▬		
	Stationary	Pumps	▬					
		Generators		▬▬▬▬				
		Compressors			▬▬▬▬			
Impact equipment		Pneumatic wrenches				▬▬▬		
		Jackhammers and rock drills			▬▬▬▬▬▬			
		Impact pile drivers, peaks					▬▬▬	
Other		Vibrator		▬▬▬▬				
		Saws		▬▬▬				

Note: Based on limited available data samples.
Source: U.S. Environmental Protection Agency, 1972, p. 2-108.

and coal car shakers), and tonal sounds (e.g., those from transformers, motors, pumps, and fans).

If noise information is unavailable for a proposed project or activity, several possible methods exist for assembling pertinent information. One approach would be to conduct a computer-based search of the literature to determine whether any emission information had been published for the source type. If this approach is unproductive, consideration could be given to the collection of existing noise emissions data for sources of similar type. Another approach is to gather noise emission information from the manufacturers of the noise-producing items. Finally, noise emission standards, if relevant and available, could be used to identify noise emission concerns.

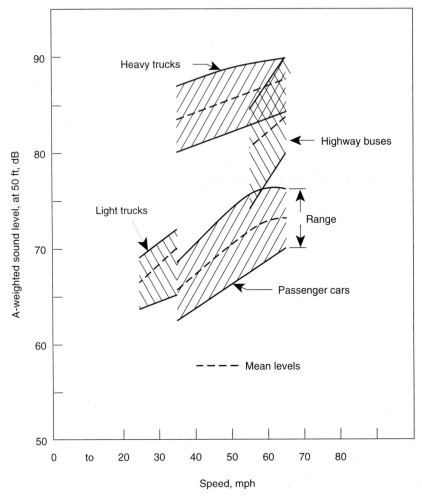

FIGURE 9.4
Single-Vehicle Noise Output as a Function of Vehicle Speed (Wyle Laboratories, 1971, p. 95).

STEP 2: DESCRIPTION OF EXISTING NOISE-ENVIRONMENT CONDITIONS

In analyzing the potential noise impacts of a proposed project (or activity), it is necessary to consider the study area (potential project area or region of influence) associated with the noise emissions. The delineation of a study area can be made based upon the boundaries of the land associated with the project, or the delineation can include a larger area by also considering the area of noise influence within the vicinity of the proposed project.

The primary information which should be accumulated in step 2 is data on existing noise levels and noise sources within the study area. Land-use and human-population-distribution maps in relation to the proposed project would also be needed. For example, for a military installation, the results of an ICUZ study, if one has been conducted, could be abstracted for this step. A properly developed noise inventory provides information concerning all source emissions and defines the location, magnitude, frequency, duration, and relative contribution of these emissions. It can be used as a baseline marker against which to judge previous and future noise emissions in the geographical study area. Additionally, it can be used in a comparative context with regard to other nearby geographical areas.

If no specific data on existing noise levels is available for the study area, it might be possible to use published noise-level information developed for projects involving similar land use. For example, Table 9.10 shows typical day-night noise levels in urban areas in the United States. Quiet suburban residential areas have an average L_{dn} of 50 dBA, while very noisy urban residential areas exhibit L_{dn} values of 70 dBA. Typical noise levels in rural settings are 30 to 35 dBA, and in wilderness locations they are on the order of 20 dBA. Seasonal and daily variations in noise levels may occur, particularly at national and state parks and recreational areas (Bowlby, Harris, and Cohn, 1990).

The sources of information on noise-monitoring data include the relevant county, regional, or state environmental agency, and private industries in the area that might be maintaining a noise-monitoring program for their particular interests. It should be noted that an extensive historical record of noise-monitoring data would be unlikely unless the study area had previous problems with excessive noise levels and resultant noise complaints. Information is available on the planning and conduct of noise surveys in Fidell (1977), U.S. Department of Housing and Urban Development (1985), and Edison Electric Institute (1984).

One method of expressing both existing noise and predicted noise levels is by using a level-weighted population value (Von Gierke et al., 1977). (This type of index approach was mentioned in Chapter 5.) A sound-level-weighted population is a single-number representation of the significance of a noise environment to the exposed population. The assumptions are that the intensity of human response is one of several consequences of average sound level, depending upon the response mode of interest (annoyance, speech interference, hearing loss) and that the impact of high noise levels on a small number of people is equivalent to the impact of lower noise levels on a larger number of people in an overall evaluation. Based on these assumptions, the "fractional impact" can be determined as the product of a sound-level-weighting value and the number of persons exposed to a specified sound level. Summing the fractional im-

TABLE 9.10

TYPICAL DAY-NIGHT NOISE LEVELS IN URBAN AREAS IN THE UNITED STATES

Description	Typical range of L_{dn}, dB	Average L_{dn}, dB	Average census tract population density, no. of people/mi^2
Quiet suburban residential	48–52	50	630
Normal suburban residential	53–57	55	2,000
Urban residential	58–62	60	6,300
Noisy urban residential	63–67	65	20,000
Very noisy urban residential	68–72	70	63,000

Source: U.S. Environmental Protection Agency, 1974, p. B-5.

pacts over the entire population provides the sound-level-weighted population (LWP). The calculation is as follows (Von Gierke et al., 1977):

$$\text{LWP} = \int P(L_{dn}) \cdot W(L_{dn}) \, d(L_{dn})$$

where $P(L_{dn})$ is the population distribution function, $W(L_{dn})$ is the day-night average sound-level-weighting function characterizing the severity of the impact as a function of sound level (its derivation is described below), and $d(L_{dn})$ is the differential change in day-night average sound level. Sufficient accuracy can be obtained by taking average values of the weighting function between equal decibel increments—say, up to 5 dB—and replacing the integrals by summations of successive increments in average sound level (Von Gierke et al., 1977).

The weighting function $W(L_{dn})$ is based on the reaction of populations to living in noise-impacted environments and other social survey data relating the fraction of sampled population expressing a high degree of annoyance to various L_{dn} values. The weighting function is normalized to unity at 75 dB; values of $W(L_{dn})$ are listed in Table 9.11.

A noise impact index (NII) can then be used for comparing the relative impact of one noise environment with that of another. It is defined as the sound-level-weighted population LWP divided by the total population P_{total} under consideration:

$$\text{NII} = \frac{\text{LWP}}{P_{total}}$$

An example calculation for this index is in Table 9.12.

The NII for existing conditions could be determined based on noise measurements, population data, and the pertinent weighting functions. Prediction of project-induced noise levels and consideration of future population data, along with the pertinent weighting functions, would enable the determination of the project-related noise index. The actual impact is related to the differences in the calculated indices. Al-

---

TABLE 9.11

SOUND-LEVEL-WEIGHTING FUNCTION FOR OVERALL IMPACT ANALYSIS

L_{dn}, in dB	$W(L_{dn})$	$\dfrac{W(L_{dn}) + W(L_{dn} + 5)}{2}$
35	0.006	0.010
40	0.013	0.021
45	0.029	0.045
50	0.061	0.093
55	0.124	0.180
60	0.235	0.324
65	0.412	0.538
70	0.664	0.832
75	1.000	1.214
80	1.428	1.697
85	1.966	2.307
90	2.647	

Note: The right-hand column is included for convenience for finding the weighting of certain 5 dB increments.
Source: Von Gierke et al., 1977, p. VII-6.

TABLE 9.12

EXAMPLE OF LEVEL-WEIGHTED POPULATION LWP AND NOISE-IMPACT-INDEX COMPUTATION

L_{dn}, in dB	Cumulative population[a]	Incremental population[a]	Weighting function[b]	Level-weighted population[a]
80	0.1	0.1	1.697	0.17
75	1.3	1.2	1.214	1.46
70	6.9	5.6	0.832	4.66
65	24.3	17.4	0.538	9.36
60	59.6	35.3	0.324	11.44
55	97.5	37.9	0.180	6.82
	Total 97.5	$NII = \dfrac{33.91}{97.5} = 0.35$		Total 33.91

[a]Population in thousands.
[b]From Table 9.11.

though not described herein, an additional weighting function at higher noise environments can be used to quantify the potential of noise-induced hearing loss and general health effects (Von Gierke et al., 1977).

STEP 3: PROCUREMENT OF RELEVANT NOISE STANDARDS AND/OR GUIDELINES

The primary sources of information on noise standards, criteria, and policies will be the relevant local, state, and federal agencies which have a mandate for overseeing the noise environment of the study area. Additional information may be available from international agencies such as the World Health Organization (WHO) or the United Nations' Environment Program. This information can be used to determine the baseline quality and the significance of noise impacts incurred during projects (or activities); it could also aid in deciding between alternative actions or in assessing the need for mitigating measures for a given alternative. Pertinent institutional information related to the noise environment has been described earlier.

STEP 4: IMPACT PREDICTION

Step 4 involves predicting the propagation of noise from a source and determining the type of affected land uses. Several approaches for predicting noise contours are outlined in the discussion of this step.

Simple Noise-Attenuation Models

Two simple models for noise-level prediction (one for point-source emissions and one for line-source emissions) will be considered. Sound travels through air in waves with the characteristics of "frequency" and "wavelength." If a sound is created at a point, a system of spherical waves propagates from that point outward through the air at a speed of 1,100 ft/sec, with the first wave making an ever-increasing sphere with time (U.S. Department of Transportation, 1972). (See Figure 9.5 for a depiction of sound propagation.) As the wave spreads, the height of the wave or the intensity of sound at any given point diminishes as the fixed amount of energy is spread over an increasing surface area of the sphere. This phenomenon is known as "geometric attenuation of sound." Point-source propagation can be defined as follows:

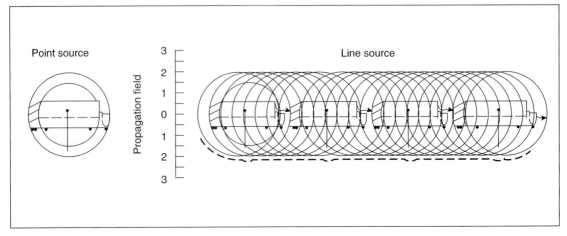

FIGURE 9.5
Sound-Propagation Comparison (U.S. Department of Transportation, 1972, p. 25).

$$\text{Sound level}_1 - \text{sound level}_2 = 20 \log \frac{r_2}{r_1}$$

Thus, the sound level at station 1 minus the sound level at station 2 is equal to 20 times the log of the ratio of the radii r_2, r_1. This means that for every doubling of distance, the sound level decreases by 6 dBA. This point-source relationship is called in the "inverse square law" and is applicable for noise emissions from aircraft and from single vehicles when sound is propagating in free air, either in a complete spherical sense (from the airplane to the ground and to the surrounding atmosphere) in the first case, or in only half a sphere in the second case. It would also apply for noise propagation from construction equipment.

Line-source propagation occurs when there is a continuous stream of noise sources. The propagation is no longer characterized by a spherical or hemispherical spreading of sound; rather, the reinforcement by the line of point sources makes the propagation field either a cylinder-shaped or a half-cylinder-shaped area. The line-source-propagation prediction model is as follows:

$$\text{Sound level}_1 - \text{sound level}_2 = 10 \log \frac{r_2}{r_1}$$

The decrease in sound level for each doubling of distance from a line source is 3 dBA. When noise levels from a busy highway are considered, it is appropriate to utilize the highway as an infinite line source and consider a 3-dBA doubling of the distance-propagation rate.

Simple Models for Specific Source Types

Simple models have been developed for specific types of sources, as well. Information on developing noise contours is presented herein for (1) impulse- and continuous-noise sources at military installations, (2) construction, and (3) airports.

Military sources of noise can be classified as either impulse or continuous. "Impulse sources" are single, discrete events where the noise levels rise with time, reach a maximum value, and then decay to the background level. The noise exposure from this type of source is assessed in terms of such events that occur throughout the day. In contrast, "continuous sources" are those for which the sound level rises to a particular value and remains there for an extended period of time. These sources are assessed in terms of the maximum level and duration of such occurrences.

Both impulse- and continuous-source propagation are explicitly expressed in the following models. Noise is measured using both day-night average sound level (L_{dn}) and energy-equivalent noise levels (L_{eq}). The basic equation for impulse sources is as follows (Goff and Novak, 1977):

$$L_{dn} = \text{SEL } 10 \log$$
$$(N_d + 10 N_n) - 49.4 \quad (9.1)$$

where L_{dn} = day-night average noise level
 SEL = maximum sound exposure level occurring for single event
 N_d = number of daytime (0700–2200 hr) operations
 N_n = number of nighttime (2200–0700 hr) operations

The basic equations for continuous sources are as follows:

$$L_{eq} = \text{AL} + 10 \log D - 35.6 \quad (9.2)$$
$$L_{dn} = \text{AL} + 10 \log (D_d + 10 D_n)$$
$$- 49.4 \quad (9.3)$$

where AL = maximum A-weighted sound level of event
 D = event duration within 1-hr period, sec
 D_d = event duration during daytime (0700–2200 hr), sec
 D_n = event duration during nighttime (2200–0700 hr), sec

$$\text{ADJ} = k \log \frac{Y}{X} \quad (9.4)$$

where ADJ = adjustment factor
 X = initial distance from source
 Y = other distances from source
 k = constant dependent on type of source

Although the equations for continuous and impulse noise are different, the contouring procedure is identical, consisting of the following substeps (Goff and Novak, 1977, pp. 28–29):

Substep 1: Determine the SEL (for impulse sources) or AL (for continuous sources) at the location of interest for each type of operation.

Substep 2: Tabulate the number of discrete events (for impulse sources) or duration (for continuous sources) for each type of operation. The following data can be used:
  a. *Average daily units*—divide total number/duration of operations in a typical month by 30.
  b. *Average hourly units*—divide total number/duration of operations in an average day by 24.
  c. *Worst case*—compute maximum possible number/duration of operations that can occur during a day/hour.

Substep 3: Determine the L_{dn} value for each type of operation using equations (9.1) through (9.3).

Substep 4: If more than one type of operation occurs at the same location, determine the total L_{dn} value for all operations by using Table 9.13 to add logarithmically the individual L_{dn}.

Substep 5: Draw contours for other noise sensitive areas (Y) using Eq. (9.4).

Noise from construction operations is different from noise from other major sources for two reasons. First, it is caused by many types of equipment. Second, the resulting adverse effects will be temporary because the operations are relatively short-term. In addition, since construction usually occurs during the day, there is a minimum of sleep interference. Thus, to develop an evaluation procedure for construction noise, the time and detail involved must be weighted carefully against its unique and temporary nature. The following substeps represent a procedure for preparing sound contours for construction noise (adapted from Goff and Novak, 1977, p. 84):

TABLE 9.13

DETERMINING THE CUMULATIVE DECIBEL SPL WHEN THE DIFFERENCES BETWEEN TWO OR MORE LEVELS ARE KNOWN

Difference between levels, dBA	No. of dBA to be added to higher level
0	3.0
1	2.6
2	2.1
3	1.8
4	1.5
5	1.2
6	1.0
7	0.8
8	0.6
10	0.4
12	0.3
14	0.2
16	0.1

Source: Chanlett, 1973, p. 526.

Substep 1: Using Table 9.8, determine the L_{eq} from the site for each phase of construction.

Substep 2: Using Equation (9.5), determine the L_{eq} for the site for the entire operation.

$$L_{eq} = 10 \log \frac{1}{T} \sum_{i=1}^{N} T_i(10)^{L_i/10} \qquad (9.5)$$

where $L_i = L_{eq}$ for ith phase (from Table 9.8)
  T_i = total time duration of ith phase (from Table 9.8)
  T = total time of operation from the beginning of initial phase ($i = 1$) until beginning of final phase ($i = N$)
  N = number of phases

Substep 3: Correct L_{eq} for the distance of interest x from the site using

$$ADJ = -20 \log (x + 250) + 48 \qquad (9.6)$$

where x is distance in feet from the boundary.

Substep 4: Draw the L_{eq} contours around the site on the appropriate map.

Airport activities generate two major noise sources: fixed-wing aircraft operations and ground operations. Fixed-wing aircraft include both jet and propeller types. Noise commonly associated with the former is from jet exhaust and turbomachinery. The noise of aircraft flyovers differs from that of ground operations. For a given distance, the maximum noise levels produced during a ground operation will be lower than those produced during a flight operation because of ground absorption, intervening buildings, and other barriers. However, because a ground operation, or run-up, may last for several minutes, and hence, is a continuous-source event, as opposed to a flyover noise signal, which is an impulse event, the run-up may produce a much higher SEL level. The total exposure from airport operations is the summation of the noise exposure from all operations of all aircraft on all flight paths (air and ground). The following substeps depict a procedure which can be used to predict airport noise contours (Goff and Novak, 1977, p. 100):

TABLE 9.14

DISTANCES TO L_{dn} CONTOURS FOR AIRPORT OPERATIONS

Effective number of operations	Distance to L_{dn} 65 contour		Distance to L_{dn}75 contour	
	1	2	1	2
0–50	500 ft 152 m	3000 ft 914 m	0	0
51–100	1000 ft 305 m	1 mi 1.6 km	0	0
101–200	1500 ft 456 m	1.5 mi 2.4 km	500 ft 125 m	3000 ft 914 m
201–400	2000 ft 609 m	2 mi 3.2 km	1000 ft 305 m	1 mi 1.6 km
401–1000	1 mi 1.6 km	2 mi 3.2 km	2000 ft 609 m	1.5 mi 2.4 km
> 1000	1 mi 1.6 km	2.5 mi 4 km	3000 ft 914 m	1.5 mi 2.4 km

Source: Goff and Novak, 1977, p. 101.

Substep 1: Determine the average number and time of flight operations.

Substep 2: Determine the effective number of operations for all aircraft operations using the following:

$$EN = d + (16.7)n \qquad (9.7)$$

where EN = effective number of operations

    d = number of daytime operations (0700–2200)

    n = number of nighttime operations (2200–0700)

Each takeoff or landing is considered an operation. The calculations assume that the number of takeoffs and landings during the day is equal to the number during the night. If takeoffs occur during the day and landings occur at night, the total number of operations should be considered as daytime operations only.

Substep 3: Determine the zone contours using Table 9.14. The columns headed by a "1" in Table 9.14 refer to the distance from the centerline

of the runway to the edge of the contours; the columns headed by a "2" refer to the distance from the end of the runway to the tip of the contour. This is illustrated in Figure 9.6.

An example problem for airfield operations is as follows:

FIGURE 9.6
Development of Noise-Zone Contours for Airports (Goff and Novak, 1977, p. 101).

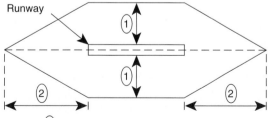

Runway

① Distance from centerline of runway

② Distance from edge of runway

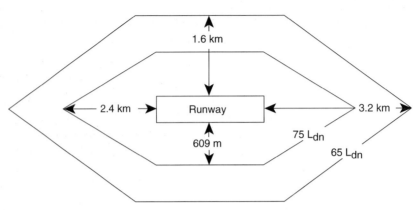

FIGURE 9.7
Airport Contours for Example Problem (Developed from Data in Goff and Novak, 1977, p. 101).

A new airfield is being proposed that will have an estimated 75 daytime operations and 20 night-time operations. Estimate the noise contours for 65 L_{dn} and 75 L_{dn}.

Substep 1: Determine the effective number EN of operations.

$$EN = d + 16.7n$$
$$= 75 + 16.7 (20)$$
$$= 409$$

Substep 2: Determine the noise contours (from Table 9.14).

65 L_{dn} distance = 1.6 km width and 3.2 km length
75 L_{dn} distance = 609 m width and 2.4 km length

Refer to Figure 9.7 for an example of how this is drawn.

Comprehensive Mathematical Models

It is beyond the scope of this chapter to derive the various mathematical models for noise predictions. This section will be oriented to the identification of several mathematical models for use in noise-impact prediction and the identification of reference sources that could be used to prepare a more detailed study for an individual project. Examples of references sources which include details on computer (or nomo-

graphic) modeling include Gordon et al. (1971), Wesler (1972), Nelson and Wolsko (1973), Magrab (1975), Lang (1986), Lipscomb and Taylor (1978), May (1978), Keast, Eldred, and Purdum (1988), FAA (1988), Fittipaldi et al. (1988), Harris (1989), and Miller and Thumann (1990).

Several examples of computer-based models will be mentioned. For example, the Federal Highway Administration uses two highway-noise-prediction approaches—the method developed by the National Cooperative Highway Research Program (Gordon et al., 1971) and the one developed by the Transportation Systems Center (Wesler, 1972). These computerized methods, which have been refined through the years, involve consideration of the characteristics of highway segments as one input variable. Pertinent characteristics include the traffic using the highway (the quantities and the speeds of both automobiles and trucks), physical dimensions of the structure (the elevation, depressions, grades, and surface types), and aspects of the environment bordering the highway that have an effect on noise levels (landscaping, structures, and barriers). Both models calculate the noise level at a particular, perpendicular distance away from a point along the highway. Once this noise level is calculated, the model "moves outward'' an incremental distance and calculates another noise level. This process is repeated until the model reaches a maximum prescribed distance away from the highway. At this point the model moves further down the highway and calculates another group of noise levels. This is repeated until the model has calculated the noise levels for the entire length of the highway section of interest. The model printouts include contour maps of noise levels over the entire length of the facility (Nelson and Wolsko, 1973).

Lang (1986) has compared quantitative prediction methods used for modeling road traffic and railway noise in Austria, Czechoslovakia, France, Germany, Hungary, The Netherlands, Scandinavian countries, Switzerland, the United Kingdom, and the United States. While specific factors included in the prediction models may vary to some extent, the basic calculation approaches are similar.

Several models are available for addressing noise levels from aircraft, including the aircraft-sound-description system, noise exposure forecast, and composite noise rating (Nelson and Wolsko, 1973). In the United States, the FAA has developed the Integrated Noise Model (INM), version 3.8. The INM program includes standard aircraft-noise and performance data for over 60 aircraft types; this data can be tailored to the characteristics of the airport in question. Use of the INM program requires the input of the physical and operational characteristics of the airport. ''Physical characteristics'' include runway coordinates, airport altitude, and temperature. ''Operational characteristics'' include aircraft mix, flight tracks, and approach profiles. Additional data that is contained within the model, or can be input specific to the airport, include departure profiles, approach parameters, and aircraft noise curves (FAA, 1988).

The U.S. Air Force has developed an overall procedure for noise-impact prediction and assessment called ''assessment system for aircraft noise'' (ASAN). ASAN incorporates procedural information and pertinent computer software models, including models for military training routes and operating areas, human effects, and structural damage (Reddingius and Smyth, 1990).

One such model is NOISEMAP, a microcomputer model for calculating noise metrics (e.g., DNL, NEF) around military bases (Lee, 1990). Noise and performance data files for both military and civilian aircraft are included. Two supporting programs are BASEOPS and NMPLOT. BASEOPS is a menu-driven, computerized air-base-operations-input program for examining different operational input scenarios (Lee and Mohlman, 1990); NMPLOT takes the output of NOISEMAP and creates the resulting noise-contour map.

A PC-compatible ''heliport noise model'' (HNM) has been developed for determining the total impact of helicopter noise in and around heliports (Keast, Eldred, and Purdum, 1988). HNM is based on the FAA's INM for modeling noise from fixed-wing aircraft.

Another example of a noise model for military installations is called ''SLICE.'' SLICE was recently developed by the Army Environmental Hygiene Office and the Institute for Water Resources. This program allows environmental coordinators to predict noise impacts of air operations, blast noise, ground transportation, and rifle-range and recreational activity. SLICE is an interactive program that enables the user to input operational and general climatic data to predict the noise annoyance factor for a proposed project. This program enables users to evaluate various scenarios to determine the most environmentally friendly alternative. For instance, possible locations of firing points can be evaluated to determine which will create the least noise annoyance. Impact prediction methods are also available for blast noise in Schomer et al. (1981).

In summary, models can be useful aids in estimating the impacts of a project on the noise environment. Also, various modifications of the proposed project can be evaluated to assess the effectiveness of impact-mitigation efforts. Selection of an appropriate model to meet a given need typically involves consideration of the technical capabilities of models and managerial issues. Typical managerial issues include economic considerations, and necessary training and experience of model users.

STEP 5: ASSESSMENT OF IMPACT SIGNIFICANCE

The next activity is impact assessment. In the terminology used herein, ''assessment'' refers to the interpretation of the significance of anticipated changes related to the proposed project. One basis for impact assessment is public input; this input could be received through a continued scoping process or through the conduction of public meetings or public participation programs, or both. The general public can often delineate important environmental resources and values for particular areas, and this should be considered in impact assessment. Professional judgment can also be useful to assess the percentage changes from baseline conditions in terms of noise levels and/or exposed human population, or a noise index; such changes could be considered for both the construction and operational phases of a project. Land-use-compatibility guidelines such as those in Table 9.5 could also be used. For some particular types of projects or noise-assessment methods, there are specific numerical standards or criteria which can be used as a basis of interpretation, and two examples will be mentioned: one relating to highways and the other to an ICUZ study.

Highway Project Example

Figures 9.8 and 9.9 include specific guidance for acceptable noise levels adjacent to highways. They could be used in assessing impacts to existing land uses and human populations and for planning land uses adjacent to rights-of-way.

ICUZ Study Example

A key output of an ICUZ study is a land-use map showing measured and/or predicted noise contours and including ICUZ zones. A land-use map should be created using a topographical map, installation map, or aerial photograph. The primary consideration is to determine the appropriate scale and level of detail for the land-use map. These map characteristics are entirely dependent on the size of the noise contours influencing the area. The larger the area of noise impact, the more general the land-use map will need to be. For instance, a land-use map for an airport would cover a large area and show only general land-use features, since the noise impacts could extend many kilometers. However, a land-use map for a construction site should show much greater detail, since the noise contours from the activity will probably extend only a relatively short distance. The following substeps are suggested for land-use map development (U.S. Department of the Army, 1990; Fittipaldi et al., 1988):

Substep 1: Determine the distance from the noise source for L_{dn} values of 65 and 75 dBA.

Substep 2: Draw the noise-contour map. Subdivide the noise contours into noise zones (see Table 9.7).

Substep 3: Procure a land-use map from local planning officials. If this is not available, use topographic maps or aerial photographs that display, at a minimum, the area of the noise contours. Then, lay a transparent sheet over the base map and determine the land-use classification of the areas surrounding the noise source (see Table 9.15). Every attempt should be made to identify all the different land-use classes within the zone III contour. However, this may be difficult if the noise source is in a urban area or the noise contours extend over a large area. If this is the case, the land-use map can be broken into grids as

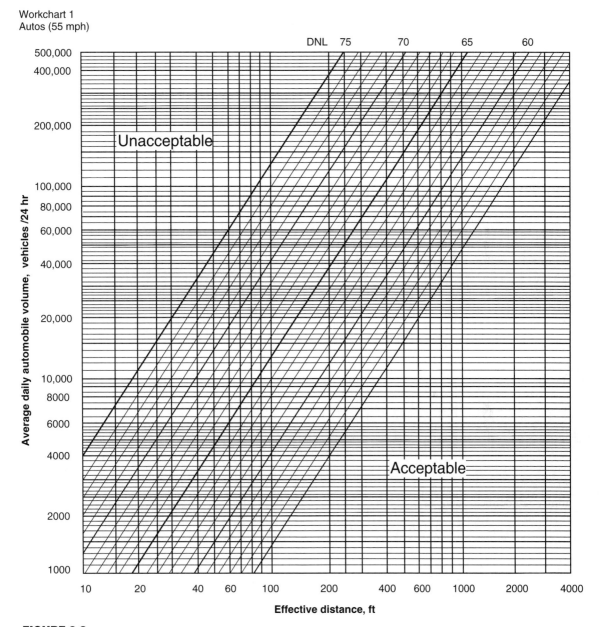

FIGURE 9.8

Noise Levels Adjacent to Highways Traveled by Automobiles (U.S. Department of Housing and Urban Development, 1985, p. 67).

Workchart 2
Heavy trucks (55 mph)

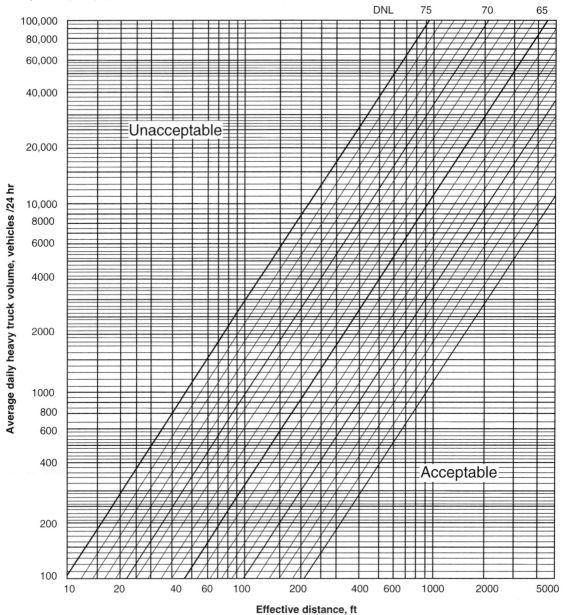

FIGURE 9.9
Noise Levels Adjacent to Highways Traveled by Heavy Trucks (U.S. Department of Housing and
Urban Development, 1985, p. 68).

TABLE 9.15

LAND-USE CLASSIFICATION

Land-use classification	Noise sensitivity	Max L_{dn} level, dBA	Land use
1	High	65	Residences, hospitals, schools, parks, retail stores, professional offices, scientific research facilities
2	Moderate	75	Most industrial facilities, parking lots, wholesale trade building-materials retail, and construction services, golf courses, water-based activities, livestock
3	Low	85+	Agriculture, utilities, rail transit, highways

Source: Compiled using data from U.S. Department of the Army, 1990.

TABLE 9.16

NOISE-CONTOUR AREA VS. LAND-USE DETAIL

Degree of detail	Area of contours	Size of grid	Source
High	9.3 km^2	60 m × 60 m	Industrial, roads
Moderate	12.8 km^2	250 m × 250 m	Pistol range, highway construction
Least	>12.8 km^2	500 m × 500 m	Aircraft, blast

Source: Goff and Novak, 1977, p. 33.

outlined in Table 9.16. If there is mixed land use within a grid, label the grid with the dominant land use.

Substep 4: The noise-contour map can easily be drafted over the land-use map at the same scale. Make sure to indicate the exact location of the noise source in relation to the noise contours.

Substep 5: Using Table 9.17, determine the environmental sensitivity of the areas impacted by the project's noise.

Substep 6: Draw a final overlay that shows the land-use classification, ICUZ zones, and areas of high to moderate environmental sensitivity. The relative desirability of alternative actions may be assessed by comparing populations in areas of high environmental sensitivity. However, any alternative that will result in pairing a zone III with a sensitive land

use (land-use classification 1) should be considered unacceptable.

Other Considerations

Noise impacts on wildlife may be an important issue for a proposed project. Considerable information has been accumulated on this subject; it ranges from specific studies on marine mammals (Richardson et al., 1991) and caribou (Harrington and Veitch, 1990), to generic studies on fish and wildlife (Gladwin, Asherin, and Manci, 1988), to general literature reviews (Fletcher and Busnel, 1978; U.S. EPA, 1973, 1974). For example, selected topics addressed in Fletcher and Busnel (1978) include the effects of high-voltage transmission-line noise on wildlife; the effects of sonic booms on reproduction; the effects of noise on hearing in fish; the effects of noise on insects; and snowmobile noise effects on wildlife. In general, such effects

TABLE 9.17

ENVIRONMENTAL SENSITIVITY TO NOISE

Noise zone	Land-use classification		
	1	2	3
I	Low	Low	Low
II	Moderate	Moderate	Low
III	High	Moderate to high	Low

Source: Compiled using data from U.S. Department of the Army, 1990.

are disruptive to normal breeding patterns and detrimental to habitat conditions. These issues can be addressed in an impact study by identifying the key species of concern (often a threatened or endangered species), searching for specific information on noise effects on the species, and applying either specific or generic techniques for impact prediction and assessment. Using this data, noise-impact-mitigation measures could also be planned during this step; for example, the construction phase of the proposed project might be timed so that it would not coincide with the breeding cycle of a nearby endangered species.

STEP 6: IDENTIFICATION AND INCORPORATION OF MITIGATION MEASURES

''Mitigation measures,'' in this context, refer to steps that can be taken to minimize the magnitude of the detrimental noise impacts. The key approach to mitigation is to reduce or control the noise expected to be emitted from the project (or activity). Mitigation can proceed along three possible courses of action, either by changing (1) the source of noise, (2) the path of noise from the source to the receiver, or (3) the receiver of noise. Some additional principles of noise control include the reduction of the number of vibrating sources, enclosure of the source, and attenuation of noise by absorption (Chanlett, 1973). Simple hand calculations and/or computer models described earlier can assist in

forecasting the relative effectiveness of various designed and/or operational-phase mitigation techniques. Examples of mitigation measures include

1. In planning military training activities, weather could be used as a decision factor. Weather conditions that are conducive to abnormal noise propagation cannot be predicted, but they can be observed, and noise-mitigation actions can be taken when bad situations occur. For example, good firing conditions for heavy weapons include days having clear skies with billowy cloud formations, especially during warm periods of the year, and a rising barometer immediately following a storm (Raspet and Bobak, 1988). Bad firing conditions include days having steady winds of 10 to 15 mi/hr with gusts of greater velocities (20 mi/hr or more) in the direction of nearby residences, clear days on which the layering of smoke or fog is observed, hazy or foggy mornings, days following a day when large temperature changes (greater than 20° C) occurred, and days on which there are generally high barometric readings with low temperatures (Raspet and Bobak, 1988). Learn to recognize thermal-inversion situations; for instance, when a plume of smoke is observed to rise only a short distance and then go horizontal, there is a low-lying inversion situation and noise focusing is likely. This is a situation where postponing activities until the inversion lifts

can be beneficial. Strong inversions are most likely in the early morning and late afternoon during clear-sky conditions, when there is a low overcast condition, or when winds are light and variable with a high-pressure cell dominating the area. Moderate to heavy winds disrupt inversions, but can enhance propagation toward (windward to) sensitive areas.

2. For military training activities, firing points should be sited near natural barriers. Although man-made barriers for blast-noise protection are prohibitively expensive, the use of valleys and ravines as firing points can affect the propagation significantly. Moving selected firing points a few hundred meters (within safety-fan boundaries) can substantially reduce the impacts. Adjustment of the firing-target points for selected operations can move the ICUZ zone II and III areas back on post (on the military installation). For instance, if a zone III is very near the installation boundary, and if it can be determined (from the contouring input or operations data) that a particular weapon and firing point is responsible for the large noise impact, then moving that firing point a few hundred meters away from the boundary might change the zone pattern significantly. Care must be taken, however, in that changing both the firing-point and the target-point might produce offsetting impacts.

3. Noise-control measures for roads or highways include construction of barriers to obstruct or dissipate sound emissions, elevated or depressed highways, and the absorption effects of landscaping (trees, bushes, and shrubs). Constructed barriers can be an effective approach for reducing noise from highways. When implementing this measure, important factors include the relative height of the barrier, the noise source and the affected area, and the horizontal distances between the source and the barrier and between the barrier and the noise-affected area (U.S.

Department of Transportation, 1972). Noise barriers around freeways can include various barrier-wall shapes and textures, along with landscaping (Farnham and Beimborn, 1990). Elevating or depressing highways in urban areas provides differences in grade, thus shielding traffic noise and reducing noise levels on adjacent properties. Plantings of trees, bushes, and shrubs adjacent to a highway generally produce little physical reduction in noise unless the plantings are very dense and have significant depth (Hendriks, 1989). Additional noise-control measures for roads or highways include limitations on allowable grades, maintenance of proper road-surface repairs, route locations planned to ensure maximum separation between roadways and existing noise-sensitive areas, and provisions for compatible land uses adjacent to highway rights-of-way.

4. The timing of the noise-generating source may be changed. For example, the time periods for construction activities could be limited to selected daytime hours. Another example, for military installations, would be to change the times of large-weapons firing. If a particular firing or target point is heavily used for night operations, then shifting those operations to a firing or target point farther away from sensitive land uses can be helpful. In general, nighttime (2200 to 0700 hr) operations should be kept to the barest minimum; some night operations may be shifted to the daytime as long as it does not impact mission requirements.

5. Purchased mobile equipment should be in compliance with noise emission standards. Information on noise-emission-control technologies for sources such as passenger vehicles, trucks, buses, motorcycles, and mopeds is in Organization for Economic Cooperation and Development (1988).

6. Noise-attenuation measures should be utilized in building design and construction; examples of such measures include (U.S.

TABLE 9.18

TRANSMISSION LOSS VALUES FOR COMMON BARRIER MATERIALS

Material	Thickness, (inches)	Transmission loss, dBA[a]
Woods		
Fir	1/2	17
	1	20
	2	24
Pine	1/2	16
	1	19
	2	23
Redwood	1/2	16
	1	19
	2	23
Cedar	1/2	15
	1	18
	2	22
Plywood	1/2	20
	1	23
Particle Board	1/2	20
Metals		
Aluminum	1/16	23
	1/8	25
	1/4	27
Steel	24 ga	18
	20 ga	22
	16 ga	15
Lead	1/16	28
Concrete, Masonry, etc.		
Light Concrete	4	38
	6	39

Department of Housing and Urban Development, 1985) reducing the total area of windows or other acoustically weaker building elements; sealing off "leaks" around windows, doors, and/or vents; improving the actual sound-attenuating properties of small building elements such as windows, doors, etc.; and improving the actual sound-attenuating properties of major building elements such as roof and wall construction. Examples of transmission loss values for common barrier materials are shown in Table 9.18. Sound-insulation methods for new and existing residences in the vicinity of airports have been developed in U.S. Department of Transportation (1992).

7. Design features can be used to reduce the noise from specific sources; for example, the mechanical noise from the gearbox of large wind turbines can be minimized by adapting specific design features (Ljunggren and Johansson, 1991).

TABLE 9.18

TRANSMISSION LOSS VALUES FOR COMMON BARRIER MATERIALS
(continued)

Material	Thickness, (inches)	Transmission loss, dBA[a]
Dense Concrete	4	40
Concrete Block	4	32
	6	36
Cinder Block (Hollow Core)	6	28
Brick	4	33
Granite	4	40
Composites		
Aluminum Faced Plywood	3/4	21–23
Aluminum Faced Particle Board	3/4	21–23
Plastic Lamina on Plywood	3/4	21–23
Plastic Lamina on Particle Board	3/4	21–23
Miscellaneous		
Glass (Safety Glass)	1/8	22
	1/4	26
Plexiglass (Shatterproof)	—	22–25
Masonite	1/2	20
Fiberglass/Resin	1/8	20
Stucco on Metal Lath	1	32
Polyester with Aggregate Surface	3	20–30

[a]A weighted TL based on generalized truck spectrum.
 Source: U.S. Department of Housing and Urban Development 1985, p. 27.

8. Noise mitigation may be facilitated by the development of a comprehensive noise-management program. For example, noise-management guidelines for airport managers are available; such guidelines include operating hours, flight-path and landing and takeoff controls, and the use of local noise reflectors or barriers (Gillen, Levesque, and Smith, 1990). Noise management and noise-impact minimization through integrated land-use planning and activity controls at military installations can also be achieved (Feather and Shekell, 1991). Noise-management programs for military airfields, base operations, training areas, and housing and recreational areas are available.

The result of step 6 should be the delineation of a pertinent mitigation plan to reduce or eliminate as many of the negative noise impacts as possible from the proposed project. The rationale for including each mitigation-plan component should be described, along with its potential effectiveness.

SUMMARY

This chapter has presented a six-step methodology for addressing the noise-environment impacts of proposed projects, plans, programs, or policies. These steps provide a general framework which can be used (1) as a guide to EIS planning and conduction; (2) as an indication of where more detailed information will be necessary; (3) to discuss a study with a contractor (or sponsor) and develop appropriate terms of reference; and (4) to review impact study work done by others.

It will be necessary to document the six-step methodology in the resulting noise-environment-impact report, which could be a section or chapter in either an EA or EIS. The report should address the potential noise impacts identified for the project or activity (step 1), the existing characteristics of the noise environment and land uses in the study area (step 2), the applicable noise standards and land-use-compatibility considerations (step 3), the quantification of the noise impacts due to the project or activity (step 4), the assessment of the predicted impacts and its rationale (step 5), and the considered and included mitigation measures (step 6).

This chapter has focused on noise impacts in relation to human health and other effects, and to effects on wildlife. Structural effects on houses, buildings, bridges, and the like, can occur as a result of vibrations generated by major noise sources. For certain types of projects, it may also be necessary to address potential vibrational impacts (Von Gierke et al., 1977); the International Standards Organization (ISO) has developed vibrational standards which can be used in the prediction and assessment process.

SELECTED REFERENCES

Benson, L. J., White, M. J., and Murphy, K. J., "Operational Noise Data for OH-58D Army Helicopters," CERL-N-92/07, U.S. Army Construction Engineering Research Laboratory, Champaign, Ill., Jan. 1992.

Berglund, B., and Berglund, U., "Loudness of Power Line Noise," SNV-3035, National Swedish Environment Protection Board, Solna, Sweden, Mar. 1986.

Bowlby, W., Harris, R. A., and Cohn, L. F., "Seasonal Measurements of Aircraft Noise in a National Park," *Journal of the Air and Waste Management Association,* vol. 40, 1990, pp. 68–76.

Chanlett, E. T., *Environmental Protection,* McGraw-Hill Book Company, New York, 1973, pp. 523–544.

Edison Electric Institute, "Electric Power Plant Environmental Noise Guide," 2 vols., Washington, D.C., 1984.

Farnham, J., and Beimborn, E., "Noise Barrier Design Guidelines," DOT-T-90-15, Center for Urban Transportation Studies, University of Wisconsin—Milwaukee, Milwaukee, July 1990.

Feather, T. D., and Shekell, T. K., "Reducing Environmental Noise Impacts: A USAREUR Noise Management Program Handbook," IWR-91-R-5, Planning and Management Consultants, Carbondale, Ill., June 1991.

Federal Aviation Administration (FAA), "Final Environmental Assessment for Max Westheimer Airpark, Norman, Oklahoma," Oklahoma City, Feb. 1988, pp. 1–47.

Federal Interagency Committee on Urban Noise, "Guidelines for Considering Noise in Land Use Planning and Control," Federal Interagency Committee on Urban Noise, Washington, D.C., June 1980.

Fidell, S., "The Urban Noise Survey," EPA 550/9-70-100, U.S. Environmental Protection Agency, Washington, D.C., Aug. 1977.

Fittipaldi, J. J., et al., "Procedures for Conducting Installation Compatible Use Zone Studies," CERL-TR-N-88/19, U.S. Army Construction Engineering Research Laboratory, Champaign, Ill., Aug. 1988.

Fletcher, J. L., and Busnel, R. G., *Effects of Noise on Wildlife,* Academic Press, New York, 1978.

Gillen, D. W., Levesque, T. J., and Smith, T. N., "Management of Airport Noise: Project Summary and Guidelines for Managers," Transportation Development Center, Montreal, 1990.

Gladwin, D. N., Asherin, D. A., and Manci, K. M., "Effects of Aircraft Noise and Sonic Booms on Fish and Wildlife: Results of a Survey of U.S. Fish and Wildlife Service Endangered Species and Ecological Services Field Offices, Refuges, Hatcheries, and Research Centers," NERC-88/30, National Ecology Research Center, U.S. Fish and Wildlife Service, Fort Collins, Colo., June 1988.

Goff, R. J., and Novak, E. W., "Environmental Noise Impact Analysis for Army Military Activities: User Manual," Tech. Rep. N-30, U.S. Army Construction Engineering Research Laboratory, Champaign, Ill., Nov. 1977.

Gordon, C. G., et al., "Highway Noise—A Design Guide for Highway Engineers," National Cooperative Highway Research Program, Rep. 117, Washington, D.C., 1971.

Haber, J. M., and Nakaki, D., "Noise and Sonic Boom Impact Technology: Sonic Boom Damage to Conventional Structures," BBN-6829, BBN Systems and Technologies Corporation, Canoga Park, Calif., Feb. 1989.

Harrington, F. H., and Veitch, A. M., "Impacts of Low-Level Jet Fighter Training on Caribou Populations in Labrador and Northern Quebec, rev. ed." Newfoundland Wildlife Division, St. John's, Newfoundland, 1990.

Harris, C. S., "Effects of Military Training Route Noise on Human Annoyance," AAMRL-TR-89-041, Harry G. Armstrong Aerospace Medical Research Laboratory, Wright-Patterson AFB, Ohio, Oct. 1989.

Hendriks, R. W., "Traffic Noise Attenuation as a Function of Ground and Vegetation (Interim Report)," REPT-65328-637327, California State Department of Transportation, Sacramento, Sept. 1989.

Keast, D., Eldred, K., and Purdum, J., "Heliport Noise Model (HNM), Version 1, User's Guide," DOT/FAA/EE-88-2, HMM Associates, Concord, Mass., Feb. 1988.

Lang, J., "Assessment of Noise Impact on the Urban Environment," Environmental Health Series No. 9, World Health Organization, Regional Office for Europe, Copenhagen, 1986.

Lee, R. A., "NOISEMAP 6.0—The USAF Microcomputer Program for Airport Noise Analysis," AAMRL-TR-90-084, Harry G. Armstrong Aerospace Medical Research Laboratory, Wright-Patterson AFB, Ohio, May 1990.

——— and Mohlman, H. T., "Air Force Procedure for Predicting Aircraft Noise Around Airbases:

Airbase Operations Program (BASEOPS) Description," AAMRL-TR-90-012, Harry G. Armstrong Aerospace Medical Research Laboratory, Wright-Patterson AFB, Ohio, Jan. 1990.

Lipscomb, D. M. and Taylor, A. C., eds., *Noise Control: Handbook of Principles and Practices,* Van Nostrand Reinhold Company, New York, 1978.

Ljunggren, S., and Johansson, M., "Measures Against Mechanical Noise from Large Wind Turbines: A Design Guide," FFA-TN-1991-26, Aeronautical Research Institute of Sweden, Stockholm, June 1991.

Magrab, E. B., *Environmental Noise Control,* John Wiley and Sons, New York, 1975.

Marsh, W. M., *Landscape Planning: Environmental Applications,* 2d ed., John Wiley and Sons, New York, 1991, p. 322.

May, D. N., *Handbook of Noise Assessment,* Van Nostrand Reinhold Company, New York, 1978.

Miller, R. K., and Thumann, A., *Fundamentals of Noise Control Engineering,* AEE Energy Books, Lilburn, Ga., 1990.

Morey, M. J., "Effects of Aircraft Noise at Williams Air Force Base Auxiliary Field on Residential Property Values," ANL/EAIS/TM-44, Argonne National Laboratory, Argonne, Ill., Nov. 1990.

Nelson, K. E., and Wolsko, T. D., "Transportation Noise: Impacts and Analysis Techniques," Rep. ANL/ES-27, prepared by Argonne National Laboratory for Illinois Institute for Environmental Quality, Chicago, Oct. 1973.

Organization for Economic Cooperation and Development (OECD), "Transport and the Environment," (OECD), Paris, 1988, pp. 110–115.

Planning and Management Consultants, "A Noise Management Handbook for USAREUR Noise Management Program," prepared for U.S. Army Corps of Engineers, Institute for Water Resources, Ft. Belvoir, Va., Mar. 1990.

Raspet, R., and Bobak, M. J., "Procedures for Estimating the Flat-Weighted Peak Level Produced by Surface and Buried Charges," Tech. Rep. CERL-N-88/07, U.S. Army Construction Engineering Research Laboratory, Champaign, Ill., 1988.

Reddingius, N. H., and Smyth, J. S., "Assessment System for Aircraft Noise (ASAN): Development of Alpha-Test Prototype System Software," HSD-TR-90-005, BBN Systems and Technologies Corporation, Canoga Park, Calif., Feb. 1990.

Richardson, W. J., Greene, C. R., Malme, C. I., and Thomson, D. H., "Effects of Noise on Marine Mammals," LGL-TA-834-1, LGL Ecological Research Associates, Bryan, Tex., Feb. 1991.

Schomer, P. D., et al., "Blast Noise for Prediction, Volume I: Data Bases and Computational Procedures," U.S. Army Construction Engineering Research Laboratory, Champaign, Ill., Mar. 1981.

U.S. Army Construction Engineering Research Laboratory, "Environmental Review Guide for USAREUR," vol. 2, U.S. Army Construction Engineering Research Laboratory, Champaign, Ill., 1989.

U.S. Department of the Army, "Environmental Protection and Enhancement," AR 200-1, U.S. Department of the Army, Washington, D.C., Apr. 1990, pp. 42–44, 72.

U.S. Department of Housing and Urban Development, "The Noise Guidebook," U.S. Department of Housing and Urban Development, Washington, D.C., 1985.

U.S. Department of Transportation, "Guidelines for the Sound Insulation of Residences Exposed to Aircraft Operations," DOT/FAA/PP-92-5, Federal Aviation Administration, Washington, D.C., Oct. 1992, pp. 1-1–1-11.

———, "Transportation Noise and Its Control," Publ. DOT P 5630.1, U.S. DOT, Washington, D.C., June 1972.

U.S. Environmental Protection Agency (EPA), "Information on Levels of Environmental Noise Requisite to Protect Public Health and Welfare with an Adequate Margin of Safety," EPA 550/9-74-004, U.S. Environmental Protection Agency, Office of Noise Abatement and Control, Washington, D.C., Mar. 1974.

———, "Noise Abatement Programs," *Code of Federal Regulations,* vol. 40, Subchap. G, Parts 201–205, July 1, 1991, pp. 24–115.

———, "Protective Noise Levels: Condensed Version of EPA Levels Document," EPA 550/9-79-100, U.S. Environmental Protection Agency, Office of Noise Abatement and Control, Washington, D.C., Nov. 1978.

———, "Public Health and Welfare Criteria for Noise," EPA 550/9-73-002, U.S. Environmental Protection Agency, Office of Noise Abatement and Control, Washington, D.C., July 1973.

———, "Report to the President and Congress on Noise," 92nd Congress, 2d Session, Doc. 92-63, Washington, D.C., Feb. 1972.

Von Gierke, H. E., et al., "Guidelines for Preparing Environmental Impact Statements on Noise," National Research Council, Washington, D.C., June 1977.

Wesler, J. E., "Manual for Highway Noise Prediction," Rep. DOT-TSC-FHWA-72-1, prepared by Transportation Systems Center for U.S. Department of Transportation, Washington, D.C., Mar. 1972.

Wyle Laboratories, "Transportation Noise and Noise from Equipment Powered by Internal Combustion Engines," Publ. NTID 300.13, prepared for Office of Noise Abatement and Control, U.S. Environmental Protection Agency, Washington, D.C., Dec. 1971.

CHAPTER 10

Prediction and Assessment of Impacts on the Biological Environment

Prediction and assessment of impacts on the biological environment entail a number of technical and professional considerations related to both the predictive aspects and the interpretation of the significance of anticipated changes. Impact prediction and assessment for the biological environment has also been called "ecological impact assessment" (Westman, 1985). The purpose of this chapter is to summarize information related to biological-impact prediction and assessment using alternatives to structured habitat-based approaches. The habitat-based approaches refer to the habitat evaluation system (HES) developed by the U.S. Army Corps of Engineers, and the habitat evaluation procedure (HEP) developed by the U.S. Fish and Wildlife Service (FWS). HES and HEP are representative of a large number of habitat-based methods and will be discussed in this chapter within the context of their applicability to impact prediction and assessment. Both HES and HEP are described in detail in Chapter 11.

Many projects (and activities) can cause undesirable impacts on terrestrial and/or aquatic ecosystems. Examples of such impacts include habitat degradation through overgrazing prac-tices; wetland drainage for agricultural, industrial, or urban development projects; habitat loss, with attendant consequences on fish and wildlife because of excessive deforestation practices; changes in habitat and associated fish and wildlife species due to the construction and operation of hydropower projects; loss of critical habitat for endangered or threatened species as a result of timber harvesting, recreational developments, and/or military training activities; multiple aquatic and terrestrial ecosystem effects from acid rain formed as a consequence of sulfur dioxide emissions from coal-fired power plants; and potential toxic effects to plants and/or animals as a result of air- or water-pollutant discharges or of waste-disposal activities of industries and municipalities.

This chapter is structured around a six-step methodology for biological-impact prediction and assessment. These six steps are analogous to the steps described for the air, surface-water, soil and groundwater, and noise environments in Chapters 6 through 9, respectively. The chapter begins with a summary of fundamental biological concepts useful in impact studies; this is followed by information on key federal laws related to the biological environment and a discussion of the six-step methodology.

BASIC INFORMATION ON BIOLOGICAL SYSTEMS

The following are important fundamental terms and concepts in biological impact studies (after Franck and Brownstone, 1992, pp. 34, 36, 48, 98–99, 146).

Biogeochemical cycles (nutrient cycles) "Biogeochemical cycles" are the series of biochemical pathways by which the earth's inorganic elements (1) are made available for use by living organisms, (2) find their way into the food chain, and (3) are later broken down to begin the cycle again. These cycles tie together the "biosphere" (the global system consisting of the totality of life) and its interaction with the nonliving environment (the geosphere). Fundamental to most of the cycles—and the most important set of biochemical reactions in the biosphere—is photosynthesis. Within the total biosphere, as within each ecosystem, organisms are categorized by their role in these cycles—that is, by how they get their energy, as follows:

- *Primary producers*—"Primary producers" are organisms, mostly green plants, that draw on the sun's energy to make the fuel for others to use.
- *Consumers*—"Consumers" are those beings who eat the food produced by plants, starting with plant-eating (herbivorous) organisms and extending into a chain of larger, animal-eating (carnivorous) organisms.
- *Decomposers*—"Decomposers" are microorganisms that break down the remains of dead plants and animals for eventual recycling within the biosphere.

Biological diversity The term "biological diversity" refers to the variety and variability of living organisms and the biological communities in which they live. Biological diversity, or "biodiversity," exists at several levels. For example, "ecosystem diversity" refers to the different types of landscapes that act as home to living organisms. "Species diversity" refers to the different types of species in an ecosystem. "Genetic diversity" refers to the range of characteristics coded in the DNA carried in the genes of the plants and animals of a species. The greater the variety of the gene pool, the more readily the species can adapt to environmental change.

Carrying capacity "Carrying capacity" is the total number of plants and animals that can be supported by a particular ecosystem, without reducing the environment's long-term ability to sustain life at the desired level and quality. It varies with the type of soil and its inherent productivity, the climate, and the usable products that grow well in that ecosystem, as well as—in the case of cultivated land—the methods used to produce them.

Ecosystem An "ecosystem" is a stable, interacting gathering of living organisms in their nonliving environment, which is unified by a circular flow of energy and nutrients. "Ecosystem" is a broad term, whose meaning imposes no size limitation, which may refer to something as small as the life in and around a fallen log to that of a huge lake or forest, even to that of the whole biosphere, depending on the focus of the observer. Each ecosystem is bound together by the biogeochemical cycles through which living organisms use energy from the sun to obtain, or "fix," nonliving (inorganic) elements (such as carbon, oxygen, and hydrogen) from the environment, and transform them into vital food, which is then used and recycled. Ecosystems are neither permanent nor unchanging. The number of organisms in a mature ecosystem, as well as their rate of growth and "lifestyle," depends on the availability of energy and key chemical elements, some of which may be in short supply and, therefore, be limiting factors, such as nitrogen. Ecosystems do not spring full-blown, but develop in stages (ecological succession); these stages vary in terms of altitude, climate, terrain, and mix of plants and animals, so wide diversity exists between a forest ecosystem and a desert ecosystem, between different types of forest ecosystems, and between ecosystems in various regions of the world.

Edge effect "Edge effect" is the influence of transitional zones on the plant communities on either side of it, such as would be exhibited by a hedge or shrubby border between forest and pasture; also, the attraction of such an area for animals.

Habitat A "habitat" is the natural home of an individual or species of plant or animal; also, it is the necessary combination of food, water, cover, and other resources the individual or species requires for life. Among these resources are other living organisms, some of them food sources, as well as conditions such as climate, water, soil, and vegetative cover, available in the right amounts and locations to be handy for feeding, escape, and breeding purposes. Each species is adapted to certain types and amounts of resources, and if these resources are not available many members of the species may be weakened and die. Habitats are classified in various ways, primarily on the basis of the dominant vegetation and its associated environmental conditions; so a habitat in the Pacific Northwest might be described as "old-growth, temperate, coniferous forest." Within the world's several general types of biological regions—"biomes," or major life zones, such as forest or grasslands—are found many varied habitats.

The biogeochemical cycle of nitrogen is shown in Figure 10.1; the environmental cycling of mercury is shown in Figure 10.2. Information on the biogeochemical cycles of carbon, oxygen, and sulfur is available in White, Mottershead, and Harrison (1984). The processes depicted in these figures indicate that the biological environment is a dynamic system which can be stressed as a result of various projects or activities. The anticipated effects of initial changes in certain portions of a biogeochemical cycle should be traced to other portions of the cycle and considered, in their entirety, in the EIA process.

An estuarine food web is shown in Figure 10.3. Three main trophic levels are depicted: T_1 = largely microscopic plants; T_2 = herbivores, a high proportion of which are saprophytic (detritus-feeding) invertebrates; and $T_{3/4}$ = carnivorous birds and fish. Energy-flow information is also available in Tivy and O'Hare (1981) and Watts and Loucks (1969). Food-web (or food-chain) relationships and energy-flow characteristics are also important indicators of the dynamic aspects of the biological environment. They can also be used in the development of qualitative-quantitative models of aquatic or terrestrial systems, with such models being useful for impact prediction. Information on model building for the biological environment is available in Starfield and Bleloch (1986).

The most-sophisticated biological models involve energy-system diagrams. An energy-system diagram for a marine community with disturbance associated with development activities is depicted in Figure 10.4. Such diagrams can be useful in the EIA process for impact identification and quantification; however, they are data- and personnel-intensive.

Another fundamental concept relevant to the EIA process is "ecological succession." "Succession" refers to normal biological changes over time that lead to alterations in community (or habitat) types and in species found within community types. Succession progresses through several seral stages until a mature, stable, and relatively unchanging stage, called the "climax stage," is reached. The following trends are characteristic of most successional processes (Goudie, 1984, p. 201):

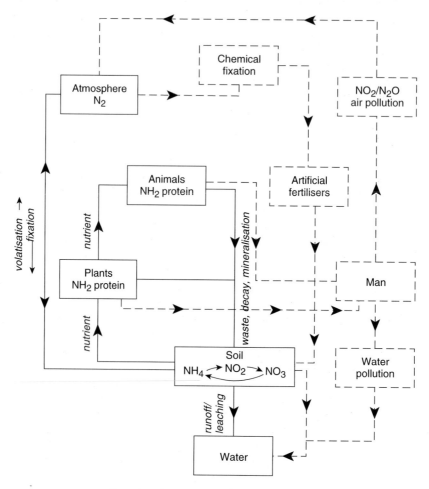

Dotted lines denote human intervention.

FIGURE 10.1
The Biogeochemical Cycle for Nitrogen (Drew, 1983, p. 15).

1. There is a progressive development of the soil, with increasing depth, increasing organic content and the differentiation of soil horizons towards the mature soil of the climax community.
2. The height of plants increases, and strata become more evident.
3. Productivity increases, as does biomass.
4. Species diversity increases from the simple communities of early succession to the richer communities of late succession.
5. As height and density of above-ground plant cover increases, the micro-climate within the community is increasingly determined by characteristics of the community itself.
6. Populations of different species rise and fall and replace one another, and the rate of this replacement tends to slow through the course of succession as smaller and shorter-lived species are replaced by larger and longer-lived ones.

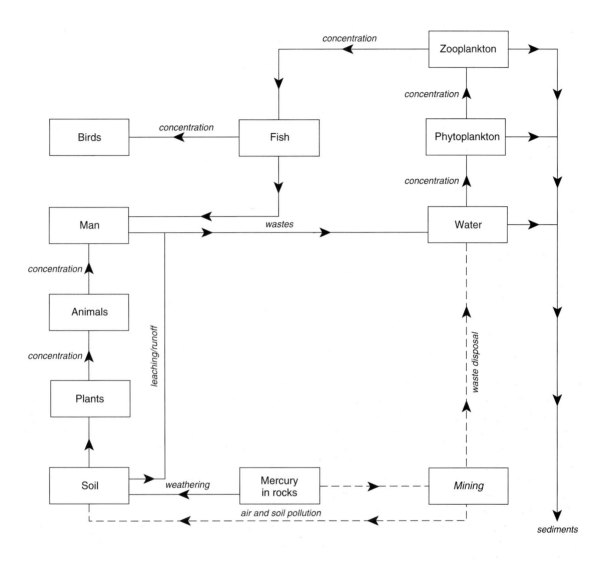

Special reference is given to biological concentration in the aquatic environment. Areas of human intervention are shown by broken lines.

FIGURE 10.2
The Biogeochemical Cycle for Mercury (Drew, 1983, p. 17).

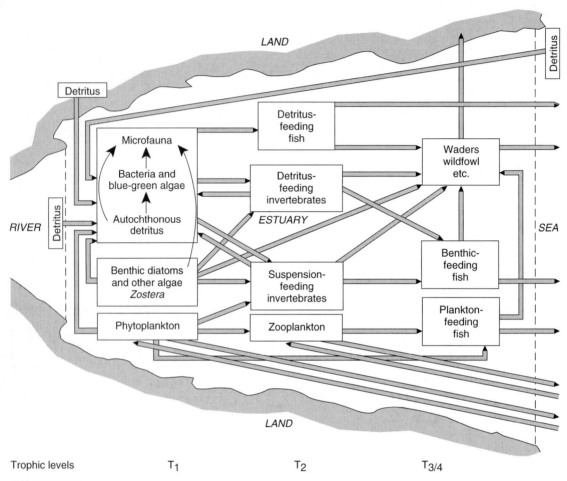

FIGURE 10.3
Estuarine-Food-Web Relationships (Tivy and O'Hare, 1981, p. 16).

7. The final community is usually more stable than the earlier communities, and there is a very tight cycling of nutrients.

KEY FEDERAL LEGISLATION

There are numerous federal laws related directly or indirectly to the biological environment. Examples include the Fish and Wildlife Coordination Act of 1966 for water resources projects, the Coastal Zone Management Act (CZMA) of 1972, the Coastal Zone Management and Im-

provement Act of 1990, and the Surface Mining Control and Reclamation Act (SMCRA) of 1977. The law which may have the greatest overall influence on the EIA process is the Endangered Species Act Amendments of 1978.

Endangered Species Act Amendments of 1978

The purposes of the Endangered Species Act (ESA) Amendments of 1978 (P.L. 95-632) are (1) to provide a means whereby the ecosystems upon which endangered species and threatened

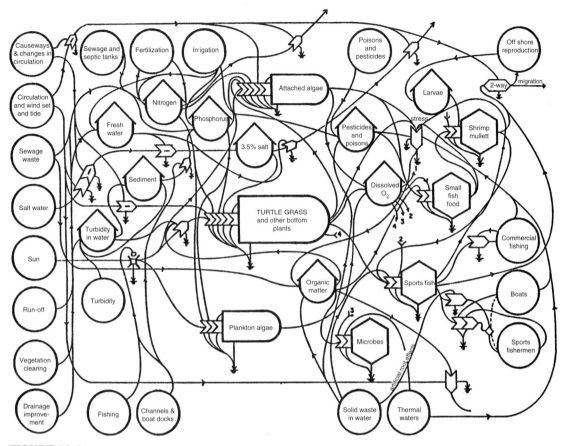

FIGURE 10.4
Energy-System Diagram of a Marine Meadows Community with Disturbances Imposed by Estuarine Development (Odum, 1972, p. 209).

species depend may be conserved, (2) to provide a program for the conservation of such endangered species and threatened species, and (3) to direct and empower appropriate agencies to take such steps as may be necessary to achieve the purposes of relevant treaties and conventions (U.S. Congress, 1978). "Endangered species" means any species which is in danger of extinction throughout all or a significant portion of its range. "Threatened species" means any species which is likely to become an endangered species within the foreseeable future throughout all or a significant portion of its range. To "conserve"

and "conservation" mean to use and the use of all methods and procedures which are necessary to bring any endangered species or threatened species to the point at which the measures provided pursuant to the ESA are no longer necessary. Such methods and procedures include, but are not limited to, all activities associated with scientific resources management such as research, census, law enforcement, habitat acquisition and maintenance, propagation, live trapping, and transplantation, and, in the extraordinary case where population pressures within a given ecosystem cannot be otherwise relieved, may include regulated taking (U.S. Congress, 1978).

The factors to be considered in determining whether any species is endangered or threatened include (U.S. Congress, 1978) (1) the present or threatened destruction, modification, or curtailment of its habitat or range; (2) overutilization for commercial, sporting, scientific, or educational purposes; (3) disease or predation; (4) the inadequacy of existing regulatory mechanisms; or (5) other natural or man-made factors affecting its continued existence.

Critical habitat may be the determining factor in species listing in the *Federal Register.* "Critical habitat" for a threatened or endangered species means the specific areas within the species range (or geographical area occupied by the species) on which are found those physical or biological features essential to the conservation of the species and which may require special management considerations or protection, and specific areas outside the range which are essential for the conservation of the species (U.S. Congress, 1978).

Endangered or threatened species lists are periodically published in the *Federal Register.* The lists are often revised by additions, deletions, or classification changes; a comprehensive review of the lists is called for on a five-year cycle. Each list refers to each species contained therein by its scientific and common name or names, if any, and must specify, with respect to each such species, over what portion of its range it is endangered or threatened, along with any critical habitat within that portion of the range (U.S. Congress, 1978).

An ESA "Section 7" consultation requires that the agency with jurisdiction over an endangered species, either the Fish and Wildlife Service (FWS) of the U.S. Department of the Interior, or the National Marine Fisheries Service (NMFS) of the U.S. Department of Commerce, must be consulted by the action agency to determine whether there is any likelihood of jeopardizing the continued existence of any endan-

gered species or threatened species or of effecting the destruction or adverse modification of the critical habitat of the given species (U.S. Congress, 1978). Details on the Section 7 process are in *U.S. Code of Federal Regulations,* Vol. 50, Sec. 230 (50 C.F.R. 230).

To facilitate compliance with the Section 7 process, the action agency, or its contractor, must conduct a biological assessment, which is separate from the EIS, for the purpose of identifying any endangered species or threatened species which is likely to be affected by such action. Typical components addressed in a biological assessment report include (1) a description of the proposed action; (2) a general description of the environmental setting; (3) biological resources in the setting; (4) any endangered and/or threatened species of concern (with reference to the federal list and state programs), any critical habitat, and supporting scientific details; (5) impacts of the proposed action on endangered or threatened species and/or critical habitat; (6) mitigation measures for impacts on endangered species, threatened species, and/or critical habitat; (7) recommendations; and (8) cited literature. The biological assessment report could be included as an appendix to an EA or EIS, and sections of it could be incorporated, as appropriate, in the EA or EIS.

The Section 7 consultation procedure is graphically depicted in Figure 10.5. Over 16,000 informal consultations and over 2,000 formal consultations were held in the five-year period from 1987 through 1991, with the following breakdown of involvement of the FWS and the NMFS: informal—FWS 96 percent and NMFS 4 percent; formal—FWS 88 percent and NMFS 12 percent. In the 2,050 formal consultations, almost 90 percent of the biological opinions by the FWS-NMFS concluded that the proposed action would not jeopardize the species. Of the total of 181 cases in which jeopardy to species was determined, 158 (or 87 percent) included the identification of reasonable and pru-

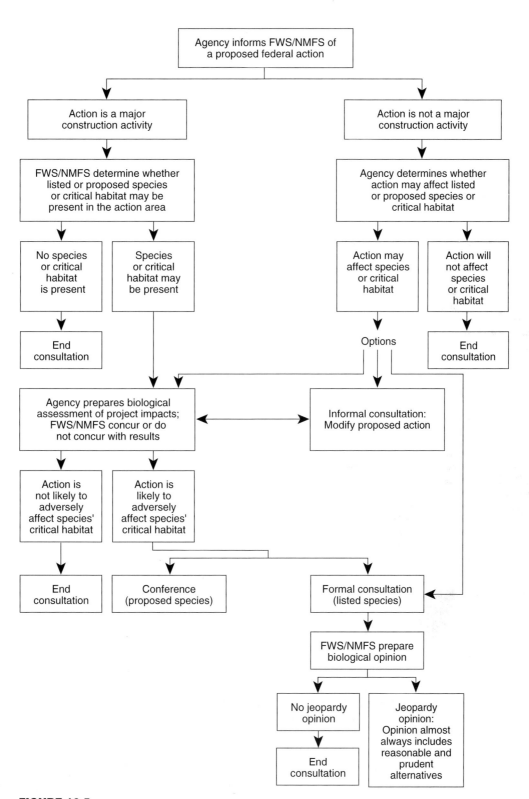

FIGURE 10.5

The Consultation Process Under Section 7 of the ESA (U.S. General Accounting Office, 1992, p. 16).

dent alternatives to the proposed action (U.S. General Accounting Office, 1992).

Detailed legal information related to the substance and procedures of the ESA is in Littell (1992).

The ESA Amendments of 1978 included an exemption procedure providing for the establishment of an endangered species committee and a review board, and the delineation of criteria for determining whether an exemption to the requirements of ESA would be justified. The Endangered Species Committee, popularly referred to as the "God Squad," consists of seven members, as a minimum: the secretaries of Agriculture, the Army, and the Interior; the administrators of the Environmental Protection Agency and the National Oceanic and Atmospheric Administration; the chairman of the Council of Economic Advisors; and one individual from each affected state, appointed by the President. The review board consists of a small group of practicing professionals appointed to review each exemption application from technical, policy, and legal perspectives. Exemptions to the requirements of the ESA can be granted by the Endangered Species Committee if (U.S. Congress, 1978) (1) it determines on the record, based on the report of the review board and on such other testimony or evidence, that there are no reasonable and prudent alternatives to the agency action; that the benefits of such action clearly outweigh the benefits of alternative courses of action consistent with conserving the species or its critical habitat, and that such action is in the public interest; and that the action is of regional or national significance; and (2) it can devise and establish reasonable mitigation and enhancement measures (including, but not limited to, live propagation, transplantation, and habitat acquisition and improvement) as necessary and appropriate, to minimize the adverse effects of the agency action upon the endangered species, threatened species, or critical habitat concerned.

Laws Related to Wetlands

A classification scheme of key federal laws related to wetlands is depicted in Figure 10.6. In addition, many states have enacted laws that are related to wetlands. More information on pertinent federal laws is in step 2 in this chapter.

Two additional federal laws related to wetlands are not listed in Figure 10.6; these are the North American Wetlands Conservation Act of 1990 and the Coastal Barrier Improvement Act of 1990. Under the North American Wetlands Conservation Act of 1990, the federal government was authorized to purchase wetlands in support of the U.S.–Canadian North American Waterfowl Management Plan (Hammond, 1991). The Coastal Barrier Improvement Act of 1990 limits federal financial assistance to development within listed coastal areas. Over 1.25 million acres (1,211 shoreline mi) are now included in the "Coastal Barrier Resource System," which comprehends coastline along the Great Lakes, the Gulf of Mexico, and the Atlantic Ocean, for examples (Hammond, 1991).

Other Related Legislation

Additional federal statutes protecting or regulating wildlife include the Lacey Act Amendments of 1988—related to illegal trafficking of fish, wildlife, or plants; the Marine Mammal Protection Act Amendments of 1984; the Bald Eagle and Golden Eagle Protection Act Amendments of 1972; the Migratory Bird Treaty Act Amendments of 1986; the Wild Free-Roaming Horses and Burros Act of 1978; and the Federal Power Act of 1986 (Littell, 1992). The Federal Insecticide, Fungicide, and Rodenticide Act of 1982, which relates to chemical applications, and the Toxic Substances Control Act of 1976, which is associated with releases of various chemicals into different environmental media, also fall into this category.

The relationship between mineral extraction laws and natural resources management are not discussed herein. This issue is addressed in

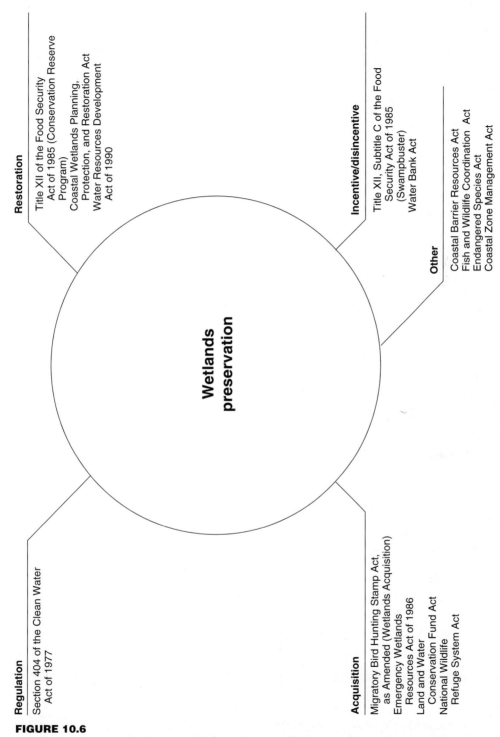

Regulation

Section 404 of the Clean Water Act of 1977

Restoration

Title XII of the Food Security Act of 1985 (Conservation Reserve Program)
Coastal Wetlands Planning, Protection, and Restoration Act
Water Resources Development Act of 1990

Acquisition

Migratory Bird Hunting Stamp Act, as Amended (Wetlands Acquisition)
Emergency Wetlands Resources Act of 1986
Land and Water Conservation Fund Act
National Wildlife Refuge System Act

Wetlands preservation

Incentive/disincentive

Title XII, Subtitle C of the Food Security Act of 1985 (Swampbuster)
Water Bank Act

Other

Coastal Barrier Resources Act
Fish and Wildlife Coordination Act
Endangered Species Act
Coastal Zone Management Act

FIGURE 10.6
Classification of Certain Major Federal Wetlands-Related Legislation by Primary Purpose (U.S. General Accounting Office, 1992, p. 20).

Government Institutes (1991), and several laws related to national forests and forestry, the national park system, the national wildlife refuge system, the wilderness preservation system, the national trails system, and the marine sanctuaries system are summarized.

CONCEPTUAL APPROACH FOR ADDRESSING BIOLOGICAL IMPACTS

To provide a basis for addressing biological environment impacts, a six-step or six-activity model is suggested for the planning and conduction of impact studies. This model is flexible and can be adapted to various project types by modification, as needed, to enable the addressing of specific concerns of specific projects in unique locations. The identified steps are typical of impact studies related to the biological environment. It should be noted that the focus in this model will be on projects and their biological system impacts; however, the model could also be applied to plans, programs, policies, and regulatory and permit actions.

The six generic steps associated with biological environment impacts are (1) identification of the potential biological impacts of the construction and/or operation of the proposed project or activity, including habitat changes or loss, chemical cycling and toxic events, and disruptions to ecological succession; (2) description of the environmental setting in terms of habitat types, selected floral and faunal species, management practices, endangered or threatened species, and special features (such as wetlands); (3) procurement of relevant laws, regulations, or criteria related to biological resources and protection of habitat or species; (4) conduction of impact prediction activities, including the use of analogies (case studies), physical modeling, and/or mathematical modeling, and based on professional judgment; (5) use of pertinent information from step 3, along with professional judgment and public input, to assess the significance of anticipated beneficial and detrimental impacts; and (6) identification, development, and incorporation of appropriate mitigation measures for the adverse impacts. Figure 10.7 delineates the relationship between the six steps or activities in the suggested conceptual approach.

The six-step model can be used to plan a study focused on biological environment impacts, to develop the scope of work for such a study, and/or to review biological-impact information in EAs or EISs.

FIGURE 10.7
Conceptual Approach for Study Focused on Biological Environment Impacts.

Step 1:   Identification of biological impacts of proposed project /activity

▼

Step 2:   Preparation of description of existing biological conditions and consideration of endangered or threatened species and critical habitat

▼

Step 3:   Procurement of relevant laws, regulations, or criteria related to impacts and/or conditions

▼

Step 4:   Impact prediction

▼

Step 5:   Assessment of impact significance

▼

Step 6:   Identification and incorporation of mitigation measures

A list of questions which can be useful in planning a study to address the biological impacts of a proposed project or activity is available in Westman (1985).

STEP 1: IDENTIFICATION OF BIOLOGICAL IMPACTS

The first step is to qualitatively identify the potential impacts of the proposed project (or activity) on biological resources, including habitats and species. For example, many projects can cause terrestrial habitat loss. Southerland (1992) suggested that the loss and degradation of terrestrial environments could be classified into eight causal categories: (1) land conversion to industrial and residential land use, (2) land conversion to agricultural use, (3) land conversion to transportation use, (4) timbering practices, (5) grazing practices, (6) mining practices, (7) water management practices, and (8) military, recreational, and other activities. These causal activities contribute to the degradation and loss of ecological values, including animal and plant species; ecosystem structure (abundance-biomass, community composition, species richness, species diversity, trophic organization, and spatial structure); and ecosystem function (energy flow, nutrient cycling, and water retention) (Southerland, 1992).

Comprehensive literature reviews on specific types of projects or activities can be useful for impact identification. For example, Darnell (1976) addressed the various potential biological impacts of construction activities in wetland areas. The conduction of computer-based literature reviews may be desirable. Case studies of similar types of projects can also aid in impact identification.

Impact-identification methods such as interaction matrices, networks, or simple and descriptive checklists can provide a systematic basis for qualitatively delineating potential impacts of concern. Chapter 3 includes information on, and examples of, these methods.

A discussion of the biological impacts from coal-fired power plants will be used to illustrate impact listing. The general adverse biological impacts from coal-fired power plants include (U.S. Fish and Wildlife Service, 1978) (1) elimination or physical disturbance of terrestrial and/or aquatic habitats and (2) additions to air, water, and soil of substances that have the potential for acute or sublethal effects on biota. Specific features of power plants and their associated impacts include (U.S. Fish and Wildlife Service, 1978, pp. 1–2):

1. *Coal slurry pipelines*—disturbance of habitat during right-of-way construction; removal of habitat for pumping stations and storage ponds; water use; accidental coal spills;
2. *Coal cleaning and coal storage*—loss of terrestrial habitat to coal storage and waste disposal; additions of acidic substances, solids, and trace elements to surface waters; difficulty in reclamation of inactive coal waste-disposal sites due to the chemical nature of the waste;
3. *Limestone storage* (applicable to power plants with flue-gas desulfurization)—loss of terrestrial habitat; additions of alkaline runoff to soils and surface waters;
4. Particulate and gaseous emissions—additions of sulfur oxides and other gases to air; additions of particulates and trace elements to air, soil, water, and vegetation; acid precipitation on soil, vegetation, and surface waters; and
5. Waste ash and flue-gas desulfurization sludge disposal—loss of terrestrial habitat; seepage of potentially toxic material to soil, ground water, and surface waters; possible toxification of water birds attracted to disposal ponds; difficulty in reclamation of disposal sites due to the physical and chemical nature of the wastes.

The study of the fate and effect of toxic agents in ecosystems has been termed ''ecotoxicology.'' Such studies may be relevant in the EIA process if the proposed project-activity has planned or potential releases of toxic agents to the atmosphere or to terrestrial or aquatic ecosystems (Bascietto et al., 1990; Cairns and Mount, 1990; Hoffman, Rattner, and Hall, 1990).

STEP 2: DESCRIPTION OF EXISTING BIOLOGICAL-ENVIRONMENT CONDITIONS

The second step in the methodology entails the preparation of a description of the flora and fauna—and other natural resources and habitats—constituting the biological-environment setting. This description should primarily focus on community types (habitat types) and their geographical distribution. It may be desirable to identify certain selected species and to include species descriptions for those selected species for each community type. There are several options for achieving this step, and four will be described: (1) use of species lists with qualitative descriptions, (2) use of structured data presentations with qualitative-quantitative descriptions, (3) use of HES, HEP, or other habitat-based methods, and (4) use of energy system diagrams.

In the early years following the passage of the NEPA, the most prevalent approach for describing the biological environmental setting consisted of preparing species lists of flora and fauna anticipated to be present in the study area, with some brief qualitative information included on the general characteristics of the area ecosystem. "Species lists" typically provide both the scientific and the common names of the flora and fauna within the study area. In more recent years, there has been a trend away from the simple reporting of species lists to using a list process that includes more-pertinent information relative to the individual biological components within the system. For example, the primary categories of flora and fauna utilized in a waterway navigation study are as follows (Canter, Risser, and Hill, 1974):

Floral Components

General vegetation patterns of entire area

Plant species for bluestem prairie

Tree species in upland forests

Tree species in lowland forests

Shrubs and vines of post-oak–blackjack forest

Shrubs and vines of floodplain forest

Herbaceous plants of post-oak–blackjack forest

Herbaceous plants of floodplain forest

Rare plant species in entire area

Faunal Components of Entire Area

Amphibians—Frogs, toads, salamanders

Reptiles—Turtles, lizards, snakes

Naiads (freshwater mussels or freshwater clams)

Fishes

Sport fisheries

Birds

Mammals

Rare faunal species

Many states have developed computerized fish and wildlife information systems (CFWISs); one example is Virginia (Cushwa and Kopf, 1985). Output from a CFWIS can be used for describing the affected environment, predicting the potential impacts of proposed projects (or activities), and assessing the significance of the anticipated changes.

Structured data presentations represent another approach which could be used for describing the biological environmental setting. Typically, structured data presentations provide a context for interpretation of information, as well as providing more detailed information than is found in species lists. For example, the ecoregion concept could be used as the basis for interpretation, where "ecoregions" refer to geographical regions of relative homogeneity in ecological systems or in relationships between organisms and their environments. Key determining factors in this context include land-surface form, soils, land use, and potential natural vegetation. [The average ecoregion in the conterminous United States is 130,000 km^2 (Omernik, 1987). Subregional delineations are available in Omernik and Gallant (1989).] In effect, the ecoregion concept is similar to that of an air quality control region (in terms of air pollution) or a drainage basin (in dealing with the surface-water environment), in that the ecoregion represents a geographical area of similar, relevant biological characteristics. Ecoregion approaches can also be used in describing existing conditions in the case of surface-water systems in such a way as to establish natural water quality and biological communities, and to provide a basis for chemical and biological water quality goals (Hughes and Larsen, 1988).

The ecoregion concept was developed by the U.S. Forest Service and represents the basis for a valid approach for identifying the general biological systems at a given project's location (Bailey, 1980). This approach could also be used to obtain more specific data, such as on the typical characteristics of the ecoregion itself—for examples land-surface form, climate, vegetation, soils, and fauna. The ecoregion concept

is endorsed and utilized by several federal agencies, including the Bureau of Land Management, Fish and Wildlife Service, Geological Survey, Soil Conservation Service, and Environmental Protection Agency. To a degree, although not exactly in the same sense, the term ''ecoregion'' may be considered a broader term to denote the general community (habitat) types present within a geographical study area.

A second approach which could be used for structured data presentations involves the assemblage of specific information on indicator floral and faunal species within the study area. For example, Table 10.1 provides a partial list of amphibians in the mid-Arkansas River Basin region, presented in a structured format. This listing includes both the scientific and the common names of the amphibians within the study area, as well as information on their occurrence within two alternative navigation routes being considered for this project. A classification code is used to denote the type of occurrence, ranging from C, indicating those amphibians which are

TABLE 10.1

PARTIAL LIST OF AMPHIBIANS OF THE MID-ARKANSAS RIVER BASIN REGION

Amphibian	Occurrence[a,b]		Habitat[c,d]	Comments
Scaphiopus hurteri (Hurter's spadefoot)	C	C	T(B)	Inhabits woodlands and grasslands
Bufo americanus (American toad)	C	C	T(B)	Inhabits moist woodlands
Bufo woodhousei (Rocky Mountain toad)	O	O	T(B-R)	Frequents moist woodlands
Acris crepitans (Blanchards cricket frog)	C	C	PR	
Rana catasbeiana (bullfrog)	C	C	PR	
Rana aerolata (northern crayfish frog)		R	T(B)	Inhabits lowland thickets, waterways; associated with crayfish burrows
Rana clamitans (green frog)	O	O	P, R, B	
Rana pipiens (leopard frog)	C	C	P, R, B	
Gastrophryne olivacea (Great Plains narrow-mouthed toad)	O		T(B)	Mainly inhabits subterranean sites

[a]The two listings under Occurrence represent two alternative routes.
[b]C = common, occurring in many localities in large numbers; O = occasional, occurs in several localities in small numbers; R = rare, highly localized, restricted by scarcity of habitat or low numbers.
[c]T = terrestrial, land areas not associated with water; R = running water, streams and springs; P = permanent, stationary bodies of water; B = temporary bodies of water.
[d]Breeding habitats are listed in parentheses.
Source: After Canter, Risser, and Hill, 1974.

commonly occurring, to an R, indicating those which are very rarely found within the study area. In addition, Table 10.1 includes information on the habitat requirements of the individual amphibians for both their normal life cycle and their breeding cycle. Finally, additional comments are included on some of the amphibians. It would not necessarily be required that every amphibian within the study area be listed, but certainly, with this approach, those that would be representative and perhaps indicative of resources of concern should be identified. This type of information can be organized for a number of different categories of floral and faunal species within a study area.

Detailed information on plant and animal indicators, community indicators, and pollution indicators is available in Spellerberg (1991). For example, plants can be used as indicators of environmental conditions and the impacts of human's activities. To serve as an illustration, Table 10.2 lists a number of site conditions in different bioclimatic regions that can be interpreted from vegetation.

To serve as another illustration of a structured data presentation, the vegetation in a study area can be classified according to several schemes such as (1) the floristic (or linnaean), which classifies individual plants according to species, genera, families, and so on, using the universally recognized system of botanical names; (2) the form and structure (or physiognomic) schemes, which classify vegetation or large assemblages of plants according to overall form (for example, forest and grassland) with special attention to dominant plants (largest and/or most abundant); and (3) various ecological schemes, which classify plants according to their habitat (for example, sand dunes, wetlands, lake shores) or some critical parameter of the environment, such as soil moisture or seasonal air temperatures (Marsh, 1991). Table 10.3 depicts a five-level scheme, with each level addressing a different classification element. Level I is based on overall structure; level II on dom-

inant plant types; level III on plant size and density; level IV on site, habitat, or associated use, and level V on significant species. Level V is included to provide for rare, endangered, protected, and highly valued species.

Another approach to a structured data presentation of flora and fauna is to include food-web relationships for the individual ecosystem of concern in the study area. Food-web relationships represent an attempt to show the interdependence of various floral and faunal components within the ecosystem upon each other (U.S. Army Corps of Engineers, 1980). This system presentation can aid in understanding that changes in one aspect of the biological environment will often lead to changes in other features of the biological system. (An illustration of a food-web relationship is shown in Figure 10.3.)

Species diversity indices or other biological indices (as described in Chapter 5) could also be used in step 2. Included in this group are various types of ecological-sensitivity ratings for aquatic or terrestrial ecosystems; such ratings often focus on system resiliency or sensitivity to various perturbations. Productivity data for both aquatic and terrestrial habitats can be summarized as a part of step 2. Finally, photosynthesis and respiration ratios could be used in a presentation of data regarding streams, lakes, and estuarine systems.

The final example approach to a structured data presentation involves the assemblage of information based on the habitat concepts, such information could be found in either HES or HEP, for example. However, rather than going into the details of the numerical system (a technique described in Chapter 11), usage in this context would involve including only qualitative information on habitat characteristics and inferring from that whether or not the habitat was of good quality in terms of its ability to support appropriate floral and faunal species. This qualitative habitat approach would be feasible in many cases and represents a logical

TABLE 10.2

VEGETATION INDICATORS OF ENVIRONMENTAL CONDITIONS

Climatic region	Absence of plant cover	Sparse herb and shrub cover	Thick herb and shrub cover	Brush and small trees	Blade and reed plants	Highly localized tree cover
Humid (Eastern North America, Pacific Northwest, South)	Bedrock at or very near surface; Active dunes; Recent human use, cultivation, etc.; Recent fire; Recent loss of water cover	Bedrock near surface; Recent or sterile soils—dunes, fill; Recently disturbed (fallow, fire, flood); Active slopes/erosion	Recently logged or burned; Too wet for trees; Managed grazing; Organic soil; Old field regrowth	Landslide/fire, flashflood scars; Old field or woodlot regrowth; Shale/clay substrate; Organic soil; Moisture deficiency	Organic soil; Standing water; High ground-watertable; Springs, seepage zones	Wet depression, organic soil, steep; Slopes in agricultural areas; Flood-prone areas
Semiarid (High Plains, S. California)	Caliche or salt pan (playa) at or very near surface; Desert pavement	Localized water sources; Eolian erosion; Overgrazing	Overgrazing; Free from burning; Too dry for trees	Channels of available moisture; Aquiferous substrate	(Same as above)	Aquiferous substrate; Seepage zone or spring; Stream valley (galleria) forest
Arid (Southwest, Great Basin)	Rock surface; Unstable ground such as dunes or rockslides; Too dry			Protected pockets; Favorable (moist) slopes; Logged/burned		Plantation
Arctic and alpine (N. Canada, Alaska, Rockies)	Rock surface; Active slopes; Semipermanent ice or snow cover; Ponded water during growing season	Above tree line; Semipermanent ice or snow cover; Active slopes; Periglacial processes active	Above tree line; Ice, snow, and wind pruning; Mildly active slopes; Wet depressions	Wind/ice pruning; Avalanche, landslide scars, fire; Recent logging near tree line; permafrost near surface	(Same as above)	Protection pockets

Source: Marsh, 1991, p. 273.

TABLE 10.3

FIVE-LEVEL VEGETATION CLASSIFICATION SCHEME

Level I (vegetative structure)	Level II (dominant plant types)	Level III (size and density)	Level IV (site or habitat or associated use)	Level V (special plant species)
Forest (trees with average height greater than 15 ft with at least 60% canopy cover)	E.g., oak, hickory, willow, cottonwood, elm, basswood, maple, beach, ash	Tree size (diameter at breast height) Density (number of average stems per acre)	E.g., upland (i.e., well-drained terrain), floodplain, slope face, woodlot, greenbelt, parkland, residential land	Rare and endangered species; often ground plants associated with certain forest types
Woodland (trees with average height greater than 15 ft with 20–60% canopy cover)	E.g., pine, spruce, balsam fir, hemlock, douglas fir, cedar	Size range (difference between largest and smallest stems)	E.g., upland (i.e., well-drained terrain), floodplain, slope face, woodlot, greenbelt, parkland, residential land	Rare and endangered species; often ground plants associated with certain forest types
Orchard or plantation (same as woodland or forest but with regular spacing)	E.g., apple, peach, cherry, spruce, pine	Tree size; density	E.g., active farmland, abandoned farmland	Species with potential in landscaping for proposed development
Brush (trees and shrubs generally less than 15 ft high with high density of stems, but variable canopy cover)	E.g., sumac, willow, lilac, hawthorn, tag alder, pin, cherry, scrub oak, juniper	Density	E.g., vacant farmland, landfill, disturbed terrain (e.g., former construction site)	Species of significance to landscaping for proposed development

method for describing the biological-environmental setting.

A third option for describing the biological setting is to solely utilize quantitative habitat-based methods, such as HES or HEP. Use of such methods requires considerable information and the development of numerical indices of habitat quality as a part of the evaluation process. The quantitative use of habitat-based methods will probably be limited to larger-scale projects which have significant concerns relative to their anticipated biological impacts. However, simplified versions of habitat-based methods have been utilized, by the U.S. Army Corps of Engineers, for example, for projects such as permit actions. Therefore, habitat-based methods can also form a basis for project evaluation, for a variety of undertakings ranging from fairly

TABLE 10.3

FIVE-LEVEL VEGETATION CLASSIFICATION SCHEME *(continued)*

Level I (vegetative structure)	Level II (dominant plant types)	Level III (size and density)	Level IV (site or habitat or associated use)	Level V (special plant species)
Fencerows (trees and shrubs of mixed forms along borders such as road, fields, yards, playgrounds)	Any trees or shrubs	Tree size; density	E.g., active farmland, road right-of-way, yards, playgrounds	Species of value as animal habitat and utility in screening
Wetland (generally low, dense plant covers in wet areas)	E.g., cattail, tag, alder, cedar, cranberry, reeds	Percent cover	E.g., floodplain, bog, tidal marsh, reservoir backwater, river delta	Species and plant communities of special importance ecologically and hydrologically; rare and endangered species
Grassland (herbs, with grasses dominant)	E.g., big blue stem, bunch grass, dune grass	Percent cover	E.g., prairie, tundra, pasture, vacant farmland	Species and communities of special ecological significance; rare and endangered species
Field (tilled or recently tilled farmland)	E.g., corn, soybeans, wheat; also weeds	Field size	E.g., sloping or flat, ditched and drained, muckland, irrigated	Special and unique crops; exceptional levels of productivity in standard crops

Source: Marsh, 1991, p. 262.

simple projects to very large-scale, complicated projects. This approach is described in Chapter 11.

The final suggested option for addressing the biological environmental setting involves the use of energy-system diagrams. Such diagrams, as briefly noted earlier, are based on accounting for the flows of all energies in the biological system, including the main components of the system—such as plants, animals, chemical processes, and so forth—as well as outside actions that may cause changes. The energy-system-

diagram approach involves the development of mathematical equations to describe energy flow from the sun to primary producers, from primary producers to secondary producers, and so forth. It is necessary to identify all of the processes of energy flow within a system and to develop information on the rates of energy transfer. Typically, successful use of energy-system diagrams requires a high degree of technical sophistication in developing the necessary information, and also requires access to extensive backup resources, such as computing capabilities. These requirements for use of energy-system diagrams basically translate into one concern: namely, the costs for this approach in an EIS are considerably greater than for the other options described.

While the information requirements of energy-system-diagram approaches are intensive, this approach does represent a scientifically valid method of dealing with the biological environment. Energy-system diagrams tend to be comprehensive, and they do provide a common measure—the energy or energy flow within the system—which can be used to quantify impacts. This common measure can be used to describe impacts between environmental factors, as well as between alternatives. In addition, it is possible to convert this common unit of measure to monetary values and thus express environmental impacts in the context of cost. At this point, the environmental impacts could be compared in a manner similar to that used for more-traditional benefit-cost analyses of projects.

In summary, several options for describing the biological environmental setting have been presented. Perhaps some of the better options are those that would be included within the general category of "structured data presentations," since such options attempt to present organized information on the biological environmental setting without becoming cost-prohibitive.

Identification of Management Practices

There is a general misconception that "man has never done anything to influence the biological environment in the past"; another way of saying this is "the biological environment has not been previously managed." This is obviously not the case, since it is difficult—within the United States, at least—to find any biological settings which have not been subject to human's activities at some point in their history. Therefore, it is important to recognize that humankind has had programs and activities related directly or indirectly to the biological environment for many years, and that the proposition of a project in a given area (whatever the project may be) is not likely to represent the first time that some description and/or management plans have been developed for the biological system.

The suggestion herein is to discuss past and current management practices as related to habitats and floral and faunal species, as well as special activities associated with protected species. Some examples of management practices which have been used include spraying for pest control, introduction of new species into an area, controlled burning of forestlands, programs to stock water bodies with fish species, and limitations on hunting seasons in terms of time and the number of animals killed. These approaches are all examples of some aspects of biological environmental management techniques being used at White Sands Missile Range in New Mexico.

Protection of habitat by means of the establishment of parks and nature preserves has been a focal point of conservation efforts in recent years. There are just under 7,000 nationally protected areas in the world, covering some 651 million hectares (ha), 4.9 percent of the earth's land surface, as shown in Table 10.4. The location of protected areas in relation to the proposed project-activity should be delineated in step 2, along with a summary history of protection efforts and requirements.

Therefore, it is desirable to identify areas where management plans are being practiced and to describe those practices, as well as any effectiveness-monitoring programs being conducted. In addition, any past practices that are no longer in operation should be described. Where plans exist for future management practices in an area that would be external but close to the project under evaluation, these should also be appropriately identified. This information would typically be obtained from the nat-

TABLE 10.4

NATIONALLY PROTECTED AREAS, SELECTED COUNTRIES AND WORLDWIDE, 1990

Country	Area of national parks and of equivalent sites[a] (thousand hectares)	Share of total land area (percent)
Venezuela	20,265	22.2
Bhutan	924	19.7
Chile	13,650	18.0
Botswana	10,025	17.2
Panama	1,326	17.2
Czechoslovakia	1,964	15.4
Namibia	10,346	12.6
United States	98,342	10.5
Indonesia	17,800	9.3
Australia	45,654	5.9
Canada	49,452	5.0
Mexico	9,420	4.8
Brazil	20,525	2.4
Madagascar	1,078	1.8
Soviet Union	24,073	1.1
World	651,468	4.9

[a]Includes protected areas over 1,000 hectares in IUCN's Categories 1–5 (Strict Nature Reserve, National Park, Natural Monument/Natural Landmark, Wildlife Sanctuary, and Protected Landscape or Seascape); does not include production-oriented areas such as timber reserves.
Source: Ryan, 1992, p. 15.

ural resources agencies at the state level as well as the federal level.

Discussion of Natural Succession

Another important aspect of describing the biological environment is to recognize the occurrence and import of natural succession. This consideration is needed primarily to counter a misconception that "the biological system is the same now as it has always been, and it will remain the same over time if the project that is being proposed is not implemented."

"Natural succession" refers to normal changes within the biological environment that lead to alteration of community types, and species within community types, over extended periods of time. This phenomena is well-known in the field of biology, and it should be taken into account in project planning and decision making. In effect, a description of the natural successional process is basic to describing what the environment will become without implementation of the proposed action. This effort will involve the usage of professional judgment on the part of biological scientists on the EIS team. Published literature can also provide information on the successional processes associated with a particular community type.

Identification of Endangered or Threatened Species

In the past several hundred years, it is estimated that about 80 species have become extinct in the continental United States and Hawaii; and an

TABLE 10.5

THREATENED AND ENDANGERED U.S. PLANT AND ANIMAL SPECIES, INCLUDING CANDIDATE SPECIES, 1990[a]

Category	Endangered	Threatened	Total	Species with recovery plans
Mammals	53	8	61	29
Birds	74	11	85	69
Reptiles	16	17	33	25
Amphibians	6	5	11	6
Fishes	54	33	87	44
Snails	3	6	9	7
Clams	37	2	39	29
Crustaceans	8	2	10	5
Insects	11	9	20	12
Arachnids	3	0	3	0
Plants	180	60	240	120
Total	445	153	597[b]	351[c]

Total endangered U.S. species	445	(265 animals, 180 plants)
Total threatened U.S. species	153	(93 animals, 60 plants)
Total listed U.S. species	597	(358 animals, 240 plants)
Total U.S. species with designated critical habitats	108	(83 animals, 25 plants)
Total candidate species	3,700	(1,600 animals, 2,100 plants)

[a]Maintained by the U.S. Department of the Interior, Fish and Wildlife Service, and the U.S. Department of Commerce, NOAA National Marine Fisheries Service, in compliance with the Endangered Species Act.
[b]Separate populations of the following species listed both as endangered and threatened are tallied twice: gray wolf, grizzly bear, bald eagle, piping plover, roseate tern, green sea turtle, and olive ridley sea turtle. For purposes of the Endangered Species Act, the term "species" can mean a species, subspecies, or distinct vertebrate population. Several entries also represent entire genera or even families.
[c]Of the 276 approved recovery plans, some cover more than one species, and a few species have separate plans covering different parts of their ranges.
Source: Council on Environmental Quality, 1991, p. 137.

additional 210 species are likely to become extinct (Reid, 1992). Table 10.5 summarizes 1990 statistics on threatened and endangered species in the United States. In addition to approximately 600 listed species, 3,700 candidate species have been identified. The top 20 threatened and endangered animal and plant species as of 1989, in order of federal and state expenditures for recovery, are shown in Table 10.6.

The original terminology referred to "rare and endangered species," and, until the Endangered Species Act of 1973, this was the dominant terminology. With the passage of the Endangered Species Act of 1973 at the federal level, the phrase was changed to "threatened or endangered species." However, many states

have programs that refer to "rare and endangered species." Therefore, to be comprehensive in this topical area, it might be appropriate to use the terms "rare, threatened, or endangered species" in any listing or documents. It should also be noted that species contained on federal lists as well as state lists should be addressed in this step.

Rare, threatened, and endangered species have caused some difficulties in project planning and decision making since the passage of the NEPA; three examples will be cited. A widely known 1970s example was the Furbish lousewart, an endangered plant, and its potential extinction by inundation caused by the Dickey Lincoln Dam, a proposed hydroelectric dam on

TABLE 10.6

TOP 20 THREATENED AND ENDANGERED ANIMAL AND PLANT SPECIES IN THE UNITED STATES, 1989, IN ORDER OF FEDERAL AND STATE EXPENDITURES FOR RECOVERY

Species	Status[a]	Historic U.S. range
Animals[b]		
1. Bald eagle	E, T	43 states
2. Brown or grizzly bears	T	48 states
3. Red-cockaded woodpecker	E	South
4. American peregrine falcon	E	Alaska, West
5. Gray wolf	E, T	47 states
6. Whooping crane	E	Rocky Mountains to North Carolina and South Carolina
7. Southern sea otter	T	Washington, Oregon, and California
8. Florida manatee	E	Southeast
9. Black-footed ferret	E	West
10. Piping plover	E	Great Lakes to Caribbean
11. Kirtland's warbler	E	West Indies
12. Least Bell's vireo	E	California
13. Florida panther	E	Louisiana–Arkansas and East to South Carolina–Florida
14. Puerto Rican parrot	E	Puerto Rico
15. California condor	E	Oregon and California
16. Humpback chub	E	West
17. Mississippi sandhill crane	E	Mississippi
18. Colorado River squawfish	E	West
19. Bonytail chub	E	West
20. Atlantic ridley sea turtle	E	East Coast
Plants		
1. Tumamoc globe-berry	E	Arizona
2. Western prairie fringed orchid	T	Midwest
3. Northern wild monkshood	T	Iowa, Wisconsin, Ohio, and New York
4. Eastern prairie fringed orchid	T	Midwest, Virginia, Pennsylvania, New Jersey, New York, and Maine
5. Prairie bush-clover	T	Iowa, Illinois, Minnesota, and Wisconsin
6. Aleutian shield-fern	E	Alaska
7. Missouri bladderpod	E	Missouri
8. Blowout penstemon	E	Nebraska
9. Arizona cliffrose	E	Arizona
10. Santa Ana River woolly-star	E	California
11. Minnesota trout lily	E	Minnesota
12. Todson's pennyroyal	E	New Mexico
13. Rhizone fleabane	T	New Mexico
14. Texas bitterweed	E	Texas
15. Mesa Verde cactus	T	Colorado and New Mexico
16. Palmate-bracted bird's beak	E	California
17. Large-flowered fiddleneck	E	California
18. Slender-horned spineflower	E	California
19. Knowlton cactus	E	Colorado and New Mexico
20. Swamp pink	T	Georgia to New York

[a]Note: E = endangered, T = threatened.
[b]For the top 20 animal species, recovery expenditures ranged from $750,000 to $3 million per species; for the top 20 plants, the range was from $19,000 to $1 million per species.
Source: Council on Environmental Quality, 1991, pp. 138–139.

the St. John River in Maine. The dam project, as proposed, was not completed.

The snail darter, a small freshwater fish, was apparently unknown until found in the Little Tennessee River in 1973, in an area to be flooded by the Tennessee Valley Authority's Tellico Dam. The species was added to the endangered species list in 1975 (Franck and Brownstone, 1992). Tellico Dam was under construction during the mid-1970s, when it became subject to litigation related to the snail darter. In 1978, the Supreme Court of the United States stopped construction on the dam, citing provisions of the Endangered Species Act of 1973, and the fact that this dam would cause the snail darter to become extinct. The Supreme Court did not comment on the merits of the snail darter as a species; it simply noted that construction of this dam would cause a violation of the act.

Subsequently, in November 1978, the U.S. Congress amended the Endangered Species Act of 1973 to include a procedure which would allow for issuance of exemptions to the requirements of the act pending an analysis of the merits of the individual species and available alternatives. (This procedure was described earlier.) The case of the snail darter was considered relative to an exemption, and it was determined that an exemption was not merited. However, in late 1979, the budget bill of the U.S. Congress was amended to include an exemption for the snail darter. When this amendment was passed, construction was restarted on the Tellico Dam, and it has been completed. However, the snail darter survived. A mid-1970s transplant of the species to a nearby river worked well, and additional snail darter populations were found during the 1980s in several other Tennessee, Georgia, and Alabama streams (Franck and Brownstone, 1992).

The final example is the spotted owl controversy in the Pacific Northwest region of the United States in the early 1990s. The controversy centered on the need of this threatened species for old-growth forest as habitat and the conflict of interest posed by the sale of timber on U.S. Bureau of Land Management lands (The Spotted Owl . . . ," 1992). A de facto compromise solution has apparently been reached in that some old-growth forest is being maintained and some timber sales in selected areas are being allowed.

Some of the difficulties described above have arisen because of the lack of thorough evaluation of the environmental requirements of the species, and of the anticipated impacts of the proposed project on the species. Accordingly, an appropriate procedural approach should consist of the following activities:

1. Determine whether any endangered, threatened, rare, or protected species is within the project or activity boundary and study area. This determination can be made based on contacts with appropriate federal and state resource agencies. Species already included in federal or state lists, as well as proposed and candidate species, should be identified.
2. Assemble pertinent information on any such listed, proposed, and candidate species. Examples of pertinent information include the species breeding and nesting requirements, life-cycle features, and other unique requirements that may be of importance in considering the impacts of alternatives. Another example is to consider the range of the species and whether or not it would be associated with the project study area during construction and/or operation. The U.S. Fish and Wildlife Service has a computerized database called "Endangered Species Information System" (ESIS) for federally listed species. A total of 66 information items are included on threatened or endangered animals and plants (Table 10.7).
3. Review information on why the species is included on the federal or state list, or why it is being proposed or considered as a candidate species.

4. If a listed species has had a critical habitat type identified, then develop information on the location and condition of such habitat in the study area. Similar consideration should be given to proposed and candidate species.

5. Determine whether or not the proposed project or activity will have an effect on each listed, proposed, and candidate plant and animal species, or on each critical habitat in the study area. If negative effects are anticipated, identify and evaluate appropriate mitigation measures (step 6 includes information on this topic).

The above five activities can be conducted as part of the ESA Section 7 consultation process described earlier.

While the emphasis in this subsection has been on federal lists and requirements, it should be recognized that most states have biological survey, natural history, or natural heritage programs which may also delineate key species and habitats. For example, the Nature Conservancy is a major conservation organization in the United States that specializes in the preservation of natural lands, critical habitats, and other natural resources areas. The Conservancy has established statewide natural heritage programs in over 20 states.

In summary, although it could be argued that it is not important to protect species that are in danger of becoming extinct because many species known to have existed at one time have now become extinct, with no grave consequences, it could also be argued that endangered plant and animal species are part of the biological system, and every effort should be made to plan projects so as not to interfere with the life cycle of these species for practical considerations. For example, such species may produce substances of value to human health or valuable with respect to other aspects of society. Perhaps the key issue is that if necessary background work is done on the requirements of the species, and then this data taken into account in project

planning and decision making, it is typically possible to proceed with projects without causing the loss of threatened or endangered species or their critical habitat.

Consideration of Wetlands—A Special Habitat

The term ''wetland'' has been difficult to define because there is no single, correct, indisputable, ecologically sound definition (Cowardin et al., 1979). The primary source of difficulty in defining wetlands is that no two wetlands exist with exactly the same characteristics. Despite their differences, all areas which are referred to as ''wetlands'' share a few vital characteristics. These characteristics have been noted in the following widely cited definition (Cowardin et al., 1979, p. 3):

> Wetlands are lands transitional between terrestrial and aquatic systems where the water table is usually at or near the surface or the land is covered by shallow water . . . . [W]etlands must have one or more of the following three attributes: (1) at least periodically, the land supports predominantly hydrophytes; (2) the substrate is predominantly undrained hydric soil; and (3) the substrate is non-soil and is saturated with water or covered by shallow water at some time during the growing season of each year.

Using these parameters, over 100,000 distinct types of wetlands may be delineated (Adamus et al., 1991). If all areas that fall within the guidelines of this definition are taken into account, wetland ecosystems constitute approximately 6 percent of the global land area and are considered to be one of the most endangered of all environmental resources (Turner and Jones, 1991).

A variation of the above definition is that wetlands are those areas that are inundated or saturated by surface water or groundwater at a frequency and a duration sufficient to support, and that under normal conditions do support, a prevalence of vegetation typically adapted for life in saturated soil conditions. Wetlands generally include swamps, marshes, bogs, and similar areas (U.S. Army Corps of Engineers, 1987). Wetland-classification schemes are typically based on types of vegetation, soil, and hydrologic characteristics. Technical guidelines for identifying wetlands and distinguishing them from aquatic habitats and nonwetland areas have been developed in U.S. Army Corps of Engineers (1987). A practical manual for identifying

wetlands is in Federal Interagency Committee for Wetland Delineation (1989).

The U.S. Fish and Wildlife Service has estimated that, in the 200-year period from the 1780s to the 1980s, the contiguous 48 states have lost an estimated 53 percent of their original 221 million acres of wetlands (U.S. General Accounting Office, 1991). An estimated 104

TABLE 10.7

ESIS INFORMATION FIELDS

Number	Field name	Description
1.	CODE	Unique identification number given to each species record
2.	KINGDOM	Species' kingdom (ANIMAL or PLANT)
3.	GROUP	Common name for species' taxonomic class (e.g., BIRD)
4.	COMMON-NAME	Species' common English name from Federal Register
5.	SCI-NAME	Species' scientific name from Federal Register
6.	NSTATUS	National level status of species (e.g., THREATENED)
7.	FACTOR	General factors contributing to species' present status
8.	CONTROL-NAME	Scientific name from standardized reference (e.g., ASC List)
9.	DIVISION	Species' taxonomic Division or Phylum
10.	CLASS	Species' taxonomic Class
11.	ORDER	Species' taxonomic Order
12.	FAMILY	Species' taxonomic Family
13.	SPECIES-TYPE	Species' general type classification (TERRESTRIAL or AQUATIC)
14.	TROPHIC-LV	Species' trophic level (e.g., HERBIVORE)
15.	HABITAT-AREA	General habitat type/area in which the species occurs
16.	FED-LAND	Federal agencies which have the species on their land
17.	FWS-REGION	USFWS Regions in which the species is known or possible
18.	JURISDICTION	Federal agency with lead responsibility for the species
19.	PSTATUS	Recovery plan status (PLAN APPROVED or PLAN NOT APPROVED)
20.	SSTATUS	State level status of the species (both federal and state)
21.	ISTATUS	International level status of the species (e.g., CITES-I)
22.	ESTATUS	Economic status of the species (e.g., COMMERCIAL)
23.	REASON	Reasons for the species' present status
24.	RECOVERY	Actions recommended for the species' recovery
25.	CONSULTS	Types of FWS consultations done involving the species
26.	LAND-TYPE	General land use types species is associated with (USGS LU/LC)
27.	FOREST	General forest types species is associated with (USFS FRES)
28.	RANGE/OPEN	General rangeland types species associated with (USFS FRES)
29.	AGRICULTURE	General agriculture types species associated with (SCS RCA)
30.	WETLAND	Wetland types the species is associated with (USFWS NWI)
31.	UNIQUE-HAB	Unique habitat types with which the species is associated
32.	SOIL	General soil types the species occurs on (if a plant)
33.	FOOD	General food types the species consumes (if an animal)
34.	CRITICAL-ST	States with designated/proposed critical habitat
35.	CRITICAL-CTY	Counties with designated/proposed critical habitat

million acres of wetlands remained in these states as of the 1980s. The annual rate of loss had declined from about 458,000 acres during the period from the mid-1950s to the mid-1970s to about 290,000 acres in 1991 (U.S. General Accounting Office, 1991).

Wetland loss and/or degradation can be caused by human activities or natural occur-

rences. Human-caused impacts can result from drainage, dredging and stream channelization, deposition of fill material, diking and damming, tilling for crop production, grazing by domesticated animals, discharge of pollutants, mining, or alteration of hydrology. Natural threats include erosion, subsidence, sea-level rise, droughts, hurricanes and other storms, and over-

TABLE 10.7

ESIS INFORMATION FIELDS *(continued)*

Number	Field name	Description
36.	PAST-ST	Species' past known/possible state level occurrence
37.	PAST-CTY	Species' past known/possible county level occurrence
38.	STATE	Species' present known/possible state level occurrence
39.	COUNTY	Species' present known/possible county level occurrence
40.	HYDROUNIT	Species' present known/possible OWDC Hydrounit occurrence
41.	ECOREGION	Species' present known/possible USFS Ecoregion occurrence
42.	OS-REGION	Species' present known/possible offshore region occurrence
43.	BIA	BIA lands on which the species is known or possible
44.	BLM	BLM lands on which the species is known or possible
45.	DOD	DOD lands on which the species is known or possible
46.	USFS	USFS lands on which the species is known or possible
47.	USFWS	USFWS lands on which the species is known or possible
48.	NPS	NPS lands on which the species is known or possible
49.	STATE-LAND	States which have the species occurring on their land
50.	COMMON-NAMES	Common name synonyms used for the species
51.	SCI-NAMES	Scientific name synonyms used for the species
52.	N-TAXONOMY	Narrative on the species' taxonomy
53.	N-STATUS	Narrative on species' status (nat./state/internat./economic)
54.	N-RECOVERY	Narrative on actions recommended for recovery and restoration
55.	N-CONSULTS	Listing of consultations involving the species
56.	N-REASONS	Narrative on reasons for the species' present status
57.	N-HABITAT	Narrative describing the species' habitat requirements
58.	N-BIOLOGY	Extensive narrative describing the species' biology
59.	N-OCCURRENCE	Narrative describing the species' distribution
60.	R-PERSONS	Resource persons for the species
61.	R-REGISTER	Federal Register citations involving the species
62.	R-BIOLOGY	References for N-TAXONOMY, N-REASONS, N-HABITAT, AND N-BIOLOGY
63.	R-OCCURRENCE	References for N-OCCURRENCE field
64.	U-ADMIN	Update information for administrative data fields
65.	U-BIOLOGY	Update information for biological data fields
66.	U-OCCURRENCE	Update information for distribution data fields

Notes: There are 15 "hidden" fields designed to permit future dataset expansion, making a total of 81 fields for ESIS.
    N = narrative text field
    R = reference text field
    U = update text field
Source: U.S. Fish and Wildlife Service, 1988, pp. 4–5.

TABLE 10.8

SUMMARY OF CAUSES OF WETLAND LOSS

Causes of loss	Estuaries	Open coasts	Floodplains	Freshwater marshes	Lakes	Peatlands	Swamp forest
Direct human actions:							
Drainage for agriculture, forestry and mosquito control	■	■	■	■	●	■	■
Dredging and stream channelization for navigation and flood protection	■	○	○	●	○	○	○
Filling for solid waste disposal, roads and commercial, residential and industrial development	■	■	■	■	●	○	○
Conversion for aquaculture/mariculture	■	●	●	●	●	○	○
Construction of dykes, dams, levees, and seawalls for flood control, water supply, irrigation and storm protection	■	■	■	■	■	○	○
Discharges of pesticides, herbicides, nutrients from domestic sewage and agricultural run-off and sediment	■	■	■	■	■	○	○
Mining of wetlands soils for peat, coal, gravel, phosphate and other materials	●	●	●	○	■	■	■
Groundwater abstraction	○	○	●	■	○	○	○
Indirect human actions:							
Sediment diversion by dams, deep channels and other structures	■	■	■	■	○	○	○
Hydrological alterations by canals, roads and other structures	■	■	■	■	■	○	○
Subsidence due to extraction of groundwater, oil, gas and other minerals	■	●	■	■	○	○	○
Natural causes:							
Subsidence	●	●	○	○	●	●	●
Sea-level rise	■	■	○	○	○	○	■
Drought	■	■	■	■	●	●	●
Hurricanes and other storms	■	■	○	○	○	●	●
Erosion	■	■	●	○	○	●	○
Biotic effects	○	○	■	■	■	○	○

Key: (○) Absent or exceptional; (●) present, but not a major cause of loss; (■) common and important cause of wetland degradation and loss
Source: Mannion and Bowlby, 1992, p. 221.

grazing by wildlife (U.S. EPA, 1988). Table 10.8 summarizes various causes of losses of different types of wetlands.

Wetlands are important resources because of the variety of functions they perform. Wetland functions are chemical, biological, or physical processes or attributes that are crucial to the integrity and stability of a wetland ecosystem (Adamus et al., 1991). The vast majority of these functions are very important to human societies. Even so, most of the valuable roles wetlands perform go unnoticed or are taken for granted. Wetlands influence (and are influenced by) hydrology and water quality, contribute to food-chain support and nutrient cycling, provide habitat for a wide variety of animals and plants, and support many economic and social pursuits. With respect to hydrology, wetlands anchor shorelines and protect upland areas from erosive forces by dissipating the impact of waves, provide flood-control capacity by absorbing excess water during storms and then slowly releasing stored water to reduce peak flows downstream, and support groundwater recharge and discharge (Adamus et al., 1991; Sather and Smith, 1984; U.S. General Accounting Office, 1991).

Wetland plants are important because they provide a substrate for bacterial growth and a media for physical filtration and absorption, and restrict algal growth and wave action (Sather and Smith, 1984). Performance of these functions results in the maintenance of water quality by filtering out pollutants, sediments, and/or toxicants to purify water before it enters streams, lakes, or oceans (U.S. General Accounting Office, 1991; Dortch, 1992). The energy and nutrients stored in wetland autotrophs (such as plants), directly or indirectly, supply the needs of heterotrophs (such as animals) in wetland-related food chains; these animals often supplement the diets of humans (Sather and Smith, 1984).

Wetlands provide habitats for a multitude of organisms, including nearly one-third of the nation's endangered and threatened wildlife species (Turner and Jones, 1991). In addition, more than half of all threatened or endangered species depend directly or indirectly on wetlands at some time during their life cycles (U.S. General Accounting Office, 1991). For examples, migratory waterfowl utilize wetlands as breeding and wintering grounds which are important to hunters, and the spawning grounds coastal wetlands sustain are crucial to the commercial industries relying on certain species of fish and shellfish (Sather and Smith, 1984; U.S. General Accounting Office, 1991).

In addition to the previously mentioned functions, wetlands also provide purely recreational and aesthetic opportunities such as fishing, hunting, and bird watching (U.S. General Accounting Office, 1991). The growing knowledge pertaining to wetlands functions, along with increasing pressure from conservation groups and various other public interest groups, has initiated a change in the attitude of governmental agencies toward wetlands from one of apathy to one of protection and conservation.

Wetland functions and relationships are summarized in Table 10.9. The value of wetlands in different locations in terms of functions, products, and attributes is shown in Table 10.10.

Key federal agencies involved either directly or indirectly in wetlands delineation and/or protection include the Army Corps of Engineers, Environmental Protection Agency, Soil Conservation Service, Fish and Wildlife Service, National Marine Fisheries Service, and Agricultural Stabilization and Conservation Service (U.S. General Accounting Office, 1991).

Federal laws that in some manner regulate development in wetlands have been growing in number and increasing in scope for the past 30 years (Figure 10.6). There are many federal regulations which can restrict development in wetlands; but the Clean Water Act of 1977—in particular, Section 404 of the act—has had the greatest impact on development in wetlands (Salvesen, 1990). This act provides the principal federal authority for the regulation of wetlands use by governmental agencies. Section 404 deals primarily with developers but also affects private landowners. It stipulates that landowners and developers must obtain permits before any dredging or filling activities are carried out in navigable waters, including wetlands. The process to acquire such permits may be lengthy and places many responsibilities on the developer or landowner. Even though its initial purpose may have been to protect navigable waters, this law is now being interpreted in such a way as to help protect and to conserve wetlands.

Prior to the passage of the Food Security Act of 1985, federal policies encouraged farmers to convert wetlands to agriculturally useful lands by providing loans, credit, and commodity-price supports (Salvesen, 1990). Because the vast majority of wetland losses were, and still are, to

TABLE 10.9

WETLAND FUNCTIONS AND RELATIONSHIPS

Wetland function	Concern	How wetlands perform function	Factors determining importance of function
Flood Conveyance	If flood flows are blocked by fills, dikes, or other structures, increased flood heights and velocities result, causing damage to adjacent upstream and downstream areas.	Some wetlands (particularly those immediately adjacent to rivers and streams) serve as floodway areas by conveying flood flows from upstream to downstream points.	Stream characteristics, wetland topography, and size, vegetation, location of wetland in relationship to river or stream, existing encroachment on floodplain (dikes, dams, levees, etc.)
Wave Barriers	Removal of vegetation increases erosion and reduces capacity to moderate wave intensity.	Wetland vegetation, with massive root and rhizome systems, bind and protect soil. Vegetation also acts as wave barriers.	Location of wetland adjacent to coastal waters, lakes, and rivers, wave intensity, type of vegetation, and soil type.
Flood Storage	Fill or dredging of wetlands reduces their flood storage capacity.	Some wetlands store and slowly release flood waters.	Wetland area relative to watershed, wetland position within watershed, surrounding topography, soil infiltration capacity in watershed, wetland size and depth, stream size and characteristics, outlets (size, depth), vegetation type, substrate type.
Sediment Control	Destruction of wetland topographic contours or vegetation decreases wetland capacity to filter surface runoff and act as sediment traps. This increases water turbidity and siltation of downstream reservoirs, storm drains, and stream channels.	Wetland vegetation binds soil particles and retards the movement of sediment in slowly flowing water.	Depth and extent of wetland, wetland vegetation (including type, condition, density, growth patterns), soil texture type and structure, normal and peak flows, wetland location relative to sediment of vegetated buffer.
Pollution Control	Destruction of wetland contours or vegetation decreases natural pollution capability, resulting in lowered water quality for downstream lakes, streams, and other waters.	Wetlands act as settling ponds and remove nutrients and other pollutants by filtering and causing chemical breakdown of pollutants.	Type and size of wetland, wetland vegetation (including type, condition, density, growth patterns), source and type of pollutants, water course, size, water volume, streamflow rate, microorganisms, etc.

TABLE 10.9

WETLAND FUNCTIONS AND RELATIONSHIPS *(continued)*

Wetland function	Concern	How wetlands perform function	Factors determining importance of function
Fish and Wildlife Habitat	Fills, dredging, damming, and other alterations destroy and damage flora and fauna and decrease productivity. Dam construction is an impediment to fish movement.	Wetlands provide water, food supply, and nesting and resting areas. Coastal wetlands contribute nutrients needed by fish and shellfish to nearby estuarine and marine waters.	Wetland type and size, dominant wetland vegetation (including diversity of life form), edge effect, location of wetland within watershed, surrounding habitat type, juxtaposition of wetlands, water chemistry, water quality, water depth, existing uses.
Recreation (water-based)	Fill, dredging or other interference with wetlands will cause loss of area for boating, swimming, bird watching, hunting, and fishing.	Wetlands provide wildlife and water for recreational uses.	Wetland vegetation, wildlife, water quality, accessibility to users, size, relative scarcity, facilities provided, surrounding land forms, vegetation, land use, degree of disturbance, availability of similar wetlands, distribution, proximity of uses, vulnerability.
Water Supply (surface)	Fills or dredging cause accelerated runoff and increase pollution.	Some wetlands store flood waters, reducing the timing and amount of surface runoff. They also filter pollutants. Some serve as sources of domestic water supply.	Precipitation, watershed runoff characteristics, wetland type, size, outlet characteristics, location of wetland in relationship to other water bodies.
Aquifer Recharge	Fills or drainage may destroy wetland aquifer recharge capability, thereby reducing base flows to streams and ground water supplies for domestic, commercial, or other uses.	Some wetlands store water and release it slowly to ground water deposits. However, many other wetlands are discharge areas for a portion or all of the year.	Location of wetland relative to water table, fluctuations in water table, geology including type and depth of substrate, permeability of substrate, size of wetland, depth, aquifer storage capacity, ground water flow, runoff retention measures.

Source: Kusler, Harwood, and Newton, 1983, p. 7.

TABLE 10.10

VALUE OF DIFFERENT TYPES OF WETLANDS

Value of wetlands	Estuaries (without mangroves)	Mangroves	Open coasts	Floodplains	Freshwater marshes	Lakes	Peatlands	Swamp forest
Functions								
Groundwater recharge	○	○	○	■	■	■	●	●
Groundwater discharge	●	●	●	●	■	●	●	■
Flood control	●	■	○	■	■	■	●	■
Shoreline stabilization erosion control	●	■	●	●	■	○	○	○
Sediment/toxicant retention	●	■	●	■	■	●	■	■
Nutrient retention	●	■	●	■	■	●	■	■
Biomass export	●	■	●	■	●	●	○	●
Storm protection/windbreak	●	■	●	○	○	○	○	●
Micro-climate stabilization	○	●	○	●	●	●	○	●
Water transport	●	●	○	●	○	●	○	○
Recreation/tourism	●	●	■	●	●	●	●	●
Products								
Forest resources	○	■	○	●	○	○	○	■
Wildlife resources	■	●	●	■	■	●	●	●
Fisheries	■	■	●	■	■	■	○	●
Forage resources	●	●	○	●	●	○	○	○
Agricultural resources	○	○	○	■	●	●	●	○
Water supply	○	○	○	●	●	■	●	●
Attributes								
Biological diversity	■	●	●	■	●	■	●	●
Uniqueness to culture or heritage	●	●	●	●	●	●	●	●

Key: (○) Absent or exceptional; (●) present; (■) common and important value of that wetland type.
Source: Mannion and Bowlby, 1992, p. 219.

agriculture, the Food Security Act of 1985 is an important mechanism by which wetlands can be protected. Even though farmers can still convert wetlands if they have a Section 404 permit, farmers who do so disqualify themselves from receiving any federal aid or benefits for commodities grown or raised on such converted lands (U.S. General Accounting Office, 1991; Salvesen, 1990).

The U.S. Fish and Wildlife Service, U.S. Environmental Protection Agency, and U.S. Army Corps of Engineers are key governmental agencies concerned with wetlands and the granting of regulatory permits to projects or activities

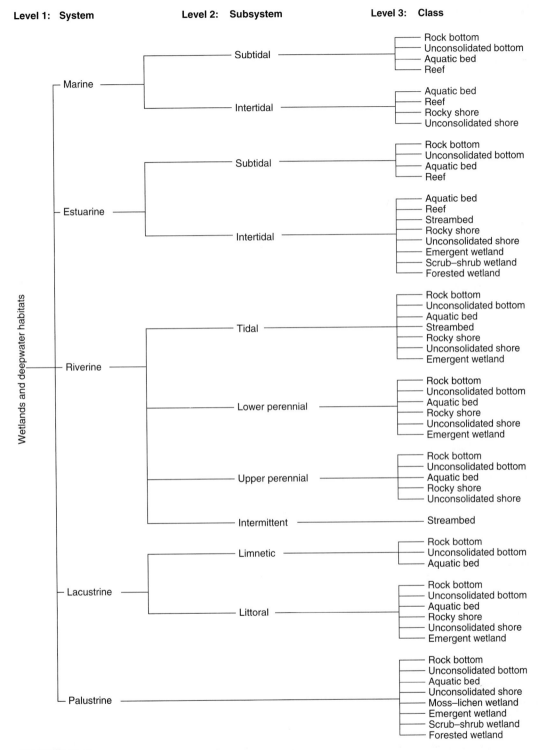

FIGURE 10.8
Three-level Classification of Wetlands and Deepwater Habitats (Marsh, 1991, p. 289).

which may exhibit wetland impacts. To facilitate regulatory review, the wetland-classification system shown in Figure 10.8 has been developed. The "marine system" consists of the deepwater habitat of the open ocean and the adjacent marine wetlands of the intertidal areas. The "estuarine system" is associated with coastal embayments and drowned river mouths. The "riverine system" is limited to freshwater stream channels, and the "lacustrine system" is limited to standing water bodies—mainly lakes, ponds, and reservoirs. Both the riverine system and the lacustrine system include deepwater habitat. The fifth system, the "palustrine system," includes only wetland habitat; it encompasses the vast majority of North America's wetlands—namely, inland marshes, swamps, and bogs (Marsh, 1991). For the palustrine wetlands, the three major classes are (1) emergent wetland, (2) scrub-shrub wetland, and (3) forested wetland (Marsh, 1991). Three criteria are basic to wetland mapping—vegetation, soils, and hydrology (flooding).

The locations of all wetlands at the site of the proposed project-activity and in the study area should be identified. Information should then be procured on the size and functions of the wetlands, including whether or not they serve as habitat for endangered or threatened plant or animal species. This information should be available from one or more federal and state environmental and resource agencies. The potential impacts of the proposed project-activity should then be delineated and classified, with the possibilities ranging from no effects, to loss or curtailment of one or more wetland functions, to loss of a portion to all of the wetland area. Systematic habitat methods such as the wetland evaluation technique (WET), described in Chapter 11, could be used to both describe existing conditions and delineate impacts.

If negative impacts are anticipated, it may be necessary to develop mitigation plans; an example list of guidelines is included in the discussion of step 6. Mitigation may include restoration, creation, and/or management efforts. Detailed information on planning wetland-restoration or -creation projects is available in Kentula et al. (1992). Payne (1992) summarizes various techniques for wildlife habitat management in wetland areas.

STEP 3: PROCUREMENT OF RELEVANT LEGISLATION AND REGULATIONS

The primary sources of information on pertinent legislation, regulations, criteria, or guidelines related to the biological environment include environmental and/or natural resource agencies at the federal and state levels. Local agencies and/or conservation groups may also provide pertinent information. (Key federal legislation has been described earlier.) Procurement of this information will facilitate the evaluation of baseline conditions and the data obtained can serve as a basis for impact-significance determination (step 5).

Most of the biological-environment legislation, regulations, criteria, or guidelines are qualitative in terms of specific requirements. This is in contrast to the substantive areas of air quality, surface water and groundwater quality, soil quality, and environmental noise, for which numerical standards are available. Sound professional judgment must be exercised in applying the qualitative requirements for the biological environment in this step.

STEP 4: IMPACT PREDICTION

The most technically demanding step in addressing the biological environment is the prediction of the impacts of the project-activity, and various alternatives, on the biological-environmental setting. As a general principle, the impacts should be quantified where possible, with qualitative descriptions provided for those impacts which cannot be quantified. From an historical perspective, impact prediction for the biological environment has focused on land-use or habitat changes and the associated implications of those factors relative to the biological system. Several options are available for impact-prediction approaches, including qualitative descriptions of impacts, the use of habitat-based methods or ecosystem models, and the use of physical models or simulations. Broader impact issues of increasing importance are biological diversity and sustainable development.

Qualitative Approaches

Qualitative descriptions could be associated with a discussion of land use or habitat changes. One tool which can be helpful in identifying the types of impacts (effects) that might take place on the biological system is the list of 52 effects found in Table 10.11. The general approach would be to consider each of these factors and determine its applicability to the project and the environmental setting. If deemed applicable, then either specific quantitative information could be assembled, or at least qualitative discussions prepared, on the implications of the project relative to the particular biological items identified. In using this approach, the considerable exercise of professional judgment would be required.

If structured data presentations, discussed earlier, have been utilized to describe the biological environmental setting, it would also be possible to apply the list of effects in Table 10.11 to the results of the structured data approaches. Again, it would be necessary to exercise professional judgment in the interpretation phase. One additional technique in the area of structured data presentations is to consider adding impact-prediction information to the structured data presentation itself. For example, in considering Table 10.1, additional information could be added in a series of columns at the right-hand side of the table. Five such ad-

ditional columns could be address the following issues

1. The likelihood of impact, with this likelihood shown in terms of a relative scale of high, medium, and low.
2. The duration of the impact in terms of whether it would be associated with the short-term construction phase of a project versus the long-term operational phase. In addition, this column could include information on the actual anticipated duration of the impact.
3. The reversibility of the impacts. Perhaps two codes could be used in this column, with one denoting items that are irreversible and the other denoting those particular impacts that might be recoverable. This column could also relate to the possibility of success in trying to reduce the impact and to potentially reverse it through various developed programs.
4. The relative resiliency of individual plant or animal species within the study area. (It is quite well-known that some species are more tolerant of change than others.)
5. Potential mitigation measures for a given project type.

Habitat-Based Methods or Models Approaches

If habitat-based methods have been used to describe the biological environmental setting, these can also form the basis for impact prediction. As indicated in Chapter 11 (relative to HES and HEP), there are detailed requirements in predicting future changes of the acreages of habitat types, as well as in predicting changes in the individual parameters used to describe the habitat for individual species (as in the case of the HEP) or for the general biotic and abiotic components of the system (as in HES). It should be noted that usage of habitat-based methods does require predictions of changes in acreage

TABLE 10.11

LIST OF POTENTIAL EFFECTS ON THE BIOLOGICAL SYSTEM

1. Resiliency and fitness of ecosystem types; for example, lowland forest, upland forest, grassland, marsh, bog, and streams
2. Total standing crop of organic matter
3. Annual plant productivity
4. Mulch or litter removal as related to top-soil stripping
5. Animal production
6. Sediment load carried by streams
7. Aquatic macroinvertebrate populations
8. Drift rate of aquatic macroinvertebrates
9. Population density of fish
10. Sediment-load effects on fish growth
11. Sediment-load effects on fish spawning
12. Species diversity of the aquatic biota
13. Undesirable proliferations of biota
14. Localized survival of rare plant and animal species
15. Habitat carrying capacity of both aquatic and terrestial systems
16. Abandonment of habitat
17. Endemic populations of plants and animals
18. Wildlife breeding and nesting sites
19. Endangered plant and animal species
20. Vegetation communities of denuded areas
21. Wildlife refuges and sanctuaries
22. Scientific and educational areas of biological interest
23. Vegetation recovery rates
24. Forage areas for both upland and lowland game species
25. Migratory game bird species
26. Terrestrial microbial communities
27. Amount of forest removed
28. Population density of past species
29. Domestic animal species
30. Amount of grassland removed
31. Natural drainage systems
32. Natural animal corridors
33. Eutrophication
34. Expansion of population range for both plant and animal species
35. Cropland removal
36. Potential for wildlife management
37. Food-web index, including herbivores, omnivores, and carnivores
38. Species diversity of the terrestrial biota
39. Nutrient supply available to terrestrial biota
40. Sport fishing and hunting
41. Resultant air pollution effects on crop yield
42. Relict vegetation types
43. Responses of sensitive native plants to air pollutants, both particulates and gases
44. Unnatural dispersion and subsequent overutilization of habitats
45. Noise level effects on reproductive inhibition of small mammals
46. Air pollutant effects on tree canopy
47. Noise level effects on broodiness of upland and lowland game birds
48. Water-temperature stability
49. Areas of high brush-fire potential
50. Water quality and dependent biota
51. Noise level effects on insect maturation and reproduction
52. Natural biological character loss

Source: Adapted from Hill, 1975.

as well as changes in several habitat parameters. The ability to accurately predict over long periods of time is questionable.

If an energy-system diagram has been used to describe the biological environmental setting, it is simple to substitute new conditions into the developed diagrams for purposes of impact prediction. The usage of this approach for describing the biological setting is fairly limited; however, it is considered that this approach for impact prediction will probably become of greater use within the next 5 to 10 years.

Both qualitative and quantitative ecosystem models are available, with such models typically based on materials-cycling considerations and mass-balance calculations. Computer-simulation modeling may also be used; such models incorporate population dynamics and food-chain considerations (Eberhardt, 1976). Specific ecosystem models are available for predicting lake eutrophication (Hornberger, Kelly, and Lederman, 1975), the impacts of dredging and dredged-material disposal (Hall, Westerdahl, and Eley, 1976), and for specific locations, such as Chesapeake Bay (Green, 1978). The model for Chesapeake Bay focuses on the ecological functions that produce the resources of commercial and recreational fisheries, habitat for migratory birds and other wildlife, waste disposal, and aesthetic water quality. Physical conditions (light, turbidity, mixing, transport, and sedimentation) and chemical conditions (sediment-water interaction, presence of pollutants) in the environment modify the rates of biological processes such as primary production, nutrient regeneration, and larval survival. The detailed ecosystem model combines wetlands, plankton, seagrasses, other benthos, and fish-trophic-dynamics submodels to demonstrate the importance of material transfer and interactions between subsystems. Some information on aquatic-ecosystem models is in Chapter 7.

Models are also available for addressing chemical uptake and transport; for example, information is available on concentration factors

and transport models for radionuclides in aquatic environments in Patzer (1976). Many other examples of both aquatic and terrestrial models could be cited (Starfield and Bleloch, 1986).

Physical-Modeling Approaches

In addition to mathematical models, physical models could also be used for biological impact prediction and assessment. Examples include bioassay, chronic-toxicity testing, microcosms, and scaled ecosystem models. Bioassays and chronic-toxicity testing are focused on the potential toxic effects on terrestrial or aquatic plant or animal species of discharges or releases of residuals by the proposed project-activity.

"Microcosms" refer to small-scale biological systems which can be used to study fundamental processes such as the effects of nutrient additions or chemical cycling between biological compartments. Scaled ecosystem models may be on the order of hundreds of square meters in size; they are used to simulate all of the processes expected to occur in a biological system. For example, the U.S. Army Corps of Engineers has constructed several scaled ecosystem models of reservoirs and surrounding terrestrial habitats.

Biodiversity and Sustainable-Development Considerations

"Biodiversity," or the variety of life and its processes, is a basic property of nature that provides enormous ecological, economic, and aesthetic benefits. Its loss is recognized as a major national as well as global concern, with potentially profound ecological and economic consequences (CEQ, 1993). Table 10.12 summarizes some of the components of biological diversity (or "biodiversity"). An international convention on biodiversity was adopted at the 1992 United Nations Conference on Environment and Development, held in Rio de Janeiro, Brazil. A status report on global biodiversity is available in World Conservation Monitoring

TABLE 10.12

COMPONENTS OF BIOLOGICAL DIVERSITY

- **Regional ecosystem diversity:** The pattern of local ecosystems across the landscape, sometimes referred to as "landscape diversity" or "large ecosystem diversity".
- **Local ecosystem diversity:** The diversity of all living and non-living components within a given area and their interrelationships. Ecosystems are the critical biological/ecological operating units in nature. A related term is "community diversity" which refers to the variety of unique assemblages of plants and animals (communities). Individual species and plant communities exist as elements of local ecosystems, linked by processes such as succession and predation.
- **Species diversity:** The variety of individual species, including animals, plants, fungi, and microorganisms.
- **Genetic diversity:** Variation within species. Genetic diversity enables species to survive in a variety of different environments, and allows them to evolve in response to changing environmental conditions.

The **hierarchical nature** of these components is an important concept. Regional ecosystem patterns form the basic matrix for, and thus have important influences on, local ecosystems. Local ecosystems, in turn, form the matrix for species and genetic diversity, which can in turn affect ecosystem and regional patterns.

Relationships and interactions are critical components as well. Plants, animals, communities, and other elements exist in complex webs, which determine their ecological significance.

Source: Council on Environmental Quality, 1993, p. 1.

Centre (1992). Scientific and policy options for maintaining biological diversity in the United States have been delineated in Office of Technology Assessment (1987).

A number of factors have contributed or are contributing to the decline of biodiversity in the United States. This decline can be seen in the loss of ecosystems, wetlands, and habitat for threatened or endangered animal species. Factors contributing to the decline of biodiversity include physical alterations to the geography due to resource exploitation and changing land usage; pollution; overharvesting; introduction of exotic (nonnative) species and elimination of native species through predation, competition, genetic modification, and disease transmission; disruption of natural processes; and global climate change (CEQ, 1993).

Biodiversity considerations can be important in environmental management. The basic goal of biodiversity conservation is to maintain naturally occurring ecosystems, communities, and native species. The basic goals when considering biodiversity in management are to identify and locate activities in less sensitive areas; to

minimize the impacts of such activities, where possible; and to restore lost diversity, where practical (CEQ, 1993). Certain principles (not rules) can be enumerated for incorporating consideration of biodiversity into environmental management; these principles include (CEQ, 1993)

1. Take a "big picture," or ecosystem view.
2. Protect communities and ecosystems.
3. Minimize fragmentation, and promote the natural pattern and connectivity of habitats.
4. Promote native species, and avoid introducing nonnative species.
5. Protect rare and ecologically important species.
6. Protect unique or sensitive environments.
7. Maintain or mimic natural ecosystem processes.
8. Maintain or mimic naturally occurring structural diversity.
9. Protect genetic diversity.
10. Restore ecosystems, communities, and species.
11. Monitor biodiversity impacts, acknowledge uncertainty, and be flexible.

TABLE 10.13

RECOMMENDATIONS FOR AGENCY IMPROVEMENTS RELATED TO BIODIVERSITY CONSIDERATIONS IN THE NEPA PROCESS

Recommendation	Comments
1. Acknowledge the conservation of biodiversity as national policy and incorporate its consideration in the NEPA process.	Agencies should insure that both staff responsible for conducting environmental impact analyses and decision makers responsible for considering the findings of those analyses are familiar with the importance of the biodiversity issue and its relevance to their work.
	Agency-sponsored environmental training courses should discuss biodiversity and how best to consider it in the NEPA process and in all planning, design, and management.
2. Encourage and seek out opportunities to participate in efforts to develop regional ecosystem plans.	Regional ecosystem frameworks are a critical element of conserving biological diversity. Such regional efforts can provide an ecosystem framework for evaluating the impacts of individual projects on biodiversity, and provide a common basis for describing the affected environment. Both will save time and financial resources in preparing NEPA documents.
	Agencies should investigate and consider participation in efforts that may be already in progress in areas where they have jurisdiction or interest.
	Some regional frameworks exist that do not explicitly address biodiversity. In such cases, agencies should consider establishing specific goals and objectives for the conservation of biodiversity, within those frameworks.
	Finally, where such efforts are lacking entirely, agencies should consider initiating them.
3. Actively seek relevant information from sources both within and outside government agencies.	While information on the status and distribution of biota is incomplete, a great deal of information is available from a wide variety of sources. Agencies should look to each other, to state agencies, and to academic and other nongovernmental entities. By doing so, agencies can benefit from the resources and technical capabilities of others and reduce the costs associated with collecting and managing information on which ecosystem and biodiversity analyses depend.
4. Encourage and participate in efforts to improve communication, cooperation, and collaboration between and among governmental and nongovernmental entities.	Improved communication, cooperation, and collaboration will enormously increase the prospects for overcoming the barriers described earlier. Working with others can help to identify common interests and overlapping or complementary missions, and can lead to mutual sharing of information, technical capabilitites, and expertise. Efforts to do so will require support at the management and policy-making levels within agencies, as well as at the level of the staff responsible for carrying out NEPA analyses.

TABLE 10.13

RECOMMENDATIONS FOR AGENCY IMPROVEMENTS RELATED TO BIODIVERSITY CONSIDERATIONS IN THE NEPA PROCESS
(continued)

Recommendation	Comments
5. Improve the availability of information on the status and distribution of biodiversity, and on techniques for managing and restoring it.	Agencies that support or sponsor research and development efforts that will improve our ability to evaluate and manage for biodiversity should ensure that the information they obtain is maintained in a format that is useful and readily accessible.
	Agencies should consider opportunities to cooperate with and benefit from the National Biodiversity Center, presently in the planning and design stages. A key role of the center will be to identify existing ecological information and make it more readily available for use in environmental planning and assessment.
6. Expand the information base on which biodiversity analyses and management decisions are based.	Basic research is needed into a host of issues relating to both ecosystem management and biodiversity conservation. These include ecosystem functioning; selection of indicators; prediction of the effects of change on ecosystems; and establishment of spatial and temporal boundaries for impacts and analyses.
	Agencies should recognize the research opportunities afforded by various projects and consider sponsoring or cooperating with academic institutions, private industry, and others on research to advance ecological understanding.

Source: Compiled from data in Council on Environmental Quality, 1993, pp. 23–24.

Biodiversity considerations should be incorporated in the NEPA process. Examples of some current weaknesses in the NEPA process in relation to biodiversity include the following (CEQ, 1993): (1) inadequate consideration of "nonlisted" species (reliance on listed threatened and endangered species only is likely to address only a small portion of the nation's imperiled biodiversity), (2) inadequate consideration of "nonprotected" areas, (3) inadequate consideration of "noneconomically important" species, and (4) inadequate consideration of cumulative impacts.

The NEPA provides a mandate for federal agencies to consider all reasonably foreseeable environmental effects of their actions; it also provides a framework for doing so. To the extent that federal actions affect biodiversity, and that such effects may be both anticipated and evaluated, the NEPA requires federal agencies to do so. Accordingly, the CEQ has developed six recommendations for agency-driven improvements in the consideration of biodiversity in NEPA analyses; these recommendations and associated comments are in Table 10.13.

"Sustainable development" is development that meets the needs of the present without compromising the ability of future generations to meet their own needs. The term embodies two key concepts (World Commission on Environment and Development, 1987, often referred to as the "Brundtland Report"): (1) the concept of "needs," in particular the essential needs of the world's poor, to which overriding priority should be given and (2) the idea of limitations imposed by the state of technology and social organization on the environment's ability to meet present and future needs. For large-scale projects or activities, it would be appropriate to consider the anticipated impacts in terms of their implications for sustainable development. Pragmatic information is needed on how to address sustainable development at the project or activity level.

STEP 5: ASSESSMENT OF IMPACT SIGNIFICANCE

Interpretation of the anticipated impacts of a proposed project (or activity) should be considered not only in terms of individual species, but also relative to the general characteristics of the affected habitat(s) and overall ecosystem. One basis for significance determination is to apply the institutional information described earlier, including relevant laws, regulations, criteria, and guidelines (Camougis, 1981).

Another basis for impact interpretation is the biological-science or professional interpretation approach. This involves the application of professional judgment and knowledge of biological-ecological principles, and it demonstrates why it is necessary for a biological scientist to be a part of an interdisciplinary study team. Examples of some biological-ecological principles and considerations which could be applied in impact interpretation include the following:

1. The role of the individual species in the food-web relationship, with this interpretation based on recognizing the biological environment as a system.
2. An analysis of the carrying capacity of the biological setting relative to individual species of concern within the project area.
3. An evaluation of the resiliency of plant and animal species, and the interpretation of that resiliency relative to the anticipated changes caused by the project.
4. An evaluation of the implications of the project relative to species diversity within the terrestrial and aquatic habitats in the study area. In general, there is a lesser ability to resist change when the species diversity is lower; therefore, this evaluation could also serve as a basis for interpreting the overall fragility of the biological environmental setting.
5. Consideration of natural succession and the implications of the project in terms of disruptions that might occur in this successional process.

6. A review of species that exhibit the ability to reconcentrate particular chemical constituents through natural environmental processes.
7. An evaluation of the implications of the proposed project on species of economic importance within the study area (these include species that might be of interest from the perspective of hunting or fishing activities).
8. Any anticipated changes that might occur in threatened or endangered species or critical habitat within the study area.

STEP 6: IDENTIFICATION AND INCORPORATION OF MITIGATION MEASURES

Mitigation measures for biological impacts can include avoidance, minimization, rectification, preservation, and/or compensation. As noted earlier, several environmental-quality and natural-resources-protection laws have been promulgated within the United States; many include policies and implementation requirements for biological-impact-mitigation measures. Table 10.14 is a listing of some of these laws. Two notable laws in addition to the NEPA are the Surface Mining Control and Reclamation Act of 1977 (SMCRA), which delineates mitigation requirements directed toward restoring habitat disturbed by surface-mining operations; and the Endangered Species Act (ESA), discussed earlier. Surface-mining activities may cause numerous undesirable impacts on the biological environment. Section 501(b) of the SMCRA requires that a permanent regulatory program be developed and implemented, and that this regulatory program delineate actions which must be undertaken as a part of mining activities. These actions may be considered as mitigation measures in that they avoid, reduce, or compensate for adverse impacts. A summary of some miti-

---

TABLE 10.14

EXAMPLE LISTING OF U.S. LAWS WHICH MAY DIRECTLY OR INDIRECTLY ADDRESS BIOLOGICAL IMPACT-MITIGATION MEASURES

Anadromous Fish Conservation Act
Bald Eagle Protection Act
Clean Air Act
Clean Water Act
Coastal Zone Management Act
Comprehensive Environmental Response, Compensation and Liability Act (Superfund)
Endangered Species Act
Federal Insecticide, Fungicide and Rodenticide Act
Fish and Wildlife Coordination Act
Golden Eagle Protection Act
Marine Mammal Protection Act
Marine Protection, Research and Sanctuaries Act
Migratory Bird Conservation Act
National Environmental Policy Act
Resource Conservation and Recovery Act
Superfund Amendments and Reauthorization Act
Surface Mining Control and Reclamation Act
Toxic Substances Control Act
Wild and Scenic Rivers Act
Wild Horses and Burros Protection Act

Source: Canter, Robertson, and Westcott, 1991, p. 37.

TABLE 10.15

SUMMARY OF BIOLOGICAL MITIGATION MEASURES AS MANDATED BY SEC. 501(b) IN SMCRA

Biological impact	Possible mitigation measures and regulatory program requirements
Loss of wildlife and wildlife habitat	A wildlife-protection plan is required as part of any mining permit application. Wildlife agencies must be consulted. Timing, shaping, and sizing operations must be conducted to avoid breeding or nesting season and trees, protecting key food, cover, and water resources. Fencing will keep large mammals from direct contact with toxic chemicals in sedimentation ponds and from roadways to reduce the number of roadkills. Revegetation must use species with high nutritional or cover value. Topsoil handling and replacement prior to revegetation must be conducive to wildlife. Topsoil storage must be covered with vegetation, thus providing cover for wildlife. A 30-m buffer zone on each side of streams must be undisturbed.
Disturbance of aquatic organisms and aquatic habitats	A regulatory program designed for restoration, protection, enhancement, and maintenance of aquatic life must be implemented. Surface and underground mine openings must be cased and sealed to prevent escape of acid and toxic discharge. Buffer strips must be left between mining operations and waterways. All streams restoration is to include alternating patterns of riffles, pools, and drops. All diversions must be removed.
Erosion and sedimentation	Surface runoff must be collected in sediment ponds. Disturbed soils must be revegetated.
Destruction of vegetation	Affected land must be restored to premining productive capacity. Topsoil must be removed, segregated, stored, and redistributed with minimum loss or contamination. Topsoil and subsoil may be removed separately and replaced in sequence. Native vegetation or appropriate substitutes after mining must be established. Agricultural lands must be returned to the same or greater productive capacity obtained under premining conditions.

Source: Developed from data in U.S. Department of the Interior, 1979.

gation measures which relate to biological systems or biological impacts is presented in Table 10.15.

Natural-resources-management plans can be useful for describing existing and historical conditions and providing a contextual framework for assessing anticipated impacts and identifying mitigation measures. Such plans are typically available for military installations and federal lands managed by agencies such as the U.S. Forest Service and U.S. Bureau of Land Management.

Even though land usage may be extensive at military and civil works facilities, and in man-

aged natural forests and other resource areas, typically, the majority of the lands are used only periodically, if at all. Such "nonusage" areas could include buffer zones between the facilities and adjacent private land uses, or between specific facility activities. Therefore, the opportunity exists for the development and implementation of effective "natural resources management plans" (NRMPs) at military bases, civil-works-project sites, and managed government-land areas. NRMPs can focus on single issues such as forestry, or they can address a wide range of environmental concerns. Table 10.16 contains a list of some substantive issues

TABLE 10.16

SUBSTANTIVE ISSUES WHICH COULD BE A PART OF A NATURAL-RESOURCES-MANAGEMENT PLAN

Chemical control of vegetation
Soil-erosion control
Drainage
Prescribed burning
Protection against unwanted natural or human-induced fires
Tree-planting program
Shrubs- and grasses- planting program
Programs to introduce new plant or animal species
Irrigation
Landscaping
Litter control and policing
Minerals extraction
Energy-resource extraction
Land-reclamation practices
Timber management, sales, and cutting cycles
Silviculture practices
Wetlands protection
Water resources development
Waste-disposal programs
Programs to protect threatened or endangered species
Unique-habitat-protection programs
Wildlife-habitat improvement
Administration of hunting and fishing seasons
Bird sanctuaries and flyways development
Grazing leases
Wildlife management
Range management
Fisheries management

Source: Compiled from data in Jahn, Cook, and Hughes, 1984.

which could be addressed in an NRMP. More-detailed information is available on all of the issues listed in Table 10.16; for example, extensive information is available on impact mitigation and management measures for wildlife resources in Martin (1986).

Mitigation strategies for wetland impacts can be divided into four categories: (1) avoidance or minimization, (2) restoration, (3) enhancement, and (4) creation (Salvesen, 1990). Mitigation plans for wetland impacts should include the following components (Salvesen, 1990):

1. A clear statement of the objectives of the mitigation
2. An assessment of the wetlands values or resources that will be lost as a result of the fill and of those that will be replaced
3. A statement of the location, elevation, and hydrology of the new site
4. A description of what will be planted where and when
5. A monitoring and maintenance plan
6. A contingency plan for problem issues, including unwanted weeds, human disturbance, predation, and plant availability

7. A guarantee that the work will be performed as planned and approved

It should be noted that duplication of naturally occurring wetlands by wetland creation or restoration of degraded wetlands is typically limited by a lack of established goals and limited scientific information and understanding (Kusler and Kentula, 1990).

SUMMARY

In summary, the biological environment can be systematically approached in terms of impact prediction and assessment. This chapter has presented several options for dealing with the biological environmental setting other than the usage of formalized, habitat-based methods. The emphasis has been on a six-step methodological framework for planning and conducting impact studies focused on the biological environment. Key points of conclusion include

1. The approach to be used should be based on considering the type of project or activity and its associated impacts.
2. Many technical tools are available, including structured data presentations, biological indices, and mathematical and physical models.
3. In addressing biological-environment impacts, it is important to recognize the biological-ecological environment as a dynamic system.

SELECTED REFERENCES

Adamus, P. R., et al., "Wetland Evaluation Technique (WET); Volume I: Literature Review and Evaluation Rationale," WRP-DE-2, U.S. Army Engineer Waterways Experiment Station, Vicksburg, Miss., Oct. 1991.

Bailey, R. G., "Description of the Ecoregions of the United States," Misc. Publ. no. 1391, U.S. Forest Service, Washington, D.C., Oct. 1980.

Bascietto, J., Hinckley, D., Plafkin, J., and Slimak, M., "Ecotoxicity and Ecological Risk Assessment: Regulatory Applications at EPA," *Environmental Science and Technology,* vol. 24, no. 1, 1990, pp. 10–15.

Cairns, J., Jr., and Mount, D. I., "Aquatic Toxicology," *Environmental Science and Technology,* vol. 24, no. 2, 1990, pp. 154–161.

Camougis, G., *Environmental Biology for Engineers,* McGraw-Hill Book Company, New York, 1981.

Canter, L. W., Risser, P. G., and Hill, L. G., "Effects Assessment of Alternative Navigation Routes from Tulsa, Oklahoma, to Vicinity of Wichita, Kansas," University of Oklahoma, Norman, June 1974.

———, Robertson, J. M., and Westcott, R. M., "Identification and Evaluation of Biological Impact Mitigation Measures," *Journal of Environmental Management,* vol. 33, no. 1, July 1991, pp. 35–50.

Council on Environmental Quality (CEQ), "Environmental Quality," Twenty-first Annual Report, U.S. Government Printing Office, Washington, D.C., 1991, pp. 136–140.

———, "Incorporating Biodiversity Considerations into Environmental Impact Analysis Under the National Environmental Policy Act," CEQ, Washington, D.C., Jan. 1993.

Cowardin, L. M., et al., "Classification of Wetlands and Deepwater Habitats of the United States," FWS/OBS-79/31, U.S. Fish and Wildlife Service, Washington, D.C., Dec. 1979.

Cushwa, C. T., and Kopf, V. E., "State Computerized Fish and Wildlife Information Systems (CFWIS) for Resource Planning and Management," SR85-6, Department of Fisheries and Wildlife Sciences, Virginia Polytechnic Institute and State University, Blacksburg, Sept. 1985.

Darnell, R. M., "Impacts of Construction Activities in Wetlands of the United States," EPA 600/3-76-045, U.S. Environmental Protection Agency, Washington, D.C., Apr. 1976.

Dortch, M. S., "Literature Analysis Addresses the Functional Ability of Wetlands to Improve Water Quality," *The Wetland Research Program Bulletin,* vol. 2, no. 4, U.S. Army Corps of Engineers Waterways Experiment Station, Vicksburg, Miss., Dec. 1992, pp. 1–4.

Drew, D., *Man-Environment Processes,* George Allen and Unwin, London, 1983, pp. 15–17.

Eberhardt, L. L., "Applied Systems Ecology: Models, Data, and Statistical Methods," CONF-760703-7, Battelle Pacific Northwest Laboratories, Richland, Wash., 1976.

Federal Interagency Committee for Wetland Delineation, "Federal Manual for Identifying and Delineating Jurisdictional Wetlands," U.S. Army Corps of Engineers, U.S. Environmental Protection Agency, U.S. Fish and Wildlife Service, and U.S. Soil Conservation Service, Washington, D.C., 1989.

Franck, I., and Brownstone, D., *The Green Encyclopedia,* Prentice-Hall General Reference, New York, 1992, pp. 34–39, 48–49, 98–100, 156–157, 285.

Goudie, A., *The Nature of the Environment: An Advanced Physical Geography,* Basil Blackwell, Oxford, England, 1984, p. 201.

Government Institutes, *Natural Resources Law Handbook,* Rockville, Md., 1991.

Green, K. A., "A Conceptual Ecological Model for Chesapeake Bay," FWS/OBS-78/69, U.S. Fish and Wildlife Service, Washington, D.C., Sept. 1978.

Hall, R. W., Westerdahl, H. E., and Eley, R. L., "Application of Ecosystem Modeling Methodologies to Dredged Material Research," WES-TR-76-3, U.S. Army Engineer Waterways Experiment Station, Vicksburg, Miss., June 1976.

Hammond, A., ed., *The 1992 Information Please Environmental Almanac,* Houghton Mifflin Company, New York, 1991, p. 87.

Hill, L. G., personal communication to author, Oklahoma Biological Station, University of Oklahoma, Norman, 1975.

Hoffman, D. J., Rattner, B. A., and Hall, R. J., "Wildlife Toxicology," *Environmental Science and Technology,* vol. 24, no. 3, 1990, pp. 276–283.

Hornberger, G. M., Kelly, M. G., and Lederman, T. C., "Evaluating a Mathematical Model for Predicting Lake Eutrophication," VPI-WRRC-Bull.-82, Water Resources Research Center, Virginia Polytechnic Institute and State University, Blacksburg, Sept. 1975.

Hughes, R. M., and Larsen, D. P., "Ecoregions: An Approach to Surface Water Protection," *Journal of Water Pollution Control Federation,* Apr. 1988, pp. 486–493.

Jahn, L. R., Cook, C. W., and Hughes, J. D., "An Evaluation of U.S. Army Natural Resource Management Programs on Selected Military Installations and Civil Works Projects," Rep. to Secretary of the Army, Washington, D.C., Oct. 1984.

Kentula, M. E., Brooks, R. P., Gwin, S. E., Holland, C. C., Sherman, A. D., and Sifneos, J. C., *An Approach to Improving Decision Making in Wetland Restoration and Creation,* Island Press, Washington, D.C., 1992.

Kusler, J. A., Harwood, C. C., and Newton, R. B., *Our National Wetland Heritage: A Protection Guidebook,* Environmental Law Institute, Washington, D.C., 1983, p. 7.

—— and Kentula, M. E., eds., *Wetland Creation and Restoration: The Status of the Science,* Island Press, Washington, D.C., 1990, pp. xvii–xxv.

Littell, R., *Endangered and Other Protected Species: Federal Law and Regulation,* Bureau of National Affairs, Washington, D.C., 1992.

Mannion, A. M., and Bowlby, S. R., *Environmental Issues in the 1990s,* John Wiley and Sons, West Sussex, England, 1992, pp. 219–221.

Marsh, W. M., *Landscape Planning: Environmental Applications,* 2d ed., John Wiley and Sons, New York, 1991, pp. 260–262, 271–273, 287–290.

Martin, C. O., "U.S. Army Corps of Engineers Wildlife Resources Management Manual," EL-86-25, U.S. Army Engineer Waterways Experiment Station, Vicksburg, Miss., July 1986.

Odum, H. T., "Use of Energy System Diagrams for Environmental Impact Statements," *Proceedings of the Conference on Tools for Coastal Zone Management,* Marine Technology Society, Washington, D.C., Feb. 1972, pp. 197–213.

Office of Technology Assessment, "Technologies to Maintain Biological Diversity," OTA-F-330, U.S. Congress, Washington, D.C., Mar. 1987.

Omernik, J. M., "Ecoregions of the Conterminous United States," *Annals of the Association of American Geographers,* vol. 77, no. 1, 1987, pp. 118–125.

—— and Gallant, A. L., "Defining Regions for Evaluating Environmental Resources," EPA 600/D-89-265, U.S. Environmental Protection Agency, Corvallis, Ore., 1989.

Patzer, R. G., "Concentration Factors and Transport Models for Radionuclides in Aquatic Environments—A Literature Report," EPA 600/3-76-054, Environmental Monitoring and Support Laboratory, U.S. Environmental Protection Agency, Las Vegas, May 1976.

Payne, N. F., *Techniques for Wildlife Habitat Management of Wetlands,* McGraw-Hill Book Company, New York, 1992.

Reid, W. V., "Conserving Life's Diversity: Can the Extinction Crisis Be Stopped?" *Environmental Science and Technology,* vol. 26, no. 6, 1992, pp. 1090–1095.

Ryan, J. C., "Conserving Biological Diversity," Chap. 2 in *State of the World, 1992,* L. Starke, ed., W. W. Norton and Company, New York, 1992, pp. 9–26.

Salvesen, D., *Wetlands: Mitigating and Regulating Development Impacts,* Urban Land Institute, Washington, D.C., 1990, pp. 70, 106–107.

Sather, J. H., and Smith, R. D., "An Overview of Major Wetland Functions," FWS/OBS-84/18, U.S. Fish and Wildlife Service, Washington, D.C., Sept. 1984.

Southerland, M. T., "Consideration of Terrestrial Environment in the Review of Environmental Impact Statements," *The Environmental Professional,* vol. 14, no. 1, 1992, pp. 1–9.

Spellerberg, I. F., *Monitoring Ecological Change,* Cambridge University Press, Cambridge, 1991, pp. 93–108.

"The Spotted Owl—Whooo Will Prevail?" *National Association of Environmental Professionals Newsletter,* vol. 17, no. 3, July 1992, p. 4.

Starfield, A. M., and Bleloch, A. L., *Building Models for Conservation and Wildlife Management,* Macmillan Publishing Company, New York, 1986.

Tivy, J., and O'Hare, G., *Human Impact on the Ecosystem,* Oliver and Boyd, Edinburgh, 1981, pp. 21–22, 166.

Turner, K., and Jones, T., *Wetlands: Market and Intervention Failures—Four Case Studies,* Earthscan Publications, London, 1991.

U.S. Army Corps of Engineers, "Corps of Engineers Wetlands Delineation Manual," Y-87-1, U.S. Army Engineer Waterways Experiment Station, Vicksburg, Miss., Jan. 1987.

——, "Long-Range Maintenance Dredging Program—Grays Harbor and Chehalis River Navigation Project, Operation and Maintenance," Final Environmental Impact Statement Supplement no. 2, Seattle District, Seattle, Oct. 1980.

U.S. Congress, "The Endangered Species Act Amendments of 1978," P.L. 95-632, 95th Congress, Washington, D.C., Nov. 10, 1978.

U.S. Department of the Interior, "Final Environmental Impact Statement: Permanent Regulatory Program Implementing Sec. 501(b) of the Surface Mining Control and Reclamation Act of 1977," Washington, D.C., pp. BIII-71–BIII-81, 1979.

U.S. Environmental Protection Agency (EPA), "America's Wetlands: Our Vital Link Between Land and Water," OPA-87-016, U.S. Environmental Protection Agency, Office of Wetlands Protection, Washington, D.C., Feb. 1988, p. 6.

U.S. Fish and Wildlife Service, "Endangered Species Information System—Project Brief," Division of Endangered Species and Habitat Conservation, Washington, D.C., May 1988.

——, "Impacts of Coal-Fired Power Plants on Fish, Wildlife, and Their Habitats," Executive Summary, Ann Arbor, Mich., 1978.

U.S. General Accounting Office, "Endangered Species Act: Types and Numbers of Implementing Actions," GAO/RCED-92-131BR, Washington, D.C., May 1992.

——, "Wetlands Overview: Federal and State Policies, Legislation, and Programs," GAO/RCED-92-79FS, Washington, D.C., Nov. 1991.

Watts, D. C., and Loucks, O. L., "Models for Describing Exchange within Ecosystems," Institute for Environmental Studies, University of Wisconsin, Madison, 1969.

Westman, W. E., *Ecology, Impact Assessment, and Environmental Planning,* John Wiley and Sons, New York, 1985, pp. 4–14.

White, I. D., Mottershead, D. N., and Harrison, S. J., *Environmental Systems,* George Allen and Unwin, London, 1984, p. 163.

World Commission on Environment and Development, *Our Common Future,* Oxford University Press, Oxford, England, 1987, pp. 43–66.

World Conservation Monitoring Centre, *Global Biodiversity—Status of the Earth's Living Resources,* Chapman and Hall, London, 1992.

CHAPTER 11

Habitat-Based Methods for Biological-Impact Prediction and Assessment

Several habitat-based methods have been developed since the mid-1970s to enable a structured approach for biological-impact prediction and assessment. The purpose of this chapter is to review two of these methods, HES and HEP, and discuss their advantages and limitations. Brief information on several other methods is also included, along with summary information on mitigation banking.

HABITAT EVALUATION SYSTEM

In 1976, the Lower Mississippi Valley Division of the U.S. Army Corps of Engineers developed a habitat-based approach, called the "habitat evaluation system" (HES), for evaluation of water resources projects in the lower Mississippi Valley area (U.S. Army Corps of Engineers, 1976). The approach focused on seven habitat types: freshwater stream, freshwater lake, bottomland hardwood forest, upland hardwood forest, open (nonforest) lands, freshwater river swamp, and freshwater nonriver swamp. A refined version of the HES was developed in 1980 and is currently being used in several Corps district offices in the planning and formulation of water resource

developments and in the regulatory program under Section 404 of the Clean Water Act of 1977 (U.S. Army Corps of Engineers, 1980).

The fundamental assumption underlying HES is that the presence or absence, abundance, and diversity of animal populations in a habitat or community are determined by basic biotic and abiotic factors that can be readily quantified. The carrying capacity of a habitat for a given species or group of species is correlated with basic chemical, physical, and biotic characteristics of the habitat (U.S. Army Corps of Engineers, 1980). Although complex biological interactions such as predation, competition, and diseases also affect fish and wildlife populations, HES assumes that if necessary habitat requirements for a species are present, then a viable population of the species will be, or could be, supported by the habitat. HES does not treat individual species, although the techniques can be modified to evaluate habitats for specific species. Instead, general habitat characteristics are used that indicate quality for fish and wildlife populations as a whole.

The 1980 version of HES includes information for addressing two aquatic habitats (streams

TABLE 11.1

HABITAT TYPES INCORPORATED IN THE HABITAT EVALUATION SYSTEM (HES)

HES version	Habitat types included
HES (1976)	Freshwater stream Freshwater lake Bottomland hardwood forest Upland hardwood forest Open (nonforest) land Freshwater river swamp Freshwater nonriver swamp
HES-80 (1980)	Stream Lake Wooded swamp Upland forest Bottomland hardwood forest Open land Terrestrial wildlife value of aquatic habitat

Source: Adapted from U.S. Army Corps of Engineers (1976, 1980).

TABLE 11.2

AQUATIC-ECOSYSTEM KEY VARIABLE WEIGHTS

Streams		Lakes	
Variables	Weights[a]	Variables	Weights
1. Fish Species Association	30	1. Total Dissolved Solids	30
2. Sinuosity Index	20	2. Spring Flooding Index	20
3. Total Dissolved Solids	20	3. Mean Depth	15
4. Turbidity	10	4. Chemical Type	15
5. Chemical Type	10	5. Turbidity	15
6. Benthic Diversity	10	6. Shoreline Development Index	5
		7. Total Fish Standing Crop	[b]
		8. Sport Fish Standing Crop	[b]

[a]Weights for stream variables may be reassigned based on knowledge of local area or other considerations. A paired comparison technique may be used to adjust weights (Dean and Nishry, 1965).
[b]If fish data are available, weights may be reassigned or the fish data may be used solely to determine habitat quality. The paired comparison technique may be used to reassign weights (Dean and Nishry, 1965).
Source: U.S. Army Corps of Engineers, 1980, p. 25.

and lakes) and five terrestrial habitats (wooded swamps, upland forests, bottomland hardwood forests, open lands, and terrestrial wildlife value of aquatic habitats), as shown in Table 11.1. The HES approach consists of determining the quality of a habitat type using functional curves relating habitat quality to quantitative biotic and abiotic characteristics of the habitat. Habitat size and quality considerations are combined in assessing project impacts. As shown in Figure 11.1, the HES procedure involves six steps for evaluating impacts of a development project. These steps are (1) obtaining habitat-type or land-use acreage data, (2) deriving habitat quality index

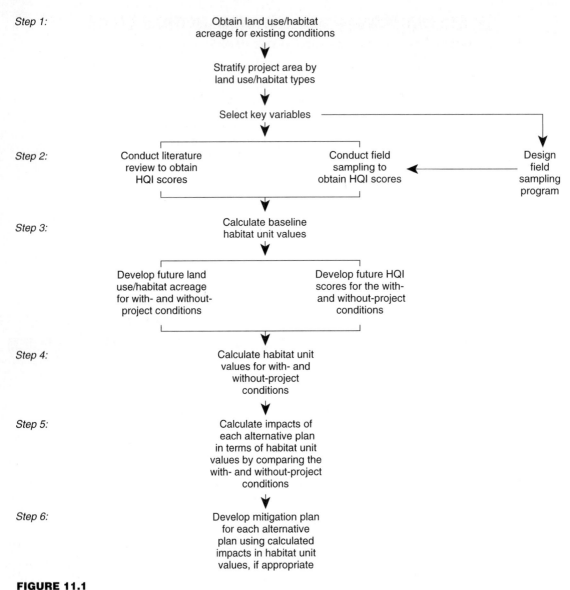

Step 1: Obtain land use/habitat acreage for existing conditions

Stratify project area by land use/habitat types

Select key variables

Step 2: Conduct literature review to obtain HQI scores   Conduct field sampling to obtain HQI scores   Design field sampling program

Step 3: Calculate baseline habitat unit values

Develop future land use/habitat acreage for with- and without-project conditions   Develop future HQI scores for the with- and without-project conditions

Step 4: Calculate habitat unit values for with- and without-project conditions

Step 5: Calculate impacts of each alternative plan in terms of habitat unit values by comparing the with- and without-project conditions

Step 6: Develop mitigation plan for each alternative plan using calculated impacts in habitat unit values, if appropriate

FIGURE 11.1
Basic Steps in Conducting an HES Evaluation (Modified from U.S. Army Corps of Engineers, 1980, p. 10).

(HQI) scores, (3) deriving habitat unit values (HUVs), (4) projecting HUVs for future with- and without-project conditions, (5) using HUVs to assess impacts of project alternatives, and (6) determining mitigation requirements, if any.

The first step in using HES is to delineate the acreage or hectares of each aquatic and terrestrial habitat type in the project area. The land-use or habitat-acreage data must be derived for existing conditions and must be projected for

TABLE 11.3

TERRESTRIAL-ECOSYSTEM KEY VARIABLE WEIGHTS

Wooded swamps

1. Species Associations	14
2. Coverage	13
3. Percent Inundation	13
4. Ground Cover—Understory Coverage	13
5. Mast Proximity	13
6. Number Trees ≥ 16 inches DBH	12
7. Tract Size	12
8. Number of Snags	10
	100

Upland forests

1. Species Associations	17
2. Mast Bearing Trees	16
3. Percent Understory	14
4. Percent Groundcover	15
5. Large Trees	14
6. Tract Size	13
7. Number of Snags	11
	100

Bottomland hardwood forests

1. Species Associations	17
2. Mast Bearing Trees	16
3. Percent Understory	14
4. Percent Groundcover	14
5. Large Trees	14
6. Tract Size	14
7. Number of Snags	11
	100

Open lands

1. Land Use Type	15
2. Land Use Diversity	15
3. Distance to Cover	15
3. Distance to Woods	14
5. Frequency of Winter Flooding	14
6. Tract Size	13
7. Perimeter Development Index	14
	100

Terrestrial wildlife value of aquatic habitats

1. Percent of Waterbody ≤ 12 inches Deep from July through February	11
2. Coverage by Aquatic Plants	12
3. Distance to Road or Other Disturbance	9
4. Water Depth in August	9
5. Distance to River	10
6. Snags/Brush Coverage	8
7. Flooding Frequency	11
8. Winter Overflow	11
9. Distance from Woods	8
10. Size of Waterbody	11
	100

Source: U.S. Corps of Army Engineers, 1980, pp. 37–38.

future without-project and with-project conditions for each alternative plan. The second step in using HES consists of deriving HQI scores for each land-use category or habitat type. Data are obtained on several key variables for each habitat type from field-measurement programs, literature, and historical information. Tables 11.2 and 11.3 display the variables and their relative weights for each habitat type. The measurements for each variable, such as total dissolved solids for a lake, or number of mast-bearing trees per acre for a forest type, are converted into an HQI score using a specific functional curve for that key variable and habitat type. The HQI score is based on a scale of 0 to 1.0, with 1.0 being the maximum value or rating. For example, Figure 11.2 shows the functional curve for total dissolved solids in streams. The HQI is a function of the general value of a habitat for fish or wildlife populations.

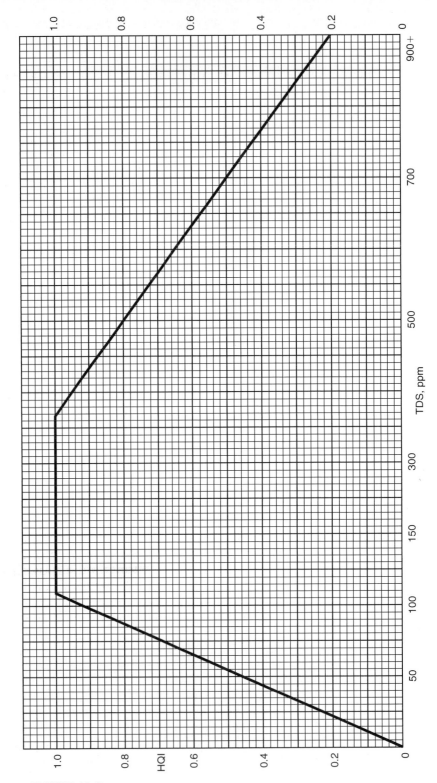

FIGURE 11.2
Functional Curve for Total Dissolved Solids in a Stream (U.S. Army Corps of Engineers, 1980, p. 101).

Each HQI score for a given habitat type is assigned a weight between 0 and 100, which reflects the relative importance of that key variable in describing overall habitat quality. (Tables 11.2 and 11.3). The weights were initially assigned through use of an informal nominal-group-process technique by about 20 biologists from the Corps of Engineers. The refined HES (the 1980 version) had the weights assigned through use of a paired-comparison technique (Dean and Nishry, 1965). The product of the HQI score for the variable and the associated weight yields a weighted HQI score for that variable. The weighted HQI scores for the array of key variables for a particular habitat type are summed, and this total is divided by 100 to yield an aggregate HQI score for the habitat type (U.S. Army Corps of Engineers, 1980). Table 11.4 contains an illustration of these calculations.

The third step in conducting HES evaluations is to combine the habitat-type or land-use size data (acreage or hectares) and the associated HQI scores to obtain an overall-habitat-value rating. Both the size or quantity and the quality of a particular habitat are considered important in evaluating the overall value of a habitat for fish and wildlife populations. The size of a given habitat type is multiplied by the aggregate HQI score to produce a habitat unit value

---

TABLE 11.4

HES FORM SHOWING TYPICAL FIELD DATA FOR A BOTTOMLAND-HARDWOOD-FOREST PLOT AND CALCULATION OF TOTAL HQI SCORE

Project: Bottomland Hardwood Habitat Type   **Date:**

Site No.   1A

Location: 20 Chains West of Intersection of   **HQI Score:** 0.76
Hwy 302 & 31, Warren Co, MS

Parameter	Data	HQI score	Curve wt.	Wt. HQI score
1. Species Association	Hackberry-elm-ash	0.96	17	16.3
2. Number of Mast Trees	3 (1 R. Oak & 2 W. Oak)	1.00	16	16.0
3. Percent Cover–understory	20% Palatable & in reach. 6 spp.	0.32	14	4.5
4. Percent Cover—groundcover	35% Desirable, 3 + spp.	0.46	14	6.4
5. No. trees 18″ DBH & Greater	2—one is over 24″	0.80	14	11.2
6. Tract Size	1,900 acres, 100% wooded	0.80	14	11.2
7. No. Snags	6 snags	0.92	11	10.2
	Total			75.8

Note: Deer browse evident—tract appears to have been logged about 3 years ago.

Understory species		Overstory	Groundcover
Swamp Privet		Hackberry	Dewberry
Swamp Dogwood		Elm	Greenbrier
Hackberry	Palatable, in reach	Ash	Panicum
Am. Elm		Nuttall Red Oak	Carex
Poison Ivy		Overcup Oak	Green Ash
Greenbrier		Gum	

$$\text{Total HQI} = \frac{75.8}{100} = 0.76$$

Source: U.S. Army Corps of Engineers, 1980, p. 13.

(HUV) for the habitat (U.S. Army Corps of Engineers, 1980). Thus,

$$\text{HUV} = \text{habitat quality index (HQI)} \quad (11.1)$$
$$\times \text{ habitat size (acreage or hectares)}$$

The fourth step in the HES entails making projections of HUVs over the project life and for various alternative plans (U.S. Army Corps of Engineers, 1980). HUVs must be developed for future conditions without the project and for future conditions corresponding to each project alternative. In addition, HUVs must be derived for each time increment, usually 10-year intervals, over the project life for without-project conditions and each alternative plan. Projected HUVs are based on estimated changes in land use or habitat size caused by such influences as land-use changes resulting from clearing. Estimated changes in land use or habitat acreage can be reasonably developed based on engineering and related planning studies. Changes in the HQI scores over time require projections of changes in all variables used to describe habitat types; these projections may be difficult to achieve with any scientific certainty. Table 11.5 shows an example HES data set and HUV calculations for a bottomland-forest habitat.

After completing step 4, the necessary data are available for assessing impacts (step 5). First, total and/or annualized HUVs are calculated for each habitat type for the without-project conditions and the with-project conditions for each alternative. As shown in Figure 11.3, total HUVs for a given condition are calculated by integrating the functions derived by plotting HUVs versus project time intervals. The area under the curve is the integral of this function. The area may simply be computed geometrically by dividing the area under the curve into trapezoids and/or rectangles, calculating the area of each subdivision, and summing the area values. For complex projects with several alternatives and habitat types, computer software is available for making HES computations for steps 1 through 5 (U.S. Army Corps of Engi-

neers, 1980). Inputs to this software program are the raw data on land use or habitat size and the HQI scores. The program calculates both total and annualized HUVs for each habitat type for each alternative plan.

Impacts of each alternative plan on each habitat type are estimated by subtracting the without-project HUV from the with-project HUV for that habitat type. Either total or annualized HUVs may be used. The impact of the alternative plan is positive, or beneficial, if the difference between the without- and with-project HUVs is positive. The impact is negative, or there is a project-induced environmental loss, if the difference is negative. Thus,

$$\text{Impact} = \text{with-project HUV}$$
$$- \text{without-project HUV} \quad (11.2)$$

Gains or losses in HUVs for each plan may be summed for all terrestrial habitats and all aquatic habitats to depict the total impacts of each plan on these two major types of environment. Table 11.6 shows a summary of project impacts for the example in Table 11.5 and Figure 11.3. Trade-off analyses and comparisons of the impacts of each alternative plan can be readily made by comparing the impacts expressed as total or annualized HUVs.

The sixth step in an HES evaluation is to determine mitigation requirements for alternative plans, if appropriate, using the HUV data derived in steps 1 through 5. In HES, "mitigation" is defined as any measures taken to return the with-project environmental quality of the area to the same level as the without-project condition. This may be accomplished by incorporation of environmental quality considerations into the formulation of each plan to minimize impacts ("good planning") and by developing or incorporating separate project features, such as purchase of lands or building of water-level-control structures to compensate for project-induced losses to the environment. HES can be readily applied to determine the amount and type of mitigation lands required to

TABLE 11.5

EXAMPLE HES DATA SET AND HUV CALCULATIONS FOR BOTTOMLAND-FOREST HABITAT

Project life (yr)	Future without			Alternative plan A			Alternative plan B			Alternative plan C		
	Acres	HQI	HUV	Acres	HQI	HUV	Acres	HQI	HUV	Acres	HQI	HUV
0	1,000	0.80	800	1,000	0.80	800	1,000	0.80	800	1,000	0.80	800
10	900	0.80	720	725	0.75	544	900	0.75	675	1,000	0.85	850
25	850	0.80	680	600	0.75	450	900	0.70	630	1,000	0.85	850
50	800	0.80	640	500	0.70	350	900	0.65	585	1,000	0.90	900
TOTAL HUV[a]			34,600			24,175			32,349			42,875
ANNUALIZED HUV[a]			692			483			647			857

[a]See Figure 11.3 for example of calculation of annualized and total HUV for plan A and Future Without-Project Conditions.
Source: U.S. Army Corps of Engineers, 1980, p. 15.

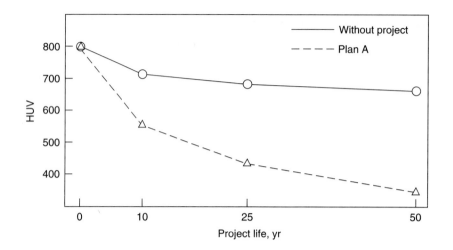

1. Total HUV without project = 1/2 (800 + 720) 10 + 1/2 (720 + 680) 15 +
   1/2 (680 + 640) 25 = 34,600

$$\text{Annualized HUV without project} = \frac{34{,}600}{50 \text{ yr}} = 692$$

2. Total HUV with plan A = 1/2 (800 + 544) 10 + 1/2 (544 + 450) 15 +
   1/2 (450 + 350) 25 = 24,175

$$\text{Annualized HUV with plan A} = \frac{24{,}175}{50 \text{ yr}} = 483$$

3. Bottomland forest losses (total HUV) = 24,175 − 34,600 = −10,425 HUV
   Bottomland forest losses (annualized HUV) = 483 − 692 = −209 HUV

FIGURE 11.3
Example Calculation of HUV Losses Using Data on Bottomland Forests from Table 11.5 (U.S. Army Corps of Engineers, 1980, p. 17).

offset project losses. The size of mitigation lands required to compensate for a particular type of wildlife or terrestrial loss expressed in terms of HUVs can be estimated from the following (U.S. Army Corps of Engineers, 1980):

$$\frac{\text{Size}}{\text{(acreage)}} = \frac{\text{annualized HUV lost}}{\substack{\text{annualized gain in HQI} \\ \text{for mitigation lands}}} \quad (11.3)$$

Assigning of the HQI score to the types of terrestrial habitat to be used for mitigation en-

tails special considerations (U.S. Army Corps of Engineers, 1980). If, for each HUV or acre of habitat lost as a result of implementing an alternative plan, a like acreage (or amount) of habitat with an equivalent HUV (acreage) is purchased, no gain in terrestrial habitat value is achieved, since no gain in habitat value is made by direct replacement. To obtain gains in HUVs for mitigation from an acre of habitat, that acre must be managed to increase its existing value as terrestrial habitat. Only the increase in the HQI score of the habitat due to management and

TABLE 11.6

SUMMARY OF HABITAT IMPACTS ON BOTTOMLAND FOREST IN TOTAL AND ANNUALIZED HUVs FOR EARLIER EXAMPLE

	Alternative plan		
	A	B	C
Total HUV	$-10,425^a$	$-2,251$	$+8,275$
Annualized HUV	-209^a	-45	$+165$
Annualized mitigation[b,c] land required	697	150	d

[a]Computed by subtracting the without-project HUV from the with-project HUV.
[b]Derived using a management potential of 0.3 HQI (Annualized HQI of lands to be bought with management minus Annualized HQI score of proposed mitigation lands without management equals $0.9 - 0.6 = 0.3$).
[c]Acreage calculated using Eq. (11.3):

$$\text{Acres} = \frac{\text{Annualized HUV lost}}{\text{Management Potential (HQI)}} \qquad (11.4)$$

[d]Since Plan C results in net benefits to bottomland forest, no mitigation is required.
Source: U.S. Army Corps of Engineers, 1980, p. 19.

development can be used in the equation for determining required acreage of mitigation lands. The gain in HQI score derived from management of the mitigation land is termed "management potential" and is expressed in HQIs. Management potential is determined by comparing the with- and without-management annualized HQIs of the type of habitat to be purchased for mitigation. Thus, management potential of a given habitat type can be calculated as follows (U.S. Army Corps of Engineers, 1980):

Management potential (HQI)
= with-management annualized HQI
− without-management annualized HQI
(11.5)

In this case, the annualized HQI score used in the equation for calculating required mitigation acreage is the management potential in HQI. Calculation of mitigation-land requirements for the example data set is shown in Table 11.6. It should be noted that some types of terrestrial habitat cannot be feasibly managed, in which case another mitigation approach would be necessary.

Another factor to be considered in determining the amount of necessary mitigation lands is the effect of habitat preservation. If land-use and environmental studies project that the specific habitat or land proposed to be purchased for mitigation will be degraded in environmental quality due to land clearing or other perturbations, credit may be claimed for preserving the habitat by acquisition. This "preservation credit," in terms of HQI scores, is calculated in the same general way as the management potential; for example (U.S. Army Corps of Engineers, 1980):

Preservation credit (HQI)
= with-purchase annualized HQI
− without-purchase annualized HQI
(11.6)

The preservation credit in HQIs can be used in the equation for calculating required mitigation acreage or, if appropriate, added to the management-potential HQI score for insertion in the equation.

HES exhibits certain characteristic advantages and disadvantages. The disadvantages are expected to be minimized over time and with further refinement of the HES. The advantages are (U.S. Army Corps of Engineers, 1980, pp. 82–83)

1. Highly dissimilar environmental characteristics are quantifiable in standardized terms (Habitat Quality Index) which can be easily understood and compared over time and among habitat types. Adverse and beneficial effects of each project alternative are clearly identified in comparable terms (Habitat Unit Values) for impact evaluation and trade-off analyses.
2. The system provides an objective method for comparing the environmental effects of project alternatives. HES is applicable to major ecosystems within the Lower Mississippi Valley Region, and with revision to curves, weights, and variables, can be applied to many other areas of the United States, and even internationally. Therefore, the concepts of HES can be used to develop a habitat-based methodology for habitat types other than those included in HES. The approach used in the development should include the identification of pertinent factors (parameters, or variables), assignment of relative importance weights to the factors, and the delineation of functional relationships for each factor.
3. HES results are reproducible. Functional curves are based on quantitative measurements of key variables which are converted to an HQI score using a functional curve. Thus, once the key variable is quantified, the HQI is fixed. Also, the system was developed to include consideration of previously unquantifiable environmental amenities, such as non-game species and esthetics.
4. Use of the HES is rapid and efficient and requires a minimum of field and laboratory data on terrestrial habitats; many of the variable measurements can be quickly determined by visual estimates. Data for most aquatic functional curves can be obtained from historical data sources.
5. The HES is a flexible method. If biologists in a specific geographical area feel that the functional curves do not correctly represent conditions of that area, curves can be altered and reweighted and other curves can be added. If weights for HQIs are questionable for technical reasons in a particular locale, they can be eliminated or revised without greatly affecting the overall results.

The disadvantages of HES are

1. "The curves and the weights assigned to each variable are subjective to some extent. The data presented represent the consensus of professionals consulted during the development of this system and are generally supported by the literature. Further refinement of the system and modification to address local or regional needs and unusual habitats are continuing. Research is needed to provide data for verifying functional curves and correlating biotic and abiotic variables to habitat quality" (U.S. Army Corps of Engineers, 1980, p. 81).
2. "The HES describes habitat quality for a broad range of species, rather than attempting to predict the density for a particular species (e.g., whitetail deer). There was a conscious effort made during the development of HES to avoid assuming that all species in a habitat are or are not of equal value, since neither of these premises is presently defensible. It is stressed, however, that if population projections for one or more species is desired, this information can be obtained by analysis of the data generated by HES. For example, a wildlife biologist could assess potential carrying capacity of habitat for deer from the detailed analysis of browse, cover vegetation, and other data that HES provides" (U.S. Army Corps of Engineers, 1980, p. 81).
3. Information needs to be developed on collecting the necessary field measurements and data for the pertinent variables in HES.
4. Information needs to be developed on the methods or approaches which can be used to predict temporal changes in habitat size and also for each variable in the HQI for each habitat type.

HABITAT EVALUATION PROCEDURE

The most frequently used habitat-based approach in environmental impact studies in the United States is the "habitat evaluation procedure" (HEP). HEP was conceived in 1972 and promulgated in 1976 by the U.S. Fish and Wildlife Service for use in evaluating major federal water projects. The objectives of HEP were (Schamberger and Farmer, 1978)

1. To develop methodologies to quantitatively assess baseline habitat conditions for fish and wildlife in nonmonetary terms
2. To provide a uniform system for predicting impacts on fish and wildlife resources
3. To display and compare the beneficial and adverse impacts of project alternatives on fish and wildlife resources
4. To provide a basis for recommending project alterations that would compensate for or mitigate adverse effects on fish and wildlife resources
5. To provide data to decision makers and the public from which sound resource decisions can be made

HEP has been applied to numerous types of projects; its use is not limited to water-resource project applications. The HEP-80 version was released by the U.S. Fish and Wildlife Service in 1980 (U.S. Fish and Wildlife Service, 1980).

HEP is a method which can be used to document the quality and quantity of available habitat for selected wildlife species. ''Wildlife'' refers to both aquatic and terrestrial animal species. HEP provides information for two general types of wildlife habitat comparisons: (1) the relative value of different areas at the same point in time and (2) the relative value of the same area at future points in time. By combining the two types of comparisons, the impact of proposed or anticipated land- and water-use changes on wildlife habitat can be quantified (U.S. Fish and Wildlife Service, 1980). HEP is based on the assumption that habitat for selected wildlife species can be described by a ''habitat suitability index'' (HSI) value. This value (which ranges from 0.0 to 1.0) is multiplied by the area of available habitat to obtain ''habitat units (HUs),'' which are used in the comparisons described above. The reliability of HEP and the significance of HUs are directly dependent on the ability of the user to assign a well-defined and accurate HSI to the selected evaluation species. Figure 11.4 shows the generalized evaluation process involved when using HEP.

The first step of a HEP application consists of (1) defining the study area, (2) delineating cover types, and (3) selecting evaluation species. The study area should include those areas where biological changes related to the land- or water-use proposal under study are expected to occur. This area should include areas that will be affected, either directly (for example, by engineering structures) or indirectly (for example, by human-use trends), by the proposed use. Additionally, the study area should include contiguous areas with significant biological linkages to the site where actual physical impacts are expected to occur (U.S. Fish and Wildlife Service, 1980).

Delineation of cover types (or habitat types) is also required as part of defining study limits. Use of cover types serves three basic functions in HEP. First, cover types facilitate the selection of evaluation species. Second, extrapolation of data from sampled areas to unsampled areas can be done with some confidence if the study area is divided into relatively homogeneous areas, by cover type, thus reducing the amount of sampling necessary. Finally, separation of the study area into cover types facilitates the treatment of HEP data (U.S. Fish and Wildlife Service, 1980). Examples of cover types used in HEP include deciduous forest, coniferous forest, grassland, residential woodland, and medium-sized warmwater stream. The concept of cover types used in HEP is analogous to that of habitat types used in HES.

''Evaluation species'' (or indicator species) are used in HEP to quantify HUs (U.S. Fish and Wildlife Service, 1980). A typical HEP study will incorporate four to six species. A method incorporating the following technical considerations and practical approaches should be used in selecting the number and types of species to be evaluated:

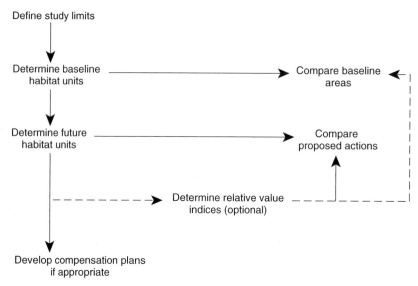

FIGURE 11.4
Generalized Evaluation Process Using HEP (After U.S. Fish and Wildlife Service, 1980, p. 102 ESMI).

1. Species known to be sensitive to specific land-use actions related to the proposed project. The species selected with this approach serve as "early warning" or indicator species for the affected wildlife community.
2. Species that perform a key function in an ecological community because of their roles in nutrient cycling or energy flows.
3. Species that represent groups of species which utilize a common environmental resource (guilds). A representative species is selected from each guild, and predicted environmental impacts for the selected species can be extended with some degree of confidence to other guild members. The recommended procedure for selecting terrestrial species involves categorizing vertebrate species in an ecological community according to their feeding and reproductive guilds. "Feeding guilds" are defined in terms of feeding mode (e.g., carnivore, herbivore, or omnivore) and strata locations in the ecosys-

tem where the foods are obtained (e.g., canopy, shrub layer, or surface). "Reproductive guilds" are defined only in terms of strata locations where reproduction occurs. "Aquatic guilds" can be developed to aggregate species into groups with similar habitat requirements. The aquatic-guild structure can have several levels, and the number of descriptors within a level can vary. Aquatic guilds may be based on (1) feeding habits, (2) reproductive habits, (3) tolerance and response to temperature, (4) preferred habitat, or (5) tolerance to the results of a potential habitat alteration, such as turbidity-siltation. After the descriptors have been established, the aquatic species are listed and categorized by guild descriptor. If more than one species has been entered into any terrestrial or aquatic guild, at least one should be selected to represent the guild. This within-guild selection can be arbitrary or according to a ranking scheme. Suggested ranking criteria include anticipated sensitivity to proposed land-use impacts, community role in nutrient cycling or energy flow, geographic range, cover-type utilization, and the availability of habitat data.

4. Species with high public interest, economic value, or both.
5. Species for which HSI models have been developed (these models, to be described later, are tailored to evaluations of individual species). In addition, environmental data necessary for the use of the model should be available or easily obtainable.
6. Species recommended for inclusion by the pertinent office of the U.S. Fish and Wildlife Service.

A HEP analysis is structured around the calculation of HUs for each evaluation species in the study area. The number of habitat units is defined as the product of the habitat suitability index (quality) and the total area of available habitat (quantity) (U.S. Fish and Wildlife Service, 1980):

$$HU = (HSI) (habitat\ size) \qquad (11.7)$$

The total area of available habitat for an evaluation species includes all areas that can be expected to provide some support to the evaluation species. The total area of available habitat is calculated by summing the areas of all cover types likely to be used by the evaluation species. If the study area is not subdivided by cover type, the total area of available habitat is identical to the entire study area. The objective of defining the total area of available habitat is to delineate only those areas that require HSI determinations. The total area of available habitat will vary between evaluation species if cover-type-use patterns are different; therefore, HSIs for each evaluation species may apply to different subareas (i.e., available habitat).

HSI values are determined through the use of HSI models. The HSI models are usually pre-sented in three basic formats: (1) graphic, (2) word, and (3) mathematical. The graphic format is a representation of the structure of the model and displays the sequential aggregation of variables into an HSI. Following this, the model relationships are discussed and the assumed relationships between variables, components, and HSIs documented. This discussion of model relationships provides a working version of the model and is, in effect, a word model. Finally, the model relationships are described in mathematical language, mimicking as closely and as simply as possible the preceding word descriptions (Schamberger, Farmer, and Terrell, 1982). The basic perspective toward HSI models is that they represent hypotheses of species-habitat relationships, rather than statements of proven cause and effect relationships.

The HSI model used in a HEP analysis must be in index form, with the index representing a ratio between some value of interest and a standard of comparison. For HEP purposes, the value of interest is an estimate of habitat conditions in the study area, and the standard of comparison is the combination of optimum habitat conditions for the same evaluation species. Therefore,

$$HSI = \frac{study\ area\ habitat\ conditions}{optimum\ habitat\ conditions} \qquad (11.8)$$

where the numerator and denominator have the same units of measure; thus, the HSI ranges between 0.0 and 1.0. The ideal goal of an HSI model is to produce an index with a proven, quantified, positive relationship to carrying capacity (i.e., units of biomass per unit area or units of biomass production per unit area).

HSI models are typically developed by a single expert, or perhaps a small group of experts, to suit the particular species. The models represent the collective best judgment of the critical habitat requirements of the species. The following summary of one such model—developed for the marten *(Martes americana),* a small animal (Allen, 1982)—will serve as an example of an HSI model.

Background

The marten inhabits late successional forest communities throughout North America. The species is most abundant in association with mature coniferous forests, but also inhabits forests of mixed deciduous and coniferous species. The marten is mostly carnivorous, generally nocturnal, and active throughout the year. Marten consume a wide variety of food items throughout the year. Invertebrates, berries, and passerine birds were the most frequent food items recorded from spring through fall in a Montana study. However, mammals were the most important food item on an annual basis, with the highest utilization of mammalian prey occurring during the winter months. Food availability is probably the most important factor affecting the distribution of marten. The reproductive requirements of the marten are assumed to be comparable to cover requirements.

HSI Model Applicability

The HSI model described below was developed for application in boreal coniferous forests of the western United States. It was developed to evaluate the potential quality of winter habitat in evergreen forests (EF) for marten. The winter cover requirements of this species are more restrictive than cover requirements for marten during other seasons of the year. It is assumed that, if adequate winter cover is available, habitat requirements throughout the balance of the year will not be limiting. "Minimum habitat area" is defined as the minimum amount of contiguous habitat that is required before an area will be occupied by a species. Information on the minimum habitat area for the marten was not reported in the literature, but home ranges in the western United States are approximately

2.38 km^2 (0.92 mi^2) for males. Based on this information, it is assumed that at least 2.59 km^2 (1 mi^2) of suitable habitat must be available before an area will be occupied by this species. If less than 2.59 km^2 (1 mi^2) of suitable habitat is present, the HSI is assumed to be 0.0.

HSI Model Description

All winter habitat requirements of the marten can be satisfied within boreal evergreen forests. The marten is, therefore, treated as utilizing evergreen forests only, and habitat evaluation using this model only considers the quality-of-life requisites provided by evergreen forests. It is assumed that food availability will not be limiting for the marten if adequate cover is present. Figure 11.5 illustrates the relationships of habitat variables, life requisites, and cover types for the marten.

The marten may range through various forested and nonforested cover types throughout the spring, summer, and fall. Based on the literature, mature stands of evergreen trees, particularly spruce and fir, are required during the winter months in order to provide adequate protective and thermal cover. Suitable winter cover is a function of the successional stage of the stand, the percent of the stand which is composed of spruce or fir, the total percent canopy closure of the stand, and the amount of downfall in the stand. Stands of mature to overmature coniferous forest, composed of 40 percent fir or spruce, with a total canopy closure greater than 50 percent, are assumed to provide near-optimal winter habitat. Forest stands which contain an abundance of downfall or windthrow are assumed to have a higher winter cover value, because such materials provide refuge sites for the marten and accessibility to small mammals active under the snowpack. Although small-diameter woody debris on the forest floor will provide cover for rodents, marten require the presence of fallen

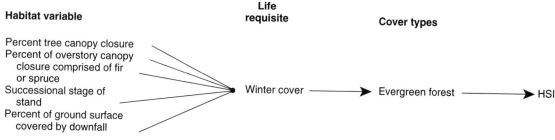

FIGURE 11.5
Relationships of Habitat Variables, Life Requisites, and Cover Types in the Marten HSI Model (Allen, 1982, p. 4).

snags or large logs on the ground surface to provide access points for foraging under the snow's surface.

Sparse forest stands are assumed to provide marginal cover for marten; therefore, a tree canopy closure of less than 25 percent will indicate no value as winter cover for the species. It is also assumed that any tree species present within a forest stand will have some value as winter cover for marten. Therefore, the lowest value which may be obtained for this variable is 0.1. Forest stands dominated by shrubs or seedling-sized trees are assumed to provide no value as winter cover for marten. Pole-sized and young stands of trees provide some cover, while mature or old-growth stands provide optimum cover. A ground surface covered by downfall ranging from 20 to 50 percent is assumed to have optimum value. However, the absence of downfall or the presence of a high density of these materials will not severely limit the cover value for marten.

The percent tree canopy closure and successional stage of the stand are the two most limiting variables for determining the suitability of marten winter habitat. When either of these variables is outside the suitable ranges defined above, marten habitat will not be present. The presence of little or no spruce or fir in a forest stand will lower the value of the habitat for marten. However, the absence of these species will not exclude the area as a potential marten habitat. Although the percent of the ground surface covered by downfall has the least influence on the determination of marten winter-habitat suitability, such material is essential to provide optimal winter habitat. An excessive amount of downfall (50 percent) is assumed to decrease the availability and accessibility of prey for marten. It is assumed that mature or old-growth forest stands will provide a sufficient number of snags and partially fallen trees to allow entry points under the snow's surface.

HSI Model

The HSI model for winter cover for the marten is as follows:

$$\text{HSI} = [(\text{SI}_{v_1})(\text{SI}_{v_2})(\text{SI}_{v_3})(\text{SI}_{v_4})]^{1/2} \quad (11.9)$$

where HSI = habitat suitability index for winter cover in evergreen forest for marten

V_1 = tree canopy closure, %; percent of ground surface that is shaded by vertical projection of canopies of all woody vegetation taller than 5.0 m (16.5 ft), measured by line intercept or remote sensing

V_2 = overstory canopy closure comprising fir or spruce, %; percent canopy closure of spruce or fir trees in overstory divided by total canopy closure of all overstory trees, measured by line intercept or remote sensing

V_3 = successional stage of stand; structural condition of forest community which occurs during its development. Four recognized stages—shrub-seedling (stage A), pole sapling (stage B), young (stage C), and mature or old growth (stage D)—are measured by on-site inspection or remote sensing

V_4 = ground surface covered by downfall which is $\geq$ 7.6 cm (3 in) in diameter, %; ground surface covered by dead, woody material which may include tree boles, stumps, root wads, or limbs, %; measured by line intercept or quadrant

Suitability index (SI) graphs for each of the four variables are shown in Figure 11.6. The SI graphs display the relationships between various conditions of habitat variables and habitat suitability for the marten. It should be noted that the SI graphs are conceptually similar to the functional relationships used in HES.

After an HSI model is obtained, it must be plugged into the HEP analysis to obtain an HSI for the available habitat. The HSI for available habitat is a function of the suitability of all cover types used by the evaluation species. The HSI for available habitat is calculated in one of several ways; the choice depends on the structure of the model. Figure 11.7 displays the various routes to calculating an HSI for available habitat. These routes are dependent on the structure of the model and can be determined by answering three questions about the model structure: (1) does use of the model produce suitability indices (SIs) for the available habitat from individual cover-type suitability indices? (2) if cover-type suitability indices are calculated, does the available habitat for the species consist of more than one cover type? and (3) if the available habitat consists of more than one cover type, is interspersion between cover types important for the species?

Habitat assessments involve measurement and description of habitat conditions for baseline (present) conditions and future with- and without-action conditions. Baseline assessments are used to describe existing ecological conditions. The results of baseline assessments provide a reference point from which resource planners can (1) compare existing conditions in two or more areas in order to define management capabilities or as a guide to future land-use planning; (2) predict and compare changes that may occur without the proposed action, with the proposed action, or with compensation measures; and (3) design monitoring studies.

Impact assessments are based on future with- and without-project conditions and are performed by quantifying habitat conditions at several points in time throughout a defined period of analysis. The assessment of land-use impacts is facilitated by dividing the study area into impact segments. An ''impact segment'' is defined as an area in which the nature and intensity of the future land use can be considered homogenous, such as the flood-pool area in a reservoir project, a recreational area, or the area of a particular agricultural practice. The advantage of dividing the study area into impact segments is that only one condition need be considered for each cover type within each impact segment. The effects of a particular project or action may be analyzed over a large area by assuming that the same condition exists throughout each impact-segment–cover-type zone.

Habitat units must be calculated for the evaluation species at each of the points in time for future with- and without-project conditions; this process includes predicting total available hab-

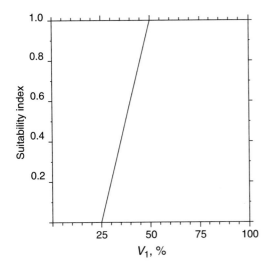

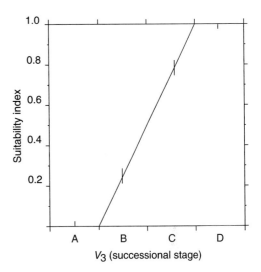

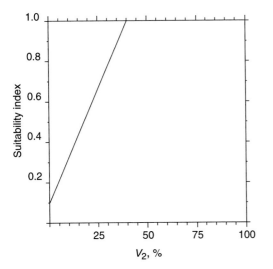

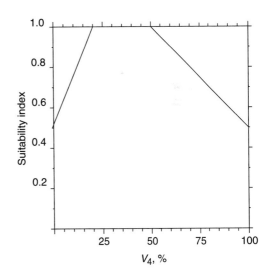

FIGURE 11.6
Suitability Indices in Marten HSI Model (Allen, 1982, pp. 5–6).

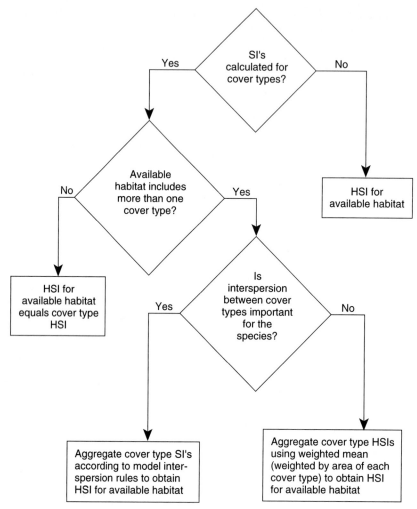

FIGURE 11.7
Options for Calculating HSI for Available Habitat (U.S. Fish and Wildlife Service, 1980, p. 102 ESM 4.2C).

itat and HSIs for each evaluation species. The impact assessment can be simplified by selecting target years (TYs) for which habitat conditions can be reasonably defined. At a minimum, target years should be selected for points in time when the rates of loss or gain in HSI or area are expected to change (if known). Rates of loss or gain in HSI or area can be assumed to occur

linearly between target years (U.S. Fish and Wildlife Service, 1980).

For each proposed action, the area of available habitat must be estimated for future years. Some cover types may increase in total area, others may decrease, and, in some cases, new cover types will be created or existing ones totally lost under forecasted with- or without-project conditions. The recommended method for determining the future area of cover types is through the development and use of cover-type maps for future years. In this method, impact-segment boundaries are superimposed on a baseline cover map. Baseline cover types will either be unaltered, altered (i.e., variables such as percent vegetation cover may change), or converted to new cover types, depending on such factors as land use within the impact segment, vegetation successional trends, and management. Areas converted to new cover types through succession or impacts should be given a new cover-type designation. An altered cover type is designated a "subtype" (e.g., deciduous forest altered by flooding). An overlay of impact-segment boundaries may be required for each target year. Figure 11.8 illustrates how a baseline cover-type map could be used in conjunction with impact segments to produce cover-type maps for future conditions.

FIGURE 11.8
Example of Cover-Type Maps (a) Existing habitat conditions and (b) Predicted conditions for target
year 20 with a proposed action (U.S. Fish and Wildlife Service, 1980).

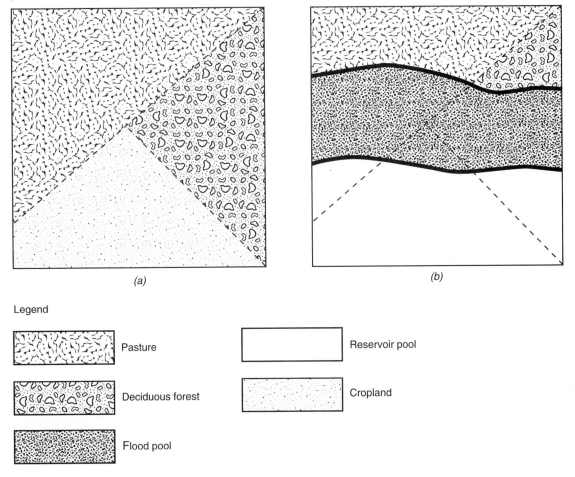

(a)   (b)

Legend

Pasture

Deciduous forest

Flood pool

Reservoir pool

Cropland

The same models that were used to determine baseline HSI values must be used to determine future HSI values. Estimating HSI values for future years requires predictions of changes in the physical, vegetative, and chemical variables of each cover type. Impact-segment overlays can be used as aids in estimating these variables. For example, seasonal flooding might alter the forest understory, but not the canopy closure. Changes in interspersion relationships because of the creation of new cover types or the conversion of existing cover types also can affect HSI-model output; these changes can be easily measured on future-cover-type maps (impact-segment overlays).

Proposed projects should be evaluated over a given period of time that can be referred to as the "life of the project," which is defined as that period between the time that the project becomes operational and the end of the project's economic life (as determined by the construction, or lead, agency). However, in many cases, gains or losses in wildlife habitat may occur before the project becomes operational, and these changes should also be considered in the impact analysis. Examples of such changes include construction impacts, implementation of a compensation plan, or other land-use changes. The habitat assessment should incorporate these changes by the use of a "period of analysis" that includes prestart impacts. Figure 11.9 displays the relationship between the life of the project and the period of analysis. If no prestart changes are expected, then the life of the project and the period of analysis are the same.

HU gains or losses should be annualized by summing HUs across all years in the period of analysis and dividing the total (cumulative HU) by the number of years in the life of the project. In this manner, prestart changes can be considered in the analysis. This calculation results in "average annual habitat units" (AAHUs). For example, assume that Table 11.7 represents the results of a HEP study with white-tailed deer as the species of interest, and Figure 11.10 contains a graphical display of the information; the cumulative HUs in a time interval can be calculated as described earlier for the HES (in Figure 11.3). The AAHUs can also be calculated in a manner similar to that described for the

FIGURE 11.9
Time Relationships for a Project (U.S. Fish and Wildlife Service, 1980, p. 102 ESM 5.2D) (Note: TY = target year).

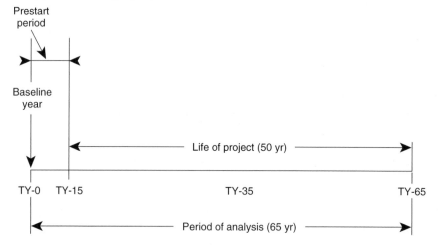

TABLE 11.7

TARGET-YEAR HABITAT FOR WHITETAIL DEER FOR BOTH FUTURE-WITH- AND WITHOUT-PROPOSED-ACTION CONDITIONS CALCULATED USING HEP

Condition	Target year	Area (acres)	HSI value	Total HU
With proposed action	Baseline	1000	0.75	750
	1	500	0.70	350
	20	500	0.20	100
	100	500	0.20	100
Without proposed action	Baseline	1000	0.75	750
	1	1000	0.75	750
	20	900	0.60	540
	100	600	0.60	360

Source: U.S. Fish and Wildlife Service, 1980, p. 5.2D.

HES (in Figure 11.3). The net annual impact of the proposed action on white-tailed deer is calculated as the difference between with- and without-project conditions over time:

$$\text{Net impact} = \text{AAHU}_{\text{with}} - \text{AAHU}_{\text{without}}$$
(11.10)

The net impact figure reflects in AAHUs the difference between the future-with and the future-without proposed action conditions.

An optional feature of HEP is the conduction of trade-off analyses among alternatives through the use of "relative value indices" (RVIs) to document value judgments made during a resource-planning effort (U.S. Fish and Wildlife Service, 1980). Trade-off analyses may be needed when proposed actions which are expected to alter habitat conditions would result in both gains and losses of different wildlife resources. In practice, RVIs are applied as weighting values to the HUs or AAHUs calculated for each evaluation species. These weighting values are determined by a user-defined set of socioeconomic and ecological criteria, and they include value judgments. The calculation of RVI values is performed in three steps: (1) defining the perceived significance of RVI criteria, (2) rating each evaluation species against each criterion, and (3) transforming the perceived sig-

nificance of each criterion and each evaluation species' rating into an RVI. The first step in the RVI calculation involves the application of relative weights to each criterion to numerically define its perceived importance to the user. The suggested weighting technique is to use pairwise comparisons, in which each criterion is compared to every other criterion, and a decision is made about which criterion of each pair is more important (Dean and Nishry, 1965).

Table 11.8 shows an illustration of the results of an RVI effort. The adjusted values can be used to compare baseline areas and proposed actions to determine where the greatest impact would occur. They can also be used to develop alternative compensation plans. Detailed information on the RVI approach is contained in U.S. Fish and Wildlife Service (1980). While this approach can be useful, it has not been frequently applied in HEP studies.

The final step in the assessment process using HEP involves the development of compensation (mitigation) plans (see Figure 11.4), if appropriate. Compensation studies identify measures that would offset unavoidable HU losses due to a proposed action. Compensation occurs by applying specified management measures to existing habitat to effect a net increase in HUs. The existing habitat may or may not be located in

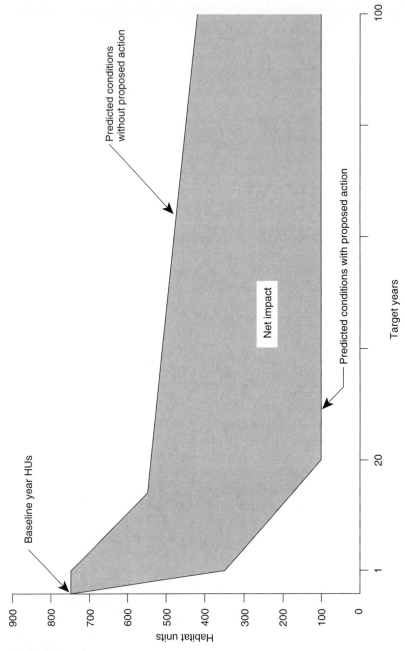

FIGURE 11.10
Relationship Between Baseline, Conditions without a Proposed Action, Conditions with a Proposed Action, and Net Impact (U.S. Fish and Wildlife Service, 1980, p. 102 ESM 5.2E).

TABLE 11.8

AGGREGATION OF HABITAT UNIT DATA USING RVIs

Evaluation species	Change in average annual habitat units	Relative value index	Adjusted value (HU × RVI)
Whitetail deer	−722	0.60	−433
Ruffed grouse	−400	0.78	−312
Red squirrel	−300	0.10	−30
Red fox	−120	0.35	−42
Yellow-rumped warbler	−550	1.00	−550
Total			−1,367

Source: U.S. Fish and Wildlife Service, 1980, p. 6.1.

the impact study area. In order to accomplish compensation, the HU losses due to the proposed action must be fully offset by the specified acquisition and/or management measures. The compensation process is depicted in Figure 11.11. Table 11.9 delineates some generic mitigation-planning objectives which could be used in the analysis. Three possible compensation goals expressed in a more specific fashion are (U.S. Fish and Wildlife Service, 1980, pp. 102 ESM 7.1 and 7.3):

1. *In-kind (no trade-off)*—This compensation goal is to precisely offset the HU loss for each evaluation species. Therefore, the list of target species must be identical to the list of negatively impacted species. The ideal compensation plan will provide, for each individual species, an increase in HUs equal in magnitude to the HU losses.
2. *Equal replacement (equal trade-off)*—This compensation goal is to precisely offset the HU losses through a gain of an equal number of HUs. With this goal, a gain of one HU for any target species can be used to offset the loss of one HU for any evaluation species. The list of target species may or may not be identical to the list of impacted species.
3. *Relative replacement (relative trade-off)*—With this goal, a gain of one HU for a target species is used to offset the loss of one HU for an evaluation species at a differential rate depending on the species involved. The trade-off

rates can be defined by RVI values for each species. For example, if the RVI values for whitetail deer and ruffed grouse are 1.0 and 0.5 respectively, one whitetail deer HU can be used to offset two ruffed grouse HUs. The lists of target and evaluation species can differ.

After the compensation goals are set, the compensation analysis is the same as that used to identify project impacts. The key steps in the process, as depicted in Figure 11.11, are to (U.S. Fish and Wildlife Service, 1980, pp. 102 ESM 7.3–7.4)

1. Select a candidate compensation study area. The area can be of any size but must be at least large enough to be a manageable unit for the target species. Develop a cover type map and determine the area of each cover type.
2. Conduct a baseline habitat assessment for each target species. Baseline data for individual species in the "impact" area may be used if the candidate compensation area is similar in terms of HSI values. If this is not the case, additional field work to determine HSIs will be necessary in the compensation study area.
3. Determine the AAHUs for the compensation study area assuming no future proposed action.
4. Identify a proposed management action that will achieve specified goals. Specify the management measures (e.g., prescribed burning, selective timber cutting, and others) that will be used to increase the HUs for target species in the candidate compensation area.

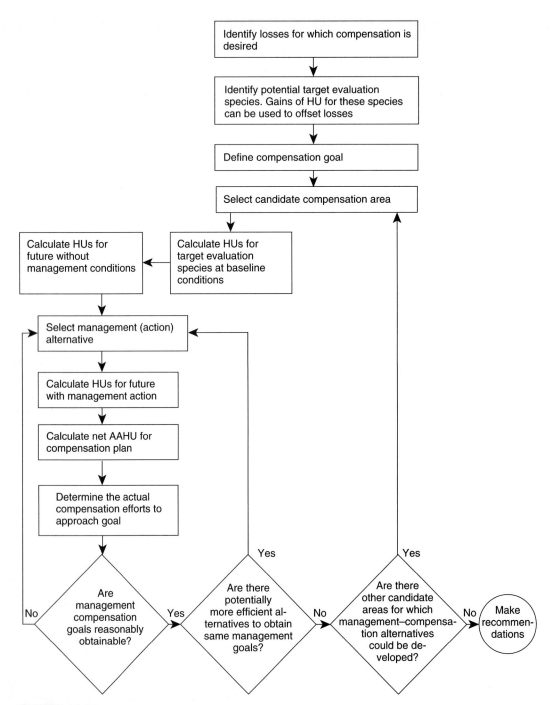

FIGURE 11.11
The Compensation Process (U.S. Fish and Wildlife Service, 1980, p. 102 ESM 7).

TABLE 11.9

RESOURCE CATEGORIES AND MITIGATION-PLANNING OBJECTIVES

Resource category and designation	Mitigation-planning objective
1. High value for evaluation species; "unique" and "irreplaceable"	Ensure no loss of existing habitat value.
2. High value for evaluation species; "scarce" or "becoming scarce"	Ensure no net loss of in-kind habitat values.
3. High to medium value for evaluation species; "abundant"	Ensure no net loss of habitat value while minimizing loss of in-kind habitat value.
4. Medium to low value for evaluation species; no designation	Minimize loss of habitat value.

5. On the compensation area, contrast the HUs without management to the HUs with proposed management measures and determine the net increase in HUs.

The process delineated above is identical to the process used to assess the net impacts of any proposed action (i.e., to estimate the net AAHU changes for a specified future action). The final step in the process involves calculating the actual size of the management area that will be required to fully offset losses.

CURRENT ISSUES RELATED TO HEP METHODS

The purpose of this section is to highlight several current issues related to the development and usage of HEP. As noted earlier, HSI models are fundamental to the use of HEP. The U.S. Fish and Wildlife Service continues to develop and test models related to various terrestrial and aquatic species (Christmas et al., 1982; Roberts, O'Neil, and Jabour, 1987). It is estimated that as of mid-1993, about 250 HSI models had been developed by the U.S. Fish and Wildlife Service.

The original versions of some HSI models were complicated from both a conceptual and a mathematical perspective. Accordingly, efforts have been made to simplify certain models and to develop community-related models. For ex-

ample, most published HSI models for warmwater fishes have too many variables, many of which are time-consuming and difficult to measure. In addition, suitability index (SI) scores are calculated using nonuniform methods that often yield different values for closely related species (Miller et al., 1987). One such community model was developed for certain warmwater fishes. Based upon the literature and the results of field studies in the Little Cypress Basin, Texas, and the Yazoo River, Mississippi, closely related species of warmwater fishes were placed into groups. For each group, a community model was developed based on only five variables: percent pools, average current velocity, percent cover, pH, and dissolved oxygen.

Optimizing the Use of HEP

The HEP approach can be both time-consuming and costly, particularly if detailed fieldwork is required. Accordingly, Wakeley and O'Neil (1988) proposed some techniques and approaches which can be used to increase efficiency and reduce the effort in applying HEP. There are two general approaches to reducing the effort in a HEP analysis. One is to use only those portions of HEP (defined broadly to include the use of HSI models) that are appropriate to the objectives of the study. This approach may be appropriate when the study objectives are limited; examples of suitable applications include habitat inventories and monitoring programs, management plans for refuges and recreation areas, and impact assessments that do not involve mitigation plans (Wakeley and O'Neil, 1988). The second approach is to simplify the procedure to achieve a low-resolution outcome that, although possibly less reliable than that obtained using a typical HEP, will still be sufficient to meet the objectives of the study or to accommodate the level of decision making. "Low resolution" is defined as that level of resolution producing a result that may be less reliable than that obtained with a typical HEP analysis (Wakeley and O'Neil, 1988). Low-resolution habitat analyses may be appropriate when (1) study areas are either very large or very small; (2) personnel, time, and funds are limited; (3) anticipated impacts are minimal; or (4) the resources involved are ubiquitous or of low priority.

The first step in tailoring HEP to a particular application (the first approach, above) is for the various agencies involved, and other interested parties, to specify priorities for cover types and evaluation elements, establish objectives, and determine the level of resolution required for the study. The next step is to identify and use only those portions of the procedure that are appropriate to the project. The objectives of a management-oriented study, for example, might be satisfied by determining HSI values for selected species without using the HEP accounting system at all. An estimate of available HUs may be all that is required for a general habitat inventory, and a simple impact analysis does not necessarily involve RVIs, management plans, or compensation needs.

One example of the second approach is the development of the Pennsylvania modified HEP, called "PAM HEP," which simplified HEP by reducing the time and personnel resources required for application and analysis, particularly regarding field-sampling procedures (Greeley-Polhemus Group, 1991).

Use of HEP in Natural-Resources Damage Assessment

In the decade of the 1980s, a great deal of attention was given to the cleanup of hazardous-waste sites. The focal point for these efforts was related to the requirements of the ''Superfund,'' and, more specifically, to the requirements of the Comprehensive Environmental Response, Compensation, and Liability Act (CERCLA) of 1980, and the Superfund Amendments and Reauthorization Act (SARA) of 1986. In addition, continuing this trend, hazardous-waste sites at governmental facilities such as those of the Department of Defense (DOD) and Department of Energy (DOE) are now being subjected to cleanup-remediation activities. As part of the Superfund requirements, and presumably applicable also to governmental facilities, ''natural-resources damage assessments'' may need to be conducted.

Section 301(c) of CERCLA requires that the President, acting through federal officials, promulgate regulations to assess damages for injury to, destruction of, or loss of natural resources. Executive Order 12316 Section 8(c), August 14, 1981, delegated this responsibility to the Secretary of the Interior. These regulations are to be used to assess damages for injuries to natural resources due to release of a hazardous substance or discharge of oil covered by CERCLA or Section 311(f) (4) and (5) of the Clean Water Act (Desvousges and Skahen, 1987). As part of the development of the regulations, it has been determined that HEP and pertinent HSI models can be used to (U.S. Fish and Wildlife Service, 1987) (1) establish that assessment- and control-area habitats are similar to prerelease habitat conditions in the assessment area and that observed species changes (e.g., diminished population numbers) are not likely due to habitat differences, (2) quantify changes in habitat resulting from a release of oil or hazardous substances, (3) determine changes in HU availability caused by a discharge or release, (4) provide a replicable and quantitative basis for determining the cost of restoring sites to attain habitat conditions present prior to release of oil or hazardous substances, and (5) provide a replicable and quantitative basis for determining the cost of achieving appropriate replacement for the lost habitat value of affected areas that cannot be restored to achieve in-kind, equal, or relative habitat replacement as described in U.S. Fish and Wildlife Service (1980).

OTHER HABITAT-BASED METHODS

As noted earlier, numerous habitat-based evaluation methods have been developed and used in natural-resources and environmental impact studies. It is beyond the scope of this chapter to review such methods, which total over 50. However, brief mention will be made of two comparative studies. Schuytema (1982) summarized and compared approximately 30 aquatic habitat assessment techniques. Table 11.10 lists the major habitat parameters and related factors used in or evaluated by the methods. The groups (surrounding area, riparian zone, general descriptors, stream banks, stream bottom, fish habitat, obstructions, and biology) are each presented with a percentage indicating the relative number of reviewed methods using the parameters within a group. Individual parameters are listed according to predominance of use. Parameters most frequently considered in the reviewed methods included flow, temperature, water surface, width, turbidity, gradient, velocity, depth, bank-stability measures, bottom size distribution, siltation, cover, pool size, attached vegetation, fish and invertebrate types, riparian-zone vegetation and shade, and obstructing factors such as waterfalls, dams, and culverts (Schuytema, 1982). Schuytema (1982) noted that a universal-habitat technique is probably not realistic

because of the diversity of watershed and stream types, but a number of methods have the potential, with regional adaptations, to be used over wide areas.

Atkinson (1985) reviewed 32 terrestrial- and/or aquatic-habitat-based evaluation methods which could be (or have been) used in EISs. These methods were reviewed in relation to the following issues:

1. Did the method provide information on delineating the study boundary in relation to time or space?
2. Did the method provide instructions on classifying the study area according to habitat type(s)?
3. Did the method incorporate appropriate information to enable an inventory and evaluation of biological resources? (Such information could be focused on specific species, the guild concept, and/or wildlife as a whole.)
4. Did the method provide information on impact assessment, including predictions of land-use changes, environmental quality changes, and/or the analysis of multiple scenarios?
5. Did the method address mitigation requirements and appropriate mitigation measures?

Table 11.11 provides a brief, graphic comparison of the 32 methods analyzed. An *x* denotes a positive response to the pertinent question (see above) for each method. It can be seen that few of the techniques go into much detail about setting the time and space boundaries of the region of influence. Almost every technique uses some classification scheme, while a few of them can be considered stand-alone classification systems (e.g., no inventory, evaluation, assessment, components). Nearly all of these techniques have an inventory and evaluation component. Only some of the techniques address how to make predictions about future habitat conditions and thereby assess the impacts of

---
TABLE 11.10

HABITAT PARAMETERS AND RELATED FACTORS EVALUATED IN THE REVIEWED METHODS[a]

Surrounding Area (26%)[b]
surrounding land use
topography/geography
upstream land use
historical land use
flood plain condition
urbanization

Riparian Zone (78%)
vegetation species/type
percent shade
streamside cover
vegetation size
vegetation density
width of zone
ungulate grazing/damage
flood plain width
vegetation successional stage

General Descriptors (10%)
flow
water temperature
water surface width
color/turbidity/transparency
gradient
velocity
average depth
air temperature
channel width
length of segment
elevation
pool/riffle ratio
stream order
stage/level
stream length
channel type/configuration
tributaries/tributary of sinuosity
pollution sources
bottom composition—general
valley bottom width

valley type/configuration
weather
drainage area
watershed type
water source(s)
water use
percent channelized
stream area
direction of flow

Stream Banks (57%)
bank stability
landform slope
mass wasting
debris jam potential
vegetative bank protection
channel capacity
bank rock content
obstructions
cutting
deposition
percent erosion/bare soil
height banks
percent damage
percent grazing

Stream Bottom (86%)
bottom size distribution
siltation/sedimentation
consolidation/particle packing
rock angularity
brightness
scouring/depositon
inbeddedness
percent channel movement
roughness coefficient

Fish Habitat (75%)
instream cover
pool length/width

pools number/percent
riffle width
pool depth
spawning gravel
  abundance/volume
pool area
spawning gravel quality
riffle depth
riffles percent
spawning gravel size
runs percent
nursery habitat
riffle velocity
runs width
runs depth
runs velocity

Biology (86%)
attached algae/macrophytes
fish species
invertebrate type/species
invertebrate abundance/rank
fish size/weight
fish abundance
invertebrate diversity

Obstructions (45%)
waterfalls
beaver dams/dams
culverts
debris piles/slides
log jams
channelization
dredging
impoundments
levies/dikes
riprap

---

[a]Rated according to predominance of use within each group.
[b]Percent of methods using parameters in each group.
Source: Schuytema, 1982, p. 3.

specified projects. Even fewer of the techniques have a component for estimating mitigation measures for offsetting adverse habitat impacts (Atkinson, 1985). Reference citations for each of the 32 methods are in Atkinson (1985).

An estuarine-habitat assessment (EHA) protocol has been developed to quantitatively assess the function of estuarine wetlands and associated nearshore areas for use as habitat by fish and wildlife species. The EHA protocol re-

TABLE 11.11

COMPARISON OF HABITAT-BASED METHODS

Method (brief title)	Boundaries	Classification	Inventory and evaluation[a] Species	Guild	Wildlife	Assessment	Mitigation
Rapid-assessment methodologies (RAMs)		X			X	X	X
Numerical classification	X	X			X	X	
Dynamic analytic silviculture (DYNAST)		X	X	X		X	X
Digitized classification		X					
Waterfowl habitat quality		X		X		X	
Trophic relationship classifications		X		X			
Habitat scarcity		X	X	X			
Habitat analysis and evaluation		X	X		X	X	
Evaluating fish and wildlife habitat		X	X		X	X	X
Habitat type classification system		X					
Estimating habitat variables			X	X	X		
Wildlife assessment		X		X	X		
Physiographic—photomorphic habitat		X					
Ecosystem impacts of urbanization		X	X	X	X		
Numerical habitat evaluation		X	X			X	X
Wildlife and fish habitat relationships		X	X			X	X
Natural-area classification		X	X		X		
Ecological systems component	X	X			X	X	
Wildlife Management Information System (WILDMIS)		X	X		X	X	X
RPA-80 Land Classification		X					
Biochemical and endocrine indicators			X				
Habitat gradient model (HGM)		X		X		X	X
Systems approach to ecological studies	X	X			X	X	
Wildlife habitats in the Blue Mountains		X		X		X	
Vertebrate or plant relationships		X		X		X	
Impact-risk mapping		X	X				
Habitat evaluation system (HES)	X	X	X		X	X	X
Habitat evaluation procedure (HEP)	X	X	X			X	X
Assessing quality of habitat		X			X	X	
Pattern recognition (PATRIC)		X	X			X	X
Habitat information systems		X	X			X	
Assessing open spaces in urban areas	X	X			X		

[a]Inventory and evaluation can be based on species-specific requirements, requirements of guilds of species, and/or requirements of wildlife as a whole.
Source: Adapted from Atkinson, 1985, p. 279.

TABLE 11.12

PREDICTORS OF WETLAND FUNCTIONAL VALUE

1. Contiguity	39. Basin alterations
2. Constriction	40. Pool-riffle ratio
3. Shape of basin	41. Basin's vegetation density
4. Fetch and exposure	42. Wetland's vegetation density
5. Basin surface	43. Sheet vs. Channel flow
6. Wetland surface area	44. Wetland-water edge
7. Basin area/watershed area ratio	45. Gradient of edge
8. Basin area/subwatershed area ratio	46. Shoreline vegetation density
9. Location in watershed	47. Shoreline soils
10. Stream order	48. Disturbance
11. Gradient of subwatershed	49. Plants: form richness
12. Gradient of tributaries	50. Plants: waterfowl value
13. Gradient of basin	51. Plants: anchoring value
14. Perched condition	52. Plants: productivity
15. Land cover of subwatershed	53. Invertebrate density: freshwater
16. Land cover trends	54. Invertebrate density: tidal flat
17. Soils of subwatershed	55. Shore erosion measurements
18. Lithologic diversity	56. Ground water measurements
19. Delta environment	57. Suspended solids
20. Evaporation-precipitation balance	58. Alkalinity
21. Wetland system	59. Eutrophic condition
22. Vegetation form	60. Water quality correlates
23. Substrate type	61. Water quality anomalies
24. Salinity and conductivity	62. Water temperature anomalies
25. pH	63. Bottom water temperature
26. Hydroperiod	64. Dissolved oxygen
27. Flooding duration and extent	65. Underlying strata
28. Artificial water level fluctuations	66. Discharge differential
29. Natural water level fluctuations	67. TSS differential
30. Tidal range	68. Nutrient differential
31. Scouring	69. Recharge effectiveness
32. Flow velocity	70. Discharge effectiveness
33. Water depth (maximum)	71. Flood storage effectiveness
34. Water depth (minimum)	72. Shoreline anchoring opportunity
35. Width	73. Shoreline anchoring effectiveness
36. Oxygenation of sediments	74. Sediment trapping opportunity
37. Morphology of wetland	75. Sediment trapping effectiveness
38. Flow blockage	

Source: Greeley-Polhemus Group, 1991, p. 30.

quires the measurement of attributes (characteristics) of estuarine habitats that promote fish and wildlife utilization and fitness; it supplements HEP and the "wetlands evaluation technique" (WET) (Simenstad et al., 1991). It can be used in impact prediction and assessment, and in the planning and implementation of baseline and post-EIS monitoring programs.

WET can be used to provide a broad overview of potential project impacts on several wetland-habitat functions. A project team implements WET by first identifying the physical characteristics of a wetland through the use of predictors; "predictors" are species or characteristics within the habitat which become representative of the study area, as shown in Table

11.12. A series of questions is asked for each predictor to more precisely define its relationship to the habitat and to determine the social significance of the wetland area. The answers for each predictor are then evaluated to determine each function's effectiveness as well as opportunity, based on interpretation keys; the evaluation ratings are "high," "moderate," or "low." Similar ratings are also used for significance. The ratings for each predictor are then combined to give a final rating of functional significance.

A threshold analysis based on 38 questions is then used to determine an overall impact rating for a proposed project or activity. If this rating is "low," the project is deemed to have negligible impact, and no further analysis is required. If the impact rating is "moderate" or "high," the earlier, question-based analysis is repeated for the postconstruction condition. The effectiveness, opportunity, and functional ratings are determined using the interpretation keys. An analysis of mitigation measures and their potential effectiveness can also be made with the same question-based approach.

COMPARISON OF FOUR METHODS

Comparative information on four habitat-based methods (HEP, PAM HEP, HES, and WET) is in Table 11.13. The level of professional acceptance of the four methods is summarized in Table 11.14. Acceptance was determined through interviews with 41 professionals, mainly biologists, associated with federal and state agencies, universities, and private groups.

MITIGATION BANKING—AN OUTGROWTH OF HES AND HEP

As described earlier, the final component of HES and HEP, and other habitat-based methods, is the determination of appropriate mitigation measures. Mitigation (including compensation, when appropriate) is often a major issue for many development projects, particularly for

those involving potential losses of wetlands. In concept, a mitigation program for a project with wetland impacts may incorporate restoration, creation, or enhancement of wetlands, and it could include "mitigation banking." Table 11.15 lists pertinent definitions of these and related terms.

Since the early 1980s, mitigation banking has been considered as one of the tools available for achieving compensation for unavoidable project-related resource losses. "Mitigation banking" has been defined as ". . . habitat protection or improvement actions taken expressly for the purpose of compensating for unavoidable, necessary losses from specific future development actions" (Short, 1988, p. 1). It has also been described by Short (1988, p.1), as follows:

> . . . mitigation banking is similar to maintaining a bank account. A developer undertakes measures to create, restore, or preserve fish and wildlife habitat in advance of an anticipated need for mitigation for project construction impacts. The benefits attributable to these measures are quantified, and the developer receives mitigation credits from the appropriate regulatory and/or planning agencies. These credits are placed in a mitigation bank account from which withdrawals can be made. When the developer proposes a project involving unavoidable losses of fish and wildlife resources, the losses (debits) are quantified using the same method that was used to determine credits, and a withdrawal equal to that amount is deducted (debited) from the bank. This can be repeated as long as mitigation credits remain available in the bank.

The concept of mitigation banking was developed in response to a number of requests to consider "banking" of management credits for future use in mitigating fish and wildlife losses. It was thought that, properly implemented, mitigation banking could be an innovative mechanism to obtain compensation for unavoidable habitat losses primarily associated with wetland-resource development projects (Short, 1988). Specifically, in the United States, mitigation banks have been designed to provide permit ap-

TABLE 11.13

COMPARISON OF FOUR HABITAT-BASED EVALUATION METHODS

Criteria	HEP	PAM HEP	HES	WET
	Habitat-based evaluation method			
1. Compatibility with accepted economic/ecologic evaluation principles				
a. Habitat-based procedures	Species-based habitat evaluation approach.	Species-based habitat evaluation approach. Modification of HEP.	Community-based habitat evaluation approach.	Community-based habitat evaluation approach.
b. Incremental analysis	Compatible with incremental analysis.	Compatible with incremental analysis.	Compatible with incremental analysis.	Not compatible with incremental analysis.
c. Cost-effectiveness	Compatible with cost-effectiveness.	Compatible with cost-effectiveness.	Compatible with cost-effectiveness.	Compatible with cost-effectiveness.
2. Professional acceptance				
a. Conceptual acceptance	Generally accepted. Some concern regarding the ability of selected species to represent entire habitat.	Varying degrees of acceptance. Some feel modifications have compromised HEP, others feel they have not affected final results.	High level of conceptual acceptance.	Not highly accepted for use in fish and wildlife applications.
b. Implementable	Procedures widely accepted.	Well developed procedures for implementation. Accounting procedures have been streamlined.	Well developed procedures which could be used nationwide if biologic models were available.	Implementation too subjective.
3. Cost of application				
a. Major factors relevant to cost	Dependent on number of species selected for evaluation; size of study area; availability of HSI's; number of team members.	Number of species selected for evaluation; size of study area; availability of HSI's; number of team members.	Size of study area; availability of functional models (HQI's); number of team members.	Size of study area; biologic knowledge of planner about area and amount of previously acquired information on area.
b. Average cost per acre	Not available.	Not available.	Not available.	Not available.
c. Average number of man years per acre	Not available.	Not available.	Not available.	Not available.
d. Most costly components	Data collection; data analysis.	Mapping cover areas; data collection and analysis.	Data analysis.	Field collection when necessary.
4. Time required for application				
a. Major factors relevant to time	Number of species selected for evaluation; size of study area; availability of HSI's; number of team members.	Number of species selected for evaluation; size of study area; availability of HSI's; number of team members; use of computer.	Size of study area; availability of HQI's; number of individuals involved with study.	Size of study area; the need for data collection (dependent on available historical information).
b. Average man-years per acre	Not available.	Not available.	Not available.	Not available.

c. Estimate of average time for application	6 months to 1 year for completion.	1 month to 3 months for completion.	2 months to 4 months for completion.	2 days to 1 week.
d. Relative time requirements	Most labor intensive approach.	Two to three times quicker than HEP.	Takes about half as much time as HEP.	
5. Biological validity				
a. Biologic measurement tool	HSI models and HU's. (HSI value × acreage = HU).	HSI models and HU's. More subjective measurement than HEP. PAM HEP recently updated with computer software.	Use of functional curves (HQI's) to determine Habitat Unit Values.	No biologic measurement tool.
b. Status of research	HSI's currently being verified. Research will then resume in developing new models. Models developed on field level when needed.	No specific research to refine procedures.	Recently refined and updated. No additional research in progress.	WES currently refining procedures.
6. Comprehensiveness				
a. National/regional application	Used nationwide; however, HSI's must be developed and modified for specific projects.	Used in Pennsylvania and some use in a couple of neighboring states with similar habitat.	Regional application to bottomlands of Lower Mississippi Valley region.	Nationwide application.
b. Design limitations	Limited to fish and wildlife applications.	Limited to fish and wildlife applications.	Biologic models designed for one specific habitat.	Limited to wetland functions.
c. Use in COE planning projects	Flood protection; navigation; water supply studies; permit applications.	Flood protection; navigation; water supply studies; permit applications.	Flood protection; navigation.	Used in preliminary phases of water supply studies; permit applications.
7. Actual application				
a. Team v. individual implementation	Inter-agency team approach required.	Inter-agency team approach required.	Not specified.	Not specified.
b. Availability of software package	Computer software aids in data analysis and mitigation evaluation.	Computer software (GDI/HEP) aids in data analysis and mitigation evaluation.	Software available to aid in data analysis.	Most recent version of technique is computer generated.
c. Use of previously collected data	HSI's available for use. Previously collected data not recommended.	HSI models used. Previously recorded useful.	HQI's available for bottomland habitat. Previously collected data useful in determining subjective judgments.	Subjective judgments based on previous knowledge of study area.
d. Use of aerial photography	Aerial photography useful in defining cover types.	Aerial photography useful in defining cover types.	Aerial photography useful.	Aerial photography not used.
8. Data requirements				
a. Methods of data collection	Field data collection required for selected species. Team approach utilized.	Field data collection; use of professional judgment. Team approach utilized.	Field data collection; use of professional judgment.	Site visit required. Field data not usually required.
b. Time demands of data collection	Average of 2–4 months for data collection efforts.	Average of 1 week required for data collection.	Average of 1–2 weeks required for data collection.	Not applicable.
c. Time demands of data analysis	Average of 6–8 months for data analysis.	Average of 1–2 months needed for data analysis.	Average of 1–3 months for data analysis.	One day needed for subjective data analysis.

TABLE 11.13

COMPARISON OF FOUR HABITAT-BASED EVALUATION METHODS (continued)

| Criteria | Habitat-based evaluation method | | | |
	HEP	PAM HEP	HES	WET
9. Potential for application to mitigation planning				
a. Designed for development of mitigation alternatives?	Yes. HU's indicate impacts to habitat and the establishment of mitigation objectives guides in the level of mitigation.	Yes. HU's indicate impacts to habitat and the establishment of mitigation objectives guides in the level of mitigation.	Yes. HUV's indicate impacts to habitat and the establishment of mitigation objectives guides in the level of mitigation.	WET useful in mitigation plan development especially for wetlands creation.
b. Useful in mitigation evaluation?	Yes. Procedure assists in evaluating alternatives by assessing impact of mitigation in terms of HU's compensated.	Yes. Procedure assists in evaluating alternatives by assessing impact of mitigation in terms of HU's compensated.	Yes. Procedure assists in evaluating alternatives by assessing impact of mitigation in terms of HUV's compensated.	Provides subjective evaluation of mitigation plans.
10. Integrability of data with socioeconomic analysis techniques				
a. Social concerns incorporated in model	Relative Value Indices (RVI) designed for HEP. Considers social importance of species.	Consideration of socially significant species during species selection.	No specific reference to social concerns in procedures.	Predictor keys provide for consideration of social values and concerns.
b. Integration with economic analysis techniques	Has been integrated with incremental analysis and cost-effectiveness techniques in the analysis of mitigation alternatives.	Has been integrated with incremental analysis and cost-effectiveness techniques in the analysis of mitigation alternatives.	Has been integrated with incremental analysis and cost-effectiveness techniques in the analysis of mitigation alternatives.	Can be integrated for use in the selection of most cost-effective alternative.

COE = U.S. Army Corps of Engineers
Source: Greeley-Polhemus Group, 1991, pp. 43–46.

TABLE 11.14

SUMMARY TABLE OF PROFESSIONAL'S REVIEW OF HABITAT EVALUATION METHODS

Criteria	General comments	HEP	PAM HEP	HES	WET
1. Compatibility with accepted economic-ecologic principles	All techniques similarly compatible except for WET. Economic principles and ecologic principles cannot be meshed. Not familiar enough with economic principles to comment.	Species-based habitat evaluation technique. Compatible with incremental analysis and cost-effectiveness techniques.	Species-based habitat evaluation technique. Applicable to incremental analysis and cost-effectiveness.	Community-based habitat evaluation technique. Compatible with incremental analysis and cost-effectiveness.	Community-based habitat evaluation technique. Not designed for economic evaluation. Compatible with cost-effectiveness principles.
2. Professional acceptance	Must be aware of only accepting techniques for the purpose they were designed: as a planning tool, not a decision-maker.	Most widely accepted. Broad support largely due to HEP being FWS supported model, not necessarily because it is the best technique. Approach is acceptable, but models must often be developed for specific projects.	Many of those contacted not familiar with method. Widely accepted on regional basis. Accepted equally with HEP by some, others accept hesitantly due to modifications.	Widely accepted on regional basis in bottomland forest habitat. Increasing confidence in technique due to better model documentation. However, HES not widely implemented due to regionalized models.	Accepted as an overview or preliminary approach to determining mitigation needs. Too subjective. Technical limitations. Addresses wetland functions that the other methods do not include.
3. Cost of application	No general comments.	Most expensive technique to implement. Cost primarily depends on the number of species selected for evaluation. Full effort of 8–10 species per cover type is expensive. Any "short-cut" will be less expensive.	Less costly than HEP. Costs are moderate if appropriate biologic models are available. Computerized approach (GDI/HEP) less labor intensive and therefore less costly.	Moderately expensive. Many have not used HES enough to assess costs.	Least expensive technique. Tries to get a "million dollar study for a buck and a half." Many not familiar enough with WET to assess.
4. Time required for application	Some of the methods conceptually different, therefore you can't really compare time requirements.	Extensive time demands. Labor intensive. Time requirements largely depend on number of species being evaluated. Field work could average 4–6 weeks distributed throughout all seasons. Field work and data analysis most time-consuming procedures. New computer model makes HEP less labor intensive.	"Streamlined" HEP. Moderate time demands. Saves an estimated 50–70 percent of the time required to implement the HEP. Field work may take up to 1 week. Cover mapping and data analysis most time consuming tasks. Computer model makes approach even less demanding.	Moderately labor intensive. Takes approximately 1/3–1/2 the time needed to apply a full HEP analysis. Can't "streamline" HES as it must evaluate the entire ecologic community.	Low to moderate time demands.

TABLE 11.14

SUMMARY TABLE OF PROFESSIONAL'S REVIEW OF HABITAT EVALUATION METHODS (continued)

Criteria	General comments	HEP	PAM HEP	HES	WET
5. Biological validity	Modelling approaches too different to effectively compare.	HSI's are the most sophisticated biologic models. Biologic models currently being verified. Some models too generalized, or too regionalized. Modifications or development of new models often required.	HSI models used when they are available. Biologic validity has never been challenged. More subjective than HEP. Some of HEP's accuracy has been taken away.	Functional relationship curves relatively accurate. Documentation has increased confidence in validity. Many not familiar enough with HES to comment.	No biologic models used. Only functional capability.
6. Comprehensiveness	Techniques constantly becoming more refined and comprehensive. However, planners must keep their use in perspective. They are still a tool to assist the planner.	Very comprehensive for fish and wildlife resources. Almost universal in its application to all habitats. However modifications to models still necessary. HSI's lacking for many species and regions. Limited to fish and wildlife habitats.	Procedures fairly comprehensive. Some modifications would be necessary for application in other regions. Biologic models would need to be developed when HSI's not available. Many not familiar enough with PAM HEP to comment.	Most comprehensive community-based technique. Developed for Lower Mississippi Valley. Use currently limited to that region. Conceptually comprehensive. Could be modified for use in other regions. Not as comprehensive as HEP, but differences are minimal.	Developed for wetland functions. Fish and wildlife component not as sophisticated as other techniques. More subjective, not data driven. Usefulness in identifying and prioritizing wetland functions.
7. Actual application	None of the methods are difficult to apply when implemented by a trained team. Team approach checks individuals who may attempt to accommodate a bias.	Inter-agency team approach. Team approach balances individual preferences. Requires extensive field data collection. "By the book" approach allows for little professional judgment. Most teams "modify" technique to a degree by using professional judgment. Number crunching and data analysis becoming less burdensome with use of computers.	Inter-agency team approach. Work towards consensus between team members. Uses more professional judgment and visual observations than HEP. Computer software assists in data analysis.	Team or individual approach. Fairly simple to train a biologist to use and apply technique. Moderate field data required. Utilizes professional knowledge and judgment.	Application of technique fairly simple for planner. Does not require extensive field data, or a biologist, but a sound biologic understanding of the study area. Technique uses a series of subjective interpretation keys. Many not familiar enough with WET to comment.

8. Data requirements	No general comments.	Extensive biologic data required through field collection efforts. Development of HSI models requires detailed information about species.	Moderate data collection efforts required. Relies more on professional knowledge and ocular techniques than HEP. When HSI models are not available they must be developed. This requires more detailed information.	Field data required to develop functional curves. Moderate field data collection efforts required when implementing HES. Many not familiar enough with HES to comment.	Little data required. Requires basic biologic knowledge of the habitat being evaluated.
9. Potential for application to mitigation evaluation	Techniques are tools to be used by planners in mitigation planning. These tools force planners to look more closely at what is being evaluated. Planners can then make more accurate recommendations.	HEP designed to be used in mitigation planning and evaluation. Potential for good mitigation plans is dependent on the reliability of individual applications. Has ability to focus planner on mitigation needs. It quantifies needs to assist planner in developing mitigation alternatives.	Great potential for mitigation planning and evaluation. Computer model aids in the development and assessment of mitigation alternatives.	Primary purpose of HES is to assess impacts to wildlife and to evaluate mitigation alternatives. Potential for mitigation comparable to HEP. Many not familiar enough with HES to comment.	WET's utility is primarily to assess impacts to wetlands, not to create and evaluate mitigation plans. Not really designed for this. Must be cautious not to force a function that it wasn't designed to perform. Good potential for wetlands creation.
10. Integrability of data with socioeconomic analysis methods	Impossible to put biological data in straight economic terms. Incremental analysis asks for values. In order to incrementally justify, one must first quantify. Techniques are tools to assist in quantifying.	Has been applied to incremental analysis. Used with cost-effectiveness techniques. HMEM developed based on HEP. Specific mitigation goals must be established in order to integrate with socioeconomic principles. Decision-making up to judgment of the team.	Used with incremental analysis. Capability for tradeoff analysis worked into PAM HEP technique. Team members must understand socioeconomic considerations for appropriate integration.	Has been used with incremental analysis and cost-effectiveness techniques. Many contacts not familiar enough with HES to comment on technique.	Not applicable to economic analysis. Social values which are built into model seem to bias results. Many not familiar enough with technique to comment.

Source: Greeley-Polhemus Group, 1991, pp. 36–39.

TABLE 11.15

PERTINENT DEFINITIONS FOR WETLAND MITIGATION

Mitigation—The actual restoration, creation, or enhancement of wetlands to compensate for permitted wetland losses. The use of the word "mitigation" here is limited to the above cases and is not used in the general manner as outlined in the Council on Environmental Quality regulations (40 C.F.R. 1508.20)

Mitigation Banking—Wetland restoration, creation, or enhancement undertaken expressly for the purpose of providing compensation for wetland losses from future development activities. It includes only actual wetland restoration, creation, or enhancement occurring prior to elimination of another wetland as part of a credit program. Credits may then be withdrawn from the bank to compensate for an individual wetland's destruction. Each bank will probably have its own unique credit system based upon the functional values of the wetlands unique to the area. As defined here, mitigation banking does not involve any exchange of money for permits. However, some mitigation programs, such as those in California, do accept money in lieu of actual wetland restoration, creation, or enhancement.

Restoration—Returned from a disturbed or totally altered condition to a previously existing natural or altered condition by some action of man. "Restoration" refers to the return to a pre-existing condition. For restoration to occur, it is not necessary that a system be returned to a pristine condition. It is, therefore, important to define the goals of a restoration project in order to properly measure its success.

Creation—The conversion of a persistent nonwetland area into a wetland through some activity of man. This definition presumes the site has not been a wetland within recent times (100–200 years) and thus restoration is not occurring. Created wetlands are subdivided into two types: artificial and man-induced. An artificial created wetland exists only as long as some continuous or persistent activity of man (i.e., irrigation, weeding) continues. Without attention from man, artificial wetlands revert to their original habitat type. Man-induced created wetlands generally result from a one-time action of man and persist on their own. The one-time action might be intentional (i.e., earthmoving to lower elevations) or unintentional (i.e., dam building). Wetlands created as a result of dredged material deposition may have subsequent periods during which additional deposits occur.

Enhancement—The increase in one or more values of all or a portion of an existing wetland by man's activities, often with the accompanying decline in other wetland values. Enhancement and restoration are often confused. For these purposes, the intentional alteration of an existing wetland to provide conditions which previously did not exist and which by consensus increase one or more values is enhancement. The diking of emergent wetlands to create persistent open-water duck habitat is an example; the creation of a littoral shelf from open water habitat is another example.

Success—Achieving established goals. Success in wetlands restoration, creation, and enhancement ideally requires that criteria, preferably measurable as quantitative values, be established prior to commencement of these activities. However, it is important to note that a project may not succeed in achieving its goals, yet provide some other values deemed acceptable when evaluated. In other words, the project failed, but the wetland was a "success." This may result in changing the success criteria for future projects. It is important, however, to acknowledge the nonattainment of previously established goals (the unsuccessful project) in order to improve goal setting.

Source: Adapted from Lewis, 1990, pp. 418–419.

---

plicants and permitting agencies with a simpler, more-effective process for complying with mitigation requirements, thus improving the resource value of mitigation projects. As such, mitigation banks present a number of potential advantages and disadvantages over more-traditional approaches. Table 11.16 contains a summary listing of these advantages and disadvantages.

Since the early 1980s initiation of mitigation banking in the United States, and according to an early 1992 survey, a total of 37 mitigation banks have been established and 63 additional banks have been proposed (Reppert, 1992). There is a widespread distribution of existing and proposed banks, with the largest number being in coastal states. Private developers and state transportation departments have shown particu-

TABLE 11.16

EVALUATION OF MITIGATION BANKING

Advantages

Can provide consolidation of mitigation for small wetland losses.
Can achieve mitigation in advance.
Requires increased planning effort which should lead to a more environmentally compatible project.
Can be used to resolve conflicts over proposed projects.
Can provide funding for monitoring and evaluation of mitigation compliance.
Can reduce federal permit processing time.
Can lead to positive public recognition for wetland mitigation actions.

Disdvantages

May focus on mitigation banking and thus not do good project planning and mitigation, other than compensation.
May still have net loss in wetlands, due to inability to create, restore, or enhance wetlands.[a]
Off-site mitigation may not ensure no net loss of species at the project site.
Conflicts can arise over the development of a mitigation banking agreement; areas where lack of agreement can delay or hinder implementation or use of a bank include the following:

(1) Selection of an evaluation methodology for use in establishing bank credits and debits associated with development projects.
(2) Comparisons of the bank and project areas to determine if "equivalent" habitat values are available within the bank habitat credits.
(3) Obtaining a long-term commitment from the bank sponsor or other involved entity for the continued operation and maintenance of the bank throughout its dedicated life.
(4) Where no impending important development project is present to act as a catalyst, agencies may be hesitant to participate in efforts to establish a bank.
(5) Because habitat improvement measures are not always as successful as predicted in increasing habitat value, some agreement needs to be reached on a later adjustment in bank credits if necessary.

[a]See Kusler and Kentula (1990) for a discussion of man's ability (or inability) to properly create, restore, or enhance wetlands.
Source: Developed from information in Short, 1988, pp. 2–7.

lar interest in the development and usage of mitigation banks.

There are eight major steps (or issues) typically involved in the development of a bank: (1) establishment of program goals and objectives; (2) identification and selection of bank site(s); (3) creation of bank operator or interagency agreements; (4) establishment of a policy for the use of credits and currency; (5) establishment of criteria for mitigation-bank use; (6) development of mitigation options; (7) construction of the mitigation-bank site(s), as well as development and implementation of maintenance, monitoring, and reporting operations; and (8) development and implementation of a long-term management plan.

The establishment of goals and objectives helps to determine the success of the bank-planning process. According to Castelle et al. (1992), the most common current goals of mitigation-banking programs are to achieve zero net loss of wetland acreage or function and to gain greater regional acceptance. The next step, the identification and selection of a bank site, is crucial when mitigating wetland functions that would be or possibly have been impacted. For example, a bottomland hardwood wetland should not be used for the compensation of coastal wetland impacts, because the functions of each are different, as are the habitats and species involved.

The "bank operator" is the legal entity specified in a "memorandum of agreement" (MOA); the MOA is a legally binding agreement between the interested parties involved in the bank. Mitigation-bank operators may include (1) resource agencies (federal, state, and/or local), (2) local governments, (3) private non-profit sector entities, (4) development associations, and/or (5) coalitions of parties. The operator is responsible for the day-to-day management of the bank site as well as the long-term integrity of the replacement wetlands (Castelle et al., 1992).

Establishment of credits can be based upon single or multiple wetland functions. Habitat value is often used to assign credit values. Because of its familiarity and wide usage throughout the United States, the HEP model is often used to determine habitat value. The use of mitigation banking is typical for projects which will impact a wetland and for which on-site mitigation is infeasible or, from a wetlands-resource perspective, undesirable. To use mitigation banking, Section 404 permit applicants must show that alternatives to avoid, minimize, rectify, and reduce the impact have been considered; only then is mitigation banking an option.

Because of the lack of ecological success of some wetland mitigation banks, several mitigation options have been developed and proposed. Restoration, enhancement, and creation are three compensatory wetland-mitigation options available. Castelle et al. (1992) define these options as follows: (1) ''restoration'' entails returning a wetland from a disturbed or totally altered condition to a relatively pristine or pre-disturbance condition, (2) ''enhancement'' means to increase one or more functions or values that an existing wetland possesses, and (3) ''creation'' involves the conversion of a persistent upland area into a wetland.

The construction, maintenance, monitoring, and reporting phases for a specific bank differ as a function of the type of bank that is being established. However, a consistent, periodic regime of monitoring and reporting is an excellent protocol to develop. For example, if the objective were to create bird habitat, then the monitoring program should focus upon habitat features and bird populations; for migratory birds, reports could be issued at the end of a migratory period. For resident bird populations, reports could be issued during spring and fall stating whether or not mating has been a success (Castelle et al., 1992).

Once all credits have been withdrawn from a bank, it enters the long-term-management phase. The primary functions of this phase are monitoring, maintenance, and remediation (Short, 1988). The purpose of these activities is to ensure that the mitigation bank remains a functioning wetland and that it is conserved in perpetuity, or for at least as long as any project impacts last. The following recommendations have been made for long-term mitigation-bank operation in Castelle et al. (1992): (1) public agencies, as well as conservation groups and/or private nonprofit groups, should be involved with mitigation-bank management; (2) public agency involvement in the operations should decrease after all mitigation credits are withdrawn, but the agencies should always be responsible for overseeing regulatory compliance; and (3) conservation-group involvement should increase as the banking program enters the long-term-management phase, and monitoring, maintenance, and remediation should be the conservation group's responsibility from the outset of the program.

The specific long-term-management responsibilities of the bank must be agreed upon initially and documented in the MOA. This helps to ensure a smooth transition to long-term management. Usually, when a bank enters the long-term-management phase, bank operation is transferred to another entity. This is advantageous in that the original operator of the bank may not be able to provide the funding or personnel to assure continued successful management of the bank (Castelle et al., 1992). Also, the continued success of the bank may not be a top priority for the original operator. Retaining the original operator may be useful though, since that agency or group is more familiar with the site, its past difficulties, and the corrective measures taken. Successful long-term management hinges upon a secure source of funds. If one is not found and agreed upon in the MOA, a successful mitigation bank may fail during long-term management because of a lack of funds for monitoring and remediation. One method to secure funds is to have the developers that utilize the bank establish long-term interest-bearing accounts that will be used for monitoring and remediation purposes (Short, 1988).

Once credits have been established for a bank, they may be considered accessible, but restrictions should be placed upon their withdrawal. One of the underlying purposes of a mitigation bank is to prevent a temporal loss of functioning wetlands. So, even though the credits may have been estimated and deposited in the bank, they should not be withdrawn until the actual wetland values have been documented. One method for accomplishing this is to allow only a certain number of credits to be withdrawn each year, based upon the anticipated or actual values realized. This prevents ''front-end loading'' of the bank, where a developer withdraws all the credits that will be generated over the life of the bank within the first few years (Short, 1988).

Another way to control the withdrawal of credits is to delineate specific wetland types that can be replaced with bank credits. Included in this specification are the geographic boundaries where credits are valid. Replacement ratios, which usually range from 1:1 to 6:1, also need to be established by policy and carefully defined (Short, 1988). As an additional safeguard, all signees of the MOA must agree to each withdrawal.

Based upon a recent review of case studies of mitigation-banking practices throughout the United States, a set of guidelines for use in mitigation-banking proposals and operation has been developed. The guidelines are as follows (Weems, 1993, p. xi):

1. The goal of resource agencies and developers should be to provide an efficient and effective means of providing mitigation while at the same time decreasing the time involved in the permitting process.

2. Site selection and construction should take place prior to wetland impacts from a project to allow the wetland functions of the bank to become established.

3. The bank operator should be the developer and/or a combination of resource/regulatory agencies whose responsibilities are to carry out administrative duties and develop long-term management plans.

4. Establishment of bank credits/debits should be based on a habitat method such as HEP, or an equivalent habitat method.

5. The preferred order of mitigation option usage in developing a bank site are restoration, enhancement, creation, and preservation.

6. Those projects that are eligible for bank use must have proven to the applicable regulatory agencies that all components of the mitigation hierarchy are not possible and the project is unavoidable.

7. It is recommended that earmarked accounts, trust funds, or long-term interest bearing accounts be used to guarantee funds for long-term management. Contingency plans, conservation easements, and transferral of the bank site to public ownership are additional ways to guarantee conservation of the bank.

8. During bank construction, best management practices of erosion and fugitive dust control should be implemented to protect adjacent ecosystems. Construction timing should be scheduled so as to not disturb migratory or breeding characteristics of area species.

9. Postmonitoring should be conducted annually for at least ten years to determine the success or failure of a bank site and to provide information on establishing a maintenance program.

10. Bank account statements should be sent semiannually to bank users and after each transaction to signees of the MOU/MOA for their response and approval.

SUMMARY

Habitat-based methods for impact prediction and assessment for the biological environment are being increasingly used. The most widely used method in the United States is the ''habitat evaluation procedure'' (HEP) of the U.S. Fish and Wildlife Service. This chapter reviews HEP and the U.S. Army Corps of Engineers' "habitat evaluation system" (HES). The following list summarizes the points of similarity and difference for HEP and HES:

1. Both methods require baseline information on the sizes (acreage) of habitat types in the study area.

2. The HES method involves an analysis of habitat quality without reference to specific faunal species; the HEP method involves the selection of representative faunal species (called "evaluation species") and the subsequent evaluation of habitat quality relative to the species. The HES method involves the calculation of a "habitat quality index" (HQI) based on a structured evaluation of selected variables indicative of habitat quality. The HEP method is based on specific "habitat suitability index" (HSI) models for the selected evaluation species; these models also require a structured evaluation of pertinent variables indicative of habitat quality for the evaluation species.

3. The HES method involves the calculation of "habitat unit values" (HUVs) based on the products of habitat size (acreage) and HQIs; the HEP method involves the calculation of "habitat units" (HUs) based on the products of habitat size (acreage) and HSIs.

4. Neither method provides sufficient guidance for the field sampling and evaluation necessary to determine HQIs or HSIs.

5. Both methods require predictions of future sizes (acreage) of different habitat types. The HES method requires prediction of future HQIs, and the HEP method requires prediction of future HSIs. The predictions of acreage can be based on engineering and planning studies. Neither the HES or the HEP method provides sufficient guidance for predictions of the variables associated with HQIs or HSIs, respectively.

6. Both methods express impacts on an annualized basis by considering the changes over the project lifetime. The HES method defines "impact" as the difference between with-project and without-project HUVs over the same given period of time. The HEP method defines "impact" in a similar manner based on HUs.

7. Both methods involve the use of a pairwise-comparison technique for importance-weight assignments. The HES method incorporates the technique in assigning weights to the variables which constitute given HQIs. The HEP method incorporates weight assignments in determining "relative value indices" (RVIs) for different evaluation species.

8. Both methods have mitigation-evaluation components to aid in identifying and evaluating mitigation measures for undesirable biological impacts.

9. Both methods are primarily useful in rural or undeveloped areas.

10. Future refinements of both HEP and HES can be anticipated. The chief improvements for HEP are expected to be in the continuing development and validation of HSI models. Field testing for both methods is needed and expected to continue.

In addition to HEP and HES, over 50 other habitat-based methods have been developed for addressing terrestrial and/or aquatic environmental settings. HEP and HES have been used in prospective studies (environmental impact studies) with regard to both settings, and HEP has been used in retrospective studies (evaluation of natural-resources damages at Superfund sites). Finally, mitigation banking appears to be an outgrowth of the development and usage of HEP and/or HES. Mitigation banking, although in its infancy, appears to have major implications for future environmental impact studies.

SELECTED REFERENCES

Allen, A. W., "Habitat Suitability Index Models: Marten," FWS/OBS-82/10.11, U.S. Fish and Wildlife Service, Fort Collins, Colo., Feb. 1982.

Atkinson, S. F., "Habitat-Based Methods for Biological Impact Assessment," *The Environmental Professional,* vol. 7, 1985, pp. 265–282.

Castelle, A. J., et al., "Wetlands Mitigation Banking," Rep. no. 92-12, Shorelands and Coastal Zone Management Program, Washington State Department of Ecology, Olympia, Mar. 1992.

Christmas, J. Y., et al., "Habitat Suitability Index Models: Gulf Menhaden," FWS/OBS-82/10.23, U.S. Fish and Wildlife Service, Gulf Coast Research Laboratory, Ocean Springs, Miss., July 1982.

Dean, B. V., and Nishry, J. J., "Scoring and Profitability Models for Evaluating and Selecting Engineering Products," *Journal of the Operations Research Society of America,* vol. 13, no. 4, July–Aug. 1965, pp. 550–569.

Desvousges, W. H., and Skahen, V. A., "Techniques to Measure Damages to Natural Resources," U.S. Department of the Interior, Washington, D.C., June 1987.

The Greeley-Polhemus Group, "Economic and Environmental Considerations for Incremental Cost Analysis in Mitigation Planning," IWR Rep. 91-R-1, Institute for Water Resources, U.S. Army Corps of Engineers, Ft. Belvoir, Va., Mar. 1991, pp. 22–58.

Kusler, J. A., and Kentula, M. E., "Executive Summary," *Wetland Creation and Restoration: The Status of the Science,* J. A. Kusler and M. E. Kentula, eds., Island Press, Washington, D.C., 1990, pp. xvii–xxi.

Lewis, R. R., "Wetlands Restoration/Creation/Enhancement Terminology: Suggestions for Standardization," *Wetland Creation and Restoration: The Status of the Science,* J. A. Kusler and M. E. Kentula, eds., Island Press, Washington, D.C., 1990, pp. 417–419.

Miller, A. C., et al., "Environmental Impact Research Program—Community Habitat Suitability Models for Warmwater Fishes," WES/MP/EL-87-14, U.S. Army Engineer Waterways Experiment Station, Vicksburg, Miss., Nov. 1987.

Reppert, R., "Wetlands Mitigation Banking Concepts," IWR Rep. 92-WMB-1, Institute for Water

Resources, U.S. Army Corps of Engineers, Ft. Belvoir, Va., July 1992.

Roberts, T. H., O'Neil, L. J., and Jabour, W. E., "Status and Sources of Habitat Models and Literature Reviews: December 1984," Misc. Paper EL-85-1, rev. by L. J. O'Neil and H. K. Gray, U.S. Army Engineer Waterways Experiment Station, Vicksburg, Miss., Aug. 1987.

Schamberger, M., and Farmer, A. H., "The Habitat Evaluation Procedures: Their Application in Project Planning and Impact Evaluation," *Transactions of the 43rd North American Wildlife and Natural Resources Conference,* Wildlife Management Institute, Washington, D.C., 1978, pp. 274–283.

Schamberger, M., Farmer, A. H., and Terrell, J. W., "Habitat Suitability Index Models: Introduction," FWS/OBS-82/10, U.S. Fish and Wildlife Service, Ft. Collins, Colo., Feb. 1982.

Schuytema, G. S., "A Review of Aquatic Habitat Assessment Methods," EPA 600/S3-82-002, U.S. Environmental Protection Agency, Environmental Research Laboratory, Corvallis, Ore., Aug. 1982.

Short, C., "Mitigation Banking," Biological Report 88(41), U.S. Fish and Wildlife Service, Ft. Collins, Colo., July 1988, pp. 1–7.

Simenstad, C. A., Tanner, C. D., Thom, R. M., and Conquest, L. L., "Estuarine Habitat Assessment Protocol," EPA 910/9-91-037, U.S. Environmental Protection Agency, Region 10, Seattle, Wash., Sept. 1991.

U.S. Army Corps of Engineers, "A Habitat Evaluation System for Water Resources Planning," Lower Mississippi Valley Division, Vicksburg, Miss., Aug. 1980.

———, "A Tentative Habitat Evaluation System (HES) for Water Resources Planning," Lower Mississippi Valley Division, Vicksburg, Miss., Nov. 1976.

U.S. Fish and Wildlife Service, "Habitat Evaluation Procedures (HEP)," ESM 102, U.S. Fish and Wildlife Service, Washington, D.C., Mar. 1980.

———, "Type B Technical Information Document: Guidance on Use of Habitat Evaluation Procedures and Suitability Index Models for CERCLA Application," Western Energy and Land Use Team, Ft. Collins, Colo., June 1987.

Wakeley, J. S., and O'Neil, L. J., "Techniques to Increase Efficiency and Reduce Effort in Applications of the Habitat Evaluation Procedures (HEP)," Tech. Rep. EL-88-13, U.S. Army Engineer Waterways Experiment Station, Vicksburg, Miss., Sept. 1988.

Weems, W. A., "Wetland Mitigation Banking: Development of Guidelines," MES Thesis, University of Oklahoma, Norman, 1993, pp. x–xi, 20–38, 112–115.

CHAPTER 12

Prediction and Assessment of Impacts on the Cultural (Historical and Archaeological) Environment

A possible major concern for many actions is their potential impact on cultural resources, which include architectural, historical, and archaeological sites, as well as areas of unique importance because of their ecological, scientific, or geological information. The sphere of cultural resources includes not only the precise limits of the project area, but also all surrounding lands on which the project may have a reasonably direct impact by modifying land-use patterns or by opening areas for agriculture or for public use, thus increasing potential vandalism (McGimsey, 1973). Possible impacts on cultural resources include inundation, destruction, disruption, or disturbance. This chapter addresses data needs and approaches for predicting and assessing impacts of proposed actions on cultural resources, including a six-step methodological approach for impact prediction and assessment. Information is also included on relevant federal legislation and associated procedures.

The National Environmental Policy Act (NEPA) of 1969 addressed potential impacts on the cultural environment by indicating that it is

the responsibility of the federal government to (NEPA, Section 101, b, 4)

. . . preserve important historic, cultural and natural aspects of our national heritage. . . .

Historical and archaeological resources are often referred to as "cultural resources." In a broader sense, the culture of a society is a composite of its history, traditions, arts, architecture, religious beliefs, sciences, and educational systems, among other things (King, Hickman, and Berg, 1977). Cultural resources management may involve consideration of the potential impacts of proposed projects or activities on various components constituting a society's culture, including its historic and archaeological resources. An even broader term is "social-impact assessment," with the "social environment" including all organized behavior by groups of human beings, and the products of such behavior—economic systems, religions, governments, schools, and housing. Social-impact assessments are very general studies that tend to deal with broad patterns of social and economic behavior that occur in the area where a project is planned

435

(King, Hickman, and Berg, 1977). Socioeconomic impacts are addressed in Chapter 14.

BASIC INFORMATION ON CULTURAL RESOURCES

The issue of cultural resources is becoming more important in relation to EISs, given the growing realization that the environment and civilization are the products of history. Cultural resources are nonrenewable, and this feature in itself is one reason that these resources are important. In addition, information on cultural resources, particularly archaeological (historical) resources, can yield important environmental data, since past ecological conditions often are reflected in archaeological sites (McGimsey, 1973).

Archaeological sites may exhibit evidence of different occupations over different periods of time. For example, in a study of nine sites along the James River in the Stutsman and LaMoure counties and three sites around Kraft slough in Dickey County, North Dakota, it was determined that the floodplain areas were recurrently occupied from the Middle Plains Woodland period (100 B.C. to A.D. 600) through the Late Plains Woodland period (A.D. 600 to 900), Early Plains Village period (A.D. 900 to 1400), and Late Plains Village period (A.D. 1400 to 1750) (Gregg, Kordecki, and Picha, 1985).

Section 106 of the National Historic Preservation Act of 1966 requires that federal agencies take into account the effects of their undertakings on properties included in or eligible for listing in the *National Register of Historic Places,* and that such agencies afford the Advisory Council on Historic Preservation the opportunity to comment on such undertakings (Advisory Council on Historic Preservation, 1980). "Archaeological properties" are those properties included in, eligible for, or potentially eligible for, inclusion in the *National Register,* and whose significance lies wholly or partly in the archaeological data they contain. Archaeological data are data embodied in material remains (artifacts, structures, refuse, and the like) utilized purposely or accidentally by human beings, in the spatial relationship among such remains, and in the environmental context of such remains.

Examples of some pertinent questions and resultant responses related to historical and archaeological resources in EISs include

1. How should "unknown" resources in the study area be addressed?

   Depending on the type of project and its proposed location, it may be necessary to plan and conduct historical and archaeological surveys in the study area.

2. Do all findings in a survey have significance; or, are all findings eligible for inclusion in the *National Register of Historic Places?*

   Survey findings need professional interpretation in relation to the significance criteria for eligibility for the *National Register.*

3. Can historical-archaeological surveys be planned and conducted by anyone or do they need to be approved or recognized by certain professional agencies and/or organizations?

   Survey work must be planned and conducted by one or more qualified professionals. Davis (1982) described a qualified professional archaeologist as *(a)* any person accredited by the Society of Professional Archaeologists (SOPA) with appropriate SOPA archaeological emphases and who is carried on the active list of persons accredited by SOPA; or *(b)* any archaeologist who has been identified as a principal investigator in a contract awarded by a federal agency where the guidelines contained in 36 *Code of Federal Regulations* (C.F.R.) 66 (or equivalent federal guidelines) have been applied to the evaluation of the principal investigator.

4. Are there limitations of historical-archaeological surveys conducted by visual observations and/or surface-soil shovel testing?

There are limitations, since deeper properties may not be found. Therefore, projects involving subsurface excavation should include plans for addressing potential historical and archaeological properties found during the construction phase. It may be necessary to include permit conditions for handling construction-phase information.

5. Who has ownership of historic and archaeological properties on private land?

Private ownership of significant cultural properties can become an issue of concern. The Advisory Council on Historic Preservation (1984) indicated that, before any property or district may be included on the *National Register* or designated a "National Historic Landmark," the owner or owners of such property, or a majority of the owners of the properties within the district in the case of an historic district, must be given the opportunity (within a reasonable period of time) to concur in, or object to, the nomination of the property or district for such inclusion or designation. If the concerned party or parties object to such inclusion or designation, the property will not be included on the *National Register* nor will it be designated a landmark until the objection is withdrawn.

6. Is it necessary for every historical-archaeological survey to provide complete coverage of the potential area of effect?

Field efforts in a survey can be optimized through professional knowledge of known resources and the application of professional judgment. Predictive modeling can be used as a basis for targeted surveys.

7. When existing and new information on historic properties is summarized in a report prepared as a part of an EIS, will such information encourage property vandalism and scavenging?

The information should be provided to the state historic preservation officer for handling in a professional, confidential manner. Summary information, which does not include site- and property-specific information, could be included in the EIS documentation.

FEDERAL LAWS, REGULATIONS, AND EXECUTIVE ORDERS

Federal laws, regulations, and executive orders related to cultural resources impacts are listed in Table 12.1, and brief information on each will be presented.

Antiquities Act of 1906 (P.L. 59-209)

The Antiquities Act of 1906 provided for the protection of all historic and prehistoric ruins of monuments on federal lands (U.S. Congress, 1906). It prohibited any excavation or destruction of such antiquities without permission of the secretary of the executive department having jurisdiction. It authorized the secretaries of the Interior, Agriculture, and War to give permission (grant permits) for excavation to reputable institutions for increasing knowledge and for permanent preservation in public museums. It also authorized the President to declare areas of public lands "national monuments" and to reserve lands for that purpose (Neal, 1975). In practice this law has largely been superseded by the Archaeological Resources Protection Act (ARPA) of 1979 (P.L. 96-95) (Limp, 1989).

Historic Sites, Buildings, and Antiquities Act of 1935 (P.L. 74-292)

The Historic Sites, Buildings, and Antiquities Act of 1935 established a national policy to preserve for public use historic sites, buildings, and objects of national significance (U.S. Congress, 1935). It transferred to the National Park Service in the U.S. Department of the Interior the responsibility for implementing this policy and establishing the *National Register of Historic Places*. In effect, this act charged the National Park Service with performing any required fed-

TABLE 12.1

MAJOR FEDERAL LAWS RELATED TO CULTURAL RESOURCES IMPACTS

Public law no.	Title	Date
59–209	Antiquities Act	1906
74–292	Historic Sites, Buildings, and Antiquities Act	1935
81–408	National Trust Act	1949
86–523	Reservoir Salvage Act	1960
89–665	National Historic Preservation Act	1966
93–291	Archaeological and Historic Preservation Act	1974
95–341	American Indian Religious Freedom Act	1978
96–95	Archaeological Resources Protection Act	1979
96–515	National Historic Preservation Act Amendments	1980
100–298	Abandoned Shipwreck Act	1987
100–555	Archaeological Resources Protection Act Amendment	1988
100–588	Archaeological Resources Protection Act Amendment	1988
101–601	Native American Graves Protection and Repatriation Act	1990
102–575	National Historic Preservation Act Amendments	1992

Source: After Limp, 1989.

eral archaeological investigations (Limp, 1989). This act also led to the establishment of the *Historic Sites Survey,* the *Historic American Building Survey,* and the *Historic American Engineering Record.* The National Historic Landmarks program and its advisory board were also established to designate properties having exceptional value as commemorating or illustrating the history of the United States. The National Historic Landmarks program was the beginning of the *National Register* program (Neal, 1975).

National Trust Act of 1949 (P.L. 81-408)

The National Trust Act of 1949 was designed to improve public participation in the preservation of archaeological and historic sites, buildings, and objects. It created the National Trust for Historic Preservation and defined its responsibilities and powers (Limp, 1989).

Reservoir Salvage Act of 1960 (P.L. 86-523)

The Reservoir Salvage Act of 1960 provided for the preservation of historic and/or archaeological resources threatened by construction of a federal or a federally licensed dam-reservoir project. The act also required that before undertaking the construction of a dam or issuing a license for construction of a dam (greater than 5,000 acre-ft or 40 surface acres of capacity), an agency must provide written notice to the Secretary of the Interior. The provisions of the act apply regardless of the size of the reservoir if the constructing agency finds or is presented with evidence that archaeological resources are affected (McGimsey, 1973). Should an agency find that the project may cause irreparable damage to cultural resources, it must notify the Secretary of the Interior and request him or her to undertake recovery-protection activities. Alternatively, the agency is to itself engage in such efforts.

National Historic Preservation Act of 1966 (P.L. 89-665)

The National Historic Preservation Act of 1966 established the Advisory Council on Historic Preservation to advise the President and Congress on matters related to historic preservation; it also established state historic preservation officers. As amended, the act specified in Section

106 that before a federal agency approved expenditure of any funds on an undertaking or issuance of a license or permit, the agency was to take into account the effect of the project on any property listed on or eligible for the *National Register of Historic Places*. It was to consult with the advisory council to find ways to mitigate or avoid adverse effects. The act also provided grants for historic preservation and gave the federal government authority to acquire significant sites by power of eminent domain. Major revision of this act occurred with the passage of the 1980 amendments (P.L. 96-515) discussed below (Limp, 1989). Additional revisions were included in the 1992 amendments (P.L. 102-575).

Executive Order 11593 of 1971: Protection and Enhancement of the Cultural Environment

Executive Order 11593 of 1971 established a policy requiring the federal government to provide leadership in preserving, restoring, and maintaining the historic and cultural environment of the nation. Further, the order indicated that federal agencies must (1) administer the cultural properties under their control in a spirit of stewardship and trusteeship for future generations; (2) initiate measures necessary to direct their policies, plans, and programs in such a way that federally owned sites, structures, and objects of historical, architectural, or archaeological significance are preserved, restored, and maintained for the inspiration and benefit of the people; and (3) in consultation with the Advisory Council on Historic Preservation, institute procedures to assure that federal plans and programs contribute to the preservation and enhancement of nonfederally owned sites, structures, and objects of historical, architectural, or archaeological significance (the President, 1971).

Two examples of the implementation of Executive Order 11593 can be cited. First, for all property under federal jurisdiction or control, the responsible agencies must survey and nominate all historic properties to the *National Register*. These historic properties must also be maintained and preserved by the responsible agency. Second, for every action funded, licensed, or executed by the federal government, the agency involved must ask the Secretary of the Interior to determine if any property in the environmental-impact area is eligible for the *National Register*. The determination-of-eligibility process is faster than the nomination process and gives the same protection. If the federal action will substantially alter or destroy a historic property, the agency must have the property recorded in the *Historic American Buildings Survey* or the *Historic American Engineering Record* (Neal, 1975).

Also included in the executive order was a directive to conduct studies to locate and nominate properties under federal control that might qualify for inclusion in the *National Register*. The order required that agencies exercise caution in the meantime to make certain that they did not unnecessarily damage eligible properties on federal lands. The order also directed the National Park Service to develop criteria and policies for evaluation of important properties and determination of their eligibility for inclusion in the *National Register*. A major portion of Executive Order 11593 was codified in Sections 110 and 206 of the 1980 amendments to the National Historic Preservation Act of 1966 (Limp, 1989).

Archaeological and Historic Preservation Act of 1974 (P.L. 93-291)

The Archaeological and Historic Preservation Act of 1974, also referred to as the "Moss-Bennett bill," amended the Reservoir Salvage Act of 1960 and provided for the preservation of historical and archaeological data which otherwise might be irreparably lost or destroyed by construction or any alteration of the terrain as a result of any federal construction project or federally licensed project, activity, or program. The act also authorized the Secretary of the Interior, or the agency itself, to undertake recovery, protection, and preservation of such data. Where the federal government financially aids in activities that may cause irreparable damage, the Secretary of the Interior may survey the data and undertake recovery and preservation. Archaeological salvage or recording by the *Historic American Buildings Survey* or the *Historic American Engineering Record* are among the alternatives available. When the activity takes place on private land, the secretary must compensate the owner for any resultant delays or loss of use of the land. This act presents two innovations in terms of previous laws: (1) only dams were covered in the Reservoir Salvage Act of 1960, now all federal projects are covered; and (2) up to 1 percent of project funds may be used for data recovery, protection, and preservation (Neal, 1975). The 1-percent level was not strictly required, nor were necessary expenditures to be limited to 1 percent.

The act further provided for the preservation of historical and archaeological data (including relics and specimens) which might otherwise be irreparably lost or destroyed as the result of (1) flooding, the building of access roads, the erection of workers' communities, the relocation of railroads and highways, and other alterations of the terrain caused by the construction of a dam by an agency; or (2) any alteration of the terrain resulting from any federal construction project or federally licensed activity or program. Before any agency undertakes the construction of a dam or issues a license to any private individual or corporation for the construction of a dam, it must give written notice to the Secretary of the Interior setting forth the site of the proposed dam and the approximate area to be flooded and otherwise changed—provided that, with respect to any floodwater-retarding dam which provides less than 5,000 acre-ft of detention capacity and with respect to any other type of dam which creates a reservoir of less than 40 surface acres, the requirements apply only when the construct-

ing agency, in its preliminary surveys, finds or is presented with evidence that historical or archaeological materials exist or may be present in the proposed reservoir area (U.S. Congress, 1974).

The federal agency involved could either transfer up to 1 percent of the project funds for cultural-resources investigation or conduct the activity itself. This limitation could be waived pursuant to Section 208(3) of the 1980 amendments to the National Historic Preservation Act (P.L. 96-515) (Limp, 1989).

American Indian Religious Freedom Act of 1978 (P.L. 95-341)

The American Indian Religious Freedom Act of 1978 established a federal policy, the protection and preservation of the right of Native Americans to exercise their traditional religions. It addressed questions of access to sites and possession of sacred objects. The act recognized the legitimacy of Native American religious and cultural activities, and directed agencies to review their policies and programs to ensure that such activities are properly considered and respected. Federal managers were directed to administer their relevant laws in consultation with traditional religious leaders (Limp, 1989).

Archaeological Resources Protection Act of 1979 (P.L. 96-95)

The Archaeological Resources Protection Act of 1979 (ARPA) was enacted to provide a comprehensive framework for protecting and regulating the use of archaeological resources on public and Native American lands. It mandated that all excavation and removal of archaeological resources on public land be done pursuant to a permit issued by the federal manager of the land involved. Penalties involving fines and/or imprisonment were authorized for those guilty of excavating or destroying such resources without a permit. These areas were formerly dealt with by the Antiquities Act of 1906. It should be noted that the Antiquities Act has not been repealed and is also currently in force (Limp, 1989).

The purpose of the ARPA was stated as follows (U.S. Congress, 1979, p. 1):

> to secure, for the present and future benefit of the American people, the protection of archaeological resources and sites which are on public lands and Native American lands, and to foster increased cooperation and exchange of information between governmental authorities, the professional archaeological community, and private individuals having collections of archaeological resources and data which were obtained before the date of the enactment of the ARPA.

Definitions of some key terms ("archaeological resources," "public lands," "Indian lands," and "federal land manager") are included in Table 12.2, along with other pertinent terms relevant to other, related legislation.

Implementing regulations for the ARPA are in 32 C.F.R. 229. A key element is the issuance of permits by the pertinent federal land manager for excavating or removing any archaeological resource located on public lands or Native American lands. Several federal agencies have adopted their own regulations for implementing the requirements of the ARPA; examples include the departments of the interior, agriculture, and defense, and the Tennessee Valley Authority. Permit requirements or conditions are typically included in these implementing regulations.

An issue which may arise following the issuance of a permit is related to the custody of identified archaeological resources; the regulations in U.S. Department of Defense (1984) indicated that (1) archaeological resources excavated or removed from the public lands remain the property of the United States and (2) archaeological resources excavated or removed from Native American lands remain the property of the Native American individual or tribe having rights of ownership over such resources.

National Historic Preservation Act Amendments of 1980 (P.L. 96-515)

The National Historic Preservation Act Amendments of 1980, which amended the act of 1966 (P.L. 89-665), established a separate historic preservation fund and gave the Secretary of the Interior discretionary authority to increase the matching ratio of grants to states for preparation of preservation plans. It also established the Advisory Council on Historic Preservation as an independent agency with authority to issue rules and regulations and codified the substance of Executive Order 11593 (Limp, 1989).

Abandoned Shipwreck Act of 1987 (P.L. 100-298)

Shipwrecks, objects lost underwater, and submerged prehistoric sites are a part of the cultural diversity of the United States. "Underwater archaeology" refers to the study of the remains of prehistoric and historic human activities found underwater (Office of Technology Assessment, 1987). Potential locational technologies for underwater cultural resources include side-scan sonar, subbottom profilers, magnetometers, and remotely operated vehicles.

The Abandoned Shipwreck Act of 1987 has as its purposes (U.S. Congress, 1988) (1) to protect natural resources and habitat areas, (2) to guarantee recreational exploration of shipwreck sites, and (3) to allow for appropriate public and private sector recovery of shipwrecks consistent with the protection of the historical values and environmental integrity of the shipwrecks and the sites. In managing the resources subject to the provisions of the act, states are encouraged to create underwater parks or areas and thus provide additional protection for such resources.

Prior to the act, the status of underwater shipwrecks was different from that of all other underwater cultural resources. All shipwrecks previously came under marine law, and under such law salvage rules overrode cultural-resource regulations. Passage of this law corrected the

TABLE 12.2

PERTINENT DEFINITIONS RELATED TO HISTORIC AND/OR ARCHAEOLOGICAL RESOURCES

Archaeological resources: Any material remains of past human life or activities which are of archaeological interest, as determined under uniform regulations promulgated pursuant to the Archaeological Resources Protection Act of 1979 are "archaeological resources." Regulations containing such determination apply to pottery, basketry, bottles, weapons, weapon projectiles, tools, structures or portions of structures, pit houses, rock paintings, rock carvings, intaglios, graves, human skeletal materials, any portion or piece of any of the foregoing items, as well as items not stated here. Nonfossilized and fossilized paleontological specimens, or any portion or piece thereof, are not considered archaeological resources unless they are found in an archaeological context. No item less than 100 years old is to be treated as an archaeological resource.

Archaeology, historic: The term "historic archaeology" refers to the scientific study of the life and culture in the New World after the advent of written records.

Archaeology, prehistoric: The scientific study of the life and culture of indigenous peoples before the advent of written records constitutes "prehistoric archaeology."

Architecture: The term "architecture" refers to the style and construction of buildings and structures.

Area of potential effects: The geographic area or areas within which an undertaking may cause changes in the character or use of historic properties, if any such properties exist, is referred to as the "area of potential effect."

Building: A "building" refers to something—such as a house, barn, church, hotel, or similar construction—that has been created to shelter any form of human activity. "Building" may also be used to refer to an historically and functionally related unit, such as a courthouse and jail or a house and barn.

District: A "district" possesses a significant concentration, linkage, or continuity of sites, buildings, structures, or objects united historically or aesthetically by plan or physical development.

Federal land manager: The term "federal land manager" means, with respect to any public lands, the Secretary of the department, or the head of any other agency or instrumentality of the United States, having primary management authority over such lands. In the case of any public lands or Indian lands with respect to which no department, agency, or instrumentality has primary management authority, such term means the Secretary of the Interior. If the Secretary of the Interior consents, the responsibilities (in whole or in part) under the Archaeological Resources Protection Act of 1979 of the Secretary of any department (other than the Department of the Interior) or the head of any other agency or instrumentality may be delegated to the Secretary of the Interior with respect to any land managed by such other secretary or agency head, and, in any such case, the term "federal land manager" means the Secretary of the Interior.

Historic conservation district: The term "historic conservation district" refers to an urban area of one or more neighborhoods which contains (1) historic properties, (2) buildings having similar or related architectural characteristics, (3) cultural cohesiveness, or (4) any combination of the foregoing.

Historic resource or historic property: Any prehistoric or historic district, site, building, structure, or object included in, or eligible for inclusion in, the *National Register of Historic Places* is a "historic resource" or "historic property"; such term includes artifacts, records, and remains which are related to such a district, site, building, structure, or object. The *National Register,* in turn, defines an historic property as a district, site, building, structure, or object significant in American history, architecture, engineering, archaeology, or culture.

Indian lands: All lands under the jurisdiction or control of an Indian tribe are referred to as "Indian lands." Lands of Indian tribes, or Indian individuals, which are either held in trust by the United States or subject to a restriction against alienation imposed by the United States, except for any subsurface interests in lands not owned or controlled by an Indian tribe or an Indian individual.

Indian tribe: The term "Indian tribe" refers to the governing body of any Indian band, nation, or other group that is recognized as an Indian tribe by the Secretary of the Interior and for which the United States holds land in trust or restricted status for that entity or its members. The term also includes any Native village corporation, regional corporation, or Native group established pursuant to the Alaska Native Claims Settlement Act.

TABLE 12.2

PERTINENT DEFINITIONS RELATED TO HISTORIC AND/OR ARCHAEOLOGICAL RESOURCES
(continued)

National historic landmark: A "National Historic Landmark" is an historic property that the Secretary of the Interior has so designated.

Object: The term "object" is used to distinguish from buildings and structures those constructions that are primarily artistic in nature or are relatively small in scale and simply constructed. Although it may be, by nature or design, movable, an "object" is associated with a specific setting or environment, such as statuary in a designed landscape.

Preservation or historic preservation: The definition of "preservation" or "historic preservation" includes the identification, evaluation, recordation, documentation, curation, acquisition, protection, management, rehabilitation, restoration, stabilization, maintenance and reconstruction, or any combination of the foregoing activities.

Public lands: The term "public lands" refers to lands which are owned and administered by the United States as part of the national park system, the national wildlife refuge system, or the national forest system; and to all other lands the fee title to which is held by the United States, other than lands on the Outer Continental Shelf and lands which are under the jurisdiction of the Smithsonian Institution.

Site: A "site" is the location of a significant event, a prehistoric or historic occupation or activity, or a building or structure, whether standing, ruined, or vanished, where the location itself possesses historical, cultural, or archaeological values regardless of the value of any existing structure.

Structure: The term "structure" is used to distinguish from buildings those functional constructions made usually for purposes other than creating shelter.

Undertaking: The term "undertaking" denotes any federal, federally assisted, or federally licensed action, activity, or program or the approval, sanction, assistance, or support of any nonfederal action, activity, or program. Undertakings include new and continuing projects and program activities (or elements of such activities not previously considered under Section 106 or Executive Order 11593) that are (1) directly undertaken by federal agencies; (2) supported in whole or in part through federal contracts, grants, subsidies, loans, loan guarantees, or other forms of direct and indirect funding assistance; (3) carried out pursuant to a federal lease, permit, license, certificate, approval, or other form of entitlement or permission; or (4) proposed by a federal agency for Congressional authorization or appropriation. Site-specific undertakings affect areas and properties that are capable of being identified at the time of approval by the federal agency. Although nonsite-specific undertakings have effects that can be anticipated on properties listed in the *National Register* and properties eligible for listing, the effects cannot be identified in terms of specific geographical areas or properties at the time of federal approval. Nonsite-specific undertakings include federal approval of state plans pursuant to federal legislation, development of comprehensive or areawide plans, agency recommendations for legislation, and the establishment or modification of regulations and planning guidelines.

Sources: Adapted from Parker, 1985; Advisory Council on Historic Preservation, 1977, 1979, 1984, 1986a, 1986b; U.S. Congress, 1979.

situation and gave the responsibility for management of shipwrecks located in state waters or submerged lands to the states, paralleling their terrestrial-cultural-resources management role. "Submerged lands" are those normally extending 3 nautical mi from shore, except in the cases of Texas, Florida, and Puerto Rico, which have lands extending beyond 3 nautical mi; riverbeds and natural lake floors are also

included in the definition. This law also required the Secretary of the Interior to promulgate guidelines (Limp, 1989).

Archaeological Resources Protection Act Amendments of 1988 (P.L. 100-555 and 100-588)

Two amendments to the ARPA were passed and signed into law (P.L. 100-555 and 100-588) dur-

TABLE 12.3

KEY TERMS IN THE NATIVE AMERICAN GRAVES PROTECTION AND REPATRIATION ACT

1. "Burial site" means any natural or prepared physical location, whether originally below, on, or above the surface of the earth into which, as a part of the death rite or ceremony of a culture, individual human remains are deposited.

2. "Cultural affiliation" refers to a relationship of shared group identity which can be reasonably traced historically or prehistorically between a present-day Native American tribe or Native Hawaiian organization and an identifiable earlier group.

3. The term "cultural items" means human remains, as well as the following items:

   a. "Associated funerary objects," or objects that, as a part of the death rite or ceremony of a culture, are reasonably believed to have been placed with individual human remains either at the time of death or later. Both the human remains and associated funerary objects must be presently in the possession or control of a federal agency or museum. However, other items exclusively made for burial purposes or to contain human remains are also considered associated funerary objects.

   b. "Unassociated funerary objects," or objects that, as a part of the death rite or ceremony of a culture, are reasonably believed to have been placed with individual human remains either at the time of death or later. The remains must not be in the possession or control of the federal agency or museum, and the objects must be identified by a preponderance of evidence related to specific individuals or families or to known human remains or, by a preponderance of evidence, showing them to have been removed from a specific burial site of an individual culturally affiliated with a particular Native American tribe.

   c. "Sacred objects," or specific ceremonial objects which are needed by traditional Native American religious leaders for the practice of traditional Native American religions by their presnt-day adherents.

   d. "Cultural patrimony," or an object having vital, ongoing historical, traditional, or cultural importance to the Native American group or culture itself. The term does not refer to property owned by an individual Native American. Cultural patrimony cannot be alienated, appropriated or conveyed by any individual even should that individual be a member of the pertinent tribe or Native Hawaiian organization. The object must have been considered inalienable by the affected Native American group at the time it was separated from the group.

4. The term "museum" means any institution or state or local government agency (including any institution of higher learning) that receives federal funds and has possession of, or control over, Native American cultural items. This term does not refer to the Smithsonian Institution or any other federal agency.

5. "Native American" means of, or relating to, a tribe, people, or culture that is indigenous to the United States.

6. "Native Hawaiian" means any individual who is a descendant of the aboriginal people who, prior to 1778, occupied and exercised sovereignty in the area that now constitutes the state of Hawaii.

7. "Right of possession" means possession obtained with the voluntary consent of an individual or group that had authority of alienation. The original acquisition of an unassociated funerary object, sacred object, or object of cultural patrimony from a Native American tribe or Native Hawaiian organization with the voluntary consent of an individual or group with authority to alienate such object is deemed to give right of possession of that object, unless the phrase so defined would result in a Fifth Amendment taking by the United States, as determined by the United States Claims Court (pursuant to 28 U.S.C. 1491). In the later case, "right of possession" is provided under otherwise applicable property law. The original acquisition of Native American human remains and associated funerary objects which were excavated, exhumed, or otherwise obtained with full knowledge and consent of the next of kin or the official governing body of the appropriate culturally affiliated Native American tribe or Native Hawaiian organization is deemed to give right to possession to those remains.

Source: Compiled using data from U.S. Congress, 1990.

ing the 100th Congress (1988). The first required federal agencies to develop plans to inventory all lands, including those not scheduled for development. This amendment added force to the Section 110 requirements of the NHPA. It also required development of a uniform system for reporting archaeological violations. The second amendment made attempting to loot or vandalize an archaeological site on federal lands a crime. It also required federal land managers to develop outreach programs for increasing public awareness of cultural resources (Limp, 1989).

Native American Graves Protection and Repatriation Act of 1990 (P.L. 101-601)

The Native American Graves Protection and Repatriation Act (NAGPRA) of 1990 (P.L. 101-601) provides for the protection of Native American graves and similar sites. Some key terms associated with NAGPRA are defined in Table 12.3. The focus of selected key sections of the NAGPRA includes (U.S. Congress, 1990)

1. *Section 3(a)*—Ownership of Native American human remains and objects
2. *Section 3(b)*—Ownership of unclaimed Native American human remains and objects
3. *Section 3(c)*—Intentional excavation and removal of Native American human remains and objects
4. *Section 3(d)*—Inadvertent discovery of Native American remains and objects
5. *Section 4*—Illegal trafficking in Native American human remains and cultural items
6. *Section 5*—Inventory of human remains and associated funerary objects
7. *Section 6*—Summary of unassociated funerary objects, sacred objects, and cultural patrimony
8. *Section 7(a)*—Repatriation of Native American human remains and objects possessed or controlled by federal agencies and museums

National Historic Preservation Act Amendments of 1992 (P.L. 102-575)

The National Historic Preservation Act Amendments of 1992 provide for a significant expansion of the historic preservation programs of Native American tribes, establish standards for professionals, and refine several definitions. One refined definition is that for ''undertaking,'' which now means (U.S. Congress, 1992)

> a project, activity, or program funded in whole or in part under the direct or indirect jurisdiction of a federal agency, including those carried out by or on behalf of the agency; those carried out with federal financial assistance; those requiring a federal permit license, or approval; and those subject to state or local regulation administered pursuant to a delegation or approval by a federal agency.

Also established is a National Center for Preservation Technology to coordinate preservation research, disseminate information, and provide training on preservation technologies (U.S. Congress, 1992).

STATE LAWS, REGULATIONS, AND EXECUTIVE ORDERS

Most states have legislation directed toward historical, archaeological, and cultural resources. Laws dealing with cultural resources regulate their disturbance on state lands and, in some instances and to different degrees, on private lands. State environmental policy acts usually include consideration of cultural resources. In some states, aggressive programs for the investigation, protection, and recovery of archaeological resources have been established; one example is the program implemented by the state of Arkansas. Details on state laws, regulations, and executive orders can be procured from the state historic preservation officer. State laws dealing with cultural resources vary considerably from state to state, and even inquiry at the local level may be necessary to determine pertinent laws and regulations.

NATIONAL HISTORIC PRESERVATION ACT PROVISIONS

The National Historic Preservation Act of 1966 (P.L. 89-665), as amended in 1980 and 1992, provides for expansion of the scope of the *National Register of Historic Places* for districts, sites, buildings, structures, and objects significant in American history, architecture, archaeology, and culture. It provides for a funding program to the states for historical surveys and planning and for preservation, acquisition, restoration, and development projects. The act also established the Advisory Council on Historic Preservation, appointed by the President, to advise the President and the Congress on matters relating to historic preservation (U.S. Congress, 1966).

Advisory Council on Historic Preservation

The Advisory Council consists of the following members (Advisory Council on Historic Preservation, 1984): (1) a chairman selected from the general public; (2) the Secretary of the Interior; (3) the Architect of the Capitol; (4) the Secretary of Agriculture and the heads of four other agencies of the United States, the activities of which affect historic preservation; (5) one governor; (6) one mayor; (7) the president of the National Conference of State Historic Preservation Officers; (8) the chairman of the National Trust for Historic Preservation; (9) four experts in the field of historic preservation from the disciplines of architecture, history, archaeology, and another appropriate discipline; and (10) three at-large members from the general public.

State Historic Preservation Officers

State historic preservation officers (SHPOs)—who are appointed by either the governors of the states, the chief executive of the territories, or, in the case of the District of Columbia, the mayor—carry out the historic preservation programs of their jurisdictions and are given the following responsibilities by the NHPA and other federal authorities (Parker, 1985, p. 6):

1. Carrying out a comprehensive statewide survey of historic properties and maintaining inventories of such properties.
2. Nominating properties to the National Register.
3. Preparing and implementing a statewide historic preservation planning process.
4. Administering Historic Preservation Fund grants.
5. Advising and assisting federal and state agencies and local governments in historic preservation matters.
6. Working with the Department of the Interior, the Advisory Council on Historic Preservation, and others to ensure that historic properties are taken into account in planning.
7. Providing public information, education, and training in historic preservation.
8. Cooperating with local governments in developing preservation programs and assisting them in becoming certified to manage Historic Preservation Fund grants and otherwise participate actively in the national program.
9. Reviewing requests for historic preservation certification and making recommendations to the National Park Service, as part of the federal tax incentives program.

Section 106 Requirements

The head of any federal agency having direct or indirect jurisdiction over a proposed federal or federally assisted undertaking in any state and the head of any federal department or independent agency having authority to license any undertaking must take into account, prior to the approval of the expenditure of any federal funds on the undertaking or prior to the issuance of any license, as the case may be, the effect of the undertaking on any district, site, building, structure, or object that is included in, or eligible for inclusion in, the *National Register*. The head of any such federal agency must afford the Advisory Council on Historic Preservation a reasonable opportunity to comment with regard to such an undertaking (Advisory Council on Historic Preservation, 1986a). Section 106 applies to all properties already listed in the *National Register,* to properties formally determined to be eligible for listing, and to properties not formally determined to be eligible but that meet specified eligibility criteria. This means that properties that have not yet been listed, and even properties that have not yet been discovered, can be eligible for consideration under Section 106 (Advisory Council on Historic Preservation, 1986a).

The Section 106 requirements refer to undertakings. The word "undertaking" was used deliberately in the National Historic Preservation Act and in council regulations to connote a broad range of federal actions (Advisory Council on Historic Preservation, 1986b). The statutory language refers specifically to undertakings over which federal agencies have either "di-

---

TABLE 12.4

CRITERIA FOR *NATIONAL REGISTER* ELIGIBILITY

The quality of significance in American history, architecture, archaeology, and culture is present in districts, sites, buildings, structures, and objects that possess integrity of location, design, setting, materials, workmanship, feeling, and association, and:

(A) that are associated with events that have made a significant contribution to the broad patterns of our history; or

(B) that are associated with the lives of persons significant in our past; or

(C) that embody the distinctive characteristics of types, period or method of construction, or that represent the work of a master, or that possess high artistic values, or that represent a significant and distinguishable entity whose components may lack individual distinction; or

(D) that have yielded, or may be likely to yield, information important in prehistory or history.

Ordinarily cemeteries, birthplaces, or graves of historical figures, properties owned by religious institutions or used for religious purposes, structures that have been moved from their original locations, reconstructed historic buildings, properties primarily commemorative in nature, and properties that have achieved significance within the past 50 years shall not be considered eligible for the National Register. However, such properties will qualify if they are integral parts of districts that do meet the criteria or if they fall within the following categories:

(A) a religious property deriving primary significance from architectural or artistic distinction or historical importance; or

(B) a building or structure removed from its original location but which is significant primarily for architectural value, or which is the surviving structure more importantly associated with an historic person or event; or

(C) a birthplace or grave of an historical figure of outstanding importance if there is no other appropriate site or building directly associated with his productive life; or

(D) a cemetery that derives its primary significance from graves of persons of transcendent importance, from age, from distinctive design features, or from association with historic events; or

(E) a reconstructed building when accurately executed in a suitable environment and presented in a dignified manner as part of a restoration master plan, and when no other building or structure with the same association has survived; or

(F) a property primarily commemorative in intent if design, age, tradition, or symbolic value has invested it with its own historical significance; or

(G) a property achieving significance within the past 50 years if it is of exceptional importance.

Source: Advisory Council on Historic Preservation, 1986b, p. 8.

rect" or "indirect" jurisdiction. Three kinds of undertakings are alluded to: federal undertakings (actions undertaken directly by a federal agency); federally assisted undertakings (for example, activities receiving direct federal financial assistance or such indirect assistance as loan guarantees and mortgage insurance); and federally licensed undertakings (undertakings requiring permits or other entitlements from federal agencies). Other definitions related to Section 106 requirements are in Table 12.2.

Criteria for *National Register*

The U.S. Department of the Interior criteria for listing of historical-archaeological properties in the *National Register of Historic Places* are in 36 C.F.R. Section 60.4; these criteria, which have been designed to guide the states, federal agencies, and the Secretary of the Interior in evaluating potential entries (other than areas of the National Park System and National Historic Landmarks), are in Table 12.4.

Properties can be nominated to the *National Register* by state historic preservation officers through certified local-government historic preservation programs and by federal agencies. As of 1986, there were more than 45,000 listings, many of which were districts, and more are being nominated on a routine basis (Advisory Council on Historic Preservation, 1986b). Detailed procedural aspects for requesting determinations of eligibility and documenting the process are in 36 C.F.R. 63 (Advisory Council on Historic Preservation, 1977).

Section 106 Process

The process for complying with Section 106 requirements has been developed by the Advisory Council on Historic Preservation; the process entitled "Protection of Historic Properties" is in 36 C.F.R. 800. (The effective date of the most recent version was October 1, 1986.) The basic steps in the Section 106 process applied to impact studies are in Figure 12.1. Steps 1 and 2 are self-explanatory. Consultation and council

comment (steps 3 and 4) always involve the agency proposing the undertaking and the SHPO (unless the SHPO declines to participate), and often includes other interested persons. Typically, such consultation results in a memorandum of agreement (MOA), which sets out specific steps for avoiding or reducing harm to historic properties. When an MOA has been accepted by the council, it serves as the council's comment (step 4). In those cases in which the consulting parties cannot reach agreement, consultation may be terminated and the agency may request council comments directly (Advisory Council on Historic Preservation, 1986b).

Section 110 Requirements

All federal agencies must carry out their programs in accordance with, and in furtherance of, national historic preservation policy; designate historic preservation officers to coordinate the agencies' activities under the act; identify and preserve historic properties under their ownership or control; and make efforts to minimize harm to national historic landmarks (Advisory Council on Historic Preservation, 1986a). One example of an agency program developed in response to Section 110 requirements is contained in 33 C.F.R. 325, App. C; these are the procedures to be followed by the U.S. Army Corps of Engineers in its regulatory (permit) program (U.S. Department of the Army, 1990). The procedures are compatible with the Section 106 process.

BASIC STEPS FOR CULTURAL-IMPACT PREDICTION AND ASSESSMENT

Based upon both general principles and the Section 106 process, the fundamental steps associated with prediction of changes in the cultural environment and assessment of the impact of these changes are as follows:

1. Identify known cultural resources in the area of interest. These resources could include

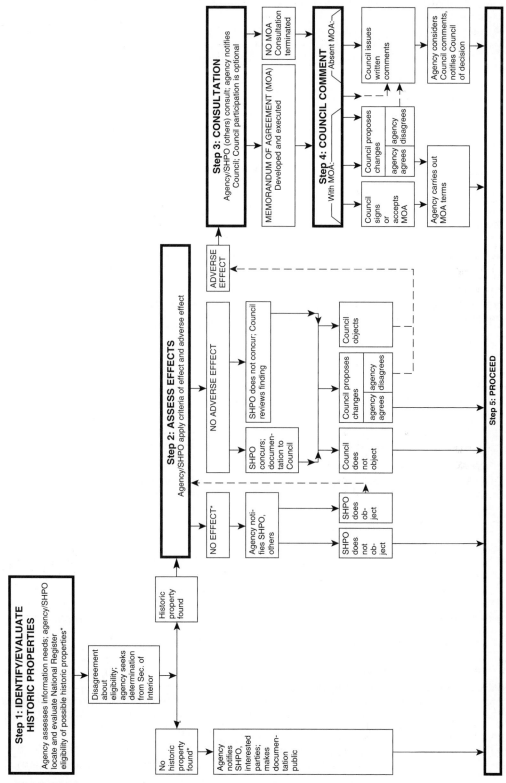

FIGURE 12.1
The Basic Steps of Section 106 Review (Advisory Council on Historic Preservation, 1986b, p. 16.).

449

historical and archaeological sites; areas of ecological, scientific, or geological significance; and areas of ethnic importance. Consider the cultural overview of the area, including prehistorical and historical patterns in the area.

2. Identify potential cultural resources in the area of interest.

3. Determine significance of known and potential cultural resources relative to local, regional, and national concerns.

4. Delineate possible impacts of alternatives on known and potential cultural resources in the area of interest. Impacts should be determined for preconstruction, construction, operation, and postoperation phases.

5. Depending upon the findings of steps 3 and 4 above, either (a) proceed with the selection of one proposed action from the alternatives or (b) eliminate one or more alternatives and then proceed with selection of the proposed action. Following selection of the proposed action, conduct a detailed reconnaissance of the pertinent project area and develop mitigation measures for impact minimization and cultural-resources preservation. If appropriate, develop historic-resources-management plans for the study area.

6. Develop procedures that will be used during the construction phase in the case of discovery of previously unidentified cultural resources.

STEP 1: IDENTIFICATION OF KNOWN CULTURAL RESOURCES

The first step consists of identifying known cultural resources in the area of interest (also known as the "area of potential effects"). "Archaeological resources" can be defined as objects and areas made or modified by humans (see Table 12.2), as well as the data associated with these artifacts and features (McGimsey, 1973). These objects can include such artifacts as Native American arrowheads, and the like

(see Table 12.2). Areas made or modified by humans include hunting stations, temporary camps, and permanent settlements. Careful excavation of these areas can reveal the location and dimensions of postholes and storage pits and can allow recovery of pollen, seeds, bones, cracked and flaked stones, and other debris indicative of food habits, manufacturing practices, and other details of the life patterns of inhabitants of the sites. Contextual information consists of the location (horizontal and vertical) in or on the ground of the artifacts or features and observations about the natural or constructed strata in or on which they are found. Persons trained in archaeology are needed to interpret the contextual information relative to the archaeological objects or areas (McGimsey, 1973).

Cultural resources related to areas of ecological, scientific, or geological importance are of recent interest. Examples include wildlife refuges, caves, and unique areas such as the Painted Desert in Arizona. Local professional societies, as well as universities and colleges, can provide information on such resources in the area of interest.

Local resources of importance to ethnic groups, such as burial grounds and cemeteries or areas of unique religious importance, can be identified through contact with local government officials as well as organized ethnic groups in the area of interest or having responsibility therein.

One aspect of identifying and evaluating historic properties involves determining the area of potential effects for an undertaking; the "area of potential effects" refers to the geographic area or areas within which an undertaking may cause changes in the character or use of historic properties, if any such properties exist. It is not necessary to know that the area in question contains historic properties, or even to suspect that such properties exist, in order to recognize the area as the area of potential effect (Advisory Council on Historic Preservation, 1986b). The area of potential effects need not be a contiguous area; it can include multiple alternative project sites or multiple areas in which possible changes are anticipated.

The identification of known historic properties can be accomplished by conducting a records search (or literature review) and by contacting or referring to potential sources of information. One such source is the *National Register of Historic Places,* published periodically in complete form by the U.S. Department of the Interior. Annual updates containing new listings are published in the *Federal Register* each February or March. In addition, listings of properties already determined by the Secretary of the Interior to be eligible for the National Register are published periodically in the *Federal Register.* Information on these listings may also be obtained from the SHPO (Advisory Council on Historic Preservation, 1986b).

The agency proposing the undertaking should also consult other sources. The SHPO can advise on the existence or availability of previous studies pertinent to the area and previously recorded historic properties not listed in the *National Register,* and on the likelihood that undiscovered or unrecorded properties exist in the area. The SHPO can provide information on properties being nominated and on state registers or state inventories. Other sources of information include local academic institutions and museums, historical and archaeological societies, historical units on military installations, local governments, Native American tribes, and published or unpublished background studies pertinent to the area (Advisory Council on Historic Preservation, 1986b). Typical state-level sources of information on known historic and/or archaeological sites include (Davis, 1982) (1) state archaeological surveys; (2) state historic preservation programs; and (3) state land offices (or equivalent agencies). The files of state archaeological surveys and state historic preservation programs are being added to constantly. It is likely that a records check will be out of

date in six months. If field surveys or project planning is delayed more than this amount of time, these two offices should be contacted for an update (Davis, 1982).

Finally, local cities, towns, and communities may have planned and conducted an historic resources survey and inventory. Such surveys, and the resulting survey data and inventories, can be used to identify properties that contribute to the community's character (or that of its neighborhoods) or that illustrate its historical and architectural development, and, as a result, such studies deserve consideration in planning (Parker, 1985).

STEP 2: IDENTIFICATION OF POTENTIAL CULTURAL RESOURCES

Even though the concept of a national register was initiated with the Historic Sites Act of 1935, there has never been a complete nationwide survey conducted. Most of the United States has not been thoroughly or professionally surveyed for cultural resources, and in many areas known resources include only those previously discovered by cursory studies or accident. Lack of information is the reason that step 1 is not sufficient, in and of itself, to identify cultural resources. Potential historic sites in the area of interest can be determined through contact with the state historic preservation officer, as well as state and local historical commissions, societies, or clubs. The SHPO can also provide information regarding the eligibility of certain sites for inclusion in the *National Register,* as well as the status of sites that have been previously recommended or are in the process of being recommended for inclusion.

To identify previously unknown archaeological resources in the area of interest, a preliminary archaeological reconnaissance is necessary (McGimsey, 1973). This reconnaissance should involve surface examination of selected portions of the area to be affected. It should be adequate to describe the general nature of cultural resources present, to assess the probable impacts of each alternative, and to estimate the costs of mitigation measures. Field research associated with preliminary archaeological reconnaissance will probably constitute no more than 25 percent of the total research time required to make an adequate assessment. Review of available information associated with step 1 and analysis of field data gathered in this step, along with report preparation, will account for the major portions of the total cultural resources efforts and budget.

As shown in the Section 106 review process chart (Figure 12.1), there is a potential need for identifying and evaluating unknown properties in the study area. The proposing agency can request the SHPO's views about whether further actions are needed to identify historic properties—for example, whether field surveys or additional background research ought to be conducted. Based on this assessment of existing information and further needs, the agency then may make a reasonable effort to actually locate historic properties that may be affected by the undertaking and to gather enough information to evaluate their eligibility for inclusion in the *National Register*. Useful references for such surveys and evaluations include "Standards and Guidelines for Archaeology and Historic Preservation" (U.S. Department of the Interior, 1983a) and "Standards and Guidelines for Identification" (U.S. Department of the Interior, 1983b).

If it is determined that a survey is needed, the SHPO can be contacted for state or regional guidelines for surveys and for appropriate forms for recording survey methods and properties discovered. The SHPO may also identify individuals, institutions, and firms that can do survey work under contract. Some SHPOs or related state agencies conduct survey work themselves on behalf of federal agencies. Reports of completed surveys, as well as of any other original identification research, should always be filed with the SHPO so that the results can be incorporated into the statewide inventory of historic

properties (Advisory Council on Historic Preservation, 1986b).

Once the proposing agency has completed the identification tasks described above, the agency may find that there are no historic properties that may be affected by its proposed action. In that event, the agency (1) must provide documentation to the SHPO verifying that it has found no historic properties; (2) should notify other interested parties, such as those with whom the agency has consulted during identification, of the same thing; and (3) must make pertinent documentation available to the public by means of an environmental assessment (EA), a finding of no significant impact, an environmental impact statement, and/or a record of decision. If the agency finds one or more historic properties that its undertaking could affect, it then proceeds to step 2 in the process, assessing effects (see Figure 12.1) (Advisory Council on Historic Preservation, 1986b).

Types of Surveys

An "archaeological survey" can be defined as a planned effort to determine the distribution of past human activity across the landscape, observe and record information about those activities, and evaluate the potential for each location to provide further information about human behavior and adaptations in the past (Davis, 1982). There are several categories of surveys; for example, Davis (1982) noted two such categories: "reconnaissance surveys" and "intensive surveys." The difference between these categories is in the percent of an area actually surveyed, the kinds of data obtained, and the degree of detail of information recorded.

King (1978) described four types of archaeological surveys: (1) the uncontrolled-exclusive survey, (2) the controlled-exclusive survey, (3) the nonexclusive survey, and (4) the predictive survey. In an "uncontrolled-exclusive survey," certain areas are excluded from inspection because it is believed that they will not contain archaeological sites; the decision to exclude such areas is made on the basis of uncontrolled—that is, unverified—assumptions. One of the most common uncontrolled assumptions used to structure archaeological surveys is the assumption that people typically live near water (King, 1978). At the opposite end of the spectrum of survey efficiency is the "controlled-exclusive survey." In such a survey, sufficient information exists on an area to make solid and defensible assumptions and judgments about where archaeological sites may and may not be (King, 1978).

In a "nonexclusive survey," no portion of the study area is excluded from inspection; coverage is complete. Coverage may be complete at a number of levels of intensity, however, and the level will naturally affect the probability of identifying all archaeological sites. In addition, this type of survey can be classified into four subtypes. The most obvious distinction among nonexclusive-survey types is that between a "nonexclusive surface survey" and a "nonexclusive survey with subsurface exploration." In conducting a nonexclusive surface survey, an inspection of the surface of the ground, wherever this surface is visible, is made, with no substantial attempt to clear brush, turf, deadfall, leaves, or other material that may cover the surface and with no attempt to look beneath the surface beyond the inspection of rodent burrows, cut banks, and other exposures that are observed by accident. A nonexclusive survey with subsurface exploration involves some definite effort to expose obscured surface conditions and/or to monitor subsurface conditions in a planned fashion (King, 1978).

The second subtype distinction is between a "nonexclusive survey with background research" and a "nonexclusive survey without background research." Background study of environmental data, historical sources, and ethnographies will generally result in special attention being given to particular portions of the study area wherein special types of sites are expected to occur; it may result in the employment of special detection techniques in such portions of the area. In the absence of such research, uniform inspection techniques should be employed throughout the study area insofar as possible (King, 1978).

The third subtype distinction is between a "nonexclusive deployed survey" and a "nonexclusive gang survey." In the former, field-crew members are deployed over the landscape in accordance with a plan to ensure total inspection of the land surface. In a survey with subsurface exploration, subsurface-testing efforts would be conducted. In a gang survey, the field crew moves through the area as a group, or gang, spreading out in some places, bunching up in others, and splitting and segmenting to check spots on either side. The final subtype distinction is between a "nonexclusive comprehensive survey" and a "nonexclusive special-purpose survey." In the former, the survey is conducted in order to determine all types of archaeological sites present in the study area; the latter is a survey to identify some particular class of sites. In the absence of sufficient data to justify a controlled-exclusive survey, a nonexclusive comprehensive deployed survey with background research and subsurface testing is desirable (King, 1978).

Finally, a "predictive survey," also called a "predictive model," involves building a database to enable sensitive, responsible historic preservation planning without conducting a 100 percent nonexclusive survey (King, 1978). In a predictive survey, physical inspections of only a fraction of the actual area of concern are conducted, and from these inspections—supplemented with background research—extrapolations are made to the entire area. Based on the results of a predictive survey that has been subjected to a number of carefully planned and executed tests, it should be possible to conduct controlled-exclusive surveys (instead of nonexclusive surveys) in advance of project implementation, thereby saving time and expense (King, 1978).

TABLE 12.5

POTENTIAL OUTLINE FOR AN ARCHAEOLOGICAL SURVEY REPORT FOR EIA STUDY PURPOSES

Title page

    A. Report title
    B. Type of investigation (e.g., intensive survey and testing)
    C. Location of investigation (e.g., county and state)
    D. Contracting sponsor, permit, and contract number
    E. Principal investigator and research organization
    F. Author(s)
    G. Date of report

Abstract (should not exceed one page)

    A. Identify specific type of project (e.g., construction) and purpose of investigation.
    B. Provide concise summary of report's contents including location, research orientation (which includes methodology), conclusions, number and nature of resources located, and any new information that may have resulted from this work.
    C. Provide reference to significance and *National Register* eligibility of site(s).

Table of contents

    A. Arrange and paginate appropriately.
    B. Include list of tables, maps, and figures.

Management or executive summary (one to five pages)

    A. Purpose and scope of investigation
    B. Constraints on the investigation (nonbudgetary)
    C. Results

        1. Significance (e.g., research, sociocultural, *National Register* eligibility information)
        2. Identification of potential impacts
        3. Recommendations for eligibility of cultural resources to the National Register of Historic Places

Report

   I. Introduction

    A. Sponsor and contract number and/or permit number, expiration date, and other appropriate agency-specific information
    B. Geographical limits of project area
    C. Description of proposed project, and of the nature and extent of ground disturbance anticipated
    D. Purpose of study
    E. Discussion of management objectives (e.g., scope of work, agency's program authority, and applicable implementing regulations)
    F. Dates of investigation
    G. Personnel and work organization
    H. Disposition of field notes and artifacts
    I. Locational project map

   II. Affected environment (general history)

    A. Physiographic province (e.g., topography, drainage)
    B. Microenvironment of the project area
    C. Relevant climatic history
    D. Historic land-use patterns
    E. Current condition of land within project area (e.g., type of crops, pasture, timber, swamps)
    F. Prehistoric resource-utilization potential (e.g., availability of raw lithic resources)

TABLE 12.5

POTENTIAL OUTLINE FOR AN ARCHAEOLOGICAL SURVEY REPORT FOR EIA STUDY PURPOSES (*continued*)

III. Existing data and literature review (step 1 activities)

A. Dates, purposes, intensity, and results of previous studies.

B. Historic documents and records (e.g., deeds to historic sites, records, and files of federal, state, and local governments, research institution files, maps, and published material).

C. Informants and their addresses (when used), both amateur and professional, as well as procedures used to locate these persons.

D. Location and nature of field notes, unpublished manuscripts, and collected materials from previous investigations.

E. Relevant ethnographic and/or ethnohistoric data of the project area.

F. The project area should be placed in its regional setting in regard to the known cultural history.

IV. Study problems (step 2 activities)

A. Study objectives

B. Problem orientation

C. Justification of problem selection

D. The definitions of "site" and "isolated find" used for purposes of the survey

E. An explicit discussion of the survey and/or testing strategy, expected results, and relationship to study objectives

F. Hypotheses to be tested, test implications, and analytical technique required to test hypotheses

V. Field methods (step 2 activities)

A. Surface survey techniques, both site-specific and general

B. Subsurface testing techniques, both site-specific and general

C. Description of any nondisruptive techniques used for conducting surveys, or testing

D. Description of data collection techniques (e.g., surface collection techniques, artifact provenance recording techniques, site of screens, recovery of soil samples) and measuring devices and circumstances, indicating when used and when not used

E. Constraints on investigation (e.g., limitations of access, poor ground visibility, or other environmental limitations such as adverse weather conditions)

F. Controls for personal bias

G. Justification of any in-field modifications of study strategy

H. Maps which indicate areas surveyed at different levels of intensity

VI. Laboratory methods and analysis (step 2 activities)

A. Classificatory typological scheme(s) used in artifact description and analysis

B. Method of chronological determination

C. Description of all natural material observed

D. Description of other potential paleological data

E. Description of assemblage(s) with illustrations, distribution tables, weights, and other measures

F. Scaled photographs and/or line drawings of all diagnostics or a representative sample of each type and class of artifacts

VII. Inventory of cultural properties (step 1 and 2 activities)

A. Physical characteristics of site.

    1. Location (legal description and UTM).

    2. Site extent must be clearly delineated on a specific site map.

    3. Site distribution within project area must be plotted on a USGS 7.5-min topographic map, if available.

TABLE 12.5

POTENTIAL OUTLINE FOR AN ARCHAEOLOGICAL SURVEY REPORT FOR EIA STUDY PURPOSES
(continued)

4. Site size (both vertical and horizontal).
5. Appropriate photographs of the adjacent environment.
6. Topographic setting.
7. Vegetation.
8. Proximity to water.
9. Elevation.
10. Soil descriptions with appropriate graphics (Munsell soil-color chart designation).

B. **Cultural and temporal characteristics of the site.**

1. Material culture.
2. Site type, with supporting evidence.
3. Site function, with supporting evidence.
4. Intrasite structure.
5. Artifact-provenance data.
6. Distribution of artifacts by type (horizontal and vertical).
7. Cultural and temporal placement within regional chronology.
8. Appropriate photographs of features.

C. Nature and extent of previous disturbance.
D. Relationship between site location and probable project impacts.
E. Site-specific activities conducted and results. For example, for testing, there must be a full description of the size, number, and location of test pits relative to the site data. All test pits—or a representative sample—should be graphically displayed and accompanied with appropriate description or interpretation.

VIII. **Evaluation of data (step 3 activities)**

A. Reliability of data (e.g., potential for unlocated or unidentified resources within project area)
B. Relationship of changes in study goals
C. Identification of changes in study goals
D. Synthesis and comparison of results of analysis
E. Integration of ancillary data
F. Identification and discussion of perceived patterns and relevant and/or other substantive concerns

IX. **Recommendations (steps 3, 4, 5, and 6 activities)**

The *National Register* criteria for significance evaluations must be explicitly addressed. Concise discussion of known and potential contributions of site or district to current regional archaeological study problems must be provided. The rationale for significance as well as nonsignificance evaluations should be clearly stated. Specifically discuss effects determinations, mitigation plans, and measures for handling construction-phase findings.

X. **References**

Use *American Antiquity's Style Guide* (44:193–205) or agency-specific instructions.

XI. **Appendices (as appropriate)**

A. Peer reviews
B. Supporting data (e.g., computer printouts)
C. Scope of work
D. Ancillary studies (e.g., palynological report)
E. Proposal
F. Other

Fieldwork in Surveys

Fieldwork is associated with archaeological surveys; and five basic principles related to planning such work are as follows (King, 1978): (1) the fieldwork should make maximum use of background information; (2) the field team should include persons trained to recognize all of the types of archaeological phenomena that are likely to occur; (3) it is often most effective to conduct the fieldwork in several stages of increasing intensity; (4) field methods should be planned carefully to allow for environmental diversity; and (5) within reason, all ground surfaces should be inspected and subsurface exploration should be done if the surface is obscured or if buried sites are thought to be present.

Shovel testing has been, and continues to be, a fundamental technique used in fieldwork. However, shovel testing may be slow, expensive, and frustrating, and it is often only marginally effective in locating archaeological sites (King, 1978). Small and/or deeper phenomena can still escape notice. Further, the technique tends to discourage team members from closely inspecting their surroundings and forces them instead to concentrate on pacing and digging. Other methods of subsurface exploration include the use of power and hand-driven posthole diggers, backhoes, tractor-drawn plows, road graders, hand-driven and power cultivators, and remote-sensing devices such as ground-penetrating radar and resistivity monitoring (King, 1978). Photographs should be made of test pits, profiles should be drawn of at least one wall of all test pits, the soil matrix should be described, and artifacts should be described and analyzed by stratigraphic levels. Placement of excavated test pits should be in relation to at least one datum, so that the pit can be relocated if necessary (Davis, 1982).

Survey Reporting

It may be necessary to contract with a professional archaeologist, state archaeological survey, or archaeological consulting firm to assemble a records search and to plan and conduct an appropriate archaeological survey for a proposed project. A ''scope of work'' (terms of reference) would be desirable for such a contract, and should include guidelines for the resultant report. Table 12.5 is an example of a potential detailed outline for an archaeological report prepared for EIA purposes. The topical issues identified could be used in negotiations with prospective contractors, and they could be used to prepare a budget and time schedule, as well.

Volunteer Digs

Professionally planned and directed surveys can be facilitated by the use of volunteers. ''Volunteer digs'' refer to surveys which have been planned by one or more professional archaeologists, and which are then implemented wholly or in part by volunteers following their receipt of appropriate instruction and continuing field supervision. Many volunteers are ''amateur archaeologists'' with personal interests in the identification and preservation of archaeological resources.

STEP 3: DETERMINATION OF SIGNIFICANCE OF CULTURAL RESOURCES

Every archaeological site is of importance to a complete understanding of human history (McGimsey, 1973). Other cultural resources are important from past as well as future historical perspectives. However, not every cultural resource can be preserved or carefully and completely excavated. Decisions must often be made with regard to which sites should be preserved, which sites should be investigated (and the nature and intensity of such investigation), and necessary mitigation efforts. Proper assessment is required to identify those sites with sufficient significance to warrant preservation, since it has been estimated that over 5,000 sites are lost on an annual basis within the United

States (Neal, 1975). Significance determinations for archaeological and other cultural resources require the professional judgment of trained specialists.

During a survey, when properties are found that may be historic but have never actually been evaluated, it is the agency's responsibility to ascertain whether the properties are eligible for inclusion in the *National Register.* The agency should review the property with reference to the criteria in Table 12.4. Agencies should follow "Standards and Guidelines for Evaluation" (U.S. Department of the Interior, 1983c). The regulations also require that the agency's determination be made in consultation with the SHPO, but if the SHPO does not provide views on the eligibility of properties, then he or she is presumed to agree with the agency's determination. If the agency and SHPO agree that a property is eligible, it is treated as eligible for purposes of Section 106; if they agree that a property is not eligible, it is treated as not eligible for purposes of Section 106. If the agency and the SHPO cannot agree about *National Register* eligibility, the agency must obtain a formal determination of eligibility from the Keeper of the *National Register,* who acts on behalf of the Secretary of the Interior, in accordance with the applicable significance regulations (Advisory Council on Historic Preservation, 1986b).

STEP 4: DETERMINATION OF IMPACTS ON CULTURAL RESOURCES

Impacts on cultural resources can include inundation, destruction, damage, and/or disruption. Impacts can directly result from construction-phase disturbances, or they can indirectly occur from activities such as treasure hunting by persons during any phase of the project. Preconstruction impacts occur primarily through acts of vandalism on known cultural resource sites in the area of potential effects. Indirect impacts primarily include those that occur as a result of land-use changes and secondary growth and development. These secondary impacts can also be a result of direct construction activities or indirect acts of vandalism and disturbance. Attempts should be made to quantify the nature and extent of impacts during the various temporal phases of a project, with qualitative discussion provided where quantification is impossible. A basic approach that can be used involves overlay mapping, in which a base map showing known and potential sites is overlaid by maps identifying the nature and extent of the impacts from various alternatives under study to meet project needs.

There are several ways of classifying potential effects of federal agency projects or activities on historic and archaeological properties. Five types of effects have been defined by King, Hickman, and Berg (1977) as follows:

1. "Direct effects" can occur; for example, an agency bulldozes through a prehistoric site while constructing a project.

2. "Permitted effects" are similar to direct effects, but there is an administrative distinction. For example, when an agency permits extraction of water from a navigable waterway, any damage to historic properties resulting directly from the construction of extraction facilities or facilities to convey the extracted water is a permitted effect, because it could not occur in the absence of the permit.

3. "Managerial effects" occur as regular results of on-going agency management activities. For example, if prehistoric sites are destroyed by erosion around a project maintained by an agency, the destruction is a managerial effect.

4. "Contingent effects" are those arising from the actions of nonfederal entities which could not occur without a direct action by a federal agency, although no formal permit is involved. For example, assume that an agency guarantees flood insurance for housing con-

struction downstream from a federal dam. The damage to historic properties resulting from the housing construction so insured is a contingent effect of the insurance authorization and the dam.

5. "Infrastructural effects" occur when a federal action alters the course of development on nonfederal lands; the resulting damage to historic properties is a special kind of contingent effect. Infrastructural effects are more difficult to place boundaries upon than are other kinds of contingent effects; they are less predictable and yet may be more far-

reaching. For example, a new interstate highway is constructed through an area previously accessible only by barge. The area is hence opened for residential and industrial development. Any resulting damage to historic properties is an infrastructural effect of the highway construction.

An important element of the Section 106 review process is the assessment of the effects of the undertaking on previously known or currently identified historic properties. The proponent agency should consult with the SHPO on

---

TABLE 12.6

CRITERIA OF EFFECT AND ADVERSE EFFECT

Criterion of Effect:

An undertaking has an effect on an historic property when the undertaking may alter characteristics of the property that may qualify the property for inclusion in the National Register. For the purpose of determining effect, alteration to features of a property's location, setting, or use may be relevant depending on a property's significant characteristics and should be considered.

Criteria of Adverse Effect:

An undertaking is considered to have an adverse effect when the effect on an historic property may diminish the integrity of the property's location, design, setting, materials, workmanship, feeling, or association. Adverse effects on historic properties include, but are not limited to:

(1) Physical destruction, damage, or alteration of all or part of the property;
(2) Isolation of the property from or alteration of the character of the property's setting when that character contributes to the property's qualification for the National Register;
(3) Introduction of visual, audible, or atmospheric elements that are out of character with the property or alter its setting;
(4) Neglect of a property resulting in its deterioration or destruction; and
(5) Transfer, lease, or sale of the property.

Exceptions to the Criteria of Adverse Effect:

Effects of an undertaking that would otherwise be found to be adverse may be considered as being not adverse for the following:

(1) When the historic property is of value only for its potential contribution to archaelogical, historical, or architectural research, and when such value can be substantially preserved through the conduct of appropriate research, and such research is conducted in accordance with applicable professional standards and guidelines;
(2) When the undertaking is limited to the rehabilitation of buildings and structures and is conducted in a manner that preserves the historical and architectural value of affected property through conformance with the Secretary's "Standards for Rehabilitation and Guidelines for Rehabilitating Historic Buildings;" or
(3) When the undertaking is limited to the transfer, lease, or sale of an historic property, and adequate restrictions or conditions are included to ensure preservation of the property's significant historic features.

Source: Advisory Council on Historic Preservation, 1986a.

such an assessment, and should take into account the views of any interested persons. The agency's judgment about whether there could be an effect is based on the criteria of effect and adverse effect, which are listed in Table 12.6. Effects may occur at the same time and place as the undertaking, or they may occur later than or at a distance from the location of the undertaking. When applying the criteria of effect and adverse effect, there are three possible findings (Advisory Council on Historic Preservation, 1986a):

1. *No effect*—There is no effect of any kind (that is, neither harmful or beneficial) on the historic properties.
2. *No adverse effect*—There could be an effect, but the effect would not be harmful to those characteristics that qualify the property for inclusion in the *National Register.*
3. *Adverse effect*—There could be an effect, and that effect could diminish the integrity of such characteristics.

If the agency determines there is no effect on historic properties, the agency must notify the SHPO and any interested persons who have made their concerns known to the agency, compile the documentation that supports the finding, and make that documentation available for public inspection. If the agency determines there is an effect but that it is not adverse, the agency may either obtain the SHPO's concurrence with the finding of no adverse effect and then notify the council with summary documentation, which it must also make available for public inspection; or it may submit the finding of no adverse effect directly to the council for a 30-day review period and notify the SHPO of its action (Advisory Council on Historic Preservation, 1986a). If there is an adverse effect, the agency proceeds with step 3 of the Section 106 process (Figure 12.1): consultation. The primary purpose of consultation is to bring together the principal parties (such as the agency, SHPO, council, and others) to consider ways to

mitigate the adverse effects of the proposed undertaking.

STEP 5: SELECTION OF PROPOSED ACTION AND IMPACT MITIGATION

Selection of the proposed action from alternatives can be based upon elimination of one or more alternatives due to their potential adverse effects on cultural resources, or selection of the alternative least likely to cause adverse effects on potential cultural resources. Once the proposed action is selected, it may then be desirable to conduct an intensive archaeological reconnaissance if it has not been previously done in conjunction with step 2 (McGimsey, 1973). This survey might be appropriate as a final step prior to selection of the proposed action, with the reconnaissance being conducted for two or more alternatives. An intensive archaeological reconnaissance typically involves an on-the-ground surface survey and testing of an area sufficient to permit a more detailed determination of the number and extent of the resources present, their scientific importance, and the time factors and cost of preserving them or otherwise mitigating any adverse affects on them. This level of investigation is appropriate once a specific region or area to be affected has been determined or the choice has been narrowed to one or a few prime alternatives.

Following an intensive reconnaissance, the archaeologist should be able to indicate which sites are particularly worth preservation and/or investigation, suggest areas where caution should be exercised because of possible buried sites, and provide an estimate of the cost of protecting or recovering an adequate amount of information from the area to be affected. Salvage may be determined to be appropriate for one or more sites in the area of interest. Archaeological excavation associated with scientifically controlled recovery or salvage can be designed to yield maximum information about the life of the inhabitants, their ways of solving human problems, and their ability to adjust to and modify their natural environment. Archaeological excavations should be programmed during the final planning stages or, at the latest, during early stages of project construction.

Mitigation Measures

The results of the consultation process (step 3 in Figure 12.1) may be required in order to accomplish step 5. Consultation typically gives first consideration to alternative ways of accomplishing the agency's goals without unacceptably damaging historic properties. Alternative sites, alternative undertakings, and alternative designs are typically addressed in an agency's planning process as well as during consultation. The alternative of not carrying out the undertaking at all (the no-project condition) should also always be considered (Advisory Council on Historic Preservation, 1986a). "Mitigation" is the term for actions that reduce or compensate for the damage an undertaking does to historic properties. Typical mitigation measures include (Advisory Council on Historic Preservation, 1986b)

1. Limiting the magnitude of the undertaking
2. Modifying the undertaking through redesign, reorientation of construction on the project site, or other similar changes
3. Repair, rehabilitation, or restoration of an affected historic property (as opposed, for instance, to demolition)
4. Preservation and maintenance operations for involved historic properties
5. Documentation (drawings, photographs, histories) of buildings or structures that must be destroyed or substantially altered
6. Relocation of historic properties
7. Salvage of archaeological or architectural information and materials

As an example, Davis (1982, p. 71) has suggested that mitigation of adverse effects can be accomplished by means of one or more of the following suggestions:

1. Preplan the project to avoid the eligible property, thereby having no adverse effect upon it. This is not an active protective measure, however, since the avoided property may be impacted by nonproject-related events.
2. Preserve and protect the eligible property in situ without adverse effect by project construction (for example, a housing development can provide interpretation of the site for those in the housing project). This is a positive and active protective mechanism.
3. The eligible property can be neither avoided nor protected, but data recovery can be accomplished so that the significant information in the site is removed from the project area and the project will, therefore, have no adverse impact upon it.

Archaeological Considerations in Route Selection

Archaeological resources can be considered in the screening and elimination of alternative routes. For example, in a study of eight alternative waterway-navigation routes from Tulsa, Oklahoma, to Wichita, Kansas, 19 factors were used to assess the significance of cultural resources (Canter, Risser, and Hill, 1974). Each parameter represented a matter of importance for each archaeological site, and each waterway route was evaluated in terms of these parameters. The considerations associated with each of the 19 factors can be generally applied; they were as follows:

1. *Age or occupation period of the site*—Site age or occupation period concerns the *approximate* age and period of the individual site. Some period sites are more rare than others, and archaeological work should make a serious effort to gather data from the entire range of time periods. Consequently, some sites from each period should be examined. As the number of sites available for each period increases, the priority in terms of losses decreases.
2. *Concern by local population*—The local inhabitants are often concerned about de-

struction or conservation of cultural resources.

3. *Cost of conducting on-foot archaeological surveys*—The cost of archaeological surveys can be used as a determinant in the selection of alternatives, since these costs are proportional to the area involved for each alternative.

4. *Depth of the occupational area*—The depth of the deposit is important in estimating excavation costs. It is also important in terms of potentially recoverable information for occupation, duration, and stratigraphy. In general, shallow deposits are less likely to be undisturbed and therefore will produce less cultural data than deep deposits will. On the other hand, shallow occupations may provide activity areas or data that can be confusing, in the case of thicker deposits.

5. *Ecological setting*—This factor is concerned with the ecological setting with reference to river valleys—that is, whether the site occupies bottomland, terrace ridge, hilltop, or rock shelter.

6. *Eligibility of the site for state or national register*—Known sites on state and national registers are of importance. Sites eligible for inclusion on either state or national registers also need to be considered in evaluating alternatives.

7. *Estimated number of sites to be obtained by survey*—Reasonable estimates can be made for the number of potential archaeological sites by considering published data for similar types of projects or land areas.

8. *Importance in terms of local, state, and national level*—The importance of archaeological sites varies in terms of local, regional, state, and national interest. Professional judgment can be used to estimate relative importance for these geographical concerns.

9. *Minimum salvage costs for estimated sites*—This factor includes an estimate of the archaeological salvage costs for the total

probable number of sites for each alternative, including both known and potential sites.

10. *Nature of the site*—The "nature of the site" factor is concerned with the character of the individual site and whether it represents a hunting camp, village site, special-activity area, burial ground, quarry, workshop area, or other use. Sites likely to contain greater diversity in the cultural activities represented will have greater value in reconstructing the culture involved.

11. *Number of known sites to be damaged*—This factor refers to the number of already identified sites to be affected by the proposed action.

12. *Presence of single or multiple occupations*—The existence of single or multiple occupations at a site will affect excavation costs and requirements for sampling. Multiple-component sites may provide data for chronology and are normally of greater value as an information source.

13. *Preservation of archaeological data*—This parameter is concerned with the preservation of materials at each archaeological site. Some sites are more likely to have materials preserved than others because of soil conditions (acid or alkaline), protection from the weather (as in a rock shelter), and other factors. Consequently, sites more likely to have preserved various archaeological materials have greater potential importance than those where this is not the case.

14. *Previous knowledge of the area*—The previous-knowledge factor takes into account the overall knowledge of the archaeological resources in the entire area. In general, a well-known region would require less investigation than unknown regions.

15. *Site frequency within the area concerned*—Site frequency here deals with the general nature of each site and the relative number of similar sites in the area to be affected. Some sites are more rare, sometimes

unique, while others may be well-represented by numerous examples. Obviously, sites that are rare are of greater importance than those that are plentiful.

16. *Site importance in terms of geographic area and problems*—Some archaeological sites are more important than others in terms of providing answers to problems of concern not only locally, but also possibly on a statewide or even regional level.

17. *Site preservation from damages*—Some archaeological sites have been damaged by vandalism, contour plowing, deep plowing, construction, erosion, and other factors. Others may be relatively intact and without obvious disturbance. Consequently, sites that are better preserved are likely to be of greater importance than those that have been disturbed.

18. *Size of occupational area*—The total area covered by the occupation will partly determine the cost of excavation; also, larger sites are more likely to provide greater amounts of information.

19. *Value of site for nonarchaeological fields*—Information collected by archaeological research may be of value to a number of other disciplines. Materials such as animal bones, plant material, tree-ring specimens, pollen samples, soil samples and profiles, baked clay, shells, and charcoal may provide raw data for numerous fields concerned with past events.

Historic-Properties Management and Preservation Plans

There is an increasing emphasis on the development of historic-properties management plans and historic preservation plans for federal lands, and particularly for military installations. For example, such a plan has been developed for the U.S. Military Academy at West Point, New York (U.S. Military Academy, undated). Such plans are shifting the focus from "compliance," or Section 106–based crisis management, to the needed long-term-management orientation required by Section 110 of the National Historic Preservation Act of 1966 (Limp, 1989). While the plans would differ in detail from installation to installation, some common sections might be those related to the legislative and regulatory bases, interagency coordination requirements, basics of the management strategies and property treatments, and similar data for the archaeological sites and other data. Each plan must be well integrated with the "installation master plan" (Limp, 1989).

Operation and maintenance plans for federal projects are also beginning to address historic-property protection and preservation. For example, to facilitate such planning for U.S. Army Corps of Engineers projects, Grosser (1991) conducted a study to assist Corps historic-property managers in identifying site impacts and selecting site-protection and -preservation strategies at operation and maintenance projects. Additionally, the study addressed means by which the Corps' internal structure and its assigned missions could integrate in the historic-preservation process. Each Corps district was contacted to obtain information on the types of projects in the district, the major impacts to historic properties at these projects, and past and ongoing protective techniques used to mitigate these impacts (Grosser, 1991). Guidelines designed to identify means for evaluating archaeological-site-preservation technical options and to establish a procedure for selecting the proper options to be employed in specific situations have been developed by the U.S. Army Engineer Waterways Experiment Station (Thorne, 1988).

STEP 6: PROCEDURES FOR CONSTRUCTION-PHASE FINDINGS

Sometimes, even after an agency has fully complied with the Section 106 review process, new or additional historic properties are discovered after work has begun on a project. This often happens in the case of projects that involve excavation or ground-disturbing activities, when previously undiscovered archaeological resources may be uncovered during the process of construction or excavation. When an agency's action is of a type likely to uncover historic properties after work has begun, the agency is encouraged to develop a plan for treating such newly discovered historic properties. Often, agencies will realize as they complete the identification step of Section 106 review that discovery of additional properties is likely later on.

Plans for handling such discoveries should be included in the documentation developed during the assessment-of-effects and the consultation steps of the Section 106 process. When an agency has developed such a plan and then discovers historic properties after completing Section 106 requirements, the agency should follow the plan that was approved during the consultation and council-comment steps of the review. If an agency has not prepared a plan in anticipation of newly discovered historic properties, the procedure is more complex. In this case, the agency must afford the council an opportunity to comment on effects to these newly discovered historic properties, in accordance with procedures delineated in 36 C.F.R. 800 (Advisory Council on Historic Preservation, 1986a).

If historic properties are discovered after work has begun, council regulations do not require agencies to stop work on the undertaking. However, depending on the nature of the property and the undertaking's apparent effects on it, agencies should try to avoid or minimize harm to any historic properties until the Section 106 requirements have been met (Advisory Council on Historic Preservation, 1986b). It may be desirable, prior to initiation of construction, for the proposing agency to develop agreements (either MOAs or "Memoranda of Understanding") with appropriate state and local archaeological agencies regarding procedures to be utilized should sites be identified during the construction phase. Such stipulations could also be included in permit conditions if the federal action involves granting such a permit. These agreements should aid in precluding lengthy delays in project construction.

SUMMARY

Proper attention should be given to cultural resources in conjunction with project planning and decision making. The geographical area to be considered for cultural resources should not be limited to the specific area where project construction will occur, but rather should encompass an area sufficient to include both primary and secondary consequences. Numerous federal, state, and local laws and regulations exist for the protection of cultural resources. If an adequate assessment of cultural resources is accomplished at each step in the planning process and if proper funding is provided, it is generally possible to provide for protection or, where necessary, preservation, or other mitigation for all significant sites, with minimal or no effect on project time schedules.

SELECTED REFERENCES

Advisory Council on Historic Preservation, "National Historic Preservation Act of 1966, as Amended," Washington, D.C., 1984, pp. 1–18.

———, "Procedures for Requesting Determinations of Eligibility," 36 *Code of Federal Regulations* Part 63, Washington, D.C., 1977.

———, "Protection of Historic and Cultural Properties," 36 *Code of Federal Regulations* Part 800, Washington, D.C., 1979, pp. 467–484.

———, "Protection of Historic Properties: Regulations of the Advisory Council on Historic Preservation Governing the Section 106 Review Process," 36 *Code of Federal Regulations* Part 800, Washington, D.C., Oct. 1986a, pp. 1–19.

———, "Section 106, Step-by-Step," Washington, D.C., Oct. 1986b, pp. 5–51.

————, "Treatment of Archaeological Properties—A Handbook," U.S. Government Printing Office, Washington, D.C., Nov. 1980, pp. 5–14.

Canter, L. W., Risser, P. G., and Hill, L. G., "Effects Assessment of Alternate Navigation Routes from Tulsa, Oklahoma, to Vicinity of Wichita, Kansas," University of Oklahoma, Norman, June 1974, pp. 295–341.

Davis, H. A., ed., "A State Plan for the Conservation of Archaeological Resources in Arkansas," Arkansas Archaeological Survey Research Series 21, *Arkansas Archaeological Survey,* Fayetteville, 1982, pp. 71, B3–B11, E3.

Gregg, M. L., Kordecki, C., and Picha, P. R., "Archaeological Investigations in the Southern Section of the Garrison Diversion Unit in North Dakota, 1984," Contribution 219, Department of Anthropology, University of North Dakota, Grand Forks, Dec. 1985.

Grosser, R. D., "Historic Property Protection and Preservation at U.S. Army Corps of Engineers Projects," WES/TR/EL-91-11, U.S. Army Engineer Waterways Experiment Station, Vicksburg, Miss., Aug. 1991.

King, T. F., "The Archaeological Survey: Methods and Uses," Heritage Conservation and Recreation Service, U.S. Department of the Interior, Washington, D.C., 1978, pp. 13, 31–37, 48–52, 73–74.

————, Hickman, P. P., and Berg, G., *Anthropology in Historic Preservation, Caring for Culture's Clutter,* Academic Press, New York, 1977, pp. 8–10, 53–59, 199–223, 276–291.

Limp, W. F., ed., "Guidelines for Historic Properties Management—Southwestern Division Management Plan," Final Report Submitted to the U.S. Army Corps of Engineers, Southwestern Division, DACW63-84-C-0149, *Arkansas Archaeological Survey,* Fayetteville, 1989, pp. 33–38, 143–144.

McGimsey, C. R., "Archaeology and Archaeological Resources," Society for American Archaeology, Washington, D.C., 1973.

Neal, L., personal communication to author, *Oklahoma Archaeological Survey,* University of Oklahoma, Norman, 1975.

Office of Technology Assessment, "Technologies for Underwater Archaeology and Maritime Preservation—Background Paper," OTA-BP-E-37, U.S. Government Printing Office, Washington, D.C., Sept. 1987, pp. 1–3.

Parker, P. L., "Guidelines for Local Surveys: A Basis for Preservation Planning," *National Register Bulletin* 24, U.S. Department of the Interior, 1985, pp. 1–2, 6–8.

President, Executive Order 11593, "Protection and Enhancement of the Cultural Environment," 1971.

Thorne, R. M., "Guidelines for the Organization of Archaeological Site Stabilization Projects: A Modeled Approach, Environmental Impact Research Program," WES/TR/EL-88-8, U.S. Army Engineer Waterways Experiment Station, Vicksburg, Miss., June 1988.

U.S. Congress, "Abandoned Shipwreck Act of 1987," P.L. 100-298, 100th Congress, Washington, D.C., Apr. 28, 1988.

————, "Antiquities Act of 1906," P.L. 59-209, 59th Congress, Washington, D.C., June 8, 1906.

————, "Archaeological and Historic Preservation Act of 1974," P.L. 93-291, 93rd Congress, Washington, D.C., May 24, 1974.

————, "Archaeological Resources Protection Act of 1979," P.L. 96-95, 96th Congress, Washington, D.C., Oct. 31, 1979.

————, "Historic Sites, Buildings, and Antiquities Act of 1935," P.L. 74-292, 74th Congress, Washington, D.C., Aug. 21, 1935.

————, "National Historic Preservation Act of 1966," P.L. 89-665 (as amended by P.L. 91-243, P.L. 93-54, P.L. 94-422, and P.L. 94-458) 89th Congress, Washington, D.C., Oct. 15, 1966.

————, "National Historic Preservation Act Amendments," P.L. 102-575, 102nd Congress, Washington, D.C., Oct. 30, 1992.

————, "Native American Graves Protection and Repatriation Act," P.L. 101-601, 101st Congress, Washington, D.C., Nov. 16, 1990.

U.S. Department of the Army, "Part 325—Processing of Department of the Army Permits—Appendix C—Procedures for the Protection of Historic Properties," *Federal Register,* vol. 55, no. 126, June 29, 1990, pp. 27003–27007.

U.S. Department of Defense, "Protection of Archaeological Resources: Uniform Regulations," 32 *Code of Federal Regulations* Part 229, *Federal Register,* vol. 49, no. 4, Jan. 6, 1984, pp. 1027–1034.

U.S. Department of the Interior, "Standards and Guidelines for Preservation Planning," *Federal Register,* vol. 48, Sept. 29, 1983a, pp. 44716–44720.

——, "Standards and Guidelines for Identification," *Federal Register,* vol. 48, Sept. 29, 1983b, pp. 44720–44723.

——, "Standards and Guidelines for Evaluation," *Federal Register,* vol. 48, Sept. 29, 1983c, pp. 44723–44726.

U.S. Military Academy, "Scope of Work for Historic Resources Management," West Point, N.Y., undated.

CHAPTER 13

Prediction and Assessment of Visual Impacts

"**A**esthetics" are often mentioned in environmental impact studies, with the meaning typically related to visual quality and potential project impacts. Reference is made to such aesthetics-related issues as environmental-design arts and visual quality in the NEPA and the CEQ regulations (CEQ, 1978). Many federal agency guidelines or regulations address visual-quality issues. This chapter addresses basic concepts and methodological approaches for conducting a scientifically based analysis of the potential aesthetic, or visual, effects of proposed projects, plans, programs, or policies. The chapter begins with a discussion of pertinent definitions and legislation; this is followed by detailed information on a six-step methodological approach.

BASIC DEFINITIONS AND CONCEPTS

Aesthetics should be considered in evaluating the potential environmental impacts of proposed actions. "Aesthetics" can be defined as that which is concerned with the characteristics of objects and of the human being perceiving them that make the object pleasing or displeasing to

the senses (U.S. Army Construction Engineering Research Laboratory, 1989, vol. 3). Table 13.1 includes definitions for several terms related to potential aesthetic impacts of proposed projects or activities. Essentially all of the terms, with the exception of "aesthetics," are related to *visual* quality and resultant impacts. Visual-impact assessment aims to predict and assess the significance and magnitude of visual impacts from proposed developments, at specific sites (Heape, 1991). However, even though numerous definitions for "aesthetics" (and related terms) have been developed, there is no uniform agreement among professionals or the public related to any one of these definitions. Furthermore, in applying a given conception of aesthetics, what is particularly pleasing in terms of visual quality to one individual may not necessarily be pleasing to another individual. There is the adage that "beauty is in the eye of the beholder." Beer (1990) noted that there is no generally accepted rule as to what constitutes beauty. Landscape architects have searched for some common measure of aesthetic quality and in doing so have realized that complicated issues are involved (Beer, 1990).

TABLE 13.1

DEFINITIONS OF SELECTED TERMS RELATED TO AESTHETIC-IMPACT STUDIES

Aesthetics: Evaluations of aesthetics and considerations regarding the sensory quality of resources (sight, sound, smell, taste, and touch) and especially with respect to judgment about their pleasurable qualities. Pertaining to the quality of human perceptual experience (including sight, sound, smell, touch, taste, and movement) evoked by environmental phenomena, elements, or configurations of elements.

Aesthetic resource: Those natural and cultural features of the environment which elicit one or more sensory reactions and evaluations by the observer, particularly in regard to pleasurable effects.

Landscape: Landform and land cover forming a distance visual pattern. Land cover comprises water, vegetation, and man-made development, including cities. "Landscape" refers to an expanse of natural scenery seen by the eye in one view, or to the sum total of the characteristcs that distinguish a certain area on the earth's surface from other areas. These characteristics are a result not only of natural forces but of human occupancy and use of the land.

Landscape character: The arrangement of a particular landscape as formed by the variety and intensity of the landscape features and the four basic elements of form, line, color, and texture. These six factors give the area a quality which distinguishes it from its immediate surroundings.

Scenic area: A place which has been designated by the U.S. Forest Service as containing outstanding or matchless beauty, which requires special management to preserve these qualities. Areas of this type and all other special interest areas are identified and formally classified primarily because of their recreational value. An area preserved primarily because of its present beauty, such as cliffs, streams, vistas, vegetation, and wildlife. A place which has been designated by the U.S. Bureau of Land Management as having outstanding scenic quality and which requires special management to preserve or enhance this quality. An area whose landscape character exhibits a high degree of variety, harmony, and contrast among the basic visual elements which results in a pleasant landscape to view.

Visual character: The visual character of a landscape is formed by the order of the patterns composing it. The elements of these patterns are the form, line, color, and texture of the landscape's visual resources. Their interrelationships can be objectively described in terms of dominance, diversity, continuity, and so on.

Basic elements: The four major elements (form, line, color, and texture) which determine how the character of a landscape is perceived. "Form" refers to the perceived aggregation of elements in which there is a consciousness of the distinction and relation of a whole to its parts. "Line" refers to a very thin, threadlike mark, such as a border or boundary, a division between conditions, and so forth, limit, demarcation. "Color" is the third of the four basic elements of visual patterns; the hue (for example, red or blue) and value (for example, light or dark) of the light reflected or emitted by an object. Finally, "texture" refers to the arrangement of the particles or constituent parts of any material (such as wood, metal, and so forth), its structure and composition.

Aerial perspective: Concerning the effects of distance from the viewer upon the color and distinctness of objects—especially as a result of the transparency of the intervening air. Typically, objects become bluer, greyer, their edges less distinct, and there is less contrast of light and shade as distance from the viewer increases.

Background lighting: The distance in the landscape where elements lose detail distinctions is its "background lighting." Emphasis is on the outline or edge of one land mass against another with a strong skyline element.

Some potential problems which might be encountered when addressing visual impacts include (1) a lack of agreement on visual-quality definition and criteria; (2) difficulties (due to item 1, above) in achieving effective visual, verbal, and written communications between various professionals, between project-activity proponents and environmental staffs, and between professionals and the general public concerning aesthetics; (3) the need to address seasonal variations in the quality of various viewscapes (viewsheds); (4) insufficient numbers of experienced professionals such as landscape architects, planners with expertise in city planning, urban and regional planning, and/or rural planning, and geographers; and (5) the large

TABLE 13.1

DEFINITIONS OF SELECTED TERMS RELATED TO AESTHETIC-IMPACT STUDIES *(continued)*

Observer position: The placement and relationship of a viewer to the landscape which is being perceived. A term employed to describe the observer's elevational relationship between himself or herself and the landscape he or she sees. It is used to indicate whether the observer is essentially below, at the same level, or above the visual objective. Three specific terms are used: (1) "observer inferior"—viewer below object; (2) "observer normal"—viewer on level of object; or (3) "observer superior"—viewer above object.

Seen area: That portion of the landscape which can be viewed from one or more observer positions. The extent of area that can be viewed is normally limited by landform, vegetation, or distance.

Simulation: The realistic visual portrayal which demonstrates the perceivable changes in the landscape features of a proposed management activity through the use of photography, artwork, computer graphics, and other such techniques.

View: Something, especially a broad landscape or panorama, that is looked toward or kept in sight; the act of looking toward this object or scene.

Viewsheds: All of the surface areas visible from an observer's viewpoint are termed "viewshed." This particularly refers to surface areas from which a critical object or viewpoint is seen. There are two types of viewshed: (1) existing viewshed—the area normally visible from an observer's viewpoint, including the screening effects of intermediate vegetation and structures; and (2) topographic viewshed—the area which would be visible from a viewpoint based on landform alone, without the screening effect of vegetation and structures.

Viewsheds composite: Composite of overlapping areas visible from a continuous sequence of viewpoints along a road; or a network of viewpoints surrounding a road (or object).

Visibility: The geographic extent of a resource and legibility of its features which can be seen by an observer(s), determined by his or her location.

Visual compatibility: The degree to which development with specific visual characteristics is visually unified with its setting.

Visual impact: The significance and/or severity of visual-resource quality change as a result of anticipated activities or land uses that are to take place (or have taken place) on or adjacent to the landscape. A contrasting intrusion in the unified order of landscape, seen and appreciated as a "misfit" in appearance or function. A visual impact contributes to a reduction in scenic values. The degree of change in visual resources and viewer response to those resources.

Adverse visual impact: Any impact on the land or water form, vegetation, or any introduction of a structure which adversely changes or interrupts the visual character of the landscape and disrupts the harmony of the natural elements.

Source: Compiled using data from Smardon, Palmer, and Felleman, 1986, pp. 310–333.

number of persons having diverse opinions on what is aesthetically pleasing (while individuals may recognize their limitations in other substantive areas, they often are willing to voice an opinion on visual quality).

LEGISLATION RELATED TO AESTHETIC RESOURCES

While there is no single federal law that relates to visual impacts, there are numerous laws, regulations, and executive-administrative orders at the federal, state, and local levels that either directly or indirectly address such impacts. Examples of pertinent specific laws or programs at the federal level, in addition to the NEPA, include the National Historic Preservation Act of 1966 (as amended in 1980 and 1992), the Federal Highway Beautification Act of 1965 (as amended), the Wilderness Act of 1964, the National Wilderness Preservation System, various statutes creating specific wilderness areas, the

National Natural Landmarks Program, the Wild and Scenic Rivers Act of 1968, the National Trails System Act of 1968, Section 4(f) of the Federal Aid Highway Act of 1966, and the Surface Mining Control and Reclamation Act of 1977 (Smardon and Karp, 1993). For example, Section 4(f) of the Federal Aid Highway Act stipulates that no highway project requiring the use of publicly owned land located in or comprising parks, recreation areas, and wildlife and waterfowl refuges, or any land located in or comprising historical sites of national, state, or local significance (all such lands are called "Section 4(f) lands") can be approved unless there are no feasible and prudent alternatives to the proposed use of such land, and unless all possible measures to minimize harm to such land are taken (Smardon and Karp, 1993). Additional federal laws related either directly or indirectly to visual quality include the Coastal Zone Management Act of 1972 (as amended), wetland executive orders, Section 404 of the Clean Water Act of 1987, the Clean Air Act Amendments of 1990, the Endangered Species Act of 1978, and the Marine Mammal Protection Act of 1972. Key federal agencies involved with administering these laws include, but are not limited to, the Department of Transportation, Forest Service, Soil Conservation Service, National Park Service, Fish and Wildlife Service, Bureau of Land Management, Department of Energy, Environmental Protection Agency, Army Corps of Engineers, and Bureau of Reclamation.

Relevant institutional information at the state level includes state environmental policy acts or related administrative-executive orders, land-use planning controls, highway beautification regulations, shoreline management acts, wild and scenic rivers acts, scenic trails systems acts, scenic areas acts, wetlands acts, and surface-mining acts. Local-level institutional information includes ordinances on signs and billboards, junkyard regulations, setback provisions around significant viewscapes, scenic easements, special districts and zoning regulations, architectural review procedures, wetlands-zoning regulations, and wetlands-protection ordinances. Architectural controls involving preservation of or changes related to individual structures or groups of structures in terms of the structures' height, mass, design materials, color, and other physical attributes can also be utilized (Smardon and Karp, 1993).

FIGURE 13.1

Conceptual Approach for Study Focused on Visual Impact Assessment.

Step 1: Identification of types of visual impacts from proposed project/activity

Step 2: Preparation of description of existing visual resources for the study area

Step 3: Procurement of relevant laws, regulations, or criteria related to impacts and/or conditions

Step 4: Prediction of the impacts of the proposed project/activity on existing visual resources

Step 5: Assessment of the significance of the predicted impacts

Step 6: Identification and incorporation of mitigation measures

CONCEPTUAL APPROACH FOR VISUAL-IMPACT PREDICTION AND ASSESSMENT

A six-step methodology for visual-impact quantification is introduced in this section, and each of the steps is discussed in greater detail in subsequent sections. These steps are based on typical visual-impact studies conducted in both Europe and the United States (Martin, 1984; Coughlin et al., 1982). While there may be specific issues that arise in conjunction with a project that would require still additional considerations, it is considered that these steps will suffice for the majority of aesthetic-impact prediction and assessment. The steps are delineated in Figure 13.1.

STEP 1: DELINEATION OF THE TYPES OF POTENTIAL VISUAL IMPACTS

The first step involves identifying the potential impacts of the proposed project (or activity) on visual resources, with the intent being to identify such impacts by making use of readily available information. To serve as an example, Table 13.2 contains a listing of some potential visual impacts, organized by industrial or public works project type. Some examples of military

TABLE 13.2

EXAMPLES OF VISUAL IMPACTS RESULTING FROM SELECTED PROJECT TYPES

Project type	Examples of potential visual impacts
Power generation: power plants	1. Plant operation Scale dominance to existing landscape Introduction of stack plume 2. Building site cuts and fills, fences, and bulk-fuel loading
Power transmission: overhead transmission	1. Transmission route Visible poles and lines over streams, rivers, lakes, coastal areas Cleared swaths across landscape Marred natural landform and vegetation patterns 2. Site preparation for field office and storage yard 3. Installation of overhead transmission line The interruption of lines-of-sight to a focal point by a transmission line Spatial interruption Disparity in relative size of transmission structures and landscape elements (houses, barns) accentuated by proximity Observable cut or fill necessitated by transmission-line installation 4. Transmission towers Rigid, unnatural appearance, medium contrast to the form, and lines expressed in natural landscape
Access structures: highways, roads	1. Road alignments, cuts, fills, retaining walls, cribs, rivetted embankments Contrast between natural landforms and engineering features of highway significant if visible from public recreation area, residential areas, or scenic highways

projects, components, or activities which can cause aesthetic impacts include (1) new physical facilities, troop movements, or military training activities which may partially or completely obstruct views (i.e., cause disruptions in viewscape) of mountains, scenic valleys, waterfalls, waterways, oceans, and/or historic sites; (2) new physical facilities which are out of character with existing architectural features in built-up areas on military installations or in residential or urban areas; (3) projects which create visual intrusions in the existing landscape character (examples of such physical intrusions include radar towers, transmission towers, and power lines, and oil or gas exploration and production activities); (4) new physical facilities or military training activities which obstruct views of open spaces such as greenbelts; (5) new physical facilities or military training activities which may require the removal of natural or built barriers or buffers, thus enabling lower-quality viewscapes to be seen; and (6) with regard to historic or archaeological properties, projects which may isolate the property or alter the character of the property's setting when that character contributes to the property's qualification as a historical or archaeological resource, or projects which may create visual, audible, or atmospheric elements that are out of character with the property or alter its set-

---

TABLE 13.2

EXAMPLES OF VISUAL IMPACTS RESULTING FROM SELECTED PROJECT TYPES *(continued)*

Project type	Examples of potential visual impacts
	2. Embankments (highway above grade), berms, elevated highway (on structures, fences, and barriers landscaping) Blocked viewlines along visual corridors (valleys, stream courses, streets) Severing of visual continuity of open-space network Decreased residential and commercial property values and rents Elevated or above-grade highway out of scale with adjacent urban development
	3. Fill slopes, grading cut slopes and faces, vegetation clearing Highly visible erosion and/or bare earth or rock scars Significant if visible from public recreation area, residential areas, or scenic highway
Water-resource development: impoundment	1. Impoundment Severing of visual continuity of open-space network Introduction of water as a visual element
	2. Grading, flooding, draining, filling, clearing Creation of permanent, highly visible landscape (drawdown rim, shoreline clearing, cut and fill faces) that vividly contrast with surrounding landscape Creation of areas of highly visible dead, dying, decaying, or unhealthy vegetation Exposure of stumps and vegetation debris Engineering feature of the project out of scale with landscape

Source: Compiled using data from Smardon et al., 1988.

ting (Advisory Council on Historic Preservation, 1986).

Sources of information for delineating the types of visual impacts associated with particular projects include site visits to existing projects of similar type (analogs), computer-based searches of published literature, review of EAs or EISs for similar types of projects, and discussions with pertinent professionals. Relative to step 1, information should be aggregated by construction and operational phases of the proposed project or activity, and this compilation of data should be used to prepare a descriptive list of the visual impacts of concern.

STEP 2: DESCRIPTION OF EXISTING VISUAL RESOURCES

In analyzing potential visual impacts of a proposed project or activity, it is necessary to define a study area associated with the potential visual intrusions. The study area should include the boundaries of the land associated with the project, and the viewsheds or viewscapes within the vicinity.

Two approaches for describing existing visual resources include a simple view-scoring technique (Beer, 1990) and the use of a checklist (Smardon, Palmer, and Felleman, 1986). Beer (1990) noted that it is necessary at an early stage in the site-planning process to determine which views from the site should be preserved as they are, which could be enhanced, and which should be screened. Input from local people and/or regular visitors to the area could be included in this stage. A relatively simple but systematic way to record subjective judgments of visual quality is to give each view a relative score. As an illustration, one would give the first view a score of 10, whatever the view is like, then score up or down for each of the following views. For instance, the next view is poorer than the reference view and is given a score of 5; the one evaluated after that is slightly better than the first, so its score is 11; and the next is much

better, so its score is 15. The actual numbers used do not matter. This process enables consideration of the relative merits of all the views from the site—it does no more than allow a ranking of the views from and to the specific site. It is normally best not to take the ranking too literally, as to do so would give an undue sense of scientific exactness. As a subsequent step, it is best to sort the views into no more than five quality categories, such as the following (Beer, 1990): (1) very good views, which must be kept open; (2) good views, which ought to be kept open, if at all possible; (3) moderately good views, which could be used to advantage; (4) poor views, which ought to be screened, if at all possible; and (5) very poor views, which must be screened. This information is useful, as it assists the site planner in deciding where visitors might like to sit or walk to enjoy views. It is also helpful in identifying views that detract from the scene, thus indicating where screens should be placed (Beer, 1990).

Checklists or questionnaire checklists can also provide a useful, documentable approach for addressing the visual impacts of projects. An example based on New York's State Environmental Quality Review Act (called SEQR) will be cited. The SEQR process consists of three basic stages* (Tables 13.3 through 13.5 offer suggestions or example forms for stages 1 through 3, respectively, in the process). The three stages in an SEQR-based visual-impact study are (Smardon, Palmer, and Felleman, 1986)

> *Stage 1:* Conduct an inventory of visual resources to establish or clarify community values, policies, and priorities related to existing visual resources before projects of controversy arise.

---

*Note: The three stages incorporate elements of all 6 steps in Figure 13.1.

TABLE 13.3

SUGGESTED VISUAL-INVENTORY PROCESS—STAGE 1 OF VISUAL-IMPACT STUDY

A. Identify Community Visual Resource Values

1. Describe and define the general character of the existing area.
2. Document visual resources and/or visually sensitive land including:
   a. state parks or state forest preserves, municipal parks;
   b. wild, scenic, or recreational water bodies designated by a state government agency;
   c. publicly or privately operated recreation areas;
   d. publicly or privately operated areas (including areas used for recreation) primarily devoted to conservation or the preservation of natural environmental features;
   e. hiking or ski-touring trails designated as such by a state or municipal government agency;
   f. architectural structures and sites of traditional importance;
   g. historic or archaeological sites designated as such by the National Register or State Register of Historic Places;
   h. parkways, highways, or scenic overlooks and vistas designated as such by the National Register or State Register of Historic Places;
   i. important urban landscapes including visual corridors, monuments, sculpture, landscape plantings, and urban "green space";
   j. important architectural elements and structures representing community style and neighborhood character.

B. Public Participation

1. Notify the public of the proposed inventory process and its purpose.
2. Conduct a survey of local resident/viewer perceptions:
   a. identify positive visual attractions;
   b. identify visual detractions or "misfits" (car dumps, gravel pits, waste disposal areas, and so forth).
   Results of the survey should indicate a preliminary consensus of the public's perceptions and values regarding its visual resources.
3. Conduct public meeting(s) to inform residents of the public's perceptions and values regarding its visual resources.
4. Adopt the municipal visual resource inventory.
5. Formalize community visual standards through creation of sign ordinances, architectural board of review's adopted standards, or other appropriate techniques.

C. Establish "Critical Areas of Environment Concern"
Special visual resources that are considered highly valued by the community and are sensitive to change may be established as Critical Areas of Environmental Concern. Thereafter, any action that takes place within, partially within, or adjacent to the critical areas would receive a fully coordinated environmental review process.

Source: Smardon, Palmer, and Felleman, 1986, pp. 152–153.

Stage 2: Establish practical visual criteria to guide decisions related to proposed projects so that the visual character and quality of an environmental setting are protected.

Stage 3: Use a questionnaire checklist form and focus on the project's potential visual impacts. The checklist-based approach is an orderly method that can be used to support a determination of nonsignificance.

STEP 3: PROCUREMENT OF RELEVANT INSTITUTIONAL INFORMATION

Step 3 should include the identification of pertinent federal, state, or local laws or regulations related either directly or indirectly to visual re-

TABLE 13.4

ESTABLISHMENT OF PRACTICAL VISUAL CRITERIA—STAGE 2 OF VISUAL-IMPACT STUDY

Agency decision makers can protect the visual character and quality of a project and its environmental setting by early consideration of the general siting and design criteria listed below. Municipalities and agencies may wish to use these suggested criteria as a base, adding their own criteria to reflect community values.

1. Locate new facilities where they are intrinsically suitable to their visual environment.

2. Insure that agency decisions prevent the exposure or creation of visual misfits (such as car dumps or waste disposal areas adjacent to scenic vistas) unless visual mitigation measures are adequate.

3. Whenever possible protect the visual privacy of residential sites.

4. Actively preserve future access to public viewing points.

5. Emphasize shared infrastructure space for public utilities.

6. In areas of high scenic quality, avoid commercial advertising, overhead utility service, and other man-made distractions.

7. Avoid development on steep slopes.

8. Take special care to enhance the visual quality of the physical entranceway to a community. The entranceway, usually a public roadway, sets the tone for the perceived visual expectation of the community.

9. Protect the integrity of visually important building facades by utilizing legal means to transfer development rights.

10. Promptly remove, refurbish, or replace abandoned facilities.

11. Be aware that visual spaces can be as important as physical objects. In this sense, air pollution can affect the visual quality of important spaces by obscuring or diminishing views.

12. Insure that transmission line corridors are not silhouetted against the skyline and traverse slopes on a diagonal rather than perpendicular basis.

13. As appropriate, either remove existing vegetation along travel corridors in order to create or enhance views or vistas, or retain existing vegetation along travel corridors to enhance the natural character.

14. Consider all possible mitigation measures. Use vegetation, landforms, or structural techniques to screen visually intrusive characteristics of a proposed development.

15. Enhance views to bodies of water.

16. Avoid adverse visual effects caused by the introduction of materials, colors, and/or forms incompatible with the surrounding landscape.

Source: Smardon, Palmer, and Felleman, 1986, p. 153.

sources. In addition, local master plans, planning guides or criteria, and/or ordinances should be considered in relation to visual resources. Pertinent information from this step can be used in assessing existing visual quality and the predicted impacts of the proposed project or activity (step 4).

STEP 4: PREDICTION OF IMPACTS ON EXISTING VISUAL RESOURCES

Prediction of the impacts of a project (or activity) on visual resources can be accomplished by any one of a wide range of approaches. Some examples include

TABLE 13.5

IDENTIFICATION OF VISUAL IMPACTS—STAGE 3 OF VISUAL-IMPACT STUDY

Part 1

1. Is the project within or adjacent to a Critical Area of Environmental Concern?

Yes ☐   No ☐

Description of Existing Visual Environment

2. Area surrounding project site can be identified by one or more of the following terms:

	Within	
	1/4 mile*	1 mile*
Essentially undeveloped	Yes ☐	No ☐
Forested	Yes ☐	No ☐
Agricultural	Yes ☐	No ☐
Suburban residential	Yes ☐	No ☐
Industrial	Yes ☐	No ☐
Commercial	Yes ☐	No ☐
Urban	Yes ☐	No ☐
River, Lake, Pond	Yes ☐	No ☐
Cliffs, Overlooks	Yes ☐	No ☐
Designated open space	Yes ☐	No ☐
Flat	Yes ☐	No ☐
Hilly	Yes ☐	No ☐
Mountains	Yes ☐	No ☐
Other _____	Yes ☐	No ☐

3. Are there visually similar projects within:

*One Mile	Yes ☐	No ☐
*Two Miles	Yes ☐	No ☐
*Three Miles	Yes ☐	No ☐
Adjacent	Yes ☐	No ☐

*Distances from project site are provided for assistance. Substitute other distances as appropriate.

Degree of Project Visibility

4. Will the project be visible from outside the limits of the project site?

Yes ☐   No ☐

5. The project may be visible from

Site or structure on the National Register or State Register of Historic Places	☐
Palisades	☐
State or County Park	☐
Parkway	☐
Interstate Route	☐
State Highway	☐
County Road	☐
Local Road	☐
Bridge	☐
Railroad	☐
Existing Residences	☐
Existing Public Facility	☐
Adjacent Property Owner(s)	☐
Designated Scenic Vistas	☐
Other _____	☐

6. Will the project eliminate, block, partially screen, or detract from views or vistas known to be important to the area?

Yes ☐   No ☐

TABLE 13.5

IDENTIFICATION OF VISUAL IMPACTS—STAGE 3 OF VISUAL-IMPACT STUDY *(continued)*

7. Is the visibility of the project seasonal? For example, screened by summer foliage, etc. but visible Fall/Winter/ Spring?

Yes ☐　　No ☐

If yes, which season(s) is project visible:

Summer　☐
Winter　☐
Spring　☐
Fall　☐

8. How many linear feet of frontage along a public thoroughfare does the project occupy?

_____feet

9. Will project open new access to or create new scenic views or vistas?

Yes ☐　　No ☐

10. Does proposed project or action plan to:
a. maintain existing natural screening

Yes ☐　　No ☐

b. introduce new screening to minimize project visibility

Yes ☐　　No ☐

If yes, is screening:
1. vegetative　☐
2. structural　☐

Viewing Context

11. Viewers will likely be in which of the following situations when the project is visible to them?

Activity	Frequency			
	Daily	Weekly	Holidays, Weekends	Seasonally
Travel to and from work	☐	☐	☐	☐
Involved in recreational activities	☐	☐	☐	☐
Routine travel by residents	☐	☐	☐	☐
At a residence	☐	☐	☐	☐
At worksite	☐	☐	☐	☐
Other _____	☐	☐	☐	☐

Visual Compatibility

12. Are the visual characteristics of the project obviously different from those of the surrounding area?

Yes ☐　　No ☐

If yes, the visual difference is because of:

Type of project	☐
Design style	☐
Size (including length, width, height, number of structures, etc.)	☐
Coloration	☐
Condition of surroundings	☐
Construction material	☐
Other _____	☐

TABLE 13.5

IDENTIFICATION OF VISUAL IMPACTS—STAGE 3 OF VISUAL-IMPACT STUDY *(continued)*

13. Is there local opposition to the project entirely, or in part, because of visual aspects?

Yes ☐   No ☐

14. Is there public support for the project because of its visual qualities?

Yes ☐   No ☐

Part 2

Apply the following series of questions to help determine the importance of each visual impact. These include:

1. What is the probability of the (visual) effect occurring?
2. What will the duration of the (visual) impact be?
3. Is the nature of the (visual) impact irreversible and will the (visual) character of the community be permanently altered?
4. Can the (visual) impact be controlled?
5. Is there a regional or statewide consequence to this (visual) impact?
6. Will the potential (visual) impact be detrimental to local goals and values?

The answers to these questions will indicate whether or not the potential impact is important.

Source: Smardon, Palmer, and Felleman, 1986, pp. 154–155.

1. The use of a descriptive approach, supplemented by photographs, wherein the visual quality of the study area is described under both baseline conditions and with the project or activity in place. (Consideration should be given to seasonal variations in the visual quality, and the influence of weather phenomena, such as fog, on the viewscape.)
2. The use of scale models to depict the project in place in the study area, coupled with a descriptive approach, as noted in (1) above.
3. The use of seasonally varied photographs of different viewscapes in the study area, with another set of the same photographs with the proposed project superimposed, coupled with a descriptive approach, as noted in (1) above.
4. The use of seasonally varied photomontages with companion photomontages on which the proposed project has been superimposed, coupled with a descriptive approach, as noted in (1) above.
5. The use of computer simulations of the viewscapes in the study area both under baseline conditions and with the project or activity in place, coupled with a descriptive approach, as noted in (1) above.
6. The use of a quantitative index method for depicting the visual quality in the study area under baseline conditions and with the project or activity superimposed, coupled with a discussion of the method, the results, and the implications of visual-quality changes which might occur as a result of the project.

To illustrate some of these approaches, Monbailliu (1984) has related the visual impacts of projects to potential changes in landscape characteristics and landscape beauty. In order to examine such impacts, it is necessary to utilize a landscape-analysis technique. Most classical landscape-analysis techniques—typically developed for specific projects, such as the siting of power lines, pipelines, roads, and other large projects—involve the use of perspective drawings and sketches, models, and photomontages. Photomontage techniques may include superimposition of photography or projected images—for example, as in an architect's conception—in order to enable a better visual appreciation of and a more accurate presentation of the development scheme. More recently, computer mapping and graphics tools have been employed in order to ensure the best integration of the proposed development in the landscape (Monbailliu, 1984).

The descriptive approaches described in steps 1 and 3, as well as in this step (4), can be facilitated by the use of the simple view-scoring technique or the questionnaire-checklist technique (shown in Table 13.5) used for step 2. Either method can provide a structure and consistency for the prediction step.

Simple Scoring Methodologies

Several comparative studies of simple scoring methodologies have been conducted. One example is a study conducted for the U.S. Environmental Protection Agency (Bagley, Kroll, and Clark, 1973). This study suggested that the methodologies designed to measure (or quantify) aesthetics can be grouped into two basic categories: visual analysis and user analysis. Those in the first category, visual-analysis methodologies, can best be described as tools to be used by a planning staff or decision maker to identify aesthetic attributes and forecast changes in the aesthetic characteristics of the environment, and to describe the implications of changes in terms of potential uses of environmental resources and environmental quality standards. Methods in the second category, user-analysis methodologies, are designed specifically for evaluating individual preferences for various aesthetic stimuli.

The methods in each of the two categories (visual analysis and user analysis) can be further divided into subcategories: those that assign numerical (quantitative) values to aesthetic characteristics and those that rank (measure) aesthetic attributes but are nonnumerical. One additional distinction is made under numerical methods; some of the methods reviewed attempt to relate aesthetic considerations to other envi-

ronmental considerations (e.g., impacts resulting from a change in air quality, landform, or water quality; economic impacts; or social impacts). These methods are called ''comprehensive environmental-analysis methods'' because visual-quality changes are evaluated concurrently with other environmental changes and weighted or ranked accordingly. Trade-offs can be made among various impacts during the evaluation process, thus providing the decision maker with a synthesis of the impact study. Methods that do not interrelate environmental components are designed to assess visual impacts as an independent environmental consideration, therefore providing the decision maker with information that must then be weighed against other impacts. Methods falling under this category can be called ''independent aesthetic-assessment methods'' (Bagley, Kroll, and Clark, 1973).

Monbailliu (1983) suggested that landscape-evaluation techniques can be used in visual-impact assessment, and that such techniques can be divided into three categories: (1) intuitive, field-based appreciation of landscape scenery, (2) grading or scaling methods, or (3) statistical techniques involving user analysis and multiple-regression analyses of user responses. In the field-based-appreciation category, the assessment of visual resources is based on the allocation of qualitative scores to landscape units using descriptions such as ''spectacular'' or ''beautiful,'' ''unsightly,'' and so on. Grading or scaling methods consist of attributing a quality rating to landscape scenery depending on topography, vegetation, and land use. Here, the terminology of landscape appreciation is replaced by a nominal-value system (for example, from -10 to $+10$), and the appraisal is typically undertaken by one person or, at most, a group comprising only a small number of persons. Statistical techniques are more objective and reliable techniques, since they establish a relationship between the overall attractiveness of a site and the value of the elements that compose the landscape scenery.

In impact studies for certain types of development (for example, power stations), simple assessment techniques are often used to justify the choice of the site of least impact. These techniques are often limited to a comparative description of a few potential construction sites. Perspective sketches and photomontages of "before-and-after" scenes often accompany this type of visual appraisal. Impact studies for linear infrastructure projects (highways, for example) will mostly require a regional assessment, which includes a listing of landscape criteria, in order to justify the corridors and their variants of least impact. In such situations, landscape-assessment techniques should be designed in accordance with the particular needs of the proposed infrastructure project (Monbailliu, 1983).

Monbailliu (1984) also distinguished between subjective and objective landscape-evaluation techniques. In the first case, a study would be conducted and would tend to use a descriptive, rather than a quantitative approach. Bagley, Kroll, and Clark (1973) referred to this as a "visual-analysis method." Objective evaluation techniques attempt to avoid the subjectivity of the observer and assessor by using a representative sample of observers viewing and statistically evaluating the landscape. Bagley, Kroll, and Clark (1973) called these approaches "user-analysis methods."

An objective landscape-evaluation technique used by the Warwickshire County Council in the United Kingdom is described in Monbailliu (1984). Twenty-three parameters (landscape elements) were used; these included relief, agriculture, woodland, parks, moorland, water, urbanized areas, industry, wasteland, and a series of linear elements such as hedgerows, trees, watercourses, roads, power lines, railway tracks, farms, listed buildings, churches, and windmills. The second stage involved the evaluation of 1-km-grid squares in terms of beauty using a ranking system of 26 scores as signed by one observer. After the identification of landform, land use and land features per square kilometer, and the ranking of each grid in terms of beauty, a regression analysis was applied in order to determine the contribution of each factor to visual quality. In addition, an index of intervisibility was also used for each grid square.

Martin (1984) has described several visual-impact assessment techniques for appraising proposed development sites. One category of techniques are "visibility studies," which are concerned only with the extent to which a proposed facility is visible. The "isovist" method involves plotting "limits-of-vision" lines on a map, relative to a focal point, or "visual center." The focal point is visited, and intermediate horizons, ultimate horizons, and dead ground (or "vision shadows") for the 360° rotation are observed and noted. The "zone-of-visual-influence" (ZVI) technique (also known as the "viewshed" technique) is quite similar to the isovist method, but works directly from topographic maps. It plots the area within which a man-made structure can be seen, making corrections for the curvature of the earth and the refraction of light.

The use of graphic representations is also a useful technique in visual-impact studies. Simple forms of graphic representation—such as plans, perspective sketches, and models—are used as a matter of course by project developers. The various "simulation" techniques developed for visual assessment perform the same function, but are quicker and more flexible. One particularly useful form of simulation is photomontage, discussed earlier. Some factors which should be considered in developing photomontages for visual-impact studies are (Heape, 1991) (1) choice of view or location—show typical views which can be experienced by significant numbers of people at regular or important times; (2) timing—time of day, time of year, and age or stage of the project, as well as light and weather conditions; (3) accuracy—the photomontage should be as accurate as possible (i.e., in terms of color, size, screening), because greater accuracy permits greater flexibility in the range of applications of the simulation; and (4) validity—the integrity of all of the above factors affects the validity of the photomontage data-gathering process, the skill and intent of the creator, the reproduction of the simulation, and the presentation.

Computer programs now exist for constructing photomontages. The most sophisticated visual simulation, however, is the color-videotape montage. This developed from methods used in the filming of architectural models. Latest techniques use videotapes of selected views of the site, and mix the images with perspectives of the development generated by a computer program. Video imaging has been recently used at Camp Shelby, Mississippi, to demonstrate to U.S. Army decision makers the potential impacts of several proposed changes in land usage (Marlatt, Hale, and Sullivan, 1993). In this technique, an existing site is captured through photographs or video footage. The picture is then digitized and edited electronically using specialized equipment to show how the site can be expected to look after a proposed land use is implemented.

Based upon the techniques discussed above, a very simple scoring methodology could be used for identifying and describing the existing visual resources in the study area (stage 3 of the visual-impact study, outlined in Table 13.5). This method is compatible with the state of the art of simple visual-impact methodologies. Specifically, it is suggested that the following scoring system be used for questions 2 and 3 (shown in Table 13.5):

1. *For question* 2—Assign two points to each identified land area within 1/4 mi of the proposed project, and one point to each identified land area within 1 mi (but not also within 1/4 mi) of the proposed project.
2. *For question* 3—Assign one point to visually similar projects adjacent to the site of the proposed project, two points if 1 mi away,

three points if 2 mi away, and four points if 3 or more mi away (choose one category only).

The above point assignments are then totaled and the following classes of existing visual quality used to designate the study area:

Class 1: Good visual quality, based on a score of greater than 15 points for questions 2 and 3.

Class 2: Modest visual quality, based on a score of from 8 to 15 points for questions 2 and 3.

Class 3: Minor visual quality, based on a score of less than 8 points for questions 2 and 3.

Documentation of the rationale used for each point assignment for questions 2 and 3 in Table 13.5 should be provided. Photographs and/or photomontages of the study area could be used as the basis for the point assignments.

A simple scoring methodology can also be used for predicting the visual impacts of the proposed project. Again, this methodology is compatible with the state of the art of simple visual-impact methodologies. Specifically, it is suggested that the following point scoring system be used for questions 4 through 7, 10, 12, and 13 in Table 13.5:

1. *Question 4*—Minus two points if answer is yes
2. *Question 5*—Minus one point for each listed location if project is visible from location
3. *Question 6*—Minus five points if answer is yes
4. *Question 7*—Minus two points if answer is yes (for one to four seasons)
5. *Question 10*—Minus two points if answer to *(a)* or *(b)* is no
6. *Question 12*—Minus five points if answer is yes
7. *Question 13*—Minus five points if answer is yes

The above negative points are then totaled and the following impact-significance categories and criteria used to designate the study area:

1. *Category A*—The cumulative negative scores based on the point assignments to the above questions are greater than 18 points; this category suggests potentially significant impacts.
2. *Category B*—The cumulative negative score based on the point assignments to the above questions is in the range of from 9 through 17 points; this category suggests possible significant impacts.
3. *Category C*—The cumulative negative score based on the point assignments to the above questions is less than 9 points; this category suggests no significant visual impacts.

Documentation of the rationale used for each negative point assignment should be provided. Photographs and/or photomontages of the study area with and without the project or activity in place could be used as the basis for the point assignments.

Systematic Scoring Methodologies for Existing Visual Resources, and for Impact Prediction and Assessment

Sophisticated techniques are available for quantifying impacts on visual resources. For example, this analysis can involve the application of a systematic, more-complex scoring methodology for describing existing visual resources and for predicting and assessing the potential visual impacts of the proposed project or activity. Background information on systematic scoring methodologies is presented here: the purpose is to summarize the range of sophisticated scoring methodologies.

Visual-resource management (VRM) systems deal with three classes of activities: (1) visual inventory and analysis for large landscape areas needing landscape planning, (2) scoping of potential visual impacts or determining thresholds, and (3) detailed evaluation of visual impacts

(Smardon, Palmer, and Felleman, 1986). The systems developed by three agencies will be mentioned and one will be briefly described. The first method was developed by the Forest Service in U.S. Department of Agriculture (1974); the second, by the Bureau of Land Management in U.S. Department of the Interior (1978a, 1978b, 1980a); and the third, by the Soil Conservation Service (SCS) in U.S. Department of Agriculture (1978). The first two are called "VRM systems," while the third is called a "landscape management system" (LMS). All three systems contain common elements, including (1) subsystems for physically based–landscape visual-quality inventory and evaluation; (2) subsystems for assessing people's use of or attitude toward the landscape and the visibility of the landscape; and (3) geographic mapping of these factors (in items 1 and 2) and the visual-quality inventory to yield classified areas for certain management objectives, visual-quality maintenance levels, or priorities for a professional landscape architect's attention (Smardon, Palmer, and Felleman, 1986).

The generic steps in the two VRM systems and one LMS system are listed below (Smardon, Palmer, and Felleman, 1986, p. 166):

1. Inventory and simultaneously evaluate visual landscape quality based on primarily physical landscape factors with aesthetic modifiers (form, line, color, texture).
2. Inventory amount of use of the landscape, travel through the landscape, or attitudes towards the landscape, indicating the degree of sensitivity.
3. Map degree of visibility or distance zones from which the landscape can be seen.
4. Combine this information to establish appropriate levels of management of visual quality. Under these management levels, certain intensities and types of activities are allowed or not allowed. In SCS's case, priorities for an appropriate level of professional involvement are established.
5. Assess whether significant visual impact may occur.

6. Assess visual impact absorption limits or thresholds to severity of visual impact allowed for specific landscape sites and provide guidance for ameliorative design or change in location of the impacting activity.
7. Integrate all of the above into appropriate levels and times of environmental decision making.

U.S. Forest Service's Visual Management System

Figure 13.2 depicts the overall elements of the U.S. Forest Service's Visual Management System (VMS). Table 13.6 identifies five physical landscape attributes and a three-level classification scheme, based on categories called "variety classes," for each. The attributes are primarily reflective of those associated with the heavily vegetated mountain landscapes in the Pacific Northwest and Rocky Mountains. Other types of systematic scoring methodologies exist for different types of landscapes.

The volume of usage of the area is considered (shown in Table 13.7) in determining the sensitivity level. The "sensitivity level" of the visual resource in the VMS is an indicator of the sensitivity of the landscape to the viewer as expressed by its visibility (i.e., can it be seen by many people?), its importance or intensity of use, or interpretations of how people actually feel about the landscape in question (Smardon, Palmer, and Felleman, 1986). A final sensitivity-level classification is then assigned, as shown in Table 13.8.

The VMS also includes the delineation of distance zones on maps. These zones show how far landscapes are from convenient viewpoints or frequently traveled viewing corridors. Dis-

FIGURE 13.2

Overview of Visual Management System of U.S. Forest Service (U.S. Department of Agriculture, 1974, p. 9).

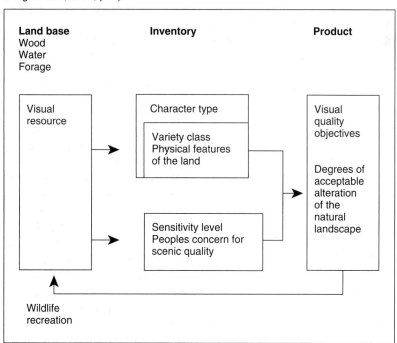

TABLE 13.6

LANDSCAPE ATTRIBUTES (VARIETY CLASSES) IN THE U.S. FOREST SERVICE'S VISUAL MANAGEMENT SYSTEM

	Class A	Class B	Class C
	Distinctive	Common	Minimal
Landform	Over 60 percent slopes which are dissected, uneven, sharp exposed ridges or large dominant features.	30–60 percent slopes which are moderately dissected or rolling.	0–30 percent slopes which have little variety. No dissection and no dominant features.
Rock form	Features stand out on landform. Unusual or outstanding avalanche chutes, talus slopes, outcrops, etc., in size, shape, and location.	Features obvious but do not stand out. Common but not outstanding, avalanche chutes, talus slopes, boulders and rock outcrops.	Small to nonexistent features. No avalanche chutes, talus slopes, boulders and rock outcrops.
Vegetation	High degree of patterns in vegetation. Large old-growth timber. Unusual or outstanding diversity in plant species.	Continuous vegetative cover with interspersed patterns. Mature but not outstanding old growth. Common diversity in plant species.	Continuous vegetative cover with little or no pattern. No understory, over story or ground cover.
Water forms— lakes	50 acres or larger. Those smaller than 50 acres with one or more of the following: (1) unusual or outstanding shoreline configuration, (2) reflects major features, (3) islands, (4) Class A shoreline vegetation or rock forms.	5 to 50 acres. Some shoreline irregularity. Minor reflections only. Class B shoreline vegetation.	Less than 5 acres. No irregularity or reflection.
Water forms— streams	Drainage with numerous or unusual changing flow characteristics, falls, rapids, pools and meanders of large volume.	Drainage, with common meandering and flow characteristics.	Intermittent streams or small perennial streams with little or no fluctuation in flow or falls, rapids, or meandering.

Source: U.S. Department of Agriculture, 1974, p. 13.

tance zones of "foreground," "middle-ground," and "background" are used (Smardon, Palmer, and Felleman, 1986).

The VMS then combines the information on landscape-attribute classes, variety classes, sensitivity levels, and distance zones into a matrix which depicts a management classification; this matrix is shown in Table 13.9. The codes within the matrix correspond to different visual-quality objectives. These objectives can be used in evaluating the visual impacts of proposed projects, and delineating appropriate mitigation meas-

TABLE 13.7

U.S. FOREST SERVICE'S TRAVEL-ROUTE, USE-AREA, AND WATER-BODY IMPORTANCE CRITERIA

	Primary importance	Secondary importance
Travel route	National importance High use volume Long use duration Forest land access roads	Local importance Low use volume Short use duration Project roads
Use areas	National importance High use volume Long use duration Large size	Local importance Low use volume Short use duration Small size
Water bodies	National importance High fishing use High boating use High swimming use	Local importance Low fishing use Low boating use Low swimming use

Source: U.S. Department of Agriculture, 1974, p. 18.

TABLE 13.8

U.S. FOREST SERVICE'S SUMMARY TABLE FOR SENSITIVITY LEVELS

Use	Sensitivity level		
	1	2	3
Primary Travel Routes, Use Areas, and Water Bodies	At least 1/8 of users have MAJOR concern for scenic qualities	Less than 1/4 of users have MAJOR concern for scenic qualities	
Secondary Travel Routes, Use Areas, and Water Bodies	At least 1/4 of users have MAJOR concern for scenic qualities	At least 1/8 and not more than 1/4 of users have MAJOR concern for scenic qualities	Less than 1/8 of users have MAJOR concern for scenic qualities

Source: U.S. Department of Agriculture, 1974, p. 21.

ures; they also determine the different degrees of modification or change allowed for a specific landscape area.

The final component of the VMS provides a means of assessing the severity of visual impacts, or the capability of specific landscape sites to absorb visual impacts. The U.S. Forest Service refers to this as "determining the visual absorption capability (VAC)." The VAC is used to calculate how much can be done to a landscape site before its visual absorption capability is exceeded by the proposed project or activity. The VAC combines physical factors of the existing landscape, highly changeable perceptual factors, existing visual-quality factors (form, line, color, texture), and proposed-activities factors (scale, configuration, duration, frequency, and so forth) as variables to determine

TABLE 13.9

U.S. FOREST SERVICE'S MATRIX FOR VRM VARIETY CLASSES[a]

	Sensitivity level [b,c]						
	fg1	mga	bg1	fg2	mg2	bg2	3
Class A	R	R	R	PR	PR	PR	PR
Class B	R	PR	PR	PR	M	M	M MM
Class C	PR	PR	M	M	M	MM	MM

[a]"Variety Class" refers to the landscape attributes and their classification as depicted in Table 13.6
[b]"Sensitivity level" refers to the sensitivity levels (such as 1, 2, or 3) shown in Table 13.8.
[c]Key: fg = foreground-distance zone
mg = middle-ground-distance zone
bg = background-distance zone
R   = retention
PR = partial retention
M   = modification
MM = maximum modification
Source: U.S. Department of Agriculture, 1974, p. 43.

the VAC score for that particular landscape. Figure 13.3 depicts the rating system which has been devised for developing the VAC. A low VAC score indicates that the range of allowable activity is very restrictive, a high score means much more activity can be allowed. The VAC score range can be then compared, by means of a matrix, to the existing visual-quality management objective(s) already determined for that area; this matrix is depicted in Table 13.10.

U.S. Army Corps of Engineers' Visual-Resources-Assessment Methodologies

The St. Paul District of the U.S. Army Corps of Engineers has developed a ''visual resource evaluation methodology'' (VREM) for use in assessing and managing the visual quality of lands in the upper Mississippi River basin area (U.S. Army Corps of Engineers, 1982). The VREM outlines a process to inventory, document, and evaluate existing viewsheds and to ascertain potential visual impacts of proposed activities within those viewsheds. The process has been organized in an objective manner in order to minimize discrepancies resulting from

variations reflecting a given observer's background, education, experiences, and expectations. A consideration of visual arts issues, physical calculations, dominant impressions received by observers, public exposure, and user expectations are all part of the process (U.S. Army Corps of Engineers, 1982).

The U.S. Army Corps of Engineers has also recently developed a ''visual resources assessment procedure'' (VRAP) for usage on typical civil works projects (Henderson and Peyman-Dove, 1988). VRAP is a systematic method which can be used to (1) evaluate and classify existing aesthetic or visual quality; (2) assess and measure visual impacts caused by various Corps water-resources projects; (3) evaluate the beneficial or adverse nature of the visual impacts; and (4) make recommendations for changes in plans, designs, and operations of water-resources projects (Smardon et al., 1988). The visual resources considered are water resources, landform, vegetation, land use, and user activities. VRAP is composed of two parts, the ''management classification system'' (MCS) and the ''visual impact assessment'' (VIA) pro-

Factors	Variables		Ratings	Viewpoints V1	V2	V3
Observer position	Superior	+300'−+500'	1			
		+100'−+300'	2	2	2	
	Normal	±100'	3			
		−100'−−300'	4			4
	Inferior	−300'−+500'	5			
Observer distance	Foreground	0 − ¼ mi	1			
		¼ − ½ mi	2			
	Middleground	½ − 1 mi	3	3	3	
		1 − 2 mi	4			4
	Background	2+				
View duration	Long	30+ sec	1		1	
		10 − 30 sec	2		2	
	Short	5 − 10 sec	3			3
		3 − 5 sec	4	4		
	Glimpse	0 − 3 sec	5			
Landscape description		Feature	1			
		Focal	2			2
		Enclosed	3	3	3	
		Panoramic	4			
		Other	5			
Slope	Very steep	45 + %	1			
	Steep	00 − 45%	2			
	Moderate	20 − 30%	3			
	Gentle	10 − 20%	4			
	Very gentle	0 − 10%	5	5	5	5
Lowest rating is the key viewpoint				17	16	18
				Summary		

Visual absorption capability

5 − 13 Low

14 − 16 Intermediate

17 − 23 High

FIGURE 13.3

Rating System Devised for Development of Visual Absorption Capability (U.S. Department of Agriculture, 1974).

TABLE 13.10

MATRIX FOR IDENTIFICATION OF APPROPRIATE LANDSCAPE-MANAGEMENT GUIDES
(Scale Rating: I—Most restrictive, V—Least Restrictive)

Visual absorption capability	Visual quality objective			
	Retention	Partial retention	Modification	Maximum modification
Low	I	II	III	V
Intermediate	I	III	IV	V
High	II	III	IV	V

Source: Smardon, Palmer, and Felleman, 1986, p. 166.

cedures. MCS is used to establish an assessment framework for a project area by setting the visual-resource criteria that are to be used throughout the visual assessment. The existing visual quality of an area is determined by taking an inventory of the visual resources and comparing the inventory with the assessment framework. Using data for all proposed alternatives, VRAP can be used to measure the change in the visual resources and determine the compatibility or acceptability of the changes in the visual resources—that is, the visual impact for the given alternative. Visual simulations of the with- and without-project conditions are used to determine changes in visual resources. The MCS criteria are used to determine the acceptability or compatibility of the visual impact (Smardon et al., 1988) with the existing visual quality or environmental setting.

Ten forms are utilized in the VRAP process; these forms are explained in Table 13.11. In concept, the VRAP process can be applied as a "general" procedure, a "basic" procedure, or a "detailed" procedure. Table 13.12 summarizes the application at the general-procedure level; this level is associated with reconnaissance or basinwide studies of water-resources projects. The basic procedure is used for typical projects with low to moderate visual-impact potential and relatively little controversy. The detailed procedure is used for projects that are

unique, controversial, and likely to cause a significant visual impact. It is a more sensitive and extensive process than the basic procedure. Detailed information on the nuances of the three procedures is in Smardon et al. (1988).

Case Studies of Methods

Five case studies in which visual resources and visual impacts were studied will be briefly described. Leopold (1969) discussed a visual-analysis method developed to compare 12 potential sites for hydroelectric power development in the Snake River area in Idaho. This particular region is one of the most scenic areas in the northwestern part of the United States. A method incorporating two approaches was developed for considering aesthetics in the site-selection process. The basic concept was that the hydroelectric power dam should be constructed and operated at the site which, out of the 12, had the least visual-quality value to society.

One approach which was used was based on the viewpoint that landscape which is different or unique has value in society. The basic assumption is that a unique landscape has more significance to society than one which is common (Leopold, 1969). For example, if every state had a Grand Canyon, then the unique visual quality represented by the Grand Canyon would be reduced. A second approach involved

TABLE 13.11

EXPLANATIONS OF THE 10 FORMS USED IN THE VISUAL RESOURCE ASSESSMENT PROCEDURE

FORM 1

FORM 1—VISUAL RESOURCE SUMMARY/DESCRIPTION FORM is used in the MCS and all VIA Procedures to describe the visual resources and aesthetic characteristics of the study area in a holistic manner. The total visual impression and unified perceptions of the landscape are recorded. Visual resource components (e.g., landform, vegetation, water resources, or structures) that are prominent in the landscape are identified. MCS Similarity Zones and designated study areas are inventoried with General VIA Procedures, and future study area conditions are forecasted on this Form. Basic and Detailed VIA Procedures inventory the existing study area conditions from each viewpoint. FORM 1 is used to record forecasting information for each viewpoint for the with- and without-plan conditions. Space is available for a written description and photographs.

FORM 2

FORM 1—VISUAL RESOURCE INVENTORY/FORECAST is used in the MCS and all VIA Procedures. This Form is a list or summary of the various characteristics and types of resources used to assess the visual quality of the study area. Whereas the VISUAL RESOURCE SUMMARY/DESCRIPTION FORM examines the landscape from an overall holistic standpoint, the VISUAL RESOURCE INVENTORY/FORECAST FORM focuses on specific visual resource components. In the MCS, the resources of the Similarity Zones are inventoried. The study area in the General Procedure and each viewpoint chosen in the Basic or Detailed Procedures are inventoried for existing conditions and assessed for future with- and without-plan conditions.

FORM 3

FORM 3—ASSESSMENT FRAMEWORK is used in the MCS to record the determinations of Distinct, Average, and Minimal resource characteristics for each Regional Landscape. ["Definitions of Distinct, Average, and Minimal" are explained in *Part II: Management Classification System* (Smardon et al., 1988, pp. 15–37).] The characteristics are determined for water resources, landform, vegetation, land use, and user activities. The framework determinations provide consistent criteria for the assessment of existing and forecasted visual quality in Similarity Zones, study areas, and viewpoints. The Assessment Framework is initially developed by environmental resource professionals. This evaluation may be combined with public information to form a composite framework.

FORM 4

FORM 4—ASSESSMENT SUMMARY uses information from FORM 1—VISUAL RESOURCE SUMMARY/ DESCRIPTION, FORM 2—VISUAL RESOURCE INVENTORY/FORECAST, and FORM 3—ASSESSMENT FRAMEWORK to produce a numerical Total Assessment Value for each Similarity Zone or study area. Each resource included on FORM 2 (water resources, landform, vegetation, land use, user activity, special considerations) is rated: Distinct = 3, Average = 2, or Minimal = 1 Total Assessment Values range from 0 to 17. FORM 4 is used in the MCS to assess existing visual quality in each Similarity Zone. In the General VIA Procedure, this Form can be used to assess existing and forecasted visual quality in the study area.

FORM 5

FORM 5—MANAGEMENT CLASSIFICATION SUMMARY is used in the MCS or General VIA Procedures to record the classification of existing visual resources of each Similarity Zone in a Regional Landscape or study area. Each zone is classified depending on its numerical Assessment Value as determined in FORM 4—ASSESSMENT SUMMARY. Management classes and Total Assessment Values include Preservation (17 or greater), Retention (14 to 16), Partial Retention (11 to 13), Modification (8 to 10), and Rehabilitation (less than 8).

specific data comparisons and the development of what could be termed an "aesthetics index." The analyses resulting from these two approaches could then be compared, and any sites that were classified in the lower grouping according to both approaches should be the ones seriously considered for hydroelectric power dam development.

In the relative-uniqueness approach, 46 factors were used to assess value (Leopold, 1969).

TABLE 13.11

EXPLANATIONS OF THE 10 FORMS USED IN THE VISUAL RESOURCE ASSESSMENT PROCEDURE *(continued)*

FORM 6
FORM 6—VIEWPOINT ASSESSMENT is used in the Basic and Detailed VIA Procedures. Each evaluator uses this Form to assess the forecasted conditions of representative viewpoints for each alternative plan. For each viewpoint, the water resources, landform, vegetation, land use, user activities, and special considerations are rated from the ASSESSMENT FRAMEWORK FORM as Distinct = 3, Average = 2, or Minimal = 1, and assessed for with- and without-plan conditions. The Viewpoint Value, a numerical difference between the with- and without-plan conditions, is calculated for each resource. The level of compatibility, scale contrast, and spatial dominance of the project to the study area is also assessed on this Form. The landscape composition of the with- and without-plan conditions are rated as inconspicuous, significant, or prominent.

FORM 7
FORM 7—SUMMARY VIEWPOINT ASSESSMENT is used in the Basic and Detailed VIA Procedures to combine the assessments of the different viewpoints into a Summary Assessment Value. A separate assessment is completed for each evaluator, each forecast period, and each alternative. The information from each viewpoint is transferred to this Form and averaged to get a Summary Viewpoint Value for each of the visual resource components, e.g., water resources or vegetation. For the modifier ratings, a majority rating is determined for compatibility, scale contrast, spatial dominance, and landscape composition.

FORM 8
FORM 8—VISUAL IMPACT ASSESSMENT SUMMARY is used to compute a VIA Value for each forecast period and alternative plan considered in the Basic or Detailed VIA Procedure. The VIA Value is determined by combining the Composite Viewpoint Values of all the evaluators. The VIA Value is the measure of visual impact caused by the project, comparing with- and without-plan conditions, and is used for comparison with the project's MCS classification of the study area. The modifier ratings of all the evaluators are averaged to give a majority rating. The landscape composition ratings are also averaged.

FORM 9
FORM 9—DESIGN ELEMENT INVENTORY/FORECAST—DETAILED is used in the Detailed VIA Procedure. The Form is used to inventory and forecast the viewpoints in terms of the design elements of line, form, color, texture, and scale. This inventory is completed along with FORM 6—VIEWPOINT ASSESSMENT during the Detailed VIA Procedure. The design elements are described for existing and forecasted conditions. The information from this Form is used to identify elements that can be changed to minimize or modify the visual impacts or to reformulate alternative plans.

FORM 10
FORM 10—DESIGN ELEMENT ASSESSMENT—DETAILED is used in the detailed VIA Procedure. This Form is used to document, in narrative, the changes in the design elements for the representative viewpoints of each forecast period and alternative plan. The assessment of differences in with- and without-plan conditions is described in reference to water, landform, vegetation, and structures.

Source: Smardon et al., 1988, pp. 9–11.

The first 14 factors were related to physical issues associated with the environment, whereas factors 15 through 28 were associated with biological and water quality issues. Factors 29 through 46 were descriptive of human use and interest issues; it could be argued that these factors are the most indicative of visual quality. The next step involved considering the range of information which might be collected for each of the 46 factors, and then dividing these 46 factors into five descriptive categories based on the data to be obtained. This was done to enable

TABLE 13.12

GENERAL PLANNING PROCESS AND VRAP PROCEDURE

Planning process	VRAP procedure	Forms
Specify problems and opportunities	Define the study area	Visual Resource Summary/Description
	Identify the regional landscape Determine MCS class Establish what method is to be used for the study (general, basic, or detailed)	Assessment Framework
Inventory and forecast	Inventory existing visual resources Forecast without-plan conditions to assess any changes from existing visual-resource conditions	Visual Resource Inventory/Forecast Visual Resource Inventory/Forecast
	Forecast with-plan conditions	Visual Resource Inventory/Forecast
Formulate alternative plans	Use simulations to show designs of alternatives	
Evaluate alternative plans	Assess visual impacts by calculating the difference between future with- and without-plan conditions for each landscape component, for each viewpoint	Visual Impact Assessment—Viewpoint
	Combine viewpoint assessments from each evaluator to calculate VIA values for the landscape modifiers	Visual Impact Assessment—Viewpoint Summary
	Combine the evaluators' VIAs to calculate a VIA value	Visual Impact Assessment—Assessment Summary
(This process is performed only if public input is available)	(Combine public and professional VIA values to calculate a total VIA value)	(Composite Project Assessment)
Compare alternative plans	Compare VIA values with MCS criteria	

Source: Smardon et al., 1988, p. 8.

information sorting and analysis. The third step involved site visits by study-team members. Each site was evaluated by each team member relative to the 46 factors; composite evaluations were then prepared and used in subsequent steps. The next step involved the calculation of a uniqueness ratio. The "uniqueness ratio," by definition, is equal to 1 divided by the number of sites having the same category value as the site being evaluated. On this basis, the minimum uniqueness ratio would be obtained where all 12 sites had the same category value, hence the

minimum uniqueness ratio would be 0.08. The maximum uniqueness ratio would be obtained where only 1 site out of the 12 had a particular category value. Hence, the maximum uniqueness ratio would be equal to 1 divided by 1, or a value of 1.00. The final step was to sum the uniqueness ratios for each of the 12 sites and the 46 factors. The sites with the lowest sums would be the best candidates for hydroelectric power development. It should be noted that the analysis discussed so far has not considered the impact of the proposed project on visual quality, it has only considered the existing visual quality. Also, it would be possible for a site to be unique, but not of good visual quality. That is, a site might be the only one of a kind, but its unique visual quality might not be very desirable.

The second approach entailed modeling specific data comparisons into aesthetic indices. In this technique, two features were identified: one called ''valley character'' and the other called ''river character.'' Valley character was considered to be a function of four parameters—namely, width of valley, height of nearby hills, scenic outlook, and degree of urbanization. River character was considered to be a function of three parameters: river depth, river width, and prevalence of rapids. Information was assembled on each of these seven factors and then aggregated using a nomographic approach (Leopold, 1969). The final step involved plotting the scale of river character versus the scale of valley character. The basic assumption was that those sites with higher scores would have greater aesthetic value than those with lower scores. The objective would again be to identify those sites that would have the greatest aesthetic value, and to consider development at those sites that would have the least aesthetic value.

The following general comments are offered on the visual-quality methodology developed by Leopold (1969):

1. This overall methodology, which involves the systematic comparison of a series of sites relative to a series of decision factors, is quite useful for relative comparisons. It should be noted that, in the example case, no specific analysis was made of the impact of the anticipated project on the visual quality; however, the 12 sites could be compared based on their relative visual quality by the use of either of the approaches.

2. Both approaches are in reality types of aesthetics-indices methods.

3. While this methodology was developed and applied by an interdisciplinary team, it would be possible to develop a methodology such as this that could include public input on the factors and categories for consideration, as well as specific data comparisons.

4. This methodology was primarily developed for a setting which includes valleys and nearby hills. There are many locations in the United States that do not have these characteristics and, therefore, different factors would need to be considered for different terrain.

5. Usage of either or both approaches would provide a structured procedure that would serve to delineate how visual quality was evaluated as a part of project planning, although there might be debate as to whether or not this is the best approach.

Brief information will be presented on the remaining four case studies. For example, Robertson and Blair (1980) described a visual-assessment study of a number of development alternatives proposed for a 128-acre waterfront property in Seattle, Washington. The various steps in the visual-analysis method used in the study, along with the results of the study, are presented. Particular emphasis is placed on the advantages which the method offers both the developer, in this case the Port of Seattle, and the public affected by the proposals.

A visual-impact-assessment methodology for industrial-development-site review in Spain has been described by Smardon, Palmer, and Felleman (1986). A logical methodology for addressing the potential visual-quality impacts of an industrial plant would typically include accomplishment of the following steps (Smardon, Palmer, and Felleman, 1986):

1. Assemble information about the project, including built volumes, spatial organization,

building materials, auxiliary elements, and so forth. Illustrated documents of the installation would be helpful.

2. Prepare a description of the surrounding landscape. Identify its principal uses. Also identify all particularly interesting elements, and so forth.

3. Prepare a definition of the area visually affected by the installation.

4. Select appropriate observation points—for example, places which will probably concentrate observers under with-project conditions. Identify probable visual elements to be affected.

5. Delineate the new viewsheds of the chosen spots, and evaluate any intrusion and dominance problems.

6. Prepare a simulation from each of the chosen spots.

7. Define the probable impacts, based on the simulation. Consider contrast between the visual elements of the installation and the environment: lines, color, shape, and texture. Also consider scale dominance and disturbance of the visual parameters.

8. Evaluate the impacts.

As another case example, the visual impact of electricity transmission towers is influenced by the following factors (Hull and Bishop, 1988): (1) the distance of an observer from the tower, (2) the environment surrounding the tower, (3) the physical characteristics of the tower, (4) the visibility of the tower, and (5) the disposition and visual preference of people viewing the tower. The first two factors were examined in a user-analysis type of study conducted in Australia. The scenic impact of 500-kV electricity transmission towers was assessed at distances ranging from 100 m to 5 km for three landscape types: (1) a rural landscape with rolling topography, minimal buildings, and mixed agricultural uses (mainly grazing); (2) a suburban landscape on a very flat topography,

with 1/4-acre blocks, minimal vegetation, and wide paved roads; and (3) a flat expanse of agricultural fields with no buildings and minimal vegetation (recently harvested or recently burned crops). The scenic impact of a tower in a landscape scene was assessed by subtracting the scenic beauty of the scene with the tower from the scenic beauty of very similar scenes without towers. The scenic beauty of landscape scenes was assessed using a 10-point rating scale; 196 female members of suburban parent-teacher associations in Melbourne, Australia, rated slides of these landscapes for scenic beauty. The results indicated that scenic impact decreases rapidly as distance increases. A reciprocal relationship between distance and scenic impact best explained the data. Landscape type did not have a statistically significant impact on this relationship. However, in general, the largest impacts occurred in the rural landscapes which were high in scenic beauty, medium impacts occurred in residential areas, and the smallest impacts occurred in flat, barren, not very scenic landscapes (Hull and Bishop, 1988).

Finally, Swihart and Petrich (1988) described a method for assessing the aesthetic impacts of small hydropower-development projects proposed by the Federal Energy Regulatory Commission. The method involved a combined study of (1) the physical factors contributing to aesthetic quality and (2) the cognitive, perceptual factors that relate to aesthetic quality and contribute to preferences for particular landscapes. Evaluations of the physical dimensions of the aesthetic resource were based on an analysis of the actual physical components of the landscape—such as water, vegetation, and landform—which influence the ability of each streamscape to visually absorb modifications and contribute to each streamscape's relative uniqueness. The users' perceptions of the attractiveness of the resource guided the evaluations of the cognitive dimensions of the aesthetic resource. Users' perceptions were elicited during public meetings, and by means of site visits and interviews with resource managers. A photoquestionnaire was used to gather this information (Swihart and Petrich, 1988).

STEP 5: ASSESSMENT OF SIGNIFICANCE OF PREDICTED IMPACTS

The assessment of predicted changes in visual quality in the study area will require the careful exercise of professional judgment. This is necessitated because of the absence of quantitative standards; however, as noted earlier, there may be multiple federal, state, or local laws, regulations, and/or executive-administrative orders which should be considered in relation to their requirements for assessing visual impacts. Several examples will be mentioned. One example of pertinent considerations for a preliminary assessment would be to apply the practical visual criteria or guidelines as delineated in Table 13.4. A second example would be to apply the six questions listed in Part 2 of Table 13.5.

The following represent still further questions which could be used singly or in various combinations as a basis for interpreting the potential significance of visual impacts:

1. Is the proposed project (or activity) compatible or in compliance with land-use zoning requirements of the local community or urban area?
2. Is the proposed project compatible with sign ordinances or regulations of the local community or urban area?
3. Is the proposed project compatible with design guides of the local community or urban area?
4. Does the proposed project meet the buffer-zone (or greenbelt-zone) requirements of the local community or urban area?
5. What is the recommended interpretation of the changes in visual quality in the study area

based on the utilized quantitative methodology for describing baseline conditions and with-project conditions? (Note: This question may be applicable for methodologies used for step 4 only).
6. What is the recommended interpretation of the changes in visual quality in the study area based on conversations with appropriate local-area community planners and/or urban planners?

STEP 6: IDENTIFICATION AND INCORPORATION OF MITIGATION MEASURES

An appropriate consideration in assessing visual impacts is related to the possibility of implementing selected mitigation measures. "Mitigation measures," in this context, refer to steps that can be taken to minimize the visual intrusion or negative impacts of the proposed project or activity, in the study area. Examples of generic mitigation measures which can be considered for visual impacts include (Heape, 1991) (1) the use of scale, shape, color, and tone; (2) the use of vegetative screening; (3) restoration procedures; and (4) siting of the proposed development. Mitigation measures for negative aesthetic impacts can include the following, either singly or in combination, as appropriate for the proposed undertaking:

1. Selection of color(s) of paint for facility exterior, and incorporation of painting patterns, so as to enable the facility to blend appropriately with the viewscape
2. Selection of construction materials for facility exterior so as to enable the facility to blend appropriately with the viewscape
3. Use of architectural features for proposed facilities so as to enable the facilities to blend appropriately with the architectural features of existing buildings and the viewscape

4. Tinting of concrete in facility or project structures so as to achieve color compatibility with existing buildings or structures and the viewscape
5. Incorporation of design features and colors so as to achieve compliance with pertinent regulatory design guides and/or professional architectural or landscape-architectural design guides
6. Reuse of materials from former facilities (or structures) in new facilities, particularly with regard to materials from demolished facilities with historical significance
7. Incorporation of underground utilities (electricity, water, sewer, and gas) in project planning
8. Provisions of appropriate visual screens or barriers in the viewscape to preclude unsightly intrusions from the project
9. Provision of greenbelts around the project
10. Planning and implementation of an appropriate landscaping (vegetative-screening)

TABLE 13.13

DECISION FRAMEWORK FOR THE RESULTS OF A VISUAL-IMPACT ASSESSMENT (VIA)

Existing visual resources[a]	Category of impact[b]	Mitigation possible	Recommended decision
1, good	A, potentially significant	Yes No	Complete study. Modify or abandon proposed project or activity.
	B, possibly significant	Yes No	Complete study. Modify proposed project or activity.
	C, not significant	Yes No	Complete study. Complete study.
2, modest	A, potentially significant	Yes No	Complete study. Modify proposed project or activity.
	B, possibly significant	Yes No	Complete study. Complete study with monitoring.
	C, not significant	Yes No	Complete study. Complete study.
3, minor	A, potentially significant	Yes No	Complete study. Complete study with monitoring.
	B, possibly significant	Yes No	Complete study. Complete study with monitoring.
	C, not significant	Yes No	Complete study. Complete study.

[a]Terminology and bases for existing visual resources are described in text.
[b]Terminology and bases for impact categories are described in text.

program for the project including consideration of initial versus full-growth characteristics of grass, plants, bushes, trees, and flowers in terms of colors, heights, and density

While painting of facilities to promote their visual blending with the landscape is often proposed and utilized, conflicts can arise over this mitigation approach. For example, the Bonneville Power Administration (BPA) built a 500-kV line across Rock Creek, a class 1 trout stream about 20 mi east of Missoula, Montana (BPA, 1988). Two 190-ft towers rise on either side of the Rock Creek Valley, and the line between is suspended 600 ft over the valley floor. The crossing poses a hazard to passing airplanes and disrupts the natural landscape. The area where the line crosses Rock Creek is prized for its scenic beauty. In response to public demand that BPA protect the visual beauty of this area, the agency painted the towers gray so that they would blend in with their natural surroundings. The issue then became one of deciding between two gray towers, two orange and white towers, or a combination. The underlying need is to resolve the conflict of pilot safety against scenic intrusion. The proposed action was to paint one gray tower aeronautical orange and white.

Impact-mitigation considerations can prompt an overall review of the information generated in the analysis and the arrival at a decision relative to the proposed project (or activity). It is possible that a modification of the originally planned project could be suggested and that an analysis of this modified project will be required. Table 13.13 summarizes a decision framework which could be used.

A review of mitigation measures in relation to the potential aesthetics impacts of a project can aid in identifying pertinent approaches which could be incorporated to minimize negative impacts. In several of the potential decisions in Table 13.13 monitoring is indicated.

"Monitoring" refers to making appropriate site visits, or visual inspections and/or visual surveys, so as to enable real-time decisions to be made relative to scheduling project construction or operational features to minimize negative visual impacts. Monitoring of visual impacts can be valuable if it provides assurances that deleterious impacts are not occurring. Visual-impact monitoring should be targeted to specific impacts of concern. An additional benefit of monitoring is that, should the impacts be of concern, the monitoring will provide early detection and allow mitigation or control actions to be taken.

It is anticipated that the decision to abandon the proposed undertaking (or any modifications thereof) will occur only when significant visual impacts can be shown to be inevitable. The severity of this decision illustrates the need for a systematic, rational, documentable methodology for justifying such a decision.

SUMMARY

This chapter has provided a summary of the principles and practices for addressing the visual impacts of proposed projects or activities. The focus is on a six-step methodology which can be used to delineate existing visual resources and predict and assess the potential impacts on such resources. Detailed analysis would require all six steps, while preliminary or intermediate analysis would include subsets of the steps. Various types of scoring or index methods provide the technical basis for the methodology. In addition, it should be noted that many undesirable visual impacts could probably be mitigated by the incorporation of simple cost-effective measures during project planning, construction, and operation.

SELECTED REFERENCES

Advisory Council on Historic Preservation, "Section 106, Step-by-Step," Advisory Council on Historic Preservation, Washington, D.C., Oct. 1986, p. 25.

Bagley, M. D., Kroll, C. A., and Clark, C., "Aesthetics in Environmental Planning," EPA 600/5-73-009, U.S. Environmental Protection Agency, Washington, D.C., Nov. 1973.

Beer, A. R., *Environmental Planning for Site Development,* Chapman and Hall, London, 1990, pp. 133–140.

Bonneville Power Administration, "Rock Creek Tower Painting Project: Environmental Assessment," Bonneville Power Administration, Portland, Ore., Oct. 1988.

Coughlin, R. E., et al., "Assessing Aesthetic Attributes in Planning Water Resources Projects," *Environmental Impact Assessment Review,* vol. 3, no. 4, 1982, pp. 406–417.

Council on Environmental Quality (CEQ), "National Environmental Policy Act—Regulations," *Federal Register,* vol. 43, no. 230, Nov. 29, 1978, pp. 55978–56007.

Heape, M., "Visual Impact Assessment," paper presented at *Twelfth International Seminar on Environmental Assessment and Management,* University of Aberdeen, Aberdeen, Scotland, July 7–20, 1991.

Henderson, J. E., and Peyman-Dove, L. D., "Environmental Impact Research Program: Visual Resources Assessment Procedure for U.S. Army Corps of Engineers," WES/IR/EL-88-1, U.S. Army Engineer Waterways Experiment Station, Vicksburg, Miss., Mar. 1988.

Hull, R. B., and Bishop, I. D., "Scenic Impacts of Electricity Transmission Towers: The Influence of Landscape Type and Observer Distance," *Journal of Environmental Management,* vol. 27, 1988, pp. 99–108.

Leopold, L. B., "Quantitative Comparison of Some Aesthetic Factors Among Rivers," *Geological Survey Circular 620, U.S. Geological Survey,* Washington, D.C., 1969.

Marlatt, R. M., Hale, T. A., and Sullivan, R. G., "Video Simulation as Part of Army Environmental Decision-Making: Observations from Camp Shelby, Mississippi," *Environmental Impact Assessment Review,* vol. 13, 1993, pp. 75–88.

Martin, J., "Visual Impact Assessment Techniques for Single Site Appraisal," paper presented at *Fifth International Seminar on Environmental Impact Assessment,* University of Aberdeen, Aberdeen, Scotland, July 8–21, 1984.

Monbailliu, X., "Assessment of Visual Impact," in *Perspectives on Environmental Impact Assessment,* B. D. Clark and A. D. Gilad, eds., D. Reidel Publishing Company, Dordrecht, The Netherlands, 1984, pp. 265–271.

———, "Current Trends in Landscape Evaluation," in *Environmental Impact Assessment,* PADC Environmental Impact Assessment and Planning Unit, eds., Martinus Nijhoff Publishers, The Hague, The Netherlands, 1983, pp. 321–326.

Robertson, I. M., and Blair, W. G., "Visual Impacts of Port Development in Seattle," *Proceedings of Second Symposium on Coastal and Ocean Management,* vol. 2, American Society of Civil Engineers, New York, 1980, pp. 917–930.

Smardon, R. C., et al., "Visual Resources Assessment Procedure for U.S. Army Corps of Engineers," Instruction Rep. EL-88-1, U.S. Army Waterways Experiment Station, Vicksburg, Miss., Mar. 1988.

———, and Karp, J. P., *The Legal Landscape—Guidelines for Regulating Environmental and Aesthetic Quality,* Van Nostrand Reinhold, New York, 1993.

———, Palmer, J. F., and Felleman, J. P., *Foundations for Visual Project Analysis,* John Wiley and Sons, New York, 1986, pp. 141–166, 295–299, 310–333.

Swihart, M. M., and Petrich, C. H., "Assessing the Aesthetic Impacts of Small Hydropower Development," *The Environmental Professional,* vol. 10, 1988, pp. 198–210.

U.S. Army Construction Engineering Research Laboratory, "Environmental Review Guide for USAREUR," 5 vols., Champaign, Ill., 1989.

U.S. Army Corps of Engineers, "Visual Resource Evaluation Methodology," St. Paul District, St. Paul, Minn., Mar. 1982.

U.S. Department of Agriculture, U.S. Forest Service, "Landscape Management Visual Display Techniques," FSH6/77, U.S. Department of Agriculture, Washington, D.C., 1977.

———, U.S. Forest Service, "National Forest Landscape Management, vol. 2, Chap. 1, The Visual Management System," Handbook 462, U.S. Department of Agriculture, Washington, D.C., Apr. 1974.

———, Soil Conservation Service, "Procedure to Establish Priorities in Landscape Architecture," Tech. Release no. 65, Soil Conservation Service, Washington, D.C., Oct. 1978.

U.S. Department of the Interior, Bureau of Land Management, "Upland Visual Resource Inventory and Evaluation," BLM Manual Section 8411, U.S. Department of the Interior, Washington, D.C., Aug. 1978b.

———, "Visual Resource Contrast Rating," BLM Manual Section 8431, U.S. Department of the Interior, Washington, D.C., Aug. 1978a.

———, "Visual Resource Management Program," U.S. Department of the Interior, Washington, D.C., 1980a.

———, "Visual Simulation Techniques," U.S. Department of the Interior, Washington, D.C., 1980b.

CHAPTER 14

Prediction and Assessment of Impacts on the Socioeconomic Environment

Governmental or private programs, policies, and projects can cause potentially significant changes in many features of the socioeconomic environment. In some cases the changes may be beneficial, in others they may be detrimental. Accordingly, environmental impact studies must systematically identify, quantify, where possible, and appropriately interpret the significance of these anticipated changes. This chapter provides an organized and systematic approach for addressing the socioeconomic impacts of major projects.

Several illustrations of socioeconomic impacts can be noted. For example, large-scale relocations of people may be required for major water resources projects. Ghana's Volta Dam required the evacuation of some 78,000 people from more than 700 towns and villages; Lake Kainji in Nigeria displaced 42,000; the Aswan High Dam, in Egypt, 120,000; Turkey's Keban Dam, 30,000; Thailand's Ubolratana Dam, 30,000; the Pa Mong project in Vietnam uprooted 450,000 people; and, in China, the vast Three Gorges Dam scheme will displace 1,400,000 people (Goldsmith and Hildyard, 1984).

Proposed projects involving the decommissioning and closure of major governmental installations or industrial sector developments can have significant socioeconomic consequences in terms of local and/or areawide decreases in jobs and revenue to the economy, declines in human population, and leftover societal debts for local infrastructure and educational facilities (Grady et al., 1987).

Major development projects can include significant requirements for associated infrastructure such as streets, highways, or railroads; water supply; sanitary sewers; storm-water drainage; erosion control, sediment control, and grading; electrical systems; gas systems; and telephone communication systems. The provision of such needed infrastructures can also generate environmental impacts. Design elements and standards for these infrastructure needs and available guidelines can be used for impact interpretation (Urban Land Institute, 1989). Similar information is available for business and industrial park developments (Urban Land Institute, 1988). Finally, socioeconomic impacts may be an important concern related to the cumulative impacts of a proposed project or activity (Hundloe et al., 1990).

This chapter includes background information on the socioeconomic impacts of projects (or activities). A conceptual approach for predicting and assessing impacts on a wide range of socioeconomic factors is presented, including more in-depth information on prediction approaches and several bases for impact-significance determination. Examples of applications of specific steps or methods useful for addressing impacts on educational services, local traffic and transportation systems, and human health are then presented. Detailed information on a much broader range of impacts is available in Canter, Atkinson, and Leistritz (1985).

BACKGROUND INFORMATION

Following the January 1, 1970, effective date of the NEPA, the initial emphasis in EISs was given to the biophysical and cultural environments. Socioeconomic concerns were introduced into the NEPA process in 1973 with the issuance of guidelines by the CEQ. These concerns were raised in relation to secondary impacts, with the relevant wording as follows (CEQ, 1973, p. 20556):

> Secondary or indirect, as well as primary or direct, consequences for the environment should be included in the analysis. Many major federal actions, in particular those that involve the construction or licensing of infrastructure investments (e.g., highways, airports, sewer systems, water resource projects, etc.), stimulate or induce secondary effects in the form of associated investments and changed patterns of social and economic activities. Such secondary effects, through their impacts on existing community facilities and activities, through inducing new facilities and activities, or through changes in natural conditions, may often be even more substantial than the primary effects of the original action itself. For example, the effects of the proposed action on population and growth may be among the more significant secondary effects. Such population and growth im-

pacts should be estimated if expected to be significant and an assessment made of the effect of any possible change in population patterns or growth upon the resource base, including land use, water, and public services, of the area in question.

Consideration of impacts to the socioeconomic environment was also stressed in the CEQ regulations which became effective in 1979 (CEQ, 1978). Specifically, the definition of "human environment" was expanded to include socioeconomic concerns. The definition is as follows (CEQ, 1978, p. 56004):

> "Human Environment" shall be interpreted comprehensively to include the natural and physical environment and the relationship of people with that environment. This means that economic or social effects are not intended by themselves to require preparation of an environmental impact statement. When an environmental impact statement is prepared and economic or social and natural or physical environmental effects are interrelated, then the environmental impact statement will discuss all of these effects on the human environment.

Factors that describe the socioeconomic environment represent a composite of numerous interrelated and nonrelated items. On the one hand, this category represents a catchall group, since it includes factors not associated with the physical-chemical, biological, or cultural environment. On the other hand, this category is the one most descriptive of human relationships and interactions. Table 14.1 includes examples of socioeconomic factors that have been used in EISs. Also included are typical corresponding changes following construction and operation of proposed undertakings. Many of the changes described in Table 14.1 represent impacts of unique significance or importance to the human population. The first four factors in Table 14.1 address demographic concerns, and the next three are associated with economic and employment concerns. Land use, values, and taxes are the focus of the next three factors in Table 14.1.

TABLE 14.1

EXAMPLES OF SOCIOECONOMIC FACTORS AND THEIR POTENTIAL CHANGES RESULTING FROM PROJECT IMPLEMENTATION

Factor	Potential change
General characteristics and trends in population for state, substate region, county, and city	Increase or decrease in population
Migrational trends in study area (The study area is a function of the alternatives being considered and the available database.)	Increase or decrease in migrational trends
Population characteristics in study area, including distributions by age, sex, ethnic group, educational level, and family size	Increase or decrease in various population distributions; people relocations
Distinct settlements of ethnic groups or deprived economic or minority groups in study area	Disruption of settlement patterns; people relocations
Economic history for state, substate region, county, and city	Increase or decrease in economic activities; change in economic patterns
Employment and unemployment patterns in study area, including occupational distribution and location and availability of workforce.	Increase or decrease in overall employment or unemployment levels; change in occupational distribution
Income levels and trends for study area	Increase or decrease in income levels
Land-use patterns and controls for study area	Change in land usage; project may or may not be in compliance with existing land-use plans
Land values in study area	Increase or decrease in land values
Tax levels and patterns in study area, including land taxes, sales taxes, and income taxes	Changes in tax levels and patterns resulting from changes in land usage and income levels
Housing characteristics in study area, including types of housing occupancy levels, and age and condition of housing	Changes in types of housing and occupancy levels
Health and social services in study area, including health manpower, law enforcement, fire protection, water supply, wastewater-treatment facilities, solid-waste collection and disposal, and utilities	Changes in demand for health and social services
Public and private educational resources in study area, including grades K–12 schools, junior colleges, and universities	Changes in demand for educational resources
Transportation systems in study area, including highway, rail, air, and waterway systems	Changes in demand for transportation systems; relocations of highways and railroads
Community attitudes and lifestyles, including history of area voting patterns	Changes in attitudes and lifestyles
Community cohesion, including organized community groups	Disruption of cohesion
Tourism and recreational opportunities in study area	Increase or decrease in tourism and recreational potential
Religious patterns and characteristics in study area	Disruption of religious patterns; change in characteristics
Areas of unique significance, such as cemeteries or religious camps	Disruption of activities in or changes to unique areas

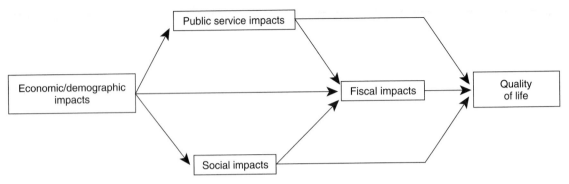

FIGURE 14.1
Example of Information Flow in Prediction and Assessment of Socioeconomic Impacts.

The next four factors (housing, health and social services, education, and transportation) are related to public service and social concerns. Community attitudes and cohesion are the subject of the next two factors, and three miscellaneous factors comprise the final group in Table 14.1. Some of the issues related to socioeconomic impacts are addressed from an economic perspective in traditional cost-benefit analyses (Hundloe et al., 1990).

Figure 14.1 displays the general information flow relative to prediction and assessment of socioeconomic impacts. Modeling of economic-demographic impacts provides basic information for addressing public service impacts (education, health services, police and fire protection, utilities, and solid-waste management), social impacts (housing, transportation, urban land use, and land ownership), and fiscal impacts. Fiscal impacts are themselves dependent upon many public services and social impacts. The quality of life (QOL) represents a composite indication of economic, demographic, public service, social, and fiscal impacts, along with impacts caused by still other factors related to a sense of well-being at a given time and location. Some QOL indices are described in Chapter 5. The impact-assessment (interpretation) approaches described subsequently provide a basis for interpreting existing and future socioeconomic conditions, both with and without the project.

"Social-impact assessment" (SIA) is a phrase which is used interchangeably with "socioeconomic-impact considerations." SIA includes health impacts, recreational activities, aesthetic interests, land and housing values, job opportunities, community cohesion, lifestyles, governmental activities, psychological well-being, and behavioral response on the part of individuals, groups, and communities (Hundloe et al., 1990). Social impacts may be reflected by heightened public concerns over the perceived risks of projects—for example, the siting of nuclear waste projects (Nieves et al., 1990).

SIA involves the systematic, advanced appraisal of the impacts of development projects or policy changes on the day-to-day quality of life of persons and communities (Burdge, 1987). Suggested SIA variables to be addressed in an impact study are listed in Table 14.2. Public involvement is required at various stages in an SIA process for a public works project, including scoping, formulation of alternatives, profiling, projection, assessment, evaluation, mitigation, and monitoring (Burdge and Robertson, 1990). Public involvement techniques are addressed in Chapter 16.

TABLE 14.2

SIA VARIABLES

Population impacts

Population change
Influx or outflux of temporary workers
Presence of seasonal (leisure) residents
Relocation of individuals and families
Dissimilarity in age, gender, racial or ethnic composition

Community/institutional arrangements

Formation of attitudes toward the project
Interest group activity
Alteration in size and structure of local government
Presence of planning and zoning activity
Industrial diversification
Enhanced economic inequities
Change in employment equity of minority groups
Change in occupational opportunities

Conflicts between local residents and newcomers

Presence of an outside agency
Introduction of new social classes
Change in the commercial/industrial focus of the
  community
Presence of weekend residents (recreational)

Individual and family level impacts

Disruption in daily living and movement patterns
Dissimilarity in religious practices
Alteration in family structure
Disruption in social networks
Perceptions of public health and safety
Change in leisure opportunities

Community infrastructure needs

Change in community infrastructure
Land acquisition and disposal
Effects on known cultural, historical, and archaeological
  resources

Source: Burdge, 1987, p. 147.

Specific advantages and features of the SIA process include the following (Burdge and Robertson, 1990, p. 83):

1. SIA is a systematic effort to identify, analyze, and evaluate social impacts of a proposed project or policy change on the individuals and so-

cial groups within a community or on an entire community in advance of the decision-making process in order that the information derived from the SIA can actually influence decisions.
2. SIA is a means for developing alternatives to the proposed course of action and determining the full range of consequences for each alternative.
3. SIA increases knowledge on the part of the project proponent and the impacted community.
4. SIA raises consciousness and the level of understanding of the community and puts the residents in a better position to understand the broader implication of the proposed action.
5. SIA includes within it a process to mitigate or alleviate the social impacts likely to occur, if that action is desired by the impacted community.

CONCEPTUAL APPROACH FOR ADDRESSING SOCIOECONOMIC IMPACTS

A conceptual approach for addressing socioeconomic impacts is shown in Figure 14.2. The basic steps are similar to the six-step methodologies for other substantive area topics addressed in Chapters 6 through 13. Identification of potential impacts represents the initial step, followed by the preparation of a description of existing conditions for the selected factors. Relevant standards, criteria, or guidelines should then be procured and utilized to assess existing conditions. Impact prediction for both the future-without-project and future-with-project conditions is the emphasis of step 4. Step 5 involves the assessment of the predicted impacts in relation to existing conditions and through the use of relevant standards, criteria, or guidelines. The final step consists of the identification and incorporation of mitigation measures to minimize the negative consequences of the proposed project. Steps 1, 2, 4, and 5 will be described in subsequent sections.

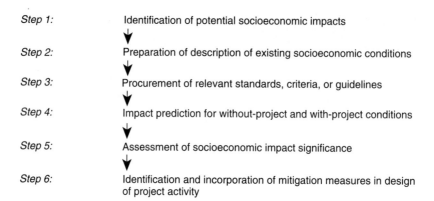

Step 1:   Identification of potential socioeconomic impacts

Step 2:   Preparation of description of existing socioeconomic conditions

Step 3:   Procurement of relevant standards, criteria, or guidelines

Step 4:   Impact prediction for without-project and with-project conditions

Step 5:   Assessment of socioeconomic impact significance

Step 6:   Identification and incorporation of mitigation measures in design of project activity

FIGURE 14.2
Conceptual Framework for Prediction and Assessment of Socioeconomic Impacts.

In contrast to Figure 14.2, a project-category-based approach for the SIA procedure for projects in developing countries has been promulgated by the Asian Development Bank; the following project categories are used:

1. *Category AP*—Those projects whose major objective is to have a direct positive impact in the form of poverty alleviation; poverty projects almost always require the active participation of beneficiaries.
2. *Category A*—Projects which are expected to have a direct, positive social impact and which, in order to be successfully executed and/or sustained, require the active participation of the intended beneficiaries; this category includes most projects in the agriculture and social infrastructure sectors.
3. *Category B*—Projects that rarely have an immediate, direct (positive or negative) social impact and/or projects that can be executed and sustained without beneficiary participation; this category includes most projects in the energy, transport, communications, and industrial (except small-scale or cottage-type) sectors.
4. *Category C*—Projects which have the potential for direct, negative impact on the lives of a significant number of people. Category

C includes all projects which may require the relocation of significant numbers of people, for example, large dams, highways, airports, and so on. Also included in this category are a range of projects (e.g., mining, plantation crops, highways) to be located in remote areas predominantly inhabited by tribal or ethnic communities or any other group having a significantly different sociocultural tradition from that of the ruling or dominant communities (Asian Development Bank, 1991b).

Specific methods for addressing the direct social impacts of projects in developing countries are also available in Finsterbusch, Ingersoll, and Llewellyn (1990).

STEP 1: IDENTIFICATION OF SOCIOECONOMIC IMPACTS

Potential socioeconomic impacts can be identified through the use of interaction matrices, networks, simple checklists, and/or descriptive checklists, as described in Chapter 3. Case studies of similar project types can also be helpful, and three examples will be mentioned herein.

The basic impact area associated with predicting and assessing impacts on the socioeconomic environment is called the "region of influence" (ROI). This represents the geographical area, or region, wherein the project-induced changes to the socioeconomic environment will occur. In an analogous sense, the ROI for addressing socioeconomic impacts would be comparable to (1) an air-quality control region for addressing air quality impacts, (2) a watershed for addressing surface-water quantity and quality impacts, or (3) an ecoregion or habitat type for addressing biological impacts.

There are no fixed rules or criteria for determining the geographical location of the ROI. The most frequent practice is to define the ROI in accordance with county boundaries, since the majority of socioeconomic information and data is available on a county basis. A reasonable beginning approach for defining the ROI is to consider the county wherein the project is located plus the contiguous counties. If a project is located in two or more counties, then each of those counties should be considered in the ROI plus all the contiguous counties to the project-counties. For projects involving large-scale construction efforts, temporary construction workers may commute from long distances. Therefore, for some large-scale projects, the ROI should be expanded to include areas beyond the project-counties and their contiguous counties, as appropriate.

A questionnaire checklist which can be used to address the socioeconomic impacts of irrigation projects in developing countries has been developed by Asian Development Bank (1991a). An emphasis is on encouraging public participation in project planning. This checklist could be modified for different types of projects.

The second example is related to socioeconomic impacts associated with wastewater treatment plants and related areawide water-quality planning efforts. Some impact concerns for these types of projects are in Table 14.3. A conceptual example of direct and indirect socioeconomic impacts of a small wastewater treatment plant, and their interrelationship, is shown in Figure 14.3; the description for the example follows (McMahon, 1982, p. 4):

Background: Smithville is a small bedroom community of 5,000 located along Interstate 24 about 20 miles south of Center City. Because of poor soils that limit development with on-site wastewater systems, the town is considering its first wastewater (sewage) treatment facility. About 2,000 of the town's residents will be served by the first phase of the project. Local capital costs for the project will be financed by long-term general obligation bonds that will be paid off by property taxes, benefit assessments, and user charges. Local capital costs will total $3 million. The only two industries in town, two leather tanning firms, presently discharge to the North River. The proposed plan calls for them to discontinue their present direct discharges of tanning wastes and to pretreat their wastes and tie into the new advanced wastewater treatment plant. Because of a shortage of sites along the North River, the plant will be built in an area characterized by large-lot single-family homes. The following numbers refer to Figure 14.3. (1) Sensory = increased visual and noise nuisances. (2) Recreation = loss of public access to boating and canoe launching area. (3) Employment = increase in construction-related and operational employment associated with the plant. (4) Land use = increase in developable residential land in the southern part of town and commercial land around Interstate 24 highway interchange. (5) Public fiscal = increase in capital and operational costs associated with public sewage. (6) Individual land values = increases in property values of proposed sewered land. (7) Individual land values = decrease in land values in the area surrounding the treatment plant. (8) Recreation = increase in demand on other riverfront public boat and canoe areas. (9) Employment = increase in multiplier service-related employment. (10) Housing = increased number of housing units; changes in single-family/multi-family mix in town; and increased rate of development. (11) Population = increased population and rate of growth. (12) Employment = increase in commercial and construction employment. (13) Employment = increase in multiplier service-related employment. (14) Public

TABLE 14.3

REPRESENTATIVE SOCIOECONOMIC ISSUES IN WATER-QUALITY-MANAGEMENT PLANNING EFFORTS

Employment and economic growth

- Increase in construction-related employment for pollution control facilities
- Locational shift of businesses
- Increase in employment for operation and maintenance of pollution control facilities
- Increase in employment for administration, planning, and management of pollution controls

Public fiscal costs

- Increase in capital, operation, and management costs related to public pollution controls
- Increase in revenue from pollution control charges and fees
- Increase in regulatory costs related to private pollution controls

Land use

- Pre-emption of land for pollution control facilities
- Changes in site design
- Changes in use of existing built environment and land uses
- Changes in growth pattern (timing, amount, locations, and type of growth

Public health/safety

- Impacts associated with operation of pollution control facilities involving hazardous wastes
- Impacts associated with improved water quality for drinking and recreational uses
- Impacts associated with strategies that offer multiple benefits in terms of flood prevention, erosion control, environmental sanitation
- Impacts associated with malfunctioning of pollution control facilities

Private cost and benefit incidence

- Increase in pollution abatement costs for firms, developers
- Increase in homeowner user charges
- Increase in property taxes
- Increase in special assessments
- Increase in real estate values

Other public services

- Change in water consumption demands
- Change in water supply availability
- Change in solid waste management
- Change in storm drainage management
- Change in street maintenance

Visual

- Conflicts in fit-with-setting
- Conflicts with visual identity
- Visual nuisances
- Conflicts with views and vistas
- Changes in natural elements

Historic resources

- Changes in the number, type, location, use, and character of historic, archaeological, and architectural resources

Recreation

- Changes in recreational opportunities
- Changes in recreational demand

Source: McMahon, 1982, p. 2.

services = changes in public services demand. (15) Private firm costs = changes in wastewater costs and tax burdens for firms connected to municipal systems. (16) Employment = decreases in local manufacturing employment. (17) Public services = change in water consumption demand. (18) Individual costs = changes in wastewater costs and taxes. (19) Public services = changes in proposed water supply facilities construction.

The final case involves consideration of the secondary (or indirect) effects of development projects causing changes in urban land use. For example, urban land-use changes can cause hydrologic effects in terms of changes in the timing and peak flows of urban runoff, and associated changes in pollutant loadings from nonpoint storm-water runoff (Leopold, 1968).

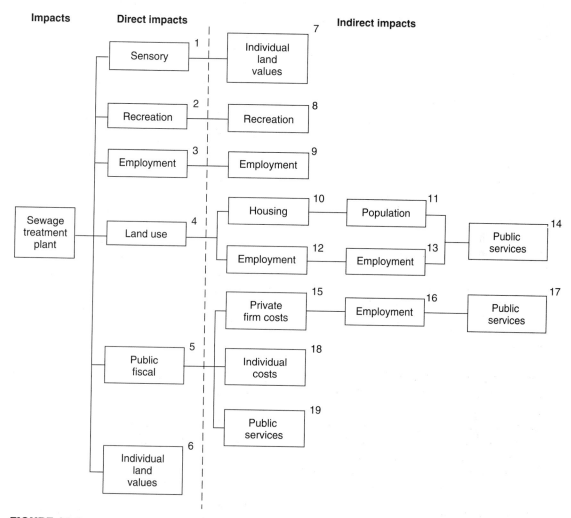

FIGURE 14.3
Direct and Indirect Socioeconomic Impacts Associated with a Wastewater-Treatment Plant in Hypothetical Town of Smithville (McMahon, 1982).

The effects can be addressed relative to the surface-water environment (Chapter 7) or the soil and groundwater environment (Chapter 8).

STEP 2: DESCRIPTION OF EXISTING SOCIOECONOMIC CONDITIONS

The key approach in describing existing conditions is to focus on selected socioeconomic factors expected to be changed by the proposed project. Sources of socioeconomic information can provide primary data (data from specific field studies on selected factors) or secondary data (data from published reports). For example, secondary data can be obtained from the U.S. Bureau of the Census; regional planning agencies such as councils of government; local planning departments; local chambers of commerce; and special studies related to transportation planning, water supply development, water quality management involving wastewater treatment, and solid-waste management.

In addition to the QOL indices mentioned in Chapter 5, other socioeconomic indices could be used as a basis for describing baseline conditions and could serve as a point of reference for delineating project-activity impacts. For example, for projects in developing countries, a human development index (HDI) could provide a relatively simple means of monitoring the overall level of socioeconomic development. The HDI combines the three factors considered to best represent the human condition—that is, life expectancy, literacy, and income (Asian Development Bank, 1991a).

STEP 4: PREDICTION OF SOCIOECONOMIC IMPACTS

The most important technical activity of the SIA process is the prediction of impacts for each of the alternatives being evaluated, including the one that will become the proposed action. Four approaches can be used for predicting impacts: (1) qualitative description, (2) quantitative description, (3) the use of application-specific prediction techniques, and (4) the use of relative comparisons of the effects of alternatives. Application of these approaches for impact prediction will vary depending upon the socioeconomic category and specific factors within the category that might be impacted. "Qualitative description" refers to the approach used by individual professionals or interdisciplinary teams to describe the effects of alternatives based on general knowledge of generic types of impacts and case studies. No quantitative presentations are made.

The approach using "quantitative description" involves use of a numerical technique by either an individual professional or an interdisciplinary team to project impacts based on an understanding of the existing situation and unit-impact information. This approach requires knowledge of effects that have occurred in similar situations and the use of quantitative impact factors.

A "descriptive-checklist" methodology for addressing selected socioeconomic impacts of housing projects in urban or suburban areas is summarized in Table 14.4. As the next phase in this approach, recommended standards or guidelines from pertinent professional organizations or governmental agencies, or temporal or geographical comparisons, could be used to assess the significance of the findings and measurements listed in Table 14.4.

Methodologies have been formulated which can be used to address the social, economic (employment, housing, and property values), fiscal, demographic, and public services impacts of land-development projects. Seven typical social-impact areas (Christensen, 1976) are (1) recreation patterns at public facilities, (2) recreational use of informal outdoor space, (3) shopping opportunities, (4) pedestrian dependency and mobility, (5) perceived quality of the natural environment, (6) personal safety and privacy, and (7) aesthetic and cultural values. Methods for addressing economic-impact factors such as area employment patterns, housing needs by types of units, and resultant impacts on property values are available in Muller (1976). A fundamental issue to consider when addressing fiscal impacts is whether the development will generate revenues in excess of the public expenditures, or vice versa. Fiscal-impact assessments undertaken by developers tend to emphasize the anticipated public gains, while those conducted under the auspices of conservation and citizen groups are more apt to stress adverse public effects (Muller, 1975). Differences in findings may often be the result of locational factors, such as the availability of underutilized services. For example, variations in the state-local fiscal structures often explain why identical growth patterns can lead to fiscal gains for some jurisdictions while causing fiscal losses elsewhere. Muller (1975) analyzed the various data sources and methodologies used in

TABLE 14.4

DESCRIPTIVE-CHECKLIST METHODOLOGY FOR ADDRESSING SOCIOECONOMIC IMPACTS

Category	Questions to identify	Necessary information	Methodology	Findings and measurements
Educational facilities	Can projected enrollments be properly handled in existing or proposed facilities with proper spacing for all activities (including classrooms, recreational areas, and staffing needs)? Will the project impact the pupil/teacher ratio so as to impede the learning process? Is the school located such that it presents a hardship for students' enrollment in terms of too great a travel time or distance or through the existence of safety hazards?	Number and type of housing units; existing pupil/teacher ratio, in local schools; size (number of pupils) of existing school facilities; miles to nearest schools with available capacity; average local bus speed	Estimate numbers, by school age, of students generated by project; compare number of students generated with existing pupil/teacher ratios; compare number of pupils in and size of present schools with number of pupils generated by project; measure distance; calculate travel time, as follows: $$\frac{\text{Distance from project to school}}{\text{Average local bus speed}}$$	Pupil population, by school age, generated by project; impact on quality of education as judged by pupil/teacher ratios; impact on quality of schools as judged by size (number of additional pupils); walking distance to schools; travel time on buses.
Employment	Will the project have an adequate supply of workers and access to workers for various types of jobs? Will the project destroy or relocate any existing jobs?	Number and types of workers generated by project; distance from project to employment opportunities; transit availability and other transportation alternatives; car ownership profiles of future residents; existing jobs on project site	Compare number and type of workers in project to available jobs (number and type) if not already employed; calculate travel time, as follows: $$\frac{\text{Distance from project to employment opportunities}}{\text{Average speed by mode}}$$	Potential unemployment in project; number and type of jobs replaced by project, if any; travel time from project to employment opportunities.
Commercial facilities	Will there be an adequate supply of and access to commercial facilities (existing or proposed) for the project?	Location, type, and size of commercial facilities; income of future residents; transit availability and other transportation alternatives; car ownership profiles of future residents	Compare retail market demand of future residents to available type and size of commercial facilities; calculate travel time, as follows: $$\frac{\text{Distance between project and stores}}{\text{Average speed by mode}}$$	Travel time to commercial facilities; adequacy of existing facilities; projection of need for new facilities

TABLE 14.4

DESCRIPTIVE-CHECKLIST METHODOLOGY FOR ADDRESSING SOCIOECONOMIC IMPACTS (continued)

Category	Questions to identify	Necessary information	Methodology	Findings and measurements
Health care and social services	Are provisions for and access to quality health care and social services adequate to meet the needs of the residents of the project area?	Local type and size of existing facilities; beds or population; socioeconomic characteristics of future population in project area; transportation availability; waiting time for existing services	Project health, hospital-bed, and day-care needs of future project population to compare with those existing facilities; calculate travel time, as follows: Distance to facilities from project ————————————— Average speed by mode	Adequacy of existing facilities—need for new facilities or expansion; travel time; emergency travel time.
Liquid-waste disposal	Is provision for sewage capacity adequate to meet the needs of the project without exceeding water quality standards? Will the project be exposed to nuisances and odors associated with wastewater-treatment plants?	Number of people or dwelling units in project; type and size of other facilities, e.g., commercial sites; offices in project; lot size and soil conditions; location of treatment facilities; Section 208 water-quality management plan	Compare future sewage needs with existing availability of public services or lot and soil suitability (percolation test) for septic tanks.	Potential water quality degradation; capacity (gal. per capita per day).
Solid-waste disposal	Is there provision for environmentally sound disposal of solid wastes generated by the project?	Number of people in project; type and size of other facilities, e.g., commercial, office, health care, etc.; existing disposal methods and capacities	Compare quantity (lb per day) and type of solid waste generated with available capacity for disposal.	Potential overloads on existing solid-waste disposal mechanisms— need for new facility or additional capacity (lb. per person per day); distance to sanitary landfill.
Water supply	Are there provisions for adequate quantity and quality of water supply to meet the needs of the project?	Number of people or residential units in project; type and size of other facilities (commercial, office, etc.); capacity of local public water system; existing water quality	Compare water supply needs of project (gal. per day) to available excess capacity of local systems.	Gal. per day
Storm-water drainage	Will provision for storm-water drainage be adequate to prevent downstream flooding and to meet U.S. HUD, state, or local standards?	Local runoff intensity (in per hr); runoff coefficient applicable to local situation; area (acres, mi², etc.) of catch basin; existing storm-water drainage system	Calculate rate of runoff; prepare unit hydrograph. Rate of runoff = runoff intensity × area of catch basin × runoff coefficient	Design flow rate (millions of gal. per day) to estimate need for and size of drainage system.

Police	Will the present or proposed police system adequately protect the additional population and facilities generated by the project?	Number and type of people in project; existing police/population ratios; existing police response time (min)	Compare police needs for new population to existing resources.	Impact on existing police service; need for additional force to maintain or improve existing service.
Fire protection	Will the project be provided with adequate fire protection services?	Location of nearest fire station; type of equipment; staff; response time; type, number, and density of buildings in project; existing available fire flow (gal. per min)	Determine distance from project to fire station; divide by average speed to obtain required fire flow (estimated).	Response time to fire; impact of project on available fire flow and of existing fire flow impact on project; need for better fire protection.
Recreation	Will the project have access to adequate facilities to meet the recreational needs of the residents?	Location, size, type, and capacity of recreational facilities; number and socioeconomic characteristics of future population in project	Determine distance to facilities with necessary capacity; compare available types and capacity of recreation facilities with needs of population.	Potential impact on available facilities and impact on project if needs not met; need for additional facilities.
Transportation	Are the transportation facilities which serve the project part of a well-integrated multimodel system and are they adequate to accommodate the project's travel demands?	Location of travelway; width and type of travelway; frequency of bus service; socioeconomic characteristics of resident population	Locate transportation facilities in relation to project; document size—maps or field measurements, e.g., size, number of lanes; estimate optimal peak-hour flows; estimate trips by mode, based on type and density of land use and on socioeconomic characteristics of resident population; compare needs to existing capacity for all modes, compare needs to existing levels of service.	Potential impact on existing facilities; need for additional or expanded facilities—all modes.
Cultural facilities	Will there be cultural facilities available to the project residents?	Location, size, type, and capacity of facilities; number and socioeconomic characteristics of project population	Determine distance to facilities with capacity; compare available types and capacity with needs of population.	Impact on available facilities; impact on residents of project if facilities inadequate.

Source: Compiled using data from Voorhees and Associates, 1975.

fiscal-impact studies and indicated which ones are preferable from the community perspective.

An economic- and demographic-impact assessment methodology for water-resources projects is available in Chalmers and Anderson (1977). A comprehensive methodology for addressing 389 socioeconomic factors related to large water-resources development projects has been developed in Fitzsimmons, Stuart, and Wolff (1975). Table 14.5 depicts a qualitative-descriptive approach for delineating existing conditions and the anticipated socioeconomic impacts of a project or activity; the approach could also be used for describing physical-chemical, biological, and cultural aspects of the environment.

''Use of application-specific or model-based techniques'' involves predicting likely effects by using an explicit and predefined relationship, such as a mathematical model. In mathematical models, for example, mathematical relationships between system variables are used to describe the way a socioeconomic system will react to an external influence. In the SIA context, mathematical models can be divided into those models which are empirical, or ''black box'' models—that is, where the relationships between inputs and outputs are established from analysis of observations in the socioeconomic environment, and those models which are ''internally descriptive''—that is, where the mathematical relationships within the model are based on some understanding of the mechanisms of the processes occurring in the socioeconomic environment (Canter, 1983).

Demographic-impact prediction involves the use of techniques for determining the number, distribution, and characteristics of the people moving into or away from the impact area, and is needed to assess public services demands, fiscal impacts, and social impacts. Population-projection techniques can be divided into five types: (1) extrapolative, curve-fitting, and regression-based techniques; (2) ratio-based techniques; (3) land-use-based techniques; (4) economic-based techniques; and (5) cohort component techniques. No matter which technique is used, a series of key steps is involved: (1) delineation of the impact area; (2) projection of baseline population for the impacted region and subregional areas; (3) determination of directly and indirectly project-related in-migrations of workers, including consideration of *(a)* characteristics of required employment and of available labor (unemployment, underemployment, skill levels, and commuting patterns), *(b)* indirect/direct worker ratios, and *(c)* local/nonlocal worker ratios; (4) projections of the geographical distribution (settlement patterns) of in-migrating workers; and (5) determination of the

---

TABLE 14.5

QUALITATIVE-DESCRIPTIVE APPROACH FOR ADDRESSING SOCIOECONOMIC IMPACTS

Socioeconomic factor	Existing conditions[a]	Nature of future conditions			
		Impact of no action[b]	Resultant conditions[a]	Impact of project[b]	Resultant conditions[a]
a	A	−	BA	+	AA
b	BA	−	BA	+	A
c	AA	−	AA	−	A
d	BA	+	A	0	A
⋮	⋮	⋮			

[a]BA = below average in terms of relevant standard, guideline, temporal average, or spatial average; A = average; AA = above average
[b]++ = very positive; + = positive; 0 = neutral; − = negative; − − = very negative.

demographic characteristics of in-migrating workers and their dependents to establish project-related population impacts by site. A series of projections that bracket the likely range of impacts should be included in whatever technique is used.

Two main types of regional-forecast economic models are available: export-base models and input-output techniques. Additional methods include intersectional flows or from-to models, econometric models, and multiregional models. Some comprehensive models have been developed to provide linkages between categories of socioeconomic impacts—such as demographic, economic, social, fiscal, and public services. Export-base models use employment or income as major indicators of economic activity and are relatively simple and low-cost. These models can be implemented using only secondary data, and the calculations involved do not necessarily require a computer. However, these models provide only limited aggregate information on the economic impacts of a project and often oversimplify employment and differences among industries.

Regional input-output models have been widely used for regional impact studies and for the assessment of economic impacts at the local and state levels. These types of models provide more-detailed impact estimates than export-base models; however, both types share many of the same basic assumptions. The major limitation to the use of input-output models is the need for an extensive amount of data collection and analysis.

One example of such an input-output model is the Economic Impact Forecast System (EIFS) developed by the U.S. Army Construction Engineering Research Laboratory. EIFS has been used by federal agencies, private consultants, and local governments (Knaap, 1992). It provides users with selected statistics regarding the socioeconomic characteristics of any county or multicounty area in the United States, and affords them an analytical process for assessing the magnitude and significance of potential socioeconomic impacts of proposed projects or activities (Defillo, 1990).

The basis of the analytical capabilities in EIFS is the calculation of multipliers which estimate impacts in terms of employment, income, or sales resulting from a change in local expenditures and/or employment. The system calculates both income and employment multipliers, which are used to estimate the effects of an action on the local economy (Vester, 1990). In calculating the multipliers, EIFS uses an economic-base model and its ratios of basic (or export) activity and nonbasic (or service) activity. "Export" in this context is defined as production consumed outside the region, while "service" indicates production consumed locally. Goods must be imported and, to pay for these goods, the region must provide exports. According to the export-base theory, a region's export industries are its economic foundation, while other industries service these industries. The theory holds that the ratio of total employment or income to export employment or income is measurable and sufficiently stable so that future changes in totals can be forecast from changes in exports (Vester, 1990).

The multipliers can also be interpreted as indices of total impacts on the regional economy resulting from a unit change in its export sector—for example, the increase in total employment resulting from an increase in a certain number of export jobs. EIFS adjusts multipliers using a location-quotient approach based on the concentration of industries within the region (Vester, 1990).

EIFS also makes available the "rational threshold value" (RTV), which facilitates the interpretation of the significance of the socioeconomic impacts of proposed projects. The RTV represents deviations of each indicator, such as income, from its average rate of change over the period of time for which data are available, usually 15 to 20 years (Defillo, 1990).

Specific models have also been developed for predicting particular impacts. For example, as noted earlier, for large construction projects, the supply of workers may extend well beyond the local area. A county-based model of construction labor markets has been developed to allow for prediction of the commuting and moving patterns of employees on large construction projects (Blair, Garey, and Stevenson, 1982).

Impact prediction may also involve "relative comparisons of the effects of alternatives."

Quantitative impact information is not necessary in this approach.

Prediction of socioeconomic impacts will probably require the use of each of the four approaches, depending upon the particular socioeconomic category and factor being addressed. For a given category or factor, several prediction approaches and techniques could be available for usage. Therefore, it may be necessary to select the most appropriate approach (or technique) for a specific usage. Table 14.6 identifies eight criteria for consideration in the se-

---

TABLE 14.6

CRITERIA FOR CHOOSING SOCIOECONOMIC-IMPACT-PREDICTION APPROACHES AND/OR TECHNIQUES

Criteria	Definition, measurement
I. Practicality	
A. Substantive relevancy	Appropriateness for the program, policy, and/or project. Has the approach (or technique) been used previously to assess impacts of similar programs, policies, and/or projects?
B. Policy relevancy	Does the approach (or technique) provide information which can be useful, particularly with respect to avoiding or mitigating impacts? Is it feasible to develop policy strategies to address these impacts? Approaches would need to address impacts of categories which can be mitigated, and be capable of producing timely predictions, etc.
C. Acceptability	Is the model acceptable to those publics likely to be impacted? Does it include the substantive areas of concern to local populations? Can be determined by previous experience, public hearings, or survey techniques.
D. Face validity	Is the approach credible in the professional research community or with others with experience in assessing socioeconomic impacts? Can validity be determined by the use of advisory groups or external review panels.
E. Applicability	Ease of using or implementing this approach. Do the data exist; are analysis routines easy to use, etc.? Must be determined by professional judgment.
F. Flexibility	Can the approach/technique be used for different substantive impact areas, for different geographic areas, different economic and social conditions, etc.?
II. Technical quality	
A. Accuracy	Is the approach likely to provide results within acceptable error ranges? Has it been the subject of previous reliability and validity studies? Has it presented significant problems in previous uses?
B. Completeness	Does the approach include a complete set of impacts? Can it be easily combined with other approaches to provide a comprehensive picture?

lection of socioeconomic-impact-prediction approaches. These criteria include both practical and technical-quality considerations. On the practical side, the approach must first be credible, demonstrating (1) substantive relevancy to the program, policy, and/or project; (2) policy relevancy in terms of providing useful information regarding impacts that can be acted upon (i.e., mitigable); (3) acceptability to affected publics; and (4) face validity to experts or professionals. If the approach is deemed credible according to these four criteria, additional desirable characteristics include how easily it can be used (applicability) and whether it can be used for different conditions and geographic areas (flexibility). Technical-quality criteria include both accuracy and completeness. Approaches should be able to provide results within acceptable error ranges, and they should provide a relatively comprehensive assessment of socioeconomic impacts.

The criteria listed in Table 14.6 can be applied in a generic sense for the selection of socioeconomic-impact-prediction approaches (or techniques), with the following possible results:

1. *Economic-demographic impacts*—Use mathematical models.
2. *Public services impacts*—Use quantitative description based on unit-impact information.
3. *Social impacts*—Use quantitative description based on unit-impact information.
4. *Fiscal impacts*—Use quantitative description based on unit-impact information.
5. *QOL changes*—Use qualitative description, quantitative description based on unit-impact information, and/or relative comparisons of the effects of alternatives.

STEP 5: ASSESSMENT OF SOCIOECONOMIC IMPACTS

Assessment of the significance of predicted changes in the socioeconomic environment requires the considerable exercise of professional judgment. Every attempt should be made to use systematic and scientific rationale for significance assessment. The conceptual framework for a systematic procedure is illustrated in Figure 14.4. The first consideration relates to the application of screening criteria. Following this, the interpretation of changes in socioeconomic features can be made based on several approaches: one would be to consider the resultant impact information relative to recommended professional and institutional standards and criteria for various socioeconomic features. Another approach would be to compare the resultant information to geographical averages or temporal trends (ranging from local or county-

FIGURE 14.4
Procedure for Assessing Socioeconomic Impact Significance.

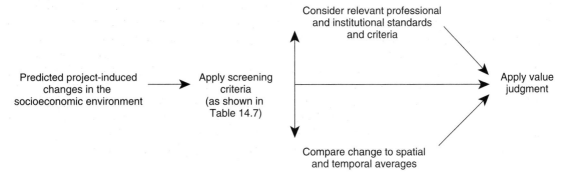

level, to regional, to state, to multistate and national averages or trends) for the same feature. Another approach for considering the significance of predicted socioeconomic changes is to examine the relationship of the information to original design standards for the socioeconomic resource being evaluated. Finally, professional and public inputs can be used to establish value judgments. Documentation of the rationale to be used for assessing the significance of impacts should be developed early in the EIA process.

Application of Screening Criteria

The predicted project-induced changes could first be subjected to screening criteria shown in Table 14.7. These general concerns (sets of characteristics) need to be addressed in order to assess impact significance: (1) the nature of the impact, (2) the impact's absolute and perceived severity, and (3) the potential for mitigation. The first set is to adequately characterize the impact according to its probability of occurrence, the number of people or groups affected, its geographic pervasiveness, and its duration. Thus, the first four screening criteria help to determine if the impact is likely, who will be affected, where, and for how long. For most impact assessments, each of these criteria is likely to be of equal importance. If a particular impact rates low on each of the first four criteria, it is not likely to occur, it would affect few people, its effects would be very isolated, and it would not last long; then it is highly unlikely that evaluation in terms of subsequent criteria will show the impact to be significant. However, if an impact rates high on any of these four screening criteria, then additional questions about its severity and potential for mitigation will need to be used.

When impacts have been determined to be likely, pervasive across populations or an area, or long-lasting, it is desirable to determine their severity. This area of concern includes a perceptual component or "local sensitivity" criteria—that is, the extent to which the local community is aware of the impact and perceives it to be threatening to its cultural, social, or economic well-being. "Severity" also addresses absolute magnitude—for example, whether the impact creates a rapid or unpredictable change, creates a change significant for its size alone, or results in a change that exceeds a commonly accepted standard or threshold value.

Third, impacts should be evaluated for mitigation potential (also related to step 6). That is, impacts which may be relatively moderate or minor in some contexts could be major in other contexts in which the existing system for mitigation already approaches or exceeds capacity. For example, relatively small economic costs can create significant financial difficulties in areas which have experienced other recent demands for services. Whereas in other communities, existing infrastructure capacity (for example, water and sewage systems, wastewater treatment plants, roads, etc.) may greatly exceed demand; thus, new populations could be more easily accommodated. The availability of relocation assistance for persons being displaced by projects needs to be considered along with provisions for impact funds to aid local communities in meeting infrastructure needs. Public projects typically have both types of funding, and private development projects can be encouraged to provide such financial assistance. Finally, some impacts are significant because they are essentially irreversible—for example, the loss of small-town value systems. A key criterion for assessing such impacts is the relevant institutional capacity for responding to the determined impacts.

Consideration of Relevant Standards and Criteria

Standards and criteria from professional groups and governmental institutions should be considered in assessing socioeconomic-impact significance (also related to step 3). Numerous professional organizations publish standards and guidelines for various social activities. The Ur-

TABLE 14.7

SCREENING CRITERIA FOR ASSESSING IMPACT SIGNIFICANCE

Criteria	Definition, measurement
I. Nature of the impact	
A. Probability of occurrence	Likelihood that a given impact will occur as a result of a program, policy, and/or project. For many socioeconomic impacts, qualitative assessments would be appropriate (high, medium, low).
B. People affected	How pervasive will the impact be across the population? This criteria should be used to assess both the percentage of the population affected and the extent to which it will affect different demographic groups.
C. Geographic pervasiveness	The extent to which the impact is experienced across a widespread area. Can frequently be addressed by mapping or use of data sources which are geographically specific (census data).
D. Duration	How long the impact will last, assuming no direct public or private sector attempts to mitigate. Can be addressed by identifying short-term, long-term, and permanent impacts.
II. Severity	
A. Local sensitivity	To what extent is the local population aware of the impact? Is it perceived to be significant? Has it been a source of previous concern in the community? Are there any organized interest groups likely to be mobilized by the impact?
B. Magnitude	How serious is the impact? Does it cause a large change over baseline conditions (e.g., will crime rates double)? Does it cause a rapid rate of change (large changes over a short time period)? Will these changes exceed local capacity to address or incorporate the change? Does it create a change which is unacceptable? Does it exceed a recognized threshold value?
III. Potential for mitigation	
A. Reversibility	How long will it take to mitigate the impact by natural or man-induced means? Is it reversible; if so, can it be reversed in the short term or the long term?
B. Economic costs	How much will it cost to mitigate this impact? How soon will finances be needed to address this impact?
C. Institutional capacity	What is the current institutional capacity for addressing the impact? Is there an existing legal, regulatory, or service structure? Is there excess capacity, or is the capacity already overloaded? Can the primary level of government (e.g., local government) deal with the impact or does it require other levels or the private sector?

ban Land Institute (ULI), for example, regularly publishes handbooks which address community development needs and suggested levels of service. These standards, commonly used in regional and city planning, are based on experience, professional judgment, and accepted land-use models. An impact could be judged significant if it is not in compliance with them.

Some examples of professional standards or criteria will be presented to illustrate their availability. In some cases, local or regional standards or criteria would be more appropriate for usage than national information. For public education impacts, general recommendations on pupil/teacher ratios, school size, and travel times are in Table 14.8 (Voorhees and Associates, 1975).

Despite the fact that there are numerous professional standards and criteria, they are not always appropriate to apply as indicators of significant impact. For example, the ULI recommends that every neighborhood of 2,500 to 40,000 persons should have a shopping center of between 30,000 and 100,000 ft^2 within a 6-mi area; a community of 40,000 to 150,000 should have one of between 100,000 and 300,000 ft^2, and so on (McKeever, Griffen, and Spink, Jr., 1977). However, larger centers attract shoppers from wide substate regions. Thus, an influx of people into an area may necessitate not only the development of neighborhood shopping centers, but the expansion of these centers into regional centers. Further, the expansion of a shopping center would require other changes, such as new parking facilities and access roads. The use of planning standards in isolation may not enable an assessment of the full impact of a proposed change. They must be used in conjunction with broader regional and city perspectives and plans.

Further, planning standards are limited to extremely basic ideas of localized physical needs—which may be incompatible with a community's prevailing philosophy. Assume, for instance, that the ULI published maximum suggested sizes for grade school classrooms. If this were applied to a community which operated its school system as an open-classroom system with no formal partitions between classes, the standard would make little sense. Therefore, such a standard could not be used to judge accurately the effects of increased grade school enrollment on the community's educational system.

Numerous local or national legal and regulatory standards from various governmental institutions have been used to assess the significance of socioeconomic impacts. In this context, an impact should be judged significant if it violates legal or regulatory standards.

TABLE 14.8

RECOMMENDED STANDARDS AND GUIDELINES FOR EDUCATIONAL FACILITIES

Issue	Standard or guideline and source		
Pupil/teacher ratio	22:1, for elementary schools, is recommended as best. The more-specialized subject areas at the jr. and sr. high school levels make standardized class size recommendations difficult to determine.		
School size		No. of students	
		Minimum	Maximum
	Elementary	200–300	500–700
	Jr. high	300–500	1,000–1,200
	Sr. high	1,000–1,500	1,500–2,000
Travel time		Access	
		Minimum walking distance	Maximum travel time, bus
	Elementary	3/4 mi	1/2 hr
	Jr. high	1-1/2 mi	1 hr
	Sr. high	2 mi	1 hr

Source: Compiled using data from Voorhees and Associates, 1975.

Comparison with Spatial and Temporal Averages

Regional and national averages also have been used as threshold values for determining socioeconomic-impact significance. An impact is judged significant if it causes a predetermined indicator—physicians per 1,000 people, for example—to deviate from the national or regional average. The problem with using these thresholds is that they do not account for local differences. Frequently, a national average will be completely inappropriate to a small rural community, which is likely to be unrepresentative of the nation as a whole. Further, numerous communities are likely to be below a national or regional development average to begin with. Significant impacts may move them closer to the average but still keep them below it. Thus, they might have undergone major negative changes which the use of a national average measure would not reveal.

Many studies have used measures of previous community-growth rates and absorption capacities to assess the significance of proposed changes. That is, the community's previous ability to deal with a particular rate of change is examined. If the proposed change exceeds this rate and absorption capacity, it is judged to be significant. Threshold levels based on historic rates of growth are, of course, highly site-specific. They can only be applied to regions or communities similar to the one for which they were developed. The major problem with this approach to assessing significance is that it does not take into consideration the wishes of local communities. For example, change may produce rates of growth unparalleled in the history of a particular community, but to a growth-oriented community the negative significance of this may be minimal.

Value Judgment

The final aspect of assessing socioeconomic-impact significance is to apply an overall value judgment. (Professional judgment has to be used in all aspects of the procedure, including applying screening criteria, considering relevant professional and institutional standards and criteria, and comparing project-induced changes to spatial and temporal averages.) Formalized approaches for generating significance-threshold values can be based on judgments made by either experts (using, for example, Delphi techniques) or affected publics (by survey research or the nominal-group technique). In each case, a group familiar with the proposed change is asked to determine a level beyond which the impact would be significant. These thresholds can serve as useful standards in determining the significance of impacts on a community's QOL. However, such approaches are not frequently used and care must be taken to ensure that the thresholds do not simply reflect the agenda of a small, vocal minority of activists.

EDUCATION SERVICES IMPACTS

The school systems within the ROI of a project can be subjected to specific, significant socioeconomic impacts. Education services, in this context, include elementary and secondary educational programs provided by public school systems. Since participation in private schools, preschools, and adult education programs is an option (i.e., not mandated by law), these institutions are not addressed in this section. However, it may be desirable to consider the impacts upon these options for certain projects or activities, using the same approach as described for grades kindergarten (K) through 12 in the public school system.

The specific method described, an application of the six-step methodology, is based primarily on the work done by Henningson, Durham, and Richardson (1982). Some modifications have been made to clarify the method used by their computer model. An example problem has been included for illustrative purposes. The ratios (or multipliers) used to convert predicted population changes into demands on the educational system are based on U.S. Department of Energy (1978) and Wisconsin State Planning Office (1975). The impacts on the education system should be calculated on a yearly basis for each school district within the ROI for each alternative.

Step 1: Identification of Potential Impacts on the Education System

Assume that potential impacts have been identified based on the number of school-age children moving into the study area during the construction and operational phases of a project and the size of the local school system.

Step 2: Description of the Existing Condition of the Education System

The first activity is to procure pertinent data and information that will enable the completion of subsequent steps. There are many possible sources of information; the ones actually used will, of necessity, depend on the circumstances encountered while conducting the study. The possible sources of data are as follows: (1) primary data sources—local school districts and postsecondary school officials; and (2) secondary data sources—the state office of education, U.S. Department of Education, state planning departments, and state research bureaus. National and regional standards can be procured from the U.S. Department of Education and the appropriate state office of education.

Data from other socioeconomic-impact categories which would be relevant for education system impacts include the breakdowns of predicted project-induced or -affected households by employment category, year, community (school district), county, and ROI.

Existing conditions should be discussed not only in current terms, but also in historical-trend terms. Any baseline critical or stressed educational categories should be highlighted. Selected indicators which could be included in the description of baseline educational conditions are as follows:

1. *Level*—Number of instructional personnel (teachers, principals, supervisors of instruction, librarians, teachers aids, library aids, and guidance and psychological personnel)
2. *Accessibility*—Geographic location of available personnel and school facility space in relation to the location of project demands
3. *Quality*—Condition of facilities, school capacity (square footage; by grade category, that is, K through 6, 7 through 9, 10 through 12), capability of the school system to meet current needs, plans, and projections for the future

Steps 3 and 4: Assembly of Pertinent Standards or Guidelines and Prediction of Impacts

The required information for these steps includes (1) the projected number of baseline households and (2) the number of dependents under 21 years old per household. The method which should be used to predict future conditions without the project for each school district within the ROI for each year of construction and operation is as follows (Canter, Atkinson, and Leistritz, 1985, p. 87):

1. Multiply projected number of baseline households by number of dependents under 21 years old per household to determine total number of dependents under 21 years of age.
2. Multiply total number of dependents by 0.65 (the assumption is that there is an equal distribution of dependents by age cohort) to determine the total number of school age children. Note: It may be necessary to use something other than an equal distribution of dependents if an area has a substantial demographic bulge in some portion of the under 21-year old population. If a bulge is taken into account, it must be projected forward for each year, so that the future dependent population will reflect the correct distribution.

3. Subdivide school age children into the following categories (see Note 1) unless a bulge in the distribution is used:

$$
\left.\begin{array}{ll}
\text{K– 6:} & 54\% \\
\text{7– 9:} & 23\% \\
\text{10–12:} & 23\%
\end{array}\right\} \quad \text{Relationship 1}
$$

4. Determine the number of instructional personnel required by applying the following pupil/teacher ratios (see Note 1):

$$
\left.\begin{array}{ll}
\text{K– 6:} & 22{:}1 \\
\text{7– 9:} & 20{:}1 \\
\text{10–12:} & 19{:}1
\end{array}\right\} \quad \text{Relationship 2}
$$

5. Determine the school facility needs based on the following per pupil space needs (see Note 1):

$$
\left.\begin{array}{ll}
\text{K– 6:} & 90 \text{ ft}^2 \\
\text{7– 9:} & 120 \text{ ft}^2 \\
\text{10–12:} & 150 \text{ ft}^2
\end{array}\right\} \quad \text{Relationship 3}
$$

The required information to predict future conditions with the project includes (1) the number of project-induced households in each employment category (for example, construction workers, operations personnel, and indirect workers) and (2) the number of dependents under 21 years old per household. The prediction method for future-with-project conditions which should be used to calculate impacts for each school district within the ROI for each year of construction and operation is as follows (Canter, Atkinson, and Leistritz, 1985, p. 88)

1. Determine the total number of dependents in each household by multiplying the number of households in each category by the following national ratios:

Construction workers:	1.60 dependents/household
Operations workers:	1.40 dependents/household
Indirect workers:	1.40 dependents/household

2. Determine the total number of school age children by multiplying the total number of dependents by 0.65.

3. Subdivide school-age children into the following categories (see Note 1):

K– 6:	54%
7– 9:	23%
10–12:	23%

4. Determine the number of instructional personnel required by applying the following pupil-to-teacher ratios (see Note 1):

K– 6:	22:1
7– 9:	20:1
10–12:	19:1

5. Determine the school facility needs based on the following per pupil space needs (see Note 1):

K– 6:	90 ft^2
7– 9:	120 ft^2
10–12:	150 ft^2

Step 5: Assessment of Predicted Impacts

Conducting an assessment of predicted impacts on the education system involves two phases. First, the predicted impacts should be converted into appropriate measures of impact; second, the measures and/or other ratios or data must be assessed in terms of the significance upon the community under consideration. Four measures

---

Note 1: Values for these parameters can be selected on the basis of (1) current local educational service levels or (2) general planning standards. The use of local service standards, as the name implies, assumes that new persons moving to an area demand and use services at the same level as existing residents. The use of general standards, such as those for the nation, a region, or a state, is based on the premise that new residents are likely to bring demands typical of populations in other areas rather than those typical of the population of the local area. In large part, then, the utility of employing general standards versus local standards depends on the disparity between local residents' and new residents' service demands and on the likely influence of new residents in changing local service demands (Leistritz and Murdock, 1981, pp. 87–88).

The advantages of local standards are that they are clearly applicable to the ROI and have high credibility with local decision makers. The disadvantages are that they may reflect already insufficient service levels and may not accurately reflect the impacts of new service demands. The advantages of general standards lie in their widespread use and acceptance, their ready availability and ease of application, and their grounding in analytical analyses. The disadvantages are that they would be inapplicable to some areas, particularly rural areas, that are not similar to the area for which they were developed (Leistritz and Murdock, 1981, pp. 87–88).

In most cases local standards are preferred, unless the growth induced by the project would transform the nature of the school system or an assessment of communities with different characteristics is being made.

of impacts can be used (Canter, Atkinson, and Leistritz, 1985):

Measure 1
The first measure expresses the percent change in requirements over baseline condition for the peak year and long term:

$$\text{Percent change} = \frac{(B - A)}{A} (100) \quad (14.1)$$

where A = level (or quality) required for baseline and B = level (or quality) required for project plus baseline.

Equation (14.1) can be used to determine (1) percent change in personnel (teacher) requirements during peak year, (2) percent change in personnel (teacher) requirements during long term (operational phase), (3) percent change in facility requirements during peak year, and (4) percent change in facility requirements during long term (operational phase).

Measure 2
The second measure expresses the annual rate of change in requirements from initial year to peak year:

$$\text{Rate of change} = \frac{\text{percent change to peak year}}{\text{years from initiation to peak}} \quad (14.2)$$

Equation (14.2) can be used to determine (1) the rate of change in personnel requirements and (2) the rate of change in facility requirements.

Measure 3
The third measure expresses temporary requirements. "Temporary need" is the percent difference between peak-year needs and long-term needs. This can be addressed by determining the temporary need for personnel requirements and the temporary need for facility requirements.

Measure 4
The fourth measure is a comparison between demand for services and supply of services. An estimate of the supply of personnel (teachers) and facilities during the peak year and the long term is needed. Local school district officials can be consulted to obtain the estimates. The following comparisons should be made, as appropriate:

a. Determine the percent difference between supply and demand for personnel requirements during the peak year.
b. Determine the percent difference between supply and demand for personnel requirements during the long term (operational phase).
c. Determine the percent difference between supply and demand for facilities during the peak year.
d. Determine the percent difference between supply and demand for facilities during the long term.

The measures of impacts should be described in a quantitative manner which represents the overall effect of the proposed project on the educational system.

The final phase of step 5 involves determining the significance of the above-calculated measures (or other appropriate ratios or data). A basic approach is to apply the professional judgment of experts and other qualified personnel in order to determine the significance of a project's impacts on a community's educational system. Comparisons of existing and predicted conditions to the norms, as suggested by relationships 2 and 3 (calculated in step 4), can be useful.

The following example is presented to demonstrate the quantification of the impact of a proposed project on an educational system:

Background
Assume that a project is planned for an area which contains school district A. Once the project is ready for the operational phase,

there will be 150 jobs to be filled by in-migrating workers (in excess of those filled by current residents). These 150 long-term jobs will be a combination of operations-related and indirect jobs and will result in a total of $150 \times 1.4 = 210$ total dependents. Of the 210 total dependents, 65 percent, or 137 dependents, will be of school age. Therefore, the long-term impact of the project will be an additional 137 children in the school system.

During the construction phase of the project, 650 in-migrating construction workers will be required during the peak year, which will occur 3 yr after initiation. Of the 650 workers, 25 percent are assumed to be unaccompanied by dependents. Therefore, 75 percent of the workers, or 488, will bring $488 \times 1.6 = 781$ dependents into the area. Of the 781 dependents, 65 percent, or 508 students, will impact the school system during the peak year. This project is typical of major energy development or military development projects near smaller communities.

The local school officials have estimated that the baseline number of students in school district A will be 1,000 during the peak year of construction. There will be 26 elementary school teachers, 13 junior high teachers, and 13 high school teachers available. Expansion of school facilities is not planned during the first 3 yr of project construction; therefore, the current facility supply will be the same as peak-year supply; the supply is 50,000 ft^2 of elementary school space, 30,000 ft^2 of junior high school space, and 40,000 ft^2 of high school space.

The project is to be completed and ready for full operation 6 yr after initiation. The local school officials have estimated that the baseline student population in 6 yr will be 1,120. The elementary school and the junior high school will have an additional 10,000 ft^2 each

added to them. Also, there will be 28 elementary school teachers, 15 junior high teachers, and 13 high school teachers available in 6 yr.

This information and the associated calculations are summarized in Table 14.9. The hypothetical project is estimated to produce the following measures of impacts on school district A:

Measure 1
a. Percent changes in personnel requirements over baseline [Equation (14.1)] are

> Peak-year: 46.9%
> Long-term: 11.1%

b. Percent changes in facility requirements over baseline [Equation (14.1)] are

> Peak-year: 50.8%
> Long-term: 12.1%

Measure 2
Annual rate of change to peak [Equation (14.2)] is

> Personnel: 15.6%
> Facility: 16.9%

Measure 3
Temporary requirements expressed as percent difference between peak-year and long-term needs are

> Personnel: 20.0%
> Facility: 20.0%

Measure 4
a. Percent personnel supply-demand differences are

> Peak-year: 38.5%
> Long-term: 9.1%

b. Percent facility supply-demand differences are

TABLE 14.9

HYPOTHETICAL SCHOOL DISTRICT A, AND THE EFFECTS OF A HYPOTHETICAL PROJECT

Peak year[a]

Category	Students[c]	Personnel Demand[d]	Personnel Supply[e]	Facility, ft² Demand[f]	Facility, ft² Supply[e]
			Baseline		
K–6	540	25	26	48,600	50,000
7–9	230	12	13	27,600	30,000
10–12	230	12	13	34,500	40,000
Total	1,000	49	52	110,700	120,000
		Project-induced plus baseline			
K–6	814	37	26	73,260	50,000
7–9	347	17	13	41,640	30,000
10–12	347	18	13	52,050	40,000
Total	1,508	72	52	166,950	120,000

Long-term[b]

Students[e]	Personnel Demand[d]	Personnel Supply[e]	Facility, ft² Reqts[f]	Facility, ft² Supply[e]
		Baseline		
604	28	28	54,450	60,000
258	13	15	30,960	40,000
258	13	13	38,700	40,000
1,120	54	56	124,110	140,000
	Project-induced plus baseline			
679	31	28	61,110	60,000
289	14	15	34,680	40,000
289	15	13	43,350	40,000
1,257	60	56	139,140	140,000

[a]Time from initiation of project construction to peak year = 3 yr.
[b]Refers to 6 yrs after initiation of construction; project will be in operation at this time.
[c]Relationship 1.
[d]Relationship 2.
[e]Estimated by local school officials.
[f]Relationship 3.

Peak-year: 39 .1%
Long-term: −0.7%

Relative to impact significance, the above measures can be considered in terms of absolute changes during both the construction and operational phases. The pupil-to-teacher ratios and facility sizes shown in Table 14.9 can be compared to the norms in relationships 2 and 3.

Step 6: Identification and Incorporation of Mitigation Measures

Examples of mitigation measures, used singly or in combination, for construction-phase impacts on school systems include the following: (1) accelerated construction of new school facilities and hiring of new teachers, (2) year-round school program (phased with construction workers and project workers), (3) operation of two shifts in school, (4) distribution of related impacts by planning for construction workers to live in other towns, (5) utilization of temporary buildings, (6) utilization of videotapes in a self-paced instruction mode, (7) making better use of the empty spaces in school system or other community facilities, (8) coordination of plans with adjacent school districts, and (9) provision for financial incentives to encourage construction workers not to bring families. One or more of these measures could be used in the example data in Table 14.9.

TRAFFIC AND TRANSPORTATION-SYSTEM IMPACTS

Development projects or other activities frequently have impacts on local and regional traffic patterns and transportation systems. The conceptual approach depicted in Figure 14.2 can be applied to these specific impacts.

Step 1: Identification of Potential Traffic and Transportation-System Impacts

The first step is to determine the potential impacts of the proposed project on local traffic and/or the transportation system in the ROI. Examples of the key transportation impacts which might occur include (1) increases or decreases in local-area or regional traffic situations, (2) temporal changes in local-area or regional traffic situations (daily, weekly, monthly, and/or seasonally), (3) construction-phase disruptions of existing local-area or regional traffic patterns, and (4) increases or decreases in commuting times and congestion in the local area and/or region.

Quantitative information should be aggregated on expected local and regional traffic changes (increases or decreases) which might occur as a result of the construction and/or operation of the proposed project. Particular attention should be given to the timing (daily, weekly, monthly, and/or seasonally) of the expected changes. It is anticipated that the project proponent (or contracted proponent) would have such information, or, if no such data exists, this information could be developed during discussions with the project proponent.

Step 2: Documentation of Baseline Traffic Information

Certain basic information on the traffic and the transportation system in the vicinity of a proposed project or activity is necessary for describing the affected environment or baseline conditions. Key information includes the following (U.S. Army Construction Engineering Research Laboratory, 1989, vol. 3): (1) the type of transportation network, its conditions, and frequency of its use; (2) the type and purpose of traffic using the network; and (3) the character of traffic flow—for example, periods of maximum and minimum use. This information can be assembled for the majority of cases by

1. Procuring from the appropriate governmental engineering staff the necessary maps showing the locations of all paved and unpaved roads in the study area. In addition, traffic

count information, if available, should be procured from the appropriate governmental engineering staff and local, regional, or national transportation agencies.

2. Making site visits to the study area and collecting ad hoc data on traffic counts for pertinent roads, streets, and highways; such counts should be focused on the peak and minimum periods of usage of the network.

The fundamental information necessary to accomplish step 2 is assumed to be readily available or easily obtainable.

Step 3: Procurement of Pertinent Standards or Criteria

Table 14.10 summarizes information on a six-category "level of service" (LOS) delineation used by the U.S. Transportation Research Board. The LOS for a highway, for example, is a qualitative measure of the effect of a number of factors, including speed and travel time, traffic interruptions, freedom to maneuver, safety, driving comfort and convenience, and operating costs. If impacts on local or regional highways are anticipated, it would be appropriate to determine the LOS classifications for the highways in the study area. In addition to the LOS system, local roads and streets in the study area may have been classified by local or regional traffic or transportation authorities, or even by the engineering section of a military facility if the project is on a military installation. The delineation of these classifications would also be appropriate in step 3.

TABLE 14.10

LEVELS OF SERVICE

The level of service concept

The definition of "level of service" is "a qualitative measure of the effect of a number of factors, which include speed and travel time, traffic interruptions, freedom to maneuver, safety, driving comfort and convenience, and operating costs." It goes on to indicate that "in practice, selected specific levels are defined in terms of particular limiting values for certain of these factors." Service levels A through F represent the best through the worst operating conditions, respectively.

Level of service A represents virtually free-flow conditions, in which the speed of individual vehicles is controlled only by the driver's desire and by prevailing conditions, not by the presence of interference of other vehicles. Ability to maneuver within the traffic stream is unrestricted.

Levels of service B, C, and D represent increasing levels of flow rate with correspondingly more interference from other vehicles in the traffic stream. Average running speed of the stream remains relatively constant through a portion of this range, but the ability of individual drivers to freely select their speed becomes increasingly restricted as the level of service worsens (goes from B to C to D).

Level of service E is representative of operation at or near capacity conditions. Few gaps in traffic are available, the ability to maneuver within the traffic stream is severely limited, and speeds are low (in the range of 30 mi/hr). Operations at this level are unstable, and a minor disruption may cause rapid deterioration of flow to level of service F.

Level of service F represents forced, or breakdown, flow. At this level, stop-and-go patterns and waves have already been set up in the traffic stream, and operations at a given point may vary widely from minute to minute, as will operations in short, adjacent highway segments, as congestion waves propagate through the traffic stream. Operations at this level are highly unstable and unpredictable.

Source: Adapted from Transportation Research Board, 1980 (pp. 163–164), 1985.

Steps 4 and 5: Prediction of Traffic and Transportation-System Impacts and Assessment of Impact Significance

Step 4 requires the consideration of the changes (in terms of increase or decrease and timing) in the baseline traffic conditions in the ROI as a result of the construction and operational phases of the proposed project. The basic mathematical relationship for this step is as follows:

$$\frac{\text{Percentage change}}{\text{in baseline conditions}} = \frac{\text{step 1 information (100)}}{\text{step 2 information}}$$

Percentage changes can be calculated for each pertinent local or regional road or highway and for each project or activity phase. For example, assume a local road has a baseline average daily traffic (ADT) of 1,000 vehicles, with the peak-hour traffic being 250 vehicles. Further assume that the project-construction phase of 6 mo will add 200 (vehicles) to the ADT, with 150 being associated with the peak hour. The project operational phase will add 75 to the ADT, and none of these vehicles will be associated with the peak hour. The percentage changes would be calculated as follows:

Construction Phase

$$\text{Percent change in ADT} = \frac{(200)(100)}{1,000} = 20\%$$

$$\text{Percent change in peak hour} = \frac{(150)(100)}{250} = 60\%$$

Operational Phase

$$\text{Percent change in ADT} = \frac{75}{1,000}(100) = 7.5\%$$

$$\text{Percent change in peak hour} = \frac{(0)(100)}{250} = 0\%$$

Since the basic output for step 4 is the percentage change information in relation to baseline traffic conditions, the next step is focused upon how to interpret this percentage change information (step 5). No transportation criteria or standards provide a delineation of an appro-

priate interpretation method; however, the absolute changes and the LOS should be given consideration.

Burchell et al. (1993) has described a five-component traffic-impact-analysis methodology for development projects. The components are (1) introduction, (2) analysis of existing conditions, (3) traffic characteristics of the development site, (4) future demands on the transportation network, and (5) impact analysis and mitigation recommendations.

The introduction of the traffic-impact analysis should contain a complete project description, including the proposed land use (or uses), the size of the development, proposed-site-access points, and phasing plans (Burchell et al., 1993).

The next component, the analysis of existing conditions, should address the current volume of traffic using the roadways and the LOS provided to current traffic. Consideration should be given to specific analysis periods—such as existing peak-traffic times and the times of peak traffic generated by the development project. Recent traffic counts should be procured from pertinent agencies. New traffic counts should be taken to fill in those critical locations without acceptable historical counts. New roadway counts are typically taken with an "automatic traffic recorder" (ATR) for a one-week period (including the weekend). The data is typically tabulated in an hourly fashion (by direction of travel) with a 24-hr volume, or "average daily traffic" (ADT), shown for each roadway in the study area (Burchell et al., 1993). In addition to ADT, turning-movement counts at key intersections may need to be taken and monitored.

The next aspect of the analysis of existing conditions is to determine the capacities and LOS within the study area. "Capacity" is defined as the maximum number of vehicles that can be expected to travel over a given section of roadway, or a specific lane, during a given time period under prevailing roadway and traffic conditions (Transportation Research Board,

TABLE 14.11

CAPACITY OF TWO-LINE ROAD IN RELATION TO LEVEL OF SERVICE

Terrain	Level of service[a]		
	C	D	E
Level	7,900	13,500	22,900
Rolling	5,200	8,000	14,800
Mountainous	2,400	3,700	8,100

[a]Assumes: Peak hour traffic = 10%; 60:40 split; 14% trucks; 4% recreational vehicles; 25 percent no passing (level terrain); 40% no passing (rolling terrain); 60% no passing (mountainous terrain).
Source: Burchell et al., 1993, p. 282.

1980, 1985). Details on measures of effectiveness for uninterrupted and interrupted flows are described in Burchell et al. (1993). This information can then be integrated; for example, Table 14.11 summarizes average capacity of a two-lane road expressed as maximum ADT volumes for three levels of service.

Addressing the traffic characteristics of the development site, the third component in the methodology, basically involves developing answers for two questions (Burchell et al., 1993): (1) How much traffic will the proposed site produce (i.e., what is the trip generation)? and (2) which roadways will site-generated traffic use (i.e., what is the trip distribution)?

Three approaches for aggregating the trip-generation information entail (1) the use of local rates, (2) the use of estimates based on the type and characteristics of the project or activity, or (3) the use of national rates, expressed in typical units such as those shown in Table 14.12. Detailed considerations involving the use, interpretation, and adjustment of information from Table 14.12 is in Burchell et al. (1993).

After the site-generated (or project- or activity-induced) traffic is estimated, the next activity is to determine the directional distribution of the traffic. For small sites, it is reasonable to assume the traffic will arrive and depart in a manner similar to existing travel patterns. Calculations for large sites often require the formation of a detailed distribution model combining elements of population, employment, travel times, highway network characteristics, and competing uses (Burchell et al., 1993).

The three most typically used methods for estimating trip distribution are either based on the use of (1) existing data, (2) origin-destination data, or (3) a trip-distribution model. The first two methods are self-explanatory. A "trip-distribution model" (also referred to as a "gravity model") assumes that the number of trips between two zones is proportional to the size of the zones and inversely proportional to the square of the distance between the two zones. Details on the development and calibration of a gravity model are contained in Burchell et al. (1993).

Determination of future demands on the transportation network is the fourth component in the traffic-impact-analysis methodology. A "horizon year" must be determined for each phase of proposed development, as well as the subsequent completion, or "buildout," year. The determination of future volumes without the site development (or project or activity) is typically calculated through the use of (1) growth rates (or trends), (2) the buildup method, or (3) the area-transportation plan (Burchell et al., 1993). The growth-rate method is the simplest to use and, as such, is most often utilized for relatively small developments or for developments with a buildout no more than five years into the future. Growth rates (or trends) are de-

TABLE 14.12

TYPICAL TRAFFIC-GENERATION BASE UNITS

| Types of land use | Traffic generation unit | |
	Preferred	Alternate
Retail		
Shopping center	1000 GLA[a]	—
Discount store	1000 GFA[b]	—
Garden center	Per employee	1000 GFA
Restaurant	Per seat	1,000 GFA
Convenience market	1000 GFA	—
Residential		
Single family	Per dwelling unit	persons
Apartment	Per dwelling unit	persons
Condominium	Per dwelling unit	persons
Retirement community	Per dwelling unit	—
Office		
General	Per employee	1000 GFA
Medical	Per employee	1000 GFA
Office park	Per employee	1000 GFA
Industrial		
General light	Per employee	1000 GFA
Industrial park	Per employee	1000 GFA
Manufacturing	Per employee	1000 GFA
Warehousing	Per employee	1000 GFA
Hospital	1000 GFA	Per employee
Hotel	Per room	Per employee
Movie theater	Per screens	Per seat
Drive-in bank	Per window	1000 GFA

[a]Per 1000 square feet of gross leasable area of building.
[b]Per 1000 square feet of gross floor area at building.
Source: After data in Institute of Transportation Engineers, 1987, various pages.

termined from historical traffic counts maintained by the appropriate traffic or transportation agencies. In the absence of specific historical traffic counts, growth rates are often indexed to area population growth. For each phase of the development (including final buildout), the existing base volumes are factored upwards by the appropriate growth rate to determine future without-site traffic volumes (Burchell et al., 1993).

The "buildup" method is most appropriately used in an area experiencing moderate to rapid growth. The buildup method combines elements of the growth-rate method with a detailed analysis of approved and anticipated developments within the study area. For each horizon year, the existing volumes are increased by the applicable growth rate. Furthermore, the trip-generation and -distribution characteristics of approved and anticipated development are estimated and added to the future base volumes to provide a total of future nonsite-traffic volumes (Burchell et al., 1993). The area transportation plan usually projects traffic volumes on major streets 20 years into the future (this is analagous to the without-project condition). If the proposed development is on one of these roadways, future volumes may be interpolated to the horizon years.

The next activity involves assigning the site-generated traffic to the study-area roadway and intersections—that is, with-project conditions. Finally, for each analysis period being studied, totals for future nonsite and site-related traffic volumes are calculated for the study area. Separate graphics and tabulations of the various components of total future traffic are useful in illustrating site-related changes (Burchell et al., 1993).

The final component in the methodology entails the actual impact analysis and the development of appropriate mitigation recommendations. This component should focus on the LOS with and without the site development. The first activity involves a calculation of the future without-project LOS for the analysis periods and horizon years described previously. After this calculation, a comparison of the results with the "acceptable standard" of the community is made. For those developments not expected to meet the extant community standard, a determination of recommended improvements necessary to achieve the desired LOS should be developed (Burchell et al., 1993).

The second activity involves the calculation of the future LOS with the development-site traffic. The results should once again be compared with the community standard and with the results of the future-without-project analysis to identify changes in the LOS caused by the development and additional improvements that may be required. As an alternative to additional capacity improvements, demand-reduction strategies (mitigation measures) may need to be seriously considered. Examples of these strategies include utilization or development of public transportation, car pools, and van pools; implementation of modified work schedules (flextime or staggered working hours); and parking limitations (Burchell et al., 1993).

Several recent references provide detailed approaches for traffic- or transportation-system-impact identification, prediction, and assessment (American Planning Association, 1984; Transportation Research Board, 1985; AASHTO, 1984; and Institute of Transportation Engineers, 1987, 1989). For example, the Transportation Research Board (1985) report, entitled "Highway Capacity Manual," includes the results of capacity calculations at signalized intersections; these are now expressed in average delay terms (shown as seconds per vehicle). In a traffic-impact study, this enables the preparer to compare intersection delay under existing conditions with that expected for future with-project conditions. The 14 chapters of the "Highway Capacity Manual" include discussion of the "traffic characteristics" variables observed during research (Chapter 2); uninterrupted-flow facilities (freeways—Chapters 3, 4, and 6; multilane highways—Chapter 7); two-lane rural highways (Chapter 8); and interrupted-flow facilities and factors, including signalized intersections, unsignalized intersections, urban and suburban arterials, transit capacity, pedestrians, and bicycles (Chapters 9 through 14, respectively) (Transportation Research Board, 1985).

The empirical basis for trip-generation information is included in an expanded form in Institute of Transportation Engineers (1987). This fourth edition is based upon 1,950 local studies, whereas the original edition, published in 1976, contained only 80 documented sources. This edition contains data on 90 individual land uses, grouped into the following categories: (1) ports and terminals, (2) industrial-agricultural, (3) residential, (4) lodging, (5) recreational, (6) institutional, (7) medical, (8) office, (9) retail, and (10) services. In addition, nine time periods are addressed, as follows (Institute of Transportation Engineers, 1987): (1) weekday (24 hr); (2) weekday, peak hour of adjacent street traffic (1 hr between 7 and 9 a.m.); (3) weekday, peak of adjacent street traffic (1 hr, between 4 and 6 p.m.); (4) weekday, a.m. peak hour of generator;

(5) weekday, p.m. peak hour of generator; (6) Saturday (24 hr); (7) Saturday, peak hour of generator; (8) Sunday (24 hr); and (9) Sunday, peak hour of generator.

In 1984, the American Planning Association published a report which delineates a methodology that can be used to determine whether the roadway network in the area of a proposed major development will be able to handle the existing through-traffic plus the additional traffic that the development will generate. Part I deals with the planning process of land development and transportation and includes chapters on urban development, site planning, and traffic analysis. Part II deals with related design considerations and includes chapters on functional circulation systems, intersection designs, access and site circulation, parking and service facilities, and drive-in facilities (American Planning Association, 1984).

Finally, AASHTO (1984) is a policy manual that provides guidelines for the design of local roads and streets, collector roads and streets, rural and urban arterials, freeways, at-grade intersections, and grade separations and interchanges. The policy manual includes chapters on highway functions, design controls and criteria, elements of design, and cross-section elements. In addition, the Institute of Transportation Engineers published a 1989 report on traffic access and impact studies for site development. The report includes chapters on planning for studies, existing conditions, non-site-traffic forecasting, trip generation and distribution, capacity analyses, site access, site planning, and off-site improvements (Institute of Transportation Engineers, 989).

In some cases, development projects may generate the need for new highways and/or transportation systems. One methodology to address such a case was developed in U.S. Department of Transportation (1975), a six-volume notebook series. The methodology was created for use by state highway departments and Federal Highway Administration field offices who

are responsible for conducting transportation-planning and -impact studies. Brief information on all six notebooks is in Table 14.13 (U.S. Department of Transportation, 1975). Thirteen impact categories are addressed in notebooks 2 through 4, as follows (U.S. Department of Transportation, 1975):

1. *Notebook 2, Social Impacts*—community cohesion, accessibility of facilities and services, and displacement of people
2. *Notebook 3, Economic Impacts*—employment, income and business activity; residential activity; effects on property taxes; regional and community plans and growth; and resources
3. *Notebook 4, Physical Impacts*—environmental design, aesthetics and historic values; terrestrial ecosystems; aquatic ecosystems; air quality; and noise and vibration

Step 6: Identification and Incorporation of Traffic and Transportation-System Impact-Mitigation Measures

''Mitigation measures,'' in this context, are steps that can be taken to minimize the magnitude of the increases in traffic in the ROI. The key approach is either to reduce the traffic or to change the timing of the traffic anticipated to be emitted from the project (or activity). Extensive information on mitigation measures is included in several of the earlier-mentioned reports. An additional source is the *Environmental Review Guide* (U.S. Army Construction Engineering Research Laboratory, 1989). This report provides examples of simple mitigation measures which can be used to reduce the traffic or change the traffic pattern in the study area of a military project or training activity, including (1) the use of car or van pooling or buses from residential areas for travel to and from military installations, (2) scheduling construction-equipment movement during nonpeak periods in the local area, (3) scheduling troop movements related to training exercises during nonpeak traffic

TABLE 14.13

U.S. DEPARTMENT OF TRANSPORTATION'S NOTEBOOK SERIES ON EIA FOR TRANSPORTATION PROJECTS

Notebook 1, *Identification of Transportation Alternatives* discusses the principal transportation planning considerations which should be incorporated in all phases of the highway planning process—from initial transportation systems studies, through project location and design phases, to the actual construction of a new roadway improvement. Notebook 1 emphasizes the continuing and evolving nature of the process of generating alternatives. This process involves successive cycles of problem identification, definition of alternatives, testing of each alternative (in terms of transportation service as well as environmental impacts), and refinement of alternatives in order to improve transportation service and to avoid or minimize environmental harm. Five case studies are included in the Notebook as actual illustrations of the alternatives development and evaluation process.

Notebook 2, *Social Impacts;* **Notebook 3,** *Economic Impacts;* **and Notebook 4,** *Physical Impacts* provide a comprehensive list of potential impacts of highway projects, together with workable state-of-the-art methods and techniques for impact identification, data collection, analysis and evaluation. Their purpose is to expedite and improve the quality and effectiveness of the environmental assessment process regardless of whether an Environmental Impact Statement (EIS) or other formal documentation is ultimately required.

Notebook 5, *Organization and Content of Environmental Assessment Materials* describes techniques for recording, organizing and communicating pertinent findings of the transportation planning and impact assessment process. Environmental assessment reports often require considerable synthesis and condensation in order to effectively communicate the large volumes of data and findings for purposes of public review, comment and decisionmaking. Notebook 5 describes examples which have been used by others to prepare both detailed and summary report materials.

Notebook 6, *Environmental Assessment Reference Book* expands on the bibliographic references contained in the preceding Notebooks, and lists other data and information which may be helpful to transportation planners and engineers responsible for environmental impact assessment.

Source: U.S. Department of Transportation, 1975, pp. 2–3.

periods in the local area, (4) implementation of traffic-volume restrictions or controls for certain times of the day, and (5) raising the pertinent LOS category by providing roadway improvements.

HUMAN HEALTH IMPACTS

The importance of health-impact considerations in project planning have been stressed by the World Health Organization [WHO (1985, 1986)]. Two examples of methodologies for incorporating such considerations will be presented. For projects financed by a development bank, the main tasks of a health impact assessment process are shown in Figure 14.5. The tasks are (Asian Development Bank, 1992, pp.

3, 5) *Definition of Project Type and Location*—Project title, location, department, executing agency and major project components are defined as part of the screening process and project classification.

2. *Health Hazard Identification*—This is the primary screen. It is based on existing experience and the screening tools provided. The output is a long list of health hazards.

3. *Initial Health Examination (IHE)*—This is the secondary screen. It uses rapid appraisal, secondary data and a fact-finding mission (if necessary). It is part of the initial environmental evaluation (IEE) and should normally be undertaken at the prefeasibility stage. The outputs are: a short list of the health hazards which may carry the most significant health risks. The identification of a hazard short list is part of the scoping process.

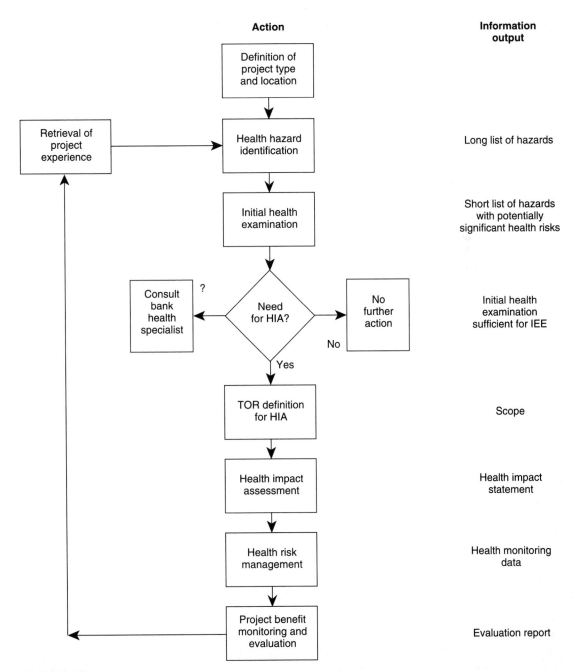

FIGURE 14.5
Overview of Health Impact Assessment (HIA) Process (Asian Development Bank, 1992).

4. *Requirement for Health Impact Assessment (HIA)*—A decision is made based on the experience of previous projects and the need to obtain further experience.

5. *Terms of Reference (TOR) Definition for HIA*—A TOR is prepared for an HIA which specifies the scope of the assessment. It includes, but is not limited by, the short list of health hazards identified by the IHE.

6. *Health Impact Assessment*—The assessment is undertaken by a specialist consultant. The output is a Health Impact Statement. The HIA may be a stand-alone study, but more typically, it will be part of an EIA.

7. *Health Risk Management*—The Health Impact Statement recommends health risk management actions including environmental management and health monitoring. Health monitoring data is an output.

8. *Benefit Monitoring and Evaluation*—The project may be evaluated by an appropriate Bank unit. The output should include a health impact evaluation report which can be used in future projects.

A generic, descriptive health-impact-prediction and -assessment methodology has been developed which aggregates principles of risk assessment (RA) methods and of traditional approaches used in EIA studies. The main concern was that it should be integrated into the unified analytical process that is basic in an EIA study. Consequently, the generic methodology was organized to parallel the structure of activities conducted in a typical impact study (Arquiaga, 1991). Figure 14.6 depicts a flow diagram of the methodology. It consists of a sequence of 10 operational activities, or components; the information derived in each component serves as the input for the next. The three activities representing the fundamental functions of the EIA process are shaded to indicate their key role. With the exception of scoping and written documentation, the activities revolve around the three key activities and, therefore, are enclosed in the larger box. The scoping and written-documentation activities are shown outside the larger box because they represent the initial and final stages, respectively, in the methodological process and are not directly connected to the fundamental functions.

Scoping

The need for an HIA should be determined based on input from regulatory agencies, other pertinent organizations, and the general public, and on the professional knowledge and judgment of the EIA study preparers, as a part of the EIA scoping process. In general, a health-impact focus should be included in an EIA study if the answer to any of the following questions is affirmative:

1. Does the nature of the proposed project (or activity) involve the handling of or emissions to the environment of materials such that their physical, chemical, radiological, or biological nature may be harmful to human health?

2. Is the location of the proposed project, together with its nature, likely to give rise to conditions that would alter the occurrence of natural hazards in the study area?

3. Could the implementation of the proposed action eventually give rise to conditions that would reduce or increase the number of adverse health-impact-causing factors?

Review and Analysis of Pertinent Institutional Information

The institutional information pertaining to the project should be reviewed and analyzed prior to any intensive effort in the EIA process. For the purpose of assessing human health impacts, any specific law, regulation, executive order, or guideline which directly or indirectly relates to human health should be identified for each pertinent administrative grouping (federal, state, regional, and/or local). In general, the institutional information will probably pertain to specific levels (or concentrations) of given health-impact-causing agents. Thus, the institutionally set level

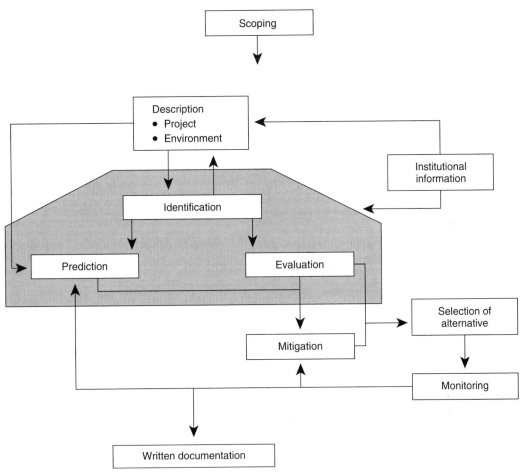

FIGURE 14.6
Flow Diagram of Generic Health-Impact-Prediction and -Assessment Methodology.

guidelines may be used for (1) determining reference doses without having to go through a dose-response assessment process and/or (2) interpreting health impacts by comparing the institutionally set levels against predicted exposure levels. In both cases, it is important to analyze the institutional information in order to identify the specific health effects and conditions for which the levels were established and to determine whether they should be applied to the conditions of the project.

Description of Project and Affected Environment

The description of the project and the affected environment is also needed for the assessment of other impacts; therefore, the collection effort can be minimized by coordinating the informational needs of all potential impacts to be addressed. The procedure for the procurement of health-related information, in particular, needed for the description of the project and that of the environment should consist of (1) structuring the description process in organizational units according to the health-related characteristics of the project and (2) collecting appropriate information for each organization unit.

Correctly organizing the description process in an EIA study in order to ensure thoroughness is important because of the great number of alternatives, phases, activities, processes, and environments that may be pertinent. Organizational units for the project and the environment may be delineated as follows:

1. Project
   - Identify the main components or activities of the project and its alternatives. Do not repeat components or activities common to two or more alternatives.
   - For each component or activity, identify phases (for example, construction, operation and closure).
   - For each phase of each main component or activity, identify subcomponents or subactivities.
2. Environment
   - Identify subcomponents or subactivities that may affect one or more environments; if this is not applicable, proceed with the following identification step.
   - Identify each environment affected by each main component or activity of the project and its alternatives.

In the context of an HIA, project information needed for each organizational unit includes the size and characteristics of the area affected, time schedules, labor-force characteristics, methods and equipment used, sources and levels of physical phenomena that may have the potential to harm human health, and sources and characteristics of any hazardous materials (chemical, biological, or radiological) that are to be used in, disposed of, or emitted to the environment.

The description of the affected environment provides the basis for determining the ways in which humans may become exposed to health-impact-causing agents. It also may lead to the identification of naturally occurring health-impact-causing agents that, if expected to be affected by the project, should be addressed in the HIA. Each environment affected by the components or activities of the project should be described in terms of three components: (1) the physical-chemical environment, including information pertaining to the meteorology and the geologic and hydrologic settings of the affected area; (2) the biological environment, including information on pertinent food chains, pathogenic organisms, and disease vectors; and (3) the human environment, including information on population, disease, land usage, health care systems, and pollution.

Identification of Potential Health Impacts

The identification of health impacts consists of three steps: (1) identifying the sources of potential health effects within each organizational unit defined during the description process, (2) defining those circumstances or scenarios under which health impacts from those sources may occur, and (3) identifying health-impact-causing agents associated with the sources and their health effects. The identification of health-impact sources requires professional knowledge, experience, and the exercise of professional judgment. In general, however, the following categories can be used to delineate potential sources:

- Processes or activities that involve the use, production, and/or handling of materials that may contain radioactive, chemical, and/or biological hazards
- Processes or activities that involve the generation of physical hazards
- Processes or activities that involve the generation of conditions that increase (or decrease) the occurrence in the environment of natural hazards

The scenarios under which health impacts from the identified sources occur can include (1) a "routine scenario," where emissions to the environment of potential health-impact-causing agents occur on a regular basis; (2) an "extraordinary scenario," which considers health impacts resulting from low-occurrence but foreseeable events; and (3) a "maximum scenario," often used to describe a more realistic scenario in which routine emissions are combined with maximum levels of emissions that may occur periodically.

In the extraordinary scenario, it is necessary to estimate the probability of occurrence of the extraordinary events that will be considered in the assessment. Methods to identify extraordinary events include hazard and operability studies, technical audits, and examination of historical records (Arquiaga, 1991). The probability of occurrence can be determined either through analysis of existing statistical data for the same or similar types of project components or activities, or through the systematic usage of event and fault trees (Arquiaga, 1991).

The task of identifying health-impact-causing agents and their associated health effects is equivalent to the hazard-identification step in conventional RA studies. In general, and for any type of health-impact-causing agent, the process involves an extensive review of studies and statistical records to determine whether exposure to the elements in or from the sources previously identified, and which are known or suspected to have some effect on human health, is likely to cause a change in the incidence of a health condition. The outcome of this process should be one or more lists of health-impact-causing agents, indicating their sources, levels (and the degree of uncertainty regarding these levels), associated health effects, and the nature and extent of evidence that the agent causes health effects in humans.

Prediction of Health Impacts

The prediction methods included in this generic methodology are based on existing RA techniques; these techniques include exposure assessment, dose-response assessment, and health-impact characterization. Additional information on RA as related to impact studies is available in Canter (1993).

EIA studies often involve the analysis of several alternatives that may encompass multiple components, activities, and/or environmental settings. Therefore, the use of a structured approach for conducting the exposure assessment and the subsequent prediction steps is needed. The generic structured approach basically consists of conducting separate exposure assessments for all combinations of project components that can be formed with the organizational units defined earlier. For each combination of project components, the exposure assessment should include several levels of analysis; the highest level of analysis is represented by the routine, extraordinary, and no-action scenarios, and a more detailed level of analysis corresponds to each subcomponent (or subactivity) in each project component (or activity). Each subcomponent should be analyzed within each pertinent phase of the project component, and each project phase should be analyzed with regard to the worker's population and the general population.

The purpose of the no-action scenario is to delineate the baseline conditions. The exposure analysis for the no-action scenario is important because it can serve a predictive function by providing background exposure levels and an evaluation function by providing the basis against which the health effects resulting from the project action can be compared.

An "exposure assessment" involves characterizing exposure pathways and quantifying exposure levels. The characterization of exposure pathways should be conducted for each project component and its associated environment. The aim is to identify and quantify individual exposure pathways so that, when the project (or any alternative) is considered as a whole, they can be aggregated, where appropriate, into exposure scenarios. "Exposure scenarios" refer to situations in which the same people are likely to be exposed to health-impact-causing agents through several exposure pathways. A key objective of the exposure assessment as conducted for the EIA process is to coordinate efforts in the prediction of impacts. That is, the exposure assessment should make all possible use of the predictive techniques used in the overall EIA process. This approach not only minimizes work, but it also fosters consistency in the prediction of impacts, since the same assumptions and criteria are used for all comparable impact calculations.

''Dose-response assessment'' consists of describing the relationship between the dose of the health-impact-causing agent and the predicted occurrence of a health effect in an exposed population. Quantitative methods that have been developed or used for each type of health-impact-causing agent should be used in this methodology (Arquiaga, 1991). The results from the dose-response and exposure assessments should then be integrated into quantitative expressions (if possible) of anticipated health-impact occurrences. In the case of health effects for which quantitative data could not be derived in the dose-response assessment, a qualitative characterization should be used.

The characterization of anticipated health impacts involves (1) matching the estimated exposure doses with the appropriate dose-response values, (2) quantifying the health-effect incidence, (3) assessing the degree of uncertainty, and (4) summarizing the results. Quantitative characterization of measurable health effects involves a two-stage process. In the first step, the probable occurrence of the health effect in an individual who is likely to actually be exposed to the expected level or concentration of the health-impact-causing agent is estimated. Then, the probability of occurrence of the exposure event is calculated and combined with the results of the first stage.

In the context of the EIA process, the evaluation of uncertainties associated with the prediction of health impacts is of utmost importance for the rational interpretation of these impacts. The most sophisticated methods for estimating uncertainty include numerical methods such as the Monte Carlo simulation method, and analytical methods such as the first-order Taylor series approximation (Arquiaga, 1991). Use of these methods, however, requires a substantial input of data that often cannot be obtained because of resource constraints. When available data are sufficient to describe the potential range the parameters might exhibit, a sensitivity analysis can be used to identify influential variables and to develop bounds on the distribution of health effects. The most practical approach to characterize uncertainty in most EIA studies will be a qualitative approach, in which a quantitative or qualitative description of the uncertainty for each parameter is developed, followed by a qualitative indication of the possible influence of these uncertainties on the numerical estimate of the health effects.

The qualitative and quantitative results of the characterized health impacts can be organized according to the flowchart shown in Figure 14.7. In this figure, only the chemical health impacts associated with a given project component, on the general population, for a routine scenario and a given alternative, are fully depicted. The same method of organization could be used for each category of health impact, project component, type of exposed population, type of project scenario, and project alternative.

Evaluation of Health Impacts

The method discussed herein for evaluating the significance of anticipated health impacts should be used for assessing those health effects associated with each of the exposure pathways or exposure scenarios delineated in the health-impact-characterization step. The evaluation method consists of two main steps and several substeps such that when the significance could not be established in the previous step or substep, a new step or substep is undertaken.

The first step involves establishing the significance of each impact by determining whether any health-impact-causing agent is sub-

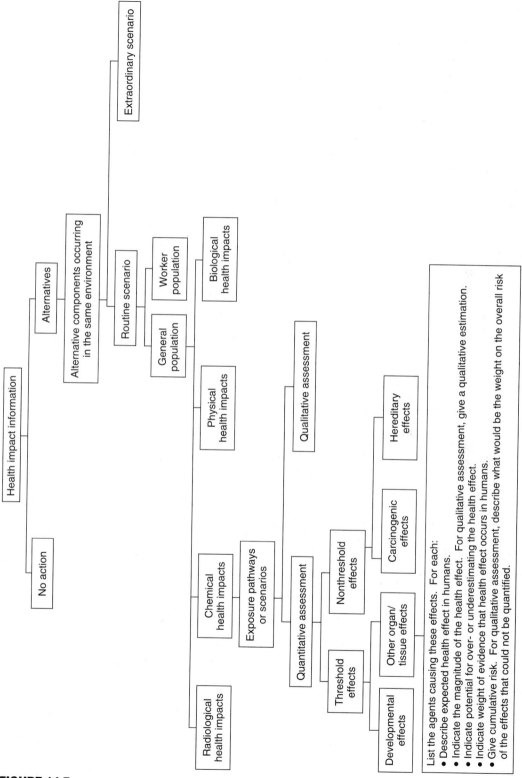

FIGURE 14.7
Organizational Flowchart for Characterized Health Effects.

ject to health-related regulatory limits. The second step involves determining the significance of health effects, first in terms of magnitude and regulatory criteria, then in terms of uncertainty, then in terms of the extent of evidence, and, finally, in terms of other evaluation factors such as cumulative effects, institutional information that might be especially relevant to the health effects or health-impact-causing agents being considered, the repercussions of the health effects at the individual and populational levels, and the perceptions related to the health effects among the human population.

One of the purposes of the evaluation process is to identify those health effects that need to be considered during the impact-mitigation planning process. These health effects, therefore, need to be included in the alternative selection process, and included in a way that is consistent, not only within each alternative, but across alternatives. Thus, in order to ensure this consistency of use, these factors should not be influenced in any way by variations that may exist among alternatives. The influence of the alternatives on the evaluation factors should be considered by the decision makers in the alternative-selection process.

Identification and Evaluation of Mitigation Measures

In most EIA studies, mitigation measures for undesirable health effects can fall into one or more of the following three categories (Arquiaga, 1991): (1) mitigation through control of source, (2) mitigation through control of exposure, and (3) mitigation through health-services development. Mitigation measures to control sources act by preventing or limiting the introduction into the environment of the health-impact-causing agents. The control of sources can be achieved through engineering techniques such as modifications in the design of the project (e.g., introducing water or product recycling) or through management methods which involve adjustments of project components according to changes in environmental conditions (e.g., re-

ducing the activity of a processing plant during thermal inversions). Control of exposure is usually achieved by preventing or limiting the access of individuals to the contaminated or potentially contaminated medium, by preventing the contaminant from reaching individuals, by stopping or reducing contact with the health-impact-causing agents, by warning individuals of potential hazards, or by devising contingency plans for both workers and the general public.

Mitigation through health-services development may involve the implementation of health education programs or the development of health-impact-prevention and health care systems. Preventive-health-oriented measures (such as vaccination programs) are aimed at protecting individuals from acquiring a disease. Health care approaches might include planning systems for the treatment of diseases or other effects of physical, chemical, or radiological exposures should such symptoms develop in the affected individuals.

Two major considerations in identifying health-effect mitigation measures are that not all the identified measures are necessarily technically or economically feasible, and that public participation is important for selecting acceptable measures. In identifying any quantifiable mitigation measure, sufficient information should be aggregated to allow quantification of the reduction of the impact. For mitigation measures which do not allow the quantification of the mitigation effect or for health effects that were characterized on a qualitative basis, a qualitative estimation of the mitigation effect should be developed using professional knowledge and judgment.

Selection of Proposed-Action Alternative

In the context of this methodology, it is not the purpose of the selection of proposed action stage to carry out the selection process itself, but to assist in the process by organizing and presenting the information on health impacts in a way which is most useful to the decision makers. The suggested approach for organizing and presenting health-impact information consists of the following steps:

1. Classify the health effects that were found to be significant, using the following categories—carcinogenic effects, hereditary effects, teratogenic effects, organ-tissue effects (including traumatic effects), and infections from biological health-impact-causing agents.
2. For each alternative and for each project scenario, estimate the number of people that would be affected by each category of health effects given above.
3. Assign a rating value to each category of health effects for which population numbers were estimated. An example of a potential rating scale is in Table 14.14.
4. Present the results obtained in steps 1, 2, and 3. These results should be accompanied by evaluation factors that are affected differently for each alternative.

Monitoring of Health Impacts

Normally, and in contrast with other environmental impacts, health impacts are not easy to detect, and, if they are detected, it may be difficult to establish a clear relationship between the health effect and the project. This is espe-cially true for carcinogenic and hereditary effects, which may take a long time to develop, be influenced by several confounding factors, and be difficult to discern from background-level effects. Thus, while one of the purposes of environmental-monitoring programs is to provide an early warning of unanticipated adverse impacts, sudden changes in impact trends, or approaches to preselected critical levels, it has to be recognized that as of today, with the techniques available to detect environmental health effects, health-impact monitoring may not always be useful as a warning system. Therefore, for long-term health effects, the main purpose of monitoring is to provide information which can be used to (1) document health impacts that might result from a proposed action, (2) review and validate impact-prediction techniques, (3) evaluate the effectiveness of implemented mitigation measures, and (4) enable a more accurate prediction, in the future, of health impacts associated with similar actions. For short-term health effects, however, monitoring could possibly also serve as a warning system.

Because the ultimate purpose of health monitoring is to determine whether a cause-response relationship exists, and to quantify the relationship, if possible, the monitoring activities should

---

TABLE 14.14

POTENTIAL SCALE FOR RATING HEALTH EFFECTS

Rating	Signification
0	No significant impacts
−1	Nondisabling, reversible adverse health effects affecting a limited number of people
−2	Nondisabling, reversible adverse health effects affecting a large number of people
−3	Disabling (not leading to death), reversible adverse health effects affecting a limited number of people
−4	Disabling (not leading to death), reversible adverse health effects affecting a large number of people
−5	Disabling (leading to death), reversible adverse health effects, either short-term or long-term, affecting a limited number of people
−6	Irreversible, long-term adverse health effects affecting a limited number of people
−7	Disabling (leading to death), reversible adverse health effects, either short-term or long-term, affecting a large number of people
−8	Irreversible, long-term adverse health effects affecting a large number of people

Note: Use same system for beneficial effects, but change − sign to + sign.

be established at two levels. One level would account for the potential causes of the predicted health effects and, therefore, would consist of monitoring the levels of the health-impact-causing agents at their sources, as well as predicted exposure points. The other monitoring level would involve the detection and recording of the health effects that develop in the nonexposed and potentially exposed populations. This monitoring level necessarily involves the health-care community of the affected area.

Finally, in terms of data collection, the monitoring of health-impact-causing agents would probably require the use of monitoring stations and systems developed specifically for the project. These monitoring stations and systems should be integrated, whenever possible, into the monitoring systems for other environmental impacts so as to minimize expenditure of resources and avoid duplication of efforts.

Preparation of Written Documentation

The findings of the environmental impact study, and in particular the information and findings corresponding to the selected alternative, need to be summarized and organized in a written EIS. The information generated and used in predicting and assessing health impacts is normally extensive. Therefore, it is recommended, as part of this methodology, to document each of the previous activities. This information can then be incorporated such that succinct information pertinent to the prediction of health impacts is contained in the appropriate sections of the EIS; the most-extensive material should be presented in appendices, and the external supporting data and literature should be adequately referenced.

SUMMARY

A systematic methodology for prediction and assessment of impacts on the socioeconomic environment and selected components is presented in this chapter. Background information is pre-

sented on the factors included in the socioeconomic environment and their relationships to components of the biophysical and cultural environments. Socioeconomic-impact prediction may involved the use of one to several approaches such as (1) qualitative description; (2) quantitative description; (3) use of application-specific or model-based techniques; and (4) use of relative comparisons of the effects of alternatives. Assessment of predicted socioeconomic impacts involves consideration of relevant professional and institutional standards and criteria; spatial and temporal averages; and the application of professional judgment.

Key summary points related to the prediction and assessment of socioeconomic impacts are as follows:

1. There are many factors and impacts to consider.
2. Baseline data is typically available from several information sources.
3. Many impacts can be quantified using multiplier factors.
4. Logical bases for assessment, derived from institutional guidelines, can be used. (There are many planning standards and criteria.)
5. Impact-mitigation measures are often straightforward; however, institutional capacity can be a limiting factor in implementing or selecting mitigation options.
6. It may be necessary to make trade-offs between positive socioeconomic impacts and negative natural environment impacts.

SELECTED REFERENCES

American Association of State Highway and Transportation Officials (AASHTO), ''A Policy on Geometric Design of Highways and Streets,'' AASHTO, Washington, D.C., 1984.

American Planning Association, ''Traffic Impact Analysis,'' American Planning Association, Washington, D.C., 1984.

Arquiaga, M. C., ''Generic Health Impact Prediction and Assessment Methodology for Environmental

Impact Studies,'' Ph.D. dissertation, University of Oklahoma, Norman, 1991.

Asian Development Bank, ''Guidelines for Social Analysis of Development Projects,'' Manila, The Philippines, June 1991a, pp. 33–35, 106–112.

———, ''Guidelines for Social Analysis of Development Projects—Operational Summary,'' Manila, The Philippines, June 1991b, pp. 2–14.

———, ''Guidelines for the Health Impact Assessment of Development Projects,'' Manila, The Philippines, Nov. 1992, pp. 3–5.

Blair, L. M., Garey, R. B., and Stevenson, W., ''An Approach to Assessing the Supply of Workers to Large Construction Projects: An Economic Model,'' *Environmental Impact Assessment Review,* vol. 3, no. 4, 1982, pp. 417–420.

Burchell, R. W., et al., ''Traffic Impact Analysis,'' Chap. 9 in *Development Impact Assessment—Handbook and Model,* Urban Land Institute, Washington, D.C., draft ed., 1993.

Burdge, R. J., ''The Social Impact Assessment Model and the Planning Process,'' *Environmental Impact Assessment Review,* vol. 7, 1987, pp. 141–150.

——— and Robertson, R. A., ''Social Impact Assessment and the Public Involvement Process,'' *Environmental Impact Assessment Review,* vol. 10, 1990, pp. 81–90.

Canter, L. W., ''Pragmatic Suggestions for Incorporating Risk Assessment Principles in EIA Studies,'' *The Environmental Professional,* vol. 15, no. 1, 1993, pp. 125–138.

———, ''Systematic Studies of Prediction Techniques,'' paper presented at the *Annual Meeting of the International Association* for *Impact Assessment,* Detroit, May 25–26, 1983.

———, Atkinson, S. F., and Leistritz, F. L., *Impact of Growth,* Lewis Publishers, Chelsea, Mich., 1985.

Chalmers, J. A., and Anderson, E. J., ''Economic/Demographic Assessment Manual: Current Practices, Procedural Recommendations, and a Test Case,'' Engineering and Research Center, U.S. Bureau of Reclamation, Denver, Colo., Nov. 1977.

Christensen, K., ''Social Impacts and Land Development,'' URI 15700, Urban Land Institute, Washington, D.C., Sept. 1976.

Council on Environmental Quality (CEQ), ''Council on Environmental Quality Guidelines,'' *Federal Register,* vol. 38, no. 147, Aug. 1, 1973, pp. 20550–20562.

———, ''National Environmental Policy Act—Regulations,'' *Federal Register,* vol. 43, no. 230, Nov. 29, 1978, pp. 55978—56007.

Defillo, R., ''Economic Impact Forecast System (EIFS): Some Insights,'' Department of Urban and Regional Planning, University of Illinois, Urbana, Dec. 1990, p. 1.

Finsterbusch, K., Ingersoll, J., and Llewellyn, L., eds., *Methods for Social Analysis in Developing Countries,* Westview Press, Boulder, Colo., 1990.

Fitzsimmons, S. J., Stuart, L. I., and Wolff, P. C., ''Social Assessment Manual—A Guide to the Preparation of the Social Well-Being Account,'' U.S. Bureau of Reclamation, Denver, Colo., July 1975.

Goldsmith, E., and Hildyard, N., *The Social and Environmental Effects of Large Dams,* Sierra Club, San Francisco, 1984, pp. 17–18.

Grady, S., Braid, R., Bradbury, J., and Kerley, C., ''Socioeconomic Assessment of Plant Closure: Three Case Studies of Large Manufacturing Facilities,'' *Environmental Impact Assessment Review,* vol. 7, 1987, pp. 151–165.

Henningson, Durham, and Richardson, ''Guidelines for the Application of the EIAP to the ICBM Modernization Program—Draft,'' prepared for the U.S. Air Force Ballistic Missile Office, Norton Air Force Base, Santa Barbara, Calif., May 1982.

Hundloe, T., McDonald, G. T., Ware, J., and Wilks, L., ''Cost-Benefit Analysis and Environmental Impact Assessment,'' *Environmental Impact Assessment Review,* vol. 10, 1990, pp. 55–68.

Institute of Transportation Engineers, ''Traffic Access and Impact Studies for Site Development,'' Institute of Transportation Engineers, Washington, D.C., 1989.

———, ''Trip Generation Manual,'' 4th ed., Institute of Transportation Engineers, Washington, D.C., Oct., 1987.

Knaap, G., ''Consultants and Local Governments Use EIFS,'' *ETIS Quarterly,* Department of Urban and Regional Planning, University of Illinois, Urbana, summer 1992, pp. 1, 3.

Leistritz, F. L., and Murdock, S. H., *Socioeconomic Impact of Resource Development: Methods for Assessment,* Westview Press, Boulder, Colo., 1981.

Leopold, L. B., ''Hydrology for Urban Land Planning—A Guidebook on the Hydrologic Effects of Urban Land Use,'' *U.S. Geological Survey Cir-*

cular 554, U.S. Geological Survey, Washington, D.C., 1968.

McKeever, R. J., Griffin, N. M., and Spink, Jr., F. H., "Shopping Center Development Handbook," Urban Land Institute, Washington, D.C., 1977.

McMahon, R. F., "Project Summary: Socioeconomic Impacts of Water Quality Strategies," EPA 600/S5-82-001, U.S. Environmental Protection Agency, Cincinnati, Ohio, Aug. 1982.

Muller, T., "Economic Impacts of Land Development: Employment, Housing, and Property Values," URI 15800, Urban Land Institute, Washington, D.C., Sept. 1976.

———, "Fiscal Impacts of Land Development," URI 98000, Urban Land Institute, Washington, D.C., 1975.

Nieves, L. A., et al., "Identification and Estimation of Socioeconomic Impacts Resulting from Perceived Risks and Changing Images: An Annotated Bibliography," ANL/EAIS/TM-24, Environmental Assessment and Information Sciences Division, Argonne National Laboratory, Argonne, Ill., Feb. 1990.

Transportation Research Board, "Highway Capacity Manual," Transportation Research Board, Washington, D.C., 1985.

———, "Interim Materials on Highway Capacity," *Transportation Research Circular* no. 212, Washington, D.C., Jan. 1980, pp. 163–164.

Urban Land Institute, *Business and Industrial Park Development Handbook,* Urban Land Institute, Washington, D.C., 1988.

———, *Project Infrastructure Development Handbook,* Urban Land Institute, Washington, D.C., 1989.

U.S. Army Construction Engineering Research Laboratory, "Environmental Review Guide For USAREUR," Vol. 3., Champaign, Ill., 1989.

U.S. Department of Energy, "Socioeconomic Impact Assessment: A Methodology Applied to Synthetic Fuels," U.S. Department of Energy, Washington, D.C., 1978.

U.S. Department of Transportation, "Environmental Assessment Notebook Series," 6 vols., U.S. Department of Transportation, Washington, D.C., 1975.

Vester, J., "EIFS Forecasting Model," *ETIS Quarterly,* Department of Urban and Regional Planning, University of Illinois, Urbana, Mar. 1990, p. 2.

Voorhees and Associates, "Interim Guide for Environmental Assessment," prepared for U.S. Department of Housing and Urban Development, Washington, D.C., June 1975.

Wisconsin State Planning Office, Department of Administration, "Public Service Costs and Development," Wisconsin State Planning Office, Madison, Wisc., 1975.

World Health Organization (WHO), "Environmental Health Impact Assessment of Urban Development Projects," WHO Regional Office for Europe, Copenhagen, June 1985.

———, "Health and Safety Component of Environmental Impact Assessment," report on a WHO Meeting, Regional Office for Europe, Copenhagen, Feb. 24–28, 1986.

CHAPTER 15

Decision Methods for Evaluation of Alternatives

Environmental impact studies typically address a minimum of two alternatives, and they can include upwards of fifty alternatives. Typical studies focus on three to five alternatives. A two-alternative study usually represents a choice between construction and operation of a project versus project nonapproval. The alternatives to be addressed may encompass a wide range of considerations. For example, alternatives for dams and hydroelectric projects may include project construction and operation at different sites; differences in design and operational procedures, including the incorporation of various mitigation measures; and timing options for the construction and operational phases. Depending upon the project need, still other alternatives for flood control, water supply, recreation, and energy supply could be included in the analysis. The categories of alternatives for projects, expressed generically, may include (1) site-location alternatives; (2) design alternatives for a site; (3) construction, operation, and decommissioning alternatives for a design; (4) project-size alternatives; (5) phasing alternatives for size groupings; (6) no-project or no-action alternatives; and (7) timing alternatives relative to

project construction, operation, and decommissioning.

The CEQ regulations highlight the importance of the assessment of alternatives by noting that this represents the "heart of the environmental impact statement" (CEQ, 1978). The regulations indicate that information on the environmental impacts of the proposed action and the alternatives should be presented in comparative form, thus sharply defining the issues and providing a clear basis for choice among options by the decision maker and the public.

The focus of this chapter is on systematic methods which can be used for comparing and evaluating alternatives. Weighting-scaling (or -ranking or -rating) checklists can be used in such comparisons and evaluations. "Scaling" refers to the assignment of algebraic scales or letter scales to the impact of each alternative being evaluated on each identified environmental factor; functional relationships typically serve as the basis for these assignments. "Ranking" checklists are where alternatives are ranked from best to worst in terms of their potential impacts on identified environmental factors, while "rating" involves the use of a predefined rating scheme. These types of checklists are useful for comparative evaluations of alternatives, thus they provide a basis for selection of the preferred alternative. "Weighting-scaling," or "-rating" checklists refer to methodologies that embody the assignment of relative-importance weights to environmental factors and impact scales or ratings for each alternative relative to each factor. "Weighting-ranking" checklists involve importance-weight assignments and the relative ranking of the alternatives from best to worst in terms of their impacts on each environmental factor. Numerous weighting-scaling (or -rating) and weighting-ranking checklist methodologies have been developed for EISs (Canter, 1979). These methodologies represent adaptations of multiple-criterion or multiple-attribute decision-making techniques used in other fields; such techniques are also called "decision-analysis techniques" and "decision-support systems." Additional in-

formation on other, related types of methodologies is in Chapter 3.

CONCEPTUAL BASIS FOR TRADE-OFF ANALYSIS

To achieve a systematic approach to deciding among alternatives, it is desirable to use trade-off analysis. Trade-off analysis typically involves the comparison of a set of alternatives relative to a series of decision factors. Table 15.1 displays a trade-off matrix for systematically comparing the groups of alternatives or specific alternatives within a group relative to a series of decision factors. The following approaches can be used to complete the trade-off matrix:

1. A qualitative approach, in which descriptive synthesized and integrated information on each alternative relative to each decision factor is presented in the matrix

TABLE 15.1

EXAMPLE OF TRADE-OFF ANALYSIS MATRIX FOR DECISION MAKING

Decision factors	Alternative plan			
	1	2	3	4
Success in meeting defined needs and identified objectives				
Economic efficiency				
Benefits, costs				
Excess benefits				
Internal rate of return				
Environmental cost-benefit analysis				
Environmental impacts				
Air quality				
Surface-water quantity, quality				
Soil quality and groundwater quantity, quality				
Noise				
Ecosystems				
Habitat quantity, quality				
Threatened or endangered species				
Historical-archaeological resources				
Socioeconomic characteristics				
Human health risks				
Public preference				

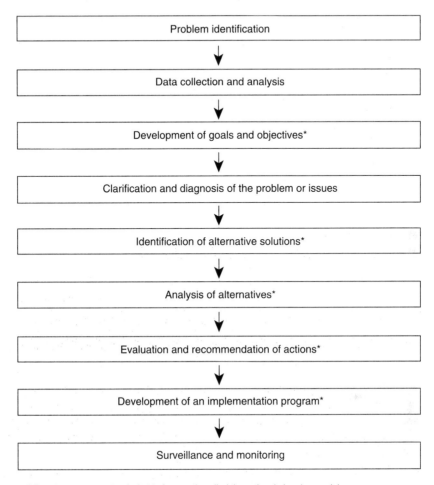

* Denotes components of what is frequently called the rational planning model.

FIGURE 15.1
Steps in Rational Planning Model (Dzurik, 1990).

2. A quantitative approach, in which quantitative synthesized and integrated information on each alternative relative to each decision factor is displayed in the matrix; or a combination qualitative-quantitative approach

3. A ranking, rating, or scaling approach, in which the qualitative or quantitative information on each alternative is summarized by using the assignment of a rank, rating, or scale value relative to each decision factor

(the rank, rating, or scale value is presented in the matrix)

4. A weighting approach, in which the importance weight of each decision factor relative to each other decision factor is considered, and the resultant discussion of the information on each alternative (qualitative, quantitative, or ranking, rating, or scaling) is presented in terms of the relative importance of the decision factors

5. A weighting–ranking, -rating, or -scaling approach, in which the importance weight for each decision factor is multiplied by the ranking, rating, or scale of each alternative, and the resulting products for each alternative are then totaled to develop an overall composite index or score for each alternative; the index may take the form of

$$\text{Index}_j = \sum_{i=1}^{n} \text{IW}_i \, R_{ij}$$

where Index_j = composited index for jth alternative
n = number of decision factors
IW_i = importance weight of ith decision factor
R_{ij} = ranking, rating or scale of jth alternative for ith decision factor

Decision making in relation to selecting the proposed action from alternatives which have been analyzed and compared should take place in relation to an overall planning model, which is also called the "rational planning model," as shown in Figure 15.1. An illustration of the application of this model to the selection of a "best practicable environmental option" (BPEO) (in this case, for pollution control) is shown in Figure 15.2. Decision-focused checklists can be used in the "Analysis of alternatives" step in Figure 15.1, and the "Select preferred option" step in Figure 15.2. Finally, McAllister (1986) has suggested that evaluation of an alternative can be divided into two phases: analysis, in which the whole is divided into parts, and synthesis, in which the parts are reformed into a whole. These phases are portrayed graphically in Figure 15.3.

At a minimum, a table containing both qualitative and quantitative comparative information (similar in structure to the matrix shown in Table 15.1) should be assembled and completed. (Table 17.5 is an illustration of a completed matrix.) This information could be used to prepare a trade-off analysis and select the proposed ac-

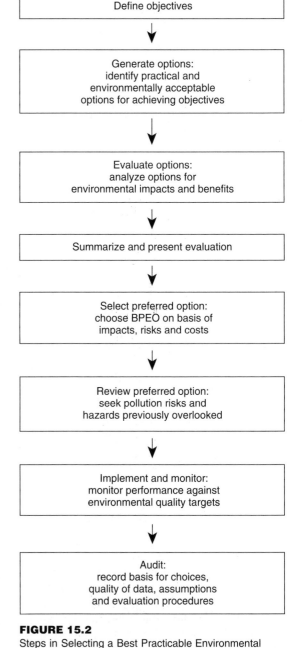

FIGURE 15.2
Steps in Selecting a Best Practicable Environmental Option (BPEO) Using the Rational Planning Model (Selman, 1992).

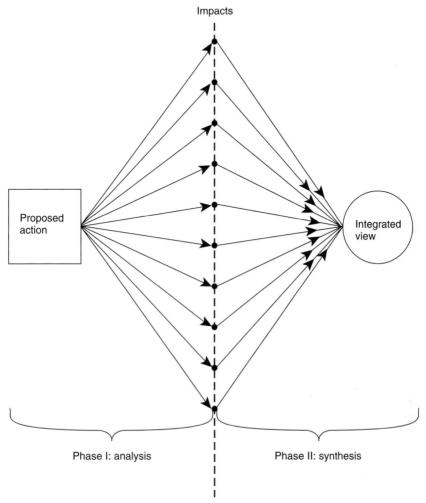

FIGURE 15.3
The Two Phases of the Alternative Evaluation Process (McAllister, 1986).

tion. If the qualitative and/or quantitative approach is used for completion of the matrix, information for this approach relative to the environmental impacts should be based on impact prediction. This information would also be needed for impact ranking, rating, or scaling. The remainder of this chapter delineates different techniques which can be used to assign importance weights and accomplish impact scaling, rating, or ranking. Examples of such

decision-focused checklists are included, along with summary observations on the usage of such approaches.

IMPORTANCE WEIGHTING OF DECISION FACTORS

If the importance-weighting approach is used in decision making, the critical issue is the use of an effective method to assign importance

TABLE 15.2

**EXAMPLES OF TYPES OF IMPORTANCE-WEIGHTING TECHNIQUES
USED IN ENVIRONMENTAL IMPACT STUDIES**

Ranking
Nominal-group process
Rating
Predefined importance scale
Multiattribute (or multicriterion) utility measurement
Unranked pairwise comparison
Ranked pairwise comparison
Delphi study

weights to the individual decision factors or, at least, to arrange the factors in a rank ordering of importance. Table 15.2 lists some structured importance-weighting or ranking techniques which could be used. These techniques have been used in numerous EIS decision-making efforts. Brief descriptions of several techniques listed in Table 15.2 will be presented as illustrations. In addition to the structured techniques, less formal approaches, such as reliance on the scoping process, can also be used as a basis for importance weighting.

"Ranking techniques" involve the rank ordering of decision factors in their relative order of importance. If there are n decision factors, rank ordering would involve assigning the value of 1 to the most important factor, 2 to the second-most-important factor, and so forth, until n is assigned to the least important factor. It should be noted that the rank order numbers could be reversed; that is, n could be assigned to the most important factor, $n - 1$ to the second-most-important factor, and so forth, until 1 is assigned to the least important factor. One illustration of a structured ranking technique, the nominal-group process (Voelker, 1977), will be briefly described herein.

The nominal-group-process (NGP) technique, an interactive group technique, was developed in 1968 (Voelker, 1977). It was derived from social-psychological studies of decision conferences, management-science studies of group-judgment aggregation, and social-work studies of problems surrounding citizen participation in program planning. The following four steps are involved in the use of NGP for importance weighting: (1) nominal (silent and independent) generation of ideas in writing by a panel of participants, (2) round-robin listing of ideas generated by participants in a serial discussion, using a flip chart, (3) group discussion of each recorded idea for the purpose of clarification and evaluations, and (4) independent voting on priority ideas, with the group decision based on mathematical rank ordering.

Rating techniques for importance weighting basically involve the assignment of importance numbers to a series of decision factors and possibly, although not always, their subsequent normalization using a mathematical procedure. Two examples of rating techniques will be presented herein: (1) the use of a predefined importance scale (Linstone and Turoff, 1975) and (2) the use of the multiattribute utility-measurement technique (Edwards, 1976). The NGP technique could also be used for *rating* the importance of decision factors. In this approach, decision factors are assigned numerical values based on predefined importance scales. Table 15.3 delineates a five-level scale of designations and corresponding definitions to be considered in the assignment of numerical values to decision factors. Usage of predefined scales can aid in systematizing importance-weight assign-

TABLE 15.3

EXAMPLE OF PREDEFINED IMPORTANCE SCALE

Scale reference[a]	Definition
1. Very important	A most relevant point First-order priority Has direct bearing on major issues Must be resolved, dealt with, or treated
2. Important	Is relevant to the issue Second-order priority Significant impact, but not until other items are treated Does not have to be fully resolved
3. Moderately important	May be relevant to the issue Third-order priority May have impact May be a determining factor to major issue
4. Unimportant	Insignificantly relevant Low priority Has little impact Not a determining factor to major issue
5. Most unimportant	No priority No relevance No measurable effect Should be dropped as an item to consider

[a]Could use numbers or letter codes in the application; the pertinent rationale for the assigned importance weight should be specified in the study; finally, one to several decision factors, or possibly no decision factors, could be assigned to each scale reference.
Source: Linstone and Turoff, 1975, p. 137.

ments. Such assignments can be made by individuals or an interdisciplinary study team.

Edwards (1976) described the multiattribute utility-measurement (MAUM) technique developed for use in decision-making processes involving different publics. The MAUM technique can be used to delineate the particular values of each participant in the process (decision maker, expert, pressure group, government, etc.) and to show how much these values differ; in doing so, the extent of such differences can frequently be reduced. The basic assumption is that the values of the participants are reflected in the importance weights they assign to individual factors. It should also be noted that the MAUM technique can be used by an individual, a small group of persons, or multiple publics.

The 10 basic steps in the MAUM technique are described as follows; steps 1 through 7 relate to the importance weighting of decision factors (Edwards, 1976):

Step 1—Identify the person or organization whose utilities are to be maximized. If, as is often the case, several organizations have stakes and voices in the decision, they must all be identified. ("Utilities" refers to general goals or objectives in the terminology used herein.)

Step 2—Identify the issue or issues to which the utilities needed are relevant. ("Issues" are, in this context, needs being addressed.)

Step 3—Identify the entities to be evaluated. Formally, they are outcomes of possible actions. But in a sense, the distinction between an outcome and the opportunity for further actions is usually fictitious. ("Entities," in the terminology used herein, are the alternatives; "outcomes" would reflect the evaluation of each entity relative to the decision factors.)

Step 4—Identify the relevant dimensions of value for evaluation of the entities. As has often been noted, goals ordinarily come in hierarchies. But it is often practical and useful to ignore their hierarchical structure, and instead to specify a simple list of goals that seem important for the purpose at hand. It is important not to be too expansive at this stage. The number of relevant dimensions of value should be modest, for reasons that will be apparent shortly. (It should be noted that "dimensions of value" refers to the decision factors for evaluation of the alternatives.)

Step 5—Rank the dimensions in order of importance. This ranking job, like step 4, can be performed either by an individual, by an interdisciplinary study team, by representatives of parties having conflicting interest (and thus, conflicting values) acting separately or as a group.

Step 6—Rate the dimensions in terms of importance, preserving existing ratios. To do this, start by assigning the least-important factor, or dimension, an importance value of 10. Now consider the next-least-important dimension. How much more important (if at all) is it than the least-important dimension? Assign it a number that reflects that ratio. Continue up the list, checking each set of implied ratios as each new judgment is made. Thus, if a dimension is assigned a weight of 20, while another is assigned a weight of 80, it means that the 20 dimension is 1/4 as important as the 80 dimension, and so on. By the time the most important factors are reached, there will be many checks to perform; typically, respondents will need to review previous judgments to make them consistent with present ones.

Step 7—Sum the importance weights, divide each by the sum, and multiply by 100. This is a purely computational step which converts importance weights into numbers that, mathematically, are like fractions. The choice of a 0-to-100 scale is, of course, completely arbitrary.

Step 8—"Measure" the location for each entity being evaluated on each dimensional "yardstick." The word "measure" is used rather loosely here. There are three classes of dimensions: purely subjective, partly subjective, and purely objective. The purely subjective dimensions are perhaps the easiest to measure; get an appropriate expert to estimate the position of the entity on that dimension on a 0-to-100 scale, where 0 is defined as the "minimum plausible value" and 100 is defined as the maximum plausible value. Note the use of the terms "minimum plausible" and "maximum plausible" rather than "minimum possible" and "maximum possible." The minimum plausible value often is not the same as total absence of the dimension. A partly subjective dimension is one in which the units of measurement are objective, but for which the locations of the entities must be subjectively estimated. A purely objective dimension is one that can be measured nonjudgmentally, in objective units, before the decision is made. For partly or purely objective dimensions, it is necessary to have the estimators provide not only values for each entity to be evaluated, but also minimum and maximum plausible values, expressed in the natural units of each dimension.

Step 9—Calculate utilities for entities. The equation is

$$U_j = \sum_{i=1}^{n} (W_i)(U_{ij})$$

where U_j = aggregate utility for ith entity (overall evaluation score, or index, for jth alternative)

j = number of entities (number of alternatives)

W_i = normalized importance weight of ith dimension of value (importance weight for ith decision factor); W_i values are output from step 7

n = number of dimensions of value (number of decision factors)

U_{ij} = scaled position of jth entity on ith dimension (scaled position of jth alternative on ith decision factor); U_{ij} measures are output from step 8

Step 10—Decide. If a single alternative is to be chosen, the rule is simple: maximize U_j.

Paired-comparison techniques (unranked and ranked) for importance weighting basically involve a series of comparisons between decision factors and a systematic tabulation of the numerical results of the comparisons. These techniques have been extensively used in decision-making efforts, including numerous examples related to EISs.

One of the most useful techniques is the unranked paired-comparison technique developed by Dean and Nishry (1965). This technique, which can be used by an individual or a group, involves the comparison of each decision factor to each other decision factor, in a systematic manner. The weighting technique consists of considering each factor relative to every other factor—thus, on a pairwise basis—and, assigning a value of 1 to the factor considered to be the more important and a value of 0 to the remaining factor. For example, suppose that there are four basic decision factors—F1 through F4; F1 could be degree of meeting identified needs or objectives, F2 could be economic efficiency, F3 could be social impacts, and F4 could be environmental impacts. The use of this unranked paired-comparison technique for this case is shown in Table 15.4. It should be noted that the assignment of 0 to a member of a pair does not denote no importance; it simply means that in the pair considered, it is of less importance.

A dummy factor, F5, is also included in the example shown in Table 15.4. The dummy factor is included so as to preclude the net assignment of a value of 0 to any of the basic factors (F1 through F4 in the process of each paired comparison)—i.e., the dummy factor is included as a "place keeper" to avoid skewing the process. The "dummy factor" is defined as that factor, of each pair, deemed less important of the two. If two factors are considered to be of equal importance, then a value of 0.5 is assigned to each factor in the pair.

TABLE 15.4

EXAMPLE OF PAIRED-COMPARISON TECHNIQUE FOR IMPORTANCE WEIGHT ASSIGNMENTS

Factor	Assignment of weight[a]										Sum	FIC
F1	1	1	1	1							4	0.40
F2	0				1	0	1				2	0.20
F3		0			0			0	1		1	0.10
F4			0			1		1	1		3	0.30
F5 (dummy)				0			0	0	0		0	0
Total											10	1.00

[a]It is vitally important that the rationale basic to each assignment be documented.

Following the assignment of relative importance to each factor pair—a process involving several iterations to be sure that each factor is being compared with each other factor in a consistent manner—the rationale for each decision should be documented. It cannot be overstressed that the most important aspect of using this technique is the careful delineation of the rationale basic to each 1 and 0 assignment. Following this documentation, the individual weight assignments are summed, and the factor-importance coefficient FIC is calculated. FIC is equal to the sum value for an individual factor divided by the sum for all of the factors and is expressed as a decimal fraction. The total of the sum column should equal $(n)(n - 1)/1$, where n is the number of factors included in the assignment of weights. In the example in Table 15.4, five factors were included, hence the Sum column total should equal 10; the total of the FIC column should equal 1.00.

The FIC column in Table 15.4 indicates that, of the four factors, F1 is the most important, followed by F4, F2, and F3. Whether or not, as a next step, the actual FIC fractions are used in a trade-off analysis, this unranked paired-comparison approach enables the rank ordering of the four decision factors from most important to least important. In addition, importance weighting of subfactors can be accomplished by the same method. For example, in Table 15.1, the Environmental Impacts row includes 10 subfactors. Ross (1976) described a method to check for the consistency of importance-weight assignments made by using the unranked paired-comparison technique.

The key feature of ranked pairwise-comparison techniques is that an initial rank ordering of all decision factors is required. A specific EIA example will be cited here; this example deals with importance weighting for water-resources projects (Dee et al., 1972). The general methodology, which was developed for the U.S. Bureau of Reclamation's water-resources projects, is called the "Battelle environmental evaluation system" (EES). The relative importance of the 78 predefined decision factors, or parameters, considered in EES was expressed in commensurate units, called "parameter-importance units" (PIUs), by quantifying several individuals' subjective value judgments. The weighting technique used by the method developers, Battelle-Columbus, was based on sociopsychological scaling techniques and a modified Delphi procedure (Dee et al., 1972). (The Delphi procedure itself will not be discussed here; however, it was summarized in Chapter 5.) The importance-weighting technique in EES forces a systematic consideration of the factors, minimizes individual bias, produces consistent comparisons, and aids in the convergence of judgment.

In ranked pairwise comparison, the list of decision factors to be compared is ranked according to selected criteria, and then successive paired comparisons are made between contiguous parameters to select, for each parameter pair, the degree of difference in importance. The initial EES ranking was made based on considering the following three criteria relative to each parameter: (1) inclusiveness, (2) reliability of measurements, and (3) sensitivity to changes in the environment.

The 78 EES decision factors, or environmental parameters, are grouped into four categories (ecology, environmental pollution, aesthetics, and human interest) and subdivided by 17 quantitative components (under "Ecology—species and populations, and habitats and communities; under "Environmental pollution"—water pollution, air pollution, land pollution, and noise pollution; under Aesthetics—land, air, water, biota, manufactured objects, and composition; and under "Human interest"—educational-scientific packages, historical packages, cultures, mood or atmosphere, and life patterns), as shown in Figure 15.4. The following 10 steps are used in importance weighting:

Step 1—Select a group of individuals for conducting the evaluation, and explain to them,

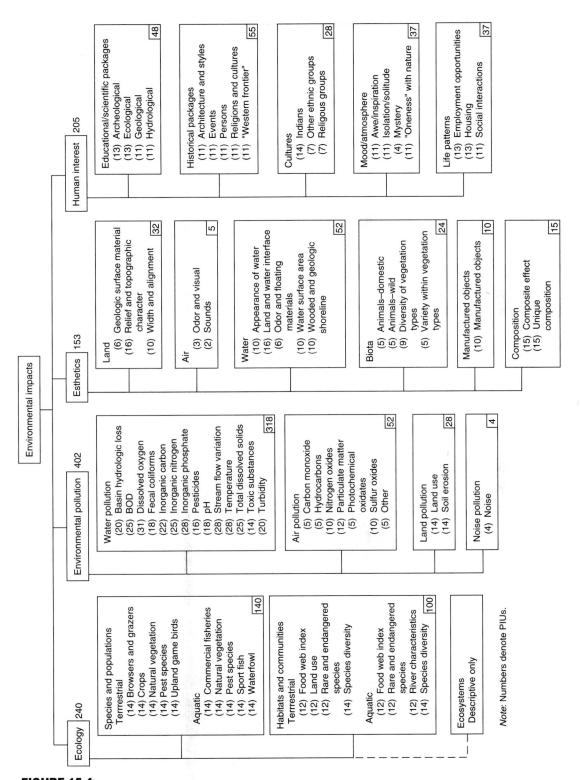

FIGURE 15.4
Environmental Evaluation System (EES) (Dee et al., 1972).

in detail, the weighting concept and the use of their rankings and weightings.

Step 2—Rank the categories, components, or parameters that are to be evaluated.

Step 3—Assign a value of 1 to the first category on the list. Then compare the second category with the first to determine how much the second is worth compared to the first. Express this value as a decimal ($0 < x \leq 1$).

Step 4—Continue with these pairwise comparisons until all items in the list have been evaluated. (Compare the third with the second, the fourth with the third, etc.)

Step 5—Multiply percentages and express over a common denominator, using the average values for all individuals in the experiment.

Step 6—In weighting the categories or components, adjust the decimal values from step 5 if the number of elements in the parameter groups being evaluated are unequal. Adjustment is made by proportioning these decimal values in relation to the number of parameters (elements) included in that grouping.*

---

*For example, the hierarchical system shown in Figure 15.4 has an unequal number of elements in each grouping. To be mathematically correct, all levels of the EES hierarchy should have an equal number of elements. However, there has not been sufficient knowledge in many of these areas to permit an equal number of elements at the same level of detail. This difference in the number of elements from group to group must be taken into consideration when assigning PIUs in the ranking and weighting. Therefore, in the ranking and weighting procedure (steps 1 through 5), the researchers were asked to assume an equal number of elements in the groupings being compared. These value judgments were then adjusted in proportion to the number of elements in each group. Because the purpose of the weighting procedure was to assign weights to parameters, if an adjustment were not made, the individual parameters grouped under "water pollution" would not receive sufficient weight as compared to those under "noise," because the total number of units available under water pollution would have to be distributed among 14 parameters whereas those for noise would all be assigned to the single noise parameter. For this reason, comparisons between components and categories should be based on average values of the PIU for the grouping, not the sum of the values.

Step 7—Multiply these averages by the number of PIUs to be distributed to the respective grouping.

Step 8—Repeat steps 2 through 7 for all categories, components, and parameters in the EES.

Step 9—Indicate to the individuals by controlled feedback the group results of the weighting procedure.

Step 10—Repeat the experiment with the same group of individuals or another group to increase the reliability of the results.

The following numerical example illustrates the use of the 10 steps (Dee et al., 1972):

Background

Consider three components *(A, B, C)* that have been selected in steps 1 and 2, with these components consisting of 8 parameters, 4 in *A*, 2 in *B*, and 2 in *C*.

Step 2

Ranking of components yields *B*, then *C*, and *A*.

Steps 3, 4

Assign weights based on professional judgments:

$$B = 1$$
$$C = 1/2 \text{ importance of } B$$
$$A = 1/2 \text{ importance of } C$$

Step 5

Multiply percentages and express over common denominator. Assume the average values of all individuals are given below:

$$
\begin{array}{l}
B = 1 \\
C = 0.5 \\
A = \underline{0.25} \\
1.75
\end{array}
$$

$$B = \frac{1}{1.75} = 0.57$$

$$C = \frac{0.5}{1.75} = 0.29$$

$$A = \frac{0.25}{1.75} = \underline{0.14}$$
$$1.00$$

Step 6

Adjust for unequal number of parameters in each component.

$$B = 0.57 \ (0.25) = \quad 0.14$$
$$C = 0.29 \ (0.25) = \quad 0.07$$
$$A = 0.14 \ (0.50) = \quad \underline{0.07}$$
$$0.28$$

Using the new total, the components' values are

$$B = \frac{0.14}{0.28} = 0.50$$

$$C = \frac{0.07}{0.28} = 0.25$$

$$A = \frac{0.07}{0.28} = \underline{0.25}$$
$$1.00$$

and the average values for the parameters in *B, C,* and *A* are

$$B = \frac{0.50}{2} = 0.25$$

$$C = \frac{0.25}{2} = 0.125$$

$$A = \frac{0.25}{4} = 0.0625$$

Step 7

Multiply these adjusted values by the appropriate PIU, assumed in this case to be 20.

$$B = 20 \ (0.50) = \quad 10$$
$$C = 20 \ (0.25) = \quad 5$$
$$A = 20 \ (0.25) = \quad 5$$

Steps 8 through 10 continue the process, addressing all categories, components, and par-

ameters. Then review the results and iterate the process until reliable estimates are obtained.

In the EES, a total of 1,000 PIUs were assigned to the parameters by first distributing the units to the 4 categories, then to the 17 quantitative components, and, finally, to the 78 parameters. That is, for each pair of categories—for example, aesthetics and environmental pollution—the participating group determined which of the two was more important, and then assigned appropriate weights. The process was repeated until all the units had been distributed among all parameters.

Dee et al. (1972) noted that instead of using the initial weight resulting from the importance-weighting procedure, an aggregate weight based on several iterations of the technique is preferred. After each iteration, the participants are given selected information about the group weights. This information can include the group mean and variance, or other pertinent data. In the weighting procedure employed in developing the EES, the participants' mean value was given in the feedback stage. All of the weighting and feedback was performed by means of formal feedback statements, thereby avoiding undesirable direct interchange of judgments between the individuals in the test (this is analogous to a feature in the Delphi-study procedure).

SCALING, RATING, OR RANKING OF ALTERNATIVES

Scaling, rating, or ranking of each alternative relative to each decision factor is the second major component in the use of the multiple-criterion, or multiattribute, decision-making approach. Selected rating and ranking concepts and techniques were described in the previous section. Westman (1985) described four types of "scales" used in environmental impact studies.

TABLE 15.5

FOUR TYPES OF IMPACT SCALES, AND PERMISSIBLE MATHEMATICAL AND STATISTICAL OPERATIONS

Scale	Nature of scale	Permissible mathematical transformation	Permissible statistical procedure
Nominal	Classifies impacts of alternatives	One-to-one substitution	Information statistics
Ordinal	Ranks alternatives in terms of impacts	Equivalence to another monotonically increasing or decreasing function	Nonparametric statistics
Interval	Rates the impacts of alternatives in units of equal difference	Linear transformation	Parametric statistics
Ratio	Rates the impacts of alternatives in units of equal difference and equal ratio	Multiplication or division by a constant or other ratio-scale value	Parametric statistics

Source: After Westman, 1985.

The four types of scales and selected attributes are listed in Table 15.5. "Nominal scales" can be used to classify objects, while "ordinal scales" rank objects or impacts in order. Ordinal scales do not convey how much better one object or impact is than another, but simply indicate relative order. "Interval scales" rate the quantitative degree of difference between objects or impacts. Finally, "ratio scales" indicate the quantitative degree of difference between objects in relation to some absolute starting point.

Several different techniques have been used for the comparative evaluation of alternatives relative to decision factors. Examples of techniques include the use of (1) the alternative-profile concept, (2) a reference alternative, (3) linear scaling based on the maximum change, (4) letter or number assignments designating impact categories, (5) evaluation guidelines, (6) unranked paired-comparison techniques, (7) functional curves, or (8) predefined impact-rating criteria.

Bishop et al. (1970) described the alternative-profile approach for impact scaling. This profile consists of a graphical, scaled presentation of the effects of each alternative relative to each decision factor. Each profile scale is expressed on a percentage basis, ranging from a negative to a positive 100 percent, with 100 percent being the maximum absolute value of the impact measure adopted for each decision factor; this is called "ratio scaling." The impact measure represents the maximum change, either plus or minus, associated with a given alternative being evaluated. If the decision factors are displayed along the impact scale, which is calibrated from $+100$ percent to -100 percent, a dotted line can be used to connect the plotted points for each alternative and thus describe its profile. The alternative profile concept is useful for visually displaying the relative impacts of alternatives on a series of decision factors. Some observations related to factor (alternative) profiling follow:

1. It is a useful technique for visually displaying the relative features of alternatives, particularly if unique color codes or other special notations are used for each alternative. However, the comparisons can become visually complicated if too many factors and/or too many alternatives are collectively compared. While there are no specific limits, it is suggested that more than 15 to 25 factors and 5 to 10 alternatives might lead to visual-interpretation complications.

2. Because of the relative comparisons presented, the trade-offs between alternatives can be easily identified, and appropriate descriptions can be prepared.
3. The technique can be used to identify where mitigation measures would be needed for an alternative. For example, suppose that one alternative exhibits good trade-off comparisons for every included decision factor other than particulate emissions to the atmosphere. This would suggest that incorporation of a particulate-emissions control system or other pertinent mitigation measures in the proposal would facilitate an advantageous trade-off analysis of the alternative relative to the particulate emissions factor.
4. By using the same vertical distance on the graph for each decision factor, it is implied that the factors are equally important. To avoid this, relative importance weights between decision factors can be depicted by differences in their vertical separation.
5. The approach of calculating percentage impacts (either beneficial or detrimental) by the relative comparisons of alternatives can be useful; however, the uncertainties related to predicted impacts and their relative size in terms of the resource bases or total pollutant loadings should be considered in the trade-off analysis and interpretation of results.

Salomon (1974) described a scaling technique for evaluation of cooling-system alternatives for nuclear power plants. To determine scale values, a reference cooling system was used and each alternative system compared to it. In the EIA process, the reference alternative could be the no-action alternative. In the example cited, the following scale values were assigned to the alternatives, based on the reference alternative: very superior ($+8$), superior ($+4$), moderately superior ($+2$), marginally superior ($+1$), no difference (0), marginally inferior (-1), moderately inferior (-2), inferior (-4),

and very inferior (-8). This technique involves nominal scaling.

Odum et al. (1971) utilized a scaling technique in which the actual measures of the decision factor for each alternative plan are normalized and expressed as a decimal of the largest measure for that factor. This constitutes linear or ratio scaling based on the maximum change. Additional information on this overall methodology is in a subsequent section.

A letter-scaling system is used in Voorhees and Associates (1975). This methodology incorporates 80 environmental factors oriented to the types of projects conducted by the U.S. Department of Housing and Urban Development. The scaling system consists of the assignment of a letter grade, from A+ to C−, to the impacts, with A+ representing a major beneficial impact and C− an undesirable, detrimental change. This constitutes a nominal scaling approach.

Duke et al. (1977) described a scaling-checklist methodology for the environmental quality (EQ) account for water-resources projects. (The EQ account is one of up to four "accounts" used in project planning.) Scaling is accomplished following the establishment of an evaluation guideline for each environmental factor. An "evaluation guideline" is defined as the smallest change in the highest existing quality in the region that would be considered significant. For example, assuming that the highest existing quality for dissolved oxygen in a region is 8 mg/L and that a reduction of 1.5 mg/L is considered significant, then the evaluation guideline would be 1.5 mg/L, irrespective of the existing quality in a given regional stream. Scaling is accomplished by quantifying the impact of each alternative relative to each environmental factor, and if the net change is less than the evaluation guideline, it is considered to be insignificant. If the net change is greater and moves the environmental factor toward its highest quality, then it is considered to be a beneficial impact; the reverse is true for those impacts that move the measure of the environmental factor away from its highest existing quality. This methodology is also a nominal scaling approach.

One of the most useful techniques for scaling, rating, or ranking of alternatives relative to each decision factor is the unranked paired-comparison technique described by Dean and Nishry (1965). This technique was also described earlier relative to its use for importance weighting of decision factors. Again, this technique can be used by an individual, interdisciplinary team, or other group for the scaling of alternatives. For example, suppose that the decision to be made involves four factors and that the importance weights have been assigned as shown in Table 15.4. Furthermore, suppose that there are three alternatives (A1, A2, and A3) to be evaluated relative to the four decision factors, and that the relevant qualitative and quantitative information is as shown in Table 15.6.

In this context, the unranked paired-comparison technique consists of considering each alternative relative to every other alternative relative to each decision factor, and assigning to the more-desirable (least environmentally disruptive) of the pair of alternatives a value of 1, and to the less desirable a value of 0. The use

and results of this paired-comparison technique for the three basic alternatives and four basic decision factors are shown in Tables 15.7 through 15.10, respectively. Again it should be noted that the assignment of 0 to a member of a pair does not denote total undesirability, merely less relative desirability. Therefore, this technique involves both interval and ordinal scaling.

A dummy alternative, called A4, is also included for this example (Tables 15.7 through 15.10), as in the earlier illustration. Again, if two alternatives are equally desirable relative to a decision factor, then a value of 0.5 is assigned to each alternative.

Following the assignment of the relative-desirability value to each alternative, with this process based on the qualitative and quantitative information in Table 15.6, the alternative choice coefficient (ACC) is determined. ACC is equal to the sum value for an individual alternative divided by the sum for all of the alternatives. The total of the Sum column, as shown in Tables 15.7 through 15.10, should be equal to $(M)(M - 1)/2$, where M is equal to the number of alternatives included in the assignments. In this example, four alternatives were included, hence the Sum column total in Tables 15.7

TABLE 15.6

EXAMPLE INFORMATION FOR TRADE-OFF ANALYSIS

Decision factor	Alternative		
	A1	A2	A3
F1	Achieves 95% of identified needs and objectives.	Achieves 75% of identified needs and objectives.	Achieves 85% of identified needs and objectives.
F2	Benefit-to-cost ratio is 1.3.	Benefit-to-cost ratio is 1.1.	Benefit-to-cost ratio is 1.5.
F3	Undesirable social impacts expected.	No social impacts expected.	Beneficial social impacts expected.
F4	Decreases overall environmental quality by 20%.[a]	Decreases overall environmental quality by 10%.[a]	Decreases overall environmental quality by 10%.[a]

[a]Environmental quality is reflected by joint consideration of air and water quality and available habitat quantity and quality.

TABLE 15.7

SCALING, RATING, OR RANKING OF ALTERNATIVES RELATIVE TO F1

Alternative	Assignment of desirability			Sum	ACC
A1	1 1	1		3	0.50
A2	0	0	1	1	0.17
A3		0 1	1	2	0.33
A4 (dummy)		0 0	0	0	0
Total				6	1.00

TABLE 15.8

SCALING, RATING, OR RANKING OF ALTERNATIVES RELATIVE TO F2

Alternative	Assignment of desirability			Sum	ACC
A1	1 0	1		2	0.33
A2	0	0	1	1	0.17
A3		1 1	1	3	0.50
A4 (dummy)		0 0	0	0	0
Total				6	1.00

TABLE 15.9

SCALING, RATING, OR RANKING OF ALTERNATIVES RELATIVE TO F3

Alternative	Assignment of desirability			Sum	ACC
A1	0 0	1		1	0.17
A2	1	0	1	2	0.33
A3		1 1	1	3	0.50
A4 (dummy)		0 0	0	0	0
Total				6	1.00

TABLE 15.10

SCALING, RATING, OR RANKING OF ALTERNATIVES RELATIVE TO F4

Alternative	Assignment of desirability			Sum	ACC
A1	0 0	1		1	0.16
A2	1	0.5	1	2.5	0.42
A3		1 0.5	1	2.5	0.42
A4 (dummy)		0 0	0	0	0
Total				6	1.00

through 15.10 is equal to 6; the total of the ACC column is equal to 1.00.

The ACC column in Table 15.7 indicates that alternative A1 is the most desirable relative to decision factor F1, and is followed by A3 and A2. Similar types of assessments could be made on the basis of the ACC values in Tables 15.8 through 15.10. Documentation of the rationale used for all ACC values is desirable. Whether or not the actual ACC fractions are later used in a trade-off analysis, this paired-comparison approach enables the rank ordering of the desirability of each alternative relative to each decision factor. As noted earlier, the paired-comparison technique based on Dean and Nishry (1965)—including the development of both FIC and ACC measures—has been used in a number of environmentally related studies.

"Functional curves," also called "functional relationships," "value functions," or "parameter-function graphs (or curves)" have also been used in a number of impact studies for scaling, rating, or ranking the impacts of alternatives relative to a series of decision factors. The functional curve is used to relate the objective evaluation of an environmental factor to a subjective judgment regarding its quality, based on a scale of designator values ranging from "high quality" to "low quality" (Dee et al., 1972); this constitutes ratio scaling. An example of a functional curve for rating impacts on species diversity is shown in Figure 15.5. Functional-curve information is also in Chapters 5 and 11. Dee et al. (1972, after p. 102) described the following seven steps which could be used in developing a functional curve for an environmental parameter:

Step 1—Obtain scientific information, when available, on the relationship between the parameter and the quality of the environment. Also, identify experts in the field who might aid in developing the value function.

Step 2—Order the parameter scale so that the lowest value of the parameter is 0 and it in-

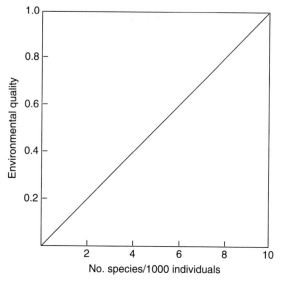

FIGURE 15.5

Functional-Relationship Graph for Species-Diversity Parameter (Dee et al., 1972).

creases in the positive direction; no negative values should be used.

Step 3—Divide the quality scale into equal intervals (typically calibrated from 0 to 1 or 0 to 100) and express the relationship between an interval and the parameter. Continue this procedure until a curve, straight-line, or step-function relationship exists.

Step 4—Average the curves over all experts in the experiment to obtain a group curve. (For parameters based solely on professional judgment, value functions should be determined by a representative population cross-section.)

Step 5—Review the group curve and expected results of using the curves in impact scaling with the experts developing the value-function estimation. If the group decides that a modification is needed, go to step 3; if not, continue.

Step 6—Repeat steps 1 through 5 until a curve exists for all parameters.

Step 7—Repeat the experiment with the same group or another group of experts to increase the reliability of the functions.

Predefined impact-rating criteria are conceptually analogous to the predefined importance scale in Table 15.3. Examples of predefined impact-rating criteria included in an EA (and, later, in an EIS) on a lignite-mine and power-plant expansion in Texas (called the "Cummins Creek project") are in Tables 15.11 (for air quality), 15.12 (for ecology), and 15.13 (for land use and aesthetics) (Wilson, 1991). These ratings were applied to various project alternatives. Usage of these ratings involves nominal scaling.

DEVELOPMENT OF DECISION MATRIX

The final step in multiple-criterion decision making is to develop a decision matrix displaying the products of the importance weights (or

---

TABLE 15.11

CUMMINS CREEK PROJECT—AIR-QUALITY-IMPACT RATING CRITERIA

Rating	Criteria
0	No potential negative impact.
1	The potential negative impacts, based on the level of emissions, would be insignificant.
2	The potential negative impacts, based on the level of emissions, would not be trivial, but would be handled by minimal controls.
3	The potential negative impacts, based on the level of emissions, would be significant but manageable.
4	The potential negative impacts, based on the level of emissions, would be serious and possibly unacceptable, but would be correctable.
5	The potential negative impacts, based on the level of emissions, would constitute a "fatal flaw"—i.e., one that is not easily mitigable.

Source: Adapted from Wilson, 1991.

TABLE 15.12

CUMMINS CREEK PROJECT—ECOLOGICAL-IMPACT RATING CRITERIA

Rating	Criteria
0	No potential negative impact to important species or habitats; no existing habitats (vegetation and/or soils) poor in quality and diversity or severely damaged.
1	The potential negative impact to important species or habitats would be minimal.
2	The potential negative impact to important species or habitats would be limited.
3	The potential negative impact to important species or habitats would be substantial.
4	The potential negative impact to important species or habitats would be only marginally acceptable.
5	The potential negative impact to important species or habitats would be excessive and unacceptable. Site is within an area containing critical habitat for endangered or threatened species.

Source: Adapted from Wilson, 1991.

TABLE 15.13

CUMMINS CREEK PROJECT—LAND-USE- AND AESTHETICS-IMPACT RATING CRITERIA

Rating	Criteria
0	No impact, no conflict with known existing or proposed land use. No alteration from assigned visual-resource-management classification. Project not visible from public access road.
1	Minimal impact, minimal conflict with known existing or proposed land use. Minimal alteration from assigned visual-resource-management classification. Minimal disturbance of existing view from public access road.
2	Limited impact, limited conflict with known existing or proposed land use. Limited alteration from assigned visual-resource-management classification. Limited disturbance of existing view from public access road.
3	Moderate impact, moderate conflict with existing or proposed land use. Moderate alteration from assigned visual-resource-management classification. Moderate disturbance of existing view from public access road.
4	Significant impact, significant conflict with known existing or proposed land use. The alteration from assigned visual-resource-management classification would be marginally acceptable. Project is highly visible from public access road. Considered marginally acceptable.
5	Major impact, major conflict with known existing or proposed land use. The alteration from assigned visual-resource-management classification would be excessive and unacceptable. Project is highly visible from public access road. Considered unacceptable. Land-use and aesthetics concerns constitute "a fatal flaw" to project development.

Source: Adapted from Wilson, 1991.

ranks) and the alternative scales (or ratings or ranks). Returning to the earlier example using the paired-comparison technique (see Table 15.4), a summary of the FIC values for the four decision factors, and the ACC values for the three alternatives (from Tables 15.7 through 15.10) is shown in Table 15.14. The final product matrix is shown in Table 15.15. Summation of the products for each alternative indicates that alternative A3 would be the best choice, followed by A1 and A2. The numerical bases for the differences in the three alternatives are indicated by the fractions shown in Table 15.15.

It should be noted that the example summarized in Tables 15.14 and 15.15 is very simple. It was presented to illustrate the concepts and mechanics of the methodology. In an actual application of the unranked paired-comparison methodology (or other multicriterion or multiattribute utility-measurement techniques), the number of decision factors would probably be greater than 4 (perhaps as many as 10 to 20 might be considered), and the number of alternatives might be greater than 3 (perhaps as many as 6 to 8 might be considered). However, the mechanics would be similar to those presented herein.

A relevant question related to the "final scores," or "indices" in Table 15.15 is "are the scores indicative of true differences in the three plans—that is, are there statistically significant differences between scores of 0.41 for A3, 0.33 for A1, and 0.26 for A2?" A nonparametric statistical test, called the "Friedman's Two-Way Analysis of Variance (Anova) by Ranks Test," can be applied to the data to obtain the answer. Detailed information on Friedman's test is contained in many nonparametric statistics books.

Nonparametric statistical testing is considered appropriate for the following reasons:

TABLE 15.14

FIC AND ACC VALUES FOR EXAMPLE DECISION PROBLEM

Decision factor	FIC values	ACC values, by alternative		
		A1	A2	A3
F1	0.40	0.50	0.17	0.33
F2	0.20	0.33	0.17	0.50
F3	0.10	0.17	0.33	0.50
F4	0.30	0.16	0.42	0.42

TABLE 15.15

PRODUCT MATRIX FOR TRADE-OFF ANALYSIS FOR EXAMPLE DECISION PROBLEM

Decision factor	FIC $\times$ ACC, by alternative		
	A1	A2	A3
F1	0.200	0.068	0.132
F2	0.066	0.034	0.100
F3	0.017	0.033	0.050
F4	0.051	0.124	0.124
Total score	0.334	0.259	0.406

1. Such testing is not based on any underlying assumptions about statistical distributions in terms of normality, variance, and the like.
2. Such testing allows the use of data ("numbers") based on either qualitative or quantitative approaches for development.
3. The computational algorithms are simple.
4. Such testing can be used with small samples (sets of information).

To serve as an example of the use of Friedman's Two-Way Anova by Ranks test, consider that the information in Table 15.15 represents the results of a comparison of alternatives. The following steps would be used to test the hypothesis that the alternatives are not statistically different:

Step 1—Assign rank-order numbers to each alternative for each decision factor (1 = worst, n = best); the numbers are shown in Table 15.16.

Step 2—Sum the rank-order numbers for each alternative, as shown in Table 15.16.

Step 3—Calculate chi-square χ_r^2, where

$$\chi_r^2 = \left[\frac{12}{n(k)\ (k+1)} \sum_1^k R_j^2 \right] - 3n(k+1)$$

and where n = number of rows (decision factors)

k = number of columns (alternatives)

R_j = sum of rank-order numbers in jth column

For the example, the calculated χ_r^2 is as follows:

$$\chi_r^2 = \left[\frac{12}{(4)(3)(4)} \sum_{i=1}^3 7^2 + 6.5^2 + 10.5^2 \right]$$
$$- (3)(4)(4)$$
$$= 2.4$$

TABLE 15.16

RANKS ASSIGNED TO ALTERNATIVES (BASED ON SCORES IN TABLE 15.15)

Decision factor	Rank order		
	A1	A2	A3
F1	3	1	2
F2	2	1	3
F3	1	2	3
F4	1	2.5	2.5
Sum (overall rank)	7	6.5	10.5

Step 4—Compare the calculated χ_r^2 with the reported χ_r^2 from a statistical table for the test, with the reported χ_r^2 representing confirmation at a 95 percent confidence level that the hypothesis stating that the alternatives are not significantly different is true. Specifically, if the calculated χ_r^2 is equal to or less than the reported χ_r^2, then the alternatives are not significantly different from each other. Conversely, if the calculated χ_r^2 exceeds the reported χ_r^2, then statistically significant differences exist between the alternatives. In this example, the reported χ_r^2 at the 95 percent confidence level and for the conditions reported in Table 15.16 is less than 2.4; thus the calculated χ_r^2 exceeds the reported value, and it can be concluded that the alternatives depicted in Tables 15.15 and 15.16 are significantly different from each other.

Many EIA practitioners are opposed to the use of numerically based decision-focused checklists. Some arguments against the use of the aggregated indices developed from these checklists relate to the following: (1) the inherent weakness in impact scaling or rating, since such decisions can be very subjective and subject to bias, (2) the reliance on differing perspectives on importance weighting of decision factors and the usage of nondocumented and potentially illogical or biased rationale in weight assignments, and (3) the likelihood of imparting

a false perception regarding the degree of sophistication, objectivity, and precision achieved in decision making (Lee, 1983). To address these issues, efforts should be made to incorporate community values and perspectives in the importance-weighting process. Finally, sensitivity analysis can be helpful in analyzing the influence of different perspectives on significant resources and impact precision.

EXAMPLES OF CHECKLISTS USED IN DECISION MAKING

Table 15.17 illustrates a simple weighting-rating checklist utilized in an environmental impact study of three proposed sites for a wastewater-treatment plant. Two groups of importance weights were used; factors assigned an importance weight of 2 were considered more important than factors assigned an importance weight of 1. Each of the three sites—A, B, and C—in the study was rated on a scale from 1 to 3, with 1 denoting the worst site and 3 the best site, relative to 13 decision factors. The best overall site based on a composite evaluation of the data presented was site C, with a rating of 40.

It is recognized that different publics can have varying perspectives on the relative importance weights of decision factors in an impact study. For example, seven potential observer weightings were used in evaluating several alternatives in an impact study for the

TABLE 15.17
WEIGHTING-RATING CHECKLIST FOR WASTEWATER-TREATMENT-PLANT SITE EVALUATION

Decision factors	Significance	Importance weight (2 = greatest)	Rating (1 = worst, 3 = best) Site A	Site B	Site C
1. Construction cost	One-time cost with federal share.	1	2	2.5	3
2. Operating cost	Ongoing cost; includes energy costs (all local share).	2	1	2	3
3. Nonpotable reuse	"Safe," economic benefit; key is proximity to users.	2	1	2	3
4. Potable reuse	More long-range than factor 3; best sites are near well field or water plant; industrial sewage should not be present.	1	3	2.5	1
5. Odor potential	Assumes good plant design and operation.	2	2	1	3
6. Other land-use conflicts	Potential to interfere with agricultural-residential land.	1	2	1	3
7. Site area available	Future expansion capability, flexibility.	2	3	2	1
8. Relationship to growth area	Assumes growth to state line (in 50-year time frame).	1	3	2	1
9. Construction impacts	Reworking of lines.	1	1	2	3
10. Health of workers	Air pollution.	1	3	3	2
11. Implementation capability	Land-acquisition problems.	1	3	1	2
12. Operability	One plant is better than two.	1	1	2	3
13. Performance reliability	Assumes equal treatment plants.	1	2	2	2
Total	Rating × importance, cumulative (highest number = best site).		34	32	40

Source: Adapted from Wilson, 1980.

Cummins Creek project discussed earlier (Wilson, 1991). The viewpoints of the seven observers (publics) are displayed in Table 15.18. The listed weights were multiplied by the ratings of each alternative for each of the eight criteria (decision factors) in order to obtain a sensitivity analysis of the selection process.

Another example of an impact-rating checklist using the method developed in Adkins and Burke (1974) is shown in Table 15.19. Comparisons of two transportation-route alternatives relative to environmental factors are depicted in Table 15.19; Table 15.20 shows summary rating comparisons relative to transportation, environmental, sociological, and economic factors. The overall evaluation for each alternative is based on the number of plus and minus ratings it received, as well as its algebraic average rating.

The ''Battelle environmental evaluation system'' (EES), as noted earlier, was an early weighting-scaling checklist methodology for water-resources projects, and it contained 78 environmental factors, as shown in Figure 15.4. Each of the elements was assigned an importance weight using the ranked pairwise-comparison technique; resultant importance-weight points (PIUs) are shown in Figure 15.4 by the numbers adjacent to the four environment categories, in the right-hand corner of the boxes representing the intermediate components, and in the parentheses in front of each environmental factor. The higher the number, the greater the relative importance. Impact scaling in the Battelle EES is accomplished through the use of functional relationships for each of the 78 factors (Dee et al., 1972).

The basic concept of the Battelle EES, in the context of this discussion, is that an index expressed in environmental impact units (EIUs) can be developed for each alternative and baseline environmental conditions. The mathematical formulation of this index is as follows:

$$EIU_j = \sum_{i=1}^{n} EQ_{ij} PIU_i$$

where EIU_j = environmental impact units for jth alternative,

EQ_{ij} = environmental-quality-scale value for ith factor and jth alternative

PIU_i = parameter importance units for ith factor

Usage of the Battelle EES consists of obtaining baseline data on the 78 environmental factors and, through use of their functional relationships, converting the data into EQ scale values. These scale values are then multiplied by the appropriate PIUs and aggregated to obtain a composite EIU score for the baseline setting. For each alternative being evaluated, it is necessary to predict the anticipated changes in the 78 factors. The predicted-factor measurements are then converted into EQ scale values using the appropriate functional relationships. Next, these values are multiplied by the PIUs and aggregated to arrive at a composite EIU score for each alternative. This numerical scaling system provides an opportunity for displaying trade-offs between the alternatives in terms of specific environmental factors, intermediate components, and categories. Professional judgment must be exercised in the interpretation of the numerical results, and the focus should be on comparative analyses, rather than on specific numerical values.

To help transform these parameter estimates into an environmental-quality scale, value-function graphs are utilized for each of the 78 parameters. Figures 15.5 through 15.8 show four value-function graphs employed in the Battelle system. Parameter values are shown on the abscissa, while the EQ scale is shown on the ordinate (Dee et al., 1972). EQ values can range from 0 to 1.0, with 0 representing poor quality, and 1.0 very good quality.

The numerical-evaluation system in the EES provides a tool that serves as a guide for impact analysis. The EES is a very highly organized methodology, and as such, it helps to ensure a

TABLE 15.18

CUMMINS CREEK PROJECT—OBSERVER WEIGHTINGS UTILIZED IN PERFORMING SENSITIVITY ANALYSIS

Impact decision factor	Equal-weight	Economic	Conservation	Pro-development	Anti-development	Casual-user	Safety and stability
				Viewpoint			
Economic	0.125	0.50	0.05	0.40	0.05	0.04	0.10
Ecological	0.125	0.07	0.25	0.033	0.15	0.18	0.033
Hydrological	0.125	0.07	0.15	0.033	0.15	0.18	0.20
Air quality	0.125	0.07	0.15	0.033	0.15	0.18	0.20
Noise	0.125	0.07	0.15	0.033	0.15	0.18	0.20
Land use, aesthetics	0.125	0.07	0.15	0.033	0.15	0.18	0.033
Geological-geotechnical	0.125	0.07	0.05	0.033	0.15	0.03	0.20
Socioeconomic	0.125	0.08	0.05	0.40	0.05	0.03	0.033

Source: Adapted from Wilson, 1991.

TABLE 15.19

EXAMPLE OF ADKINS-BURKE IMPACT-RATING CHECKLIST METHOD FOR ENVIRONMENTAL CATEGORY

Factor	Definition or explanation	Rating alternative		Comments
		1	2	
A. Community (local area)				
1. Noise pollution	Relation to present levels, Policy and Procedures Memorandum 20-8 (PPM 20-8)			
a. Adjacent to freeway		−2	−1	Relief of street traffic helps offset
b. General area		+3	+1	Improves due to relief of street traffic
2. Air pollution	PPM 20-8			
a. Adjacent to freeway		+2	+1	Relief of street traffic
b. General area		+5	+2	Relief of street traffic
3. Drainage	Effects on chances of flooding, etc.			
a. Adjacent to freeway		+1	0	Route 1 will help slightly
b. General area		0	0	
4. Water supply				
a. Water pollution	PPM 20-8, permanent or serious temporary	0	0	Little, if any, effect
b. Water quantity	Interference with movement or level of groundwater	0	0	Little, if any, effect
5. Waste disposal	PPM 20-8, access to, interference, etc.	0	0	Little, if any, effect
6. Flora effects	NEPA and PPM 20-8, irreplaceable losses, etc.	0	0	Little, if any, effect
7. Fauna effects	NEPA and PPM 20-8, breeding or nesting, etc.	0	0	Little, if any, effect
8. Parks	PPM 20-8, improvement or damage to	+5	+2	Improves access to
9. Playgrounds	PPM 20-8, improvement or damage to	+5	0	Route 1 improves access to
10. Archeological sites	NEPA and PPM 20-8, loss of or access to, etc.	0	0	None affected
11. Historical sites	PPM 20-8, loss of or access to, etc.	+2	+1	Improves access to
12. Open space		+3	+1	Opens area by removing structures, some undesirable

13. Visual aspects — PPM 20-8, community view of freeway

	Alternative 1	Alternative 2	
a. Adjacent to freeway	+3	+1	Through proper treatment areas improved
b. General area	+2	+0	Route 1 would help, route 2 not likely to help

14. Safety — PPM 20-8, any change in hazards

	Alternative 1	Alternative 2	
a. Traffic	+3	+1	Route 1 gives more relief to streets and removes railroads
b. Pedestrian	+5	+1	Route 1 gives more relief to streets and removes railroads, route 2 more persons involved
c. Other	—	—	

15. Other — PPM 20-8, e.g., other resources

PPM 20-8

B. Freeway motorist experience

		Alternative 1	Alternative 2	
1. View of freeway	Appearance and security	+3	+1	Route 1 clearer and nicer view
2. View of adjacent area	Aesthetics or special sights	0	+1	Route 2 could give special views on curves
3. Panoramic views	Vistas	+1	+3	Route 2 good, route 1 downtown area
4. Area hazards	Hazards to freeway users and vehicles	+3	−1	Route 1 would displace hazards, route 2 would expose motorists to industrial smog, etc.

Summary rating

	Alternative 1	Alternative 2
No. of plus ratings	15	12
No. of minus ratings	1	2
Ratio of plus ratings	0.94	0.86
Algebraic sum of ratings	44	14
Average of ratings	2.75	1.00

Source: Adkins and Burke, 1974, pp. 67–69.

TABLE 15.20

OVERALL COMPARISON OF RATINGS USING ADKINS-BURKE METHOD

Parameters	No. of plus ratings	No. of minus ratings	Total no. of ratings	Algebraic sum of ratings	Ratio of plus ratings	Average rating
Transportation						
Local area						
Alt. 1	7	6	13	18	0.54	1.38
Alt. 2	4	2	6	1	0.67	0.17
Metropolitan area						
Alt. 1	8	0	8	34	1.00	4.25
Alt. 2	6	1	7	7	0.86	1.00
Environmental						
Alt. 1	15	1	16	44	0.94	2.75
Alt. 2	12	2	14	14	0.86	1.00
Sociological						
Community						
Alt1	9	2	11	27	0.82	2.46
Alt. 2	6	3	9	−1	0.67	−0.11
Metropolitan						
Alt. 1	9	0	9	31	1.00	3.44
Alt. 2	6	1	7	7	0.86	1.00
Economic						
Alt. 1	15	14	29	27	0.52	0.93
Alt. 2	14	14	28	−11	0.50	−0.39
All ratings						
Alt. 1	63	23	86	181	0.73	2.10
Alt. 2	48	23	71	17	0.68	0.24

Source: Adkins and Burke, 1974, p. 77.

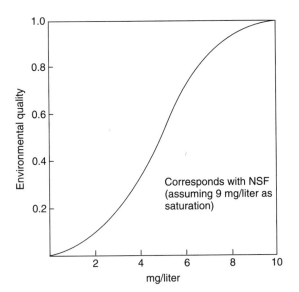

FIGURE 15.6
Functional Curve for Dissolved Oxygen (Dee et al., 1972).

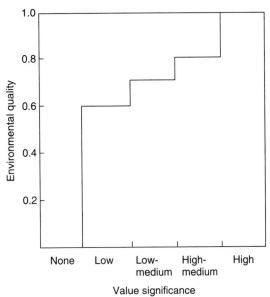

FIGURE 15.8
Functional Relationship for Historical External Package (Dee et al., 1972).

FIGURE 15.7
Functional Relationship for Appearance of Water (Dee et al., 1972).

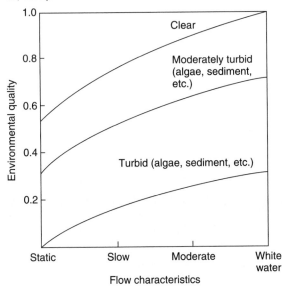

systematic, all-inclusive study approach and to identify critical changes. However, because of the time constraints under which this methodology was developed, very little emphasis is given to socioeconomic factors.

One of the key points to note is that there is no passing or failing score in the EES, since the resultant numerical evaluations must be subjected to professional interpretation. The methodology is most valuable for an analysis of trade-offs with a component, within a category, or between categories. Table 15.21 shows an example of a trade-off presentation using the Battelle environmental evaluation system.

As noted earlier, the Battelle EES has been used directly or in modified form in EISs for several water-resources projects. For example, it has been used on the Pa Mong water-resources project in south Thailand (ESCAP, 1990) and in modified form for other water-resources projects (Lohani and Halim, 1990). The general approach of the EES can be applied to other proj-

TABLE 15.21

TRADE-OFF ANALYSIS USING THE BATTELLE ENVIRONMENTAL EVALUATION SYSTEM

Category	A[a]	B	C
	Alternative		
Ecology (240)[b]	215	200	220
Environmental pollution (402)	340	350	310
Aesthetics (153)	110	120	135
Human interest (205)	180	175	175

[a]No action.
[b]The number in parentheses is the maximum for a category.
Source: Compiled using data from Dee et al., 1972.

ect types in terms of selection of pertinent environmental factors, assignment of importance weights, and the development of appropriate functional relationships for the factors. For examples, the approach used in the Battelle EES has been applied to a rapid transit system (Smith, 1974), a waterway navigation project (School of Civil Engineering and Environmental Science and Oklahoma Biological Survey, 1974), and to highway projects, pipeline projects, channel-improvement projects, and waste-water-treatment plants (Dee et al., 1973).

Another example of a scaling-weighting checklist is the optimum-pathway matrix developed by Odum et al. (1971). This methodology was applied to an evaluation of eight alternatives for completion of a section of Interstate I-75 north of Atlanta, Georgia. The first step of the analysis consisted of defining factors to be included in the evaluation. A total of 56 factors were identified and sorted into 4 general groups, as follows:

Group E: Economic and highway-engineering factors

Group L: Environmental and land-use considerations

Group R: Recreation considerations

Group S: Social and human considerations

Data were then developed for each of the 56 factors from existing engineering reports and other sources of information. The next step consisted of unitizing the information for each factor and for each alternative. The unitization (scaling) was accomplished by the use of the following mathematical formulation:

$$S_i = \frac{1}{\text{maximum of } X_1, X_2, \ldots, X_8}$$

where S_i = scaling factor for the ith factor and $X_1, X_2, \ldots, X_8$ = value of eight alternatives for the ith factor,

and
$$UV_{ij} = S_i X_{ij}$$

where UV_{ij} = unitized value for the ith factor and the jth alternative and X_{ij} = value for the ith factor and the jth alternative.

The next step consisted of an interdisciplinary effort to determine the relative importance weights of initial and long-term effects of project construction on individual factors. The range of values selected was from -20 to $+50$. After the relative weights were decided, composite weighting values were developed for each of the 56 factors. These values were calculated by assuming that the long-term effects were 10 times greater than the initial effects. The primary reason for this assumption was that the operational period for a highway project is at least 10 times longer than the construction period. Then, the composite weighting values were unitized by dividing each weight by the sum of all weights, in accordance with the following mathematical relationship:

$$N_i = \frac{W_i}{\sum\limits_{1}^{56} W_i}$$

where N_i = unitized weighting value and W_i = composite weighting factor.

The final step consisted of the calculation of environmental indices for each route in accordance with the following mathematical equation:

$$I_j = \sum\limits_{1}^{56} N_i S_i X_{ij} \pm e(N_i S_i X_{ij})$$

where I_j = environmental index and e = error term to allow for misjudgment on relative weights by ± 50 percent (stochastically selected).

The key feature of the Odum method is that an error term e is included to allow for misjudgment on relative weights. This error term e is handled by computational analysis using a packaged stochastic computer program. An environmental index was calculated 20 times for each route with the computer program (Odum, 1971).

Another example of a weighting-scaling checklist for water-resources projects is the "water-resources-assessment methodology" (WRAM) developed by the U.S. Army Corps of Engineers (Solomon et al., 1977). The steps, or elements, of the methodology are listed in Table 15.22; they are (1) the selection of an interdisciplinary team; (2) the selection and inventory of decision factors (environmental factors); (3) impact prediction, assessment, and evaluation; and (4) documentation of the results. In this

technique, the element including weighting and scaling (item C2 in Table 15.22) also involves impact prediction, assessment, and evaluation. The unranked paired-comparison technique is used to determine the factor-importance coefficient (FIC) for each decision factor. Importance-weight assignments are required for each study and should reflect the importance of the factor in terms of the given geographical location.

Impact scaling in the WRAM can be accomplished through the use of functional graphs (discussed earlier), linear proportioning (or scaling), or the development of alternative-choice coefficients (ACCs). "Linear scaling" entails determining the largest impact, which is assigned a scale value of 1.0, and calculating the relative proportions of each change for each alternative with respect to the largest impact. It should be noted that changes can be either beneficial or detrimental, and linear scaling should differentiate between them. As noted earlier, development of ACCs involves the comparative evaluation of the impacts of each alternative on each factor.

The final matrix resulting from the application of the WRAM is similar to that shown in Table 15.15. The best choice is represented by the alternative with the highest product summation. Although WRAM was developed with an orientation to water-resources projects, examination of the procedure outlined in Table 15.22 indicates that this approach could be used for any project type, as the methodology is more of a generic framework or structure than an application-specific "cookbook" approach.

The first example of the usage of WRAM was on the Tensas River Basin project (a flood-control project) in eastern Louisiana (Richardson et al., 1978). Existing data was used, along with information gathered from site visits by an interdisciplinary team. Decision factors were selected and weighted relative to their importance in assessment and evaluation within each of four typical project-evaluation accounts: national economic development (NED), environmental

TABLE 15.22

WATER-RESOURCES-ASSESSMENT METHODOLOGY (WRAM) FOR TRADE-OFF ANALYSIS INVOLVING ENVIRONMENTAL IMPACTS

Element	Delineation

A. Establish Interdisciplinary Team	1. Selection
	  a. Select members of interdisciplinary team.
	  b. Designate team leader.
	2. Review and Familiarization
	  a. Review information on potential project.
	  b. Visit locations with similar projects.

B. Select Decision Factors and Assemble Basic Information	1. Selection
	  a. Assemble preliminary list of decision factors.
	  b. Use technical questions and findings from A.2, along with professional judgment, to select additional relevant factors.
	  c. Identify any resulting interactive or cross-impact factors or categories.
	2. Environmental Inventory
	  a. Assemble extant baseline data for selected factors.
	  b. Identify factors with data deficiencies, and plan data collection effort.
	  c. Conduct field studies or assemble information on data-deficient factors.

C. Evaluate Alternatives Relative to Decision Factors	1. Prediction and Delineation
	  a. Predict changes in each factor for each alternative using available techniques and/or professional judgment.
	  b. Delinate potential impacts of alternatives.
	  c. Highlight significant impacts and "red flag" any critical issues.
	2. Weighting and Scaling
	  a. Use unranked paired comparison technique, or some other importance weighting technique, to determine importance coefficients for each factor (FIC).
	  b. Scale/rate/rank predicted impacts through development of alternative choice coefficients, or use of some other technique for evaluation of alternatives relative to decision factors (ACC).
	3. Evaluation and Interpretation of Results
	  a. Multiply FIC by ACC to obtain final coefficient matrix. Sum coefficient values for each alternative.
	  b. Use values in final coefficient matrix as basis for description of impacts of alternatives and trade-offs between alternatives.
	  c. Discuss any critical issues and predicted impacts.

D. Document Results	1. Rationale
	  a. Describe rationale for selection of decision factors.
	  b. Describe procedure for impact identification and prediction, and rationale for weighting, scaling/rating/ranking, and interpreting results.
	2. Provide Referencing of Sources of Information

Source: Solomon et al., 1977, p. 27

quality (EQ), social well-being (SWB), and regional development (RD). The projected effects on each factor were then scaled across each of the alternative-plan scenarios and the without-project condition. Next, the weights and scales were multiplied for each factor within an account and then added for each alternative plan. Another example of the usage of WRAM involved the selection of more-favorable sites for a low-flow augmentation reservoir development (Bizer and Wang, 1980).

In the late 1970s and early 1980s, the concept of "commensuration" was introduced into the water-resources-planning vocabulary. "Commensuration" was used to refer to the process of measuring different things by a single standard or measure (Lord, Deane, and Waterstone, 1979). In essence, commensuration involves the development of "common units" of measurement for evaluating a number of plans, with these units serving as the basis for trade-off analysis among the various plans. Lord, Deane, and Waterstone (1979) noted that the four essential components of commensuration are (1) identification of the factors which are to be commensurated, (2) making a determination of which participants' value judgments about those factors are to be considered, (3) discovery of what those value judgments are, and (4) amalgamation of the judgments of all of the selected individuals into a single set of collective judgments.

As a result of this emphasis on commensuration, several water-resources methodologies were developed; the methodologies were basically weighting-scaling (or -rating or -ranking) checklists. Four examples of these methodologies will be briefly noted. Brown, Quinn, and Hammond (1980) addressed impact scaling for alternative water-development plans, with particular emphasis given to environmental and social impacts. Mumpower and Bollacker (1981) developed the Evaluation and Sensitivity Analysis Program (ESAP), which is a computerized environmental-planning technique for the eval-

uation of alternative water-resource-management plans. Anderson (1981) also developed a multiple-objective, multiple-publics-oriented method for evaluating alternatives in water-resource planning; the method is called the "cascaded–trade-offs method." This method can be used to arrive at an overall ranking of planning alternatives on the basis of public values. The key feature of the method is that it provides for making trade-offs across both issue dimensions (decision factors) and publics. Finally, in terms of commensuration approaches, Brown and Valenti (1983) developed the Multi-attribute Tradeoff System (MATS). MATS is a computer program designed to help planners evaluate multiattribute alternatives in order to reach a judgment of each alternative's relative worth or desirability. MATS leads the user through a series of questions (a trade-off analysis) as a means of identifying the relative importance of various characteristics of the alternatives. The program documents the judgments which lead to the development of a policy for evaluating alternatives. Importance weighting in MATS involves the usage of the unranked paired-comparison technique, described earlier. Impact scaling is based on functional curves, which were also described earlier.

COMPARATIVE CASE STUDIES

It can be instructive to review actual impact studies to determine what methods are being used in decision making between alternatives. For example, Canter (1979) conducted a detailed review of 28 draft or final EISs prepared by the U.S. Environmental Protection Agency, or its contractors, for wastewater-facility plans. At the time of the review, the sample group represented approximately 15 percent of all EISs prepared on wastewater-treatment-plant projects. The geographic distribution included 17 states. The sample group EISs encompassed a wide variety of projects for meeting the wastewater-management needs of sites having populations ranging from as few as 1,600 persons (in Jacksonville, Oregon) to 3.6 million persons (in Los Angeles County, California). Project types included 16 involving collection or conveyance, treatment, and disposal; 4 involving treatment and disposal only; 6 involving collection or conveyance only; and 2 involving sludge disposal only. The methodologies used to select the proposed action from a series of alternatives were summarized for the 28 EISs in the sample group (Canter, 1979). Twenty EISs utilized an impact-assessment methodology in selecting the proposed action from the alternatives which were studied. Four descriptive checklists, four ranking checklists, six scaling checklists, one weighting-ranking checklist, and five weighting-scaling checklists were used. In the case of eight EISs for which no selection methodology was identified, it is possible that the methodologies were described in separate facility-plans reports and not included in the EISs.

Detailed information on the decision-making approaches used in five representative EISs prepared during the 1980s was recently summarized by Canter (1990) and indicated that qualitative-quantitative comparisons (descriptive checklists) were used on four of the EISs, and a ranking checklist was used on the other EIS. A qualitative comparison of alternatives was used in the Old Metairie Railroad EIS (a railroad rehabilitation project); and qualitative-quantitative comparisons were used in the Saddle River flood-protection EIS (a flood-control project) and a surface lignite mine EIS (resource-extraction project). Quantitative comparisons of the impacts of three route alternatives were used in a crude-oil pipeline EIS. Finally, ranking (or rating) checklists were used in the decision-making process for selecting a wastewater-treatment scheme and a sludge-disposal option in a wastewater-facilities-plan EIS.

In a recent study on actual techniques used in decision making, a study sample of 35 environmental projects was utilized; the sample consisted of 30 development projects and 5 Superfund remedial-action projects for which either an EIS or a record of decision (SROD) was prepared (Lahlou and Canter, 1993). The listing of specific projects is contained in Lahlou (1991).

The evaluation steps in the decision-making process can be viewed as the information-processing operations used by the decision maker or the project analyst to arrive at a preferred alternative. The two most common processes consist of either a single evaluation step, where a preferred alternative is identified after one evaluation of the entire set of candidate alternatives; or a two-step process, where alternatives which pass a screening phase are further refined and then undergo a final evaluation and selection step. The review of the 30 EISs in the study noted above indicated that half of the projects had adopted a multistep process, with at least one preliminary-screening phase and an intermediate step in which new alternatives identified later in the process could be addressed, or where alternatives initially screened out, but upon reexamination deemed viable (subject to improvement), were reincorporated.

To evaluate the quality of the decision techniques used in the study sample, all procedures used in both the screening and the final-evaluation steps were identified and sorted into two categories: qualitative procedures and analytical procedures. ''Qualitative procedures'' do not use analytical processes to identify the subset of preferred alternatives among the initial set. They rely mainly on the analysts' qualitative judgments in differentiating between alternatives. Three types of qualitative procedures were identified: (1) fully descriptive techniques, (2) open-matrix-based rating techniques, and (3) selection by advantages or advantages-disadvantages. In contrast, analytical procedures uniformly process information on alternative impacts in order to identify the feasible or preferred alternative(s). Three types of analytical procedures were identified: (1) conjunctive-rules techniques, (2) rating and/or ranking techniques, and (3) weighting-rating and/or -ranking techniques.

The distribution of evaluation procedures used in the study sample is presented in Table 15.23. Note, however, that a number of projects reviewed neither explicitly defined the procedure adopted—particularly in the screening step, nor clearly differentiated between the alternative-development phase, the scoping process, and the screening phase. These cases are not included in Table 15.23. The table shows that qualitative procedures, encountered in 67 percent of all evaluation steps, were most often used to tentatively select the preferred action in both EIS and SROD (Superfund record of decision) projects. Within this category, the fully descriptive procedure was used most widely, in both the screening and the final-evaluation steps, followed by the open-matrix-based rating technique and the advantages-disadvantages procedures, which were used mainly in the final-evaluation steps.

Analytical procedures were used in 33 percent of all the evaluation steps. Their distribution among the screening and the final-evaluation steps indicates that conjunctive rules are more frequently used as screening tools, and

that the comparative procedures (rating-ranking and rating-ranking-weighting) are used more often in the selection of the preferred alternative. In fact, this distribution of conjunctive and comparative procedures has been recommended as a logical model to arrive at a preferred solution in multicriterion and multiphased decision problems (Wright and Barbour, 1977).

Scales were used to describe or measure the impact intensities of each alternative with respect to a given set of decision criteria. Their expressions can be numeric, linguistic, or symbolic. The ratings (or scores) of alternatives with respect to a selected scale can be displayed in a uniformly scaled payoff matrix.

In the study sample, the ratings of the alternatives with respect to decision criteria were typically based on ordinal scales. A total of 12 numeric scales, explicitly described, were identified in 10 projects; 4 were absolute and 8 were comparative. It should be noted that unless different types of scales were used within a project, only one scale per study was counted. Mathematical operations, particularly additions, have been performed directly on these scales to derive an overall score for the alternatives. These operations on ordinally scaled data can be ar-

TABLE 15.23

EVALUATION PROCEDURES IDENTIFIED IN STUDY SAMPLE

Procedures	Screening step[a]	Final evaluation[a]	Total
Qualitative procedures	11 (55%)	26 (77%)	37 (67%)
• Descriptive	9	18	27
• Rating in an Open Matrix	1	5	6
• Advantages/Disadvantages	1	3	4
Analytic procedures	9 (45%)	9 (23%)	18 (33%)
• Conjunctive Rules	8	3	11
• Rating/Ranking	1	4	5
• Rating/Ranking/Weighting	0	2	2
Total	20	35	55

[a]Numbers refer to number of projects out of 35 in the study sample.
Source: Lahlou and Canter, 1993, p. 47.

bitrary, since different scales may produce different rankings of alternatives.

Scaling through direct ranking of the alternatives and adding the resulting ranks was also used in the study sample. This practice provided little information about the significance of impacts and the trade-offs among the alternatives. Further, its results were a function of the number of alternatives considered, rather than of the differences between the impact expressions of each alternative.

Another component of comparative procedures, particularly in environmental studies, is the determination of impact significance, the relative importance of the decision criteria, and/or the prioritization of project objectives. This significance is usually derived as importance weights of each attribute with respect to other attributes and/or with respect to higher-level objectives. Two cases in the study sample incorporated a weighting procedure in the final-evaluation and selection step (Lahlou, 1991). In one study, a rating technique using an absolute scale of 1 to 3 was used by the project technical team to derive weights for 26 decision criteria. The weight for each attribute was the average of scores assigned by seven technical team members. In the second study addressing impact significance, a paired-comparison technique was used. The method and intermediate steps were not explicitly described.

CURRENT TRENDS

A recent trend in decision making in environmental studies has been the usage of computer software. For example, Torno et al. (1988), developed a training manual for the evaluation of the environmental impacts of large-scale water-resources-development projects. The training manual is primarily related to the use of a multiobjective, multicriterion decision-analysis approach. An interactive computer program has been developed to simplify application of the methodology described in the training manual and to serve as a valuable learning aid.

In a more generic sense, a computer model has been developed which, with user input, can aid in determining the relative weights of the evaluation parameters used to evaluate projects under consideration, as well as determining the utility function for each of the attributes (Klee, 1988). A unique feature of this model is that it incorporates uncertainties of three types: (1) those dealing with the factor (parameter) weights; (2) those dealing with the worth of each project with regard to each factor; and (3) those dealing with the utilities of the attributes. This model may be applied to any evaluation of competing alternatives.

The requirements for the computerized model were that the model must (Klee, 1988) (1) produce a ranking of alternatives on at least an interval scale; (2) be based upon sound decision theory; (3) be easy to understand and to use; (4) generate easily understood output; (5) be usable on a wide variety of computers, including personal computers; and (6) be able to deal with the uncertainties of its inputs. The computerized model (called "d-SSYS," for "decision-support system") that has been developed meets all of these requirements (Klee, 1988). It is a true decision-support system; that is, it is an interactive system that provides the user with easy access to decision models and data in order to support semistructured decision-making tasks.

In a recent study conducted at the University of Oklahoma, Lahlou (1991) reviewed multiattribute decision-making techniques in order to identify and develop a comprehensive set of applicable techniques to be used in a proposed generic decision-based methodology. The set of techniques is listed in Table 15.24. The review of the 14 selected techniques was directed toward identifying a comprehensive set of applicable techniques for environmental development and/or remediation projects. Review

TABLE 15.24

EXAMPLES OF MULTIATTRIBUTE DECISION-MAKING TECHNIQUES

Objective of technique	Examples of technique
Elimination of the nonfeasible set of alternatives (also called "sequential elimination techniques")	Conjunctive screening Conjunctive ranking Disjunctive screening Lexicographic screening Compensatory screening
Elimination of the dominated set of alternatives	Noninferior curve technique Indifference map technique Reasonable-social-welfare function (RSWF) Outranking approaches
Evaluation of the nondominated set of alternatives	Utility theory and decision analysis[a] Compromise programming Displaced-ideal technique Cooperative game theory Analytic-hierarchy process (AHP)

[a]Includes four models—the additive model, the multiplicative model, the quasi-additive model, and the hierarchical additive model.

criteria which were utilized included the following (Lahlou and Canter, 1993): (1) the type and amount of information required and its availability or ease of collection; (2) the adequacy and relevancy of the information-processing methods and/or ease and validity of the mathematical operations; (3) demand on the decision maker(s) and ease of eliciting the preference structure; (4) the flexibility of decision structuring and potential for decision decomposition into smaller, homogeneous components; (5) the ease of conducting consistency and sensitivity analyses; (6) the ease of developing applicable and valid scales to represent impact values; (7) its applicability and suitability to a wide range of environmental decision situations; (8) the stability of the final ranking of the alternatives; and (9) the ease of interpretation of ranking, overall score, and/or trade-off of and/or between alternatives.

Based upon these reviews, some of the more simple and useful techniques in Table 15.24 include conjunctive screening, compensatory screening, the weighted reasonable-social-welfare function, and the analytic-hierarchy process. Summary reasons for their identification follow (Lahlou, 1991):

1. *Conjunctive screening*—Conjunctive screening is the most common screening technique in environmental impact analysis. It uses categorical constraints, as well as either quantitative or qualitative cutoffs, or both.
2. *Compensatory screening*—Compensatory screening combines several lower-level attributes according to their factual relationship and allows for constraints on higher-level objectives.
3. *Weighted reasonable-social-welfare function (RSWF)*—RSWF results in a complete ordering of the alternatives, uses flexible cutoff levels, and exerts less demand on the decision maker during the assessment of his or her preferences.
4. *Analytic-hierarchy process (AHP)*—AHP offers a logical and representative way of struc-

turing the decision problem. It is an efficient way of deriving priorities, allows for a systematic analysis of consistency, and applies to a wide range of decision situations.

SUMMARY OBSERVATIONS ON DECISION-FOCUSED CHECKLISTS

Weighting-scaling, -ranking, or -rating checklists are particularly valuable tools for displaying trade-offs between alternatives and their associated environmental impacts; thus, they are useful in the selection of a proposed action. Computerization of these decision-focused checklists is a current trend. Based upon the usage of these methods to date, the following observations can be made:

1. It is important to describe the process used for importance weighting of the individual decision factors, and the rationale utilized in determining the relative importance weights of the individual factors. It is also important to describe the rationale used for the scaling, rating, or ranking of individual alternatives relative to each decision factor. The description of the rationale is more important than the final numerical scores or classifications which might result therefrom.
2. There are several approaches available for assigning importance weights and for scaling, rating, or ranking. These approaches have relative advantages and disadvantages, which should be considered in choosing the specific approaches to be utilized in a given study.
3. The most debatable issue is associated with the assignment of importance weights to the individual decision factors. Several approaches for importance weighting involve the use of public participation. Where this type of participation can be incorporated, it is desirable to do so, in order to legitimize the overall decision-making process.

4. Decision-focused checklists can help structure the decision process for comparing alternatives and selecting the proposed action from a set of alternatives. Usage of a decision-focused checklist can in and of itself help structure the decision process and provide a basis for the comparison and evaluation of alternatives.
5. The concepts used in decision-focused checklists are the same as those used in multicriterion decision making, multiattribute utility measurement, decision analysis, and other techniques which have been used for decision making in various disciplines for more than 25 years. These concepts are only now being utilized in environmental decision making in place of other, more-established types of decision-making approaches.
6. There is a fairly wide range of computer software to aid in the decision-making process. This software is typically user-friendly and can be utilized to guide the assignment of importance weights and the accomplishment of the scaling, rating, or ranking of alternatives. The software can then calculate final index scores for each alternative. In addition, because of the availability of this software, it is becoming increasingly easy to conduct sensitivity analyses of the overall decision-making process. Usage of the software for sensitivity analysis can aid the decision-making process by indicating the relative sensitivity of the index scores to individual changes.
7. Decision-focused checklists can also be valuable for demonstrating the relative trade-offs between different alternatives.
8. It is important to keep the weighting-scaling, -rating, or -ranking checklist simple in order to facilitate the decision-making process. Numerous alternatives and decision factors do not necessarily indicate that a better overall decision will be made.

9. Decision-focused checklists can be used at several points in overall project planning. For example, they could be used early in a study to reduce the number of alternatives and thus allow for more-detailed analyses to be made. In addition, the number of decision factors can be reduced through this process so that, in the final selection, there would be a smaller number of alternatives compared in relation to the truly key decision factors.

10. Decision-focused checklists can aid in clarifying the thinking of the EIA study team and various decision makers. In addition, its use can facilitate the systematic comparisons which would be needed for different public participation activities. In that regard, its use can also help to clarify the thinking of various publics.

11. Usage of decision-focused checklists encourages decision making in context—that is, it keeps the decision maker from giving too much attention to a single issue in the decision-making process. It forces the decision maker to consider each issue and impact in relation to other issues and impacts.

TABLE 15.25

CRITERIA FOR EVALUATING EIA METHODOLOGIES

1. Assessment methods should recognize the probabilistic nature of effects. Environmental cause-effect chains are rarely deterministic because of many random factors and uncertain links between conditional human activities and states of nature.

2. Cumulative and indirect effects are important, although there are obviously limits on the extent to which they can be considered. Natural systems are highly interrelated, and a series of minor actions may have significant cumulative impact. Indirect effects may be cyclical due to positive or negative feedback.

3. A good methodology should reflect dynamic environmental effects through a capacity to distinguish between short-term and long-term effects. Impacts may vary over time in direction, magnitude, or rates of change. The larger system itself may be in ecological or social flux, and decision makers have time horizons of varying lengths.

4. Decision making necessarily encompasses multiple objectives (or multiple values). Assessment methods should include the diverse elements of environmental quality: maintenance of ecosystems and resource productivity; human health and safety; amenities and aesthetics; and historical and cultural resources. Environmental values can be divided into three types: social norms, functional values (environmental services, e.g., fisheries), and individual preferences. In addition, a good assessment method should recognize other societal objectives, such as economic efficiency, equity to individuals and regions, and social well-being.

5. Environmental assessment necessarily involves both facts and values. Values enter the process when deciding which effects to examine, whether an effect is good or bad, and how important it is relative to other effects. Methods should separate facts and values to the extent possible, and identify explicitly the source of values. Where the influence of values is obscure, the analysis itself may become a source of conflict. Under optimal conditions, results should be amenable to a sensitivity analysis where alternative value judgments are applied to a set of factors.

6. It is also important to consider whose values enter the analysis. Assessment techniques should encourage a participatory approach to incorporate the multiplicity of values provided by the public as well as by experts from varying disciplines and interest groups. Lack of participation by key actors can mitigate the usefulness of assessment results.

7. With all other things held constant, the best decision process is efficient in its requirements for time, money, and skilled labor. Increased complexity is justified only when there is a sufficient increase in the validity and decision-making utility of the analytical results.

Source: Nichols and Hyman, 1982, pp. 89–90.

12. These types of checklists can be utilized for decision making based on environmental considerations only. They can also be used when the environmental impacts and the economic characteristics of different alternatives are compared and evaluated. Finally, such approaches can also be used for systematic decision making that encompasses considering environmental-impact, economic, and engineering-feasibility issues. In other words, the range of usage can be from applications considering only the environment to those consisting of a composite consideration of the three "Es" of decision making: environment, economics, and engineering.

SELECTION OF A METHODOLOGY

There is no "universal" decision-focused methodology for meeting the EIA needs for all project types in all environmental settings. Accordingly, selection of an existing methodology (or portions thereof) or the development of a new methodology may be required in the conduction of an impact study. A reasonable, early selection should focus on methodologies developed for projects similar in type to the potential project being evaluated. For example, if the impact study is to be conducted for a highway project, methodologies which have been developed for other highway projects should be reviewed.

Several criteria which can be used in methodology (or checklist) selection have been identified. Nichols and Hyman (1982) identified seven criteria for evaluating environmental-assessment methods, and Table 15.25 summarizes these criteria. The first three reflect the complex attributes of real environmental responses to natural or man-induced changes. The remaining four represent the preferable attributes of a planning and decision-making process.

Finally, multicriterion decision-making methods have been referred to as "amalgamation methods" by Hobbs (1985), as they involve combining disparate impacts so that alternatives can be ranked. Hobbs (1985) suggested four issues to be considered in choosing an amalgamation method; these are (1) the purpose to be served, (2) the ease of use (time, money, necessary computer facilities, etc.), (3) the validity of the method, and (4) the anticipated results when compared to other methods.

SUMMARY

Checklist approaches range from simple listings of environmental factors to complex methods involving the assignment of relative-importance weights to environmental factors and impact scaling, rating, or ranking for each of a series of alternatives on the environmental factors. Weighting-scaling, weighting-ranking, or weighting-rating checklists are particularly valuable for displaying trade-offs between alternatives and their associated environmental impacts, thus they are useful in the selection of a proposed action. These decision-focused checklists can also be referred to as multicriterion decision-making techniques, multiattribute utility measurement techniques, decision-analysis techniques, and decision-support systems. User-friendly computerization of these checklists is a current trend.

SELECTED REFERENCES

Adkins, W. G., and Burke, D., "Social, Economic and Environmental Factors in Highway Decision Making," Res. Rep. 148-4, Texas Transportation Institute, Texas A&M University, College Station, Nov. 1974.

Anderson, B. F., "Cascaded Tradeoffs: A Multiple Objective, Multiple Publics Method for Alternatives Evaluation in Water Resources Planning," U.S. Bureau of Reclamation, Denver, Colo., Aug. 1981.

Bishop, A. B., et al., "Socio-Economic and Community Factors in Planning Urban Freeways," Department of Civil Engineering, Stanford University, Menlo Park, Calif., Sept. 1970.

Bizer, J. R., and Wang, L. L., "Evaluation of Alternatives in Selection of Sites for Surface Water Impoundments," *Proceedings of the Symposium on Surface Water Impoundments,* vol. 1, American Society of Civil Engineers, New York, 1980, pp. 163–171.

Brown, C. A., Quinn, R. J., and Hammond, K. R., "Scaling Impacts of Alternative Plans," U.S. Bureau of Reclamation, Denver, Colo., June 1980.

——— and Valenti T., "Multi-Attribute Tradeoff System: User's and Programmer's Manual," U.S. Bureau of Reclamation, Denver, Colo., Mar. 1983.

Canter, L. W., "Decision Making in Selecting the Proposed Action in Environmental Impact Statements—Case Studies and Approaches," Environmental and Ground Water Institute, University of Oklahoma, Norman, May 1990.

———, *Environmental Impact Statements on Municipal Wastewater Programs,* Information Resources Press, Washington, D.C., 1979.

Council on Environmental Quality (CEQ), "National Environmental Policy Act—Regulations," *Federal Register,* vol. 43, no. 230, Nov. 29, 1978, pp. 55978–56007.

Dean, B. V., and Nishry, J. J., "Scoring and Profitability Models for Evaluating and Selecting Engineering Products," *Journal Operations Research Society of America,* vol. 13, no. 4, July–Aug. 1965, pp. 550–569.

Dee, N., et al., "Environmental Evaluation System for Water Resources Planning," Final Rep., Battelle-Columbus Laboratories, Columbus, Ohio, 1972.

———, "Planning Methodology for Water Quality Management: Environmental Evaluation System," Battelle-Columbus Laboratories, Columbus, Ohio, July 1973.

Duke, K. M., et al., "Environmental Quality Assessment in Multi-Objective Planning," U.S. Bureau of Reclamation, Denver, Colo., Nov. 1977.

Dzurik, A. A., *Water Resources Planning,* Rowman and Littlefield Publishers, Savage, Md., 1990, pp. 83–92.

Economic and Social Commission for Asia and the Pacific (ESCAP), "Environmental Impact Assessment—Guidelines for Water Resources Development," ST/ESCAP/786, United Nations, New York, 1990, pp. 19–48.

Edwards, W., "How to Use Multi-Attribute Utility Measurement for Social Decision Making," SSRI Research Rep. 76-3, Social Science Research Institute, University of Southern California, Los Angeles, Aug. 1976.

Hobbs, B. F., "Choosing How to Choose: Comparing Amalgamation Methods for Environmental Impact Assessment," *Environmental Impact Assessment Review,* vol. 5, 1985, pp. 301–319.

Klee, K. J., "d-SYSS: A Computer Model for the Evaluation of Competing Alternatives," EPA 600/S2-88-038, U.S. Environmental Protection Agency, Cincinnati, Ohio, July 1988.

Lahlou, M., "Development of an Alternative Evaluation and Selection Methodology for Development and Environmental Remediation Projects," Ph.D. dissertation, University of Oklahoma, Norman, 1991.

——— and Canter, L. W., "Alternatives Evaluation and Selection in Development and Environmental Remediation Projects," *Environmental Impact Assessment Review,* vol. 13, 1993, pp. 37–61.

Lee, N., "Environmental Impact Assessment: A Review," *Applied Geography,* vol. 3, 1983, pp. 5–27.

Linstone, H. A., and Turoff, M., *The Delphi Method—Techniques and Applications,* Addison-Wesley Publishing Company, Reading, Mass., 1975.

Lohani, B. N., and Halim, N., "Environmental Impact Identification and Prediction: Methodologies and Resource Requirements," background papers for course on environmental impact assessment of hydropower and irrigation projects, International Center for Water Resources Management and Training (CEFIGRE), Bangkok, Thailand, Aug. 13–31, 1990, pp. 152–182.

Lord, W. B., Deane, D. H., and Waterstone, M., "Commensuration in Federal Water Resources Planning: Problem Analysis and Research Appraisal," Rep. no. 79-2, U.S. Bureau of Reclamation, Denver, Colo., Apr. 1979.

McAllister, D. M., *Evaluation in Environmental Planning,* The MIT Press, Cambridge, Mass., 1986, pp. 6–7.

Mumpower, J., and Bollacker, L., "User's Manual for the Evaluation and Sensitivity Analysis Program," Tech. Rep. E-81-4, U.S. Army Engineer Waterways Experiment Station, Vicksburg, Miss., Mar. 1981.

Nichols, R., and Hyman, E., "Evaluation of Environmental Assessment Methods," *Journal of the Water Resources Management and Planning Division,* American Society of Civil Engineers, vol. 108, no. WR1, Mar. 1982, pp. 87–105.

Odum, E. P., et al., "Optimum Pathway Matrix Analysis Approach to the Environmental Decision Making Process—Test Case: Relative Impact of Proposed Highway Alternates," Institute of Ecology, University of Georgia, Athens, 1971.

Richardson, S. E., et al., "Preliminary Field Test of the Water Resources Assessment Methodology (WRAM)—Tensas River, Louisiana," Misc. Paper Y-78-1, U.S. Army Engineer Waterways Experiment Station, Vicksburg, Miss., Feb. 1978.

Ross, J. H., "The Numeric Weighting of Environmental Interactions," Occasional Paper no. 10, Lands Directorate, Environment Canada, Ottawa, July 1976.

Salomon, S. N., "Cost-Benefit Methodology for the Selection of a Nuclear Power Plant Cooling System," paper presented at the *Energy Forum, Spring Meeting of the American Physics Society,* Washington, D.C., Apr. 22, 1974.

School of Civil Engineering and Environmental Science and Oklahoma Biological Survey, "Mid-Arkansas River Basin Study—Effects Assessment of Alternative Navigation Routes from Tulsa, Oklahoma to Vicinity of Wichita, Kansas," University of Oklahoma, Norman, June 1974.

Selman, P., *Environmental Planning,* Paul Chapman Publishing, London, 1992, p. 176.

Smith, M. A., "Field Test of an Environmental Impact Assessment Methodology," Rep. ERC-1574, Environmental Resources Center, Georgia Institute of Technology, Atlanta, Aug. 1974.

Solomon, R. C., et al., "Water Resources Assessment Methodology (WRAM): Impact Assessment and Alternatives Evaluation," Rep. 77-1, U.S. Army Engineer Waterways Experiment Station, Vicksburg, Miss., 1977.

Torno, H. C., et al., "Training Guidance for the Integrated Environmental Evaluation of Water Resources Development Projects," UNESCO, Paris, 1988.

Voelker, A. H., "Power Plant Siting, An Application of the Nominal Group Process Technique," ORNL/NUREG/TM-81, Oak Ridge National Laboratory, Oak Ridge, Tenn., Feb. 1977.

Voorhees, A. M., and Associates, "Interim Guide for Environmental Assessment: HUD Field Office Ed.," Washington, D.C., June 1975.

Westman, W. E., *Ecology, Impact Assessment and Environmental Planning,* John Wiley and Sons, New York, 1985, pp. 138–143.

Wilson, L. A., personal communication to author, Santa Fe, N.Mex., 1980.

——, personal communication to author, Santa Fe, N.Mex., June 10, 1991.

Wright, P., and Barbour, E., "Phased Decision Strategies: Sequels to an Initial Screening," *Multiple Criteria Decision Making,* M. K. Starr and M. Zeleny, ed., Studies in Management Sciences, vol. 6, North Holland Publishing, Amsterdam, 1977.

CHAPTER 16

Public Participation in Environmental Decision Making

The basic purpose of including public-participation programs or activities in the environmental decision-making process is to enable productive use of inputs and perceptions from governmental agencies, private citizens, and public interest groups in order to improve the quality of environmental decision making. The citizen-oriented activities are variously referred to as "citizen participation," "public participation," "public involvement," and "citizen involvement." Interest groups include those representative of industry, development, conservation, and preservation. This chapter has been prepared to highlight issues related to the planning and implementation of public participation programs. The initial sections contain some basic definitions, summarize public participation concepts from the Council on Environmental Quality (CEQ) regulations, and delineate general advantages and disadvantages of public participation. Additional issues include public participation in the EIA process (and related problems and constraints), delineation of objectives for public participation, identification of publics, techniques of public participation, conflict management and resolution, practical suggestions for implementation of a public participation program, and verbal communications in the EIA process.

BASIC DEFINITIONS

"Public participation" can be defined as a continuous, two-way communication process which involves promoting full public understanding of the processes and mechanisms through which environmental problems and needs are investigated and solved by the responsible agency; keeping the public fully informed about the status and progress of studies and implications of project, plan, program, or policy formulation and evaluation activities; and actively soliciting from all concerned citizens their opinions and perceptions of objectives and needs and their preferences regarding resource use and alternative development or management strategies and any other information and assistance relative to the decision. In essence, public participation involves both information feed-forward and feedback. "Feed-forward" is the process whereby information is communicated from public officials to citizens concerning public policy.

"Feedback," in this context, is the communication of information from citizens to public officials regarding public policy. Feedback information should be useful to decision makers in reaching timing and content decisions.

"Public participation" and "public relations" are not synonymous. Public participation is a planned effort to involve citizens in the decision-making process and to prevent or resolve citizen conflict through mutual two-way communication, while public relations is a planned effort to influence opinion through socially responsible performance based on mutually satisfactory two-way communication (Allingham and Fiber, 1990).

REGULATORY REQUIREMENTS

The CEQ regulations stipulate the need for public participation in terms of scoping, general public-involvement requirements, and the review process for draft environmental impact statements (CEQ, 1987). Therefore, the EIA process requires public participation, with the best approach for an agency to take being an active and positive one as compared to a passive approach to fulfill only the letter of the CEQ regulations. Most federal agencies have their own public participation requirements which are in consonance with, but supplementary to, the CEQ regulations.

The scoping process includes a number of public participation elements. "Scoping," as noted earlier, refers to an early and open process for determining the scope of issues to be addressed and for identifying the significant issues related to a proposed action (CEQ, 1987). Scoping requirements for the lead agency related to public participation are summarized in Chapter 1 (CEQ, 1987). General public involvement can include feedback on proposals and attendance at impact study meetings. Another instance of public participation occurs in the review process for draft environmental impact statements. The CEQ regulations indicate that after preparing a draft EIS and before preparing a final EIS the lead agency must provide opportunities for such a review (CEQ, 1987).

ADVANTAGES AND DISADVANTAGES OF PUBLIC PARTICIPATION

There are advantages and disadvantages associated with public-participation activities. Benefits accrue when affected persons likely to be unrepresented in EIA processes are provided an opportunity to present their views. Creighton, Chalmers, and Branch (1981) indicated that public participation can perform three vital functions: (1) it can serve as a mechanism for exchange of information; (2) it may provide a source of information on local values; and (3) it can aid in establishing the credibility of the planning and assessment process.

An additional benefit is that the agency, by constructing a record of decision making (draft statement, review, and final statement), provides for both judicial and public examination of the factors and considerations in the decision-making process. Thus, an added accountability is placed on political and administrative decision makers, since the process is open to public view. Openness exerts pressure on administrators to adhere to required procedures in decision making. Finally, through public participation, the agency is forced to be responsive to issues beyond those immediately related to the project.

Disadvantages, or costs, of public participation include the potential for confusion of the issues, since many new perspectives may be introduced. It is possible to receive erroneous information that results from the lack of knowledge on the part of the participants. Additional disadvantages include uncertainty of the results of the process, as well as potential project delay and increased project costs. However, a properly planned public-participation program need not represent a major funding item, and it need not cause an extensive period of delay in the process of decision making.

There is an increasing body of literature on planning and implementing public participation programs. For example, a bibliography of citizen participation in environmental issues compiled for the period 1970–1986 has been prepared by Frankena and Frankena (1988). In addition, several professional groups with expertise in public involvement have been formed. These include the International Association of Public Participation Practitioners (IAP3) and the International Environmental Negotiation Network (IENN). The IAP3 was established in 1990 as a nonprofit corporation to serve and represent public participation practitioners throughout the world. The IENN, also formed in 1990, is committed to improving the process of environmental treaty making through informal consensus building, information exchange, workshops, training sessions, and the dissemination of educational materials.

PUBLIC PARTICIPATION IN THE ENVIRONMENTAL-IMPACT-ASSESSMENT PROCESS

The EIA process can be considered to consist of seven stages: (1) identification of issues and impacts (scoping), (2) conduction of baseline studies of the environment, (3) prediction and evaluation of impacts, (4) mitigation planning, (5) comparison of alternatives, (6) decision making relative to the proposed action, and (7) study documentation through the preparation of an environmental assessment (EA) or an EIS. Public participation should be associated with all seven stages for major undertakings (projects, plans, programs, or policies) (Canter, Miller, and Fairchild, 1982).

The initial identification of issues and impacts establishes the scope of the environmental impact study. Public-participation activities at this stage are primarily devoted to informing the public about the project and determining what citizens feel about the need being addressed and the potential project. At this early stage, it will be possible to start identifying which groups see themselves as "winners" and which as "losers." The effort by the proponent agency or entity should be to establish a rapport with the involved parties and a spirit of cooperation.

The baseline study records the environmental status quo in the study area. At this stage, the information given to the public could take the form of what is being surveyed and why. Feedback to this information is often helpful in identifying existing databases. Thus, the public's response can reduce the time and cost of the baseline survey. Often, citizens can also identify areas of particular local interest which should be highlighted in the environmental impact report.

Impact evaluation consists of the prediction and interpretation of changes that would result from implementation of the alternatives under consideration. The public can assist in this process in several ways. For example, by reviewing the alternatives being considered, they can ensure that no viable alternative is inadvertently omitted. Where legal standards are not in force, comments from the public can be useful in establishing project-specific criteria or maximum tolerable levels of change. Finally, the information-feedback cycle must be maintained to hold the public's interest and prevent alienation.

Mitigation measures are planned to reduce undesirable project effects. One of the major inputs at this stage is ensuring that the mitigation measure is itself acceptable. Consider, for example, a new housing development that draws heavily on a diminishing water supply. One mitigation measure is to collect and treat wastewater from the urban area and recycle it. In many areas this measure, though technically feasible, is culturally unacceptable. As before, public review will ensure that all reasonable measures are considered.

The comparison of alternatives is done to identify the one or several preferred actions. Local values could be used to weight the importance of environmental factors at this stage. It is very important at this stage that the public have an input into what is recommended to decision makers. It is at the comparison-of-alternatives stage that the preferred project alternative is identified; therefore, at this stage any potential conflicts will come clearly into focus. (Methods of conflict resolution will be discussed in a later section.) If the public involvement program has been effective to this point, it should be possible to resolve conflicts in a spirit of cooperation.

The sixth stage in an environmental impact study is the actual decision on which alternative will be implemented. At this stage, public-involvement activities have three objectives. First, the public should be informed what the decision is and why. Ideally, the decision should be based on the recommendations arising out of the comparison of alternatives. However, this is not always the case. The second objective is the final resolution of conflicts. In this regard, it may be necessary to compensate certain publics in order to even out the distribution of benefits. Finally, if the decision makers are responsible to the public, the third objective will be the solicitation of feedback concerning the final decision.

The seventh stage is the preparation of study documentation in the form of an EA or EIS. Public involvement would consist of reviews and comments on draft documents. Stages 6 and 7 in the EIA process could actually be combined.

If public participation is to be effective in the various stages of the EIA process, the public participation program must be carefully planned. A good public participation program does not occur by accident. Planning for public participation should address the following elements:

1. Delineation of objectives of public participation during the pertinent EIA stages.
2. Identification of publics anticipated to be involved in the pertinent EIA stages.
3. Selection of public participation techniques which are most appropriate for meeting the objectives and communicating with the publics. It may be necessary to delineate techniques for conflict management and resolution.
4. Development of a practical plan for implementing the public participation program.

Each of these four major elements will be addressed in subsequent sections of this chapter. However, background information on three topics should be considered: (1) levels of citizen participation, (2) inherent problems in implementing public participation programs, and (3) usage of observations and/or general principles related to planning public-participation programs.

Levels of Citizen Participation

Levels of citizen participation can range from situations in which the citizens do not participate at all, to situations involving token citizen participation, to situations where citizens share equally in planning, to situations where citizens actually control the planning process (Schwertz, Jr., 1979). Figure 16.1 shows various stages along this continuum. The bottom two rungs of the ladder describe levels of nonparticipation that have been contrived by insincere public-participation planners to substitute for genuine participation. Their real objective is not to enable people to participate, but to enable "power holders" to "educate" or "persuade" the participants. Rungs 3 and 4 progress to levels of "tokenism" that allow the outsiders to hear and to have a voice. When these activities are proffered by power holders as the total extent of participation, citizens may indeed hear and be heard. But under these conditions they lack the power to insure that their views will be heeded. Rung 5 is simply a higher-level tokenism, because the ground rules allow have-nots to advise, but retain for the power holders the continued right to decide. Further up the ladder are levels of citizen power with increasing degrees of decision-making involvement. Citizens can enter into a "partnership" that enables them to

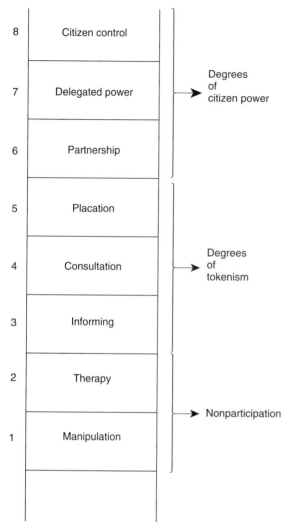

FIGURE 16.1
Levels of Citizen Participation (Arnstein, 1969, p. 217).

negotiate and engage in trade-offs with traditional power holders. At the topmost rungs, "delegated power" and "citizen control," have-not citizens obtain the majority of decision-making seats, or full decision-making power. Most public participation programs stop at rungs 3, 4, or 5. Citizen involvement will rarely reach the highest rung unless there is passage of citizen referendums. Figure 16.1 pri-

TABLE 16.1

LEVELS OF PUBLIC INVOLVEMENT

Awareness	Involvement	Participation
Monologue	Dialogue	Empowerment
Altering	Interaction	Planning
One-way	Two-way	Partnership
"Tokenism"	Engagement	Citizen Control
"Manipulation"	Consultation	
Therapy		

Source: Castensson, Falkenmark, and Gustafsson, 1990, p. 81.

marily relates to public projects; Hance, Chess, and Sandman, (1990) have developed a similar ladder for industrial projects.

The concept of different levels of public involvement and participation can be displayed on a continuum within a tabular format. For example, Table 16.1 shows characteristic words related to varying degrees of awareness, involvement, and participation.

Inherent Problems in Implementing Public Participation Programs

Delli Priscoli (1981) identified four overriding problems which continually surface in implementing public participation programs: (1) coordination, (2) control, (3) representativeness, and (4) dissonance. Although these problems are never "solved," they can and should be creatively managed. Key points about these four problems are as follows (Delli Priscoli, 1981, after pp. 173–174):

1. *Coordination*—One of the most critical problems for government today is the relationship between different governmental units and levels. Often policies and/or plans of one agency are implemented by another. Projects or facilities of one agency may even be operated or maintained by a second, third, or fourth. Furthermore, actions are rarely limited to federal agencies. State, local, and private actors may also be involved, and each agency may embody different missions and purposes. As a consequence of this mix of

purposes and actors, different citizen involvement programs frequently are developed. In some cases, these programs ameliorate interagency and citizen-government conflict; in others, they generate such conflict.

2. *Control*—When a federal agency deals with a public policy issue, its responsibility is to find and assure the federal interest. Such interest frequently takes the form of centralized control through regulation, licensing, funding, and the like. Citizen involvement, however, is by nature a decentralizing concept. Therefore, a tension always exists between the centralized needs of the agency and the decentralized interests of citizens.

3. *Representativeness*—One of the most frequent criticisms of public involvement programs is that the citizens who become involved do not represent the majority, but rather a "citizen elite" that represents special interests. This is a very serious problem for agencies that make use of citizen involvement to develop consensus and support for a policy or program. For this reason, agencies must develop multiple links in the citizen involvement process.

4. *Dissonance*—One of the facts of life for government agencies is the conflict between political interests and technical interests in decision making. The excessive practice of using technical justifications to rationalize controversial political discussions is undoubtedly one of the factors that have led to greater demands for citizen involvement. As a result, government agencies should expect that citizen involvement will increase the tension between technical and political considerations.

Observations and Principles

A review of practical observations and principles in the literature can be useful in the planning of a public participation program. One such example can be found in Canada's Federal Environmental Assessment Review Office (1988, vol. 1, p. 33), which includes the following practical observations related to public participation programs:

1. Public involvement must have two-way communication.
2. Most decision processes will benefit from some public involvement.
3. A public is any person or group of people with a distinctive interest or stake in an issue.
4. The interested public will be different for every project.
5. Use multiple techniques for public involvement.
6. Senior management need to be involved in supporting and reviewing the public involvement program.
7. For open communication to develop with the community, open communication is needed within the organization.
8. Monitor current issues of public concern, as an "early warning system."
9. If consensus is to be achieved, early public involvement is essential.

To serve as a second example, Helweg (1985, p. 279) identified the following principles for public involvement in water-resource projects:

1. The planner should seek and use input from all interested publics during all pertinent steps of the planning process.
2. Public involvement has limitations that must be understood. Well-organized minorities may overshadow silent majorities and impose an unpopular solution, or infeasible alternatives may be suggested by publics with inadequate technical knowledge.
3. The planner must be honest and must empathize with the various interest groups.
4. Media selection is important. Public meetings are not the only vehicle through which this can be accomplished. The planner should consider interviews, paid advertising, radio and television talk shows, questionnaires, newsletters, the Delphi technique, simulation games, field trips, work-study groups, advisory groups, brochures, and other methods.

OBJECTIVES OF PUBLIC PARTICIPATION

The delineation of objectives for public participation programs or activities during different stages within the EIA process is an important element in developing a public participation plan. Two basic reasons for identifying and classifying objectives are that (1) objectives change over the various stages of a study and (2) some public participation techniques are better than others for achieving particular objectives. The objectives could be general or specific. Hanchey (1981) suggested three general types of objectives which should be considered in the design of a public participation program for a specific planning situation, such as a water-resources project. These are referred to as (1) the "public relations objective," (2) the "information objective," and (3) the "conflict-resolution objective." As shown in Figure 16.2, these general objectives can be disaggregated into seven sec-

ond-order objectives which serve to clarify and to provide workable concepts for both the design and evaluation of such programs (Hanchey, 1981). Bishop (1975) delineated six objectives for public participation and tied them to the various stages of an EIA study. The six general objectives are

1. Information dissemination, education, and liaison
2. Identification of problems, needs, and important values
3. Idea generation and problem solving
4. Reaction and feedback on proposals
5. Evaluation of alternatives
6. Conflict resolution by consensus

The relationship of these six objectives, which were for water-resources project planning in the United States, to the seven stages in an EIA study is shown in Table 16.2. The first of these objectives is directed toward education of the citizenry on EISs and their purpose, and on

FIGURE 16.2
Objectives of Public Participation (Hanchey, 1981, p. 16).

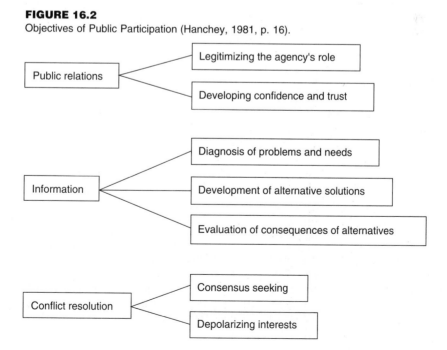

TABLE 16.2

PUBLIC INVOLVEMENT OBJECTIVES AT VARIOUS EIA PROCESS STAGES

Objective	Stage of EIA process						
	Impact identification (scoping)	Baseline study	Impact evaluation	Mitigation planning	Comparison of alternatives	Decision making	Documentation
Inform, educate	X	X	X	X	X	X	X
Identify problems, needs, values	X	X	X	X	X		
Identify problem-solving approaches			X	X	X		X
Obtain feedback		X	X	X	X	X	
Evaluate alternatives			X	X	X		
Resolve conflicts					X	X	

Source: After Bishop, 1975.

the process of citizen participation. In addition, this objective includes dissemination of information on the study progress and findings, as well as data on potential environmental impacts. The second objective—identification of problems, needs, and important values—is related to the determination of the environmental resources important to various segments of the public in an area. In addition, this objective is focused on defining areas of environmental problems and needs and the relation of potential solutions being addressed in the project study.

The third objective is directed toward identification of problem-solving approaches (alternatives) that may not have been considered in normal planning processes. In addition to designating specific alternatives for identified needs, it is possible to enumerate mitigating measures for various alternatives so as to minimize adverse environmental effects. The focus of the fourth objective is on probing public perceptions of the actions and resource interrelations. In addition, successful accomplishment of this objective can provide input useful in assessing the significance of various types of impacts. The alternatives-evaluation objective is closely related to the earlier objective—reaction and feedback on proposals. In the process of evaluating alternatives, valuable information can be received about the significance of unquantified and quantified environmental amenities. Public reaction to value trade-offs in the process of selection can also be reassessed.

The final objective is related to resolving conflicts that exist over the proposed action. Accomplishment of this objective may involve mediation of differences among various interest groups, development of mechanisms for environmental-costs compensation, and efforts directed toward arriving at a consensus of opinion on a preferred action. Successful accomplishment of this objective can prevent unnecessary and costly litigation.

As another example, the objectives for a public participation program related to five stages of a generic EIA study for a project in Canada are shown in Table 16.3. The listed stages are different from but compatible with the seven stages described earlier, and the listed objectives represent a finer delineation of the objectives suggested by Bishop (1975).

Extensive communication will be necessary to achieve the objectives of a public participation program. Bishop (1981) identified the following four communication-process models for meeting the basic objectives of public participation: (1) diffusion, (2) collection, (3) interaction, and (4) diffusion-collection. In the "diffusion process," the agency sends a message using different media to various target groups, who in turn transmit the message to still other groups or individuals. The net result is that the agency is able to reach a broader segment of the public, in terms of the total impact, than just the initial target group. The diffusion process indicates that communication is not just a single-, but a multistep activity whereby target groups become senders in transferring a message to others through media which they can access. Corollary to this is the fact that the sender cannot completely control the communication process, since intermediaries are present to influence or interrupt the process. Also, a target public can be contacted through several media, thus reinforcing and clarifying the message.

The "collection process" can be seen as "diffusion in reverse" (Bishop, 1981). It may serve to obtain feedback to complete a communication loop or to collect information. The messages may or may not return by the same media path. "Interaction" describes the situation where communication is an interchange among several groups. The agency may assume the central role in acting as a moderator and facilitator in the communication exchange between other groups, or may simply take the role of one of the communicators in the interaction. Interaction processes typically involve meetings, work groups, committees, advisory panels, and the like. The "diffusion-collection process"

TABLE 16.3

DEVELOPMENT OF OBJECTIVES FOR A PUBLIC PARTICIPATION PROGRAM

Stage	Objectives
Issues Identification/Sensing Public Interest	Obtain a complete understanding of how the issue is viewed by all significant interests. Identify the level of interest in future public participation activities on this issue.
Detailed Identification of Impacts	Undertake a scoping exercise to identify the most significant impacts.
Assessment of Impacts	Develop a complete understanding of the impacts as viewed by the public. Assess the relative merit assigned to the project by various interests, including their reasons for these evaluations. Develop understanding of possible mitigation procedures, as viewed by the public.
Decision Making	Make a decision that is technically, economically and socially feasible, and politically acceptable.
Post Approval Decision Making	Develop all requirements for public involvement following the decision.

Source: Federal Environmental Assessment Review Office, 1988, vol. 2, p. 37.

describes the situation where information is disseminated with the specific intent of eliciting some desired information in response. Usually, in addition, the mechanism or medium for response will be specified or provided in order to facilitate information collection. A simple example is a questionnaire that is sent to, say, some public groups and to a newspaper. Target publics are asked to send their responses by individual letter to the agency as the originator of the questionnaire (Bishop, 1981).

The communication-process models can be matched, as appropriate, with one or more of the six key objectives of public participation as delineated by Bishop (1975). For example, the "inform," "educate," and "liaison" objectives are all dependent on dissemination of information. The diffusion model describes this process. Identification, assessment, and feedback are objectives that are described by the reverse, the collection model. Idea generation, problem solving, conflict resolution, and consensus are generally best accomplished by interaction processes. Accomplishment of the review, reaction, and evaluation objectives requires a two-step process. An information "stimulus" is first directed to the "publics," then the publics respond with their reactions or evaluations. A comprehensive public-participation program will usually involve all four communication-process models (Bishop, 1981).

IDENTIFICATION OF PUBLICS

The identification of publics which would potentially be involved in the various stages of an environmental impact study is another basic element in the development of a public participation program. A "public" is any person, or group of people, that has a distinctive interest or stake in an issue (Federal Environmental Assessment Review Office, 1988). Information on the potential impacts of a project could be displayed according to the affected publics. Also of importance is the delineation of public participation techniques which are most effective in communicating with different publics. This section will address the recognition of types of publics, pragmatic approaches for identifying publics, and selected techniques for communicating with identified publics.

Recognition of Types of Publics

The general public cannot be considered as one body. The general public is diffuse, but at the same time highly segmented into interest groups, geographic communities, and individuals. There are sets or groups of publics that have common goals, ideals, and values. Any one person may belong to several different sets of these publics, since the groups may be professionally, socially, or politically oriented. Multiple association thus allows the opportunity for multiple access to individuals as participants, clients, or critics in a planning process (Bishop, 1981).

Schwertz, Jr., (1979) suggested that the purpose of identifying various publics is two-fold. The first consideration is to ensure that no group of persons is excluded from participating. Such exclusion, whether intentional or not, will be resented. Those who are left out will often seek redress outside of the formal public-involvement program—for example, by public and/or political pressure or by court action. The second consideration relates to the public-involvement technique to be used, since some techniques are more effective with certain groups than with others.

There are several ways of categorizing various publics that might be involved in a public participation program for an EIA study. One broad-based system for grouping publics consists of using four separate categories:

1. Persons who are immediately affected by the project and live in the vicinity of the project
2. Ecologists ranging from preservationists to those who want to ensure that development is as effectively integrated into the needs of the environment as possible
3. Business and commercial developers who would benefit from initiation of the proposed action
4. The part of the general public comprising those who enjoy a high standard of living and who do not want to sacrifice this standard in order to preserve wilderness or scenic areas or to have pollution-free air and water.

The U.S. Army Corps of Engineers (1971, p. 3) has defined the following publics in conjunction with water-resources development projects:

1. Individual citizens, including the general public and key individuals who do not express their preferences through, or participate in, any groups or organizations.
2. Sporting groups.
3. Conservation/environmental groups.
4. Farm organizations.
5. Property owners and users, representing those persons who will be or might be displaced by any alternative under study.
6. Business and industrial groups, including Chambers of Commerce and selected trade and industrial associations.
7. Professional groups and organizations, such as the American Institute of Planners, American Society of Civil Engineers, and others.
8. Educational institutions, including universities, high schools, and vocational schools. General participation is by a few key faculty members and students or student groups and organizations.
9. Service clubs and civic organizations, including service clubs in a community such as Rotary Club, Lions Club, League of Women Voters, and others.
10. Labor unions.
11. State and local governmental agencies, including planning commissions, councils of government, and individual agencies.
12. State and local elected officials.
13. Federal agencies.
14. Other groups and organizations, possibly including various urban groups, economic op-

portunity groups, political clubs and associations, minority groups, religious groups and organizations, and many others.

15. Media, including the staff of newspapers, radio, television, and various trade media.

Another example of a categorization scheme using four types of participants involves democratic societal decision making. Specifically, these participants, also called "actors" or "P-actor groups," include (after Castensson, Falkenmark, and Gustafsson, 1990, p. 79):

1. *The Public*—According to democratic doctrine and tradition, political power belongs to the people; generally speaking, the public has a low awareness of many environmental issues.
2. *The Politicians*—They are the representatives of the people; in democratic elections they receive a political mandate from the public.
3. *The Professionals*—They play a key role in the impact-communication process. They are experts in a wide range of small but well-defined areas.
4. *The Practitioners*—They belong to a category of people with a variety of practical skills. Typically a practitioner is experienced in the implementation of political decisions.

Many public groups can be considered under the broad heading of "nongovernmental organizations" (NGOs). NGOs include a diversity of groups ranging from preservationists to labor unions to "green movement" entities. This collective term has had considerable usage outside of the United States. It is important to note that NGOs can encompass a broad range of viewpoints regarding a particular proposed project.

Pragmatic Approaches for Identifying Publics

Having recognized that there are different types of publics based on several classification schemes, concern then turns to pragmatic approaches for identifying the potential publics which might be associated with a study during its various stages. One approach in targeting publics is to identify those persons who believe themselves to be affected by the study outcome. The difficulty is that the degree to which people feel affected by a proposed project is a result of their subjective perception; people the agency feels are most directly impacted may not be as concerned as those that the agency perceives as only peripherally involved. However, the starting point always entails some effort to objectively analyze the likelihood that someone will feel affected by the study. Some of the bases on which people are most likely to feel affected are (Creighton, 1981b, pp. 201–202)

1. *Proximity:* People who live in the immediate area of a project and who are likely to be affected by noise, odors, dust, or possibly even threat of dislocation are the most obvious publics to be identified.
2. *Economic:* Groups that have jobs to gain or competitive advantages to win are again an obvious starting point in any analysis of possible publics.
3. *Use:* Those people whose use of the area is likely to be affected in any way by the outcome of the study are also likely to be interested in participating. These include recreationists, hikers, fishermen, hunters, etc.
4. *Social:* Increasingly, people who see projects as a threat to the tradition and culture of the local community are likely to be interested in projects. They may perceive that a large influx of construction workers into an area may produce either a positive or negative effect on the community. Or they may perceive that the project will allow for a substantial population growth in the area which they may again view either positively or negatively.
5. *Values:* Some groups may be only peripherally affected by the first four bases but find that some of the issues raised in the study directly affect their values, their "sense of the way things ought to be." Anytime a project touches on such issues as free enterprise vs. government control, or jobs vs. environmental enhancement, there may be a number of individ-

uals who participate primarily because of the value issues involved.

Creighton (1981b) suggested three broad approaches to targeting relevant publics: (1) self-identification, (2) third-party identification, and (3) staff identification. "Self-identification" simply means that individuals or groups step forward and indicate an interest, either pro or con, in the project. The use of the news media, the preparation of brochures and newsletters, and the conduction of well-publicized public meetings are all means of encouraging self-identification. Anyone who participates by attending a meeting or writing a letter or phoning a hot line has clearly indicated an interest in being an active public in the study. As a result, it is crit-

ical that anyone who expresses an interest in any way be placed on a mailing list and kept continually informed of the EIA study progress (Creighton, 1981b).

One of the best ways to obtain information about other interest groups or individuals which should be included in the study is to ask an existing advisory committee, or representatives of known interests, to indicate anyone else who should be involved. One variation on this theme is to enclose a response form in any mailings inviting people to suggest other groups that should be included. These simple techniques of consulting with known representatives to recommend others who should be involved often prove to be among the most effective means of targeting relevant publics (Creighton, 1981b).

TABLE 16.4

CITIZEN-INVOLVEMENT-GROUP CHECKLIST

_____ Bankers associations
_____ Business associations
_____ Chambers of Commerce
_____ Civic organizations
    _____ Jaycees
    _____ League of Women Voters
    _____ Others
_____ Developer associations
_____ Elder citizens' organizations
_____ Entrepreneurs
_____ Executive and professional associations (local)
_____ Farmers' organizations
_____ Fraternal organizations
_____ Labor organizations
_____ Neighborhood associations
_____ Neighboring communities (where appropriate)
_____ Newspaper editors and owners
_____ Philanthropic organizations and/or foundations
_____ Real estate brokers' associations
_____ Religious organizations (by denomination)
_____ Retail trade associations
_____ Service organizations
    _____ Rotary Club
    _____ Others

_____ Special groups and/or organizations
    _____ Audubon Society
    _____ Conservation societies
    _____ Historic preservation societies
    _____ Sierra Club
    _____ Others
_____ Social clubs
_____ Trade organizations
_____ Utility companies
_____ Youth organizations
_____ Veterans' organizations
_____ Women's auxiliaries

Others

_____ _____

_____ _____

_____ _____

_____ _____

Source: Schwertz, Jr., 1979, p. 15.

Schwertz, Jr., (1979) noted that many communities have existing organizations or interest groups which can and should be made a part of a public involvement program. Table 16.4 contains a typical citizen-involvement-group checklist which can be useful to the public-participation-program planning staff in identifying publics which would be potentially associated with a project.

Continuously updated mailing lists can also be used in identifying and communicating with various publics. For example, Ragan, Jr., (1981b) reported on a study of mailing list usage in 13 U.S. Army Corps of Engineers district offices. He found that all 13 used a mailing list as the basis for identifying the publics which might be interested in a specific study. The mailing list was a compilation of governmental and private

TABLE 16.5

EFFECTIVENESS OF DIFFERENT COMMUNICATION TECHNIQUES ON VARIOUS "PUBLICS"

Public	Public hearings and meetings	Printed brochures	Radio programs and news	TV programs and news	Newspaper articles	Magazine articles	Direct mail and newsletters	Motion picture, film	Slide-tape presentation	Telelecture
Individual Citizens	M	L	H	H	H	L	L	M	M	L
Sportsmen Groups	M	M	M	M	M	H	H	H	H	M
Conservation-Environment Groups	M	M	M	M	M	H	H	H	H	M
Farm Organizations	M	M	M	M	M	H	H	M	M	M
Property Owners and Users	M	L	H	H	H	L	L	M	M	L
Business-Industrial	L	L	M	M	M	M	H	M	M	L
Professional Groups and Organizations	L	L	M	M	M	M	H	M	M	L
Educational Institutions	M	L	L	L	M	M	H	M	M	M
Service Clubs and Civic Organizations	L	L	M	M	M	M	L	H	H	M
Labor Unions	L	L	M	M	M	L	L	M	M	L
State-Local Agencies	H	M	L	L	L	M	H	H	H	H
State-Local Elected Officials	H	M	L	L	L	L	H	H	H	H
Federal Agencies	H	M	L	L	L	L	H	M	M	M
Other Groups and Organizations	H	M	M	M	M	M	H	H	H	M

Note: H = Highly effective; M = moderately effective; L = least effective.
Source: Bishop, 1975, p. 85.

organizations and individuals who, by virtue of their positions or prior indication of interest, need or want to be apprised of planning activities. While mailing lists may be adequate to notify parties of public meetings, there are problems in using them as the primary basis for identifying people for more-intensive public participation. First, because public meetings are "official" sessions, Ragan, Jr., (1981b) noted that fully 75 to 80 percent of most mailing lists comprise public officials and agencies. Second, mailing lists are hard to maintain, and an EIA study manager may not have the time. Third, many districts have several lists: for example, one for planning, one for design and construction, and one for various operational functions. In some cases, the environmental-resources and recreation sections may have separate lists. Fourth, mailing lists categorized by public organization, media, and all others make it difficult to identify potential interests to be contacted for specific EIA studies.

The identification of publics should also incorporate the dimension of participation through time (Bishop, 1981). At the onset of planning, such as in the scoping phase, a certain segment of the public will have an interest in participating. These are usually people or groups that (1) have participated in the past, (2) are affected by a problem, or (3) will be affected by a possible solution to a problem. As the study progresses, some of those identified will not participate, while some previously unidentified publics will identify themselves. Looking forward into time, there will always be those who, although not identified previously, may come into the process, even in the final stage. Hence, the planner must be prepared to communicate with three sets of publics: (1) those than can be identified and will participate, (2) those that will be identified as the process progresses, and (3) those that will step forward to be identified in the future. Thus, of the publics initially identified by the agency, some will follow through, others

will drop out, and some previously unidentified interests will enter the arena of participation. Continuity of participation is a problem for major EIA studies that extend over months or years. Three approaches can be used in dealing with the problem of publics changing over time: (1) at the outset of a study, actively seek out and engage a broad and representative range of public interests; (2) keep as much flexibility in the process for as long as possible, insofar as selecting a plan or recommending action; and (3) document the process and the public inputs relating to alternatives and impacts studied (Bishop, 1981).

Selected Techniques for Communicating with Identified Publics

Some public involvement techniques are better than others for communicating with different publics. Table 16.5 summarizes the effectiveness of several techniques for communicating with various publics identified earlier as associated with water-resources projects (Bishop, 1975). Table 16.6 delineates the various types of participants who might typically be associated with several public participation techniques.

SELECTION OF PUBLIC PARTICIPATION TECHNIQUES

A critical element in planning a public participation program is associated with the selection of public participation techniques to meet identified objectives and the needs of identified publics. It should be noted that there are numerous techniques, and a well-planned public-participation program will probably involve the use of multiple techniques over the lifetime of an EIS (Ragan, Jr., 1981a). For example, a survey of the usage of techniques in 13 U.S. Army Corps of Engineers district offices, noted earlier, determined that each district employed from 5 to 14 different techniques to inform and educate the public and/or obtain information on

TABLE 16.6

EXAMPLES OF PUBLIC PARTICIPATION TECHNIQUES AND MOST-LIKELY PARTICIPANTS

Type of participant favored	Public hearings	Structured workshop	Legal notices	Information meetings	Citizen advisory committees	Polls	Ombudsperson or field office	Written comment
Highly literate			X					X
Highly verbal	X	X			X		X	
Employed during the day								
Blue-collar worker			X	X		X	X	X
White-collar worker	X	X	X	X	X	X	X	X
Management						X		
Houseperson or unemployed	X	X		X	X	X	X	
Highly informed, involved	X	X	X	X	X		X	X
Degree of representativeness of multiple publics	1	3	1	2	2	3	1	2

Note: 1 = low; 2 = medium; 3 = high.
Source: Westman, 1985, p. 118.

an individual study (Ragan, Jr., 1981b). The techniques used most frequently were public meetings, informational brochures, advisory committees, media content analysis, public speeches, and newsletters (Ragan, Jr., 1981b).

The most traditional public-participation technique is the ''public hearing,'' which is a formal meeting for which written statements are received and a transcript is kept. The public hearing is generally not the most effective forum for public participation in the EIA process. Several classification schemes have been developed for public participation techniques, in accordance with their function (Schwertz, Jr., 1979), communication characteristics, and potential for meeting stated objectives (Federal Environmental Assessment Review Office, 1988; Bishop, 1975). Creighton (1981c) developed a short catalog of the advantages

and disadvantages of several additional techniques.

Techniques Classification According to Function

Schwertz, Jr., (1979) summarized a U.S. Department of Transportation (DOT) classification scheme for 37 public participation techniques, which were grouped into six functional classes, as follows: (1) information dissemination, (2) information collection, (3) initiative planning, (4) reactive planning, (5) decision making, and (6) participation-process support. Table 16.7 lists the 37 citizen-participation techniques, by functional class. Summary comments relative to the six functional classes are as follows (after Schwertz, Jr., 1979, pp. 21–23):

1. Information-dissemination techniques are used to inform the public of actions being

---

TABLE 16.7

PUBLIC PARTICIPATION TECHNIQUES CLASSIFIED BY FUNCTION

1. **Information dissemination**
   Public information programs
   Drop-in centers
   Hot lines
   Meetings—open information
2. **Information collection**
   Surveys
   Focused group discussions
   Delphi-based techniques
   Community-sponsored meetings
   Public hearings
   Ombudsman activities
3. **Initiative planning**
   Advocacy planning
   Charettes
   Community planning centers
   Computer-based techniques
   Design-in and color mapping
   Plural planning
   Task forces
   Workshops
4. **Reactive planning**
   Citizens' advisory committees

   Citizen representatives on policy-making boards
   "Fishbowl" planning
   Interactive cable TV-based participation
   Meetings—neighborhood
   Neighborhood planning councils
   Policy capturing
   Value analysis
5. **Decision making**
   Arbitrative and mediative planning
   Citizen referendum
   Citizen review board
   Media-based issue balloting
6. **Participation process support**
   Citizen employment
   Citizen honoraria
   Citizen training
   Community technical assistance
   Coordinator or coordinator-catalyst
   Game simulation
   Group dynamics

Source: Schwertz, Jr., 1979, p. 20.

taken in a program area, the opportunities for citizen input, and any proposed plans that have been submitted.

2. Information-collection techniques are used to gather both technical data and opinion-related information.

3. Initiative-planning techniques assign the responsibility for producing proposals and plans to citizen representatives.

4. Reactive-planning techniques assign the responsibility for producing proposals and plans to the local planning agency, and citizens are asked to react to those proposals and plans.

5. Decision-making techniques represent the highest rung of Arnstein's "ladder of citizen participation" (Arnstein, 1969). Accordingly, these techniques are not feasible in program areas where the responsibility for the decision is mandated to elected officials.

6. Participation-process-support techniques are not citizen-participation techniques in them-

---

TABLE 16.8

PUBLIC PARTICIPATION TECHNIQUES CLASSIFIED BY OBJECTIVES

	Public information	Information feedback	Consultation	Extended involvement	Joint planning
Public Information					
Advertising	X				
Brochures	X				
Citizen Training Programs	X				
Contests/Events	X				
Direct Mail	X				
Exhibits/Displays	X				
News Conferences	X				
Newsletters	X				
Newspaper Inserts	X				
News Releases	X				
Position Papers	X				
Political Preview	X				
Publications	X				
Publicity	X				
Public Service Announcements	X				
Reports	X				
Public Information Feedback					
Analyzing Public Involvement Data		X			
Briefs		X			
Community or Social Profiles		X			
Computer Assisted Participation		X	X		
Content Analysis		X			
Focus Groups		X	X		
Interviews		X	X		
Policy Profiling		X			
Polls		X			
Questionnaires		X			
Surveys		X			
Written Submissions		X			

selves, but mechanisms to strengthen the effectiveness of the other techniques.

Techniques Classification According to Communication Characteristics and Potential for Meeting Stated Objectives

The Federal Environmental Assessment Review Office (1988), published in Canada, has identified a number of techniques for providing public

information, soliciting information feedback, consulting, achieving extended involvement, and accomplishing joint planning; these techniques are listed in Table 16.8. Several of these techniques can be used to simultaneously satisfy more than one objective.

Bishop (1975) developed a structured classification scheme for 24 public participation techniques. Table 16.9 displays the techniques

TABLE 16.8

PUBLIC PARTICIPATION TECHNIQUES CLASSIFIED BY OBJECTIVES *(continued)*

	Public information	Information feedback	Consultation	Extended involvement	Joint planning
Consultation					
Brainstorming			X		
Coffee Klatches			X		
Conferences		X	X		
Delphi		X	X		
Dialogues			X		
Field Offices	X	X	X		
Large Meetings			X		
Nominal Group Process		X	X		
Open Houses	X	X	X		
Panels	X	X	X		
Participatory Television		X	X		
Phone Lines	X	X	X		
Public Meetings	X	X	X		
Simulation Games		X	X	X	
Technical Assistance	X	X	X	X	
Town Meetings		X	X		
Trade-off Games		X	X	X	X
Workshops			X	X	X
Extended Involvement					
Advisory Committees			X	X	
Charettes			X	X	
Task Forces			X	X	
Joint Planning					
Arbitration				X	X
Collaborative Problem Solving				X	X
Conciliation				X	X
Mediation				X	X
Negotiation				X	X
Niagara Process				X	X

Source: Federal Environmental Assessment Review Office, 1988, vol. 3, pp. 59–60.

TABLE 16.9

CAPABILITIES OF PUBLIC PARTICIPATION TECHNIQUES

Communication characteristics[a]				Impact assessment objectives					
Level of public contact achieved	Ability to handle specific interest	Degree of two-way communication	Public participation techniques	Inform/ educate	Identify problems/ values	Get ideas/ solve problems	Feedback	Evaluate	Resolve conflict/ consensus
M	L	L	Public hearings	X	X		X		
M	L	M	Public meetings	X	X		X		
L	M	H	Informal small group meetings	X	X	X	X	X	X
M	L	M	General public information meetings	X					
L	M	M	Presentations to community organization	X	X		X		
L	H	H	Information coordination seminars	X			X		
L	M	L	Operating field offices	X	X	X	X	X	
L	H	H	Local planning visits		X		X	X	
L	H	L	Planning brochures and workbooks	X		X	X	X	
M	M	L	Information brochures and pamphlets	X					
L	H	H	Field trips and site visits	X	X				
H	L	M	Public displays	X		X	X		
M	L	M	Model demonstration projects	X		X	X	X	X
H	L	L	Material for mass media	X					
L	L	M	Response to public inquiries	X			X		
H	L	L	Press releases inviting comments	X		X	X		
L	H	L	Letter requests for comments	X		X	X		
L	H	H	Workshops		X	X	X	X	X
L	H	H	Charettes			X		X	X
L	H	H	Advisory committees		X	X	X	X	
L	H	H	Task forces		X	X		X	
L	H	H	Employment of community residents		X	X			X
L	H	H	Community interest advocates			X	X	X	X
L	H	H	Ombudsman or representative		X	X	X	X	X

[a]L = low; M = medium; H = high.
Source: Bishop, 1975, p. 52.

in three groups: (1) the first six listed techniques represent public forums, (2) the following eleven listed techniques represent community contacts, and (3) the final seven listed techniques represent interactive group methods. The communication characteristics of the 24 techniques are displayed in Table 16.9 in terms of the level of public contact achieved, ability to handle specific interest, and degree of two-way communication. A relative scale of "low" (L), "medium" (M), and "high" (H) effectiveness designators is used to assess the communication characteristics of the 24 techniques. Table 16.9 also has information on the potential usefulness of each technique relative to meeting one or more of the six objectives delineated earlier (regarding the same agency) for public participation in an EIS. Specific comments on the advantages and limitations of 12 selected techniques are as follows (after Bishop, 1975, pp. 55–68):

1. *Public hearings*—Public hearings tend to be formal and rather highly structured. Appropriate records are maintained—that is, the transcript and written statements.

2. *Public meetings*—Public meetings are less formalized than are hearings and do not require a transcript. However, detailed notes should be kept on file.

3. *Informal, small group meetings*—While this type of meeting may take any of several forms and serve several purposes, the overall format is much the same as that of a public meeting. In this respect, small group meetings may function as a series of small-scale public meetings to allow more-intimate contact with publics from various geographic or interest-group areas.

4. *Information and coordination seminars*—This tool is not used to inform the general public directly, but functions to inform and coordinate with special-interest groups, specific individuals, and groups representing segments of the public.

5. *Forum of other agencies or groups*—Forums of other agencies and groups—such as civic group meetings, organization meetings, and the like—can also be used for pertinent presentations or statements.

6. *Operating field offices*—The operation of special field offices serves to establish a more or less specialized liaison between the agency office and the public. In studies necessitating close local contact and coordination this approach may be used efficiently, particularly if the offices are well publicized.

7. *Local planning visits*—Community or on-site visits are oriented toward increased understanding and coordination with cooperating agencies, knowledgeable community interest groups, and individuals.

8. *Field trips and site visits*—These excursions differ from planning visits in that they are primarily nonprofessional "show-me" trips. These visits can be used to accurately inform public groups, local officials, and the media about the specifics of a plan. Trips can be combined with or considered as field press conferences.

9. *Public displays and model demonstrations*—Under appropriate conditions, displays and demonstrations can provide an overview of a project, quick appraisal of alternatives, description of impacts of the project, and information on any number of project-related issues.

10. *Workshops*—The success of workshops depends, in large part, on the degree of advance preparation; therefore, this should be as comprehensive as possible. Advance preparation for workshops might include distribution of the various types of brochures, planning visits, coverage by the media and direct contact of interested parties. Workshops can be of several types depending on the planning activity and stage of the study, the publics to be contacted, and the subject matter for discussion. These types

include an open public workshop, an invitational workshop, and a combination of the two.

11. *Charettes*—The charette functions as a highly intense resolution-oriented meeting. It can be thought of as a miniworkshop or small, select group meeting with the express purpose of reaching a decision or resolving a conflict.

12. *Special committees*—Several types of citizens' committees have been used in planning studies. Committees, as representative public bodies, can serve a very useful purpose in EIA studies. Types of committees that have functioned in such studies include citizens' advisory committees for specific projects, ad hoc committees, and citizens' committees.

Creighton and Delli Priscoli (1981) indicated that a citizens' committee can perform a number of useful functions in a planning task; for examples, the committee may (1) assist in setting planning priorities, (2) review technical data and make recommendations on its adequacy, (3) help resolve conflicts among various interests, (4) assist in the design and evaluation of the public involvement program, (5) serve as a communication link to other groups and agencies and report reactions back to the agency, (6) review and make recommendations on the planning process, (7) assist in developing and evaluating alternatives, (8) help select consultants and review contracts, (9) review and make recommendations on the program budget, (10) review written material prior to its release to the general public, (11) participate in and help host public meetings, and (12) assist in educating the public about the project and the planning process.

The biggest single problem in establishing a citizens' committee is to select members in such a way that the committee represents the community (Creighton and Delli Priscoli, 1981). There are five basic strategies by which member

of a citizens' committee can be selected (after Creighton and Delli Priscoli, 1981, pp. 303–304):

1. Members may be selected by the agency (or proponent), with an effort being made to balance the different interests.

2. The agency may turn over the selection of the citizens' committee to some third party or group.

3. As an alternative, the agency may merely identify the interests it wishes to have represented and allow the various groups within those interests to select their own representatives.

4. It is also possible to use any of the three methods above to form the core of the committee and then to augment the membership with the addition of volunteers.

5. In a few cases, membership on a citizens' committee has been determined by popular election (of volunteers or local-group nominees).

Additional Techniques

As noted earlier, Creighton (1981c) prepared a short catalog of public involvement techniques, including a brief description of each technique plus a discussion of its advantages and disadvantages. Summary information related to some techniques not previously mentioned is as follows (after Creighton, 1981c, pp. 279, 281, 285–286, 293):

1. *Interviews*—One technique for quickly assessing public sentiment is to conduct a series of interviews with key individuals representing the range of publics most likely to be interested or affected by the study.

2. *Hot line*—A hot line is a telephone number, usually easy to remember, which is publicized through repetition in brochures, reports, newsstands, paid advertising, and so on, as a single source that citizens can call to ask questions or make comments about project-related issues.

3. *Surveys*—Surveys are tools to determine public attitudes, values, and perceptions on various issues; the survey takes a rigorous methodology to insure that the findings of the survey in fact represent the sentiment of the community being sampled. Surveys can be conducted by phone, by mail, by individual interviews, or in small group interviews.

4. *Participatory television*—Because of the number of people reached by television, it holds considerable potential as a tool for both informing the public and soliciting participation.

5. *Simulation games*—There have been a number of simulation games which have been designed to allow people to simulate the effects of making particular policy choices and decisions and, in the process, learn more about the impact of decisions and the interrelatedness of various features of an environmental or economic system. Simulation-game playing provides an opportunity for people to try out their positions, and to see what the consequences would be and how other groups would react.

TECHNIQUES FOR CONFLICT MANAGEMENT AND DISPUTE RESOLUTION

Conflicts and disputes related to environmental impact issues are becoming commonplace in the United States. Conflicts can arise over resource management strategies, anticipated environmental impacts of new projects and/or economic development plans, operation of existing projects (such as dams), and environmental restoration efforts associated with cleaning up degraded environmental resources. Issues subject to conflicts may often be scientifically complex, technically complicated, and highly uncertain (Bidol, 1986).

Traditional approaches for conflict management have included the use of litigation, legislation and/or regulation, administrative procedures, and arbitration. Newer techniques for the management and resolution of environmental conflicts involve mediation and negotiation between the parties in conflict. A variety of terms have been used for these techniques, including "environmental mediation," "environmental negotiation," "environmental bargaining," "environmental conciliation," "conflict management," "consensus building," "alternative dispute resolution," "alternative environmental-conflict management," "environmental-dispute resolution," and "environmental-dispute settlement." A key feature of all of these techniques is "collaborative problem solving."

In order for collaborative problem solving to be effective in resolving environmental conflicts, the approaches used must be voluntary, encourage face-to-face interaction of parties involved in the dispute, and seek a consensus among the parties; such approaches are often assisted by an impartial third-party mediator, or facilitator (Dunning, 1987). Mediation processes are generally perceived to be faster and less expensive than litigation (Bingham, 1986; Amy, 1987). Additional advantages of using conflict-resolution, mediation, and collaborative problem-solving techniques include the following: (Delli Priscoli, 1988): (1) expensive adversarial battles can be avoided; (2) durable agreements can be reached among seemingly irreconcilable adversaries; (3) productive relationships can be built as a result of the conflict; and (4) mutual interests can be discovered among environmental, development, industrial, federal, private, and public interests.

Causes of Environmental Conflict

There are numerous possible causes of environmental conflicts which can arise over particular situations in given geographical locations. The causes can be relatively simple—for example, related to land ownership or land-use rights—or they can involve complex social issues and interrelationships. Four types (or causes) of conflict have been delineated by Creighton (1981a, after pp. 454–455) as follows:

1. *Cognitive conflict*—Cognitive conflict occurs when people have different understandings or judgments as to the facts of a case.
2. *Values conflict*—Values conflict is a dispute over goals—for example, whether an action or outcome is desirable (or undesirable) or should (or should not) occur.
3. *Interest conflict*—Since the costs and benefits resulting from an action are rarely distributed equally, some people will have a greater interest in an action than others. Some may have an interest in assuring it does not occur. As a result, it is possible to have agreement on facts, and on values, and still have conflict based on interest.
4. *Relationship conflict*—There are several psychologically oriented bases for conflict. Every time people communicate, they communicate both content (information, facts) and relationship (how much someone is valued, accepted, etc.). Decision-making processes can also communicate relationships—decision-making processes may, for example, favor those groups which are well enough financed and organized to present scientific supporting data over those which primarily argue from a values base. The result is that there are a number of emotional motivations that may lead to conflict on grounds other than facts, values, or interests.

Examples of Techniques Used for Conflict Resolution

A review of 161 cases involving environmental-dispute resolution (EDR) in the United States has been prepared by Bingham (1986). The first documented case occurred in 1973 and involved a proposed flood-control dam on the Snoqualmie River in the state of Washington. Several different parties were involved in the 161 cases. Environmental-interest groups and private companies were involved in direct negotiations in

21 percent of the cases; in some cases, the mediated environmental disputes involved only public agencies. In total, environmental groups were at the negotiating table in 35 percent of the cases, private corporations were involved in 34 percent, and federal and state agencies and units of local government were involved in 82 percent (Bingham, 1986). (These percentages exceed 100 percent, since three or more parties were involved in some cases.)

The success of the mediation-based EDR processes can be determined based on comparisons of the outcomes of the 161 cases against the objectives established for the EDR processes. Three main objectives were noted—to reach a decision, to agree on recommendations to a decision-making body not directly represented in the dispute-resolution process, or to improve communications. In 29 of the 161 cases, the parties' primary objective was to improve communications. In 132 of the cases, the parties' objective was to reach some form of agreement with one another. Of these cases, 99 involved site-specific issues and 33 involved policy issues. Overall, agreements were reached in 103, or 78 percent, of these cases; no agreements were reached in the remaining 29 cases (Bingham, 1986).

One approach to EDR can be illustrated with two cases involving the U.S. Army Corps of Engineers. The Corps is charged with the responsibility of issuing permits for projects or actions that have the potential for negatively impacting the navigable waters of the United States. Such permits are issued under the requirements of Section 404 of the Water Pollution Control Act of 1972 (now called the "Clean Water Act"). Permit actions are often subject to conflicts among environmentalists, developers (proponents of projects or actions), and governmental agencies. Delli Priscoli (1988) has described how the Corps of Engineers used a methodology to resolve such conflicts in the case of two general permits (GPs)—one for a wetland fill on Sanibel Island in

Florida, and the other for hydrocarbon exploration drilling throughout the states of Louisiana and Mississippi. In each case, the Corps suggested that the parties in conflict over the permit applications get together and write the technical specifications for a GP. The Corps told environmentalists, citizens, contractors, industrialists, developers, and representatives of government agencies that if they could agree upon the specifications of a permit within the broad legal constraints of the 404 law, the Corps would confirm the agreement, call it a GP, and enforce its provisions.

To implement the conflict-resolution and consensus-building process, selected representatives of the pertinent Corps offices, following the receipt of appropriate training, facilitated four workshops, attended by representatives of the groups in conflict, on each project. The first workshop was focused on the identification of issues to be included in the GP. This was accomplished by randomly assigning each of the approximately 50 attendees to one of six groups. At the second workshop, the individuals focused on their special concerns. Therefore, participants were not randomly assigned to groups but rather chose appropriate groups based on the individual's area of expertise. The second workshop refined the large range of identified issues and began the detailed work of outlining the language for special conditions. The third workshop completed the writing of the special conditions. Therefore, the Corps used a format whereby each of the small groups had addressed all of the issues and language under consideration. Consensus was reached by the end of the day. The final workshop consisted primarily of presenting an explanation of the permit to all interested parties. Because some citizens had not previously participated in the workshops, the Corps asked those who had emerged as leaders in the first three workshops to serve on a panel to answer questions from the audience. Since the citizens' representatives themselves had developed these special conditions, they

were the ones to explain their work before others. In this way, the Corps was not in a defensive posture and was also able to demonstrate the joint ownership of the permit (Delli Priscoli, 1988).

Finally, in another approach, the EIA report for a proposed project has been used as a negotiating tool to reach agreement on resource management. Specifically, a paper company from the United States used the EIA report for a proposed eucalyptus plantation in Indonesia to negotiate environmental regulatory requirements and to win the support of pertinent governmental, environmental, and community groups (Stern, 1991). The various groups were participants in the planning and preparation of the EIA report.

Conditions Requisite to Using Conflict-Resolution Techniques

Practical experience suggests that there are certain key conditions that must exist—or must be within the power of the third party to create—if the intervention is to be a success (Delli Priscoli, 1990; Wehr, 1979). Six key conditions are (after Creighton, 1981a, pp. 460–462)

1. *Motivation towards resolution*—The first requisite is that all critical parties must have motivations that make resolution desirable.
2. *Roughly equal power*—Neither side is likely to compromise if it thinks it has the political or legal power to "win" outright.
3. *Acceptable, minimal risk of failure*—Sometimes the consequences of a failure at third-party intervention may be that a controllable conflict may become totally unmanageable.
4. *Organizational authority*—To be effective, a conciliator (intervenor or mediator) must usually speak for an organization that possesses authority and credibility. Even when an individual is hired as a mediator, he or she is placed in that position with the authority and credibility of the organization do-

ing the hiring. If the organization the conciliator represents is not credible to the antagonists, the conciliator will not be accepted.

5. *Negotiability of issues*—One of the tactics engaged in by a conciliator is to attempt to enlarge the number of issues which are negotiable. The more issues which are negotiable, the more likelihood exists that a "positive-sum" solution can be found.

6. *Control over the process*—Experienced conciliators stress the importance of the conciliator's control over the communication process. This is particularly true in formal mediation, or at critical junctures in negotiations.

Three additional characteristics for dispute resolution are that the focus needs to be on problem solving, the negotiation must be interest-based, and an impartial third party can be of assistance (Dunning, 1987). For problem solving, techniques should be used to bring parties together in face-to-face meetings to share information. The approaches should stress the importance of trying to focus on the issues of concern as problems to be solved, and not as contests to be won. Interest-based negotiation is focused on the underlying interests of the parties rather than the positions that parties have. By identifying the underlying interests of all parties involved in the dispute, negotiations can proceed which are focused on trying to find ways to accommodate all the interests, rather than trying to deal with essentially incompatible positions regarding the policy issue, project, or plan (Dunning, 1987). The third-party negotiator must be perceived by all parties as an impartial "process keeper" rather than as someone having a stake in a particular outcome. The negotiator must confine his or her efforts to a role of keeping disputants working together (Dunning, 1987).

If it is decided that third-party intervention is appropriate, the next step is selection of a conciliator (intervenor or mediator). There are two qualities which are essential in the intervenor: credibility and neutrality. The conflicting parties must believe that the intervenor is sincere in doing his or her job. He or she should be a person who has a good reputation, has successfully resolved disputes in the past, and has some authority. Second, he or she must have no vested interest in the dispute except the desire to see it settled (Frost and Wilmot, 1978). Baldwin (1978) has identified some possible roles the intervenor can play in conflict management and resolution: (1) creating a climate of trust and a willingness to negotiate on the disputants' parts, (2) ensuring fair and adequate representation, (3) bringing the best available environmental information and expertise to the discussions, (4) breaking deadlocks by setting goals and deadlines, (5) suggesting solutions, or alternative solutions, and/or (6) outlining implementation plans and helping create mechanisms for implementation and enforcement of the agreement.

Meetings of Disputants

At some point during the conflict-resolution process, the disputants must meet face-to-face to negotiate a mutually acceptable solution. This meeting must be carefully controlled, or it can degenerate into a shouting match. If negotiations are to be successful, all affected parties should be included. However, it is sometimes necessary to exclude all others. The presence of observers may retard progress because negotiators may assume extreme positions in the presence of witnesses. It is vital that the representatives of each party be persons with authority who can make commitments on behalf of their group. Table 16.10 includes some suggested guidelines for achieving consensus on an environmental issue through the formation of a working group composed of members with differing viewpoints (Federal Environmental Assessment Review Office, 1988). Creighton (1981a, after pp. 467–469) has summarized the

---

TABLE 16.10

GUIDELINES FOR CONSENSUS BUILDING IN A WORKING GROUP

- Take time in the beginning of the group's life for members to meet other members as colleagues, rather than as antagonists across a negotiation table. This may require field trips, guest speakers, or other common activities where resource issues are looked at in general.

- Have each member write a short statement of what they wish to achieve through the group's work, then discuss these expectations in an information sharing meeting.

- Ensure that each member understands the group's final terms of reference, and the kinds of results consistent with those terms.

- Before trying to address issues, make sure that there is a clear and satisfactory statement of what the issues are.

- When proposals for a decision are being made, seek a number of options, rather than trying to draft a single statement. The options must embody the differing viewpoints of participants. They can then be examined by everyone for pros and cons.

- When different viewpoints or options are presented, take the time to hear each participant, without judging or condemning the proposal at first hearing. Most communication gaps begin with problems in listening, and unwillingness to let a speaker make his point.

- Everyone must be willling to re-open issues or concerns already decided, if a new compromise changes the way a member perceives the balance of interests taking shape. Tolerance for this review must be balanced by a respect for the group's work by each member.

- Consensus cannot survive if any members of the group are working to a hidden agenda or manipulative strategy. Concerns have to be explicit and all cards must be placed on the table.

- Sometimes, compromise is not acceptable to one or more of the participants in a working group. In these cases, it may not be possible to achieve consensus on a single proposal. The remedy for this situation is to present either a minority report, or to present a range of alternatives. With either approach, it is essential to give reasons for the alternatives, so that they can be assessed by the agency. It should be remembered that the closer to consensus a working group can get, the more likely it is that their recommendations will be accepted and implemented.

Source: Federal Environmental Assessment Review Office, 1988, vol. 3, p. 54.

---

usual negotiating procedure as a four-step strategy:

1. *Areas of agreement*—Parties enumerate all areas on which there is agreement. These are then eliminated from discussion. This step saves time later on and, most important, establishes common ground and fosters a feeling of mutual trust.

2. *Areas of disagreement*—Parties clearly define all areas of disagreement. Each party must state its position on each point of conflict, giving the underlying reasons for its position. This gives negotiators an idea of the magnitude of the problem. It also ranks points of conflict in a rough order of importance.

3. *Conflict-resolution procedure*—If possible, a procedure for resolving disagreements should be agreed upon. Doing this establishes a suitable climate in which agreements can be made.

4. *Issue-by-issue negotiation*—It is not usually possible to resolve all outstanding disagreements at once. A more realistic approach is to try to solve points of conflict one at a time. It may be advisable to negotiate minor issues first and then progress to major ones. In this way, the negotiators will address the more difficult problems after having established a

record of successful negotiations on less difficult issues.

Lessons Learned

Because of the growing body of knowledge and experience on the use of conflict-resolution techniques, some fundamental hypotheses can now be drawn. For example, Susskind and McMahon (1985) have noted the following 13 lessons from the alternative dispute-resolution field:

1. People bargain as long as they believe negotiations will produce an outcome as good as or better than outcomes that would result from other conflict-resolution methods.
2. Issues of concern must be readily apparent and parties must be ready to address them.
3. Success depends on having a large enough range of issues or options to either trade off options or to create new options.
4. Agreement is unlikely if parties must compromise fundamental values.
5. If power is unequal, the parties will not negotiate.
6. There is a rough practical limit in the numbers of participants—around 15 parties.
7. The pressure for deadline must be present.
8. Some means of implementing the final agreement must be available and acceptable to the parties.
9. Successful negotiation depends on mutual education.
10. Durable agreements depend on procedural, psychological, and substantive satisfaction.
11. Process can make a difference.
12. Facilitators must be perceived as neutral parties.
13. Conflict can be positive.

While the emphasis herein has been on positive outcomes, there are costs associated with resolving disputes. Some potential disadvantages of dispute-resolution processes from a citizens' group perspective include the following (after Crowfoot and Wondolleck, 1990, pp. 258–259):

1. Unlike many traditional arenas where problems are handed over to lawyers and judges or agency officials, if a citizens' group becomes involved in an environmental-dispute-settlement (EDS) process for resolving a conflict, the members will be directly involved in the resolution of the dispute. In the end, citizens' issues may be addressed more promptly and more effectively than they would be in the traditional forums; however, participants may have to devote more personal and organizational time to the process.
2. Citizens' groups may have to challenge well-financed, generously staffed government agencies and industries. The representatives from government and industry usually are able to participate in the course of performing their jobs; for citizens, participation is often an additional commitment outside of work and family responsibilities, and they usually lack the support staffs that government and industry have.
3. Citizens' organizations may not have the necessary skilled manpower to participate effectively in an EDS process.
4. Part of the organizational cost of an EDS process is the time and energy spent to maintain interaction with constituents. Representatives from citizens' groups may have to devote their energies not only to participating in the actual problem-solving sessions, but also to assuming additional responsibilities pertaining to actions within their own organizations to maintain group cohesiveness and support for negotiated agreements.
5. Successful participation of citizens' organizations in EDS processes often demands the formation of strong coalitions. Managing the coalitions places additional demands on an organization's time, energy, and financial resources.
6. An alternative process may limit the types and timing of issues that can be addressed. In some cases, the more complex and controversial issues may be left unaddressed. The easier issues are purposely handled first

in order to reinforce a sense of success for the group; more-difficult issues may repeatedly get pushed back on the agenda.

7. The pressure to reach consensus in an EDS process may cause some participants to lose sight of their original objectives. Citizens may be co-opted by more-powerful groups and end up endorsing the activities of adversaries. Citizens may be frustrated by their lack of leverage and political naivete and may lack access to critical decision networks.

In summary, numerous techniques for addressing environmental disputes have been successfully developed and utilized over the last two decades. Usage of such techniques requires the training of key individuals and the careful planning of the process. With increasing environmental awareness, additional environmental-protection and resource-management legislation, and greater societal pressures for responsible en-

vironmental management, there is every reason to believe that dispute resolution will remain a part of and even increase in importance to the EIA process.

PRACTICAL CONSIDERATIONS FOR IMPLEMENTING A PUBLIC PARTICIPATION PROGRAM

Actual implementation of a public participation program involves a number of considerations. The following list of items represents some practical ideas and suggestions that can be useful in organizing a public participation program:

1. Coordinate the various federal, state, and local agencies that have interests and responsibilities in the same geographic or technical areas of the study. Develop formal agreements or informal relationships.

TABLE 16.11

SUGGESTIONS FOR ACTIVITIES ON DAY OF A PUBLIC MEETING

- Take all handout materials to the meeting site. Do not have them bound or assembled in packages; let people choose which ones they care to take.
- Set up the registration table next to the entrance, and place handout materials in easy view.
- If the meeting is a formal event to solicit public comment, have people sign in with name, affiliation, address, and phone number. Most importantly, have them indicate whether or not they wish to speak. Attendees may then be called in order during the discussion period.
- Arrange for panel members (if applicable) to meet at the site at least one hour before the meeting is to begin.
- Meet with media representatives who arrive early, and provide them with any necessary background information.
- Personally greet local officials and residents who have had an active role in past activities.
- Ensure that the court reporter (if applicable) has all necessary information, including acronyms that may be used, names and titles of presenters, and a list of persons to whom a transcript should be sent.
- Start the meeting on time, if at all possible. The only reason why the meeting should not begin on time is if there are many people still waiting to find seats.
- During the opening remarks, the speaker should ask people to hold their questions until the question-and-answer period, but it should be done in a manner to encourage participation.
- Allow sufficient time for questions, answers, and discussion.

Source: After Allingham and Fiber, 1990, p. 44.

2. Assemble a newspaper-clipping file on project needs and a previous history of the project or study.

3. Disseminate study information through the news media (newspapers, radio, and television) and through regular publication of a planning newsletter. The mailing list should encompass all state and federal interests, as well as local groups and individuals who have participated in previous meetings or shown interest in the study.

4. Every third or fourth issue of the newsletter should contain a mail-in coupon for persons wanting to continue to receive the newsletter. Each issue should contain a coupon form for reader suggestions of other persons or groups that should receive the newsletter.

Some practical suggestions for actually conducting a public meeting are as follows:

1. Be concerned about the details of the meeting. Some practical suggestions for activities on the day of a public meeting are in Table 16.11.

2. Carefully plan the meeting agenda. Some guidelines for chairing a public meeting are in Table 16.12.

3. Keep data presentations simple. The purpose of data presentations is to inform, not to confuse or disillusion. Use simple visual aids. Slides of the actual area prove very beneficial.

4. Only those persons who have the ability to speak on general matters, commingled with technical expertise, should be considered for meeting with the public. It is well known that not everybody has the ability to speak, answer questions, and perhaps debate while holding a specific image in front of the public. The ability to speak well is not the only trait on which the selection should be based. A person who can answer questions quickly and confidently from an audience in which the sentiment is mixed or opposed can establish a profound positive relation with the audience.

5. Be familiar with the area.

6. Be earnest, sincere, and willing to work on problems with individual groups.

TABLE 16.12

GUIDELINES FOR CHAIRING A PUBLIC MEETING

- Always check out the agenda with participants before starting (remain open to altering the agenda).

- Limit speakers only when necessary to give fair time to others (present time limits with this proviso).

- Never put down or ridicule a speaker who has annoyed or challenged you (courtesy is always necessary, especially when limiting a speaker or ruling someone out of order).

- When soliciting comments, look around the hall systematically and leave enough time for hesitant people to come forward before moving on.

- Treat all points of view as valid, and do not editorialize on what people present (you are a facilitator, not a judge).

- If people seem uncomfortable with the way things are going, ask for comments and deal with them directly. Sometimes a "straw vote" helps clarify how an audience feels about an issue. For example, you may ask: "How many people feel we have given enough time for questioning the last speaker?"

- If the meeting is running long, you may have to choose between extending the time or re-convening another meeting. Sometimes an extended meeting is essential for a resolution; at other times it becomes a frustrating squabble as people become increasingly tired and distracted. It is preferable if the chairman makes the decision, explains the reasons, and asks for assent; avoid calling for a vote, as this can act as a precedent in other areas.

Source: Federal Environmental Assessment Review Office, 1988, vol. 3, p. 41.

INCORPORATION OF RESULTS IN DECISION MAKING

The feedback loop from public participation must be used in the EIA process or the purpose of public participation will not have been fully satisfied. Public participation results can be useful in defining project need, describing unique features of the environmental setting, and identifying environmental impacts, potential alternatives, and mitigation measures. Results can also be utilized by the proposing agency in assigning significance (importance) values to both environmental items and impacts. Finally, selection of the most desirable alternative for meeting project needs can be aided by public participation.

Two levels are suggested for incorporation of public participation information in an environmental impact study. First, all public meetings and the entirety of the planned and accomplished public participation program should be summarized. Any information relative to the objectives outlined earlier and obtained through questionnaire surveys or other public participation techniques should be summarized. Second, public preference can be used as one basis for the selection of the alternative to become the proposed action for meeting a project need. Such public-preference results can be incorporated with other data for use in multiple-criterion decision-making techniques.

VERBAL COMMUNICATIONS IN ENVIRONMENTAL IMPACT STUDIES

During the planning and conduct of an environmental impact study, numerous occasions arise wherein informal verbal information needs to be communicated, and/or more-formal verbal presentations need to be made, on study plans, preliminary and resultant findings, and study difficulties. Such communications can be required before a variety of audiences ranging from private audiences consisting of single individuals internal to the proponent organization or group conducting the study to public audiences with several hundred or more individuals.

Variety of Audiences

A number of audiences can be associated with verbal presentations prior to, during, and/or following a given environmental impact study. Table 16.13 includes examples of meetings and/or points of contact wherein verbal communication is involved in the EIA process. During the scoping phase, there may be numerous agencies and groups contacted regarding the plans for the study and in order to solicit information and feedback relative to key issues and impacts of concern. Verbal communication in scoping may be done by means of informal meetings and short briefings on the study, with interwoven general discussions and question-answer sessions. During public participation programs, a range of publics with different interests may be represented, including professional organizations, elected officials, farmers, conservationists, developers, and others.

In addition to organized meetings, there may be numerous occasions wherein information is verbally provided to individuals, perhaps with follow-up involving the provision of written materials. Finally, relative to potential audiences, and depending upon the degree of controversy associated with a project, there may be numerous times where it is necessary to communicate information about the impact study to individuals representing both the written and electronic media.

A key point relative to audiences is that the communication techniques and the information actually provided may need to be appropriately tailored to meet the particular interests and needs of various audiences. In other words, it may not be possible to have one verbal briefing which will suffice for all possible audiences. Accordingly, this reinforces the need for careful initial identification of audiences (publics or clientele groups), and the planning of appropriate

TABLE 16.13	

EXAMPLES OF EIA-RELATED MEETINGS AND/OR POINTS OF CONTACT NECESSITATING VERBAL COMMUNICATION

1. Scoping by means of public meetings with a wide range of publics.

2. Scoping by means of meetings in the offices of pertinent governmental agencies, public interest groups, nongovernmental organizations, and private corporations (proponents for project and/or those with expertise and experience relevant to the type of project and/or environmental setting).

3. Scoping by means of telephonic contacts with various groups or individuals, as listed in 1 or 2 above.

4. Telephonic contacts with and/or visits to the offices of the parties listed in 2 above for the purpose of requesting and/or procuring information pertinent to the EIA study. Examples of such information include data related to describing the affected environment, predicting or assessing impacts, comparing alternatives, and developing and evaluating effective mitigation measures).

5. Conduction of public participation meetings to explain the project and solicit input on environmental issues and concerns, and to generate feedback on the potential project and alternatives.

6. Implementation of an environmental-mediation (or negotiation) program to resolve conflicts related to the proposed project and the evaluation of alternatives. This activity also includes issues-resolution conferences between the proponent agency and other pertinent governmental agencies.

7. Provision of information to the written or electronic media representatives and responding to queries from the media.

8. Conduction of meetings between the project proponent, engineering design team or consultants, and the EIA study director, team, and/or consultants.

9. Responding to telephonic and/or written queries related to the EIA study.

10. Conduction of meetings between the EIA study director, the interdisciplinary study team, and/or special consultants for purposes of study planning and coordination.

11. Negotiations between the EIA study director, the proponent agency's contracting officials, and contractors with regard to conducting targeted or comprehensive studies and producing results for the EIA study. This may involve detailed discussions relative to information gathering, monitoring, selection and use of impact-prediction techniques, and the content of written products.

12. Negotiations between the EIA study director, the project proponent, and pertinent regulatory agencies with regard to the preparation of memoranda of agreements or understandings (MOAs or MOUs).

verbal presentations for effectively communicating with specific groups.

Planning Verbal Presentations

A fundamental issue in planning verbal presentations is related to deciding what topics should specifically be addressed in a verbal briefing. In general, the following topics should be addressed, although the treatment of some could be minimized, embellished, or eliminated, depending upon the specific audience:

1. The basis for conducting the environmental impact study. This may include a citation of regulatory requirements or a delineation of the fundamental reasons the project falls within the purview of such regulatory requirements.

2. Needs to be met by the project and the purposes for planning such a project at this time and at this location. This provides stage setting in terms of helping the audience understand why the project is being considered.

3. A summary of the affected, or baseline, environment, with particular attention being given to key resources or extant environmental issues. Examples of key resources and is-

sues may include wetlands, threatened or endangered plant or animal species, and historic and cultural resources in the project study area.

4. Alternatives which have been identified for consideration, and the selection process which will be used, or which has been used, in comparing the specific alternatives. Depending upon the timing of the verbal briefing, if the alternatives have actually been compared and the proposed action selected, then summary information should be presented on the key comparative features of the alternatives considered.

5. Key information on the proposed project, including its location, general design considerations and features, timing of construction, operational patterns, and the overall costs and expected benefits resulting from the project, including, if possible, information on the environmental benefits and costs.

6. A summary of the key impacts (both beneficial and detrimental) and/or risks of the proposed project (Committee on Risk Perception and Communication, 1989), and information on pertinent mitigation measures which have been planned for implementation in project-construction and -operation phases. It might also be appropriate to briefly describe utilized impact-prediction methodologies and bases for interpretation of the anticipated impacts, and how the resulting information was used to identify appropriate mitigation measures for the subject project.

7. Any positive experiences incurred by the interdisciplinary study team during the conduction of the study, and specific study difficulties that may have been encountered, including unresolved issues that may be in the process of being addressed.

8. The identification of uniquely helpful persons, groups, and/or agencies which have provided information and participated in a positive manner in the EIS.

9. A roster of interdisciplinary-study-team members, including a summary of their qualifications and experience in conducting similar or related studies.

Use of Visual Aids

Verbal presentations at meetings, whether to small or large groups, can be enhanced by the effective use of visual aids. Presentations using visual aids can be extremely complicated, particularly with the use of detailed information on maps, figures, and tables; accordingly, any visual aids which are used should be simple and prepared in a neat and concise fashion so that information can be most effectively communicated. The eight principal types of visual aids are flip charts or easel pads, chalkboards, charts, transparencies shown on overhead projectors, slides, filmstrips or movies, videotapes, and handouts (Helweg, 1985).

Practicing for the Presentation

There is no substitute for thoroughly studying the material to be presented in a verbal communication effort, whether the briefing is to be given to one person or to several hundred people. The presenter should carefully study the material so that he or she is thoroughly familiar with what is there and where to find information if questions arise. Practice run-throughs of the presentation can be helpful in verifying the timing required for different segments and can serve as a basis for adjusting timing and giving greater emphasis to certain topics. If possible, it is desirable to have this run-through heard by persons interested in polishing the presentation and who can give positive feedback and constructive criticism. It can be useful to have both technical and nontechnical persons participate in the practice presentation.

Another preparation consideration for verbal briefings is to anticipate questions that might arise from the audience, and to practice answering these questions. An approach which the preparer can use is to formulate some questions that he or she would like answered about the project if he or she were a member of the audience, or that would typically arise from a person hearing the presentation for the first time. If a practice audience has been used in the suggested run-through of the presentation, it can also be helpful in identifying potential questions. Practicing responses to these questions can be beneficial.

Finally, there are a number of written materials which have been prepared to provide more-detailed information on making effective verbal presentations. There are many useful reference books on verbal and other means of communication, including written communication. Scott (1984) and Mills and Walter (1978) are books which include sections devoted to effective speaking, writing, meetings, and interviews.

SUMMARY

Public participation opportunities within the EIA process include scoping, provision of information related to studies, review and feedback on impact studies, and the review of environmental documentation relative to EAs or draft EISs. Public participation programs or activities need to be carefully planned and implemented. Components of planning are related to the delineation of objectives, identification of publics, and selection of relevant public-involvement techniques. Techniques are available for providing information, soliciting feedback and evaluations, and resolving environmental conflicts. Conflict-resolution techniques typically include collaborative problem solving facilitated by an environmental mediator. Many practical issues are associated with the implementation of public-involvement efforts, including coordination with various agencies, concern with any meeting details, and rehearsal of verbal presentations. Public-involvement-program results should be incorporated in study documentation. In some instances, such involvement has influenced project decision making.

SELECTED REFERENCES

Allingham, M. E., and Fiber, D. D., "Commander's Guide to Public Involvement in the Army's In-

stallation Restoration Program," CETHA-PA-CR-90182, U.S. Army Toxic and Hazardous Materials Agency, Aberdeen Proving Ground, Maryland, Nov. 1990, pp. 7–8, 44, 51–54.

Amy, D. J., *The Politics of Environmental Mediation,* Columbia University Press, New York, 1987.

Arnstein, S. R., "A Ladder of Citizen Participation," *Journal of the American Institute of Planners,* vol. 35, no. 4, July 1969, pp. 216–224.

Baldwin, P., "Environmental Mediation: An Effective Alternative?," *Proceedings of Conference Held in Reston, Virginia,* RESOLVE, Center for Environmental Conflict Resolution, Palo Alto, Calif., Jan. 11–13, 1970.

Bidol, P., "An Adaptive Conflict Management Model," *Proceedings of Engineering Foundation Conference on Social and Environmental Objectives in Water Resources Planning and Management, Santa Barbara, California,* American Society of Civil Engineers, New York, May 11–16, 1986, pp. 72–78.

Bingham, G., *Resolving Environmental Disputes: A Decade of Experience,* Conservation Foundation, Washington, D.C., 1986, pp. xv–xxviii.

Bishop, A. B., "Communication in the Planning Process," in *Public Involvement Techniques: A Reader of Ten Years Experience at the Institute of Water Resources,* J. L. Creighton, and J. D. Delli Priscoli, eds., IWR Staff Rep. 81-1, U.S. Army Engineer Institute for Water Resources, Fort Belvoir, Va., 1981.

———, "Structuring Communications Programs for Public Participation in Water Resources Planning," IWR Contract Rep. 75-2, U.S. Army Engineer Institute for Water Resources, Fort Belvoir, Va., May 1975.

Canter, L. W., *Environmental Impact Assessment,* McGraw-Hill Book Company, New York, 1977, pp. 220–232.

———, Miller, G. D., and Fairchild, D. M., "Sentry Missile Environmental Assessment Program," rep. submitted to U.S. Army Ballistic Missile Division Systems Command, Huntsville, Alabama, Environmental and Ground Water Institute, Norman, Okla., Sept. 1982, p. 2424.

Castensson, R., Falkenmark, M., and Gustafsson, J. E., "Water Awareness in Planning and Decision Making," Swedish Council for Planning and Coordination of Research, Stockholm, 1990, pp. 78–89.

Committee on Risk Perception and Communication, *Improving Risk Communication,* National Research Council, National Academy Press, Washington, D.C., 1989.

Council on Environmental Quality (CEQ) "Regulations for Implementing the National Environmental Policy Act," *40 Code of Federal Regulations,* Chap. V, July 1, 1987, pp. 929–971.

Creighton, J. L., "Acting as a Conflict Conciliator," in *Public Involvement Techniques: A Reader of Ten Years Experience at the Institute of Water Resources,* J. L. Creighton and J. D. Delli Priscoli, eds., IWR Staff Rep. 81-1, U.S. Army Engineer Institute for Water Resources, Fort Belvoir, Va., 1981a.

———, "Identifying Publics/Staff Identification Techniques," in *Public Involvement Techniques: A Reader of Ten Years Experience at the Institute of Water Resources,* J. L. Creighton and J. D. Delli Priscoli, eds., IWR Staff Rep. 81-1, U.S. Army Engineer Institute for Water Resources, Fort Belvoir, Va., 1981b.

———, "A Short Catalogue of Public Involvement Techniques," in *Public Involvement Techniques: A Reader of Ten Years Experience at the Institute of Water Resources,* J. L. Creighton and J. D. Delli Priscoli, eds., IWR Staff Rep. 81-1, U.S. Army Engineer Institute for Water Resources, Fort Belvoir, Va., 1981c.

———, Chalmers, J. A., and Branch, K., "Integrating Planning and Assessment Through Public Involvement," in *Public Involvement Techniques: A Reader of Ten Years Experience at the Institute of Water Resources,* J. L. Creighton, and J. D. Delli Priscoli, eds., IWR Staff Rep. 81-1, U.S. Army Engineer Institute for Water Resources, Fort Belvoir, Va., 1981.

——— and Delli Priscoli, J. D., "Establishing Citizens' Committees," in *Public Involvement Techniques: A Reader of Ten Years Experience at the Institute of Water Resources,* J. L. Creighton, and J. D. Delli Priscoli, eds., IWR Staff Rep. 81-1, U.S. Army Engineer Institute for Water Resources, Fort Belvoir, Va., 1981.

Crowfoot, J. E., and Wondolleck, J. M., *Environmental Disputes: Community Involvement and Conflict Resolution,* Island Press, Washington, D.C., 1990, pp. 254–263.

Delli Priscoli, J. D., "Conflict Resolution in Water Resources: Two 404 General Permits," *Journal of Water Resources Planning and Management,* vol. 114, no. 1, Jan. 1988, pp. 66–77.

———, "Implementing Public Involvement Programs in Federal Agencies," in *Public Involvement Techniques: A Reader of Ten Years Experience at the Institute of Water Resources,* J. L. Creighton, and J. D. Delli Priscoli, eds., IWR Staff Rep. 81-1, U.S. Army Engineer Institute for Water Resources, Fort Belvoir, Va., 1981.

———, "Public Involvement, Conflict Management, and Dispute Resolution in Water Resources and Environmental Decision Making," IWR-90-ADR-WP-2, U.S. Army Engineer Institute for Water Resources, Fort Belvoir, Va., Oct. 1990.

Dunning, C. M., "Alternative Dispute Resolution in Water Resources Management," *Proceedings of Engineering Foundation Conference on the Role of Social and Behavioral Sciences in Water Resources Planning and Management, Santa Barbara, California,* American Society of Civil Engineers, New York, May 3–8, 1987, pp. 99–109.

Federal Environmental Assessment Review Office, "Manual on Public Involvement in Environmental Assessment: Planning and Implementing Public Involvement Programs," 1988, Ottawa, Ontario, Canada, 3 vols., pp. 8, 11–15, 33—Vol. 1, pp. 21–22, 26, 37—Vol. 2, and pp. 2, 41, 54, 57–60—Vol. 3.

Frankena, F., and Frankena, J. K., *Citizen Participation in Environmental Affairs—1970–1986: A Bibliography,* AMS Press, New York, 1988.

Frost, J. H., and Wilmot, W. W., *Interpersonal Conflict,* Wm. C. Brown Company Publishers, Dubuque, Iowa, 1978.

Hance, B. J., Chess, C., and Sandman, P. M., *Industry Risk Communication Manual: Improving Dialogue with Communities,* Lewis Publishers, Chelsea, Mich., 1990, p. 30.

Hanchey, J. R., "Objectives of Public Participation," in *Public Involvement Techniques: A Reader of Ten Years Experience at the Institute of Water Resources,* J. L. Creighton and J. D. Delli Priscoli, eds., IWR Staff rep. 81-1, U.S. Army Engineer Institute for Water Resources, Fort Belvoir, Va., 1981.

Helweg, O. J., *Water Resources Planning and Management,* John Wiley and Sons, New York, 1985, pp. 279–283.

Mills, G. H., and Walter, J. A., *Technical Writing,* Holt, Rinehart, and Winston, Dallas, Texas, 1978.

Ragan, J. F., Jr. "Constraints on Effective Public Participation," in *Public Involvement Techniques: A Reader of Ten Years Experience at the Institute of Water Resources,* J. L. Creighton and J. D. Delli Priscoli, eds., IWR Staff Rep. 81-1, U.S. Army Engineer Institute for Water Resources, Fort Belvoir, Va., 1981a.

———, "An Evaluation of Public Participation in Corps of Engineers Field Offices," in *Public Involvement Techniques: A Reader of Ten Years Experience at the Institute of Water Resources,* J. L. Creighton and J. D. Delli Priscoli, eds., IWR Staff Rep. 81-1, U.S. Army Engineer Institute for Water Resources, Fort Belvoir, Va., 1981b.

Schwertz, E. L., Jr., "The Local Growth Management Guidebook," Center for Local Government Technology, Oklahoma State University, Stillwater, 1979.

Scott, W. P., *Communication for Professional Engineers,* Thomas Telford, London, 1984.

Stern, A. J., "Using Environmental Impact Assessments for Dispute Management," *Environmental Impact Assessment Review,* vol. 11, 1991, pp. 81–87.

Susskind, L., and McMahon, G., "The Theory and Practice of Negotiated Rulemaking," *Yale Journal on Regulation,* vol. 3, no. 133, 1985, pp. 132–165.

U.S. Army Corps of Engineers, "Public Participation in Water Resources Planning," EC 1165-2-100, U.S. Army Corps of Engineers, Washington, D.C., May 1971.

U.S. Department of Transportation, "Effective Citizen Participation in Transportation Planning: Community Involvement Processes," 2 vols., U.S. Department of Transportation, Washington, D.C., 1976.

Wehr, P., *Conflict Regulation,* Westview Press, Boulder, Colo., 1979.

Westman, W. E., *Ecology, Impact Assessment, and Environmental Planning,* John Wiley and Sons, New York, 1985, p. 118.

CHAPTER 17

Preparation of Written Documentation

Perhaps the most important activity in the environmental impact assessment (EIA) process is the preparation of one or more written reports which document the impact study findings. This written communication will be utilized by decision makers in making their final choices relative to the project (or plan, program, or policy). It will also be scrutinized by interested publics and governmental agencies in the review process. Therefore, special attention should be given to the preparation of documentation that effectively communicates information about the study findings. The resultant document or documents may be referred to as "environmental assessments" (EAs), "environmental impact statements" (EISs), "environmental impact reports," "environmental impact declarations," and/or "findings of no significant impacts" (FONSIs).

To illustrate the importance of written documentation, the Council on Environmental Quality regulations stated in paragraph 1502.8 (CEQ, 1987, p. 939):

> Environmental impact statements shall be written in plain language and may use appropriate graphics so that decision makers and the public can readily understand them. Agencies should employ writers of clear prose or editors to write, review, or edit statements, which will be based upon the analysis and supporting data from the natural and social sciences and the environmental design arts.

It is important to apply certain basic principles of technical writing in planning the document and preparing written materials. Five important principles have been delineated by Mills and Walter (1978, p. 16):

1. Always have in mind a specific reader, and always assume that this reader is intelligent, but uninformed.
2. Before you start to write, always decide what the exact purpose of your report is; and make sure that every paragraph, every sentence, and every word makes a clear contribution to that purpose, and makes it at the right time.
3. Use language that is simple, concrete, and familiar.
4. At the beginning and end of every section, check your writing according to this principle: "First, you tell your readers what you are going to tell them; then you tell them, and then you tell them what you have told them."
5. Make your report visually attractive.

It should be noted that it is easier to review and comment upon prepared EISs than it is to prepare such EISs. Accordingly, considerably more time and effort may be required for the planning and preparation of written documentation than would be required for the review of such documentation.

An important issue in impact study documentation is that the target audience typically consists of two groups: (1) a nontechnical audience, as represented by decision makers and interested members of various publics; and (2) a technical audience, as represented by professional colleagues in various governmental agencies and specific public groups who have interest in the study. Accordingly, it is necessary to prepare written documentation on impact studies to address the information needs of both nontechnical and technical audiences. Another way of considering target audiences is related to their roles or perspectives with regard to the proposed action. For example, Weiss (1989) suggested three motivated groups of EIS readers: (1) those belonging to a fairly high jurisdiction of government, who are responsible for a series of decisions, often including enforcement, and who are frequently under political pressure to approve or disapprove (in the guise of ''neutral'' review for compliance); (2) supporters of the proposed action, who are hoping that the EIS will not forecast any unavoidable consequences or identify any more-attractive alternatives, and who are often impatient to have it approved as quickly as possible; and (3) opponents of the action, who are alert to any instance in which its adverse effects are minimized or in which those of the alternatives are exaggerated, and who are typically especially skeptical of all assumptions, inferences, and secondhand or imputed data.

TABLE 17.1

SUGGESTIONS FOR OVERCOMING COMMON WRITING-RELATED ERRORS IN EISs

Strategic errors

Eliminating strategic errors calls for genuine project leadership by a project director with intellectual vigor. The various specialists on the assessment team, including subcontractors, must not begin to write their sections until their respective findings and conclusions have been discussed and evaluated. As impractical as it may sound, the individual contributors should not be turned loose on their writing assignments until they have presented written summaries of their data and interpretations for discussion by the team as a whole. In this context, only what is relevant should be featured prominently in the body of the EIS. Incidental information, tutorials and primers on environmental science, and miscellaneous exhibits should be relegated to appendices.

Structural errors

Eliminating structural errors calls for more-professional document design. The documents should be packaged and summarized in a way that lets most readers find what they need to know at once. Most important, answers to the following central questions of any EIS must be answered definitively: Are there unavoidable consequences? Are there alternative sites or technologies with less environmental hazard? etc. The most useful way to help readers follow complicated discussions is to unify the text with the associated tables, maps, and charts.

Tactical errors

The only way to eliminate or control tactical errors is with the help of a professional editor or team leader (or member) assigned the responsibility of a tactical review of the EIS. Every EIS, like every final report, needs the attention of someone who knows how to reduce the burden on the reader, to contain the effort and "overhead" needed to read and use the document. To do that job, however, takes time. Schedules must allow room for the editor or team leader to work.

Source: After Weiss, 1989, p. 240.

EISs have often been criticized relative to their communication attributes. For example, three broad classes of writing-related errors associated with EISs prepared in the United States include (Weiss, 1989) (1) "strategic errors," which are mistakes of planning and often are products of the failure to understand why the EIS is being written and for whom, (2) "structural errors," which are mistakes of organization and often products of the failure to arrange the elements in the document in a way that makes them easy to follow, and (3) "tactical errors," which are mistakes of editing and often products of the failure to test and revise the text for clarity and readability. Suggestions for eliminating these errors and overcoming such failures are included in Table 17.1.

To provide a positive approach to writing activities in the EIA process, this chapter summarizes information relative to the planning and preparation of written documentation on environmental impact studies. Three phases are suggested: (1) initial planning, (2) detailed planning, and (3) writing. Included are examples of practice and suggestions for appropriate issues to be addressed for each of these phases.

INITIAL PLANNING PHASE

An effective EIA report, like any other technical report, must be designed to fill a particular purpose in a specific situation. Accordingly, early attention to reporting requirements is desirable in an impact study. This attention should focus on the scope of the reporting requirements and the preparation of preliminary outlines for each written document, including an EIS (if one is required). A useful perspective to take in planning a report is to "put yourself in your reader's shoes" (Woolston, Robinson, and Kutzbach, 1988). Fundamentally, this entails asking yourself what you would like to see in the report if you had to read it (but not prepare it).

Following consideration of this perspective, a preliminary outline of the specific report or

EIS should be prepared. This effort will be a reminder for the team leader and every team member that one or more written products will be required during and/or at the completion of the impact study. It is also a useful guide for a single-person study effort.

Two kinds of outlines are relevant for a technical report: (1) the topic outline and (2) the sentence outline (Mills and Walter, 1978). In a "topic outline," each entry is a phrase or a single word; no entry is in the form of a complete sentence. Conversely, in a "sentence outline," every entry is a complete sentence. The sentence outline has one important advantage over the topic outline, but it also has at least one important disadvantage. The advantage is that, in making a sentence outline, the writer is forced to think out each entry to a much greater degree than for the topical form. However, the sentence outline is more difficult and time-consuming to write than the topic outline. The topic outline is easier to prepare and it is considered appropriate for the initial planning phase. The topic outline should primarily focus on main headings and merely indicate the general type of information to be subsequently included in the documentation.

To facilitate the outline preparation, it is desirable to follow the suggested outline of the specific proponent agency or sponsor for the project (or plan, program, or policy). A review of the reporting requirements of the proponent and some current environmental documents generated by the same proponent or by others for similar undertakings can be useful. For example, the broad topics to be included in an EA prepared to meet the requirements of the Council on Environmental Quality include the following (CEQ, 1987): (1) the need for the proposal, (2) a description of alternatives, (3) environmental impacts of the proposed action and any alternatives, and (4) a list of agencies and persons consulted. A generic structure for an EIS topical outline is in Table 17.2.

The initial planning phase should be accomplished early in the impact study; it is useful to

TABLE 17.2

STRUCTURE OF AN EIS TOPICAL OUTLINE

Section	Comments
Cover sheet	The cover sheet must not exceed one page. It must include a list of the responsible agencies, including the lead agency and any cooperating agencies; the title of the proposed action; the name, address, and telephone number of the person at the agency who can supply further information; a designation of the statement as a draft, final, or supplement; a one-paragraph abstract of the statement; and the date by which comments must be received.
Summary	Each EIS must contain an adequate and accurate summary. The summary must stress the major conclusions, areas of controversy (including issues raised by agencies and the public), and the issues to be resolved (including the choice among alternatives). It will normally not exceed 15 pages.
Purpose and need	The EIS shall briefly specify the underlying purpose and need to which the agency is responding in proposing the action and any alternatives.
Alternatives, including the proposed action	This section is the heart of the EIS. Based on the information and analysis presented in the sections Affected Environment and Environmental Consequences, it presents the environmental impacts of the proposal and the alternatives in comparative form, thus sharply defining the issues and providing a clear basis for choice among options by the decision maker and the public.
Affected environment	The EIS must succinctly describe the environment of the area(s) to be affected or created by the action and alternatives under consideration. The descriptions are to be no longer than is necessary to ensure understanding of the effects of the alternatives. Data and analyses in a statement must be commensurate with the importance of the impact, with less-important material summarized, consolidated, or simply referenced. Useless bulk in statements is to be avoided, and effort and attention must be concentrated on important issues. Verbosity does not enhance the adequacy of an EIS.
Environmental consequences	This section forms the scientific and analytic basis for the comparisons of the action and alternatives. The discussion is to include the environmental impacts, any adverse environmental effects which cannot be avoided should the proposal be implemented, the relationship between short-term uses of humans' environment and the maintenance and enhancement of long-term productivity, and any irreversible or irretrievable commitments of resources which would be involved in the proposal.
List of preparers	The EIS must list the names and qualifications (expertise, experience, professional disciplines) of the persons who were primarily responsible for preparing the document or significant background papers, including basic components of the statement. Where possible, the persons who are responsible for each particular analysis, including analyses in background papers, should be identified. Normally, the list will not exceed two pages.
Appendices	If an agency prepares an appendix to an EIS the appendix must consist of material prepared in connection with the EIS and material which substantiates any analysis fundamental to the statement. It must be analytic and relevant to the decision to be made and must be circulated with the EIS or be readily available on request.

Source: After Council on Environmental Quality, 1987, pp. 940–941.

consider this phase when developing an overall study strategy. This early planning can ensure that items that subsequently must be addressed in the written documentation are included in the information-gathering and -analysis segments of the study. While not necessary, it can be of value to assign individuals to work on specific aspects of the written documentation, even in the initial planning phase; this will aid in emphasizing the importance of written documentation throughout the study.

DETAILED PLANNING PHASE

The detailed planning phase for the written documentation should occur somewhere near the midpoint of the impact study. The first task should involve the preparation of a detailed outline of all required items of written documentation. This outline could be topical and also include sentences written to explain the individual topical issues. Table 17.3 shows a generic outline for a written report resulting from an en-

---

TABLE 17.3

GENERIC TOPICAL OUTLINE FOR AN ENVIRONMENTAL IMPACT REPORT

    I. Abstract or Executive Summary
   II. Chap. 1 Introduction
  III. Chap. 2 Delineation of need for project[a]
  IV. Chap. 3 Description of proposed project[a]
      A. What it is and how it will function
      B. When will it occur (timing for construction and operation)
      C. Extent of effectiveness in meeting need
   V. Chap. 4 Description of affected environment
      A. Components of baseline conditions and study area boundaries
      B. Interpretation of existing quality for components
  VI. Chap. 5 Impacts of proposed project[a]
      A. Identification and description and/or quantification of impacts on environmental components
      B. Interpretation of significance of impacts
      C. Mitigation measures for adverse impacts
 VII. Chap 6 Evaluation of alternatives
      A. Description of alternatives
      B. Selection method and results leading to proposed action
VIII. Chap. 7 Planned environmental monitoring
      A. Need for monitoring
      B. Description of monitoring program
      C. Outputs and decision points
  IX. Selected references
   X. Glossary of terms
  XI. List of abbreviations
 XII. Index
XIII. Appendices (as appropriate; the following are examples)
      A. Pertinent laws, regulations, executive orders, and policies
      B. Species lists
      C. Impact calculations
      D. Technical descriptions of project[a]
      E. Construction specifications to mitigate negative impacts
      F. Description of scoping program
      G. Description of public participation program
      H. Environmental factors considered and deemed not relevant

[a]Project is used as an example. The same outline steps may be used for a plan, program, or policy.

vironmental impact study. This generic outline addresses the primary activities in an impact study and presents them in a logical fashion relative to the decision-making process. The following comments are pertinent to the chapters and other components of the outline:

1. The report (whether an EA or EIS) should begin with an abstract or an executive summary. An "informational abstract" or "executive summary" consists of a short description, or a condensation, of the report. The executive summary presents a synopsis of the key aspects of the report, as well. It is typically longer than an abstract, with the length being from 2 to 15 pages, compared with a typical length of 1 page for the abstract.

2. The first chapter (or section) in the report should be an introduction. The primary purposes of an introduction to a technical report are to state the subject, the purpose, the scope, and the plan of development of the report (Mills and Walter, 1978).

3. The second chapter should be a description of the need for the project. This is particularly important in that it provides a framework for considering the characteristics of the project and the various economic and environmental ramifications of the project relative to meeting the identified need.

4. The third chapter should include a description of the proposed project, with this description identifying what activities are planned, as well as when these particular activities will take place. The mitigation measures to be used should be described in this chapter. The primary focus should be on those project features anticipated to have impacts on various categories of the environment. Finally, Chapter 3 should address the extent to which the proposed project is expected to meet all or a portion of the identified need described in Chapter 2.

5. Chapter 4 should provide a description of the affected environment; this description can serve as a point of departure for impact identification, prediction, and assessment. It is important to address the components of the environment which are expected to be changed by the proposed project. These components can include physical and chemical features, biological features, historic and archaeological resources, visual quality, socioeconomic characteristics, and health characteristics within the project boundaries. It is also important to clearly delineate the study area boundaries being addressed, with these boundaries primarily based upon the inclusion of those geographical areas expected to be changed by the proposed project. Chapter 4 should also include an interpretation of the existing quality of the categories of the environment. Pertinent laws, regulations, and executive orders; professional judgment; and other societal management tools can serve as the bases for this interpretation.

6. Chapter 5 should focus on the impacts of the proposed project, and it should identify mitigation measures. This is the most important technical chapter from the viewpoint of identifying and quantifying, where possible, the impacts of the proposed project. The impacts of the proposed project on each pertinent component of the baseline environment should be addressed. This portion of the chapter should primarily depend upon information from the initial identification of potential impacts, as well as on the application of impact-prediction techniques. Interpretation of the anticipated changes should be based on professional judgment, the application of pertinent institutional information, and public input. The use of a tabular approach can facilitate the presentation of information on impact prediction and assessment. Table 17.4 shows a suggested graphical presentation format.

TABLE 17.4

EXAMPLE GRAPHICAL PRESENTATION FORMAT FOR IMPACT PREDICTION AND ASSESSMENT

Environmental factor	Predicted impact	Method used for prediction	Interpretation of impact significance[a]	Basis for significance determination
a	Quantitative	Model x	S	Standard
b	Qualitative	Case study	MS	Professional judgment
c	Qualitative	Professional judgment	LS	Public imput
d	Quantitative	Unit impact factor	WNV	Qualitative law
e	.	.	.	.
	.	.	.	.
	.	.	.	.

[a]S = significant impact; MS = moderately significant impact; LS = low-signficance impact; WNV = expected change within normal variations for factor.

Impact mitigation should be considered relative to the potential for reducing any undesirable effects of the proposed project. There are numerous mitigation measures, and the emphasis in this chapter should be on the identification and evaluation of mitigation in terms of the actual selection of included measures. The chapter should also delineate the mitigation implementation responsibilities and time schedules for the measures.

7. Chapter 6 should provide a summary of the evaluation of alternatives; in the United States, this particular evaluation is considered the heart of the process (CEQ, 1987). The chapter should include a description of the alternatives, as well as a discussion of the method used for comparative evaluation of the alternatives and the results of the use of the method. It should be made clear why the proposed action was selected from among the alternatives evaluated. Table 17.5 illustrates a comparative display of information on alternative plans; this display can be useful in summarizing the features of viable plans as a part of the process lead-

ing to the selection of a proposed action (U.S. Army Corps of Engineers, 1985). If decision-focused checklists involving multicriterion decision-making techniques have been used, they should be described. It is recognized that detailed information on each alternative need not be presented in this chapter, but could be included as supplemental information to the environmental impact document, instead, or the selection could have been made primarily on the basis of a qualitative comparison. At any rate, the key is to document how alternatives were identified and evaluated as a part of the impact study.

8. Chapter 7 should summarize the planned environmental-monitoring program for the proposed project, if relevant. It is anticipated that the primary emphasis of this monitoring will be during project construction and operation. The chapter should include information to justify the recommended monitoring program and to demonstrate how it relates to the anticipated impacts of the particular project. The anticipated outputs from monitoring programs—

TABLE 17.5

EXAMPLE DISPLAY OF COMPARATIVE IMPACTS OF ALTERNATIVES

Base condition and alternatives	Blue River wetlands	Fort America historic site	Endangered yellow trout habitat	Blue River class A water quality	Marina-related employment
Base Condition	Available: 1,000 acres	Available: site on National Register of Historic Places (NRHP)	Available: 10 acres	Available: 15 river-miles	Available: 10 full-time equivalent (FTE) employees
Without Condition (No Action)	Available: 200 acres Impact: 0 acres/year	Available: site on NRHP Impact: none	Available: 10 acres Impact: 0 acres/year	Available: 10 river-miles Impact: 0 river-miles/year	Available: 10 FTE employees Impact: 0 FTE employees/year
Plan 1	Available: 0 acres Impact: −350 acres/year	Available: site on NRHP Impact: loss of base condition visual setting due to unavoidable introduction of visible elements	Available: 0 acres Impact: −10 acres/year	Available: 0 river-miles Impact: −10 river-miles/year	Available: 40 FTE employees Impact: +27.5 FTE employees/year
Plan 2	Available: 200 acres Impact: 0 acres/year	Available: site on NRHP Impact: none	Available: 0 acres Impact: −10 acres/year	Available 10 river-miles Impact: 0 river-miles/year	Available: 20 FTE employees Impact: +9.5 FTE employees/year
Plan 3	Available: 1,500 acres Impact: +775 acres/year	Available: site on NRHP Impact: none	Available: 10 acres Impact: 0 acres/year	Available: 10 river-miles Impact: 0 river-miles/year	Available: 10 FTE employees Impact: 0 FTE employees/year
Plan 4	Available: 200 acres Impact: 0 acres/year	Available: site on NRHP Impact: none	Available: 8 acres Impact: −2 acres/year	Available: 10 river-miles Impact: 0 river-miles/year	Available: 40 FTE employees Impact: +27.5 FTE employees/year

Note: Base condition year = 1990; period of analysis = 100 years.
Source: U.S. Army Corp of Engineers, 1985, pp. D-10–D-11.

such as quarterly, biannual, or annual reports—as well as decision points relative to continuation or modification of the monitoring program, should also be delineated.

9. In addition to the seven main chapters, selected references should be appropriately cited. A glossary of terms, providing definitions of technical terms that might be unfamiliar to a nontechnical audience, would also be appropriate, along with a list of abbreviations included in the document. It may also be desirable to include an index of important issues.

10. The final portion of the environmental impact report should consist of appendices, as appropriate. Table 17.3 lists eight types of appendices as examples.

As noted in the discussion of the initial planning phase, the review of proponent-report guidelines and tables of contents in actual EISs for similar types of projects can be useful in developing a detailed topical outline.

The final task of the detailed planning phase is to make writing assignments to individual members of the interdisciplinary team. These assignments should be made in view of the discipline represented by the team member, and the portion of the impact study in which he or she participated. It is also desirable to identify a time schedule for completion of chapters and appendices and to adhere as closely as possible to that schedule. In addition, written guidance on details such as numbering tables and figures, and the format for citing references, should be made available to team members.

WRITING PHASE

The third phase involves the preparation of materials for inclusion in the appropriate environmental documentation and the actual writing of the report. The practical suggestions included herein are not in any order of importance.

Organizing Relevant Information

One approach for organizing information relative to an environmental impact document is to assemble basic reference materials and information according to chapters, and sections within chapters, in one to several notebooks. The advantage of this approach is that information can be added or revised prior to the actual writing process. Pertinent materials can be placed in file folders in lieu of notebooks; thus, the files may serve as aids in organizing information. Some information might be pertinent to more than one section or chapter; therefore, it may be necessary to have photocopying capability in order to facilitate inclusion of information in multiple sections. Some individuals use index cards or computer diskettes for preparing and organizing information. These tools can be used for both the organization of topical information and the preparation of a reference list.

Initiating the Writing

Two examples of environmental-impact-report outlines have been shown; however, it is not necessary to actually prepare chapters and sections in the order shown in the developed outline. Table 17.6 presents some suggestions for generating an initial draft of an impact report (Woolston, Robinson, and Kutzbach, 1988). The most effective approach would be to prepare materials in accordance with their ease of preparation. For example, the appendices or the list of selected references might be prepared before the specific chapters are written. One of the advantages of preparing easier segments first is that it aids the writer in becoming mentally prepared for addressing more-difficult components or chapters. It is stressed that the outline prepared during the detailed planning phase can be modified during the writing phase. Major modifications would not be expected; however, modifications within subsections of chapters would be appropriate as final information is assembled.

TABLE 17.6

SUGGESTIONS FOR GENERATING THE INITIAL DRAFT OF THE REPORT

1. Start by expanding on the easiest part of the outline. Complete whole sections at a time, but take a break about every hour, so that you can maintain your efficiency.

2. Write as quickly as possible; if you are familiar with the material, you should be able to write 500–700 words per hour. Do not stop to check spelling, browse through a thesaurus, or ponder punctuation. If you are missing a few numbers but otherwise know what you want to say, keep going and look up the facts later. Put all of your mental energy into developing paragraphs from each topic in the outline.

3. After you finish a section, let it sit overnight.

4. After this cooling-off period, revise the section.

5. Give the draft to one or more colleagues for comments. Choose readers whose judgment you trust and whose suggestions will be fair and firm. Above all, have your draft reviewed by someone who is not familiar with the project on which you are writing, so that any jargon or shoptalk can be exposed for what it is—an avoidable rhetorical shortcut.

6. After a second cooling-off period, revise again, this time giving particular attention to consistency and emphasis.

Source: Woolston, Robinson, and Kutzbach, 1988, pp. 52–53.

The writing sequence which should yield the most-targeted and -concise products is as follows:

1. Prepare the basic EIA report.
2. Prepare the summary of the report. It should be perhaps 10 to 15 pages in length.
3. Prepare the executive summary (if one is included) from the summary. It should be perhaps three to five pages in length.
4. Prepare the abstract (about one paragraph in length) from the executive summary or the summary.

Some individuals use dictation devices or word processors during the writing phase, while others feel more comfortable by writing the report first in longhand, to be followed by subsequent typing. In some cases, several modes are used for one report. If dictation can be done with ease, this is an efficient way to assemble a report in that the writer can dictate materials, have them typed on a word processor, and then modify them upon review. Report formatting and layout can be facilitated by means of desktop publishing software.

Effective writing style is a product of attention to sentence structure and length, paragraph structure and length, and word precision. Good technical writing calls for reliance on natural word order, simple sentence structure, and fairly short sentences (Mills and Walter, 1978). Natural word order typically consists of the subject, a verb, and an object or complement. A "paragraph" may be defined as the compositional unit for the development of a single thought (Mills and Walter, 1978). Typically, a paragraph begins with a sentence (the topic sentence) that states the gist of the idea to be developed. The other sentences of the paragraph develop, support, and clarify this central idea. Two considerations govern paragraph length: unity of thought and eye relief for the reader. Effective technical writing also requires precision in the use of words. Precision is achieved when the technical writer demonstrates an exact knowledge of the meaning of words, avoids words that are vague, leaves out unnecessary words, uses simple words wherever possible, avoids overworked or trite words, and avoids technical jargon (Mills and Walter, 1978).

Use of Visual-Display Materials

Environmental impact documents can be enhanced by the use of visual-display materials. Woolston, Robinson, and Kutzbach (1988) have suggested that visual-display materials are useful in the following situations: (1) when words will not suffice, (2) when information is faster and easier to understand in graphic form, and/or (3) when visual-display material can be used to highlight an important point.

Visual-display material can include charts or graphs, drawings and photographs, and tabular presentations (Mills and Walter, 1978). ''Charts'' or ''graphs'' are means of presenting numerical quantities visually so that trends of and relationships between numerical quantities can be illustrated. Types of charts include the line or curve chart, the bar or column chart, the surface chart, the circle (or pie) chart, the organization (or line-of-flow) chart, and the map chart. Table 17.7 presents some guidelines for constructing several types of graphs or charts.

Drawings and diagrams can be valuable for showing principles and relationships that might be obscured in a photograph; they are sometimes used instead of photographs because they are less expensive to reproduce. However, a photograph can supply far more concreteness and realism than drawings or diagrams (Mills and Walter, 1978).

Tables offer a convenient method of presenting a large body of precise quantitative data in

---

TABLE 17.7

GUIDELINES FOR CONSTRUCTING GRAPHICAL PRESENTATIONS OF INFORMATION

Line graphs

1. Limit the number of lines on a graph to three or four.
2. Distinguish different lines by design or color.
3. Choose the range of tick marks on the scale lines so that the data fill up as much of the graph area as possible.
4. Put tick marks outside the data region, and keep them to a minimum.
5. Label each scale line (quantity and units).
6. Place the figure title and legend below the graph.

Bar graphs

1. Arrange the bars in a logical sequence.
2. Use the same width for each bar. Make the distance between bars different from the width of the bars.
3. Make the bars stand out from the white background using stippling.
4. Since each bar represents magnitude by its length, include the zero line for accurate representation and interpretation.
5. To gain vertical distance for the shorter bars, show physical break in the case of an occasional, excessively long bar above the range of the other bars.
6. Use vertical bars for comparing the magnitudes of a variable over time. Use horizontal bars for comparing the magnitudes of categories with descriptive labels.
7. Label each scale line.
8. Place the figure title and legend below the graph.

Pie graphs

1. Limit the number of segments to five or six.
2. Order from largest to smallest segment, beginning at 12 o'clock and moving to the right (clockwise).
3. Identify each sector with a label.
4. Keep labels within segments, if possible, and keep labels horizontal.

Source: After Woolston, Robinson, and Kutzbach, 1988, pp. 80, 83–84, 86.

an easily understood form. Tables are intended to be read from the top down in the first column and to the right. The first, or left, column normally lists the independent variables (time, item number, and so on); and the columns to the right typically list dependent variables (Mills and Walter, 1978). Tabular displays of materials need to be carefully planned prior to their inclusion; for example, Woolston, Robinson, and Kutzbach (1988, p. 17) suggested the following guidelines for constructing tables:

1. Place columns to be compared next to each other.
2. Make headings and data reflect an organizational principle (priority, descending order, alphabetical order).
3. Label each column and row.
4. Include units of measure in the headings.
5. Align decimals in a column.
6. Put table number and title at the top.
7. Use footnotes for more extensive explanations of data or headings.

Maps can also be useful in impact reports for communicating information on project location and specific site designs. Maps can also be utilized to demonstrate environmental features, such as the range of specific faunal species and general air and water quality within the study area. Figures and tables can be useful in communicating numerical information about project features and anticipated impacts. Photographs can be used to identify project need and any existing environmental resources in the study area that might be subject to anticipated impacts. Interaction matrices can be useful for summarizing impact information, particularly if color codes are used to display anticipated beneficial and detrimental impacts. Finally, conceptual drawings by landscape architects can serve to illustrate what the project would look like within the environs of the environmental setting. Conceptual drawings are particularly useful for analyzing the visual impacts of a project.

Use of Referencing and Numbering Systems

It is important to use a consistent referencing system for the in-text citations and the list of cited references. One approach for in-text citations is to note the author and year of the citation in parentheses at the end of the pertinent sentence. This is better than using a numbered-reference approach, since it eliminates the renumbering often necessary with the latter when references are either added or deleted late in the writing process. The list of cited references can be an alphabetical listing by author. It is also important to use a consistent reference citation system for the list of cited references. It is recognized that different professional organizations use different systems for citing references; the important issue is not the citation system chosen, but consistency in delineating references in a common format.

Another important issue is the numbering system for figures, tables, maps, and other visual-display materials. One suggested approach is to number each of these visual-display materials by chapter in a sequential manner. Therefore, the first table in Chapter 3 would be listed as "Table 3.1," the second as "Table 3.2," and so forth. This approach can aid in simplifying the subsequent assemblage of the written document, in contrast to an approach based on sequential numbering throughout the document; this is because sequential numbering is often interrupted by additions or deletions of visual-display materials near the end of the writing phase.

Coordination of Team-Writing Effort

Many environmental impact documents are written by an interdisciplinary team wherein various individuals have different writing styles. Woolston, Robinson, and Kutzbach (1988) suggested the following three-step strategy as an aid to a cohesive team-writing effort: (1) the team agrees on purpose, scope, and

TABLE 17.8

GENERAL WRITING SUGGESTIONS FOR AN ENVIRONMENTAL IMPACT REPORT

1. Do not use cliches and catchwords. The former may reduce the report's impact and effectiveness; the latter may be understandable to the writers but unfamiliar to reviewers and agency decision makers.

2. Every effort should be made to make EISs (or other such reports) succinct and clear, with minimal use of written texts and liberal use of visual-display methods.

3. Do not use vague generalities in the report. Examples of such statements include: "Developer will exercise supervision and control to prevent siltation," "Special consideration will be given to providing in-plant controls," and "Construction noise will be minimized." While statements such as these may be true and applicable, it is better to identify exact methods of implementation and enforcement, if they are known.

4. Try to avoid creating a credibility gap as a result of too many technical errors and mistakes in the document. In any scientific writing, and particularly for documents prepared under very stringent time constraints, many errors and omissions may occur. There is no foolproof way of eliminating this problem. One approach is to subject the draft statement to internal review by persons unfamiliar with details of the project.

5. Both pro and con information with regard to a proposed action should be presented. Value judgments indicating that the writer or writers are attempting to justify the project are easily identifiable in an EIS (or similar document).

6. Efforts should be made to provide as complete a document as possible within the time frame and monetary constraints associated with a given writing effort. Even though all potential impacts of a project may not be analyzed with the same degree of detail, it is useful to identify all known items of potential impacts even though they may be dealt with only in a cursory manner.

7. Care should be taken to prevent plagiarism from existing documents. This particular problem can be eliminated if a proper referencing and documentation system is used.

8. Attempts should be made to provide a document that has continuity from one section to another. Poor organization of the contents of a report will lead to criticisms by reviewers, even though appropriate substantive materials are present.

9. Since environmental impact reports are written by interdisciplinary teams, it is not uncommon to find differences in writing style within the work. This can result in contradictory information and statements in various sections of the report. One of the most important steps in EIS preparation is the internal review of the completed draft report. Conflicts of professional opinion or scientific information in substantive areas should not be omitted, but rather they should be included to provide decision makers and reviewers with more complete information and to enable better realization of the consequences of initiating a particular action. However, conflicting statements based on misinformation or errors in interpretation should be eliminated from the reports.

10. One of the frequent criticisms of EISs is that the information is so general it is of little relevance to an analysis of environmental impacts. The writers must be as specific as possible, within the bounds of data availability. If general information or information from other locales is used, it should be clearly identified, with appropriate explanations as to why this information is considered to be relevant.

Source: Compiled using data from Garing, Taylor, and Associates, 1974, pp. 19–21, 46–47.

overall outline; (2) individuals are assigned sections for which they develop detailed plans and write drafts; and (3) the team reviews individual written work frequently and provides feedback in writing, or by marking the draft with notes and specific suggestions for improvement.

In a team-writing effort, it is important for one person (either the team leader or a technical writer–professional editor) to have the responsibility of reviewing the entire written document for consistency from chapter to chapter, and for consistency in terms of project information and anticipated impacts.

The typical approach in the preparation of written documentation includes the assemblage of a draft report which is then subjected to internal review prior to release. The internal review should be done by individuals within the preparing agency or organization—preferably by both a technical person and a nontechnical individual. It is desirable to have the document reviewed by persons unfamiliar with the project, as well. If the information is communicated in an appropriate fashion, both the technical and nontechnical persons should be able to identify the key points within the document. This represents a good final check on a draft report prior to its release for public review and analysis.

Use of Reminder Checklists

It might be desirable to use a checklist to both plan the report and check the final product. Table 17.8 contains several general writing suggestions for environmental impact reports. Many other practical writing suggestions are available; one example is a series of ''don'ts and dos'' (Hellstrom, 1975).

Finally, an issue which may arise in the preparation of an EIA report is related to how to deal with incomplete or unavailable information. The Council on Environmental Quality has issued specific guidance on this issue; this guidance is summarized in paragraph 1502.22 of the CEQ regulations (CEQ, 1987).

SUMMARY

In summary, the preparation of written documentation for an environmental impact study should follow a logical process which consists of a preliminary planning phase and a detailed planning phase, followed by the actual writing phase. In the United States, it has been suggested that technical writers be included as a part of this process, particularly to convert the more-detailed technical information prepared by technical specialists into a written form that is understandable to nontechnical audiences. However, the use of technical writers does not eliminate the writing responsibilities of study team professionals. It is further suggested that a master file of the findings of the study be maintained, since questions frequently arise during the review phase of the written documentation as to the sources of information, its relevance, and its validity relative to the individual study. Therefore, it may be necessary for the preparers of the written documentation to refer to previously utilized materials during this part of the EIA process. In summary, written documentation for environmental impact studies can be prepared in a consistent manner which can aid in communicating information to both technical and nontechnical audiences.

SELECTED REFERENCES

Council on Environmental Quality (CEQ), ''Regulations for Implementing the National Environmental Policy Act,'' *40 Code of Federal Regulations,* Chap. V, July 1, 1987, pp. 929–971.

Garing, Taylor, and Associates, ''A Handbook Approach to the Environmental Impact Report,'' 2d ed., Arroyo Grande, Calif., 1974, pp. 19–46.

Hellstrom, D. I., ''A Methodology for Preparing Environmental Statements,'' AFCEC-TR-75-28, U.S. Air Force Civil Engineering Center, Tyndall Air Force Base, Fla., Aug. 1975.

Mills, G. H., and Walter, J. A., *Technical Writing,* Holt, Rinehart and Winston, Dallas, 1978.

U.S. Army Corps of Engineers, ''Organization and Content of EISs Combined with Planning Reports,'' ER 1105-2-60, App. D, U.S. Army Corps of Engineers, Washington, D.C., Nov. 1985.

Weiss, E. H., ''An Unreadable EIS Is an Environmental Hazard,'' *The Environmental Professional,* vol. 11, 1989, pp. 236–240.

Woolston, D. C., Robinson, P. A., and Kutzbach, G., *Effective Writing Strategies for Engineers and Scientists,* Lewis Publishers, Chelsea, Mich., 1988, pp. 7, 13, 52–53, 77–86.

CHAPTER 18

Environmental Monitoring

A comprehensive (or targeted) post-EIS environmental monitoring program should be required of major projects, plans, or programs as a part of their life cycle, and the resultant information should be used in environmentally responsible management and decision making. "Comprehensive environmental monitoring" refers to the set of activities which provide chemical, physical, geological, biological, and other environmental, social, or health data required by environmental managers (U.S. EPA, 1985). A "targeted monitoring program" could include elements related to environmental media (air, surface, and/or groundwater; soil; and noise), biological features (plants, animals, and habitats), visual resources, social impacts, and human health. Pertinent elements should be selected based on the project type, baseline environmental sensitivity, expected impacts, and monitoring objectives.

An integrating term being used in some countries to denote life-cycle environmental management is "post-project analysis" (PPA). "PPAs" refer to environmental studies undertaken during the implementation phase (prior to construction, during construction or operation,

and at the time of abandonment) of a given activity after the decision to proceed has been made (ECE, 1990). Such studies can include comprehensive or targeted environmental monitoring, evaluation of the collected data and information, environmentally focused decision making, as appropriate, and implementation of the management decisions.

Examples of environmentally responsible project-, plan-, or program-management decisions which can be based on monitoring data, and which can be beneficial in terms of minimizing adverse impacts and enhancing environmental management include (1) reducing power production (and resultant atmospheric emissions) at a coal-fired power plant when atmospheric dispersion conditions are limiting, (2) planning training activities at a military installation so as to not coincide with the use of certain areas for breeding or nesting by threatened or endangered faunal species, (3) planning and implementation of a metals removal system at an industrial wastewater-treatment plant so as to minimize metals uptake in aquatic food chains downstream of the wastewater discharge, and (4) changing surface-water reservoir levels

637

and water-release patterns to optimize dissolved-oxygen concentrations in the water phase during various seasons. Spellerberg (1991) has described the following three ways in which floral- and/or faunal-species-monitoring data can be used in environmental management: (1) to establish a basis for the sustainable use of populations, (2) to detect and, it is hoped, minimize the detrimental environmental impacts, and (3) to provide data which can be used as a scientific basis for conservation.

BACKGROUND INFORMATION

The Council on Environmental Quality regulations (CEQ, 1987) enunciate the principles of post-EIS environmental monitoring in sections 1505.3 and 1505.2(c). The CEQ regulations primarily focus on the use of monitoring in conjunction with the implementation of mitigation measures. Monitoring could also be used to determine the effectiveness of each of the several types of mitigation measures.

Several agencies have developed monitoring information related to mitigation measures in their EIA guidance; one example was developed for U.S. Army projects or activities (U.S. Department of the Army, 1988). In this case, monitoring is identified as an integral part of any mitigation program. Two basic types of monitoring are defined as follows (U.S. Department of the Army, 1988, p. 46355):

1. *Enforcement monitoring* Enforcement monitoring ensures that mitigation is being performed as described in the environmental document, including provisions that mitigation requirements and penalty clauses are written into any contracts. It also includes ensuring that these provisions are enforced.

2. *Effectiveness monitoring* Effectiveness monitoring measures the success of the mitigation effort and/or the environmental effect. This must be a scientifically based quantitative investigation. Generally, qualitative measurements are not acceptable. However, it is not necessary to measure everything that may be affected by the action, only enough information to judge the method's effectiveness.

Sadler and Davies (1988) have delineated three types of environmental monitoring which might be associated with the life cycle of an undertaking; these are baseline monitoring, effects or impact monitoring, and compliance monitoring. "Baseline monitoring" refers to the measurement of environmental variables during a representative preproject period to determine existing conditions, ranges of variation, and processes of change. "Effects monitoring" or "impact monitoring" involves the measurement of environmental variables during project construction and operation to determine the changes which may have occurred as a result of the project. Finally, "compliance monitoring" takes the form of periodic sampling and/or continuous measurement of levels of waste discharge, noise, or similar emissions, to ensure that conditions are observed and standards are met. Pre-EIS monitoring includes baseline monitoring, while post-EIS monitoring encompasses effects or impact monitoring, and/or compliance monitoring.

Only minimal attention has been given to comprehensive or targeted environmental monitoring in conjunction with major actions subjected to the EIA process in the United States. In contrast, other countries—such as Canada, member states of the European Community, and many developing countries—have focused attention on this subject. Some reasons environmental monitoring, and, in particular, post-EIS monitoring, have been given minimal attention in the United States are given below:

1. Environmental monitoring is not required in the current EIA process; the emphasis has been on getting the EIS completed so the project, plan, or program can be started.

2. Monitoring requirements may be included, or assumed to be included, as part of environmental-media (air, surface, or groundwater, and/or noise) or other permit conditions (e.g., Section 404, habitat, plant or animal species, and/or cultural resources).

3. There is the presumption that numerous federal, state, and even local monitoring networks could be used if necessary, and that they would meet project-monitoring needs, if any.

4. There is resistance to planning and implementing a monitoring program, since collected data could be used by regulatory agencies as a basis for notification of violations, or even the levying of fines.

5. Even if monitoring is considered a necessity, agency staffing and funding may be limited. For example, monitoring is an important element in the U.S. Bureau of Land Management rangeland-management and grazing allotment program. Monitoring *should* be used to document experienced grazing impacts and to establish pertinent grazing levels and allotments. However, BLM officials attribute their inability to perform all needed monitoring largely to staff shortages and the need to concentrate on other rangeland-management tasks (U.S. General Accounting Office, 1992).

Some reasons other countries are interested in post-EIS monitoring (whereas the United States has not emphasized this topic) follow: (1) extant environmental-monitoring programs may be minimal in scope, particularly in developing countries; (2) in the EIA process, the emphasis is on life-cycle environmental management, and not just on getting initial approval through the preparation of an EIS; (3) in many countries, there is no structured, legalistic environmental-management system which focuses attention on regulatory compliance, legislative violations, fines, and possibly lawsuits; and (4) these countries recognize that monitoring provides the opportunity to gather environmental data and to use it to increase scientific understanding of environmental-transport and -fate processes and ecological stresses.

In summary relative to background information, the General Counsel of the CEQ has identified monitoring and mitigation as a NEPA-related, topical issue which needs to be emphasized (Smith, 1989). The U.S. EPA's Office of Federal Activities has also noted the need for post-project monitoring, particularly for situations where good predictive techniques may be lacking (Smith, 1989).

PURPOSES OF ENVIRONMENTAL MONITORING

Numerous purposes (and implied benefits) can be delineated for pre- and/or post-EIS environmental monitoring. For example, Marcus (1979, after p. 2) identified the following six general purposes or uses of information gleaned from the conduction of post-EIS monitoring:

1. Environmental monitoring provides information that can be used for documentation of the impacts that result from a proposed federal action; this information enables more-accurate prediction of impacts associated with similar federal actions.

2. The monitoring system could warn agencies of unanticipated adverse impacts or sudden changes in impact trends.

3. The monitoring system could provide an immediate warning whenever a preselected impact indicator approaches a predetermined critical level.

4. Environmental monitoring provides information which could be used by agencies to control the timing, location, and level of impacts of a project. Control measures would involve preliminary planning as well as the possible implementation of regulation and enforcement measures.

5. Environmental monitoring provides information which could be used for evaluating the effectiveness of implemented mitigation measures.

6. Environmental monitoring provides information which could be used to verify predicted impacts and thus validate impact-prediction techniques. Based on these findings, the techniques—for example, mathematical

models—could be modified or adjusted, as appropriate.

Recently, a multicountry task force on EIA auditing conducted a comparative analysis of 11 case studies in order to document environmental-monitoring practices (ECE, 1990). Some generic purposes for conducting such monitoring as delineated in the case studies included the following (ECE, 1990):

1. To monitor compliance with the agreed conditions set out in construction permits and operating licenses
2. To review predicted environmental impacts for proper management of risks and uncertainties
3. To modify the activity or develop mitigation measures in case of unpredicted harmful effects on the environment
4. To determine the accuracy of past impact predictions and the effectiveness of mitigation measures in order to apply this experience to future activities of the same type
5. To review the effectiveness of environmental management for the activity
6. To use the monitoring results in order to determine the compensation required to be paid to local citizens affected by a project

Monitoring can be useful in distinguishing between natural change and those changes caused directly or indirectly by pollution and other impacts. Spellerberg (1991) has delineated six reasons biological and ecological monitoring are of value: (1) as a basis for managing biological resources for sustainable development and resource assessment; (2) as aids in the management and conservation of ecosystems and populations; (3) as tools to focus on land use and landscapes as a basis for better use of the land—that is, combining conservation with other uses; (4) as a source of data to aid in the use of organisms to monitor pollution and to indicate the quality of the environment; (5) usage to advance knowledge about the dynamics

of ecosystems; and (6) as a means of targeting insect pests of agriculture and forestry for study so as to establish effective means of control of those pests.

In the context of human health impacts, biological monitoring can be used to relate environmental-media concentrations to potential health effects (Schweitzer, 1981). One type of biological monitoring involves measuring the chemicals that accumulate in species indigenous to the local area. It can be hypothesized that chemicals not detected at significant levels in air, water, or soil might reach higher levels in biota because of the multiple routes of exposure. A second type of monitoring entails the measurement of biological responses to chemical contaminants using either indigenous biological species or species introduced into the area of concern. Considerable documented information is available on the use of different organisms as biomonitors in Ontario Ministry of the Environment (1989). Finally, the most direct approach involves medical investigations and human surveillance to identify possible health impacts on nearby populations. A variety of techniques may be applicable, ranging from routine chemical analyses of blood, urine, and breath, to investigations of impacts on responses of the nervous and immunological systems (Schweitzer, 1981; Burtan, 1991).

Five objectives for social-impact assessment and monitoring are (Krawetz, MacDonald, and Nichols, 1987) (1) to document compliance with expected performance (for example, inspection, surveillance in terms of regulatory permits, and contractual agreements); (2) to achieve impact management—that is, project control to ensure that problems do not develop which interfere with construction through delays or cost overruns; (3) to facilitate research and development, including straight documentation, enhancing technical capacity for future project planning, evaluating predictions, and testing specific hypotheses; (4) to establish credibility (public assurance); and (5) to provide evidence of change, including determination of status, trend monitoring, and early warning systems.

Environmental monitoring can also serve as a basic component of a periodic environmental regulatory auditing program for a project (Allison, 1988). In this context, "auditing" can be defined as a systematic, documented, periodic, and objective review, conducted by regulated entities, of facility operations and practices related to meeting environmental requirements (U.S. EPA, 1986). Some purposes of environmental auditing are to verify compliance with environmental requirements; to evaluate the effectiveness of in-place environmental management systems; and/or to assess risks from regulated and unregulated substances and practices. Some direct results of an auditing program include an increased environmental awareness by project employees, early detection and correction of problems and thus avoidance of environmental-agency enforcement actions, and improved management control of environmental programs (Allison, 1988). There are several available references describing protocols and experiences in auditing related to the EIA process; Canter (1985a), Munro, Bryant, and Matte-Baker (1986), PADC Environmental Impact Assessment and Planning Unit (1982), Sadler (1987), and United Nations Environment Program (1990) are some examples.

An emerging topic which often involves monitoring is the planning and conduction of site assessments. A "site assessment" refers to a study of a specific parcel of land (and the building and facilities associated therewith) for the purpose of documenting previous or current soil, groundwater, and/or building or facility contamination. This information is then used by property owners, potential buyers, and lending institutions to establish appropriate liability for the extant contamination. Site assessments are also referred to as "preacquisition site assessments" or "property-transfer assessments." Information for planning necessary environmental monitoring is available in numerous references; one example is the protocol for environmental baseline surveys at military installations in U.S. Department of the Army (1990).

The primary point to note from these delineations of different monitoring purposes is that such purposes can be wide-ranging; therefore, monitoring purposes need to be incorporated in the planning and implementation of a monitoring effort for a project, plan, or program.

CASE STUDIES OF MONITORING

To illustrate the various uses of monitoring in environmental impact work, eight case studies will be noted; Table 18.1 contains a summary of the case studies in terms of project-program type, monitoring conducted, and the uses of the monitoring information. The case studies comprise a pest control program, a wastewater-treatment facility, two lignite-extraction projects, an airport modification project, an evaluation of historical and needed waste-disposal practices at a nuclear facility, an existing multipurpose surface-water reservoir system, and a proposed multipurpose surface-water reservoir project (Canter, 1993).

As noted in the previous section, environmental monitoring can be incorporated in EISs for a variety of purposes, such as for establishing project or program need. For example, monitoring of septic-tank discharges, groundwater inflow to a lake, and lake water quality and aquatic ecology was used to establish the need for a centralized wastewater-collection and -treatment system in the environs of Crystal Lake in Michigan (U.S. EPA, 1980). As another example, at the Savannah River plant of the U.S. Department of Energy, located in South Carolina, monitoring of soil and groundwater quality at extant waste-disposal sites was used to establish the need for both a remediation program and modified practices in waste-disposal-facility siting and/or operation (U.S. Department of Energy, 1987).

TABLE 18.1

SUMMARY OF ENVIRONMENTAL MONITORING IN EIGHT CASE STUDIES

Case study (reference)	Project, program type	Monitoring[a]	Uses or purposes of monitoring
Fire ant control program (Animal and Plant Health Inspection Service, 1981)	Pest control	Pesticide effectiveness Pesticide residues in environmental compartments	To describe project effectiveness and environmental consequences
Wastewater treatment around Crystal Lake (U.S. Environmental Protection Agency, 1980)	Wastewater-treatment facility	Septic tank discharges Groundwater flow Lake water quality and aquatic ecology	To describe need for project and to determine effectiveness of project for water quality improvements
Surface lignite mine (U.S. Environmental Protection Agency, 1983)	Lignite extraction	Water quality in surface streams and lakes Discharge permit monitoring	To describe existing water quality and to establish basis for controlling potential impacts
Airport runway extension (Federal Aviation Administration, 1988)	Airport modification	Noise from aircraft and existing ambient noise levels	To describe baseline noise and to use as input to noise-prediction model
Nuclear facility waste management program (U.S. Department of Energy, 1987)	Waste disposal	Soil and groundwater quality	To establish need for project and to serve as basis for waste-disposal planning
Surface lignite mine (U.S. Environmental Protection Agency, 1990b)	Lignite extraction	Soil composition, wheat production, hydrogeological parameters, groundwater quality, baseline flows in streams, and noise	To describe baseline environmental conditions and to serve as "look-alike" information for project design and impact prediction
Reservoir system on Tennessee River (Tennessee Valley Authority, 1991)	Operation of 16 reservoirs and dams in Tennessee River system	River flow, water quality (dissolved oxygen and other constituents), and effectiveness of aeration of water releases from dams	To determine influence of reservoir operational patterns on water quality (particularly dissolved oxygen), and to improve water quality and aquatic habitat by increasing minimum flow rates and aerating releases from the TVA dams to raise dissolved oxygen levels, and to extend the recreation season on TVA lakes by delaying drawdown for other reservoir operating purposes, primarily hydropower generation
Construction and operation of Elk Creek Lake (U.S. Army Corps of Engineers, 1991)	Multipurpose reservoir, the third of three reservoirs in the river basin	Water temperature, turbidity, and suspended sediment; river flow rates; game fish; and terrestrial habitats for eight evaluation species; monitoring at two existing reservoirs and at proposed site for Elk Creek Lake	To validate extant water quality models, and to serve as basis for predicting both single-project and cumulative impacts on fisheries, water quality, and terrestrial wildlife habitat

[a]Other monitoring may have been mentioned in the EIS but not addressed herein.
Source: Canter, 1993, p. 80.

An often-cited purpose of environmental monitoring is for describing the affected environment (i.e., establishing the baseline conditions). The water-quality monitoring in surface streams and lakes in the environs of a proposed surface lignite mine in Rusk County, Texas, is an example of "baseline delineation" (U.S. EPA, 1983). Another example is the noise monitoring conducted in the area of a proposed airport runway modification in Oklahoma (FAA, 1988).

Baseline monitoring conducted as a part of the preparation of the EIS for the surface lignite mine in Titus County, Texas, included the following components (U.S. EPA, 1990b): (1) determination of soil composition in the study area, since the project was expected to alter the structure, increase the bulk density, reduce the permeability, and modify the texture of the soil; (2) measurement of wheat production over a three-year period on a 10-acre portion of post-minespoil at a nearby mine area operated by the project proponent (this monitoring information was used to predict potential wheat production on minespoil for the proposed mine); (3) measurement of hydrogeological parameters and groundwater quality by means of 43 wells drilled in the study area (this information was used to estimate the probable hydrogeologic consequences of the mining operations and the dewatering requirements for the mining area); (4) measurement of baseline flows in streams in the study area by means of seven crest-gauge monitoring stations and one continuous-level monitoring gauge; and (5) conduction of a noise survey at seven receptors—three which were located in the study area and four which were within 1.5 mi of the study area boundaries (this information documented existing conditions and was used in assessing construction and mining impacts).

Another purpose for monitoring is to provide sufficient collected information so as to enable the prediction of the undertaking's potential effectiveness and/or its potential environmental impacts. One example related to effectiveness prediction is the pesticide-effectiveness monitoring conducted as part of a program to control the imported fire ant in nine southeastern states (Animal and Plant Health Inspection Services, 1981). In this same study, monitoring data on pesticide residues in various environmental compartments was used as a basis for predicting the potential environmental consequences of pesticide usage. Another example of monitoring for impact prediction is in the airport-runway modification project (FAA, 1988). Background-noise data, as well as noise data from various types of aircraft and their operations, was used as input to a noise-prediction model. This model was then used to examine the noise impacts of various alternatives.

Environmental monitoring can also be used as an aid to project or program operation and management. In fact, specific post-EIS monitoring was addressed in six of the eight case studies (pest control program, both surface lignite mines, waste disposal at a nuclear facility, a reservoir system on the Tennessee River, and construction and operation of Elk Creek Lake). The most comprehensive illustration of environmental monitoring coupled with ongoing decision making was in conjunction with the operation of 16 extant reservoirs and dams in the Tennessee River system (TVA, 1991).

A targeted pre- and post-EIS environmental-monitoring program was described in the final EIS, Supplement no. 2, for an undertaking known as the "Elk Creek Lake project." A portion of this monitoring effort was instituted to ensure compliance with the requirements of some court decisions (U.S. Army Corps of Engineers, 1991). Monitoring for several water-quality, fishery, and terrestrial-habitat parameters was conducted at two existing reservoirs and dams (Applegate Dam and Lost Creek Dam) and the proposed site for the Elk Creek Dam in the Rogue River Basin. The Elk Creek project is a concrete dam and reservoir to be located on Elk Creek, approximately 1.7 mi upstream from its confluence with the Rogue River. Water-quality and terrestrial-habitat modeling was used for analyzing single and cumulative impacts related to the Elk Creek Lake project. Water-quality models used for evaluating temperature, turbidity, and suspended-sediment impacts included the Water Resources Engineers (WRE) model, two Corps of Engineers models (WESTEX and CE-THERM-R1), and the U.S. EPA's QUAL-IIE model. Four physical parameters (land cover, soils, slope, and stream network) were monitored in a remote-sensing–GIS (geographic information system) analysis of suspended-sediment turbidity. Fisheries-resources studies for salmon and steelhead populations assessed changes in emergence timing of fry from river gravel; the abundance, size, growth rate, and migration timing of juvenile fish; and the abundance, migration timing, pre-spawning mortality, and spawning of adult fish. Eight terrestrial species in the Rogue River Basin were studied through usage of the habitat evaluation procedures (HEP, discussed in an earlier chapter) of the U.S. Fish and Wildlife Service (U.S. Army Corps of Engineers, 1991).

It should be evident from these illustrations of the various purposes of environmental monitoring that such monitoring might be conducted prior to, during, or after environmental-impact studies and the preparation of EISs. Each monitoring program in the eight case studies was unique and a function of the project or program type and the geographical location. While essentially no information on the costs of the environmental-monitoring efforts was included in the eight case studies, it can be concluded that these efforts can be expensive. Monitoring costs for a specific study are a function of numerous factors, including availability of extant data, the number and types of parameters to be monitored, the length of the monitoring program, and data-management and -interpretation needs.

PLANNING CONSIDERATIONS FOR A MONITORING PROGRAM

Careful planning and implementation of an environmental-monitoring program is a requisite for meeting the stated purposes of monitoring. Three premises relative to monitoring programs in the United States are stated below:

1. There is an abundance of environmental-monitoring data routinely collected by various governmental agencies and the private sector. This data typically needs to be identified, aggregated, and interpreted.
2. Environmental-monitoring programs are expensive to plan and implement; therefore, every effort should be made to utilize or to modify extant monitoring programs, as appropriate.
3. Because of overlapping environmental-management and -monitoring responsibilities of many local, state, and federal government agencies, it may be necessary to carefully coordinate environmental-monitoring planning among several agencies.

Several conceptual models exist for the planning and implementation of environmental-monitoring programs, and two examples will be cited (Marcus, 1979; Spellerberg, 1991). Marcus (1979) described two phases in a conceptual model: (1) development of a monitoring system and (2) implementation and operation of a monitoring system. Figure 18.1 identifies 11 work elements associated with development of a monitoring system; key points to note follow (Marcus, 1979):

1. Work elements 1 through 3 are related to the preparation and results of an EIS.
2. Agency coordination is addressed in work elements 4, 5, and 9. Coordination with the ongoing monitoring programs of various agencies is vital in the development of a post-EIS monitoring system.
3. The monitoring objectives (work element 6) should be related to the anticipated impacts of the action. Impacts can occur on the physical-chemical, biological, cultural, and socioeconomic components of the environment. Comprehensive or targeted monitoring can be planned.

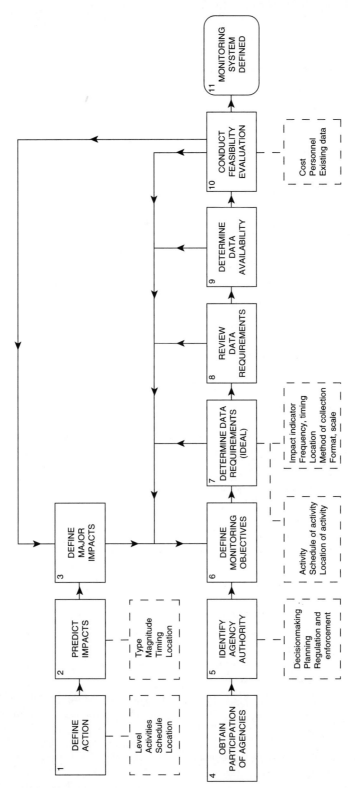

FIGURE 18.1
Monitoring Methodology Flowchart. Phase I—Development of a
Monitoring System (Marcus, 1979, plate 1 in report packet).

645

TABLE 18.2

TASKS ASSOCIATED WITH DETERMINING AND REVIEWING DATA REQUIREMENTS IN PLANNING A MONITORING PROGRAM

Work element[a]	Tasks
7—Determine ideal data requirements	• Reevaluate impacts on the basis of monitoring objectives; eliminate overlap in monitoring objectives and monitoring effort.
	• Select impact indicators. "Impact indicators" are the parameters that must be monitored to assess the magnitude of impacts; several parameters may be indicative of a particular impact. Any impact indicator should be selected on the basis of its utility for decision making, planning, regulation, and enforcement.
	• Determine the frequency and timing of data collection. Frequency of data collection should be the minimum necessary for trend analysis, enforcement of regulations, and correlation of cause and effects. For some parameters, the timing of data collection may be more important than the frequency level; for example, collection of water-quality data during a major runoff event is more important than a precise data-collection frequency. The timing of data collection should relate to the timing of activities causing the impact. Different phases of an action may produce different impacts that persist after an activity ceases.
	• Determine monitoring sites or collection areas. These should be identified based on the location of the activities causing the impacts, predictions of areas most likely to be affected, and a determination of locations where integrated measurements would assist in gaining comprehensive understanding.
	• Determine the method of data collection.
	• Determine the data type and storage format. Data format possibilities include statistical tables, charts, graphs, summaries, maps, map overlays, computer printouts, and a variety of graphic techniques. Criteria for selecting suitable format include ease and convenience of access to data by all users, intelligibility, interrelatability among formats, and ease of updating.
	• Determine the data-analysis method.
8—Review data requirements	• Review data needs for conformance with monitoring objectives.
	• Revise data needs as necessary to meet monitoring objectives.

[a]From Figure 18.1.
Source: After Marcus, 1979, pp. 31–33.

4. Work elements 7 and 8 (determining and reviewing data requirements) constitute the key technical component. They require detailed planning based on scientific rationale. Examples of specific tasks in work elements 7 and 8 are included in Table 18.2.
5. It will be necessary to adjust the post-EIS monitoring program to coincide with available budgetary resources. Several iterations may be necessary to achieve a workable monitoring system (work elements 10 and 11).

The second phase of a post-EIS monitoring program involves the implementation and operation of the monitoring system, and Figure 18.2 delineates the work elements (Marcus, 1979). Key points to note from this phase are

1. Work element 12 (implementing the monitoring system) may require considerable effort in obtaining specific interagency agreements and necessary funding.

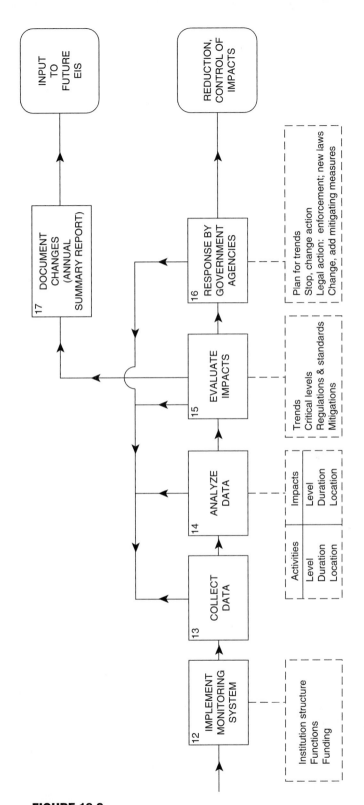

FIGURE 18.2
Monitoring Methodology Flowchart. Phase II—Implementation and
Operation of a Monitoring System (Marcus, 1979, plate 2 in report
pocket).

2. Work elements 13 through 15 involve data collection, analysis, and evaluation. Evaluation of impacts will involve the predetermination of criteria to be used for interpretation. These criteria should be based on legal or institutional limits, professional judgment, and public inputs. Four issues associated with work element 15 include (a) consideration of impact trends and rates of change (the rate at which an impact is increasing is significant because of the need to respond to impact trends in a timely fashion before critical impact levels are reached), (b) consideration of impacts that have reached critical impact levels (critical impact levels requiring immediate notification of participants should be set for each impact being monitored), (c) consideration of impacts that have exceeded legal limits, and (d) the effectiveness of mitigation measures.

3. Development of appropriate response plans to impact trends (work element 16) can be time-consuming and technically difficult, and may require considerable coordinating efforts. It is important that such plans be developed prior to implementation of the monitoring system.

4. It is vitally important that annual summary reports, or reports issued on more-frequent intervals, be prepared in order to document the findings and resultant responses to the post-EIS monitoring program (work element 17).

There are several fundamental books and articles which would be useful in the detailed planning and implementation of a monitoring program. The following noninclusive references are available for air-quality monitoring (Noll and Miller, 1977; Lodge, 1989), surface-water-quality monitoring (Canter, 1985b; Loftis et al., 1989), groundwater-quality monitoring (Aller et al., 1989), noise monitoring (Lipscomb and Taylor, 1978), species and habitat monitoring (Brown and Dycus, 1986; Gray, 1988; Horner, Richey, and Thomas, 1986; Ontario Ministry of the Environment, 1989; Roberts and Roberts, 1984; Spellerberg, 1991), social-impact assessment and monitoring (Krawetz, MacDonald, and Nichols, 1987), and health-effects monitoring (Burtan, 1991; Schweitzer, 1981). General references which encompass several types of environmental monitoring include Cheremisinoff and Manganiello (1990), Gilbert (1987), and Keith (1991).

A generic conceptual framework for developing an environmental-monitoring plan for biological and ecological monitoring has been described by Spellerberg (1991). The framework can be adapted in principle to the monitoring of environmental media, visual impacts, socioeconomic and/or social impacts, and health impacts. The first step in developing such a plan is to define the monitoring objectives. In many monitoring schemes, the objectives are either not stated or are so complex that they become meaningless (Spellerberg, 1991). The second step is to determine the places where the monitoring will take place. In considering sites, three questions can be used (Spellerberg, 1991):

1. Are the localities suitable for monitoring, particularly in terms of the objectives?

2. Will the locality be secure for the duration of the monitoring?

3. Are the localities representative?

The third step is to make sure the data collected is documented for future use. When monitoring programs extend over long periods of time, those working on the project change. So the need arises for the use of suitable methods to assure the retention of such data, which should be accessible and understandable to successive monitors. The fourth step is the arrangements for data collection and storage. The success of a monitoring program depends not only on good planning and logistical support, but also on coordination with other, related programs (Spellerberg, 1991).

The fifth step involves the process of selecting the variables. The ideal variable and process would have a wholly ecological basis, but logistic limitations (finance, time, and effort) may prevail. Methods for collection of data from the field or assemblage of data from other sources should therefore be considered, along with the choice of parameters. Indicators and composite indices can be useful in monitoring studies (Spellerberg, 1991). For example, ecological status and trend indicators for six resource categories (near-coastal and inland surface waters, wetlands, forests, arid lands, and agroecosystems) have been identified for the U.S. EPA's Environmental Monitoring and Assessment Program (EMAP) (U.S. EPA, 1990a). "Diatoms" can be used as aquatic indicators, since they respond rapidly to changes in many ecological characteristics and it is relatively easy to obtain large numbers of diverse individuals in a monitoring program (Dixit et al., 1992). Monitoring data can be aggregated into pertinent indices to reflect the composite quality of different environmental categories. Several indices for the biological environment have been developed; one example is an "index of biotic integrity" (IBI) for stream-fish assemblages, which is used in both North America and Europe (Oberdorff and Hughes, 1992).

The sixth step involves preliminary data gathering and conduction of baseline surveys. Before the planning of biological monitoring can start, biological information from published sources or preliminary field studies should be assembled. Finally, the seventh step involves the analysis and presentation of the data. Considerations of who will use the data to make recommendations should be remembered when the form of data presentation is selected (Spellerberg, 1991).

It may be possible to coordinate a project- or region-specific monitoring program with an ongoing program. For example, the near-coastal component of EMAP consists of estuaries, coastal waters, coastal and estuarine wetlands, and the Great Lakes (Paul et al., 1991). Review of findings of this EPA program can be used to ascertain its relevance to project- or region-specific biological- or ecological-monitoring needs.

Numerous special issues may be encountered in the planning and implementation of monitoring programs (Schweitzer, 1981). Selected examples include

1. Statistical aspects and representative sampling are important factors in the design of a monitoring program. A statistician on the planning team can help insure that adequate consideration is given to these aspects both in designing the program and in formatting and interpreting the data (Stevens and Olsen, 1992).

2. Access to preferred sampling sites is not always possible. Therefore, the sampling plan should be sufficiently flexible to compensate for such problems.

3. Special efforts are needed to minimize holding times between sampling and analysis. However, extended holding times (beyond two weeks) may be unavoidable. In that event, appropriate storage procedures should be used.

4. A quality assurance program involving surrogate recoveries, inter- and intralaboratory duplicates, and field and laboratory blanks is essential. The quality assurance program may account for up to 20 percent of the monitoring costs.

Potential problems which may be encountered when utilizing environmental data from extant monitoring systems include (1) the likely absence of a quality control program, particularly for older data; (2) difficulties in matching and integrating data on common resources from diverse sources of information; and (3) the general absence of information on data interpretation.

GUIDELINES AND POLICIES

The basic premise is that targeted (or comprehensive) post-EIS monitoring programs should be planned and implemented for selected projects with potentially significant negative impacts. To facilitate and institutionalize such programs, certain guidelines and policies will be needed. Some guidelines and policies for environmental monitoring and auditing in relation to the EIA process have been proposed by Sadler and Davies (1988). Examples of some policy statements underlying the development of a monitoring program include (Sadler and Davies, 1988, after pp. 3–6)

1. *Baseline monitoring*—Baseline monitoring should be planned and initiated during the scoping phase of EIA. Monitoring can then be integrated with impact prediction and assessment and readjusted as necessary to focus on key impacts as the project proceeds.
2. *Formulation of impact predictions*—Predictive statements must be expressed as verifiable impact hypotheses, so that statistical tests can be applied. Probabilities or degrees of certainty should be stated explicitly, making predictions more conducive to analysis and providing a more-precise indication of what is anticipated. Great precision in the expression of prediction should be achieved. Information should be made as explicit as possible. In cases where it is difficult to provide a systematic expression of an impact in quantified terms, sound judgment, based on experience, should be used to make qualitative assessments. Wherever possible, quantitative thresholds should be applied to the definition of minor, moderate, or major impacts. Such thresholds may be based upon statutory recommended or in-house standards or policy objectives. However, the basis of

the categorization should be explicit. Where quantitative thresholds cannot be applied, each term should be defined as clearly as possible and be based upon (a) the importance of the environment; (b) activities or interests affected; (c) public acceptability of impact; (d) whether the impact affects rare or endangered species, habitats, or sites; (e) the reversibility or irreversibility of effects; (f) the frequency, duration, and magnitude of impact; and (g) expert judgment.
3. *Effects monitoring*—Effects monitoring must be designed to establish cause-effect relationships which provide the basis for impact management through the implementation of corrective action.

Ten selected principles and recommendations associated with post-project analysis (PPA) as developed by the task force mentioned earlier are as follows (ECE, 1990, after pp. 3–5):

1. Post-project analysis should be used to complete the EIA process by providing the necessary feedback in the project-implementation phase both for proper and cost-effective management and for EIA-process development.
2. A preliminary plan for the PPA should be prepared during the environmental review of a project; the PPA framework should be fully developed when the EIA decision on the project is made.
3. The PPA should focus on important impacts about which there is insufficient information; identification of these impacts and their priorities is undertaken during the environmental review process.
4. The authority to undertake a PPA should be linked to the EIA process so that the concerns identified for inclusion in the PPA during the environmental review can be properly addressed.
5. PPAs should be done for all major projects with potentially significant impacts. In addition, for other projects, focused PPAs

may be suitable either to facilitate environmental management of the project or the acquisition of knowledge from the project.

6. The development of hypotheses to test should be a part of PPAs. The hypotheses will depend greatly on the nature of the PPA and may involve comparisons of impacts with predictions or with standards, or they may relate to how well the environmental-management system is working.

7. In order to undertake PPAs effectively, baseline data relevant to the hypotheses should be collected and should be as complete as possible.

8. Monitoring and evaluation of the data collected in the monitoring process should be an essential part of PPA. These steps are needed in order to test the hypotheses. This is equivalent to the "scope of work" or "defined responsibilities" process.

9. As a tool for managing PPAs, advisory boards consisting of representatives of industry, government, contractors, independent experts, and the public should be used. Such boards, with well-defined terms of reference, increase the credibility and quality of the PPA.

10. Public participation in the PPA should be encouraged, and PPA reports should be made public.

The CEQ regulations do not currently include an EIS section on environmental monitoring (CEQ, 1987); however, in many countries and for many groups, monitoring documentation is incorporated in many environmental impact reports, and two examples will be cited. The required format for EIS reports on water-resources projects in southeast Asian and Pacific countries includes a description of the monitoring program (ESCAP, 1990). The monitoring program must be designed so that the environmental agency receives monitoring reports which will ensure that all necessary environmental-protection measures are being carried out as listed in the approved project plan. As another example, the format for environmental-assessment reports for projects being financed by the World Bank includes a description of the monitoring plan regarding environmental impacts and performance. The plan must specify the type of monitoring, who would do it, how much it would cost, and what other inputs (for example, training) are necessary (World Bank, 1989).

SUMMARY

Comprehensive or targeted monitoring can be used as an integral component of responsible life-cycle environmental management of major projects, plans, or programs. Current EIA-process considerations in the United States are focused on the use of monitoring in conjunction with the implementation of mitigation measures. Additional valid purposes of environmental monitoring include, but are not limited to, establishing baseline conditions, documenting and managing experienced impacts, evaluating the effectiveness of mitigation measures, and validating impact-prediction techniques.

Planning and implementation of a comprehensive or targeted environmental-monitoring program should include usage of extant monitoring data and coordination with pertinent governmental monitoring systems. Program planning includes the delineation of objectives related to expected impacts, selection of pertinent indicators (variables) and determination of sampling location and frequency and analytical requirements. Implementation includes the pre-development of response strategies (management actions) and periodic reporting. "Postproject analysis" (PPA) is a term being used in some countries to denote the role of environmental monitoring in life-cycle project management. Incorporation of environmental-monitoring requirements in the EIA process in the United States would be a logical outgrowth of the EIS-focused nature of the EIA process.

SELECTED REFERENCES

Aller, L., et al., "Handbook of Suggested Practices for the Design and Installation of Ground-Water Monitoring Wells," EPA 600/4-89-034, National Water Well Association, Dublin, Ohio, 1989.

Allison, R. C., "Some Perspectives on Environmental Auditing," *The Environmental Professional,* vol. 10, 1988, pp. 185–188.

Animal and Plant Health Inspection Service, "Cooperative Imported Fire Ant Program, Final Programmatic Environmental Impact Statement," U.S. Department of Agriculture, Washington, D.C., June 1981.

Brown, R. T., and Dycus, D. L., "Characterizing the Influence of Natural Variables During Environmental Impact Analysis," *Rationale for Sampling and Interpretation of Ecological Data in the Assessment of Freshwater Ecosystems,* American Society for Testing and Materials, Philadelphia, 1986, pp. 60–75.

Burtan, R. C., "Medical Monitoring's Expanding Role," *Environmental Protection,* vol. 2, no. 6, Sept. 1991, pp. 16–18.

Canter, L. W., "Impact Prediction Auditing," *The Environmental Professional,* vol. 7, no. 3, 1985a, pp. 255–264.

———, *River Water Quality Monitoring,* Lewis Publishers, Chelsea, Mich., 1985b.

———, "The Role of Environmental Monitoring in Responsible Environmental Management," *The Environmental Professional,* vol. 15, no. 1, 1993, pp. 76–87.

Cheremisinoff, P. N., and Manganiello, B. T., *Environmental Field Sampling Manual,* Pudvan Publishing Company, Northbrook, Ill., 1990.

Council on Environmental Quality (CEQ), "Regulations for Implementing the National Environmental Policy Act," *40 Code of Federal Regulations,* Chap. V, July 1, 1987, pp. 929–971.

Dixit, S. S., Smol, J. P., Kingston, J. C., and Charles, D. F., "Diatoms: Powerful Indicators of Environmental Change," *Canadian Journal of Fisheries and Aquatic Sciences,* vol. 49, no. 1, 1992, pp. 128–141.

Economic Commission for Europe (ECE) "Post-Project Analysis in Environmental Impact Assessment," ECE/ENVWA/11, United Nations, Geneva, Switzerland, 1990, pp. 1–10, 21–38.

Economic and Social Commission for Asia and the Pacific (ESCAP), "Environmental Impact Assessment—Guidelines for Water Resources Development," ST/ESCAP/786, Bangkok, Thailand, 1990.

Federal Aviation Administration (FAA) "Final Environmental Assessment for Max Westheimer Airpark, Norman, Oklahoma," Oklahoma City, Feb. 1988.

Gilbert, R. O., *Statistical Methods for Environmental Pollution,* Van Nostrand Reinhold, New York, 1987.

Gray, R. H., "Overview of a Comprehensive Environmental Monitoring and Surveillance Program: The Role of Fish and Wildlife," PNL-SA-15922, Battelle Pacific Northwest Labs, Richland, Wash., May 1988.

Horner, R. R., Richey, J. S., and Thomas, G. L., "Conceptual Framework to Guide Aquatic Monitoring Program Design for Thermal Electric Power Plants," *Rationale for Sampling and Interpretation of Ecological Data in the Assessment of Freshwater Ecosystems,* American Society for Testing and Materials, Philadelphia, 1986, pp. 86–100.

Keith, L. H., *Environmental Sampling and Analysis: A Practical Guide,* Lewis Publishers, Chelsea, Mich., 1991.

Krawetz, N. M., MacDonald, W. R., and Nichols, P., "A Framework for Effective Monitoring," Canadian Environmental Assessment Research Council, Hull, Quebec, Canada, 1987.

Lipscomb, D. M., and Taylor, A. C., eds., *Noise Control Handbook of Principles and Practices,* Van Nostrand Reinhold, New York, 1978.

Lodge, J. P., ed, *Methods of Air Sampling and Analysis,* 3d ed., Lewis Publishers, Chelsea, Mich., 1989.

Loftis, J. C., Ward, R. C., Phillips, R. D., and Taylor, C. H., "An Evaluation of Trend Detection Techniques for Use in Water Quality Monitoring Programs," EPA 600/S3-89-037, U.S. Environmental Protection Agency, Center for Environmental Research Information, Cincinnati, Ohio, Sept. 1989.

Marcus, L. G., "A Methodology for Post-EIS (Environmental Impact Statement) Monitoring," *U.S. Geological Survey Circular* 782, *U.S. Geological Survey,* Washington, D.C., 1979.

Munro, D. A., Bryant, T. J., and Matte-Baker, A., "Learning from Experience: A State-of-the-Art Review and Evaluation of Environmental Impact

Assessment Audits," Canadian Environmental Assessment Research Council, Hull, Quebec, Canada, 1986.

Noll, K. E., and Miller, T. L., *Air Monitoring Survey Design,* Ann Arbor Science Publishers, Ann Arbor, Mich., 1977.

Oberdorff, T., and Hughes, R. M., "Modification of an Index of Biotic Integrity Based on Fish Assemblages to Characterize Rivers of the Seine Basin, France," *Hydrobiologia,* no. 228, 1992, pp. 117–130.

Ontario Ministry of the Environment, "Investigation, Evaluation, and Recommendations of Biomonitoring Organisms for Procedures Development for Environmental Monitoring," MIC-90-00873/WEP, Ontario Ministry of the Environment, Toronto, Ontario, Canada, 1989.

PADC Environmental Impact Assessment and Planning Unit, "Post-Development Audits to Test the Effectiveness of Environmental Impact Prediction Methods and Techniques," University of Aberdeen, Aberdeen, Scotland, 1982.

Paul, J. F., Holland, F., Summers, J. K., Schimmel, S. C., and Scott, J. K., "EPA's Environmental Monitoring and Assessment Program: An Ecological Status and Trends Program," EPA 600/D-91-250, U.S. Environmental Protection Agency, Narragansett, R.I., 1991.

Roberts, R. D., and Roberts, T. M., *Planning and Ecology,* Chapman and Hall, New York, 1984.

Sadler, B., ed., "Audit and Evaluation in Environmental Assessment and Management: Canadian and International Experience," 2 vols., Environment Canada, Hull, Quebec, Canada, 1987.

—— and Davies, M., "Environmental Monitoring and Audit: Guidelines for Post-Project Analysis of Development Impacts and Assessment Methodology," Centre for Environmental Management and Planning, Aberdeen University, Aberdeen, Scotland, Aug. 1988, pp. 3–6, 11–14.

Schweitzer, G. E., "Risk Assessment Near Uncontrolled Hazardous Waste Sites: Role of Monitoring Data," *Proceedings of National Conference on Management of Uncontrolled Hazardous Waste Sites,* Hazardous Materials Control Research Institute, Silver Spring, Md., Oct. 1981, pp. 238–247.

Smith, E. D., "Future Challenges of NEPA: A Panel Discussion," CONF-891098-10, Oak Ridge National Laboratory, Oak Ridge, Tenn., 1989.

Spellerberg, I. F., *Monitoring Ecological Change,* Cambridge University Press, Cambridge, England, 1991, pp. 181–182.

Stevens, D. L., and Olsen, A. R., "Statistical Issues in Environmental Monitoring and Assessment," EPA 600/R-92-073, U.S. Environmental Protection Agency, Research Triangle Park, N.C., Apr. 1992.

Tennessee Valley Authority (TVA), "Tennessee River and Reservoir System Operation and Planning Review," Final Environmental Impact Statement, TVA/RDG/EQS—91/1, Knoxville, Tenn., 1991.

United Nations Environment Program, "Environmental Auditing," Tech. Rep. Series no. 2, Paris, 1990.

U.S. Army Corps of Engineers, "Elk Creek Lake, Rogue River Basin, Oregon," Final Environmental Impact Statement, Supplement no. 2, Portland District, Portland, Ore., May 1991.

U.S. Department of Energy, "Final Environmental Impact Statement, Waste Management Activities for Ground Water Protection, Savannah River Plant, Aiken, South Carolina," 2 vols., Washington, D.C., Dec. 1987.

U.S. Department of the Army, "Environmental Effects of Army Actions," Regulation 200-2, *Federal Register,* vol. 53, no. 221, Nov. 16, 1988, pp. 46322–46361.

——, "Environmental Protection and Enhancement," AR 200-1, U.S. Department of the Army, Washington, D.C., Apr. 1990, pp. 73–78.

U.S. Environmental Protection Agency (EPA), "Draft Environmental Impact Statement: Martin Lake D Area Lignite Surface Mine, Henderson, Rusk County, Texas," EPA 906/83-003, Region 6, Dallas, Texas, Mar. 1983.

—— "Environmental Auditing Policy Statement," *Federal Register,* vol. 51, no. 131, July 9, 1986, p. 25004.

——, "Environmental Impact Statement, Alternative Waste Treatment Systems for Rural Lake Projects, Case Study Number 1, Crystal Lake Area, Sewage Disposal Authority, Benzie County, Michigan," EPA-5-MI-Benzie-Crystal Lake-LA-80, Region 5, Chicago, July 1980.

——, "Environmental Monitoring and Assessment Program: Ecological Indicators," EPA 600/3-9-060, Las Vegas, 1990a.

———, "Monticello B-2 Area Surface Lignite Mine, Titus County, Texas," Draft Environmental Impact Statement, EPA 906/04-90-003, Dallas, Texas, Apr. 1990b.

———, "Resource Document for the Ground Water Monitoring Strategy Workshop," Office of Ground Water Protection, Washington, D.C., Mar. 1985.

U.S. General Accounting Office, "Rangeland Management: Interior's Monitoring Has Fallen Short of Agency Requirements," U.S. General Accounting Office, GAO/RCED-92-51, Washington, D.C., Feb. 1992.

World Bank, "Operational Directive 4.00, Annex A: Environmental Assessment," World Bank, Washington, D.C., Sept. 1989.

Index